Sustainable Processes and Clean Energy Transition
ICSuPCET2022

International Conference on Sustainable Processes and Clean Energy Transition 2022 (ICSuPCET2022), Borneo Convention Centre Kuching, Sarawak, 1-2 December 2022

Editors

Yeong Yin Fong[1]
Lam Man Kee[2]
Norwahyu Jusoh[1]
Chuah Chong Yang[1]

[1]CO_2 Research Centre, Institute of Contaminant Management, Chemical Engineering Department, Universiti Teknologi PETRONAS

[2]Institute of Self-Sustainable Building, Chemical Engineering Department, Universiti Teknologi PETRONAS

Peer review statement

All papers published in this volume of "Materials Research Proceedings" have been peer reviewed. The process of peer review was initiated and overseen by the above proceedings editors. All reviews were conducted by expert referees in accordance to Materials Research Forum LLC high standards.

Published under License by **Materials Research Forum LLC**
Millersville, PA 17551, USA

Published as part of the proceedings series
Materials Research Proceedings
Volume 29 (2023)

ISSN 2474-3941 (Print)
ISSN 2474-395X (Online)

ISBN 978-1-64490-250-9 (Print)
ISBN 978-1-64490-251-6 (eBook)

Distributed worldwide by

Materials Research Forum LLC
105 Springdale Lane
Millersville, PA 17551
USA
https://www.mrforum.com

Manufactured in the United State of America
10 9 8 7 6 5 4 3 2 1

Table of Contents

Preface

The International Conference on Sustainable Processes and Clean Energy Transition 2022 (ICSuPCET2022) was successfully held physically in Sarawak, Malaysia from 1 – 2 December 2022. This is the 7th series of a conference organized under the umbrella of World Engineering, Science and Technology Congress (ESTCON), UTP since 2010. It was an overwhelming success, serving as a critical platform for experts from multidisciplinary science and engineering to share their insights and engage in discussions about the challenges and opportunities of technological innovations, particularly Industrial Revolution 4.0 (IR 4.0), in achieving a sustainable future. Themed as "Energy Transition for Sustainable Futures," the conference covered an extensive range of topics, including Biomass and Biofuel, Green Processes and Materials, and Safety and Energy System. The accepted papers, which underwent rigorous peer-review by experts, meet the highest international publication standards.

Over 70 papers were submitted, and 57 were selected for inclusion in the conference proceedings, representing a compilation of exciting outcomes from the conference. With four technical sessions featuring national and international presenters, the insightful presentations sparked heated debates and triggered engaging discussions among participants. Feedback from every participant was overwhelmingly positive, with many praising the conference for disseminating useful and insightful knowledge.

The organizing committee expressed their heartfelt appreciation to all individuals and institutions who supported ICSuPCET2022, as their contributions were instrumental in the success of the conference. The committee hopes that ICSuPCET2022 will inspire collaborative research, ignite new ideas, and drive sustainable development. Above all, the conference proceedings will serve as a crucial source of knowledge and reference, enabling the scientific and engineering community to discover groundbreaking findings, develop new products and technologies, and propel sustainable progress forward.

Organizing Committee of ICSuPCET2022

Organizing Committee

Chairman
Assoc. Prof Ir Dr Abdul Halim Shah Maulud

Co Chairman
Assoc. Prof Ts Dr Oh Pei Ching

Secretary
Ts Dr Khairiraihanna Johari

Treasurer
Ts Dr Nor Erniza Mohammad Rozali

UEvent
Dr Chew Thiam Leng
Dr Serene Lock Sow Mun

Technical
Assoc. Prof Ts Dr Yeong Yin Fong
Ts Dr Lam Man Kee
Dr Norwahyu Jusoh
Ts Dr Chong Yang Chuah

Publication
Dr Nurul Aini Amran
Dr Shafirah Samsuri

Logistics/Event Management
Dr Azizul Buang

Promotion & Website
Dr Nabila Farhana Jamaludin
Ts Dr Nurul Ekmi Rabat
Dr Adilla Rashidi
Mrs Siti Nur Fatimah Manap

Characteristic study of blended wastes for thermal process: Alum sludge mixed palm oil decanter cake for low slagging tendency

KUNMI Joshua Abioye[1,a], NOORFIDZA Yub Harun[1,b*], AFOLABI Haruna Kolawole[1,c] and AHMAD Hussaini Jagaba[2,d]

[1]Department of Chemical Engineering, Universiti Teknologi PETRONAS, 32610 Seri Iskandar, Perak, Malaysia

[2]Department of Civil and Environmental Engineering, Universiti Teknologi PETRONAS, 32610 Seri Iskandar, Perak, Malaysia

[a]kunmi_20001938@utp.edu.my, [b*]noorfidza.yub@utp.edu.my, [c]haruna_18002856@utp.edu.my, [d]ahmad_19001511@utp.edu.my

Keywords: Alum Sludge, Ash Fusion Test, Biomass Waste, Palm Oil Decanter Cake, Slagging

Abstract. Proper and effective utilization of biomass wastes, couple with efficient conversion into useful energy have made the study of biomass wastes and finding solution to ash deposition during thermal conversion crucial. In this work, Alum Sludge (AS) was used to regulate ash deposition in Palm Oil Decanter cake (PODC). AS and PODC were characterized using Scanning Electron Microscopy and Energy Dispersive Spectroscopy (SEM-EDS), Thermogravimetric analysis (TGA), Advance X-ray diffractometer (XRD) while Ash fusion test (AFT) was used to determine the ash fusion temperature. SEM-EDS revealed the presence of Aluminum (13.50 wt%), Potassium (1.46 wt%), Iron (2.43 wt%) and Silicon (14.95 wt%) in AS while Potassium (1.19 wt%), Silicon (1.41 wt%), Calcium (2.16 wt%) and Chlorine (0.67 wt%) in PODC. TGA revealed that fixed carbon, volatile matter and ash content in AS are 8.94 wt%, 3.60 wt% and 86.72 wt% while PODC are 7.13 wt%, 51.84 wt% and 38.15 respectively. The total weight loss of AS and PODC was 13.29% and 61.86% respectively. XRD revealed the presence of Kaolinite and Quartz/ Zeolite in AS while Whewellite, Quartz, Potassium Chloride in PODC. AFT revealed 1215°C and above 1400°C as the sintering temperature for PODC and AS respectively. At 50% and above of AS addition, low slagging tendency of PODC was attained. This study provides a new insight on how AS and PODC wastes can be properly managed and utilized in clean energy production.

Introduction

Economic and population growth have led to an increase in the utilization of biomass. Biomass are utilized on daily basis and as a result, numerous biomass wastes are generated annually all over the world. Proper handling and management of this wastes have been a challenge as most end up on landfills and flowing streams or river. In Malaysia, over 2 million tons of Alum Sludge (AS) are generated per annum [1] as AS is an unavoidable waste from water purification and waste water treatment where alum salts are used during the coagulation process [2]. AS are generally known to be dominated with Aluminum and Silicon content, low organic and neutral pH [2]. When discharge into water bodies, it causes siltation, serve as potential risk to human health and disrupt the life of river biota due to the high presence of Aluminum and other metal contents. On the other hand, palm oil decanter cake (PODC); a palm oil biomass is a solid waste generated from the purification of crude palm oil, usually separated from liquid [3]. PODC comes from the mesocarp of oil palm fibers during processing at the mill. Malaysia is the second largest producer of palm oil with total production of 19,700 million tons which is equivalent to 26% of the total world palm oil production [4]. As a result, huge amount of PODC is generated in Malaysia. According to [5], each palm oil mill process 1,065 tons of fresh fruit bunch per day and one ton of fresh fruit bunch can produce PODC as much as 4.0% or as much as 40 kg. Mills dumped PODC around oil palm plantations or air burn thereby resulting in environmental pollution and hazards [6]. According to [7,8], PODC like other oil palm biomass have high potential for biofuel or bioenergy applications.

Both AS and PODC are biomass wastes which constitute nuisance to the environment when dispose improperly. According to [9], all kinds of biomass can be transformed into energy including starchy, triglyceride and lignocellulosic biomass. Biomass wastes can be converted to biofuels and biopower via thermochemical [7,10]and biochemical processes [11,12]. Thermochemical conversion produces products like bio-methanol, biodiesel, bio-oil, bio-syngas and biohydrogen while biochemical conversion produces liquid or gaseous fuels through fermentation or anaerobic respiration. The production of biofuels through thermochemical conversion processes with a broad range of technologies has drawn the most attention in the world [9] because of its instantaneous production and high-quality end products. However, the thermochemical conversion of oil palm biomass through gasification faces ash deposition issues [13] which affects the end product quality, reduces biomass gasification efficiency and cut short equipment lifespan. Oil palm biomass i.e., PODC, oil palm fronds, oil palm trunk are good source of energy as they possess high calorific value[12,14]. According to [13], ash deposition in biomass gasification is as a result of the presence of Alkali metals and Chlorine. Several researchers have worked on solving biomass ash deposition using Aluminosilicate based additives such as Kaolin/ Kaolinite and Zeolite. [15] reported that Kaolin addition to biomass with high Potassium and Chlorine such as Olive cake makes the ash compositions viable for combustion. [16] reported that Kaolin and coal fly ash which are Aluminosilicate based additives effectively captured Potassium Chloride in water slurry fed into an entrained flow reactor. [17] reported that 16% kaolin/ Kaolinite was enough to completely capture Potassium compound in empty fruit bunch during combustion process in furnace at 900°C.

AS; a biomass waste which has abundance of Aluminum and Silicon just as Kaolin, can be likened to Aluminosilicate based additive and be used to combat ash deposition problems during gasification. Up till now, there have been no report on regulating the ash deposition problem of PODC with AS. Looking into this will be worthwhile, contribute significantly to the effective thermal conversion of PODC into biofuel and aid proper management of both AS and PODC wastes. Therefore, this study intends to characterize and examine AS and PODC mix for low slagging tendency and efficient thermal process.

Materials and Methods

AS and PODC were collected from waste water treatment plant and oil palm processing factory respectively in Perak, Malaysia. Both were oven dried at 110°C for 24 hours, grinded mechanically with biomass grinder and sieve to 250 μm.

Chemical Composition Analysis:

Chemical composition analysis of the elements presents in both AS and PODC was conducted using Scanning Electron Microscopy and Energy Dispersive Spectroscopy (SEM-EDS) [Brand: Zeiss, Model: Evo LS15 VPSEM]. The scanning electron microscopy (SEM) coupled with EDS detector has the ability to determine elemental concentrations in selected spots or areas [18]. The use of SEM-EDS techniques for the detection of elemental composition have been used widely in many research activities. Samples were prepared on circular supports of 12.5 mm diameter over carbon stickers. The working distance (WD) was approximately 10 mm and the dead time (DT%) was kept at less than 30% for the experiment [19].

Thermal Decomposition:

Thermogravimetric analysis (TGA) of AS and PODC was carried out using Thermogravimetric analysis (TGA) [Brand: Perkin Elmer, Model: STA6000] in order to examine their decomposition behavior. 5.0 mg of each samples was paved uniformly in a platinum crucible and heated linearly from ambient temperature to 800°C under ambient atmosphere at a heating rate of 25°C/min. The experiment was duplicated to ensure the reproducibility of results.

Mineral Analysis:

The XRD (X-ray Diffraction) analysis was carried out to identify the crystalline compounds present in AS and PODC using Advance X-ray diffractometer (XRD) [Brand: Panalytical, Model: Xpert3 Powder]. The scan for the samples were collected from 2° to 80° (2θ scale) using Cu Kα radiation. Operating conditions are 45 kV and 45 mA. Peak identification was performed through comparison with reference standards from X'pert HighScore software package.

Ashing and ash fusion test:

2 g of AS and PODC blends in the proportion of 0% AS, 30% AS, 50% AS, 70% AS and 100% AS was ashed in a muffle furnace in accordance with National Renewable Energy Laboratory (NREL) ashing standard for biomass [20]. Deformation temperature (DT), softening temperature (ST), hemispherical temperature (HT), and flow temperature (FT) of the ashes was determined using AF700 Ash Fusion Determinator (LECO, USA). The heat treatment of ash was carried out at a heating rate of 10°C/min [21].

Result and Discussions

Proximate and Ultimate Analysis:

Table 1 shows the proximate and ultimate analysis of AS and PODC. The proximate analysis revealed the values for moisture, volatile matter, fixed carbon and ash of AS are 0.64 wt%, 3.60 wt%, 8.94 wt% and 86.72 wt% respectively. Ultimate analysis revealed that Aluminum, Carbon, Iron, Oxygen, Potassium and Silicon were the elements present with 13.50 wt %, 18.16 wt %, 2.43 wt %, 49.49 wt %, 1.46 wt % and 14.95 wt % respectively. The low percentage of volatile indicate AS is inorganic in nature and this was further proved by its low amount of carbon content. The high presence of ash content indicates its poor combustion status and increase in burning time. The values obtained was in agreement with [2] findings where 2.66% of volatile matter and 89.78% of ash content were recorded. The dominant components were found out to be Aluminum and Silicon. This resulted from the use of alum in the coagulation process. The combination of Aluminum oxide and Silicon dioxide present in AS is approximately 70% [17-18]. In PODC, the proximate analysis obtained for moisture, volatile matter, fixed carbon and ash are 2.78 wt%, 51.84 wt%, 7.13 wt% and 38.15 wt% respectively. Carbon, Oxygen, Potassium, Silicon, Calcium and Chlorine are the elements found present with 74.80 wt %, 19.76 wt %, 1.19 wt %, 1.41 wt %, 2.16 wt % and 0.67 wt % respectively. The amount of volatile matter present indicates the presence of gaseous fuels and its suitability as energy source. 74.80% Carbon content proves a good source as biomass energy. [8] reported higher volatile matter and lower ash content of PODC as 72.13% and 14.73% respectively. This is as a result of difference in location and region of the PODC used for the experiment. [24] in a similar manner reported the value of potassium and calcium in PODC to be 1.27% and 1.18% respectively while [25] gave a very close value of 74.4% for the value of organic carbon in their work.

Thermal Decomposition:

Fig. 1 depicts the weight loss (TGA) and derivative weight loss (DTG) curves upon the degradation of AS and PODC at a heating rate of 25°C/min under ambient atmosphere. As observed, three distinct sections of thermal degradation were recorded for both AS and PODC and the weight loss of PODC is higher than that of AS. In AS, total weight loss of 13.29% was observed over three stages while 61.86% was observed in PODC. The first section indicates the removal of moisture present which occurred at 29-156°C for both AS and PODC. The devolatilization stage which is the second section has the most significant change in weight loss. Devolatilization occur at 156-370°C and 156-598°C for AS and PODC respectively. The differences in temperature range resulted from the degradation of hemicellulose followed by cellulose compound present in PODC. Hemicellulose was degraded first, followed by cellulose which has a long polymer of glucose with low branching structure [25]. Due to lesser volatile matter, AS was less reactive compare to PODC. The third section showed a slow degradation in weight loss as combustion occurred at 370-800°C and 485-800°C for AS and PODC respectively. The slow degradation experienced in PODC was as a result of degradation of carbonaceous material and compounds [26]. As reflected from the DTG curve peaks, it could be deduced that PODC is more reactive than AS. Similar thermal decomposition trend was reported by [27]for AS. However, a different PODC thermal decomposition report of four distinct sections of temperature 50-100°C, 100-250°C, 250-400°C and 400°C above was reported by [8].

Table 1: Proximate and Ultimate Analysis

	AS	PODC
	Proximate analysis (wt%)	
Moisture	0.64	2.78
Volatile Matter	3.60	51.84
Fixed Carbon	8.94	7.13
Ash	86.72	38.15
	Ultimate analysis (wt%)	
Aluminum	13.50	-
Carbon	18.16	74.80
Iron	2.43	-
Oxygen	49.49	19.76
Potassium	1.46	1.19
Silicon	14.95	1.41
Calcium	-	2.16
Chlorine	-	0.67

(a)

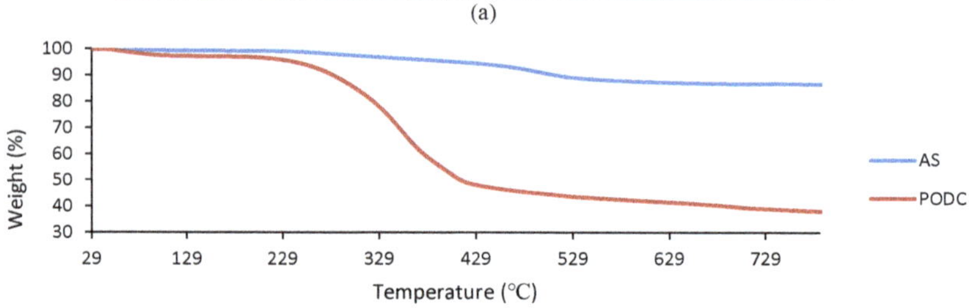

(b)

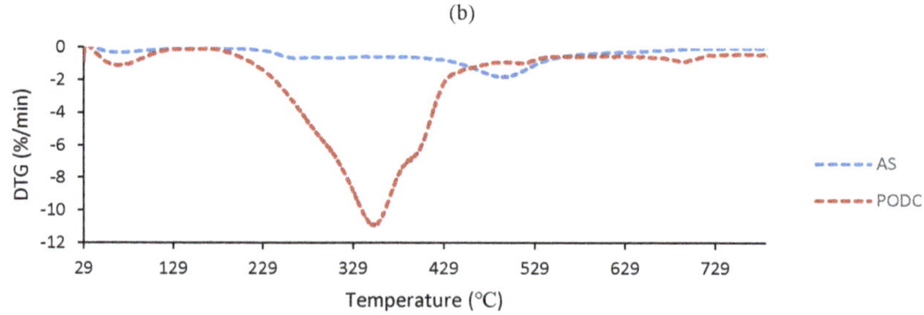

Fig. 1: Thermal decomposition curves of AS and PODC (a) TGA and (b) DTG

Mineral Composition:

The XRD patterns of raw AS and PODC are shown in Fig. 2. The patterns reveal that both AS and PODC have well defined crystalline structures as indicated by the presence of sharp peaks. Kaolinite ($Al_2H_4O_9Si_2$), Quartz/ Zeolite (SiO_2) peaks were found in the XRD structure of AS while Whewellite ($CaC_2H_2O_5$), Quartz (SiO_2) and Potassium Chloride (KCl) were found in the XRD structure of PODC. Since Aluminum Sulphate is used as coagulant in water treatment plant, the existence of Quartz/ Zeolite in AS may be due to suspended sand and clay particles in raw water [28]. Similar report about AS was given by [26]. The existence of Potassium Chloride in oil palm biomass i.e palm oil mill effluent has been reported by [29]which is basically the reason behind ash deposition.

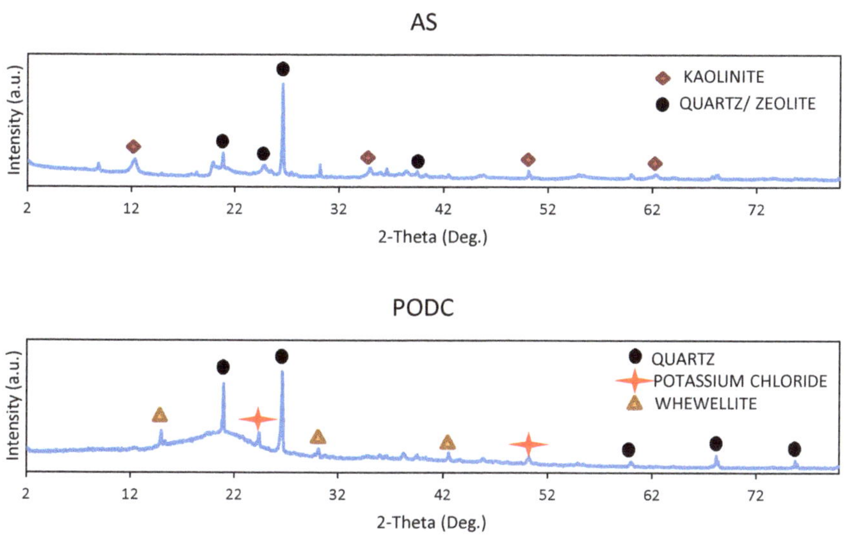

Fig. 2: XRD pattern of raw AS and PODC

Ash Fusion Test:

Table 2 shows the ash fusion temperature of AS, PODC and their blends. ST has been widely employed to predict whether a combustion process is likely to experience slagging [30]. As illustrated in Table 3, ash slagging occurs when ST is lower than 1390°C [31]. ST for PODC and 70PODC are 1215°C and 1280°C respectively, indicating that the possibilities of slagging is relatively high in both samples. When the content of PODC is 50% or less in the blends, the slagging possibilities were low. The improvement in slagging when AS in the blend is 50% and above could be attributed to kaolinite in AS capturing Potassium in PODC to form high melting point Potassium Aluminum Silicate which is in agreement with [32]shown in Eq. 1 below.

$$Al_2O_3.xSiO_2 + 2KCl + H_2O \rightarrow K_2O.Al_2O_3.xSiO_2 + 2HCl_{(g)}. \quad (1)$$

Table 2: Ash fusion Temperature of AS, PODC and their blends

	DT (°C)	ST (°C)	HT (°C)	FT (°C)
PODC	1195	1215	1285	1400
AS30+70PODC	1225	1280	1350	>1400
AS50+50PODC	>1400	>1400	>1400	>1400
AS70+30PODC	>1400	>1400	>1400	>1400
AS	>1400	>1400	>1400	>1400

Table 3: Index of slagging level

Index		**Slagging Level**	
	Low	**Medium**	**High**
ST (°C)	>1390	1390-1260	<1260

Conclusions

In this study, AS and PODC were characterized by SEM-EDS, TGA and XRD. SEM-EDS revealed the presence of Aluminum (13.50 wt%), Potassium (1.46 wt%), Iron (2.43 wt%) and Silicon (14.95 wt%) in AS while Potassium (1.19 wt%), Silicon (1.41 wt%), Calcium (2.16 wt%) and Chlorine (0.67 wt%) in PODC. TGA revealed that fixed carbon, volatile matter and ash content in AS are 8.94 wt%, 3.60 wt% and 86.72 wt% while PODC are 7.13 wt%, 51.84 wt% and 38.15 respectively. The total weight loss of AS and PODC was 13.29% and 61.86% respectively. XRD revealed the presence of Kaolinite and Quartz/ Zeolite in AS while Whewellite, Quartz, Potassium Chloride are present in PODC. AFT revealed 1215°C and above 1400°C as the ST for PODC and AS respectively. At 50% and above of AS addition, low slagging tendency of PODC was attained. In order words, ash deposition was avoided and efficient thermal process was attained.

Acknowledgment

The authors would like to acknowledge the efforts of Universiti Teknologi Petronas Malaysia for financing the project through FRGS/1/2019/TK10/UTP/02/1.

References

[1] K. M. Breesem, F. G. Faris, and I. M. Abdel-Magid, "Reuse of Alum Sludge in Construction Materials and Concrete Works: A General Overview," *Infrastructure University Kuala Lumpur Research Journal*, vol. 2, no. 1, pp. 20–30, 2014, [Online]. Available: https://www.researchgate.net/publication/321019764

[2] T. Ahmad, K. Ahmad, and M. Alam, "Characterization of Water Treatment Plant's Sludge and its Safe Disposal Options," *Procedia Environ Sci*, vol. 35, pp. 950–955, 2016, https://doi.org/10.1016/j.proenv.2016.07.088

[3] M. D. Duaja, E. Kartika, B. Buhaira, and W. P. Armita, "Optimization the effect of decanter cake with fermented fertilizer of cow urine in edamame growth and yield," in *Proceedings of the 3rd Green Development International Conference (GDIC 2020)*, 2021, pp. 498–504.

[4] USDA, "Palm Oil Explorer," 2021. https://ipad.fas.usda.gov/cropexplorer/cropview/commodityView.aspx?cropid=4243000 (accessed Jan. 25, 2022).

[5] M. D. Duaja, E. Kartika, and L. Lizawati, "Application of indigenous AMF from ex-coal mining soil combined with phosphorus fertilizers to improved oil palm seedling growth (elaeis

guineensis Jacq.)," *Biogenesis: Jurnal Ilmiah Biologi*, vol. 7, no. 1, pp. 38–43, Jun. 2019, https://doi.org/10.24252/bio.v7i1.5990

[6] E. Onoja, S. Chandren, F. I. Abdul Razak, N. A. Mahat, and R. A. Wahab, "Oil palm (elaeis guineensis) biomass in Malaysia: The present and future prospects," *Waste and Biomass Valorization*, vol. 10, no. 8. Springer Netherlands, pp. 2099–2117, Aug. 01, 2019. https://doi.org/10.1007/s12649-018-0258-1

[7] H. A. Umar, S. A. Sulaiman, R. K. Ahmad, and S. N. Tamili, "Characterisation of oil palm trunk and frond as fuel for biomass thermochemical," in *IOP Conference Series: Materials Science and Engineering*, Jun. 2020, vol. 863, no. 1. https://doi.org/10.1088/1757-899X/863/1/012011

[8] P. Shrivastava, P. Khongphakdi, A. Palamanit, A. Kumar, and P. Tekasakul, "Investigation of physicochemical properties of oil palm biomass for evaluating potential of biofuels production via pyrolysis processes," *Biomass Convers Biorefin*, vol. 11, no. 5, pp. 1987–2001, Oct. 2021, https://doi.org/10.1007/s13399-019-00596-x

[9] M. Akia, F. Yazdani, E. Motaee, D. Han, and H. Arandiyan, "A review on conversion of biomass to biofuel by nanocatalysts," 2014.

[10] T. A. Ishola, E. A. Alhassan, R. A. Busari, and J. K. Abioye, "Development of a biomass pyrolyser for sawdust and rice husks conversion to energy," *Journal of Agricultural Engineering and Technology (JAET)*, vol. 26, no. 1, pp. 77–87, 2021, [Online]. Available: www.niae.net

[11] A. H. Jagaba *et al.*, "Removal of nutrients from pulp and paper biorefinery effluent: Operation, kinetic modelling and optimization by response surface methodology," *Environ Res*, vol. 214, p. 114091, Nov. 2022, https://doi.org/10.1016/j.envres.2022.114091

[12] A. H. Jagaba *et al.*, "A systematic literature review on waste-to-resource potential of palm oil clinker for sustainable engineering and environmental applications," *Materials*, vol. 14, no. 16, p. 4456, Aug. 2021, https://doi.org/10.3390/ma14164456

[13] K. E. Anyaoha, R. Sakrabani, K. Patchigolla, and A. M. Mouazen, "Critical evaluation of oil palm fresh fruit bunch solid wastes as soil amendments: Prospects and challenges," *Resour Conserv Recycl*, vol. 136, pp. 399–409, Sep. 2018, https://doi.org/10.1016/j.resconrec.2018.04.022

[14] H. A. Umar, S. A. Sulaiman, M. A. B. M. Said, and R. K. Ahmad, "Palm kernel shell as potential fuel for syngas production," in *Lecture Notes in Mechanical Engineering*, 2020, pp. 263–273. https://doi.org/10.1007/978-981-15-5753-8_25

[15] L. J. Roberts *et al.*, "The Impact of Aluminosilicate-based Additives upon the Sintering and Melting Behaviour of Biomass Ash," *Biomass Bioenergy*, vol. 127, p. 105284, Aug. 2019, https://doi.org/10.1016/J.BIOMBIOE.2019.105284

[16] G. Wang *et al.*, "Thermal behavior and kinetic analysis of co-combustion of waste biomass/low rank coal blends," *Energy Convers Manag*, vol. 124, pp. 414–426, Sep. 2016, https://doi.org/10.1016/j.enconman.2016.07.045

[17] S. Konsomboon, S. Pipatmanomai, T. Madhiyanon, and S. Tia, "Effect of kaolin addition on ash characteristics of palm empty fruit bunch (EFB) upon combustion," *Appl Energy*, vol. 88, no. 1, pp. 298–305, 2011, https://doi.org/10.1016/j.apenergy.2010.07.008

[18] J. George, P. Arun, and C. Muraleedharan, "Experimental investigation on co-gasification of coffee husk and sawdust in a bubbling fluidised bed gasifier," *Journal of the Energy Institute*, vol. 92, no. 6, pp. 1977–1986, Dec. 2019, https://doi.org/10.1016/j.joei.2018.10.014

[19] L. Febrero, E. Granada, D. Patiño, P. Eguía, and A. Regueiro, "A comparative study of fouling and bottom ash from woody biomass combustion in a fixed-bed small-scale boiler and

evaluation of the analytical techniques used," *Sustainability (Switzerland)*, vol. 7, no. 5, pp. 5819–5837, 2015, https://doi.org/10.3390/su7055819

[20] N. Yub Harun, T. Jin Han, T. Vijayakumar, A. Saeed, and M. Afzal, "Ash Deposition Characteristics of Industrial Biomass Waste and Agricultural Residues," 2019. [Online]. Available: www.sciencedirect.comwww.materialstoday.com/proceedings2214-7853

[21] T. Wang *et al.*, "Regulation of ash slagging behavior for sewage sludge by rice husk addition: Focusing on control mechanisms," *J Clean Prod*, vol. 284, Feb. 2021, https://doi.org/10.1016/j.jclepro.2020.124677

[22] L. C. Peres, A. de C. O. Carneiro, C. G. Figueiró, L. D. F. Fialho, M. F. Gomes, and B. M. D. R. T. Valente, "Clonal Selection of Corymbia for Energy and Charcoal Production," *Advances in Forestry Science*, vol. 6, no. 3, pp. 749–753, Nov. 2019, https://doi.org/10.34062/afs.v6i3.8293

[23] Y. Zhao, B. Ren, A. O'Brien, and S. O'Toole, "Using Alum Sludge for Clay Brick: An Irish Investigation," *Environmental Science*, vol. 73, no. 5, pp. 719–730, Sep. 2016, https://doi.org/10.1080/00207233.2016.1160651

[24] S. F. Osman, M. M. Hashim, A. Sulaiman, and M. Y. Abdullah, "Hot water treatment of oil palm decanter cake (OPDC) for bio-organic media of oil palm seedlings," *Konvensyen Kebangsaan Kejuruteraan Pertanian Dan Makanan*, pp. 22–26, 2019.

[25] A. Embrandiri, P. F. Rupani, S. A. Ismail, R. P. Singh, M. H. Ibrahim, and M. O. b. Abd. Kadir, "The effect of oil palm decanter cake on the accumulation of nutrients and the stomatal opening of solanum melongena (brinjal) plants," *International Journal of Recycling of Organic Waste in Agriculture*, vol. 5, no. 2, pp. 141–147, Jun. 2016, https://doi.org/10.1007/s40093-016-0124-8

[26] J. Y. Teoh *et al.*, "Co-pyrolysis of Palm Empty Fruit Bunch and Palm Kernel Shell with Palm Oil Mill Effluent (POME) Sludge," *International Journal of Biomass & Renewables*, vol. 9, no. 1, pp. 1–9, 2020.

[27] Y. Wang, J. Jin, D. Liu, H. Yang, and S. Li, "Understanding Ash Deposition for the Combustion of Zhundong Coal: Focusing on Different Additives Effects," *Energy and Fuels*, vol. 32, no. 6, pp. 7103–7111, Jun. 2018, https://doi.org/10.1021/acs.energyfuels.8b00384

[28] M. Y. Soleha *et al.*, "Characterization of raw and thermally treated alum sludge," in *Key Engineering Materials*, 2016, vol. 701, pp. 138–142. https://doi.org/10.4028/www.scientific.net/KEM.701.138

[29] H. A. Umar *et al.*, "Assessing the implementation levels of oil palm waste conversion methods in Malaysia and the challenges of commercialisation: Towards sustainable energy production," *Biomass Bioenergy*, vol. 151, Aug. 2021, https://doi.org/10.1016/j.biombioe.2021.106179

[30] S. v. Vassilev, D. Baxter, and C. G. Vassileva, "An overview of the behaviour of biomass during combustion: Part II. Ash fusion and ash formation mechanisms of biomass types," *Fuel*, vol. 117, no. PART A, pp. 152–183, 2014, https://doi.org/10.1016/j.fuel.2013.09.024

[31] H. Rong, T. Wang, M. Zhou, H. Wang, H. Hou, and Y. Xue, "Combustion characteristics and slagging during co-combustion of rice husk and sewage sludge blends," *Energies (Basel)*, vol. 10, no. 4, Mar. 2017, https://doi.org/10.3390/en10040438

[32] R. I. Mainil and Y. Matsumura, "Effect of heating rate on gasification and phosphorus recovery for palm oil mill effluent in supercritical water," *Journal of Supercritical Fluids*, vol. 173, Jul. 2021, https://doi.org/10.1016/j.supflu.2021.105217

Absorption of acid gases (CO_2, H_2S) from natural gas using a ternary blend of N-methyldiethanolamine (MDEA), 2-amino-2-methyl-1-propanol (AMP), and Sulfolane

ABID Salam Farooqi[1,2,a*], RAIHAN Mahirah Binti Ramli[1,2,b*], SERENE Sow Mun Lock[1,3,c], AHMAD Salam Farooqi[1,d], SYED Muhammad Wajahat[1,e] and SIDRA Anis Farooqi[4,f]

[1]Department of Chemical Engineering, Universiti Teknologi PETRONAS, Bandar Seri Iskandar, 32610 Seri Iskandar, Perak, Malaysia

[2]Centre of Innovative Nanostructures & Nanodevices, Universiti Teknologi PETRONAS, Bandar Seri Iskandar, 32610 Seri Iskandar, Perak, Malaysia

[3]CO_2 Research Center (CO2RES), Universiti Teknologi PETRONAS, 32610 Seri Iskandar, Malaysia

[4]Soochow Institute for Energy and Materials Innovations, College of Energy, Soochow University, Suzhou 215006, China

[a]abid_20001276@utp.edu.my, [b]raihan.ramli@utp.edu.my, [c]sowmun.lock@utp.edu.my, [d]ahmad_1700920@utp.edu.my, [e]syed_22000103@utp.edu.my, [f]sidrafarooqi66@gmail.com

Keywords: Absorption, Simulation, Aspen HYSYS, Amines

Abstract. Natural gas (NG) must be treated to remove sulphur compounds and acid gases i.e., carbon dioxide (CO_2) and hydrogen sulfide (H_2S) to ensure that it complies with requirements for sale and transportation. More than 95% of the NG processing plants are operated through the acid gas removal unit (AGRU) using aqueous amine solvent in removing sour gas components from the hydrocarbon gas due to the availability of amine solvent at a low cost. However, the main limitation of this process is the high operating cost of providing sufficient thermal energy at the reboiler for solvent regeneration. Meanwhile, the reboiler duty requirement generally increases with the requirement of CO_2 removal efficiency as higher energy consumption is required to strip off a more significant amount of CO_2 from the rich solvent. This current study addresses the absorption performance of acid gases using ternary hybrid solvents of MDEA, AMP, and Sulfolane. A study on the effects of solubility on H_2S and CO_2 absorption was performed at varying pressure (1000-6000 kPa) and temperatures (25°C-50°C) using Aspen HYSYS®V12.1. The results revealed that the concentration of CO_2 and H_2S in sweet gas increased with the decrease in pressure, while increasing temperature increased the concentration of H_2S and CO_2 in sweet gas. The future study will look at the reboiler duty required for solvent regeneration using this ternary blend of MDEA, AMP, and Sulfolane.

Introduction

For the past few decades, fossil fuels have become the primary source of electricity generation worldwide due to their high heating value. In 2021, nearly 80% of the electricity production in Malaysia will be produced from the combustion of fossil fuels of which 43.6% is contributed by natural gas (NG), followed by 31.1% from coal [1]. NG is one of the cleanest fossil fuels compared to other hydrocarbon deposits such as coal and oil [2]. NG consists primarily of Methane (CH_4), the primary component, but it also contains heavier hydrocarbons, acid gases (CO_2, H_2S), water, mercury, and inert gases in different proportions. These impurities must be removed to obtain rich amine NG and comply with emissions regulations and sales gas specifications [3]. Among the non-hydrocarbon components in NG, CO_2 and H_2S may severely damage the environment and industrial equipment. For example, in water (H_2O), CO_2 and H_2S can form an acidic solution, contributing to pipeline corrosion problems [4].

Table 1 shows the typical NG composition in the Malaysian gas field. In Malaysia, most of the NG reserves are found to be sour gas fields whereas sour gas refers to NG that consists of a high amount of CO_2 and H_2S. Typically, the NG reserves in Malaysia are high in CO_2 content which ranges between 28% and 87%, and relatively low H_2S content of less than 1% [5]. Therefore, this shows that removing high CO_2 content from raw NG has always been a significant challenge for Malaysian oil and gas operators to meet the sales of gas specifications.

Table 1:Typical composition of NG in Malaysia.

Components	Composition (%)
Methane	40-50
Ethane	5-10
Propane	1-5
Carbon dioxide	20-30
Hydrogen Sulphide	0-1

Many technologies have been used to separate acid gases from NG, including distillation, absorption, adsorption and membrane separation. Out of the mentioned technologies, absorption using alkanolamines is the most effective and widely used on an industrial scale to remove CO_2 and H_2S. However, due to this process maturity and economic feasibility, more than 95% of the gas processing plants are operated through chemical absorption using an aqueous amine solvent [6]. The common alkanolamines used in the current industry are monoethanolamine (MEA), diglycolamine (DGA), diethanolamine (DEA), and diisopropanolamine (DIPA) and methyldiethanolamine (MDEA). Usually, the reaction between H_2S and amines is instantaneous. However, for CO_2 different amines have different reactivity and selectivity. MEA and DEA amines react with CO_2 to form carbamate while MDEA is hindered from reacting with CO_2 due to nitrogen atoms in the structure. MDEA is thus said to react selectively with H_2S in the presence of CO_2. MDEA is more favorable in industries because of its loading capacity of 1 mole of CO_2 per mole of the amine as compared to 0.5 moles of CO_2 per mole of amine for MEA and DEA, but the main disadvantage of MDEA is that its rate of reaction and reaction kinetics are very slow as compared to primary and secondary amine [7].

To overcome this problem, an activator typically primary, secondary, and hindered amines like MEA, DEA, 2-amino-2-methyl-1-propanol (AMP), piperazine (PZ), and DIPA is introduced to improve the rate of carbamate hydrolysis and dissolved CO_2 absorption kinetics of tertiary amines. Hindered amines such as AMP have high CO_2 and H_2S absorption rates and high absorption capacity. Therefore, activator AMP is added to increase the reaction rate and the loading capacity of MDEA [8]. Despite the maturity of the technology in AGRU using aqueous amine absorption since the 1960s, one of the major challenges in this process is the high reboiler duty necessary to regenerate the solvent for continuous operation. In most cases, up to 70% of the operating costs excluding the labor force result from the regeneration [9]. Contrary, physical solvents are also one of the alternatives used to reduce reboiler duty for solvent regeneration since the acid gases in AGRU can be stripped off by reducing the pressure without giving extra heat. Removal of acid gases using physical solvents is usually beneficial at a high partial pressure of acid gases while a lower pressure system favors absorption by chemical solvents. Typical physical solvents are Sulfolane, N-methyl 2- pyrrolidone (NMP), and methanol [10].

To eliminate contaminants, a hybrid solvent, a combination of physical and chemical solvents was used as an alternative solvent in this case. In physical solvents, the solubility of acid gas and

Sulphur compounds is nearly linear. Therefore, mixing chemical and physical solvents gives advantages at the same time. Sulfinol, a combination of DIPA, tetramethylene sulfone, and H_2O was the first mixed solvent employed. MDEA eventually substituted DIPA, and the solvent was renamed Sulfinol-M [11].

In the past, numerous researchers depicted MDEA-based solutions in different gas-sweetening industries. Nejat et al. [12] performed the exergy and energy analysis on the khangrian sweetening plant using MDEA as a solvent. He proposed that replacing MDEA with a hybrid solution of MDEA-sulfolane required less energy than aqueous MDEA. Ghanbarabadi et al. [13] evaluated and contrasted the practicability of sulfolane-MDEA-water, DGA, MDEA-AMP, and MDEA for CO_2, H_2S, and other components. Their findings demonstrate that a sulfolane-MDEA-water solution with a lower flow rate absorbs more than 25-35% of mercaptans and acid gases, and regeneration requires 15-20% more energy. Sarker et al. [14] discovered that employing a blend of amines for natural gas sweetening produces better results than using a single amine.

The primary objective of this investigation is to capture both CO_2 and H_2S from NG using a hybrid amine blend of MDEA, AMP, and sulfolane. Simulation software Aspen HYSYS V12.1 was used to build a gas-sweetening plant. The novelty of this work is that no study has been done so far on the high content of CO_2 (40%) and H_2S (1%) at high pressure using a hybrid solvent. Different operating parameters affecting the absorption process i.e., the effect of temperature (25-50°C), pressure (1000-6000 kPa) and solvent flow rate at different concentrations were then observed on the composition of sweet gas using an aqueous alkanolamine solvent.

Methodology

Process Description

The inlet separator was used to eliminate soluble, heavy hydrocarbons as liquid, suspended liquid, and entrained hydrocarbons from the incoming feed gas. The CO_2 was then brought into contact with the amine solvent by the evacuated gas from the top of the inlet separator entering the bottom of the absorption tower. Sour gas enters the absorber from the bottom while solvent enters from the top. The sour gas enters an absorber at 35°C with a flow rate of 25965 kgmole/h. Specification of the simulation of AGRU is given in Table 2. The lean amine solvent enters at a temperature 42°C higher than the input feed gas and departs at the bottom of the absorption tower after passing the counter-current through the packed column. The solvent temperature is at least 5°C higher than the temperature of feed gas to avoid condensate forming.

Table 2: Input specification for Aspen Hysys simulation.

Absorber Specifications	
Solvent temperature (°C)	42
Inlet gas temperature (°C)	35
Solvent pressure (kPa)	5220
inlet gas pressure (kPa)	5210
inlet gas flow rate (kgmol/h)	25965
Number of stages	23
CO_2 in inlet gas (mol.%)	45
H_2S in inlet gas (mol.%)	1
Solvent feed tray	1

After the absorption of acid gases in the absorber, the output gas becomes hotter because of the exothermic reactions inside the absorber. It exits the tower from the top as sweet gas, while rich amine containing CO_2 and H_2S goes from the bottom of the absorption column at 57.3°C. Figure 1 shows the complete flow diagram of AGRU built on Aspen Hysys Simulation software. Before entering the stripping or regenerator column, to remove soluble hydrocarbons from rich amine, in the flash tank the pressure was reduced to 656 kPa and then heated again to 95°C. The heat from a steam reboiler generates vapor in the stripping column, which removes CO_2 and H_2S from the rich amine as it passes up the column. The stream that carries the lean amine leaves the column from the bottom and exchanges heat with the exchanger. Then this lean amine is sent to a cooler where it is cooled to 40°C before returning to the absorber. The acid gas can be expelled from the top of the stripping column.

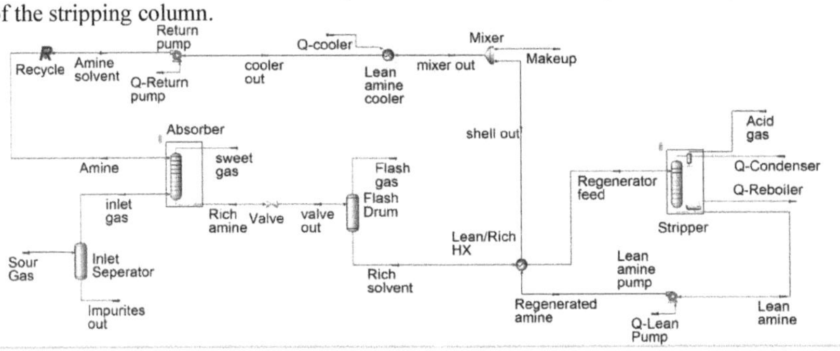

Figure 1: Aspen HYSYS flow sheet of AGRU using amine solvents.

Simulation basis

For the modeling of the column, the two most prominent methods are the rate-based model and the equilibrium model. The equilibrium model assumes that the liquid and vapors leaving at each column level are in equilibrium. Contrary, the rate bases model uses transport characteristics and concentration gradient to study heat and mass transfer happening between the contracting phase. That's why the output of the rate-based model is closer to actual plant data, allowing for estimating different parameters under different conditions. The removal of acid gases using chemical absorption is based on an acid-base reaction. CO_2 and H_2S are acid gases as they dissociate to form a weak acidic solution in an aqueous solution [15]. The reaction paths undergone by the acid gases with alkanolamines strongly depend on the structural characteristics of the alkanolamines. The reactions for CO_2 and H_2S are both exothermic where the aqueous alkanolamine solution absorbs the acid gases at a relatively low temperature with the release of heat energy.

Formation of amine hydrosulphide from the reaction between H_2S and amine solution:

$$\mathrm{RR'R''N + H_2S \leftrightarrow RR'R''N\,H^+ + HS^-} \tag{1}$$

Formation of amine carbamate via zwitterion mechanism for the primary and secondary amines:

$$\mathrm{RR'R''N + CO_2 \leftrightarrow RR'R''N\,H^+COO^-} \tag{2}$$

Hydrolysis and dissociation of CO_2 for tertiary amine:

$$\mathrm{RR'R''N + CO_2 + H_2O \leftrightarrow RR'R''N\,H^+ + HCO_3{}^-} \tag{3}$$

For the removal of H_2S, it reacts instantaneously with all types of amines regardless of the structural characteristics via a direct proton transfer reaction as given in Eq. 1. Meanwhile, CO_2 can react with the amine solution in two different reaction mechanisms depending on the amine structure. Primary and secondary amines are very reactive where they react with CO_2 to form carbamates via direct reaction as shown in Eq. 2. However, since tertiary amine has no N-H bond.

Therefore, it can only form a protonated amine and bicarbonate ion through hydrolysis and dissociation of CO_2 given by Eq. 3. Hence, compared to primary and secondary amines, tertiary amine has a lower CO_2 absorption rate due to the relatively slow reaction of carbonic acid dissociation to the bicarbonate [16]. Since the reaction of H_2S with amine occurs almost instantaneously, it can be said that the H_2S absorption process takes place in the gas phase, and a relatively slow reaction of CO_2 with amine occurs in the liquid phase [17].

Result & Discussion

Validation of actual plant data

AGRU model was developed using Aspen HYSYS® V12.1 simulation software and validated against the actual plant data from a gas processing plant in Malaysia to ensure the reliability and applicability of the results on the specific plant setup. Figure 1 shows the developed ARGU model in the Aspen HYSYS simulation software based on the actual plant data and equipment specification (Table 2). In contrast, Table 3 presents the percentage deviation between the simulation model and the real plant data for the sweet gas stream is less than 5% [18].

Table 3: Validation of actual plant data with simulation data.

Parameters	Actual plant data	Simulation model	Deviation (%)
Temperature (°C)	44.40	45.18	1.73
Pressure (kPa)	5140	5210	1.34
Flowrate (kgmol/h)	25965	25302	2.51

Effect of temperature on the absorption of CO_2 & H_2S

The sensitivity analysis was performed at six different temperatures ranging from 25°C-50°C to study the effect of absorption of acid gases. Figure 2 depicts the influence of temperature on CO_2 and H_2S concentrations in sweet gas. The amine solvent loading capacity drops considerably as the system temperature rises, implying that CO_2 and H_2S concentrations also rise. The same trends were seen in previous studies: lower temperatures always enhance absorption, and higher temperatures inhibit absorption. This is because the diffusion of gas molecules in chemical solvents is reduced at higher temperatures. Another factor could be increased temperature reduced absorption for both acid gases, namely H_2S and CO_2, and vice versa. This is due to the gas molecules being activated, which raises their average kinetic energy and causes the molecules of the absorbate to move through the absorbent more quickly, lowering the contact time. For both physical and chemical absorption, the better the absorption the longer the contact time between the absorbent and the absorbate.

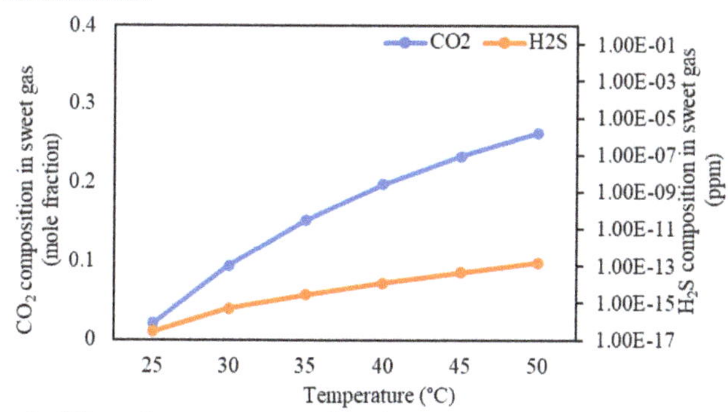

Figure 2: Effect of temperature on CO_2 & H_2S Concentration in the sweet gas.

Effect of pressure on the absorption of CO_2 & H_2S

Over a wide pressure range, the effect of operational pressure on the aqueous amine solvent absorption ability was examined (1000-6000 kPa). To establish the chemical solvent ability to absorb CO_2 gas efficiently, the absorption performance of the hybrid solvents was assessed in terms of loading capacity, which is an essential attribute in the absorption process. According to the simulation investigation, the higher pressure of the system promoted the absorption process. Figure 3 shows how pressure substantially impacts CO_2 and H_2S concentrations in sweet gas. As the pressure increases the solubility of CO_2 and H_2S increases in amine solvent. The better loading capacity could be attributable to two factors. First, increased pressure promotes absorption. Secondly, physical diffusion occurs at higher pressure. The increased pressure aids the diffusion of CO_2 gas in the amine solvent aqueous solution. Higher pressure causes more CO_2 gas molecules to diffuse into the aqueous solution, increasing the solvent loading capacity. According to a literature review, higher pressure encourages the absorption process, and chemical solvent uptake capability improves when the partial pressure of CO_2 gas increases.

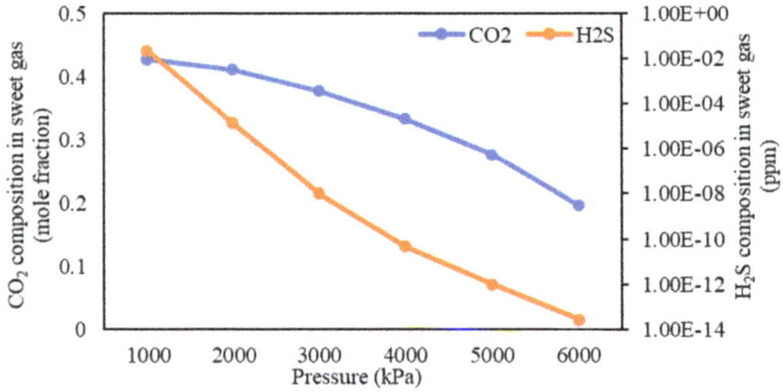

Figure 3: Effect of pressure on CO_2 & H_2S Concentration in the sweet gas.

Effect of solvent flow rate on the absorption of CO_2 and H_2S

One of the most critical factors which affect the operating cost of AGRU is the solvent flow rate. The effect of solvent flow rate was observed by keeping the chemical solvents (MDEA and AMP) composition constant while changing the concentration of sulfolane. The results reveal that by increasing the concentration of sulfolane less flow rate was required to achieve the allowable specifications of acid gases in the sweet gas stream. This is because the physical solvents are not constrained by any absorption limits, as the partial pressure of the gases is directly affected the absorption capacity of the system. The concentration of chemical amines MDEA and AMP are kept constant at 20% and 5% respectively. As shown in Figure 4, as the sulfolane concentration was increased less flow rate was required for the absorption of acid gases below 3% CO_2 and 4 ppm H_2S. The results depicted that the solvent containing 40% sulfolane requires a 35% less flow rate than the solvent containing 20% sulfolane to achieve the allowable limit for CO_2.

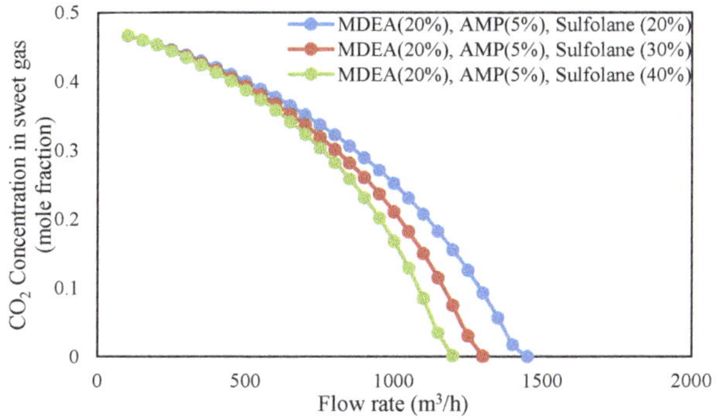

Figure 4: Effect of flow rate on the absorption of CO_2 in sweet gas.

Conclusion

The overall CO_2 and H_2S absorption in a ternary hybrid mixture of MDEA+AMP+Sulfolane was investigated using Aspen HYSYS®V12.1 simulation software. Operating parameters i.e., Pressure and temperature, and their effect on the concentration of CO_2 and H_2S in sweet gas were studied. It was investigated that the pressure positively impacts the CO_2 and H_2S concentration in the sweet gas. At the same time, temperature shows a negative effect as the temperature increases the concentration of CO_2 and H_2S increases in the sweet gas. In addition, increasing the percentage of sulfolane (40%) in a solvent requires a 35% less flow rate than Sulfolane (20%). This solubility data will be further used to study the effect of solvent circulation rate and concentration of amines for solvent regeneration in the stripper column to lessen the reboiler duty.

Acknowledgment

The authors acknowledge financial support from Universiti Teknologi PETRONAS, Malaysia, under the Yayasan Universiti Teknologi PETRONAS (YUTP) grant number 015LC0-293.

Reference

[1] "Global Energy Review: CO2 Emissions in 2021 - Analysis - IEA." https://www.iea.org/reports/global-energy-review-co2-emissions-in-2021-2.

[2] B. Obama, "The irreversible momentum of clean energy: Private-sector efforts help drive decoupling of emissions and economic growth," Science (80-.)., vol. 355, no. 6321, pp. 126-129, Jan. 2017, https://doi.org/10.1126/science.aam6284

[3] A. S. Farooqi et al., "A comprehensive review on improving the production of rich-hydrogen via combined steam and CO2 reforming of methane over Ni-based catalysts," Int. J. Hydrogen Energy, vol. 46, no. 60, pp. 31024-31040, Sep. 2021, https://doi.org/10.1016/j.ijhydene.2021.01.049

[4] A. S. Farooqi, R. M. Ramli, S. S. M. Lock, N. Hussein, M. Z. Shahid, and A. S. Farooqi, "Simulation of Natural Gas Treatment for Acid Gas Removal Using the Ternary Blend of MDEA, AEEA, and NMP," Sustain. 2022, Vol. 14, Page 10815, vol. 14, no. 17, p. 10815, Aug. 2022, https://doi.org/10.3390/su141710815

[5] L. S. Tan, K. K. Lau, M. A. Bustam, and A. M. Shariff, "Removal of high concentration CO2 from natural gas at elevated pressure via absorption process in packed column," J. Nat. Gas Chem., vol. 21, no. 1, pp. 7-10, Jan. 2012, https://doi.org/10.1016/S1003-9953(11)60325-3

[6] M. Torabi Angaji, H. Ghanbarabadi, and F. Karimi Zad Gohari, "Optimizations of Sulfolane concentration in propose Sulfinol-M solvent instead of MDEA solvent in the refineries of Sarakhs," J. Nat. Gas Sci. Eng., vol. 15, pp. 22-26, Nov. 2013, https://doi.org/10.1016/j.jngse.2013.08.003

[7] N. Dave, T. Do, G. Puxty, R. Rowland, P. H. M. Feron, and M. I. Attalla, "CO2 capture by aqueous amines and aqueous ammonia-A Comparison," Energy Procedia, vol. 1, no. 1, pp. 949-954, Feb. 2009, https://doi.org/10.1016/j.egypro.2009.01.126

[8] A. T. Zoghi, F. Feyzi, and S. Zarrinpashneh, "Experimental investigation on the effect of addition of amine activators to aqueous solutions of N-methyldiethanolamine on the rate of carbon dioxide absorption," Int. J. Greenh. Gas Control, vol. 7, pp. 12-19, Mar. 2012, https://doi.org/10.1016/j.ijggc.2011.12.001

[9] N. M. A. Al-Lagtah, S. Al-Habsi, and S. A. Onaizi, "Optimization and performance improvement of Lekhwair natural gas sweetening plant using Aspen HYSYS," J. Nat. Gas Sci. Eng., vol. 26, pp. 367-381, Sep. 2015, https://doi.org/10.1016/j.jngse.2015.06.030

[10] M. I. Stewart, Gas Sweetening. 2014. https://doi.org/10.1016/B978-0-12-382207-9.00009-3

[11] A. S. Farooqi, R. M. Ramli, S. S. M. Lock, N. Hussein, A. S. Farooqi, and S. M. Wajahat, "Simulation of acid gas removal unit using DIPA+TEA amine solvent," IOP Conf. Ser. Mater. Sci. Eng., vol. 1257, no. 1, p. 012033, Oct. 2022, https://doi.org/10.1088/1757-899X/1257/1/012033

[12] T. Nejat, A. Movasati, D. A. Wood, and H. Ghanbarabadi, "Simulated exergy and energy performance comparison of physical-chemical and chemical solvents in a sour gas treatment plant," Chem. Eng. Res. Des., vol. 133, pp. 40-54, May 2018, https://doi.org/10.1016/j.cherd.2018.02.040

[13] H. Ghanbarabadi and B. Khoshandam, "Simulation and comparison of Sulfinol solvent performance with Amine solvents in removing sulfur compounds and acid gases from natural sour gas," J. Nat. Gas Sci. Eng., vol. 22, pp. 415-420, Jan. 2015, https://doi.org/10.1016/j.jngse.2014.12.024

[14] N. K. Sarker, "Theoretical effect of concentration, circulation rate, stages, pressure and temperature of single amine and amine mixture solvents on gas sweetening performance," Egypt. J. Pet., vol. 25, no. 3, pp. 343-354, Sep. 2016, https://doi.org/10.1016/j.ejpe.2015.08.004

[15] V. Abkhiz and I. Heydari, "Comparison of amine solutions performance for gas sweetening," Asia-Pacific J. Chem. Eng., vol. 9, no. 5, pp. 656-662, Sep. 2014, https://doi.org/10.1002/apj.1795

[16] S. Mitra, "A Technical Report on Gas Sweetening by Amines," Sr. Process Eng. Petrofac Eng., no. June, pp. 1-58, 2015.

[17] P. M. M. Blauwhoff, G. F. Versteeg, and W. P. M. Van Swaaij, "A study on the reaction between CO2 and alkanolamines in aqueous solutions," Chem. Eng. Sci., vol. 38, no. 9, pp. 1411-1429, 1983, https://doi.org/10.1016/0009-2509(83)80077-3

[18] U. Zahid, F. N. Al Rowaili, M. K. Ayodeji, and U. Ahmed, "Simulation and parametric analysis of CO2 capture from natural gas using diglycolamine," Int. J. Greenh. Gas Control, vol. 57, pp. 42-51, Feb. 2017, https://doi.org/10.1016/j.ijggc.2016.12.016

Materials Research Proceedings 29 (2023) 17-27 https://doi.org/10.21741/9781644902516-3

Syngas production via combined dry and steam reforming methane over Ni-based catalyst: A review

SYED MUHAMMAD WAJAHAT ul Hasnain[1,2,a], AHMAD Salam Farooqi[1,2,b], BAMIDELE Victor Ayodele[1,c], and BAWADI Abdullah[1,2,d*]

[1]Department of Chemical Engineering, Universiti Teknologi PETRONAS, Seri Iskandar, 32610, Perak, Malaysia

[2]Centre of Contamination Control and Utilization (CenCoU), Universiti Teknologi PETRONAS, Seri Iskandar, 32610, Perak, Malaysia

[a]syed_22000103@utp.edu.my, [b]ahmad_17007920@utp.edu.my, [c]bamidele.ayodele@utp.edu.my, [d]bawadi_abdullah@utp.edu.my

Keywords: Catalyst Development, Ni-Based Catalysts, Syngas Production, CDSRM

Abstract. Global energy consumption has eventually increased as a result of the growing world population. Various problems arise as a result. The accumulation of greenhouse gases (GHGs), which led to a shift in the world's climate, is the most problematic. Combined dry and steam reforming of methane (CDRSM) is a highly advantageous method since it uses two of the most significant GHGs, CH_4 and CO_2, to produce syngas, an intermediate product to produce valuable fuels. Ni-based catalysts are inexpensive, compared to many noble metals, and exhibit good reaction activity. However, deactivation, coking, and sintering of catalysts continue to be the major obstacles to commercialization. Due to better and more stable catalytic structure, which is both coke and sintering resistant at high temperatures, bimetallic catalysts have established increased activity and prolonged durability when compared to monometallic catalysts. This review highlights recent advancements in Ni-based catalysts for CDSRM by emphasizing factors such as catalyst support, bimetallic catalyst, promoters, and strong metal-support interactions (SMSI).

Introduction

Advancement in the quality of life, which is directly related to the amount of energy consumed, has accelerated the increase in global energy consumption. The industrialization has transformed the power generation scenario over the years due to rising energy demand [1]. Many of the world's energy needs are currently met by burning fossil fuels (natural gas, coal, petroleum products, etc.). They have become a pivotal and fundamental element of modern society, relying on them more and more ever since industrialization, although the oil and gas prices are much higher than in the past [2]. Wind, geothermal, solar, and other renewable energy sources contribute only a small fraction of the total energy demand. It is also likely that energy consumption will increase further in the future since it has been increasing in recent years [3]. The current reliance on fossil fuels such as coal, natural gas (NG), and oil to meet energy demand has resulted in the production of hazardous greenhouse gases (GHGs), which has caused environmental problems [4]. GHGs with the highest abundance are methane and CO_2, and they are the key contributors to the present climate challenges. In recent decades, CH_4 emissions in the atmosphere have increased, making it a major contributor to increased global warming [5]. Earth System Research Laboratory (ESRL) [6] data show that worldwide methane and carbon dioxide concentrations rose from 1650 parts per billion and 345.36 parts per million in 1985 to 1884.7 parts per billion and 412.31 parts per million, respectively, in September 2021 [7]. These two gases boost the greenhouse effect by absorbing and radiating the energy. GHG emissions will affect the Earth's climate and degrade natural ecosystems, resulting in huge socio-economic losses [8]. Hence, reducing GHG emissions into the atmosphere and synthesizing a clean fuel is a main priority. Furthermore, efficacious technologies for converting CH_4 and CO_2 into useful products, mainly syngas, are required [9].

For the creation of clean fuel, like hydrogen, natural gas needs to undergo a catalytic process described as reforming. Reforming is the most usual method for producing syngas, which is the intermediate product to produce liquid fuels, in the industry and it involves one of three reforming processes. Steam reforming (SRM), partial oxidation (POM), and dry reforming of methane (DRM) are the most frequent methods used in industry to make syngas [10]. But every process produces a different ratio of syngas 3, 2, and 1 respectively [11]. The stoichiometric ratio required for the Fischer Tropsch synthesis to produce methanol is 2 [12]. Only, POM produces syngas with a ratio of 2. However, it has various disadvantages like hotspot formation, high explosion risk, and the requirement of cryogenic unit [13]. Among all other processes, combined dry and steam reforming methane methane, which includes both SRM and DRM, has recently sparked interest within the research and educational communities in gaining benefits while minimizing the disadvantages of each of the three fundamental methods [14].

CDSRM has ability to decrease the formation of carbon deposition, easier modification of the H_2/CO ratio by varying the feed composition and high catalytic stability in the coexistence of CO_2 and H_2O oxidizing reactants [15]. It will help to reduce the coke formation by the oxidation of the carbon deposited on the surface. CDSRM, on the other hand, is a complicated reaction since it involves two reforming reactions, secondary reactions, and coke production reactions all at the same time [16]. The reaction pathway is highly dependent on reaction variables such temperature, feed composition, and pressure [17]. Excess steam has been shown to be useful in preventing the formation of coke, but it also raises the expense of the operating process. Due to endothermicity, CDSRM requires elevated reaction temperatures and catalyst helps in reducing the reaction's activation energy and increases the conversion rate [18]. However, the deposition of carbon on the catalyst's surface deactivates it to a large extent, causing harm to the catalyst's stability . The sintering of the active catalyst phase, the production of inactive metal oxide, and the unfavorable interaction with oxide support are all factors that affect catalyst performance [19].

A lot of studies have been done for the development of an efficient catalyst for the CDSRM [20] that has good selectivity and catalytic performance and is resistant to coke formation. Noble metals (Ru, Rh, Pd, Ir, and Pt) are enormously active and resistant to the development of carbon whiskers, however, carbon deposition can be reduced most effectively by Ru and Rh, according to researchers. However, apart from their high activity, noble metals are not used industrially because of their high cost [21]. There are a number of existing catalysts, but Ni-based catalysts are currently the most commonly used in industrial processes. The CDSRM process will require a highly active catalyst that has improved metal support interactions and is resilient to carbon formation in order to be commercialized [22]. The goal of this review is to highlight recent new developments in the synthesis of appropriate catalysts for CDSRM done in the last ten years. A comprehensive analysis of numerous emerging methods and measures used to stabilize catalysts is presented, along with associated data revealing activity and stability levels.

Ni-based catalysts for CDSRM reaction

Nowadays, the use of nickel as an active site rather than noble metals is the main focus of catalyst development due to its high intrinsic activity, lower cost, and abundancy. [23]. However, the stability of Ni-based catalysts in harsh conditions is a significant problem because coke deposition and thermal sintering on the catalyst's surface encourage deactivation [24]. Laosiripojana et al. [25] suggest that poisonous chemicals in reactants or products cause a rapid decrease in reforming potential over time. There have been various investigations performed so far that have investigated the nature of the support, the methods for preparing it and the addition of catalysts[26]. It is therefore essential to develop catalysts that resist carbonation, oxidative environments, and high temperatures in order to commercialize the process.

Irrespective of the fact that Ni-based catalysts are cheap and exhibit better performance that is comparable to expensive metallic catalysts, still thermal sintering and coke formation promote the deactivation of catalysts [27]. Since, the generation of carbon deposits is dependent on the operating conditions, namely the utilization of surplus steam or not during reactor feeding, and the nature of the catalyst and CDSRM uses low steam-carbon ratios in order to achieve the required the H_2/CO ratio, hence Ni-based catalyst deactivates due to carbon deposition [28]. In high-temperature CDSRM reactions, using a nickel-based catalyst with a high oxygen capacity or oxygen mobility is a useful strategy to prevent coke formation and improve catalytic performance. A lot of research emphasizes the use of a catalyst that has strong interaction between metal and support (MSI), redox properties and small particle size (<2 nm), and excellent metal dispersion, so as to prevent coke formation and catalyst sintering due to high temperature [29].

Effect of Catalyst Support. Catalysts typically consist of multiple components, which are constructed into the desired shape and structure. Materials embedded with active metal usually support metal catalysts. There are several functions that these support materials play in the catalyst's activity [30]. Metallic compounds can disperse over a large surface area, maximizing the surface area of active sites, which allows the coarse geometry of the catalyst to be tailored to the reactor [31]. Because of their interactions with active metal sites, these supports are typically inactive on their own but participate in the whole process when involved in a reaction [32]. Using Ni-Co in combination with supports containing strong Lewis basicity can limit the accumulation of carbon on metal active sites, resulting in an enhanced ability for the catalyst to absorb CO_2, resulting in a decrease in carbon generation by reverse Boudouard reaction, thereby altering the equilibrium concentration [33]. During the reforming reaction, catalyst-support interactions significantly influence carbon deposition over metal-support catalysts. According to a study by Roh et al. [34], adding La_2O_3 to Ni/Al_2O_3 improved the dispersion of Ni particles over the CDSRM's support surface. It proved advantageous to add La_2O_3 to Ni catalysts supported by SiO_2. The inclusion of La_2O_3 improved Ni dispersion, increasing its interaction with the support, and so promoting CO_2 activation.

Because of its high surface area, which makes it possible for metallic particles to be easily and uniformly dispersed on either internal or external surfaces, alumina is frequently used as a catalytic carrier in practically all reforming reactions [35]. Additionally, despite being chemically and physically stable, Al_2O_3 has a moderately acidic pH, which tends to encourage a side-reaction called methane breakdown that is more likely to occur on acidic sites, which tends to increase carbon deposition [36]. In a study conducted by Nakoua and Naas [37], the authors tested Ni/Al_2O_3 in a detachable reactor. The author noted that during the dry reforming of methane at 700°C, 1 atm, rapid carbon deposition was seen. However, a decrease in carbon deposition on the surface was seen when CO_2 reforming was carried out in conjunction with steam reforming reaction in thinner channels deposited with a thinner layer of catalyst.

Roh et al. used various supports for the Ni catalyst including Ni/ZrO_2, Ni/MgO, $Ni/MgO\text{-}Al_2O_3$, Ni/CeO_2, and $Ni\alpha/Al_2O_3$. The authors tested these catalysts for CCSRM to produce syngas with a ratio of 2.0. The catalyst was prepared by the wetness impregnation method. The feed ratio H_2O: CO_2: CH_4 of 2.4:1.2:3 was used at the temperature of 800 °C [38]. A study on zirconia-supported nickel (Ni/ZrO_2) catalysts with less than 2% metal content demonstrated that they were exceptionally durable at the stoichiometric ratio of CO_2 to CH_4. According to research by Bae et al. [39], co-impregnating Ni and mixed $CeO_2\text{-}ZrO_2$ oxides onto $MgAl_2O_4$ surfaces is significantly more important than post-impregnating Ni on $CeO_2\text{-}ZrO_2$-coated $MgAl_2O_4$ surfaces. Rezaei [40] presented similar findings by synthesizing Ni catalyst over nano-crystalline ZrO_2 using a surfactant-based method that produced high levels of CH_4 transformation. A strong interaction between the catalyst and the support plays a significant role in determining activity, stability and performance, as evidenced in the literature presented here.

Effect of Promoters. It is reported that the anti-coking and anti-sintering ability of the catalyst can be improved by using support along with promoters. The coke-resistance properties can be increased by introducing promoters such as alkali, alkaline earth, and metal oxides of rare-earth elements, such as lanthanum, magnesium, cerium, and calcium, and also diminish the active site's thermal sintering. In their study of Boron-doped Ni/SBA-15, Siang et al found that Boron promoted CO_2 adsorption by donating electrons to nearby Ni. This results in enhanced catalytic activity [41]. They concluded that for Ni/SBA-15, 3 wt.% boron was optimum. Moreover, Ni-SiO_2-MgO catalysts showed better catalytic activity having 80% of methane and 60% of CO_2 conversion. The operating temperature was 750°C and the produced syngas ratio of 2. Stroud et al., tested Sn modified Ni/Al_2O_3 and dual oxide support Ni/CeO_2-Al_2O_3 catalyst. It showed promising activity, as it occupies the C nucleation sites near the Ni atom. As a result, it decreases the coke formation rate during reaction [42].

Li et al. [43] have found that Ni/ZrO_2 promoted by Mo_2C for CDSRM enhanced the Ni dispersion and it significantly increased the activity. Moreover, various authors develop Ni-based catalysts that have high stability and reactivity which are supported by the Ce-ZrO_2, MgO, ZrO_2, and MgO-ZrO_2 for CDSRM. They studied the effect of the support and promoters on the activity of the nickel catalyst [26]. Furthermore, a study conducted by Jabbour et al. [44], in which they used Ni/Al_2O_3 promoted by 5 wt.% Mg and Ca. And gives the conclusion that the addition of Ca and Mg as a promoter significantly increased the Ni catalyst activity. Moreover, various authors develop Ni-based catalysts that have high stability and reactivity which are supported by the Ce-ZrO_2, MgO, ZrO_2, and MgO-ZrO_2 for CCSRM. They studied the effect of the support and promoters on the activity of the nickel catalyst [45].

Farooqi et al. [46] studied the production of syngas over Ni/ZrO_2-MgO by comparing the performance of Ni/ZrO_2-MgO with the Ni-MgO. The authors observed that coke deposition varies with the type of catalyst used. Ni/MgO had significant carbon deposition on its active sites, and the carbon was in the form of sheets having a negative influence on the activity of the reaction and resulting in catalyst deactivation. However, Ni/ZrO_2-MgO generated a different type of carbon, which was in the form of filamentous carbon having structures like nanotubes and the intensity was insignificant. Moreover, Ni/ZrO_2-MgO exhibited better stability and activity for CDSRM reaction at 800 °C by using a stoichiometric feed ratio, and carbon deposition amount and type strictly depend on the catalyst [47]. Ji et al. [48] investigated Ce-promoted Ni/SBA-15 catalysts for CDSRM in an analogous manner. Based on the findings, the Ce-promoted catalyst supported on SBA-15 demonstrated a higher surface area, longer stability, higher activity, and higher oxygen storage capacity. In addition to alkaline earth metals, noble metals doping metal catalysts resulted in encouraging results that improved coke resistance and stability. Catalysts can be greatly improved with even a minute amount of noble metals[49].

In a study, Khani et al. [50] concluded that $ZnLaAlO_4$ catalyst with a 3 wt.% Ru promoter produced the best results due to increased catalytic activity and reduced coke deposition. As a result, the promoters boosted dispersion, which also lowered the temperature required in order to achieve optimal performance through synergistic interactions. Since the actual reaction mechanism pathway must be identified before selection can begin, choosing an appropriate co-metal may be challenging.

Effect of Bimetallic Catalyst. Apart from designing monometallic Ni-based catalysts with different support variations and altering their properties by adding metals with redox or basic properties, alloying a second metal (whether noble or transition) with the existing Ni metal, which acts as co-metal helps in the modification of various catalytic properties via a synergistic effect between each other [51]. As, the second metal offers various advantages including high activity, and stability [52].

Zhang et al. [53] perform a stability test for the Ni-Co bimetallic catalyst and it shows minute deactivation for 2000 hrs. TOS. Due to the technological impacts, bimetallic catalysts are receiving a lot of attention from researchers for CDSRM-based processes [54]. When these catalysts are used along with oxide support, they generate bi-functional capabilities as a result of the change in metal-oxygen bond strength, which subsequently reflects variations in their performances and catalytic properties [55]. Introducing a minute amount of noble metal modifies the Ni environment. Garcia et al. [56] studied the effect of 0.04% Rh_2O_3 and found the enhanced CDSRM performance. Furthermore, Shakouri et al. [57] studied the effect of Ni-Co bimetallic catalyst and compared its performance with monometallic Co and Ni catalyst. Authors concluded that Ni-Co bimetallic exhibits better performance than the monometallic ones as it converts 70% more CO_2 and the respective H_2/CO ratio was 1.8-2.

Therefore, from the above studies, it was observed that as compared to monometallic catalysts, the bimetallic catalyst demonstrated improved performance. As, bimetallic catalysts significantly activated both metals via a synergistic effect and a stable alloy was formed that prevented the catalyst from sintering at high temperatures. Also, the use of suitable promoters helps in better metal dispersion, reduced coke formation, increased oxygen storage capacity, and better stability. Hence, co-metal and promoters should be selected carefully as they vary the main reaction mechanism.

Table 1 Recent developments of catalyst used in CDSRM for syngas

Catalyst	T (°C)	P (bar)	Feed Ratio ($H_2O:CO_2:CH_4$)	GHSV[1] ($Lgcat^{-1}hr^{-1}$)	TOS[2] (hrs)	CH_4 %Con.	CO_2 %Con.	H_2/CO	Ref
Ni/MgO-ZrO_2	800	1	2:1:3	36	20	82.5	81.5	1.70	[46]
Ni/MgO-Al_2O_3	850	1	2:1:3	-	-	61.8	59.8	2	[58]
13XZnO-Ni-u	800	1	2:1:3	-	6	90	58	-	[59]
Ni-Co/AlMgOx	850	1	2:1:3	110	15	80.3	89.1	1.68	[57]
Ni/CeO_2-SrO_2	800	1	1.5:2.4:3	-	250	91	78	1.9	[26]
Ni-K/LDH	800	1	2:1:3	-	-	90	64.6	2	[60]
Ni-Sn/Ce-Al	700	1	1:1:1	60	24	70	40	1.6	[42]
Ni/Al_2O_3	800	1	0.8:0.4:1	69	3	71	56	2.5	[44]
Ni/MgO	830	1	0.8:0.4:1	60	160	71	73	1.92	[61]
Ni/SBA-15	800	1	2:1:3	36	-	61.6	58.9	2.14	[62]
Ni-Ce/$MgAl_2O_4$	850	1	1:0.4:1	-	20	83	52	2.2	[34]
Ni/SiO_2 MgO	750	1	2:1:3	160	140	80	60	2	[63]
Ni/SBA-15 3%B-Promoter	800	1	2:1:3	36	24	71.4	70	2.5	[14]
Ni-Mg/Al_2O_3	800	1	0.8:0.4:1	138	40	80	74	2.10	[44]
Ni-Ce-Fe/Al_2O_3	900	20	1:0.8:1	25	50	89	70	2	[2]

[1]GHSV stands for Gas hourly space velocity
[2]TOS stands for Time on stream

Conclusions and recommendations

Considering the future's increasing demand for fossil fuels, CDSRM offers an environmentally friendly alternative. The use of CO_2 and CH_4 as feedstock might also make this a useful technology for mitigating GHG emissions. Despite this, a few obstacles prevent it from becoming widely commercialized. Economic considerations force the employment of Ni-based catalysts in the majority of situations. However, these materials can degrade through the creation of coke and sintering. The commercialization of CDSRM consequently requires the creation of a commercially viable catalyst that is highly active, immune to carbon formation, resistant to sintering, and has enhanced SMSI (metal support interaction). Researchers found that adding promoters and catalyst supports to Ni-based catalysts decreased the generation of coke. Improved performance can be achieved by doping metal-based active catalysts with alkali, alkaline earths, or noble metals. It results from improved stability, oxygen storage capacity, and enhance catalytic structure. Furthermore, since the preparation of catalysts affects both their structural behavior and performance, it is vital to examine this process. Bimetallic catalysts have been discovered to operate pretty well in comparison to the monometallic catalysts at high temperatures because of stable alloy formation. However, there is still a need for a coke-resistant, highly active bimetallic catalyst that uses less energy by operating at lower temperatures and has a longer lifespan for the CDSRM process. Interactions between metals and supports, along with dispersion of metals in supports, are also important future considerations. A number of variables, including as the composition of feedstock gases, reaction temperature, and gas hourly space velocity (GHSV) also influences the catalytic activity of CDSRM. It is essential to establish a reliable relationship between the process variables and the performance of CDSRM catalysts.

Future outlook

In order to comprehend the mechanism of the CDSRM reaction, numerous experiments on Ni-based catalysts have been carried out during the past few decades. These investigations have helped to develop new strategies to stop the creation of coke in addition to identifying the causes of coke formation. To lessen the likelihood of coke generation employing Ni-based catalysts, several strategies have been used. This entails choosing the right method for effective and viable catalyst formation, picking the right metal for improved SMSI, keeping the support material alkaline, doping with the right promoter to increase oxygen storage capacity, and choosing the right bimetallic catalyst to improve mutual synergies.

Future research should concentrate on creating suitable bimetallic and trimetallic catalysts based on Ni, as this catalyst has demonstrated significant and encouraging evidence of enhanced activity. However, carbon deposition continues to be a constraint. Future studies should concentrate on metal dispersion and active metal particle size for improved catalyst durability. A major factor in the performance of the catalyst is also its preparation method. The proper technique for catalyst synthesis can enhance Ni dispersion on the support, stability, activity, SMSI, and resistance to coke formation. Future studies should pay particular attention to these factors in order to create the optimum catalyst.

Acknowledgement

The authors acknowledge financial support from Universiti Teknologi PETRONAS, Malaysia, under the Yayasan Universiti Teknologi PETRONAS (YUTP) grant number 015LC0-392

References

[1] X. Li, "Diversification and localization of energy systems for sustainable development and energy security," *Energy Policy*, vol. 33, no. 17, pp. 2237–2243, 2005, https://doi.org/10.1016/j.enpol.2004.05.002

[2] P. Li, Y. H. Park, D. J. Moon, N. C. Park, and Y. C. Kim, "Carbon deposition onto Ni-Based catalysts for combined steam/CO2reforming of methane," *J. Nanosci. Nanotechnol.*, vol. 16, no. 2, pp. 1562–1566, Feb. 2016, https://doi.org/10.1166/jnn.2016.12006

[3] N. Apergis, "Oil prices and corporate high-yield spreads: Evidence from panels of nonenergy and energy European firms," *Q. Rev. Econ. Financ.*, vol. 72, pp. 34–40, 2019, https://doi.org/10.1016/j.qref.2019.01.012

[4] S. Li and J. Gong, "Strategies for improving the performance and stability of Ni-based catalysts for reforming reactions," *Chem. Soc. Rev.*, vol. 43, no. 21, pp. 7245–7256, Oct. 2014, https://doi.org/10.1039/C4CS00223G

[5] M. Saunois, R. B. Jackson, P. Bousquet, B. Poulter, and J. G. Canadell, "The growing role of methane in anthropogenic climate change," *Environ. Res. Lett.*, vol. 11, no. 12, 2016, https://doi.org/10.1088/1748-9326/11/12/120207

[6] E. J. Dlugokencky, L. P. Steele, P. M. Lang, and K. A. Masarie, "Global Monitoring Laboratory - Carbon Cycle Greenhouse Gases," *J. Geophy. Res.*, vol. 99, no. D8, pp. 17021–17043, 1994, Accessed: Jul. 01, 2022. [Online]. Available: https://gml.noaa.gov/ccgg/trends/global.html

[7] "Inventory of U.S. Greenhouse Gas Emissions and Sinks | US EPA." https://www.epa.gov/ghgemissions/inventory-us-greenhouse-gas-emissions-and-sinks (accessed Apr. 28, 2022).

[8] B. M. Al–Swai *et al.*, "Low-temperature catalytic conversion of greenhouse gases (CO_2 and CH_4) to syngas over ceria-magnesia mixed oxide supported nickel catalysts," *Int. J. Hydrogen Energy*, vol. 46, no. 48, pp. 24768–24780, 2021, https://doi.org/10.1016/j.ijhydene.2020.04.233

[9] N. Yang and R. Wang, "Sustainable technologies for the reclamation of greenhouse gas CO2," *J. Clean. Prod.*, vol. 103, pp. 784–792, 2015, https://doi.org/10.1016/j.jclepro.2014.10.025

[10] A. S. Farooqi *et al.*, "Catalytic conversion of greenhouse gases (CO2 and CH4) to syngas over Ni-based catalyst: Effects of Ce-La promoters," *Arab. J. Chem.*, vol. 13, no. 6, pp. 5740–5749, Jun. 2020, https://doi.org/10.1016/j.arabjc.2020.04.012

[11] M. Yusuf *et al.*, "Syngas production from greenhouse gases using Ni–W bimetallic catalyst via dry methane reforming: Effect of W addition," *Int. J. Hydrogen Energy*, vol. 46, no. 53, pp. 27044–27061, 2021, https://doi.org/10.1016/j.ijhydene.2021.05.186

[12] J. S. Yu, J. M. Park, J. H. Kwon, and K. S. Park, "Roles of Al2O3 coating layer on an ordered mesoporous Ni/m-Al2O3 for combined steam and CO2 reforming with CH4," *Fuel*, vol. 3, no. July 2022, pp. 37–38, 2023, https://doi.org/10.1016/j.fuel.2022.125702

[13] Y. Zhang *et al.*, "Combined steam and CO2 reforming of methane over Co–Ce/AC-N catalyst: Effect of preparation methods on catalyst activity and stability," *Int. J. Hydrogen Energy*, vol. 47, no. 5, pp. 2914–2925, 2022, https://doi.org/10.1016/j.ijhydene.2021.10.202

[14] T. J. Siang *et al.*, "Methane bi-reforming over boron-doped Ni/SBA-15 catalyst: Longevity evaluation," *Int. J. Hydrogen Energy*, vol. 44, no. 37, pp. 20839–20850, 2019, https://doi.org/10.1016/j.ijhydene.2018.06.123

[15] G. A. Olah, A. Goeppert, M. Czaun, and G. K. S. Prakash, "Bi-reforming of methane from any source with steam and carbon dioxide exclusively to metgas (CO-2H2) for methanol

and hydrocarbon synthesis," *J. Am. Chem. Soc.*, vol. 135, no. 2, pp. 648–650, 2013, https://doi.org/10.1021/ja311796n

[16] M. Yusuf, A. S. Farooqi, M. A. Alam, L. K. Keong, K. Hellgardt, and B. Abdullah, "Response surface optimization of syngas production from greenhouse gases via DRM over high performance Ni–W catalyst," *Int. J. Hydrogen Energy*, no. xxxx, 2021, https://doi.org/10.1016/j.ijhydene.2021.05.153

[17] W. J. Jang *et al.*, "Combined steam and carbon dioxide reforming of methane and side reactions: Thermodynamic equilibrium analysis and experimental application," *Appl. Energy*, vol. 173, pp. 80–91, 2016, https://doi.org/10.1016/j.apenergy.2016.04.006

[18] G. A. Olah, G. K. S. Prakash, A. Goeppert, M. Czaun, and T. Mathew, "Self-sufficient and exclusive oxygenation of methane and its source materials with oxygen to methanol via metgas using oxidative bi-reforming," *J. Am. Chem. Soc.*, vol. 135, no. 27, pp. 10030–10031, 2013, https://doi.org/10.1021/ja405439c

[19] V. Dal Santo, A. Gallo, A. Naldoni, M. Guidotti, and R. Psaro, "Bimetallic heterogeneous catalysts for hydrogen production," *Catal. Today*, vol. 197, no. 1, pp. 190–205, Dec. 2012, https://doi.org/10.1016/j.cattod.2012.07.037

[20] S. Yu, Y. Hu, H. Cui, Z. Cheng, and Z. Zhou, "Ni-based catalysts supported on MgAl2O4 with different properties for combined steam and CO2 reforming of methane," *Chem. Eng. Sci.*, vol. 232, p. 116379, 2021, https://doi.org/10.1016/j.ces.2020.116379

[21] G. A. Olah, "Beyond oil and gas: The methanol economy," *Angew. Chemie - Int. Ed.*, vol. 44, no. 18, pp. 2636–2639, 2005, https://doi.org/10.1002/anie.200462121

[22] B. Abdullah, N. A. Abd Ghani, and D. V. N. Vo, "Recent advances in dry reforming of methane over Ni-based catalysts," *J. Clean. Prod.*, vol. 162, pp. 170–185, 2017, https://doi.org/10.1016/j.jclepro.2017.05.176

[23] K. Jabbour, "Tuning combined steam and dry reforming of methane for 'metgas' production: A thermodynamic approach and state-of-the-art catalysts," *Journal of Energy Chemistry*, vol. 48. Elsevier B.V., pp. 54–91, Sep. 01, 2020. https://doi.org/10.1016/j.jechem.2019.12.017

[24] C. C. Chong, S. N. Bukhari, Y. W. Cheng, H. D. Setiabudi, A. A. Jalil, and C. Phalakornkule, "Robust Ni/Dendritic fibrous SBA-15 (Ni/DFSBA-15) for methane dry reforming: Effect of Ni loadings," *Appl. Catal. A Gen.*, vol. 584, no. May, 2019, https://doi.org/10.1016/j.apcata.2019.117174

[25] N. Laosiripojana and S. Assabumrungrat, "Catalytic dry reforming of methane over high surface area ceria," *Appl. Catal. B Environ.*, vol. 60, no. 1–2, pp. 107–116, Sep. 2005, https://doi.org/10.1016/J.APCATB.2005.03.001

[26] G. Mallikarjun, T. V. Sagar, S. Swapna, N. Raju, P. Chandrashekar, and N. Lingaiah, "Hydrogen rich syngas production by bi-reforming of methane with CO2over Ni supported on CeO2-SrO mixed oxide catalysts," *Catal. Today*, vol. 356, no. December 2019, pp. 597–603, 2020, https://doi.org/10.1016/j.cattod.2020.01.005

[27] C. C. Chong, N. Abdullah, S. N. Bukhari, N. Ainirazali, L. P. Teh, and H. D. Setiabudi, "Hydrogen production via CO2 reforming of CH4 over low-cost Ni/SBA-15 from silica-rich palm oil fuel ash (POFA) waste," *Int. J. Hydrogen Energy*, vol. 44, no. 37, pp. 20815–20825, 2019, https://doi.org/10.1016/j.ijhydene.2018.06.169

[28] B. C. Ekeoma, M. Yusuf, K. Johari, and B. Abdullah, "Mesoporous silica supported Ni-based catalysts for methane dry reforming: A review of recent studies," *Int. J. Hydrogen*

Energy, no. xxxx, 2022, https://doi.org/10.1016/j.ijhydene.2022.05.297

[29] S. Singh *et al.*, "Metgas production from Bi-reforming of methane over lamodified santa barbara amorphous-15 supported nickel catalyst," *Chem. Eng. Trans.*, vol. 56, pp. 1573–1578, 2017, https://doi.org/10.3303/CET1756263

[30] Y. Tian, X. Ma, X. Chen, and C. Zhang, "Effect of Ni-Co bimetallic core-shell catalyst for coke resistance in CO2 reforming of biomass Tar," *J. Anal. Appl. Pyrolysis*, vol. 164, p. 105539, Jun. 2022, https://doi.org/10.1016/j.jaap.2022.105539

[31] M. A. Salaev, L. F. Liotta, and O. V. Vodyankina, "Lanthanoid-containing Ni-based catalysts for dry reforming of methane: A review," *Int. J. Hydrogen Energy*, vol. 47, no. 7, pp. 4489–4535, 2022, https://doi.org/10.1016/j.ijhydene.2021.11.086

[32] S. K. Ryi, S. W. Lee, J. W. Park, D. K. Oh, J. S. Park, and S. S. Kim, "Combined steam and CO2 reforming of methane using catalytic nickel membrane for gas to liquid (GTL) process," *Catal. Today*, vol. 236, no. PART A, pp. 49–56, Nov. 2014, https://doi.org/10.1016/j.cattod.2013.11.001

[33] A. S. Al-Fatesh *et al.*, "CO2-reforming of methane to produce syngas over Co-Ni/SBA-15 catalyst: Effect of support modifiers (Mg, la and Sc) on catalytic stability," *J. CO2 Util.*, vol. 21, no. March, pp. 395–404, 2017, https://doi.org/10.1016/j.jcou.2017.08.001

[34] S. C. Baek, J. W. Bae, J. Y. Cheon, K. W. Jun, and K. Y. Lee, "Combined steam and carbon dioxide reforming of methane on Ni/MgAl 2O4: Effect of CeO2 promoter to catalytic performance," *Catal. Letters*, vol. 141, no. 2, pp. 224–234, Nov. 2011, https://doi.org/10.1007/s10562-010-0483-0

[35] L. Zhu, Z. Lv, X. Huang, J. Ran, J. Chen, and C. Qin, "Understanding the role of support structure in methane dry reforming for syngas production," *Fuel*, vol. 327, no. June, p. 125163, 2022, https://doi.org/10.1016/j.fuel.2022.125163

[36] N. D. Charisiou *et al.*, "Syngas production via the biogas dry reforming reaction over nickel supported on modified with CeO2 and/or La2O3 alumina catalysts," *J. Nat. Gas Sci. Eng.*, vol. 31, pp. 164–183, Apr. 2016, https://doi.org/10.1016/J.JNGSE.2016.02.021

[37] M. A. Al-Nakoua and M. H. El-Naas, "Combined steam and dry reforming of methane in narrow channel reactors," *Int. J. Hydrogen Energy*, vol. 37, no. 9, pp. 7538–7544, May 2012, https://doi.org/10.1016/J.IJHYDENE.2012.02.031

[38] H. S. Roh *et al.*, "Combined reforming of methane over supported Ni catalysts," *Catal. Letters*, vol. 117, no. 1–2, pp. 85–90, Aug. 2007, https://doi.org/10.1007/s10562-007-9113-x

[39] J. W. Bae, A. R. Kim, S. C. Baek, and K. W. Jun, "The role of CeO2-ZrO2 distribution on the Ni/MgAl2O4 catalyst during the combined steam and CO 2 reforming of methane," *React. Kinet. Mech. Catal.*, vol. 104, no. 2, pp. 377–388, Dec. 2011, https://doi.org/10.1007/S11144-011-0371-7/FIGURES/5

[40] E. Rezaei and L. J. J. Catalan, "Evaluation of CO2utilization for methanol production via tri-reforming of methane," *J. CO2 Util.*, vol. 42, Dec. 2020, https://doi.org/10.1016/J.JCOU.2020.101272

[41] T. J. Siang *et al.*, "Combined steam and CO2 reforming of methane for syngas production over carbon-resistant boron-promoted Ni/SBA-15 catalysts," *Microporous Mesoporous Mater.*, vol. 262, pp. 122–132, May 2018, https://doi.org/10.1016/j.micromeso.2017.11.028

[42] T. Stroud *et al.*, "Chemical CO2 recycling via dry and bi reforming of methane using Ni-

Sn/Al2O3 and Ni-Sn/CeO2-Al2O3 catalysts," *Appl. Catal. B Environ.*, vol. 224, no. September 2017, pp. 125–135, 2018, https://doi.org/10.1016/j.apcatb.2017.10.047

[43] W. Li, Z. Zhao, P. Ren, and G. Wang, "Effect of molybdenum carbide concentration on the Ni/ZrO2 catalysts for steam-CO2 bi-reforming of methane," *RSC Adv.*, vol. 5, no. 122, pp. 100865–100872, Nov. 2015, https://doi.org/10.1039/c5ra22237k

[44] K. Jabbour, P. Massiani, A. Davidson, S. Casale, and N. El Hassan, "Ordered mesoporous 'one-pot' synthesized Ni-Mg(Ca)-Al2O3 as effective and remarkably stable catalysts for combined steam and dry reforming of methane (CSDRM)," *Appl. Catal. B Environ.*, vol. 201, pp. 527–542, 2017, https://doi.org/10.1016/j.apcatb.2016.08.009

[45] U. S. Mohanty, M. Ali, M. R. Azhar, A. Al-Yaseri, A. Keshavarz, and S. Iglauer, "Current advances in syngas (CO + H2) production through bi-reforming of methane using various catalysts: A review," *Int. J. Hydrogen Energy*, vol. 46, no. 65, pp. 32809–32845, 2021, https://doi.org/10.1016/j.ijhydene.2021.07.097

[46] A. S. Farooqi, "Hydrogen-rich syngas production from bi-reforming of greenhouse gases over zirconia modified Ni / MgO catalyst," no. September, pp. 1–17, 2021, https://doi.org/10.1002/er.7325

[47] M. Yusuf, A. S. Farooqi, L. K. Keong, K. Hellgardt, and B. Abdullah, "Contemporary trends in composite Ni-based catalysts for CO2 reforming of methane," *Chem. Eng. Sci.*, vol. 229, p. 116072, 2021, https://doi.org/10.1016/j.ces.2020.116072

[48] T. J. Siang, H. T. Danh, S. Singh, Q. D. Truong, H. D. Setiabudi, and D. V. N. Vo, "Syngas production from combined steam and carbon dioxide reforming of methane over Ce-modified silicasupported nickel catalysts," *Chem. Eng. Trans.*, vol. 56, pp. 1129–1134, 2017, https://doi.org/10.3303/CET1756189

[49] A. S. Al-Fatesh *et al.*, "Greenhouse gases utilization via catalytic reforming with Sc promoted Ni/SBA-15," *Fuel*, vol. 330, no. August, p. 125523, 2022, https://doi.org/10.1016/j.fuel.2022.125523

[50] Y. Khani, Z. Shariatinia, and F. Bahadoran, "High catalytic activity and stability of ZnLaAlO4 supported Ni, Pt and Ru nanocatalysts applied in the dry, steam and combined dry-steam reforming of methane," *Chem. Eng. J.*, vol. 299, pp. 353–366, 2016, https://doi.org/10.1016/j.cej.2016.04.108

[51] X. Wang, S. Xu, W. Yang, X. Fan, Q. Pan, and H. Chen, "Carbon Capture Science & Technology Development of Ni-Co supported on SBA-15 catalysts for non-thermal plasma assisted co-conversion of CO 2 and CH 4 : Results and lessons learnt," *Carbon Capture Sci. Technol.*, vol. 5, no. September, p. 100067, 2022, https://doi.org/10.1016/j.ccst.2022.100067

[52] S. S. Itkulova, Y. Y. Nurmakanov, S. K. Kussanova, and Y. A. Boleubayev, "Production of a hydrogen-enriched syngas by combined CO2-steam reforming of methane over Co-based catalysts supported on alumina modified with zirconia," *Catal. Today*, vol. 299, no. November 2016, pp. 272–279, 2018, https://doi.org/10.1016/j.cattod.2017.07.014

[53] J. Zhang, H. Wang, and A. K. Dalai, "Development of stable bimetallic catalysts for carbon dioxide reforming of methane," *J. Catal.*, vol. 249, no. 2, pp. 300–310, Jul. 2007, https://doi.org/10.1016/j.jcat.2007.05.004

[54] S. S. Itkulova, G. D. Zakumbaeva, Y. Y. Nurmakanov, A. A. Mukazhanova, and A. K. Yermaganbetova, "Syngas production by bireforming of methane over Co-based alumina-supported catalysts," *Catal. Today*, vol. 228, pp. 194–198, 2014, https://doi.org/10.1016/j.cattod.2014.01.013

[55] H. C. Foley, A. J. Hong, J. S. Brinen, L. F. Allard, and A. J. Garratt-Reed, "Bimetallic catalysts comprised of dissimilar metals for the reduction of carbon monoxide with hydrogen," *Appl. Catal.*, vol. 61, no. 1, pp. 351–375, May 1990, https://doi.org/10.1016/S0166-9834(00)82155-7

[56] J. D. Medrano-García, R. Ruiz-Femenia, and J. A. Caballero, "Multi-objective optimization of combined synthesis gas reforming technologies," *J. CO2 Util.*, vol. 22, pp. 355–373, 2017, https://doi.org/10.1016/j.jcou.2017.09.019

[57] M. Shakouri, Y. Hu, R. Lehoux, and H. Wang, "CO2 conversion through combined steam and CO2 reforming of methane reactions over Ni and Co catalysts," *Can. J. Chem. Eng.*, vol. 99, no. 1, pp. 153–165, 2021, https://doi.org/10.1002/cjce.23828

[58] D. He, Y. Zhang, Z. Wang, Y. Mei, and Y. Jiang, "Bi-reforming of Methane with Carbon Dioxide and Steam on Nickel-Supported Binary Mg−Al Metal Oxide Catalysts," 2020, https://doi.org/10.1021/acs.energyfuels.9b03312

[59] A. F. Cunha, S. Morales-Torres, L. M. Pastrana-Martínez, F. J. Maldonado-Hódar, and N. S. Caetano, "Syngas production by bi-reforming of methane on a bimetallic Ni-ZnO doped zeolite 13X," *Fuel*, vol. 311, Mar. 2022, https://doi.org/10.1016/j.fuel.2021.122592

[60] A. F. Cunha *et al.*, "Syngas production by bi-reforming methane on an Ni-K-promoted catalyst using hydrotalcites and filamentous carbon as a support material," *RSC Adv.*, vol. 10, no. 36, pp. 21158–21173, 2020, https://doi.org/10.1039/d0ra03264f

[61] G. A. Olah, A. Goeppert, M. Czaun, T. Mathew, R. B. May, and G. K. S. Prakash, "Single Step Bi-reforming and Oxidative Bi-reforming of Methane (Natural Gas) with Steam and Carbon Dioxide to Metgas (CO-2H2) for Methanol Synthesis: Self-Sufficient Effective and Exclusive Oxygenation of Methane to Methanol with Oxygen," *J. Am. Chem. Soc.*, vol. 137, no. 27, pp. 8720–8729, 2015, https://doi.org/10.1021/jacs.5b02029

[62] S. Singh *et al.*, "Bi-reforming of methane on Ni/SBA-15 catalyst for syngas production: Influence of feed composition," *Int. J. Hydrogen Energy*, vol. 43, no. 36, pp. 17230–17243, 2018, https://doi.org/10.1016/j.ijhydene.2018.07.136

[63] J. Ashok, Z. Bian, Z. Wang, and S. Kawi, "Ni-phyllosilicate structure derived Ni-SiO2-MgO catalysts for bi-reforming applications: Acidity, basicity and thermal stability," *Catal. Sci. Technol.*, vol. 8, no. 6, pp. 1730–1742, 2018, https://doi.org/10.1039/c7cy02475d

Materials Research Proceedings 29 (2023) 28-34 https://doi.org/10.21741/9781644902516-4

Sunlight mediated enhanced removal of metoprolol using graphitic carbon nitride ($g\text{-}C_3N_4$)

KHEW Chung Wei[1,a], LEONG Kah Hon[2,b*], SIM Lan Ching[3,c], ABD AZIZ Azrina[4,d], DAI Chaomeng[5,e], OH Wen-Da[6,f] and IBRAHIM Shaliza[7,g]

[1]Department of Environmental Engineering, Faculty of Engineering and Green Technology, Universiti Tunku Abdul Rahman, Jalan Universiti, Bandar Barat, 31900, Kampar, Perak Malaysia

[2]Department of Environmental Engineering, Faculty of Engineering and Green Technology, Universiti Tunku Abdul Rahman, Jalan Universiti, Bandar Barat, 31900, Kampar, Perak Malaysia

[3]Department of Chemical Engineering, Lee Kong Chian Faculty of Engineering and Science, Universiti Tunku Abdul Rahman, Jalan Sungai Long 9, Bandar Sungai Long, 43000, Kajang, Selangor, Malaysia

[4]Faculty of Civil Engineering Technology, Universiti Malaysia Pahang, 26300, Gambang, Kuantan, Pahang, Malaysia

[5]College of Civil Engineering, Tongji University, 1239 Siping Road, 200092, Shanghai, China

[6]School of Chemical Sciences, Universiti Sains Malaysia, 11800, Penang, Malaysia

[7]Institute of Ocean and Earth Sciences (IOES), University of Malaya, 50603, Kuala Lumpur, Malaysia

[a]angelakcw1016@1utar.my, [b]khleong@utar.edu.my, [c]simcl@utar.edu.my, [d]azrinaaziz@ump.edu.my, [e]chaomeng.dai@tongji.edu.cn, [f]ohwenda@usm.my, [g]shaliza@um.edu.my

Keywords: Sunlight, $g\text{-}C_3N_4$, Metoprolol, Photocatalyst

Abstract. Graphitic carbon nitride ($g\text{-}C_3N_4$) is a photocatalyst that has recently been given a lot of attention due to its effectiveness in wastewater and environmental treatment, solar energy utilization, biomedical applications, etc. In this study, $g\text{-}C_3N4$ was synthesized and characterized to carry out the degradation of metoprolol tartrate salt (MET), which is classified as an emerging contaminant. MET is one of the most commonly used pharmaceuticals to treat patients with cardiovascular diseases and disorders, a common disease in Malaysia. Recent discoveries of MET in surface waters and drinking water raise awareness and concerns. $g\text{-}C_3N_4$ was synthesized using solid urea by placing it in a muffle furnace of 550°C for 3 hours. The photocatalytic activities of $g\text{-}C_3N_4$ were investigated by photodegradation of MET, $g\text{-}C_3N_4$ of different dosages were added into MET-containing solution, and a dark reaction was carried out for 24 hours for complete adsorption equilibrium. Various physical and chemical analyses were conducted to elucidate the properties of $g\text{-}C_3N_4$, such as FESEM, FTIR and UV-Vis. The absorbance and reflectance graphs of $g\text{-}C_3N_4$ show that there will be higher absorption in the visible light spectra. The results show that the optimum dosage to treat 10 ppm of MET is by using 0.3 g of $g\text{-}C_3N_4$. Under sunlight irradiation of 4 hours, the degradation of MET achieved 54.6% of removal. Hence, it proves that $g\text{-}C_3N_4$ nanosheet can be applied to remove complex pollutants such as MET under sunlight irradiation. This path is an alternative removal method for MET in a sustainable approach.

Materials Research Proceedings 29 (2023) 28-34 https://doi.org/10.21741/9781644902516-4

Introduction

Pharmaceuticals and personal care products (PPCPs) pollutants are a highly recalcitrant aquatic micropollutant that severely harms the water source. Pollutants such as metoprolol, amoxicillin, ciprofloxacin, triclosan, carbamazepine are common PPCPs pollutants that are even detected at the tap water source in Kajang and Putrajaya, Malaysia [1,2]. This is an alarming and needed immediate remediation as such pollutants will pose adverse effects to human growth, allergies, lung diseases, obesity, and other waterborne diseases. Furthermore, PPCPs pollutants can generate various toxic effects on aquatic life. It will retard the reproduction of aquatic diseases. Furthermore, PPCPs pollutants can generate various toxic effects on aquatic life. It will retard the reproduction of aquatic life and increases the rate of malformation and mortality. Critically, most of these pollutants are persistent and pass through conventional biological treatment systems without any significant changes in their chemistry or concentrations. Moreover, the conventional treatment processes for water, such as filtration, coagulation and chlorination usually remove part of the PPCPs. [3] A promising treatment option for eliminating PPCPs completely from water is advanced oxidation technology (AOPs) which can degrade various active PPCPs pollutants in water based on the generation of active radicals.

In AOPs based on solar photocatalysis, a heterogeneous catalyst is employed to activate the generation of reactive oxygen species (ROS) through the free electron mobility mechanism under the presence of sunlight. The intensity of sunlight exceeded the catalyst bandgap energy will then excites the electron from the valence band to the conduction band. It will then generate an electron-hole pair that leads to a series of electrons mobility to the photodegradation of pollutants [4,5]. The movement of these electrons and holes leads to the formation of active radicals. This ROS formation is responsible for breaking the target PPCPs pollutants' and degrading them down into harmless by-products. These radicals' strong oxidation potential can destroy stable bonds and unreactive organic molecules to form organic intermediates, resulting in CO_2 and H_2O [6]. Due to the importance of generating these active radicals, sunlight mediated photocatalysts is important to maximise the absorption of sunlight for the generation of active radicals. Therefore, it can promote the massive generation of ROS and improve photocatalytic degradation efficiency. Moreover, such photocatalysts can enhance the entire solar spectrum's harvesting to promote a sustainable water treatment process.

The well-known TiO_2 semiconductor is effectively removing different organic pollutants due to its unique properties. The successful application of this catalyst in organic pollutants removal had led to the other environmental application such as PPCPs. However, despite its advantages, it possesses a major drawback of limiting its absorption ability only under the UV light region due to the wide bandgap energy of 3.2 eV. This will retard the ability of TiO_2 to fully utilise the sunlight energy to generate active radicals to promote sustainable photocatalysts. Therefore, enhancing its photosensitivity to utilise the entire solar spectrum from UV is important. Therefore, g-C_3N_4 could be the best alternative to overcome the drawbacks of TiO_2. This is due to the unique properties of g-C_3N_4 of having strong ability in harvesting the visible light. It will promote the maximise utilisation of solar spectrum to generate active radicals. Hence, this study will focus on the simple synthesis of g-C_3N_4 in removing MET under sunlight irradiation.

Material and Methods

Material

Synthesis of g-C_3N_4

Decomposition of g-C_3N_4 using thermal approach was done by using the precursor of urea. 73g of urea was dried in an oven at a temperature of 80°C for 24h. The dried sample was then placed in a crucible before transferring to a muffle furnace. The temperature is set to 550°C and heated for 3h. Then it formed a yellowish product of g-C_3N_4. To remove any impurities in the product, nitric acid and ultrapure water were used to wash the product. It then repeated the washing three times. Finally, the sample was dried in an oven at 80°C overnight [7].

Characterisation
The morphology structure of g-C_3N_4 was analysed using the Field emission scanning electron microscope, (FESEM) Jeol JSM 7601. The presence of the functional elements of g-C_3N_4 was analysed using the Fourier transform infrared (FTIR) spectrum between 400 to 4000 cm^{-1} using Perkin Elmer Spectrum 400 Spectrophotometer. The optical absorbance ability was examined by using UV-Vis diffuse reflectance spectra, (UV-DRS) with white standard reference ($BaSO_4$).

Photodegradation of MET
The sustainable g-C_3N_4 photocatalytic activity was evaluated using the removal of metoprolol (MET) under sunlight irradiation. First, 0.3 g and 0.5 g of the prepared photocatalysts (g-C_3N_4) were added to the beakers that contained 10 ppm of MET. Next, the dark reaction was carried out for 24h to achieve an adsorption and desorption equilibrium. Then, it was followed by 24h of dark reaction before it underwent sustainable photocatalysis of MET using sunlight as the light source over 4h. Finally, the beakers were relocated to a location fully exposed to sunlight, as shown in Fig. 1. Samples were collected every 15 minutes intervals. Before conducting High-Performance Liquid Chromatography (HPLC) analysis, the samples will be centrifuged and filtered using a syringe filter. The solvent was filtered using a nylon membrane filter of 0.45µm porosity before conducting the HPLC analysis.

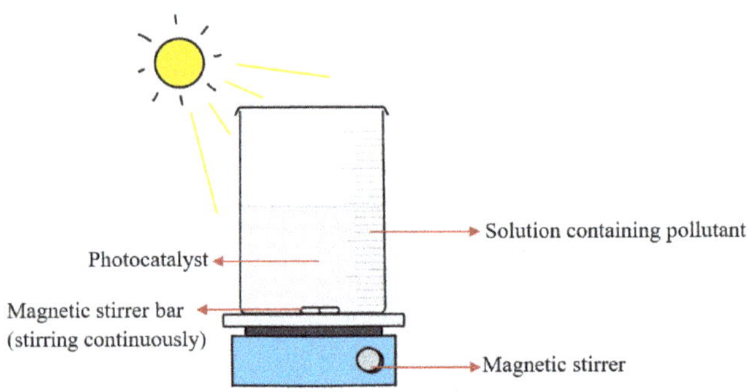

Figure 1: Photocatalysis analysis setup up

Results and Discussion
FESEM analysis is performed to study the microstructure and morphology of the as-synthesized g-C_3N_4, as shown in Fig. 2. It showed that pure g-C_3N_4 photocatalysts displayed fibre-like nanosheets. With the increased magnification, it observed that the nanostructures of g-C_3N_4 are uneven surfaces. In addition, the surfaces are rough and porous. Other researchers also observed similar structure nature and proved the successful synthesis of g-C_3N_4 [8].

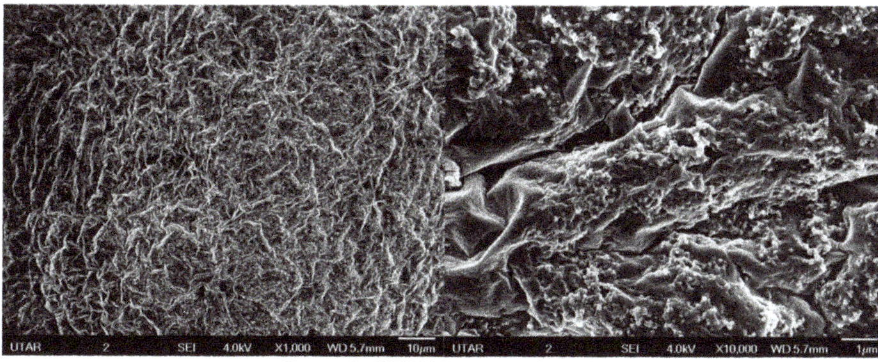

Figure 2: FESEM images of g-C_3N_4

The functional groups of g-C_3N_4 were analysed by FTIR analysis, as shown in Fig. 3. It can be seen that the band was detected at 3413 to 3234 cm^{-1}, which was caused by the stretching vibration of N-H bonds and surface adsorbed molecules. Apart from that, a few peaks from 1200 to 1700 cm^{-1} were observed, indicating the CN heterocycles stretching modes. In addition, 810 cm^{-1} peak is caused by the tri-s-triazine units breathing mode, while the consecutive peak of 822 cm^{-1} is caused by the N-H bonds deformation mode [9]. This observation confirmed the presence of successful synthesis of g-C_3N_4.

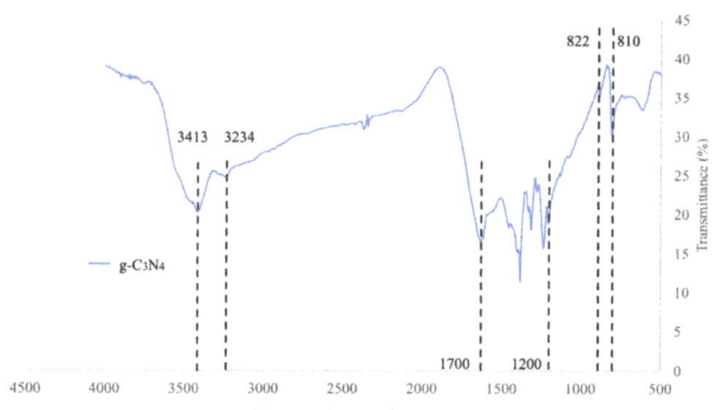

Figure 3: FTIR spectra of g-C_3N_4

The UV-DRS analysis was used to measure the absorption and reflection of g-C_3N_4. As observed in Fig. 4, the g-C_3N_4 photocatalyst possesses a very strong absorbance in UV and visible light regions. This strongly indicates the huge potential of g-C_3N_4 in absorbing the entire solar spectrum to generate active radicals. Due to the strong light g-C_3N_4 harvesting ability, it is the most promising photocatalyst compared to conventional TiO_2 and ZnO, which are limited to UV light absorbance. The harvesting of the entire solar spectrum promotes the excessive electrons and holes generation, which are responsible for the active radical's generation. The formation of these active radicals then be responsible for the removal of MET under sunlight irradiation.

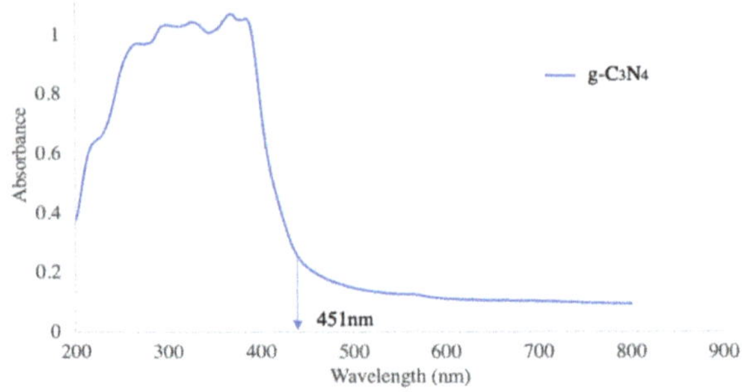

Figure 4: UV-Vis spectra of g-C_3N_4

The MET removal efficiency is shown in Fig. 5. It is observed that there is no degradation of MET during the 24h dark reaction. Thus, it can conclude that MET is not favourable to self-degradation without the presence of any catalyst. In addition, there is no significant removal of MET in the absence of a light source. This is mainly due to the non-photosensitizing properties of MET. Hence, MET degradation solely depends on the photocatalytic degradation of g-C_3N_4 and the light source. However, in Fig. 5, the blank graph increased to around 1.3. This might be due to the clinging of materials in the column contamination, causing some of the graphs to be not symmetrical due to minor errors. Upon the sunlight illumination, MET photodegradation took place rapidly and found that 0.3 g and 0.5 g of g-C_3N_4 resulted in the removal efficiency of 54.6% and 51.6%, respectively. Therefore, it can conclude that 0.3 g is the optimum catalyst dosage for MET removal under sunlight irradiation.

On the other hand, the research found that increasing the photocatalyst dosage did not drastically increase the photodegradation of MET. This is mainly contributed to the optimization of g-C_3N_4 in generating electrons and holes. Furthermore, the electron and hole pairs will recombine and form heat. This retarded the generation of active radicals.

The enhancement of MET removal by g-C_3N_4 is mainly contributed by the strong UV and visible light absorbance that led to the formation of electron and hole pairs. Once the g-C_3N_4 is illuminated with sunlight, it forms the electrons and holes at the valance and conduction band. The free electrons and holes will undergo a redox reaction to form active radicals such as hydroxyl and super anion radicals. These two active radicals will then attack the pollutant and break its component to form H_2O and CO_2. Therefore, the role of photocatalysts such as g-C_3N_4 will provide practical solutions to tackle the complex nature of the pollutants.

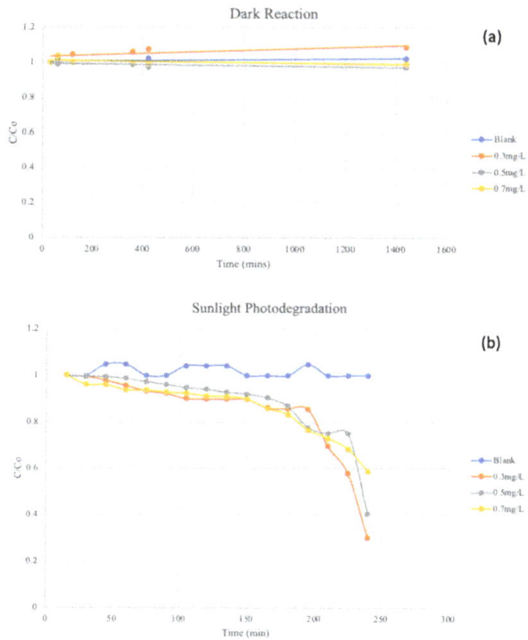

Figure 5: (a) dark reaction and (b) sunlight irradiation for removal of MET

Conclusion

In summary, the usage of g-C_3N_4 proves its ability to remove MET under sunlight irradiation effectively. The study of different dosages of photocatalyst showed that 0.3g is an optimum catalyst dosage for maximum removal of MET within 4h of sunlight irradiation. This research strongly indicates that MET will not undergo self-degradation, and the g-C_3N_4 catalyst must assist it. This is mainly due to the non-photosensitizing effect of MET. The superior g-C_3N_4 properties contribute to this research's positive achievement, that they can absorb UV and visible light due to the low band gap energy. This leads to the massive generation of electron and hole pairs. With the free electrons and holes, it then triggered the active radical's production that are responsible for degrading MET. Hence, this study proves the ability of g-C_3N_4 to degrade MET under sunlight irradiation effectively. This positive outcome will surely path a new direction in the field of photocatalysts in removing complex pollutants and promote sustainable remediation.

Acknowledge

The research was supported by the Ministry of Higher Education (MoHE), through the Fundamental Research Grant Scheme (FRGS), (FRGS/1/2021/TK0/UTAR/02/4)

References

[1] F.A.M. Nasir, S.M. Praveena, A.Z. Aris, Public awareness level and occurrence of pharmaceutical residues in drinking water with potential health risk: A study from Kajang (Malaysia). *Ecotox and Environ Safe*, 185 (2019) 109681. https://doi.org/10.1016/j.ecoenv.2019.109681

[2] S.M. Praveena, M.Z.M. Rashid, F.A.M. Nasir, W.S. Yee, A.Z. Aris, Occurrence and potential human health risk of pharmaceutical residues in drinking water from Putrajaya (Malaysia). *Ecotox and Environ Safe*, 180 (2019) 549-556. https://doi.org/10.1016/j.ecoenv.2019.05.051

[3] F. Lu, D. Astruc, Nanocatalysts and other nanomaterials for water remediation from organic pollutants. *Coordin. Chem. Rev.* 408 (2020) 213180. https://doi.org/10.1016/j.ccr.2020.213180

[4] D. Awfa, M. Ateia, M. Fujii, MS. Johnson, C. Yoshimura, Photodegradation of pharmaceuticals and personal care products in water treatment using carbonaceous-TiO_2 composite: a critical review of recent literature. *Water Res.* 142 (2018) 26-45. https://doi.org/10.1016/j.watres.2018.05.036

[5] PF. Lim, KH. Leong, LC. Sim, WD. Oh, YH. Chin, P. Saravanan, C. Dai, Mechanism insight of dual synergistic effects of plasmonic Pd-$SrTiO_3$ for enhanced solar energy photocatalysis. *Appl. Phys. A*, 126 (2020) 1-10. https://doi.org/10.1007/s00339-020-03739-4

[6] Q. Zhang, R. Du, G. Tan, P. Chen, G. Yu, S. Deng, Efficient degradation of typical pharmaceuticals in water using a novel TiO_2/ONLH nano-photocatalyst under natural sunlight. J. Hazar. Mat. 403 (2021) 123582. https://doi.org/10.1016/j.jhazmat.2020.123582

[7] J. Liu, T. Zhang, Z. Wang, G. Dawson, W. Chen, Simple Pyrolysis of Urea into Graphitic Carbon Nitride with Recyclable Adsorption and Photocatalytic Activity. *J. Mater. Chem.* 21 (2011) 14398-14401. https://doi.org/10.1039/c1jm12620b

[8] W. Aiwu, W. Chundong, F. Li, W. Winnie, L. Yucheng, Recent Advances of Graphitic Carbon Nitride-Based Structures and Applications in Catalyst, Sensing, Imaging, and LEDs. *Nano-Micro Lett.* 9 (2017) 47. https://doi.org/10.1007/s40820-017-0148-2

[9] Z. Bicheng, X. Pengfei, L. Yao, H. Wingkei, Y. Jiaguo, Fabrication and photocatalytic activity enhanced mechanism of direct Zscheme g-C_3N_4/Ag_2WO_4 photocatalyst. *Appl. Surf. Sci.* 391 (2016) 175-183.

Sustainable Processes and Clean Energy Transition - ICSuPCET2022 | Materials Research Forum LLC
Materials Research Proceedings 29 (2023) 35-42 | https://doi.org/10.21741/9781644902516-5

Fruit palm peel extract derived from supercritical CO_2 extraction as a potent source of bioactive compounds

BELLO Usman[2,a], AMRAN Nurul Aini[1,2,b*], SAMSURI Shafirah[1,2,c] and RUSLAN Muhammad Syafiq Hazwan[3,d]

[1]HICOE—Centre for Biofuels and Biochemical Research (CBBR), Institute of Self-Sustainable Building (ISB), Universiti Teknologi PETRONAS, Seri Iskandar 32610, Malaysia

[2]Chemical Engineering Department, Universiti Teknologi PETRONAS, Seri Iskandar 32610, Malaysia

[3]School of Chemical Engineering, College of Engineering, Universiti Teknologi MARA, Shah Alam 40450, Malaysia

[a]usman_20001935@utp.edu.my, [b]nurul.amran@utp.edu.my, [c]shafirah.samsuri@utp.edu.my, [d]syafiqhazwan@uitm.edu.my

Keywords: Fruit Palm Peels, Oil Extract, Bioactive Compounds, S-CO_2 Extraction

Abstract. Supercritical CO_2 extraction (S-CO_2) was used as a green technology to extract oil from the milled fruit palm peels as a sustainable feedstock. The process was carried out at five different temperatures (40, 50, 60, 70, and 80 °C), constant pressure of 25 MPa, a flow rate of 5 mL/min, and a co-solvent ratio of 5% volume. An optimum oil extract yield of 3.95% was obtained at the highest temperature (80 °C) after 150 mins of extraction. Fourier Transform Infrared (FTIR) and Scanning Electron Microscope (SEM) were explored to characterize the chemical composition of the extract and simultaneously evaluated the presence of bioactive compounds using High-Performance Liquid Chromatography (HPLC). Phytochemicals present in the extracts, namely, quercetin and beta-carotene, were quantified as 1.1929 and 2.3354 mg/g, respectively, while there was no amount of gallic acid detected in the extract.

Introduction

Oil palm fruit is a monocotyledonous plant belonging to the palm family Palmae, which comprises *E. guineensis,* and *E. oleifera*, with suitable and different morphological traits [1]. Palm oil is increasingly becoming a demanded commodity worldwide, particularly for use in food products and other applications, and as a revenue generator through export. Presently, Indonesia and Malaysia are the major palm oil producers, accounting for 56% and 31% of total global production, respectively [2]. Studies have shown that about 42 million tons of global oil production are contributed by palm oil, and this tabled it as the most traded vegetable oil that engulfed nearly 60% of the global oil exports [3,4]. The phenomenon compelled the growing rate of palm oil plantations in most palm-producing countries and is expected to multiply by 2030, considering its substantial growing demand.

Notably, the palm fruits contained carotenoids and chlorophylls as naturally occurring pigments, with the fruit containing plenty of these phytochemicals than the oil. Therefore, the utilization of fruit palm peel extract as a source of bioactive compounds such as antioxidants, antimicrobials, and as a raw material for many industrial applications in food, pharmaceuticals, body care, biofuel, and other green chemicals is a promising option. However, the potency of the recovered value-added bioproducts and their stabilities depends on the plantpart, extraction technique employed and other preparatory conditions [5, 6]. Consequently, S-CO_2 extraction is placed forward as a clean and green extraction technique, that is environmentally friendlier, non-polluting, and toxic-free. This ruled out the drawbacks constituted by the conventional methods such as time consumption, use of hazardous solvents, high temperature, and intensive energy demand.

Moreover, the non-polarity of the extracting solvent (CO2), exempted the possibility of oxidization which often affects the stability and bioactivity strength of the extracts obtained through conventional extraction methods. Hence, these are some of the merits that placed S- CO2 extraction forward as the most plausible technique for the extraction of bioactive compounds [7].

Material and methods

Materials. Fresh palm fruit was obtained as a gift from a palm tree farm in the Bota area, Perak-Malaysia. The sample was washed to remove dirt, and other particle bounded, peeled off using stainless steel knife, and subjected to an oven (Memmert, UN55BO), drying at 50 °C for 72 hours[8]. Thereafter, pulverized milling machine (FRITSCH, V2A 1.4301) was used to grind the dried sample and sieved with a 0.5 mm size mesh. Next, the milled sample was transferred to an airtight bottle and stored in a refrigerator for upward extraction using S-CO2. Figure 1 shows the various preparatory steps of the mango peels. Besides, HPLC-grade chemicals such as beta-carotene (≥ 93%), quercetin (≥ 95%), gallic acid (97.5 %), and other high-purity solvents like methanol,acetonitrile, acetic acid, and trichloromethane were used in this research.

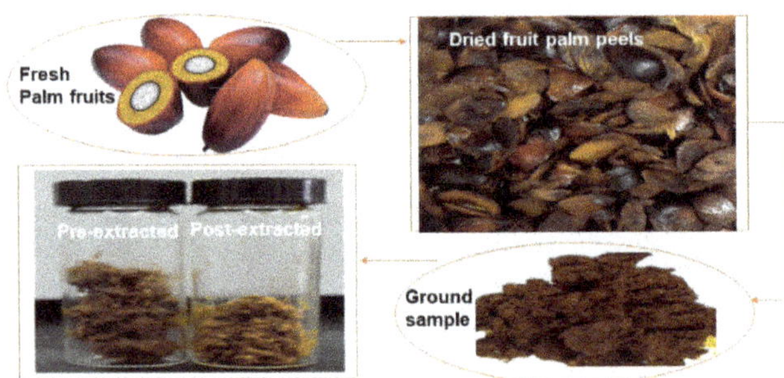

Fig. 1: Sample preparation steps.

S-CO2 Extraction. The milled fruit palm peels were extracted using S-CO2 extraction according to a procedure reported by Ruslan et al. [9], in a modified version. The extraction setup comprises the extracting solvent reservoir (CO2 supply tank) supplying liquid chilled CO2 (0 °C) via conducting pipe to the extraction vessel inside the oven. The pumping rate typically controls the flow rate, while pressure drop is controlled by the automated back pressure regulator, ABPR (Jasco, BP-2080 Plus). Extraction conditions for the significant S-CO2 extraction parameters used were a pressure of 25 MPa, a flow rate of 5 mL/min, and a co-solvent ratio of 5% vol. Next, 5.0 g of the milled peel sample was loaded into the extraction vessel (Jasco, EV- 3-50-2), and the time was noted after the first released pressure purge from the ABPR was completed. Subsequently, the oil extract was collected in an amber glass bottle immersed in an ice bath for 30, 60, 90, 120, and 150 mins. Finally, the collected extracts were diluted with 2.0 mL of n-hexane and stored diluted extracts in a refrigerator at 4 °C for upward HPLC analysis. Consequently, the percentage extract yield (XX_0 %) of the milled fruit palm was calculated using equation 1.

$$X_0\,(\%) = \frac{Weight\ of\ extract\ obtained}{Weight\ of\ milled\ sample\ used}\ \mathrm{X}\ 100 \qquad (1)$$

HPLC Analysis. SHIMADZU 20 AD HPLC profiling, integrated with an ultraviolet/diode array detector (UV/DAD), was used to isolate three bioactive compounds, namely, beta-carotene, quercetin, and gallic acid in the fruit palm oil extract derived from S-CO2 extraction. In this work, a reversed-phase (RP) separation using a poroshell column, 120 EC-C18 (4.6 mm × 100 mm, 2.7 μm) was employed for this determination. Solvent compositions of acetonitrile/acetic acid and methanol/trichloromethane with a volume ratio of (98:2, v/v) each were prepared, filtered, degassed, and used as the mobile phases for the identification of the polyphenols (quercetin & gallic acid) carotenoids (beta-carotene) respectively, [10]. Before the analysis, the oil extracts were diluted with solvents using a 1:9 ratio, v/v to reduce their concentrations. Next, 3.0 mL of each of the diluted extracts and their respective bioactive compounds' external standard solutions were filtered using a 0.45 μm nylon syringe. Finally, 20 μL of each sample was used for injection into the HPLC, maintained at 40 °C column temperature, 1.0 mL/min flow rate, and 280 and 454 nm UV detector ranges for detecting quercetin, gallic acid, and beta-carotene, respectively.

Physicochemical characterization. The superficial structure of the milled banana peels was analyzed using SEM, Hitachi TM3030, Tabletop. The dried sample was stuck to the specimen stub with conductive tape [11]. Next, the specimen stub was tied to the specimen holder and placed into the specimen chamber. The micrographs were recorded at a medium magnification of 1000 x 100 μm. The sample extract was subjected to infrared spectral analysis within the range of 4000 – 650 cm^{-1} according to a procedure reported by Bello et al. [12], using Cary 630 Agilent Technologies.

Results and Discussion

Extract Yield. The experimental yield *(X_o)* results obtained for this work, at various temperatures of 40 – 80 °C over increasing extraction time from 30 to 150 min, are shown in Table 2. According to the results, the lowest extract yield was obtained at the initial time (30 min) while the highest was recorded at the maximum extraction time. Overall, the lowest and highest yields recovered were 2.61and 3.95%, obtained at 40 °C, 30 min, and 80 °C, 150 min, respectively. This shows that the extract yield increases proportionally as the temperature rises from 40 – 80 °C, and that role played by the extraction time follows the same trend. Therefore, in kinetic terms, it can be referred to that the rate of extract recovery from the fruit palm peels using S-CO2 carried out in this work is a forward reaction [13]. However, at the initial extraction time (30min), the difference in yield between the two boundary temperatures (40 and 80 °C) is 0.21%.

On the other hand, a difference of 0.74% yield was recorded between the same temperature gap at the final extraction time of 150 min. This justified that prolonging extraction time from 30 – 150 min, has a direct impact on the recovery of the extract, where an increment in the extract yield by 0.53% was recorded. Looking at the impact of temperature on the recovery of the oil extract over time shown in Figure 2, the yield increases linearly as the temperature rises over the respective extraction times. However, at the lowest temperature (40°C), the difference in yield between the initial extraction time (30 min) and the final extraction time (150 min) is 0.6%. While the difference in yield between the same extraction time boundary (30 - 150 min) at the highest temperature (80 °C), is 1.13%. This significant difference in yield between the highest and lowest temperature over the same extraction time range can be explained based on the vital role played by temperature in reducing solvent density and enhancing solubilization of the solute by the extractant (CO2) solvent. Hence, this trend for the effects of temperature on the oil extract yield observed is in line with the conditioned theory of S-CO2 which stipulates that the yield of the extract generally increases at higher temperatures due improve solute mass transfer into the plant matrices [14]. Consequently, the obtained experimental extract yield ranging from 2.61 - 3.95% was appreciably higher, especially in the context of S-CO2 extraction. This must be connected to the greater molecular interactions between the compounds present in the fruit palm peel extracts (carotenoids) and the non-polar CO2 (extracting solvent), which utterly dissolves the compounds due to the similarity law of like dissolve like since the carotenoids were also non- polar [15].

Table 1: Results of fruit palm peel extracts derived from SCO_2 at different temperatures

		Extract yield over time (%)				
Samples	Time (min)	40 °C	50 °C	60 °C	70 °C	80 °C
Milled fruit palm peels	30	2.61	2.65	2.68	2.71	2.82
ü	60	2.72	2.78	2.85	3.00	3.18
ü	90	2.90	3.15	3.41	3.57	3.63
ü	120	3.15	3.41	3.68	3.76	3.82
ü	150	3.21	3.50	3.75	3.84	3.95

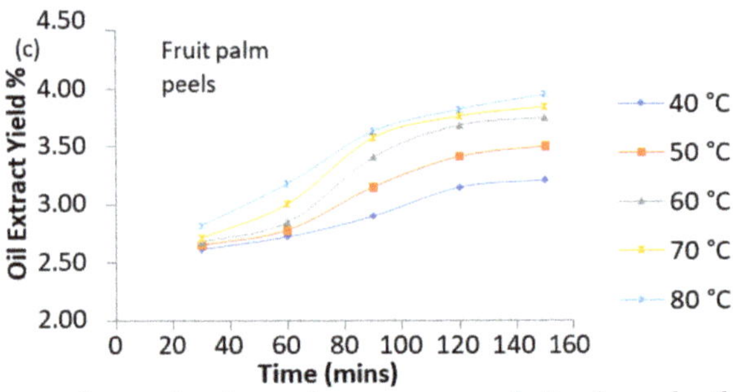

Fig. 2: Influence of extraction time and temperature on fruit palm peels oil extract yield

Quantification of bioactive compounds by HPLC. The disparity in the retention times, peak areas, chromatographic peaks of the spectra for the sample and those of the standard compounds, and the quantity of the targeted bioactive compounds was observed. However, the fitting of the regression curves shows good linearity with R2 values of 0.9754, 0.9955, and 0.9662 for quercetin, gallic acid, and beta-carotene, respectively (Table 2). The quantification of the targeted bioactive compounds, namely beta- carotene, quercetin, and gallic acid present in the oil extract, was carried out using the calibration curve determination method [16], where a graph of peak area (µV.s) on the Y-axis against concentrations (ppm) of the standard solutions on the X-axis was plotted. The generated calibration equations, peak areas, retention times, and the quantified amount of the respective target compounds are shown in Table 2. The estimated quantities of the target compounds in the oil palm extract samples calculated in milligrams per gram (mg/g) are 2.3354 and 1.1929 for the beta-carotene and quercetin, respectively. On the other hand, no amount of gallic acid was detected in this work. However, the result obtained in this work is in agreement with other previous research findings [17], establishing the presence of a higher number of carotenoids (5 – 7 mg/g) in fruit palm.

Table 2: Standard curve analysis of the external standards obtained by RP-HPLC–DAD, and quantified mass of the samples in mg/g

Standard compounds	Oil extracts	Retention time (min)	Peak area (µV.s)	Standard equation	R^2	Quantified extracts (mg/g)
Quercetin	-	2.563	16200537	$y = 16557x - 2x10^6$	0.98	-
	Fruit palm peels	2.581	194499	-	-	1.1929
Gallic acid	-	2.459	2772466	y = 45815x - 54305	0.998	-
	Fruit palm peels	ND	ND	-	-	ND
Beta-carotene	-	54.48	625881	y = 633.04x - 43379	0.97	
	Fruit palm peels	53.49	120890	-	-	2.3354

ND = not detected

Qualitative analysis using FT-IR. Absorption bands ranging between 2957.80 – 723.42 cm^{-1} were detected in the FT-IR spectrum of the fruit palm peel oil extract, as shown in Figure 3. A procedure reported by Barbara, [18] was followed in assigning the various vibrational modes. An absorption band at 1747.81 cm^{-1} could be assigned to C=C stretching consistent with Csp2 hybridized alkene that continued to overlap within the ring system. Another notable peak at 1462.51 cm^{-1} is linked to the C=C stretching in the phenyl groups attached to the two ends of the beta-carotene structure. Similarly, the appearance of low-intensity stretches at 1378.50 and 1242.83 cm^{-1} are for the methyl substituent symmetry (-CH3). Other absorption bands detected within the fingerprint region, precisely 1161.99 cm^{-1} and 1065.64 cm^{-1}, might be accorded to C-C skeletal having Csp3 hybridized. However, peaks spotted at 2957.80, 2923.42, and 2855.89 cm^{-1} are for OH stretches associated with traces of polyphenol (quercetin). Overall, the purity of the oil extract can be confirmed by looking at the excellent baseline resolution obtained in the spectrum [19].

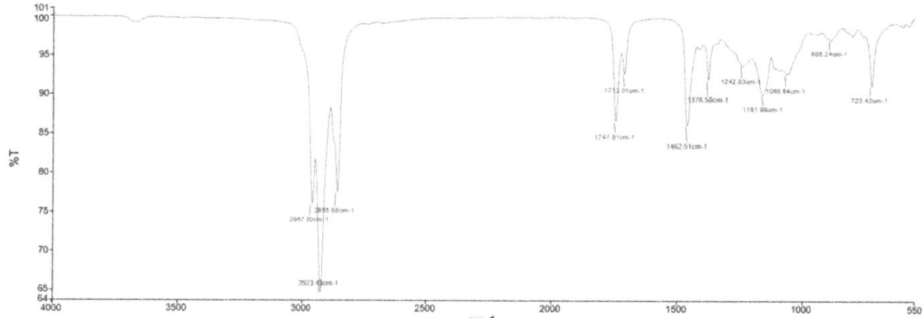

Fig. 3: FT-IR spectrum of fruit palm peel oil extract

Morphology Analysis of SEM. The surface morphology of the fruit palm peels before and after extraction studied using SEM, are presented in Figures 4 a and b, respectively. The images were viewed at a medium magnification of 1000 x 100 μm for both pre-and post-extracted samples. Comparatively, a smoother and relatively hydrated nature of the pre-extracted samples, appearing darker (Figure 1 and 4a) was due to unextracted oil surrounding the particle [20]. However, a more porous surface and lighter appearance (Figures 1 and 4b) of the post-extracted sample suggested the dissolution or rupture of the cell wall matrix and exit of oil that embedded the plant matrices [21]. To this end, a visible difference was observed concerning the physical appearance of the particle surface and colour. This feature was uniquely attributed to the selectivity of the SCO2 extraction process, where the extractable solutes broke down from inside the plant matrices due to high temperature and pressure compression leading to the withdrawal of oil extract. A similar observation was previously reported by Chai et al., [22] on the pre and post-extraction sample of papaya Linn. leaves using SCO_2.

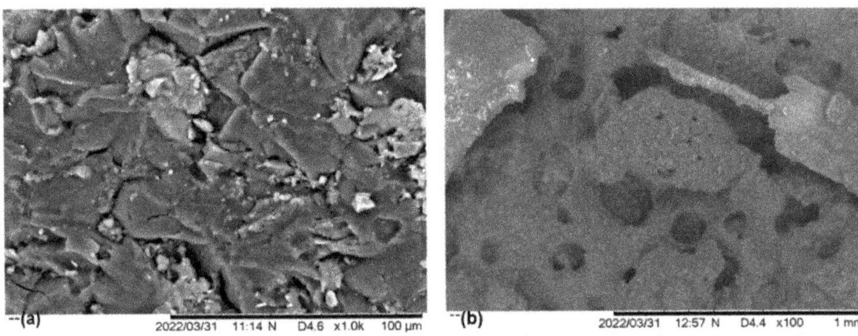

Fig. 4: Pre (a) and post-extracted (b) milled peel samples viewed at the same magnification

Conclusion

This work has explored using SCO2 as a green extraction technique to recover oil extract from fruit palm peels. Based on the results, the highest oil extract (3.95%) was recovered at the maximum temperature of 80 °C over a prolonged extraction time of 150 min. However, the lowest yield obtained was 2.61% at 40°C after 30 mins of extraction. Thus, these results established that a forward reaction is favoured considering the symmetrical increase in the oil recovery rate as the temperature increases over time. Therefore, temperature significantly increases the oil yield rate by enhancing diffusivity and solubilization of the analyte from the inner pores of the plant matrices. Quantitatively, the amount of bioactive compounds analyzed using RP-HPLC-DAD shows that the extract contained a higher quantity of beta-carotene than quercetin and gallic acid. In the qualitative characterization, the purity of the obtained oil extract as claimed by the SCO2 technique was confirmed by the FT-IR analysis considering the good baseline resolution and absence of unwanted peaks. Moreover, the visible difference in the physical appearance of the pre-and post-extracted samples further attested to the uniqueness of the SCO2 technique.

Acknowledgement

The authors would like to acknowledge the funding supports from the Ministry of Higher Education (MOHE) Malaysia through HICOE Grant (cost centre 015MA0-052/015MA0-104/15MA0-136) to CBBR.

References

[1] A. Camacho-Villalobos, F. Serna, J. Flores, H. Flores, P. Manrique, and J. Bendezu, "Morphological and molecular characterization of an Elaeis oleifera (H.B.K) Cortes germplasm collection located in Ucayali, Peru," PLoS One, vol. 16, no. 5 May, pp. 1–18, 2021, https://doi.org/10.1371/journal.pone.0250445

[2] L. M. Ayompe, M. Schaafsma, and B. N. Egoh, "Towards sustainable palm oil production: The positive and negative impacts on ecosystem services and human wellbeing," J. Clean. Prod., vol. 278, p. 123914, 2021, https://doi.org/10.1016/j.jclepro.2020.123914

[3] M. A. M. Anuar, N. A. Amran, and M. S. H. Ruslan, "Optimization of Progressive Freezing for Residual Oil Recovery from a Palm Oil-Water Mixture (POME Model)," ACS Omega, vol. 6, no. 4, pp. 2707–2716, 2021, https://doi.org/10.1021/acsomega.0c04897

[4] H. J. W. Mutsaers, "The challenge of the oil palm: Using degraded land for its cultivation," Outlook Agric., vol. 48, no. 3, pp. 190–197, 2019, https://doi.org/10.1177/0030727019858720.

[5] N. Aini, U. Bello, M. Sya, and H. Ruslan, "The role of antioxidants in improving biodiesel ' s oxidative stability, poor cold flow properties, and the effects of the duo on engine performance : A review," vol. 8, no. 3, 2022, https://doi.org/10.1016/j.heliyon.2022.e09846

[6] E. M. Coelho, M. E. A. O. De Souza, L. C. Corrêa, A. C. Viana, L. C. De Azevêdo, and M. D.S. Lima, "Bioactive compounds and antioxidant activity of mango peel liqueurs (Mangifera indica L.) produced by different methods of maceration," Antioxidants, vol. 8, no. 4, pp. 1–11, 2019, https://doi.org/10.3390/antiox8040102

[7] N. Mękus, A. Iqbal, K. Marszałek, C. Puchalski, and A. Świergiel, "Green chemistry extractions of carotenoids from daucus carota L.-Supercritical carbon dioxide and enzyme-assisted methods," Molecules, vol. 24, no. 23, pp. 1–20, 2019, https://doi.org/10.3390/molecules24234339

[8] R. R. Mphahlele, O. A. Fawole, N. P. Makunga, and U. L. Opara, "Effect of drying on the bioactive compounds, antioxidant, antibacterial and antityrosinase activities of pomegranate peel," BMC Complement. Altern. Med., vol. 16, no. 1, pp. 1–12, 2016, https://doi.org/10.1186/s12906- 016-1132-y

[9] M. S. H. Ruslan, Z. Idham, L. Nian Yian, M. A. Ahmad Zaini, and M. A. Che Yunus, "Effect of operating conditions on catechin extraction from betel nuts using supercritical CO2-methanol extraction," Sep. Sci. Technol., vol. 53, no. 4, pp. 662–670, 2018, https://doi.org/10.1080/01496395.2017.1406947

[10] Y. Dai and K. H. Row, "Isolation and Determination of Beta-Carotene in Carrots by Magnetic Chitosan Beta-Cyclodextrin Extraction and HPLC," Anal. Lett., vol. 52, no. 11, pp. 1828–1843, 2019, https://doi.org/10.1080/00032719.2019.1570245

[11] S. R. Sommano, P. Ounamornmas, M. Nisoa, S. Sriwattana, P. Page, and G. Colelli, "Characterisation and physiochemical properties of mango peel pectin extracted by conventional and phase control microwave-assisted extractions," Int. Food Res. J., vol. 25, no. 6, pp. 2657–2665, 2018.

[12] G. A, Shallangwa Bello, U., and A. A. Idris, S. O., 1 Musa, M., and 1 Aisha, "Synthesis, Spectroscopic Characterization of Cu(II) Complexes Derived from 6-hydroflavone Schiff bases.," Bayero J. Appl. Sci, vol. 11, no. 2, pp. 136–141, 2018, https://doi.org/http://dx.doi.org/10.4314/bajopas.v11i2.16

[13] S. T. Mgoma, M. Basitere, and V. V. Mshayisa, "Kinetics and thermodynamics of oil extraction from South African as avocados using hexane as a solvent," South African J. Chem. Eng., vol. 37, pp. 244–251, 2021, https://doi.org/10.1016/j.sajce.2021.06.007

[14] R. P. F. F. da Silva, T. A. P. Rocha-Santos, and A. C. Duarte, "Supercritical fluid extraction of bioactive compounds," TrAC - Trends Anal. Chem., vol. 76, pp. 40–51, 2016, https://doi.org/10.1016/j.trac.2015.11.013

[15] M. P. Garcia-Mendoza, J. T. Paula, L. C. Paviani, F. A. Cabral, and H. A. Martinez-Correa, "Extracts from mango peel by-product obtained by supercritical CO2 and pressurized solvent processes," Lwt, vol. 62, no. 1, pp. 131–137, 2015, https://doi.org/10.1016/j.lwt.2015.01.026

[16] C. Malaphong, A. Tangwanitchakul, S. Boriboon, and N. Tangtreamjitmun, "A simple and rapid HPLC method for determination of S-allyl-L-cysteine and its use in quality control of black garlic samples," Lwt, vol. 160, no. February, p. 113290, 2022, doi:10.1016/j.lwt.2022.113290

[17] M. Sundalian, A. Zainuddin, and Anita, "Analysis of carotenoid composition in oil palm fruits (Elaeis guineensis jacq.) from several varieties: A review," Biointerface Res. Appl. Chem., vol. 12, no. 2, pp. 1470–1479, 2022, https://doi.org/10.33263/BRIAC122.14701479

[18] S. Barbara, Analytical Techniques in Science.Infrared Spectroscopy; Fundamentals and Application", vol. 8. John Wiley & Sons, Inc., London New York, 2004.

[19] F. Kormin, M. Khan, and A. C. Iwansyah, "Microwave assisted extraction; phytochemical evaluation of Malaysian palm oil trunk epiphytes ferns," Int. J. Pharm. Pharm. Sci., vol. 8, no. 4, pp. 174–180, 2016.

[20] L. Baldino, G. Della Porta, L. S. Osseo, E. Reverchon, and R. Adami, "Concentrated oleuropein powder from olive leaves using alcoholic extraction and supercritical CO2 assisted extraction," J. Supercrit. Fluids, vol. 133, no. June 2017, pp. 65–69, 2018, https://doi.org/10.1016/j.supflu.2017.09.026

[21] Y. H. Tan, M. O. Abdullah, C. Nolasco-Hipolito, and Y. H. Taufiq-Yap, "Waste ostrich- and chicken-eggshells as a heterogeneous base catalyst for biodiesel production from used cooking oil: Catalyst characterization and biodiesel yield performance," Appl. Energy, vol. 160, pp. 58–70, 2015, https://doi.org/10.1016/j.apenergy.2015.09.023

[22] Y. H. Chai, S. Yusup, M. S. H. Ruslan, and B. L. F. Chin, "Supercritical fluid extraction and solubilization of Carica papaya Linn. leaves in a ternary system with CO2 + ethanol solvents," Chem. Eng. Res. Des., vol. 156, pp. 31–42, 2020, https://doi.org/10.1016/j.cherd.2020.01.025

Materials Research Proceedings 29 (2023) 43-50 https://doi.org/10.21741/9781644902516-6

Microwave degradation of carbon fiber reinforced plastics in choline chloride

KAZUTOSHI Ikenaga[1,a*], MASYU Miyakawa[1] and, KATSUKI Kusakabe[1,b]

[1]Division of Applied Chemistry, Graduate School of Engineering, Sojo University, 4-22-1 Ikeda, Nishi-ku, Kumamoto, 860-0082 Japan

[a]ikenaga@nano.sojo-u.ac.jp, [b]kusakabe@nano.sojo-u.ac.jp

Keywords: Microwave, Degradation, Carbon Fiber Reinforced Plastics, Choline Chloride, Chemical Recycling

Abstract. In spite of resin recycling of carbon fiber reinforced plastics (CFRP) is very important as well-known on the viewpoint of environmental protection and resource conservation, the development of effective resin recycling had hardly been investigated. In this study, we investigated the operating conditions for microwave decomposition of CFRP using various alcohols as solvents. In order to avoid rapid absorption of microwaves into carbon fibers (CF), choline chloride was added to the solvent for microwave degradation. Ethylene glycol monoallyl ether (EGMA) was used as a solvent in order to reuse the resin part. Double bonds could be introduced into the resin degradation products. The preparation of cured products by condensation reaction by using the resin degradation products was also successful.

Introduction

It has been well known that the use of carbon fiber reinforced plastic (CFRP) is extremely effective in reducing the weight of sports tools, such as tennis rackets, fishing rods, and golf shafts. In recent years, lightweight airplanes and automobiles made of CFRP have become popular to save petroleum-based fuels for climate change strategy. General CFRP is a composite material manufactured from epoxy resin with a carbon fiber framework and is expected to be one of the next-generation materials with hard and tough properties. However, most waste CFRP products are disposed of in landfills without any treatment due to their hardness. Carbon fiber (CF), produced by carbonizing polyacrylonitrile or pitch at about 3000°C, is a valuable carbon resource that can be reused by recovering it from CFRP. Thus, the recycling of CFRP is an important issue that needs to be solved immediately in terms of environmental protection and resource conservation [1, 2].

Recycling methods of CFRP by distillation carbonization [3, 4], supercritical fluid [5 – 7], atmospheric pressure dissolution [8], and microwave thermolysis [9] have been reported. This distillation carbonization method uses that the gas produced in the dry distillation of CFRP as a heating source burns through the dry distilled CFRP to recover CF finally [3, 4]. Therefore, the resin portion cannot be recycled in this method. In the supercritical water treatment of acid anhydride-cured CFRP, the resin decomposed at a reaction pressure of 26 MPa and a reaction temperature of 380°C, producing phenolic resin monomer as a product. When supercritical methanol was used at a reaction pressure of 10 MPa and a reaction temperature of 270°C, the products mainly contained methyl ester structures. Furthermore, supercritical acetone was also reported to be effective for amine-cured CFRP [5, 6]. In the atmospheric pressure dissolution, Benzyl alcohol solvolysis at 200°C in the presence of a tripotassium phosphate catalyst under ambient pressure conditions requires a prolong reaction time of 10 hours [8]. However, these recycling methods of CFRP have only recovered carbon fiber, and a total recycling method that includes the reuse of epoxy resin has not been established [1, 2].

For glass fiber reinforced plastics (GFRP), we recently found that the non-catalytic pressurized microwave degradation method is effective for recycling resin parts and glass fibers [10]. However, when the pressurized microwave degradation method is applied to CFRP, CF strongly absorbs microwaves to cause discharge phenomena and often the generation of sparks. At this time, it is expected that low-boiling compounds will rapidly evaporate or flammable compounds will ignite around the CF, causing an uncontrollable and rapid increase in pressure due to the gases generated.

In this study, we investigated the operating conditions for microwave decomposition of CFRP using various alcohols as solvents. Furthermore, in order to avoid rapid absorption of microwaves by CF, choline chloride as an ionic compound was added to the solvent for microwave decomposition. Ethylene glycol monoallyl ether (EGMA) was used as a solvent in order to reuse the resin part. Double bonds were introduced into the resin degradation products [10], and polymerization by condensation reaction was attempted.

Experimental section

Materials

Small pieces (4 mm x 4 mm, thickness 3mm) of carbon fiber reinforced plastic (CFRP-A, Japan) were used for microwave degradation. FT-IR measurement of CFRP-A confirmed that the resin portion is acid anhydride cured, based on the existence of a C=O absorption band at 1731.9 cm^{-1}. Resin content in CFRP-A was found to be 43.7 % based on the mass of carbon fiber after the combustion experiment. Ethylene glycol (EG), diethylene glycol (DEG), triethylene glycol (TEG), benzyl alcohol, glycerol, phenol, ethylene glycol monoallyl ether (EGMA), and deuterated chloroform were supplied by Kanto Chemical Co., Inc (Japan). Choline chloride, diethanolamine hydrochloride (DAH), and triethanolamine hydrochloride (TAH) as ionic compounds were supplied by Tokyo Chemical Industry Co., Ltd. (Japan). Tetramethylsilane (TMS, ACROS ORGNICS) and 1,4-bis(trimethylsilyl)benzene (BTMSB, Tokyo Chemical Industry Co., Ltd.) were used as internal standards for NMR. Unsaturated polyester (157BQTN) was supplied by Showa Denko (Japan), and curing agent (Kayamek M) supplied by Kayaku Akzo.

Microwave degradation of CFRP (Fig. 1)

Microwave Reactor (Biotage Initiator+8, Sweden) was used for microwave degradation of CFRP-A. A 5-mL glass vial was used for pressurized microwave degradation of CFRP-A. The vial containing solvent, stirrer, and CFRP-A chip was mounted on the Initiator+8. Microwave output was controlled manually on a continuous basis so that the prescribed temperature and pressure were maintained during microwave irradiation. The output operation was performed automatically by a 23-step execution program with reference to the appropriate manual output controls.

Extraction of the resin degradation products was performed using 20 mL of heated chloroform for 90 minutes. Suction filtration was used to separate the extract from the insoluble material. The extract was concentrated by evaporating the chloroform using an evaporator under reduced pressure. Kugelrohr distillation was performed to separate the degradation residues from the higher boiling compounds. The weight loss and the degradation ratio of the resin were determined from the weight of the resin degradation products after drying. In the degradation reaction with choline chloride, the extraction of choline chloride was repeated three times with 20 mL of water after extraction with heated chloroform.

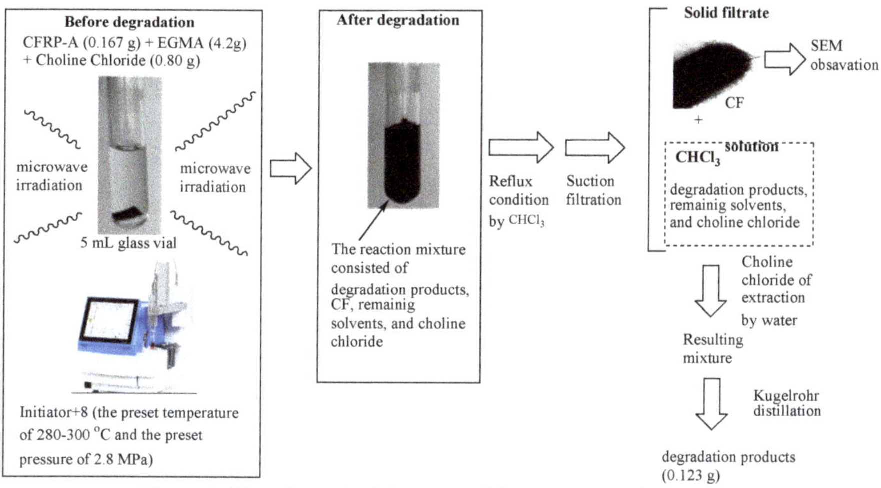

Figure 1. The schematic daiagrams of the experimental procedure

The weight reduction ratio of CFRP and the reaction ratio of the resin are defined by the following equations.

The weight reduction ratio of CFRP = 1 – (Weight of residue)/(Initial weight of CFRP) (1)
The degradation ratio of the resin =
(The weight reduction ratio of CFRP)/(Resin content of CFRP) (2)

Characterization

Fourier transform infrared (FT-IR) spectra were measured via an attenuated total reflection (ATR) method using a Perkin Elmer Spectrum 100 spectrometer (Perkin Elmer, USA). The nuclear magnetic resonance (NMR) spectra were recorded in deuterated chloroform using TMS or BTMSB by a superconducting Fourier transform NMR spectrometer ESC270 (JEOL, Japan). The scanning electron microscopy (SEM) image of CF after the degradation was observed by 3D real surface view microscope VE-8800 (Keyence Corp., Japan).

Results and discussion

Microwave Degradation of CFRP

In microwave heating of CFRP, carbon fibers in CFRP selectively absorb microwaves and are rapidly heated from the inside. Table 1 shows the results of CFRP degradation by pressurized microwave heating when various alcohols are used as solvents. For the first CFRP degradation reaction using benzyl alcohol and glycerol (Entry 1), the Initiator+8 automatically stopped after 18 minutes from the start of the reaction due to the collapse of the vial by pressurization. The next microwave heating was performed with continuous manual adjustment of microwave output to prevent rapid pressure increase. As a result, after 3 hours of microwave irradiation, the resin degradation ratios with DEG and TEG were 42.3% (Entry 2) and 47.6% (Entry 3), respectively. When DEG and GC were mixed at a mass ratio of 1:1 (Entry 4), the temperature could be maintained at 270°C after 1 hour of microwave irradiation, resulting in a resin degradation ratio

Materials Research Proceedings 29 (2023) 43-50 https://doi.org/10.21741/9781644902516-6

Table 1. Pressurized microwave degradation of CFRP-A[a].

Entry	CFRP / g	Solvent / g	Temperature / °C	Wt. Reduction ratio /%	Degradation ratio /%
1	0.142	Benzyl alcohol 2.00 + Glycerol 3.34	260	-	-
2[b]	0.259	DEG 5.00	170	18.5	42.3
3[b]	0.358	TEG 5.01	210	20.8	47.6
4[b]	0.212	DEG 3.47 + Glycerol 3.50	270	41.8	95.6
5[c]	0.190	DEG 3.50 + Glycerol 3.50	280	42.1	96.3
6[c]	0.188	Glycerol 3.50 + DEG 3.50	270	42.8	97.9

a) These reactions were carried out at 2.8 MPa for 60 min. b) Manual output control. c) Automatic output control.

Step	Irradiation time / min	Power / W	Temperature / ° C	Pressure / bar
0	0			
1	1	400	290	10
2	5	350	290	12
3	9	250	290	14
4	13	250	290	15
5	17	250	290	16
6	21	250	290	17
7	23.5	200	290	18
8	24.5	150	290	19
9	26.5	150	290	19
10	26.75	150	290	19
11	27	125	290	19
12	27.25	100	290	19
13	27.5	80	290	19
14	27.75	60	290	19
15	29.75	60	290	20
16	32.75	60	290	21
17	35.75	60	290	22
18	38.75	60	290	23
19	41.75	60	290	24
20	44.75	60	290	25
21	47.75	60	290	26
22	50.75	60	290	27
23	60	60	290	18

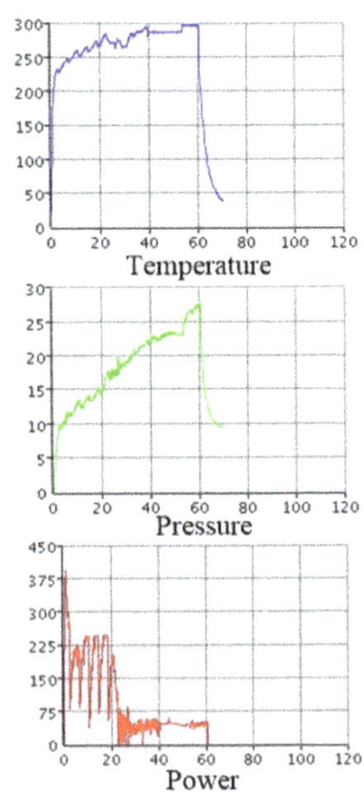

Figure 2. Setting of power, temperature, and pressure in the automatic output control program

Figure 3. SEM observations of (a) virgin CF and (b) recovered CF of Entry7.

of 96.3 %. Approximately 100 experiments in which the pressure and temperature were gradually increased by manually controlling the output were examined, and it was confirmed that the temperature and pressure could be stably and simultaneously raised and boosted even after simplifying to automatic control of the 23-step output created as shown in Fig.2 (Entry 5). The results of the experiments repeated by automatic output control were almost identical, confirming reproducibility (Entry 6).

Fig. 3 shows SEM images of CF surface (photo (b) in Fig. 3) recovered after the Entry 7 experiment. Compared to virgin CF (photo (a) in Fig. 3), there was no significant damage or dimensional change due to microwave heating, but grooves were formed along the fiber axis on the surface.

Microwave Degradation of CFRP-A with Choline Chloride

In the experiments using alcohols as solvents, complex operating conditions were obtained to obtain high resin degradation rates, but excessive microwave absorption by the carbon fibers in the CFRP formed internal hot spots and sometimes caused the breakdown of the pressure-resistant vessel. Therefore, to suppress the selective absorption of microwaves by CF, we performed microwave degradation of CFRP-A by adding choline chloride as an ionic compound with high microwave absorption capacity into EGMA (Entry7 in Table 2). Surprisingly, in the degradation

Step	Irradiation time / min	Power / W	Temperature / °C	Pressure / bar
0	0			
1	1	400	300	20
2	5	350	300	27
3	54	250	300	30

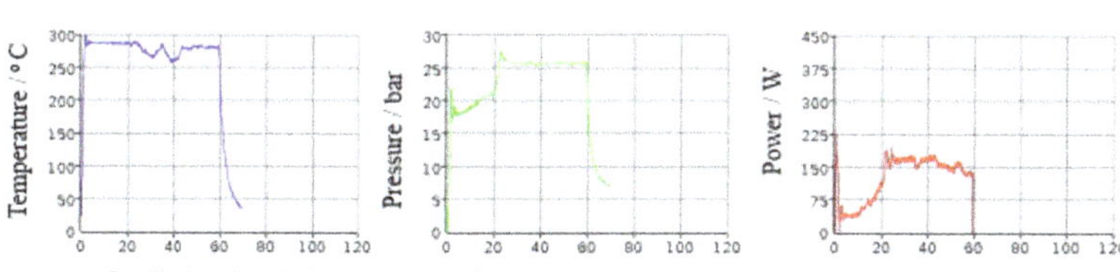

Figure 4. Setting of simple three-step program for pressure-controlled degradation reaction in Entry 7.

Table 2. Effect of choline chroride derivatives for pressurized microwave degaradation of CFRP-A[a].

Entry	CFRP-A / g	EGMA / g	Additives /g	Wt. Reduction ratio /%	Degradation ratio /%
7	0.163	4.20	Choline Chloride 0.80	40.0	91.5
8	0.116	1.25	Choline Chloride 1.59 + Phenol 2.16	39.9	91.2
9	0.134	1.25	Choline Chloride 1.78 + EG 1.97	39.6	90.2
10[b]	0.145	4.14	Choline Chloride 0.90	39.6	81.6
11	0.149	4.03	DAH 0.99	40.7	93.2
12	0.144	4.04	TAH 1.01	40.0	91.5

a) These reactions were carried out at 2.6 MPa and 290°C for 60 min. b) The reaction time was 30 min.

experiments shown in Table 2, a simple three-step program was used to perform a pressure-controlled degradation reaction (Fig. 4). The degradation of CFRP-A using the mixture of EGMA and choline chloride resulted in a resin degradation ratio of 91.5 %. Further addition of EG or phenol as an aid to miscibility in the mixture of choline chloride and EGMA was also investigated (Entry 8 and 9). However, the addition of EG and phenol did not change the resin degradation ratio. DAH and TAH were used in the microwave decomposition experiments of CFRP-A instead

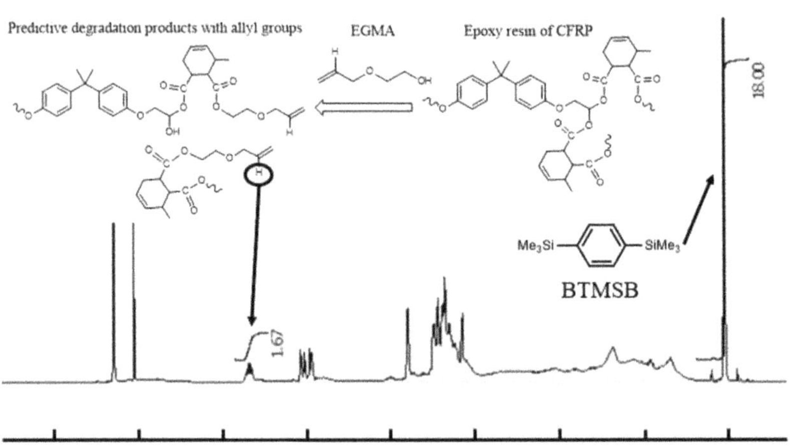

Figure 5. ^{1}H-NMR analysis for the calculation of allyl groups in the degradation products by using BTMSB.

of choline chloride (Entry 11 and 12). When DHA was used under the same conditions as Entry 7, the resin degradation ratio was 93.2%. when TAH was used, Initiator+8 stopped after 7 minutes microwave irradiation due to a sudden increase in pressure. The next reaction, in which the ratio of TAH was increased, yielded results comparable to those obtained with choline chloride. These results fully suggest that choline chloride derivatives with quaternary ammonium salt and ethanol structures and EGMA form an ionic mixture that can effectively suppress the microwave absorption of CF.

Structural Analysis of Resin Degradation Products

Fig.3 shows ^{1}H-NMR spectra of the resin degradation products. ^{1}H-NMR analysis of the resin degradation products (Entry 7) showed four chemical shifts of the crosslinking-reactive allyl groups introduced by EGMA at 3.8 (2 protons of methylene in allyl group), 4.9, 5.1, and 5.7 (three protons of vinyl moiety in allyl group) ppm. In addition, not so large chemical shifts of the methylene protons of the by-product EGMA oligomer [5] were observed around 0.5-1.5 ppm and 3.2-3.5 ppm. FT-IR analysis of the resin degradation product showed the main absorption bands of the OH group (3393.9 cm^{-1}) and the C=O group (1729.9 cm^{-1}). It is presumably suggested that a considerable amount of the EGMA oligomer was simultaneously removed by three times aqueous washing for choline chloride removal.

The use of 1,4-bis(trimethylsilyl)benzene (BTMSB, 0.0436 mmol) as an internal standard in the 1H-NMR analysis allowed the calculation of the mole concentration of allyl groups (0.0728 mmol) introduced into 0.0505g of the resin degradation product (Entry 7, Fig. 3). From the results of 1H-NMR analysis, the concentration of allyl groups in the resin degradation product was calculated to be 1.44 mmol/g.

Preparation of cured products using resin degradation products

The cured product was prepared using resin degradation products to confirm the cross-linking reactivity of the allyl group introduced by EGMA. The resin degradation product (Entry 7, 1.0 g), unsaturated polyester (157BQTN, 3.0 g), and a curing agent (Kayamek M, 0.06 g) were mixed. After 3 days of mixing and standing in the room temperature condition, the mixture was completely cured (Fig. 6). It is suggested that the allyl groups introduced into the resin degradation products are capable of cross-linking reaction with unsaturated polyesters. Furthermore, in the TG-DTA measurement of the cured product none rapid weight loss showed that the cured product had a structure with a sufficiently dense three-dimensional crosslinked structure. We will plan to carry out strength measurements in the future.

Figure 6. A cured product using resin degradation products of Entry 7 and unsaturated polyester.

Summary

(1) The degradation reaction of CFRP in various alcohols under the pressurized microwave irradiation proceeded well without the rupture of the vessel when the pressure increase was adjusted in detail by the output control.

(2) The addition of choline chloride derivatives was effective to suppress the rapid absorption of microwave on the CF, making the output control simple.
(3) When EGMA was used as a reactive solvent, the resin decomposition products with allyl groups, which have crosslinking reactivity by curing reaction, were formed.
(4) The microwave degradation reaction of CFRP is an effective recycling method for the CF and the resin by using EGMA and choline chloride derivatives.

Acknowledgments
We would like to thank Mr. Shinji Kugimiya, Mr. Wang Chen, and Mr. Keita Moriya for their support of our experiments.

References

[1] S. Pimenta, S.T. Pinho, Recycling carbon fibre reinforced polymers for structural applications: technology review and market outlook, Waste Management, 31 (2011) 378-392. https://doi.org/10.1016/j.wasman.2010.09.019

[2] H. Sukanto, W. W. Raharjo, D. Ariawan, J. Triyono, Carbon fibers recovery from CFRP recycling process and their usage: A review, 2021 IOP Conference Series Materials Science and Engineering, 1034, 012087. https://doi.org/10.1088/1757-899X/1034/1/012087

[3] H. Itazu, H. Kamiyoshi, H. Moritomi, Recycling of Carbon Fiber using Two-stage Thermal Treatment System, Haikibutu shigenjunkan gakkaishi, 24 (2013) 371-378. https://doi.org/10.3985/mcwmr.24.371

[4] H. Itazu, M. Itazu, H. Moritomi, Apparatus for regenerative treatment of carbon fiber, JP4949123.

[5] I. Okajima, K. Watanabe, S. Haramiishi, M. Nakamura, Y. Shimamura, T. Sako, Recycling of carbon fiber reinforced plastic containing amine-cured epoxy resin using supercritical and subcritical fluids Author links open overlay panel, J. Supercrit. Fluids, 119 (2017) 44-51. https://doi.org/10.1016/j.supflu.2016.08.015

[6] I. Okajima, K. Yamada, T. Sugeta, T. Sako, Decomposition of epoxy resin and recycling of CFRP with sub- and supercritical water, Kagaku Kogaku Ronbunshu, 28 (2002) 553-558. https://doi.org/10.1252/kakoronbunshu.28.553

[7] L. Weihao, H. Haihong, L. Yun, L. Lei, C. Huanbo, L. Zhifeng, Life cycle assessment and energy intensity of CFRP recycling using supercritical N-butanol, Journal of Material Cycles and Waste Management, 23 (2021) 1303-1319. https://doi.org/10.1007/s10163-021-01206-7

[8] K. Shibata, CFRP Recycling Technology using Depolymerization of Epoxy Resin under Ordinary Pressure, Hitati-Kasei Tecnical Report, 24 (2013) 358-363. https://doi.org/10.3985/mcwmr.24.358

[9] D. Jianying, X. Lei, Z. Libo, P. Jinhui, G. Shenghui, L. Jianhua, K. Sivasankar, Recycling of carbon fibers from CFRP waste by microwave thermolysis, Processes (2019), 7(4), 207. https://doi.org/10.3390/pr7040207

[10] K. Ikenaga, H. Nagamitsu, T. Kikukawa, K. Kusakabe, A new recycling process for recycled plastic production from glass fiber reinforced plastics, 2018 OP Conference Series Materials Science and Engineering, 458, 012034. https://doi.org/10.1088/1757-899X/458/1/012034

Carbon dioxide hydrate formation in pure water and highly saline water

SIRISHA Nallakukkala[1,2,a], BHAJAN Lal[1,2,b*] HANI Abulkhair[3,4,c], ABDULMOHSEN Alsaiari[3,4,d], IQBAL Ahmad[3,4,e], EYDHAH Almatrafi[3,4,f], OMAR Bamaga[3,4,g] and AZMI Mohd Shariff[1,2,h]

[1]Chemical Engineering Department, Universiti Teknologi PETRONAS, Bandar Seri Iskandar,32610 Perak Darul Ridzuan, Malaysia

[2]Research Centre for CO_2 Capture (RCCO2C), Universiti Teknologi PETRONAS, Bandar Seri Iskandar,32610 Perak Darul Ridzuan, Malaysia

[3]Center of Excellence in Desalination Technology, King Abdulaziz University, Jedah 80200, Saudi Arabia

[4]Mechanical Engineering Department, King Abdulaziz University, Jedah 80200, Saudi Arabia

[a]sirisha_18002191@utp.edu.my, [b]bhajan.lal@utp.edu.my, [c]haboalkhair@kau.edu.sa, [d]aoalsaiari@kau.edu.sa, [e]irajboot@kau.edu.sa, [f]Ealmatrafi@kau.edu.sa [g]oabamaga1@yahoo.com, [h]azmish@utp.edu.my

Keywords: Gas Hydrate-Based Desalination, Produced Water, Water Recovery, Gas Uptake, CO_2 Hydrate

Abstract. In this study, the hydrate kinetics for CO_2 gas was assessed in treating highly saline water at 3 MPa and 275.15 K to acquire perception towards water recovery and uptake of gas for desalination purpose. The experimentation was performed using a stainless-steel reactor by implementing isochoric (constant cooling) technique in treating highly saline water and related with the deionized water system. The study discloses that CO_2 hydrate forms quicker in deionised water at 75 mins achieving uptake of CO_2 gas as 0.0575 mol/mol with a recovery of 65.7% as opposed to 83.5 mins and uptake of gas of 0.0505 mol/mol and water recovery of 45.5% in 2.8wt% saline water sample. Hence it is evident that the existence of salts slightly inhibits the formation of hydrate but still produces a higher percentage of water recovery compared to conventional technologies. The results from this study are useful for the design of the efficient reactor for hydrate desalination.

Introduction

The term "produced water" in the oil and gas segment refers to the water that is mixed with the oil throughout the extraction method and is highly saline than seawater. It is one of the major watercourses of effluent produced in the oil and gas industries, with an estimated global production in 2009 of >70 billion barrels per year, of which 21 billion barrels were generated in the US alone [1]. Produced water makes up the majority of the waste stream in production of oil and gas activities at offshore, accounting for 80% of the residues and wastes generated during the natural gas production and highly saline compared to seawater. Additionally, the quantity of produced water generated upsurges as the well's span upsurges, the production of oil drops[2]. With only 2% of fossil fuel output, the volume of water produced might reach 98% in nearly exhausted fields [3]. The water to oil ratio in the course of oil production is roughly 3:1 [4].

Produced water is a complex combination of inorganics (trace metals, salts, suspended particles), radioactive materials and organics (distributed and dissolved hydrocarbons, organic acids), as well as, frequently, residual chemical additives (corrosion inhibitors and scale inhibitors) that were added during the production of hydrocarbon. Prior to being released to the marine environment, these compounds must be handled since they may have varied degrees of impact on the receiving environment. The availability of technology, cost, and local regulatory requirements all influence how produced water is treated. Produced water management is becoming increasingly difficult as the industry attempts to recover production from mature fields and develop new fields in difficult terrain due to high water cut, the existence of emulsion, and a considerable volume of contaminants like mercury (Hg), sand, and hydrogen sulphide. Due to these difficulties, the sector must investigate new or improved practical solutions for the management disposal and treatment of produced water that are more cost-effective and sustainable.

To eliminate the dissolved solids from the highly saline water, oil and gas producers have used membrane techniques, electric separation, distillation and treatment with chemicals [4-6]. Though these techniques are good enough to treat produced water but challenged with limitations like high energy requirement, large sludge generation, scaling and fouling issues. Hence there is a need to look into alternate desalination technologies like application of gas hydrate technology. The key benefits of this technology are that it theoretically requires less energy than previous methods[7] and that it has a low propensity to damage pipelines since it functions at low temperatures above the water freezing point. Gas hydrate desalination of seawater was initiated in the 1940s and acquired significant recognition from the 1960s to the present [8]. Owing to the structure of hydrates (water + guest component), salts and other pollutants of saline water can be removed during the formation of hydrate. Salt are removed from the hydrate structure owing to the cage size, varies between 0.395 to 0.571 nm [9-10]. After the hydrate has formed, it is dissociated to produce treated or pure water[8] [11-14]. The three structure types that make up hydrates' primary structures are sI, sII, and sH [15-16]. Some of the biggest issues of hydrate-based desalination are include the slow formation kinetics of hydrate, reuse of guest molecule, and hydrate-highly saline water separation. There is substantial literature available regarding the feasibility of this technology by using different hydrate formers like refrigerants and hydrocarbons[17-21]. But the various hydrate formers used for desalination purposes in literature are not biodegradable and either form a suspension leading to difficulty in the separation of hydrate and residual, hydrate former is insoluble in water and some of the hydrate formers damage the ozone layer. Hence it is desired to look into suitable guest molecules for hydrate-based desalination purpose that can enhance moles consumption and recovery of water.

In this paper the kinetics and use of hydrate-based desalination (HBD) in place of conventional techniques like distillation and reverse osmosis systems have been discussed. This study examines to contribute insights on the possible application of gas hydrates in desalination with respect to the effect of salts in treating produced water, by using CO_2 guest gas. The kinetic evaluation of moles of gas consumed uptake of gas and water recovery produced in treating highly saline water and comparatively assessed with the pure water system. The findings of the study provide insights into CO_2 hydrate formation and mechanism in presence of salts which are valuable to possibly treat saline water by producing a higher yield of treated water with CO_2 hydrates.

Methodology

The system considered in this study produced water sample of 2.8 wt%. procured from West Patricia, offshore of Sarawak region at a latitude of 3°32′22.1″N and a longitude of 112°48′38.6″. The composition of the highly saline-produced water sample is represented in Table 1. The guest component selected for the hydrate-based desalination is CO_2 to signify utilization of CO_2 in gas hydrate applications miscibility with water and ability to form hydrate at lower pressures compared to methane.

Table 1: Characterization of produced water sample

Materials Used	Supplier	Purity	Initial Ion concentration (C_{A_0}, g/l)					
Salts	Merck	99.99%	Na^+	K^+	Mg^{2+}	Ca^{2+}	Cl^-	SO_4^{2-}
			7.858	0.277	0.744	0.293	13.61501	1.955
			Gases					
Carbon dioxide (CO_2)	Gas Walker Sdn Bhd	Assay ≥ 99.995%	-					
Ultrapure water	University Teknologi PETRONAS	18.2MΩ-cm	-					

In this work, the CO_2 hydrate kinetic experiments were measured using (99.9% pure CO_2) to treat highly saline synthetic produced water. The illustrative drawing of the hydrate desalination reactor is represented in figure1 [11] The reactor is designated to function under high-pressure conditions (10MPa) and has a volume capacity of 2600 ml. In the experimental approach, the pressure and temperature discrepancies inside the reactor were supervised by means of three thermocouples (reactor, chiller, and water bath) and two pressure sensors(gas inlet and reactor pressure). The hydrate reactor is submerged in the water bath. The system temperature is controlled by the water bath which is in turn linked to a chiller. Throughout the experimentation approach, the pressure and temperature variations of the reactor were logged via a data acquisition system. To execute each experiment, the reactor was initially fed with the synthetic highly saline water sample of 500ml volume through a high suction pump into the reactor. CO_2 gas is then pressurized to the experimental conditions of 3.0 MPa at 283.15 K. When the reactor was stable at the initial settings (3.0 MPa & 283.15 K), the experimentation was initiated by lowering the temperature from 283.15 K to the experimental condition of 275.15 °K to start hydrate formation. CO_2 hydrate formation was observed by a fast drop in the reactor pressure followed by a concurrent rise in the temperature owing to its exothermic nature. The hydrate formation was examined to be complete when the temperature and pressure in the reactor were steady. Then the residual saline water is emptied and the formed hydrates were washed with wash water to remove any salts occluded on their surface. The wash water is further drained out. The reactor is depressurized to dissociate the hydrate. The treated water is further collected and further analyzed for its initial final concentrations.

The CO_2 hydrate formation induction time (t_i) was measured using the subsequent Eq. 1

$$t_i = t_s - t_h \tag{1}$$

where t_s and t_h represent the time period at the start of the experimentation and the instant when hydrates formed, respectively.

The quantity of CO_2 hydrates formed was assessed by measuring the gas uptake at the initial point (time t=0) and any time t using Eq. 2

$$(\Delta n_H) = (\frac{PV}{zRT})_0 - (\frac{PV}{zRT})_t \tag{2}$$

where Δn_H is the CO_2 gas uptake and V is volume, P is pressure, T is temperature, and R signifies the molar gas constant, z is the compressibility factor evaluated using Pitzer's correlations.

Water recovery is calculated based on the volume of hydrate produced to the initial volume of feed input by using Eq 3

$$\text{Water recovery} = \frac{\textit{Volume of dissociated water}}{\textit{Intiial volume of feed}} * 100 \quad (3)$$

Results and Discussion

In this experimental work, the kinetic behavior of carbon dioxide in the existence of highly saline synthetic produced water of concentration 2.8wt% was studied to understand how much quantity of treated water can be obtained. The experimentations were performed at 3.0 MPa and 275.15 K to estimate the gas uptake and water recovery.

Carbon dioxide hydrate formation in deionized water (DIW)

Experiments were executed with deionized water using CO_2 gas as the hydrate former. Figure 1 represents the temperature and uptake of CO_2 gas in a pure water system carried out at 275.15 K and 3.0MPa. These experimental conditions are chosen as it provides a higher driving force. The gas hydrate process is a stochastic nucleation process and is comparable to the crystallization mechanism. In a stirred reactor system, the gas dissolves in the water, which is further supersaturated forming clusters of gas-water. The amount of time it takes for these gas-water clusters to grow into stable nuclei that can continue to further expand is known as induction time. These nuclei continue to enlarge and aggregate during the growth phase. As a result, we can say that there is water present as a continuous phase constituting hydrate crystals.

Figure 1 below illustrates a typical gas uptake graph during the hydrate formation from liquid water and CO_2. Three regions can be observed from figure 1(a) gas dissolution, (b) rapid hydrate development kinetics after nucleation, and (c) a phase of slow hydrate growth. The induction period is the amount of time it takes for visible nuclei to form once the gas begins to dissolve in water. Figure 1 illustrates the first nucleation occurrence within the first 54.58 mins since the beginning of the formation of the hydrate. At the point of nucleation, gas is rapidly taken up by the hydrates. From the temperature and gas uptake versus time plot when there is an abrupt spike in the temperature and correspondingly in the gas uptake profile there is a sudden spike observed this is assumed to be the starting point of the growth phase. Further, there is an initial, quick gas intake that suggests a faster rate of growth of hydrate. It is followed by a decline in the gas uptake, and finally, a steady state is reached, maybe as a result of constraints on the mass or heat transfer or simply because hydrate equilibrium is reached. In this study, depending on the type of gaseous system, the overall kinetics experiment duration ranges from 4-8 hours. As nucleation is a probabilistic phenomenon hence, the uncertainties in induction time are reduced by repeating each of the experiments four times at the same conditions, and average readings are used to evaluate their kinetic performance. Upon nucleation the gas uptake increases significantly, demonstrating the hydrate formation within the reactor. The gas uptake profile plateaued at around 4hrs. The time for these three regions from a-c varies depending on the experimental setup.

Materials Research Proceedings 29 (2023) 51-58 https://doi.org/10.21741/9781644902516-7

(i)

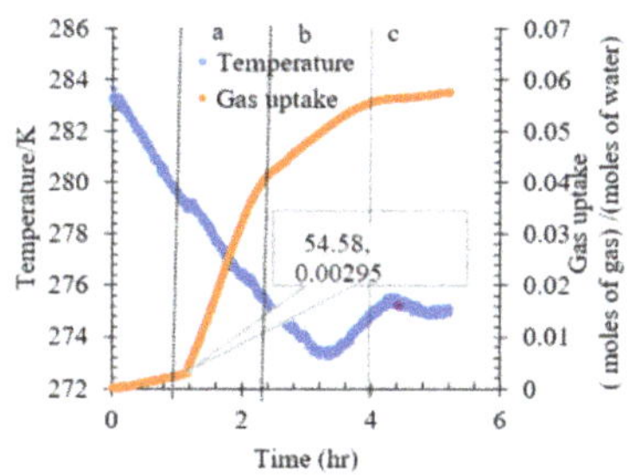

Figure 1: Temperature and gas uptake profile in pure water system

Carbon dioxide hydrate formation in the existence of highly saline water

In the hydrate-based desalination reactor, the hydrates can be produced under specific temperature and pressure parameters. It is figured that throughout the formation of hydrate inside the reactor the salinity of the residual water increases gradually as the free water is converted into hydrate. It is due to the fact that the salt ions are rejected from the surface of the hydrate. In this work, kinetic experiments were executed for 180 minutes after the initial nucleation seed. Experiments were performed with produced water with a salinity of 2.8 wt% at 3.0 MPa, 275.15 K, and a stirrer speed of 450 rpm. The time taken for the first nucleation is figured to be 86 min. for the start of hydrate formation as represented in figure 2. This displays that the existence of salts inhibits hydrate formation [22-23]. The rise in temperature peak during this period is as hydrate formation is exothermic and pressure change is determined by the degree of formation of hydrate. The effect of gas uptake and liberation of heat under the influence of pressure are opposing. The gas uptake dictates the quantity of hydrate formation. The CO_2 gas uptake improved from 1.0 % to 5.05 % with the hydrate formation. Before the nucleation stage, the solubility of CO_2 increased and hence the gas uptake increased with a decrease in solution temperature. Unusually the initial rate of gas uptake of the highly saline water system in the first 60 mins is slightly higher than that of the deionized water pure system which validates well with previous works on the promotion effect in presence of salts[24-26].

It is observed that at 100 min the gas uptake in PW system decreased by 33.6% which is highly saline compared to the deionized water system having a purity of 18MΩ-cm. At 2 hrs the gas uptake decreased by 18.6% in the saline system using CO_2 as hydrate former. At the end of 4hrs, it is further found to decrease by 11.8%. Linga's lab reported the uptake of CO_2 gas as 0.0389 mol/mol when operated at 3.1 MPa and 274.2 K [27] in a stirred tank reactor using a fixed bed. Lirio's lab reported a CO_2 uptake of 0.038 mol/mol at 3.0 MPa and 274.15 K [28]. Higher pressures of above 3.5 MPa were not carried out in this study to consider the industrial application of hydrate-based desalination. Hence for their application, it requires reducing energy consumption thus higher pressures were not considered, and also at higher pressures CO_2 gets liquified.

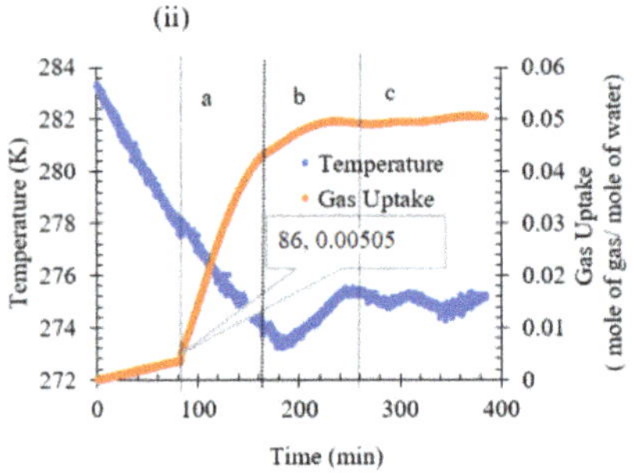

Figure 2: Temperature and gas uptake profile of PW+ carbon dioxide

Effect of salinity on the CO_2 hydrate formation

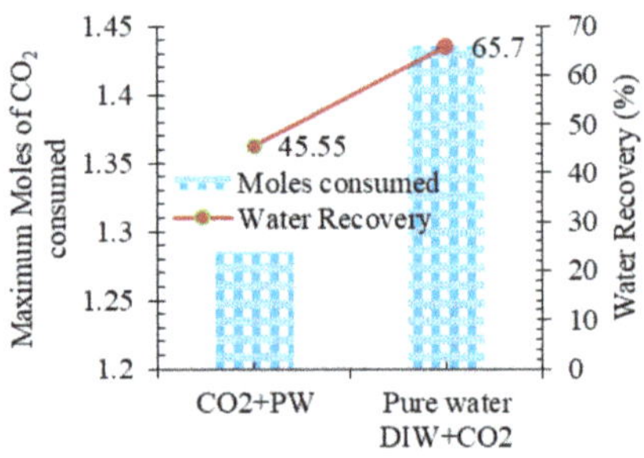

Figure 3: Comparison between moles consumed and water recovery based on the concentration

The above figure 3. shows the comparison between deionized water (DIW) 18MΩ-cm (pure water) and produced water (PW) having a salinity of 2.8wt% in terms of the amount of moles consumed and water recovery. The moles consumed and water recovery by produced water decreased by 10.43% and 30.66% compared to the pure water system. Hence, with the upsurge in salt concentration, the quantity of gas consumed declines. Since the formation of CO_2 hydrate initiates at the interface between water and gas, after some time of forming hydrate at the interface, the mass transfer from gas to liquid declines, and consumption of gas is reduced. As salts are dissolved in water the solubility of carbon dioxide gas in water is reduced and the strong electrostatic forces of interaction between ions and water molecules is established. These ions are solvated by the water molecules. As a result, the formation of hydrate crystals in highly saline water is more difficult compared to pure water.

Conclusion

Experiments were executed to study the application of the gas hydrate process for hydrate desalination of produced water, in particular, hydrate formation kinetics with respect to gas uptake and water recovery were calculated and compared with the pure deionized water. We propose to use CO_2 as its polar and hydrophilic to evaluate the water recovery of the process as the treated water produced is not harmful because of the presence of CO_2. Results revealed that water recovery of 45.5.% is produced by treating produced water and 67.5 % water recovery in presence of deionized water by using CO_2 as hydrate former at 3.0 MPa and 275.15 K. In addition, some green and biodegradable additives, can be tested to boost the kinetics of hydrate growth at the similar conditions.

Reference

[1] M. A. Al-Ghouti, M. A. Al-Kaabi, M. Y. Ashfaq, and D. A. Da'na, Produced water characteristics, treatment and reuse: A review, Journal of Water Process Engineering, 28(2019) 222-239 https://doi.org/10.1016/j.jwpe.2019.02.001

[2] T. D. Kusworo, N. Aryanti, Qudratun, and D. P. Utomo, "Oilfield produced water treatment to clean water using integrated activated carbon-bentonite adsorbent and double stages membrane process, Chemical Engineering Journal, 347(2018) 462-471 https://doi.org/10.1016/j.cej.2018.04.136

[3] P. J. Mccabe, Oil and Natural Gas: Global Resources, Encycl. Sustain. Sci. Technol. 1(2012) Springer, New York, NY https://doi.org/10.1007/978-1-4939-9763-3_71

[4] E. T. Igunnu and G. Z. Chen, Produced water treatment technologies, Int. J. Low-Carbon Technol. 9(2014)157-177 https://doi.org/10.1093/ijlct/cts049

[5] S. Nallakukkala and B. Lal, "Seawater and Produced water treatment via gas hydrate: Review, J. Environ. Chem. Eng. vol. 9 (2021) 105053 https://doi.org/10.1016/j.jece.2021.105053

[6] S. Nallakukkala, A. U. Rehman, D. B. Zaini, and B. Lal, Gas Hydrate-Based Heavy Metal Ion Removal from Industrial Wastewater: A Review, Water (Switzerland) 14(2022) 1171-1203 https://doi.org/10.3390/w14071171

[7] Y. Ghalavand, M. S. Hatamipour, and A. Rahimi, "A review on energy consumption of desalination processes, Desalin. Water Treat. 54(2015) 1526-1541

[8] K. Park et al., A new apparatus for seawater desalination by gas hydrate process and removal characteristics of dissolved minerals (Na + , Mg 2+ , Ca 2+ , K+ , B3+), Desalination. 274(2011) 91-96, https://doi.org/10.1016/j.desal.2011.01.084

[9] Z. X. Chungang Xu, Xiaosen Li , Kefeng Yan, Xuke Ruan, Zhaoyang Chen, Research progress in hydrate-based technologies and processes in China, Chinese J. Chem. Eng. 27(2019) 1998-2013 https://doi.org/10.1016/j.cjche.2018.12.002

[10] B. Lal and O. Nashed, Chemical Additives for Gas Hydrates. Springer International Publishing, 2020. https://doi.org/10.1007/978-3-030-30750-9

[11] S. Nallakukkala, B. Lal, and M. A. Shariff, "Influence of water volume on CO2 hydrate-based desalination of brine solution, Mater. Today Proc., 56(2021) 2172-2177 https://doi.org/10.1016/j.matpr.2021.11.495

[12] N. Gaikwad, R. Nakka, V. Khavala, A. Bhadani, H. Mamane, and R. Kumar, Gas Hydrate-Based Process for Desalination of Heavy Metal Ions from an Aqueous Solution : Kinetics and Rate of Recovery, ACS ES&T Water 1(2020) 134-144 https://doi.org/10.1021/acsestwater.0c00025

[13] K. Chan, P. Linga, K. Park, S. Choi, and J. Dong, Seawater desalination by gas hydrate process and removal characteristics, Desalination. 353(2014) 84-90 https://doi.org/10.1016/j.desal.2014.09.007

[14] H. Fakharian, H. Ganji, and A. Naderifar, "Desalination of high salinity produced water using natural gas hydrate, J. Taiwan Inst. Chem. Eng. 72(2017) 157-162, 2017 https://doi.org/10.1016/j.jtice.2017.01.025

[15] K. Sloan, Clathrate hydrates of natural gases., 3rd ed. CRC Press, 2007.

[16] Bhajan Lal, S. Nallakukkala, Gas Hydrate in Water Treatment Technological . economic and inductrial aspects, 1st ed. John Wiley & Sons, Ltd, 2022.

[17] A. Nambiar and P. Babu, Improved Kinetics and Water Recovery with Propane as Co-Guest Gas on the Hydrate-Based Desalination (HyDesal) Process, Chemengineering, vol. 3(2019) 31-47 https://doi.org/10.3390/chemengineering3010031

[18] H. Dong, Z. Fan, B. Wang, and S. Xue, Hydrate-based reduction of heavy metal ion from aqueous solution, Energy Procedia. 105(2017) 4706-4712, https://doi.org/10.1016/j.egypro.2017.03.1020

[19] Y. Yang, H. Zhou, F. Li, C. Shi, S. Wang, and Z. Ling, Cyclopentane hydrate-based processes for treating heavy metal containing wastewater, in International Conference on Advances in Energy and Environment Research. 04039(2019) 10-13 https://doi.org/10.1051/e3sconf/201911804039

[20] M. Karamoddin and F. Varaminian, Water desalination using R141b gas hydrate formation Water desalination using R141b gas hydrate formation, Desalin. Water Treat. 52 (2014) 2450-2456 https://doi.org/10.1080/19443994.2013.798840

[21] S. Nallakukkala et al., Suitable Binary and Ternary Thermodynamic Conditions for Hydrate Mixtures of CH 4 , CO 2 , and C 3 H 8 for Gas Hydrate-Based Applications, ACS Omega. 7(2022) 10877-10889 https://doi.org/10.1021/acsomega.1c06186

[22] S. H. B. Yang, P. Babu, S. F. S. Chua, and P. Linga, "Carbon dioxide hydrate kinetics in porous media with and without salts, Appl. Energy 162 (2016) 1131-1140 https://doi.org/10.1016/j.apenergy.2014.11.052

[23] R. Sakemoto, H. Sakamoto, K. Shiraiwa, R. Ohmura, and T. Uchida, "Clathrate hydrate crystal growth at the seawater/hydrophobic-guest-liquid interface, Cryst. Growth Des. 10(2010) 1296-1300 https://doi.org/10.1021/cg901334z

[24] A. K. Ruplai Gautam, Sanat kumar, Manisha Sahai, "Solid CO2 hydrates for sustainable environment: Application in carbon capture and desalination, Mater. Today Proc. 2022 https://doi.org/10.1016/j.matpr.2022.05.494

[25] A. K. J. Niraj Thakre, Avinash V. Palodkar, Harshal J Dongre, Microscopic Molecular Insights into Hydrate Formation andand Growth in Pure and Saline Water Environment, J. Phys. Chem.124 (2020) 4241-4252 https://doi.org/10.1021/acs.jpca.0c00621

[26] Y. L. Mingjun Yang, Jianan Zheng, Weiguo Liu, Yongchen Song, "Effects of C3H8 on hydrate formation and dissociation for integrated CO2 capture and desalination technology, Energy. 93(2015)1971-1979 https://doi.org/10.1016/j.energy.2015.10.076

[27] P. Linga, N. Daraboina, J. A. Ripmeester, and P. Englezos, Enhanced rate of gas hydrate formation in a fixed bed column filled with sand compared to a stirred vessel, Chem. Eng. Sci.68 (2012) 617-623, Jan. 2012, https://doi.org/10.1016/j.ces.2011.10.030

[28] C. F. da S. Lirio, F. L. P. Pessoa, and A. M. C. Uller, Storage capacity of carbon dioxide hydrates in the presence of sodium dodecyl sulfate (SDS) and tetrahydrofuran (THF), Chem. Eng. Sci. 96(2013) 118-123 https://doi.org/10.1016/j.ces.2012.10.022

The effect of fluoride based salt etching in the synthesis of MXene

DINESH Rao Prakas Rao[1,a*], DINESH Sivanesan[1,b] and SITI Nur Azella Zaine[1,2,c]

[1]Chemical Engineering Department, Universiti Teknologi PETRONAS, 32610 Seri Iskandar, Perak, Malaysia

[2]Centre of Innovative Nanostructure and Nanodevices, Institute of Autonomous Systems, Universiti Teknologi Petronas, 32610 Seri Iskandar, Perak, Malaysia

[a]draodsp@gmail.com, [b]dinesh_sivanesan@outlook.com, [c]nurazella.zaine@utp.edu.my

Keywords: Titanium Aluminium Carbide, MXene, Fluoride-Based Salt Etching, In-situ HF, Green Etching Process

Abstract. Here we reported the effect of fluoride-based salt etching in the synthesis of Ti_3C_2 MXene by etching Ti_3AlC_2 MAX phase precursor. Lithium fluoride (LiF) and ammonium fluoride were the fluoride-based salts were chosen as an etching agent in this study. The optimum etchant concentration and etching temperature of the MAX phase were evaluated. The presence of aluminium etched was determined by using the Inducted Couple Plasma Optical Emission Spectrometry (ICP-OES). The initial concentration of aluminium in Ti_3AlC_2 precursor was estimated based on the data from Energy Dispersive X-Ray Analysis (EDX). The study shows that the optimum etchant concentration of LiF is5M and NH_4F is 3M. Room temperature is the optimum etching temperature due to the exothermic reaction of the process. Compared to LiF, NH_4F is the preferred salt for in-situ HF fluoride-based salt etchant due to the capability of the salt to etch the maximum amount of Al at a low concentration of 3M within 24 hours at room temperature.

Introduction

Since the isolation of single-layer graphene in 2004, two-dimensional (2D) materials have gained tremendous attention because of their distinctive properties relative to their bulk form. The isolation of graphene has become a reference for all 2D materials and opened the possibility of novel discovery [1]. Due to its distinct structural and physical characteristics, a new family of two-dimensional early transition metal carbides and carbon nitrides, known as MXene has recently gained attention [2]. MXenes have important applications in the areas of catalysts [3], sensors [4], super-capacitors [5], and rechargeable batteries [6]. The MXene family comprises of transition metal carbides, carbonitrides, and nitrides with a general formula of $M_{n+1}X_n$, where M represents transition metals, and X is carbon or nitrogen [7]. The name "MXene" was given to describe the similarities between this 2D material family with graphene and to recognize the parent ternary carbide and nitrides, MAX phases, which are the precursor of MXenes [1]. MAX phases are layered ternary carbides and nitrides with a general formula $M_{n+1}AX_n$, where A represents elements from groups 13 and 14 of the periodic table.

Hydrofluoric acid (HF) containing and HF-forming etchants, which add surface functionalities like O, F, or OH, represented by Tx in this formula as $M_{n+1}X_nT_x$, are used to synthesize MXenes [2]. The first method of synthesis MXene was reported in 2011 that Ti_3C_2 MXene was synthesized by immersing Ti_3AlC_2 powders in 50% hydrofluoric acid (HF) [8]. In that method, HF is an essential etchant. However, HF has strong causticity and toxicity. Therefore, it is of great significance to find a new etchant to replace hydrofluoric acid. Since 2014, a new etchant, lithium fluoride (LiF) in hydrochloric acid (HCl), instead of HF, was used to prepare Ti_3C_2. Besides LiF, other fluoride salts in HCl were also reported to produce Ti_3C_2 [9]. Compared with HF, fluoride salts in HCl are milder and safer.

In this study, we discuss the effect of the synthesis parameter of fluoride-based salt etching namely lithium fluoride (LiF) and ammonium fluoride (NH_4F) in the synthesis of $Ti_3C_2T_x$-based MXene. Two synthesis parameters were manipulated, namely concentration and temperature. The amount of aluminium etched from the Ti_3AlC_2-based MAX precursor were measured at different etching interval by using inductively coupled plasma - optical emission spectrometry (ICP-OES).

Experimental

A) Choice of Materials

The Ti_3AlC_2 powder (Purity: 99%) was purchased from Nanoshel. The etching agent LiF and NH_4F were purchased from Sigma-Aldrich. Hydrochloric acid (HCL, 37 wt %) was purchased from Merck.

B) Varying Synthesis Temperature

To study the effect of temperature on the amount of Al etched, 1L of 6M hydrochloric acid (HCL) solution was prepared by diluting with deionized water [2]. Then, LiF salt was added to the HCl solution to prepare 5M of in-situ HF etchant solution. An oil bath was set up to heat the etchant solution to the desired temperature. The solution was soaked in the oil bath and simultaneously stirred at a rate of 130 rpm. Then, 4 g of Ti_3AlC_2 powder were slowly added into the solution under continuous stirring for 7 hours. The solution was then separated from the solid precursor and tested for the presence of Al by using ICP-OES. The experiment was repeated at different temperatures (40°C, 50°C and 60°C). The same experimental method was repeated for NH_4F salt. However, a lower concentration of 3M NH_4F-based etchant solution in 6M HCl solution was selected.

C) Varying Concentration of Etchant Solutions

To study the effect of salt concentration on the amount of Al etched, different weight of Li-based salts (LiF and NH_4F) was added to 1L of diluted HCl (6M). The solutions were left under continuous stirring at 100 rpm for 30 minutes by using a magnetic hotplate stirrer. 4 g of Ti_3AlC_2 powder were then added gradually to the etchant solution. The concentration of aluminium in the sample at different time intervals (1 hour, 7 hours, 22 hours, and 24 hours) was measured by using ICP-OES. The overall steps were repeated by using different concentrations (4M to 7M) of LiF and NH_4F salts in a diluted HCl solution.

Result and Discussion

A) EDX Analysis of MAX phase

Fig. 1 shows the Energy Dispersive X-Ray Analysis (EDX) of the Ti_3AlC_2-based MAX precursor sample. The analysis confirms the presence of titanium, aluminium and carbon elements that represent the Ti_3AlC_2 structure. The analysis shows that the sample is consist of 58.14 wt% of Ti element, 16.32 wt% of C element, 14.12 wt% of Al element and 11.42 wt% of O element. The presence of O elements could be due to the oxidation of the sample. The mass and initial concentration of aluminium was calculated as below:

$$Weight\ \% = \frac{mass\ of\ Al}{total\ mass} x\ 100\% \tag{1}$$

$$Concentration_{i(Al)} = \frac{mass\ of\ Al}{volume\ of\ solution} \tag{2}$$

Materials Research Proceedings 29 (2023) 59-65 https://doi.org/10.21741/9781644902516-8

From the calculation, the initial concentration of the aluminium presence in the MAX precursor isabout 7060 mg/L.

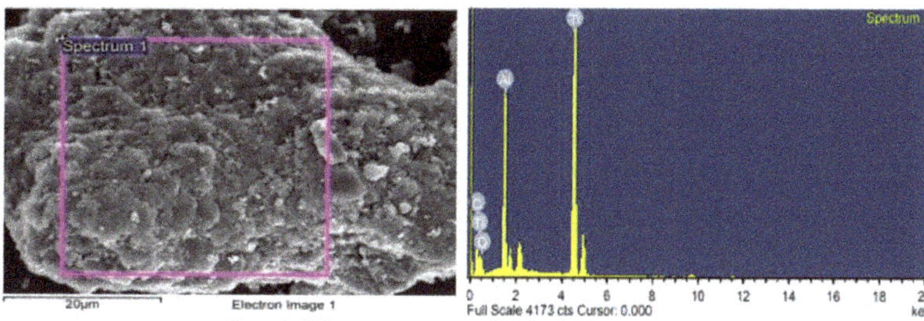

Fig. 1. The composite of Ti3AlC2 structure

B) The Effect of Synthesis Temperature on the Amount of Aluminium Etched

Fig. 2 and Fig. 3 show the amount of Al etched at different etching temperatures by using LiF and NH_4F salt, respectively. Regardless of the type of salt etchant, increase the synthesis temperature results in decreasing the amount of Al etched from the MAX precursor. Etching at room temperature results in the highest amount of AL etched. The reduce amount of Al etched could be due to the reduce the flowability of the salt solution into the lattice of MAX phase. The etching process of Al is an exothermic reaction, where the heat is release because of the chemical reaction between the acid and the MAX precursor. Introduce heat during the etching process resulting in increase the temperature of the etching solution close to the boiling point of the solution thus reduce the capability of the salt solution to penetrate the lattice of the MAX precursor. However, detail study must be conducted to prove the theory. The findings indicate that heating is not required in the synthesis of MXene by using in-situ HF fluoride-based salt etchant. Compared the amount of Al etched at room temperature, 5M of LiF solution able to extract about 26.85 mg/L while NH_4F salt solution can etch higher amount of Al (403 mg/L)even at lower concentration of 3M.

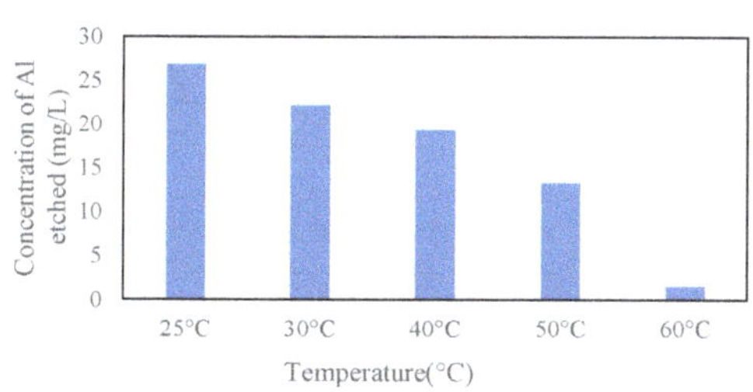

Fig. 2. Temperature vs concentration of Al etched

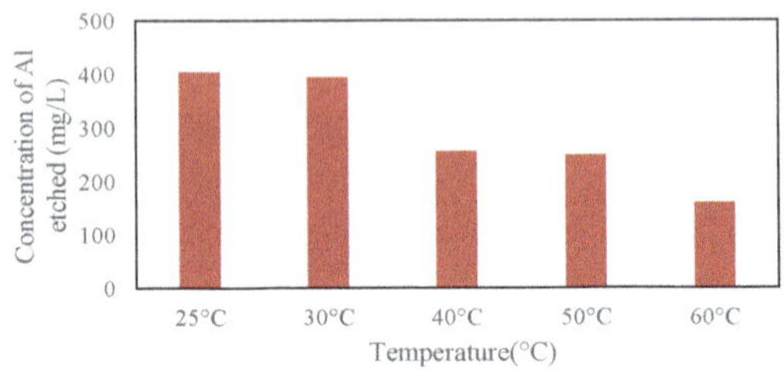

Fig. 3. Temperature vs concentration of Al etched

Table 1 shows the images of the etchant solution after 7 hours of the etching process at different etching temperatures by using 5M of LiF and 3M of NH_4F salt solutions. The images show that there is a significant change in the colour of the sample as the etching temperature increases. Regardless of the type of salt, the solution exhibit almost colourless at an etching temperature of 30°C. The colour change from pale yellow to green when the temperature increases from 40°C to 60°C. However, an obvious colour change is observed when the temperature is increased to 50°C and 60°C. It is suggested that detailed analysis must be conducted to evaluate the presence of any elements that contribute to the change in the colour of the solution.

Table 1. Change of the etchant colour at different temperature

Temp.	30°C		40°C		50°C		60°C	
Label	LiF	NH_4F	LiF	NH_4F	LiF	NH_4F	LiF	NH_4F
Sample	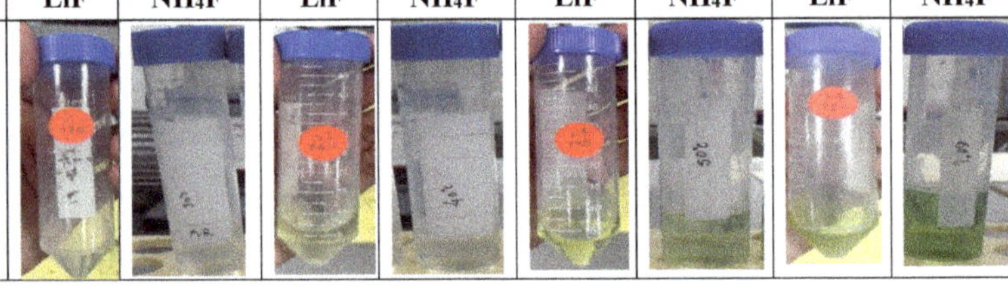							

C) The Effect of Aluminium Concentration at Different Etchant Concentrations

Fig. 4 and Fig. 5 show the amount of Al etched by using different concentrations of the salt solution measured at different time intervals at room temperature. Regardless of the type of fluoride-based salt solution, increasing the etching time results in increasing the amount of Al etched. Based on Fig 5, the concentration of salt solution has slightly affected the amount of Al etched. After 24 hours of etching time, 4M and 5M of LiF-based salt solution exhibit the highest amount of Al etched. Further increasing the concentration of LiF salt solution to 6M and 7M result in reducing the concentration of Al presence in the etched solution. The same trends were also observed for NH_4F-based salt solution, where the sample with the lowest concentration of3M exhibits the highest amount of Al etched. Further increasing the concentration results in reducing the amount of Al etched. Based on the analysis, etching process by using 3M of NH_4F solution at room temperature can etch almost all Al element presence in the MAX precursor sample.

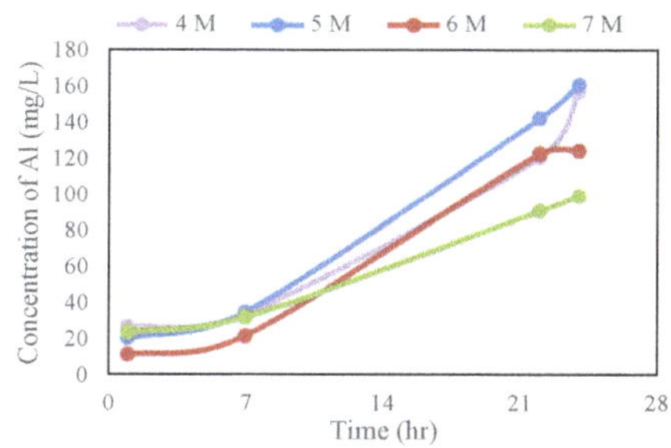

Fig. 4. Aluminium etched in different concentrations of LiF

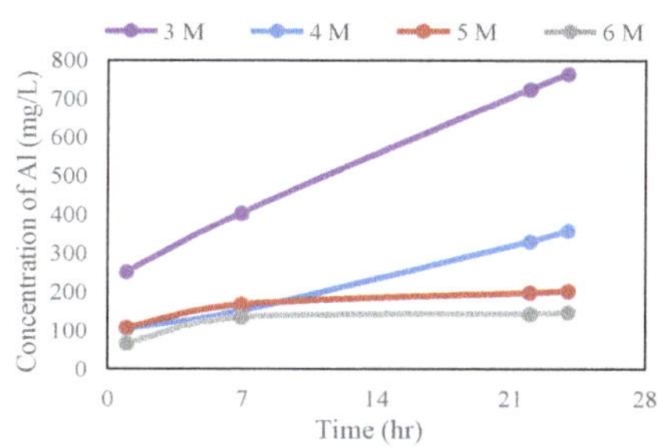

Fig. 5. Aluminium etched in different concentration of NH4F

Table 2 and Table 3 show the difference in etchant colour after etching for 7 hours by using different concentrations of LiF and NH_4F-based salt solution. The images show that the colour of the LiF-based salt etching change from colourless to green when the concentration increase from 4M to 6M. In different, NH_4F-based salt solution exhibit colourless solution at lowconcentrations of 3M to 4M. The solution change to pale yellow when the concentration increases to 5M and 6M. The greenish solution produced when the etched solution is filtered is likely to be a titanium (III) complex of fluoride and potentially chloride [10]. The greenish colouris not formed by a complex of chloride alone, because titanium produces a blue or purple compound when it reacts with chloride alone [11]. As the concentration of etchant increases, the intensity of the green colour in the sample increases which results in the increase in titanium concentration.

Table 2. Change of the etchant colour at different concentration of LiF

Description	4 M	5 M	6 M	7 M
Samples				
Colour	Colourless	Pale yellow	Pale green	Darker green

Table 3. Change of the etchant colour at different concentration of NH4F

Description	3 M	4 M	5 M	6 M
Samples			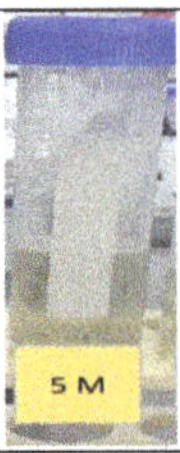	
Colour	Colourless	Colourless	Pale yellow	Pale yellow

Conclusion

In conclusion, two-dimensional MXene Ti_3C_2 can be synthesized from Ti_3AlC_2 MAX precursor via in-situ HF fluoride-based salt etching, namely LiF and NH_4F in HCl solution. Based on the study conducted, the optimum etchant concentration of LiF is 5 M and NH4F is3M, where the highest amount of aluminium is etched. Room temperature is the optimum etchingtemperature due to the exothermic reaction of the etching process. An increase in the etching temperature results in reduces the efficiency of the etching process due to the reduction of the penetration of the solution into the lattice of the MAX precursor. Moreover, NH_4F shows a better etchant compared to LiF since it etched more aluminium within 24 hours of the etching process.

Acknowledgement

The authors would like to acknowledge Yayasan Universiti Teknologi PETRONAS for funding this project under Fundamental. Research Grant (YUTP-FRG), grant number 015LC0-300.

References

[1] Alhabeb, M., Maleski, K., Anasori, B., Lelyukh, P., Clark, L., Sin, S., & Gogotsi, Y. Guidelines for Synthesis and Processing of Two-Dimensional Titanium Carbide (Ti3C2Tx MXene). Chemistry of Materials, 29(18) (2017) 7633- 7644. https://doi.org/10.1021/acs.chemmater.7b02847

[2] Liu, F., Zhou, A., Chen, J., Jia, J., Zhou, W., Wang, L., & Hu, Q. Preparation of Ti3C2 and Ti2C MXenes by fluoride salts etching and methane adsorptive properties. Applied Surface Science, 416 (2017) 781-789. https://doi.org/10.1016/j.apsusc.2017.04.239

[3] Y. Gao, L. Wang, Z. Li, A. Zhou, Q. Hu, X. Cao, Preparation of MXene-Cu2O nanocomposite and effect on thermal decomposition of ammonium perchlorate, Solid State Sci. 35 (2014) 62-64. https://doi.org/10.1016/j.solidstatesciences.2014.06.014

[4] F. Wang, C. Yang, C. Duan, D. Xiao, Y. Tang, J. Zhu, An organ-like titanium carbide material (MXene) with multilayer structure encapsulating hemoglobin for a mediator-free biosensor, J. Electrochem. Soc. 162 (2015) B16-B21. https://doi.org/10.1149/2.0371501jes

[5] M. Ghidiu, M.R. Lukatskaya, M.Q. Zhao, Y. Gogotsi, M.W. Barsoum, Conductive two-dimensional titanium carbide 'clay' with high volumetric capacitance, Nature 516 (2014) 78-81. https://doi.org/10.1038/nature13970

[6] D. Sun, M. Wang, Z. Li, G. Fan, L.-Z. Fan, A. Zhou, Two-dimensional Ti3C2 as anode material for Li-ion batteries, Electrochem. Commun. 47 (2014) 80-83. https://doi.org/10.1016/j.elecom.2014.07.026

[7] Natu, V., Pai, R., Sokol, M., Carey, M., Kalra, V., & Barsoum, M. W 2D Ti3C2Tz MXene Synthesized by Water-free Etching of Ti3Al2Cz in Polar Organic Solvents. Chem, 6(3) (2020) 616-630. https://doi.org/10.1016/j.chempr.2020.01.019

[8] M. Naguib, M. Kurtoglu, V. Presser, J. Lu, J. Niu, M. Heon, L. Hultman, Y.Gogotsi, M.W. Barsoum, Two-dimensional nanocrystals produced by exfoliation of Ti3AlC2, Adv. Mater. 23 (2011) 4248-4253. https://doi.org/10.1002/adma.201102306

[9] M. Naguib, R.R. Unocic, B.L. Armstrong, J. Nanda, Large-scale delamination of multi-layers transition metal carbides and carbonitrides 'MXenes', Dalton Trans. 44 (2015) 9353-9358. https://doi.org/10.1039/C5DT01247C

[10] Gediz, U., & Geyikçi, F. Kinetic Study on Copper Leaching in Electroplating Waste Sludge (EWS) with Ammonium Nitrate Solution (ANS). Kastamonu University Journal of Engineering and Sciences, 5(1) (2019) 1-15.

[11] Nikiforov, G. B., Roesky, H. W., & Koley, D. A survey of titanium fluoride complexes, their preparation, reactivity, and applications. Coordination Chemistry Reviews, 258-259 (2019) 16-57. https://doi.org/10.1016/j.ccr.2013.09.002

Trizane functionalized inverse vulcanized copolymer prepared using one-pot method

ALI SHAAN MANZOOR Ghumman[1,2,a], RASHID Shamsuddin[1,2,b,*], MUHAMMAD Ayoub[1,2,c] and TEH Men Qi[1,d]

[1]Chemical Engineering Department, Universiti Teknologi PETRONAS, 32610 Bandar Seri Iskandar, Perak Darul Ridzuan, Malaysia

[2]HICoE, Center for Biofuel and Biochemical Research (CBBR), Institute of Self-Sustainable Building, Universiti Teknologi PETRONAS, 32610, Seri Iskandar, Perak, Malaysia

[a]alishaan.ghumman@utp.edu.my, [b]mrashids@utp.edu.my, [c]muhammad.ayoub@utp.edu.my, [d]men_17006475@utp.edu.my

Keywords: Heavy Metal Removal, Trizane, Inverse Vulcanized Copolymers, Sulfur Enriched Polymers

Abstract. Inverse vulcanized copolymers have emerged as a promising adsorbent for mercury remediation due the presence of S-S chain in their structure. However, these adsorbents are still suffering from low selectivity, adsorption capacity due to their hydrophobic nature and less functionality. Herein we reported the synthesis of Trizane functionalized inverse vulcanized copolymer using one-pot method by reacting sulfur and 2,4,6-triallyl-1,3,5-trizane and DIB (crosslinker) at 163 °C. Fourier transform infrared spectroscopy (FTIR) and thermogravimetric analysis (TGA) were used to characterize the resultant copolymers. FTIR spectra of all copolymers shows the absence of C=C bond and C-C=C as there were no peak observed at 1660 and 3005 cm^{-1}. TGA revealed that the produced copolymers are thermally stable and only started to degrade at 200 °C. It can be concluded here that TAC and DIB can react successfully with sulfur to form trizane containing copolymer.

Introduction

Water scarcity is alarming as it is essential for living organism. Although 72 % earth is covered with water but only >1 % is available as freshwater. Freshwater is being contaminated by discarding waste in various water bodies in form of oil spill, industrial effluent, and marine dumping [1,2]. Diverse range of pollutant are found in water, but heavy metals are considered as the most lethal pollutants due to their persistent nature, bioaccumulation, toxicity, neurotoxicity, great solubility in water and mobility [3]. Fossil fuels combustion, artisanal gold mining, incineration of medical and municipal waste are major sources of mercury contamination which accounts to almost 90% of the total mercury pollution [4]. Coal combustion alone contributes about 50% of the total anthropogenic mercury contamination in ecosystem [2]. It is estimated by the UN environment agency that around 2000-2500 tons of mercury have already been released in the atmosphere in the 21st century so far, which is almost 30% of the total atmospheric emissions [1].

To overcome the mercury contamination, diverse range of methods have already been developed such as ion exchange, chemical precipitation, coagulation-flocculation, flotation, membrane filtration and adsorption [5]. Nevertheless, these technologies suffer from high operational cost, limited removal efficacy and generation of secondary pollutants. Owing to simplicity, high efficiency, availability of diverse range of adsorbents, easy operation, high reliability, promising reversibility, and low cost; adsorption is the widely considered method amidst all removal technologies [6–8]. To date various adsorbents have been utilized for the removal of heavy metals from waste water which includes geopolymers [9] activated coke [10], biochar, low dimensional carbon [11], carbon composite and polymers [2], metal-organic frameworks [2], fly ash [12], metal oxides, natural zeolites [13], biomaterials, and nanomaterials [14,15]. Nevertheless, these adsorbents are still suffering from low selectivity, high operational cost, low surface area, instable pores, and low reversibility.

Inverse vulcanized copolymers are a new class of sulfur enriched polymers prepared using facile method invented by Pyun et al. [16]. These copolymers have attracted a lot of attention and have been utilized in many applications such as slow-release fertilizers [17–22], IR materials [23], boron removal [24] and mercury remediation [25–27]. The sulfur backbone of these copolymers makes them as potential mercury adsorbents and previous studies have shown promising results for example Limjuco et al. prepared a mercury adsorbent from inverse vulcanized copolymer obtained by reacting sulfur and 2-carboxyethyl acrylate (CEA) which showed an adsorption capacity of 835 mg/g which is very high as compared to other adsorbents [25]. Diverse range of monomers have already been utilized to produce the mercury adsorbents by utilizing hydrocarbons and aromatic enriched co-monomers, but these monomers produce hydrophobic polymers making them incapable of forming hydrogen bond with aqueous feed. Moreover, these copolymers do not have appropriate functional groups to bind mercury and to functionalize these copolymers, lengthy procedure of functionalization are adopted which increases the overall cost of the copolymers [24].

Herein, we reported the synthesis of trizane functionalized inverse vulcanized copolymer by reacting sulfur with 2,4,4-triallyloxy-1,3,5-trizane (TAC) at 163 °C using diisopropenyl benzene as a crosslinker. The resultant copolymer was then characterized using Fourier transform infrared spectroscopy (FTIR) and thermogravimetric analysis (TGA) to confirm the formation of copolymer and to determine its thermal stability.

Materials

2,4,6-triallyloxy-1,3,5-trizane (TAC) and sulfur were procured from Merck, Malaysia.

Methods

Synthesis of copolymer. 3 g of sulfur was added in 30 ml glass vial and heated at 163 °C to initiate the ring-opening process which was accompanied with a color change. Upon completion of ring-opening process which was indicated by orange color of the melted sulfur, 1.5 g of TAC and 1.5 g of DIB was added and was allowed to react for 1 hour under continuous stirring. After completion of reaction, the polymer was then extracted from glass vial, into a petri dish and was allowed to cool to room temperature [18,19,28].

FTIR. FTIR Spectroscopy (PerkinElmer frontier model) is used to identify and characterize the synthesized copolymer via inverse vulcanization of sulfur and 2,4,6-triallyloxy-1,3,5- triazine (TAC). This model is equipped with the attenuated total reflectance (ATR) at the resolution of 4 cm^{-1} and scanning range of 500 to 4000cm^{-1} frequency. The sample is scanned for 8 times to have a precise result on the formation of the synthesized copolymer.

TGA. TGA (PerkinElmer STA 6000) is used to determine the thermal stability of the synthesized sulfur-based copolymer. The temperature range of this instrument is 30 to 800°C and a heating rate of 10°C/min under a nitrogenic atmosphere.

Results and Discussion

FTIR. The FTIR spectra of all copolymers synthesized, pure TAC and pure DIB are shown in Figure 1. Based on the infrared spectroscopy absorption table, the spectrum of TAC showed a high intensity peak associated with C-H stretching vibration for the alkane functional group at 2890 cm^{-1} [19]. The intensity of the peaks for the synthesized copolymers reduces as compared to TAC. It can be deduced that the TAC was consumed during the synthesis as the decrease in the intensity of the peaks was resulted by the decrease in the amount of the functional groups in the sample. The copolymer synthesized with 50 wt% sulfur shows the lowest intensity peak at 2890 cm^{-1}, indicating that the highest amount of TAC was converted in this copolymer. The spectra of the synthesized copolymer also showed less intense peaks for C-H stretching at 3005 cm^{-1} [7], [20], [21]. A peak shift was observed from 3290 cm^{-1} for DIB spectrum to 3305 cm^{-1} for the synthesized copolymers' spectra. This indicates that structural changes occurred within the molecule during synthesis [22]. The synthesized copolymers had introduced a new broad peak at 2275 cm^{-1}, which is the isocyanate with N=C=O stretching. This peak was not found in both spectra for TAC and DIB and therefore, it can be deduced that a new material had been formed via the inverse vulcanization method.

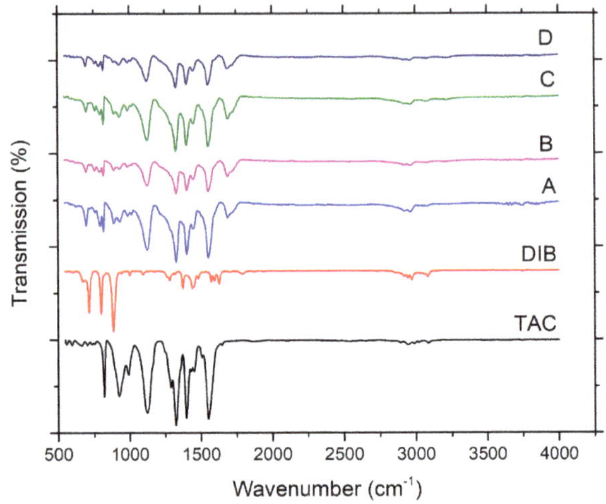

Fig. 1: FTIR Spectra of TAC, DIB, and Copolymer with (A) 50 wt% S, (B) 60 wt% S, (B) 70 wt% S (C) 80 wt% S and (D) 90 wt% S.

TGA. The TGA thermogram of elemental sulfur shows a single-phase degradation step and completely decomposes at around 345 °C. The thermogram reveals that the all the synthesized copolymers degraded in a two-step-degradation manner, except for copolymer with 40 wt% sulfur which underwent a one-step-degradation manner. The decomposition of all copolymers onset at around 200 °C, which is similar to the onsets decomposition temperature of elemental sulfur. For polymers that underwent the two-step-degradation manner, the loosely bonded sulfur chain decomposes mainly in the first step. The percentage of weight loss in the first step (200 °C to 355 °C) increases with the increase of sulfur weight percentage from 40 wt% to 70 wt%. The second step of degradation involves the decomposition of strongly bonded sulfur (C-S bond) in the sample. The copolymer with 40 wt% sulfur portrayed a single-phase degradation mainly due to the reduced sulfur content in the synthesized copolymer. It showed higher thermal stability as it retained the highest weight percentage during heating especially at temperature between 400 °C to 700 °C. In contrast, the copolymer with 70 wt% showed the most obvious manner for two-step-degradation. This is due to the highest sulfur content that is present in the synthesized copolymer. Copolymer with 70 wt% sulfur displayed the lowest thermal stability as it started to decompose at the lowest temperature, and retained the least weight percentage during heating, especially between temperature 400 °C to 700 °C [17,29–31]. Based on the TGA result, it can be concluded that the degradation of the copolymers takes place at similar degradation temperature of elemental sulfur, producing a polymeric material with high thermal stability. As can be seen from the TGA thermogram that even at 800°C, there is some char left and the pattern of the degradation is also different from monomers which confirms the succeessful structure formation of the polymer.

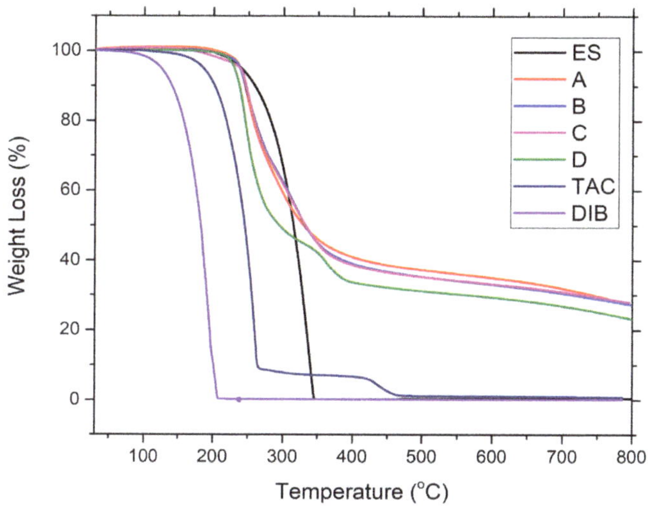

Fig. 2: TGA thermogram of TAC, DIB, and Copolymer with (A) 50 wt% S, (B) 60 wt% S, (B) 70 wt% S (C) 80 wt% S and (D) 90 wt% S.

Conclusion

A novel trizane functionalized copolymer was synthesized using inverse vulcanization process by reacting sulfur and 2,4,6-triallyl-1,3,5-trizane at 163 °C and using DIB as crosslinker. FTIR confirmed the formation of a copolymer as there was no peak observed at 1660 and 3005 cm^{-1} which indication the utilization of C=C and C-C=C bons for the formation of C-S bond. The TGA analysis demonstrated that the produced are thermally stable and onsets to degrade at 200 °C. TGA thermogram further confirmed the formation of polymer as char is left at 800 °C and the decomposition pattern is also different. This copolymer contains trizane as indicated by the FTIR, whihc makes this copolymer a potential adsorbent for mercury removal. In future this copolymer should be tested for mercury removal.

Acknowledgement

The authors would like to acknowledge the funding supports from Ministry of Higher Education (MOHE) Malaysia through HICoE grant (Cost Centre 015MA0-052/015MA0-104/015MA0-136) to CBBR and Yayasan UTP (Cost Center 015C0-271).

References

[1] M. Zunita, Graphene Oxide-Based Nanofiltration for Hg Removal from Wastewater: A Mini Review, Membranes (Basel). 11 (2021) 269. https://doi.org/10.3390/membranes11040269

[2] T. Rasheed, A.A. Hassan, M. Bilal, T. Hussain, K. Rizwan, Metal-organic frameworks based adsorbents: A review from removal perspective of various environmental contaminants from wastewater, Chemosphere. 259 (2020) 127369. https://doi.org/10.1016/j.chemosphere.2020.127369

[3] D. Jampaiah, S.J. Ippolito, Y.M. Sabri, B.M. Reddy, S.K. Bhargava, Highly efficient nanosized Mn and Fe codoped ceria-based solid solutions for elemental mercury removal at low flue gas temperatures, Catal. Sci. Technol. 5 (2015) 2913–2924. https://doi.org/10.1039/c5cy00231a

[4] K. Suresh Kumar Reddy, A. Al Shoaibi, C. Srinivasakannan, Elemental mercury

adsorption on sulfur-impregnated porous carbon-A review, Environ. Technol. (United Kingdom). 35 (2014) 18–26. https://doi.org/10.1080/21622515.2013.804589
[5] J.G. Yu, B.Y. Yue, X.W. Wu, Q. Liu, F.P. Jiao, X.Y. Jiang, X.Q. Chen, Removal of mercury by adsorption: a review, Environ. Sci. Pollut. Res. 23 (2016) 5056–5076. https://doi.org/10.1007/s11356-015-5880-x
[6] B. Pan, B. Pan, W. Zhang, L. Lv, Q. Zhang, S. Zheng, Development of polymeric and polymer-based hybrid adsorbents for pollutants removal from waters, Chem. Eng. J. 151 (2009) 19–29. https://doi.org/10.1016/j.cej.2009.02.036
[7] M. Hua, S. Zhang, B. Pan, W. Zhang, L. Lv, Q. Zhang, Heavy metal removal from water/wastewater by nanosized metal oxides: A review, J. Hazard. Mater. 211–212 (2012) 317–331. https://doi.org/10.1016/j.jhazmat.2011.10.016
[8] T. Chen, Z. Zhou, R. Han, R. Meng, H. Wang, W. Lu, Adsorption of cadmium by biochar derived from municipal sewage sludge: Impact factors and adsorption mechanism, Chemosphere. 134 (2015) 286–293. https://doi.org/10.1016/j.chemosphere.2015.04.052
[9] A.A. Siyal, M.R. Shamsuddin, M.I. Khan, N.E. Rabat, M. Zulfiqar, Z. Man, J. Siame, K.A. Azizli, A review on geopolymers as emerging materials for the adsorption of heavy metals and dyes, J. Environ. Manage. 224 (2018) 327–339. https://doi.org/10.1016/j.jenvman.2018.07.046
[10] D. Liu, C. Li, J. Wu, Y. Liu, Novel carbon-based sorbents for elemental mercury removal from gas streams: A review, Chem. Eng. J. 391 (2020) 123514. https://doi.org/10.1016/j.cej.2019.123514
[11] F. Zhu, Y.M. Zheng, B.G. Zhang, Y.R. Dai, A critical review on the electrospun nanofibrous membranes for the adsorption of heavy metals in water treatment, J. Hazard. Mater. 401 (2021) 123608. https://doi.org/10.1016/j.jhazmat.2020.123608
[12] F.O. Ochedi, Y. Liu, A. Hussain, A review on coal fly ash-based adsorbents for mercury and arsenic removal, J. Clean. Prod. 267 (2020) 122143. https://doi.org/10.1016/j.jclepro.2020.122143
[13] R. Egashira, S. Tanabe, H. Habaki, Adsorption of heavy metals in mine wastewater by Mongolian natural zeolite, Procedia Eng. 42 (2012) 49–57. https://doi.org/10.1016/j.proeng.2012.07.394
[14] M. Carballa, F. Omil, J.M. Lema, M. Llompart, C. García-Jares, I. Rodríguez, M. Gómez, T. Ternes, Behavior of pharmaceuticals, cosmetics and hormones in a sewage treatment plant, Water Res. 38 (2004) 2918–2926. https://doi.org/10.1016/j.watres.2004.03.029
[15] C. Zwiener, S.D. Richardson, D.M. DeMarini, T. Grummt, T. Glauner, F.H. Frimmel, Drowning in disinfection byproducts? Swimming pool water qualtiy, Water. 34 (2007) 42–46.
[16] W.J. Chung, J.J. Griebel, E.T. Kim, H. Yoon, A.G. Simmonds, H.J. Ji, P.T. Dirlam, R.S. Glass, J.J. Wie, N.A. Nguyen, B.W. Guralnick, P. Theato, M.E. Mackay, Y. Sung, J. Park, The use of elemental sulfur as an alternative feedstock for polymeric materials, Nat. Chem. 5 (2013). https://doi.org/10.1038/NCHEM.1624
[17] A.S.M. Ghumman, M.M. Nasef, M.R. Shamsuddin, A. Abbasi, Evaluation of properties of sulfur-based polymers obtained by inverse vulcanization: Techniques and challenges, Polym. Polym. Compos. (2020). https://doi.org/10.1177/0967391120954072
[18] A.S.M. Ghumman, M.R. Shamsuddin, M.M. Nasef, W.Z.N. Yahya, M. Ayoub, B. Cheah, A. Abbasi, Synthesis and Characterization of Sustainable Inverse Vulcanized Copolymers from Non-Edible Oil, ChemistrySelect. 6 (2021) 1180–1190. https://doi.org/https://doi.org/10.1002/slct.202004554
[19] A.S.M. Ghumman, R. Shamsuddin, M.M. Nasef, W.Z. Nisa Yahya, A. Abbasi, Optimization of synthesis of inverse vulcanized copolymers from rubber seed oil using response

surface methodology, Polymer (Guildf). 219 (2021) 123553. https://doi.org/10.1016/j.polymer.2021.123553

[20] A.S.M. Ghumman, R. Shamsuddin, M.M. Nasef, C. Maucieri, O.U. Rehman, A.A. Rosman, M.I. Haziq, A. Abbasi, Degradable Slow-Release Fertilizer Composite Prepared by Ex Situ Mixing of Inverse Vulcanized Copolymer with Urea, Agronomy. 12 (2022). https://doi.org/https://doi.org/10.3390/agronomy12010065

[21] A.S.M. Ghumman, R. Shamsuddin, M.M. Nasef, E.G. Krivoborodov, S. Ahmad, A.A. Zanin, Y.O. Mezhuev, A. Abbasi, A Degradable Inverse Vulcanized Copolymer as a Coating Material for Urea Produced under Optimized Conditions, Polymers (Basel). 13 (2021). https://doi.org/https://doi.org/10.3390/polym13224040

[22] A.S.M. Ghumman, R. Shamsuddin, M.M. Nasef, W.Z.N. Yahya, A. Abbasi, H. Almohamadi, Sulfur enriched slow-release coated urea produced from inverse vulcanized copolymer, Sci. Total Environ. 846 (2022) 157417. https://doi.org/10.1016/j.scitotenv.2022.157417

[23] M. Arslan, B. Kiskan, Y. Yagci, Combining Elemental Sulfur with Polybenzoxazines via Inverse Vulcanization, (2016). https://doi.org/10.1021/acs.macromol.5b02791

[24] A. Abbasi, W.Z.N. Yahya, M.M. Nasef, M. Moniruzzaman, A.S.M. Ghumman, H.K. Afolabi, Boron removal by glucamine-functionalized inverse vulcanized sulfur polymer, React. Funct. Polym. 177 (2022) 105311. https://doi.org/10.1016/j.reactfunctpolym.2022.105311

[25] L.A. Limjuco, G.M. Nisola, K.J. Parohinog, K.N.G. Valdehuesa, S.P. Lee, H. Kim, W.J. Chung, Water-insoluble hydrophilic polysulfides as microfibrous composites towards highly effective and practical Hg2+ capture, Chem. Eng. J. 378 (2019) 122216. https://doi.org/10.1016/j.cej.2019.122216

[26] S. Petcher, D.J. Parker, T. Hasell, Macroporous sulfur polymers from a sodium chloride porogen - A low cost, versatile remediation material, Environ. Sci. Water Res. Technol. 5 (2019) 2142–2149. https://doi.org/10.1039/c9ew00477g

[27] M.J.H. Worthington, R.L. Kucera, I.S. Albuquerque, C.T. Gibson, A. Sibley, A.D. Slattery, J.A. Campbell, S.F.K. Alboaiji, K.A. Muller, J. Young, N. Adamson, J.R. Gascooke, D. Jampaiah, Y.M. Sabri, S.K. Bhargava, S.J. Ippolito, D.A. Lewis, J.S. Quinton, A. V. Ellis, A. Johs, G.J.L. Bernardes, J.M. Chalker, Laying Waste to Mercury: Inexpensive Sorbents Made from Sulfur and Recycled Cooking Oils, Chem. - A Eur. J. 23 (2017) 16219–16230. https://doi.org/10.1002/chem.201702871

[28] A. Abbasi, M.M. Nasef, W.Z.N. Yahya, M. Moniruzzaman, A.S.M. Ghumman, Preparation and characterization of sulfur-vinylbenzyl chloride polymer under optimized reaction conditions using inverse vulcanization, Eur. Polym. J. 143 (2021) 110202. https://doi.org/10.1016/j.eurpolymj.2020.110202

[29] A.D. Smith, T. Thiounn, E.W. Lyles, E.K. Kibler, R.C. Smith, A.G. Tennyson, Combining agriculture and energy industry waste products to yield recyclable, thermally healable copolymers of elemental sulfur and oleic acid, J. Polym. Sci. Part A Polym. Chem. 57 (2019) 1704–1710. https://doi.org/10.1002/pola.29436

[30] A.D. Tikoalu, N.A. Lundquist, J.M. Chalker, Mercury Sorbents Made By Inverse Vulcanization of Sustainable Triglycerides: The Plant Oil Structure Influences the Rate of Mercury Removal from Water, Adv. Sustain. Syst. 1900111 (2020) 1–9. https://doi.org/10.1002/adsu.201900111

[31] A. Abbasi, M.M. Nasef, W.Z.N. Yahya, M. Moniruzzaman, A.S. Ghumman, Preparation and characterization of green polymer by copolymerization of corn oil and sulphur at molten state, Polym. Polym. Compos. (2020) 959536. https://doi.org/10.1177/0967391120959536

Optimization of bioethanol production from papaya waste through fermentation using response surface methodology (RSM)

TAN Ching Mun[1,a] and OH Pei Ching[2,b*]

[1] Department of Chemical Engineering, Universiti Teknologi PETRONAS, 32610 Seri Iskandar, Perak, Malaysia

[2] CO_2 Research Centre (CO2RES), R&D Building, Universiti Teknologi PETRONAS, 32610 Seri Iskandar, Perak, Malaysia

[a]hanatancm@gmail.com, [b]peiching.oh@utp.edu.my

Keywords: Biofuel, Ethanol, Optimization, Response Surface Methodology

Abstract. There is a growing pressure for the development of sustainable and environmental-friendly source of energy such as ethanol that could substitute the depleting fossil fuels. Papaya waste including papaya seed and papaya peel is one of the main fruit wastes in Southeast Asia which has great potential to be utilized as substrate for bioethanol production. In this study, papaya waste was fermented to produce bioethanol using *Saccharomyces Cerevisiae*. The effect of pH, temperature, and incubation time on bioethanol production was studied within the range of 3.0-6.0, 25-45°C and 24-96 h, respectively. These parameters were optimized using Response Surface Methodology (RSM) based Box-Behnken Design (BBD). It was found that a maximum ethanol concentration of 0.2224 g/ml was obtained from papaya waste at pH 4.5, 45°C and 24 hours. The significance of the parameters increased from incubation time, pH to temperature.

Introduction

Energy is a necessary component in life. According to the bp Statistical Review of World Energy 2020, oil (33.1%), gas (24.2%) and coal (27%) account for most of the global energy consumption, all of which are non-renewable energy sources [1]. The excessive consumption of fossil fuels has not only escalated the fuel price, but significantly elevated the environmental issues and public health concerns [2]. The issue has prompted the focus and development of biofuel technology as an alternative renewable energy, which has been developing rapidly since year 2000 [3]. Bioethanol which is mainly produced through fermentation of plant-based biomass is one of the environmentally friendly alternatives for the current energy issue.

To date, majority of bioethanol is still made from starch- or sugar-based crops which are known as first-generation biofuels such as potato and sugarcane due to high yield. However, increase in bioethanol production using first-generation biofuels has raised concerns about food security and fertile land. Thus, the exploration for second-generation biomass has mostly concentrated on cheap and abundant lignocellulosic municipal waste [4-6].

Papaya is one of the 15 most produced and marketed fruits in the world with a global production of 13.2 million metric tons in 2016 [7]. Papaya waste consisting of peels and seeds takes up approximately 20-25% of the fruit weight is one of the most abundant fruit wastes found in wet markets, main household solid municipal waste and processing industry waste [8-9]. These wastes are usually discarded in landfills or dumping site that result in land and water contamination and greenhouse gas emission [8]. Therefore, by utilizing papaya waste combining papaya peel and papaya seed to produce bioethanol through fermentation, several problems aforementioned could be overcome. However, it was found that there are limited studies on the capability of papaya waste, an abundant lignocellulosic solid food waste in bioethanol production through separated hydrolysis and fermentation method (SHF). Besides, papaya peel and papaya seed are believed to be a promising source for bioethanol production based on its fibre composition, with high cellulose (24.6%, 25.4%) and hemicellulose content (20.4%, 3.4%) which can be converted into simple sugars for fermentation through hydrolysis, and relatively low lignin content (2.7%, 1.9%) [10,11] which acts as a barrier for the digestion of hemicellulose and cellulose into pentoses and hexoses [12]. Since the lignin content for both papaya peel and papaya seed are low, the pretreatment process can be carried out together with different pretreatment steps. Moreover, native microorganism existing on papaya waste during the decay process are proven to aid the bioethanol fermentation [9].

Bioethanol production from lignocellulosic waste is still not competitive to fossil fuels till date. Most of the researchers focus on the pre-treatment step to extract glucose in hydrolysis step for bioethanol production, but barely on fermentation process. A research gap for this topic was found to be the lack of study on the effects of fermentation parameters for lignocellulosic waste to maximize the yield of bioethanol. The variables in pH, temperature and incubation time highly affect the yield.

A research has been done to produce bioethanol using papaya peels through fermentation using the one-variable-at-a-time (OVAT) approach to optimize yield [13]. OVAT is often used in bioethanol production experiments. It was found that the results obtained using OVAT is not fully reliable as it does not take the interaction among the listed parameters into account [14]. Effective optimization of several independent fermentation parameters such as pH, temperature and incubation time using effective optimization tools are significant in maximizing the yield of ethanol.

In this study, papaya seed and papaya peel were used as the substrate to produce bioethanol through separated hydrolysis and fermentation method. The effect of fermentation parameters on the bioethanol yield was studied and Response Surface Methodology (RSM) was employed to optimize the bioethanol yield. The ethanol concentration product produced from the same amount of papaya reflects the yield of bioethanol.

Materials and Methods

Materials. Fresh papaya fruits were purchased from the local market in Ipoh, Perak, Malaysia. Only papaya waste comprising papaya seed and papaya peel was used in this study. The fermenting baker's yeast (*Saccharomyces cerevisiae*) used in the study was purchased from a food manufacturer in Malaysia and stored in a refrigerator until further use.

Design of Experiment. RSM based Box-Behnken Design (BBD) was utilized to optimize the fermentation pH, temperature, and incubation time for bioethanol production from papaya waste. A three-factor three-level BBD with 17 experimental runs were carried out. The design of experiment (DOE) was generated by Design Expert Software Version 8. The numerical factors were pH, temperature (°C), and incubation time (h), while the response variable was bioethanol concentration (g/mL). The range of study for pH, temperature, and incubation time were 3.0-6.0, 25-45°C and 24-96 h, respectively.

Experimental Methodology

The experimental methodology was adapted from M. R. A. Ghani and P. C. Oh [19].

Preparation of Substrate. 400 g of papaya peel and 400 g of papaya seed were collected and washed using water. Then, the waste was dried in oven at 70°C for 24 h, and grinded into powder.

Acid Hydrolysis. 15 g of papaya waste was hydrolysed in 100 mL of 0.5 M sulphuric acid. The opening of the flask was covered with cotton plug and held at 121°C for 15 mins in an autoclave (Hirayama) for sterilization. The hydrolysate was collected using vacuum filtration through coarse filter paper. The glucose concentration of hydrolysate was analysed using refractometer (Atago).

Propagation of Yeast Cell. 7 g of dried yeast powder was added into capped bottle containing 50 mL glucose yeast extract (GYE) and placed in an incubator water bath shaker (Julabo) at 30°C and 100 rpm. After 48 h, 10% of the sample was transferred into a 1 L capped bottle with 500 mL GYE to allow cell multiplication. Incubation was repeated at 35°C and 100 rpm for 24 h. The cells were then transferred to sterilized 50 mL centrifuge tubes and centrifuged at 4°C and 10000 rpm for 10 mins using Centrifuge (Labogene). The supernatant (solution) was subsequently collected for fermentation.

Fermentation. The hydrolysate was neutralized using 1.0 M sodium hydroxide (NaOH), with 0.3% (w/v) yeast extract, and 0.2% (w/v) peptone. The flask containing the mixture was wrapped with foil and heated at 80°C and 250 rpm for 30 mins in a water bath. The fermentation was performed at pH, temperature and incubation time based on the DOE obtained from Design Expert software. The pH of the mixture was maintained using 0.5 M H_2SO_4 and 1.0 M NaOH. After each fermentation cycle, the samples were analysed for ethanol concentration using refractometer.

Results and Discussion

Pretreatment. Six samples of glucose solutions with different concentrations (0, 25, 50, 75, 100 w/v%) were used as standard calibration to find the glucose content of the papaya waste hydrolysate. The calibration curve of glucose concentration vs refractive index can be represented by Eq. 1.

$$Y = 0.00008x + 1.3381 \quad (1)$$

The average refractive index for papaya waste hydrolysate is 1.3492 after 3 trial tuns, hence the average glucose concentration can be determined as 135.25 g/L by applying Eq. 1. Since glucose is simple sugar ready for fermentation, hence the higher the glucose content in the solution, the higher the ethanol amount that can be produced through fermentation, the higher the yield of bioethanol.

Ethanol Concentration. Five ethanol solutions with different ratio of 95% ethanol to distilled water were used as standard calibration to find the response ethanol concentration for each fermentation run. The calibration curve of ethanol concentration vs refractive index can be represented by Eq. 2.

$$Y = -(8 \times 10^{-11})\,x^3 + (5 \times 10^{-8})\,x^2 + (5 \times 10^{-5})\,x + 1.3329 \quad (2)$$

Model Fitting and Analysis of Variance (ANOVA). After completing 17 runs of experiment based on the DOE using BBD, the response (i.e. ethanol concentration) is tabulated in Table 1. The ANOVA with backward elimination regression was used to determine the significant model terms by reducing insignificant terms and is shown in Table 2. Based on Table 2, the Model F-value of 32.44 implies that the model is significant after reduction and has only 0.01% probability that the large F-value could occur due to noise. The low model p-value was a positive indication that shows the predictability of the model has over 95% confidence level. After backward elimination regression was done, only significant model terms with Prob>F less than 0.05 remained. Thus, linear factors A, B and C and quadratic factors A^2 and B^2 were maintained. The effect of pH and temperature were more significant than time on yield. Quadratic effect of time was not taken into considerations, while the quadratic effect of temperature was greater than that of pH. The Lack of Fit F-value of 4.23 implies that there is a 9.09% chance that a value this large could occur due to noise. Lack of Fit p-value of 0.0909 was not significant, which indicates that the model fits well and can be used to predict the response accurately.

Materials Research Proceedings 29 (2023) 72-80 https://doi.org/10.21741/9781644902516-10

Table 1. Experimental Response According to Experimental Design

Run	Factor 1 A: pH	Factor 2 B: Temperature [°C]	Factor 3 C: Time [h]	Response Ethanol Concentration [g/mL]
1	4.5	35	64	0.1954
2	6.0	25	64	0.1756
3	4.5	35	64	0.1999
4	3.0	35	96	0.1973
5	4.5	35	64	0.1974
6	4.5	45	96	0.2174
7	4.5	35	64	0.2016
8	4.5	25	24	0.1930
9	6.0	35	96	0.1754
10	4.5	45	24	0.2240
11	6.0	35	24	0.1920
12	3.0	25	64	0.1951
13	4.5	25	96	0.1976
14	3.0	45	64	0.2247
15	4.5	35	64	0.1971
16	3.0	35	24	0.2101
17	6.0	45	64	0.2118

Table 2. ANOVA for Variables After Backward Elimination Regression

Source	Sum of Squares	df	Mean Square	F Value	p value Prob > F
Model	2.92×10^{-3}	5	5.84×10^{-4}	32.44	< 0.0001[a]
A-pH	6.57×10^{-4}	1	6.57×10^{-4}	36.48	< 0.0001[a]
B-Temperature	1.70×10^{-3}	1	1.70×10^{-3}	94.44	< 0.0001[a]
C-Time	1.25×10^{-4}	1	1.25×10^{-4}	6.95	0.0231[a]
A^2	1.22×10^{-4}	1	1.22×10^{-4}	6.77	0.0246[a]
B^2	3.36×10^{-4}	1	3.36×10^{-4}	18.67	0.0012[a]
Residual	1.98×10^{-4}	11	1.80×10^{-5}		
Lack of Fit	1.74×10^{-4}	7	2.49×10^{-5}	4.23	0.0909[b]
Pure Error	2.36×10^{-5}	4	5.89×10^{-6}		
Cor Total	3.12×10^{-3}	16			

[a] significant
[b] not significant

The quadratic equations for ethanol yield (Y) in terms of coded factors and actual factors are given in Eq. 3 and Eq. 4, respectively.

$$Y = 0.200 - (9.059\text{x}10^{-3})\,A + 0.015\,B - (3.943\text{x}10^{-3})\,C - (5.373\text{x}10^{-3})\,A^2 + (8.921\text{x}10^{-3})\,B^2 \quad (3)$$

$$Y = 0.243 + 0.016\,\text{pH} - (4.787\text{x}10^{-3})\,\text{Temperature} - (1.095\text{x}10^{-4})\,\text{Time} - (2.388\text{x}10^{-3})\,\text{pH}^2 + (8.921\text{x}10^{-5})\,\text{Temperature}^2 \quad (4)$$

The model has a high R^2 value of 0.9365 which shows that the model is robust, and the experimental data can be well-fitted. Besides, the model has a low coefficient of variance (C.V. %) of 2.1179 (<10%) indicates that the model has high reproducibility.

Mutual Effects of Parameters. The correlation between each independent factor and their effects on the corresponding ethanol yield was plotted graphically and analyzed through Fig. 1, Fig. 2, and Fig. 3.

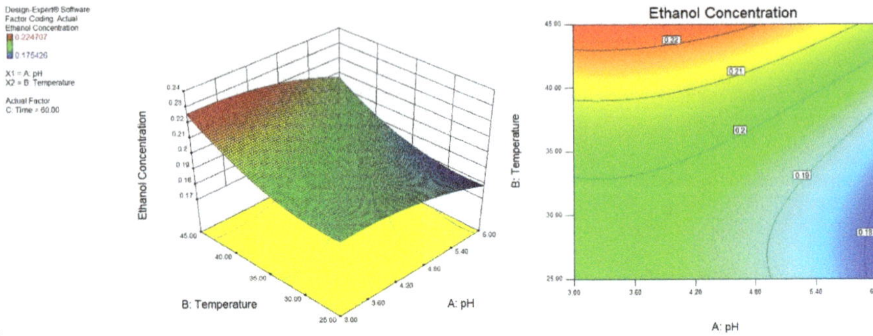

Fig. 1. (a) 3D Surface Plot (b) Contour Plot of Effect of pH & Temperature on Ethanol Yield

Fig. 1 illustrates the effect of fermentation pH and temperature on the ethanol concentration. These two parameters were varied in the range of pH 3.0 – 6.0, and 25 – 45°C, while the incubation time was kept constant at 60 h. It was observed that pH and temperature have great effect on the ethanol concentration. This was further proven by the ANOVA which shows both pH and temperature have *p*-values <0.0001.

The ethanol concentration decreased with increasing pH and decreasing temperature, which is in agreement with previous study by Abdulla et al. [13]. The maximum ethanol concentration of 0.2247 g/ml was achieved at 45°C with pH 3.0. The minimum ethanol concentration of 0.1754 g/ml was obtained at 25°C with 6.0. According to Abdulla et al. [13] , the yield of bioethanol will have an increasing trend until pH 5, which will then decrease beyond pH 5 [13, 14]. Hence, the response obtained partially agreed with the previous studies, where pH above 5 did not support fermentation process effectively. The analysis of the 3D surface plot and contour plot proved that the optimum fermentation temperature was at 45°C with mild acidic pH from 3.0 to 4.8 where the ethanol concentration was at its peak. Although pH generally has a greater effect than temperature on the production of bioethanol, in this range of study, temperature was observed to have a greater effect on the ethanol production from papaya waste since pH value within the range of 3.0 to 5.5 was suggested to be optimum for bioethanol production [14, 15] which is in line with what was observed from this work.

From the response obtained, it can be concluded that the higher the temperature, the higher the ethanol concentration produced through SHF using *S. cerevisiae*. This result was supported by G. Reed and H.J. Peppler [16] who suggested that the fermentation rate of yeast will grow with a factor of 1.5 to 2 for every 10°C increase in temperature, up until the temperature reaches 45°C [16]. Increase in fermentation temperature beyond 45°C will cause thermal deactivation of yeast and decrease the rate of ethanol production [14, 17]. Next, the increase in pH value decreased the production of bioethanol as fermentation process for biomass is favoured in slightly acidic condition. The optimum pH value to produce bioethanol from papaya waste through SHF was 3.0 to 4.8. The ethanol concentration was observed to be low at high pH value of 6.0 as this affect the shape of protein in enzymes by disturbing the bonds, which lowers the metabolic activity of the microorganisms [14]. In addition, high pH value also favoured the production of acid instead of ethanol during fermentation, hence lowered the ethanol production [17].

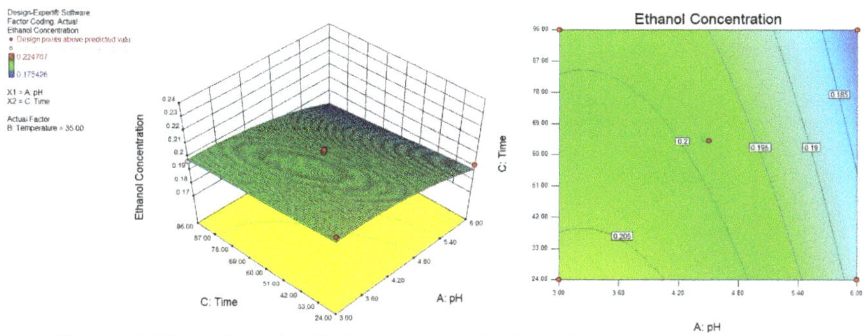

Fig. 2. (a) 3D Surface Plot (b) Contour Plot of Effect of pH & Time on Ethanol Yield

Fig. 2 shows the influence of fermentation pH and incubation time over the ethanol concentration. These two parameters were varied in the range of pH 3.0 – 6.0 and 24 – 96 h, at constant temperature of 35°C. Based on Fig. 2, pH showed a greater effect on the ethanol concentration than temperature, which was proven by the low *p*-value of pH (<0.0001) compared to time (0.0231). The lower the *p*-value, the bigger the influence of the variable over the response of the model.

The ethanol concentration decreased with increasing pH which agreed with previous findings [15, 20]. On the other hand, the ethanol production decreased with increasing time beyond 24 h, which contradicted with previous studies, implying that optimum incubation time ranged from 36 to 72 h [6, 14, 15, 21]. At constant temperature of 35°C, the maximum ethanol concentration of 0.2101 g/ml was achieved at 24 h with pH 3.0. The minimum ethanol concentration of 0.1754 g/ml was obtained at 96 h with pH 6.0. The analysis of Fig. 2 proves that the optimum fermentation time was at 24 h within pH range of 3.0 to 4.8, where the ethanol concentration was at its peak.

Ethanol concentration decreased with increasing pH as acidic environment favours the growth of yeast during fermentation. The ethanol production will increase with incubation time due to quick consumption of glucose by yeast cell if bioethanol is produced in bulk. Then, the production will remain constant at certain time, due to decrease in reducing sugar consumption by yeast cell [15]. However, prolonged fermentation may not be favoured as side products produced during the process may inhibit the yeast cell from further producing bioethanol at optimum rate [14]. In this study, the optimum incubation time was at 24 h instead of longer incubation time range from 36 to 72 h, which was suggested by other researchers [5, 14, 19, 22] . The contradiction may be due to the sample size of the lab-scale experiment was relatively small, where the available reducing sugar in the incubator was depleted and prolonged fermentation produced side products such as furfurals, via the degradation of the monosaccharides that further inhibited the ethanol fermentation process [19].

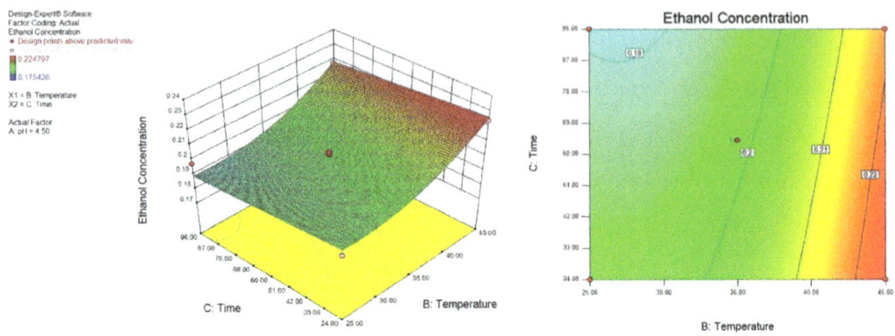

Fig. 3. (a) 3D Surface Plot (b) Contour Plot of Effect of Temperature & Time on Ethanol Yield

Fig. 3 illustrates the effect of fermentation temperature and incubation time over the ethanol concentration. These parameters were varied in the range of 25 – 45°C and 24 – 96 h, while pH was kept constant at 4.50. Based on Fig. 3, temperature has a greater effect on the ethanol concentration than time which was in agreement with a several previous researches which have proven that temperature has a greater influence on ethanol yield [14, 20]. This was also validated by the *p*-value in ANOVA where temperature and time had values of <0.0001 and 0.0231, respectively. The ethanol concentration was observed to increase with increasing temperature and decreasing time. At constant pH of 4.5, the maximum ethanol concentration of 0.2240 g/ml was achieved at 24 h and 45°C. The minimum ethanol concentration was obtained at 96 h and 25°C. The analysis of Fig. 3 proved that the optimum fermentation time was at 24 h with temperature of 45°C where the ethanol concentration was at its peak. The glucose available in the prepared solution was fully utilized by *S. Cerevisiae* at this condition.

Response Surface Optimization and Results Verification. Numerical optimization was used to determine the maximum ethanol concentration at the optimum condition of pH, temperature, and incubation time. The highest desirability shows the highest ethanol production at the optimum condition of all three varied parameters. Table 3 shows the most desirable operating condition suggested by RSM to produce bioethanol from papaya waste through fermentation.

Table 3. Numerical Optimization of Variables for Bioethanol Production

Run	pH	Temperature (°C)	Time (h)	Predicted Concentration (g/mL)	Desirability
1	4.50	45.00	24.00	0.2264	1[a]
2	3.00	45.00	64.00	0.2256	1
3	3.36	45.00	92.27	0.2226	0.987
4	5.92	45.00	24.00	0.2129	0.861

[a] selected

The higher the desirability, the higher the predicted ethanol concentration. Three confirmation experimental runs were carried out at pH 4.5, temperature of 45°C and incubation time of 24 h. The obtained responses for the confirmation runs are shown in Table 4.

Table 4. Verification of Response Obtained for Confirmation Experimental Run

Run	pH	Temperature (°C)	Time (h)	Predicted Concentration (g/mL)	Actual Concentration (g/mL)	Std Dev. (%)
1					0.2167	4.25
2	4.5	45.00	24.00	0.2263	0.2171	4.08
3					0.2224	1.75

Based on Table 4, the actual concentration of ethanol produced by the three confirmation runs was close with the predicted concentration with standard deviations below 5%. This result is satisfying and denotes that the numerical optimization is reliable to produce bioethanol with high concentration yield.

Conclusion

In this project, papaya waste including papaya seed and papaya peel was used as a substrate to produce bioethanol through fermentation process to enhance the development of green energy from lignocellulosic biomass. The fermentation conditions such as pH, temperature and incubation time were studied and optimized using RSM. This study was carried out within the fermentation condition of pH 3.0-6.0, temperature of 25-45°C and incubation time of 24-96 hours. According to ANOVA and model verification plots, the quadratic model was suitable to describe the interaction between variables with high reliability and reproducibility. The validation performed found that the predicted values fit the model. Based on the results obtained through the numerical optimization and verification run, the maximum ethanol concentration was produced at pH 4.5, 45°C and 24 h.

References

[1] "bp Statistical Review of World Energy 2020," in "Statistical Review of World Energy," bp, 2020, vol. 69. [Online]. Available: sr@bp.com

[2] Z. H. Zhao, "Design Principle on Carbon Nanomaterials Electrocatalysts for Energy Storage and Conversion," p. 134, 2017, https://doi.org/10.13140/RG.2.2.25704.16641

[3] S. Nikolic, V. Lazic, D. Veljovic, and L. Mojovic, "Production of bioethanol from pre-treated cotton fabrics and waste cotton materials," *Carbohydrate Polymers,* vol. 164, pp. 136-144, May 15 2017, https://doi.org/10.1016/j.carbpol.2017.01.090

[4] O. Oiwoh, B. V. Ayodele, N. A. Amenaghawon, and C. O. Okieimen, "Optimization of bioethanol production from simultaneous saccharification and fermentation of pineapple eels using Saccharomyces cerevisiae," *Journal of Applied Sciences and Environmental Management,* vol. 22, no. 1, 2018, https://doi.org/10.4314/jasem.v22i1.10

[5] N. Annamalai, H. Al Battashi, S. N. Anu, A. Al Azkawi, S. Al Bahry, and N. Sivakumar, "Enhanced Bioethanol Production from Waste Paper Through Separate Hydrolysis and Fermentation," *Waste and Biomass Valorization,* vol. 11, no. 1, pp. 121-131, 2018, https://doi.org/10.1007/s12649-018-0400-0

[6] M. Bhuvaneswari and N. Sivakumar, "Bioethanol Production from Fruit and Vegetable Wastes," *Bioprocessing for Biomolecules Production,* no. 1, pp. 417-427, 2020.

[7] A. Oliveira, "Market analysis for Papaya (Production and Trade)," 2018.

[8] C. Pavithra, S. S. Devi, J. S. W, and C. V. D. Rani, "Nutritional properties of papaya peel," *The Pharma Innovation Journal,* vol. 6, no. 7, pp. 170-173, 2017.

[9] G. L. Utama, Tyagita, I. Krissanti, D. W. Wira, and R. L. Balia, "Stress Tolerance Yeast Strain from Papaya Wastes for Bioethanol Production," *International Journal of GEOMATE,* vol. 17, no. 61, 2019, https://doi.org/10.21660/2019.61.4796

[10] N. A. A. Hamid and N. Z. Zulkifli, "Papaya peels as source of hydro char via hydrothermal carbonization," IOP Conf. Series: Earth and Environmental Science, vol. 765, 2021.

[11] J. J. Romero, M. A. Zarate, I. M. Ogunade, K. G. Arriola, and A. T. Adesogan, "Tropical plant supplementation effects on the performance and parasite burden of goats," Asian-Australas J Anim Sci, vol. 31, no. 2, pp. 208-217, Feb 2018, https://doi.org/10.5713/ajas.17.0321

[12] N. I. Wan Azelee, J. Md Jahim, A. Rabu, A. M. Abdul Murad, F. D. Abu Bakar, and R. Md Illias, "Efficient removal of lignin with the maintenance of hemicellulose from kenaf by two-stage pretreatment process," Carbohydr Polym, vol. 99, pp. 447-53, Jan 2014, https://doi.org/10.1016/j.carbpol.2013.08.043

[13] R. Abdulla, E. Derman, P. T. Ravintaran, and S. A. Jambo, "Fuel Ethanol Production from Papaya Waste using Immobilized Saccharomyces cerevisiae," *ASM Science Journal,* vol. 11, no. 2, pp. 112-123, 2018

[14] P. K. Dash, S. Mohaptra, M. R. Swain, and H. Thatoi, "Optimization of bioethanol production from saccharified sweet potato root flour by co-fermentation of Saccharomyces cerevisiae and Pichia sp. using OVAT and response surface methodologies," *Acta Biologica Szegediensis,* vol. 61, pp. 13-23, 2017.

[15] K. Ma, Z. Ruan, Z. Shui, Y. Wang, G. Hu, and M. He, "Open Fermentative Production of Fuel Ethanol from Food Waste by An Acid-Tolerant Mutant Strain of Zymomonas Mobilis," *Bioresour Technol,* vol. 203, pp. 295-302, Mar 2016, https://doi.org/10.1016/j.biortech.2015.12.054

[16] G. Reed and H. J. Peppler, "Yeast technology," Food, vol. 19, no. 4, 1975, https://doi.org/10.1002/food.19750190413

[17] A. K. Pai, R. K. Vikas, H. S. Neeraja, C. A. Puneeth, and C. V. Rao, "Bioethanol Production from Waste Biomass," India Patent Appl. 2373/CHE/2009 A, 2011.

[18] Y. Lugani *et al.*, "Recent advances in bioethanol production from lignocelluloses: a comprehensive review with a focus on enzyme engineering and designer biocatalysts," *Biofuel Research Journal,* vol. 7, no. 4, pp. 1267-1295, 2020, https://doi.org/10.18331/brj2020.7.4.5

[19] M. R. A. Ghani and P. C. Oh, "Optimization of Ethanol Production from Mango Peels Using Response Surface Methodology," *Applied Mechanics and Materials,* vol. 625, pp. 766-769, 2014, https://doi.org/10.4028/www.scientific.net/AMM.625.766

[20] S. Chongkhong, B. Lolharat, and P. Chetpattananondh, "Optimization of Ethanol Production from Fresh Jackfruit Seeds Using Response Surface Methodology," *Journal of Sustainable Energy & Environment,* vol. 3, pp. 97-101, 2012.

[21] D. Dahnum, S. O. Tasum, E. Triwahyuni, M. Nurdin, and H. Abimanyu, "Comparison of SHF and SSF Process Using Enzyme and Dry Yeast for Optimization of Bioethanol Production from Empty Fruit Bunch," *Energy Procedia,* vol 68, pp.107-116, 2015, https://doi.org/10.1016/j.egypro.2015.03.238

[22] A. M. Saeed, H. Ma, S. Yue, Q. Wang, and M. Tu, "Concise review on ethanol production from food waste: development and sustainability," *Environ Sci Pollut Res Int,* vol. 25, no. 29, pp. 28851-28863, Oct 2018, https://doi.org/10.1007/s11356-018-2972-4

Materials Research Proceedings 29 (2023) 81-86 https://doi.org/10.21741/9781644902516-11

Performance analysis of chicken manure coated slow-release urea fertilizer (CM-CSRUF)

ALI Shaan Manzoor Ghumman[1,2,a], RASHID Shamsuddin[1,2,b,*], YEONG Yin Fong[1,c] and HAMAD Almohamadi[3,d]

[1]Chemical Engineering Department, Universiti Teknologi PETRONAS, 32610 Bandar Seri Iskandar, Perak Darul Ridzuan, Malaysia

[2]HICoE, Center for Biofuel and Biochemical Research (CBBR), Institute of Self-Sustainable Building (ISB), Universiti Teknologi PETRONAS, 32610, Seri Iskandar, Perak, Malaysia

[3]Department of Chemical Engineering, Faculty of Engineering, Islamic University of Madinah, Madinah Saudi Arabia

[a]alishaan.ghumman@utp.edu.my, [b]mrashids@utp.edu.my, [c]yinfong.yeong@utp.edu.my, [d]hha@iu.edu.sa

Keywords: Chicken Manure, Coated Urea, Nutrient Use Efficiency, Soil Contamination

Abstract. Slow-release fertilizers (SRFs) are developed, to halt the nutrient loss to the environment, either by coating the urea or by infusing the urea in a hydrophobic material. SRFs reduces the nutrient release to that point where a single application of fertilizer can meet the nutrient demand of the plants. Diverse range of materials have already been utilized to produce SRFs, however several disadvantages such hydrophilicity, non-biodegradability and crystallinity limit their scale up. Herein, we reported the production of coating of urea using chicken manure with paraffin wax as a binder. Nitrogen release test of the coated urea revealed that only 11.8% of total nutrients were released in first 12 h whereas uncoated urea released ≥99.9 % nutrients. Ritger-Peppas model was found to best fit the nutrient release kinetic data with an R^2 value of 0.97.

Introduction

With the global population dramatically growing to reach the estimated number of 9.5 billion by 2050, the worldwide food demand has also risen and thus the anticipated apiece food requirement is likely to be doubled by that year [1,2]. Asynchrony between the release of nutrients (ammonium (NH_4^+) and nitrate (NO_3^-)) from urea and the crop demand has resulted in losing surplus nutrients due to the leaching, volatilization, and nitrous emission, provoking a ineffective plant nutrient uptake efficiency (NUE) and also serious environmental problems [3,4]. Only 30% of the applied urea applied is absorbed by the plants [5–7].

Slow-release fertilizers (SRFs), a purposefully engineered manure that delays the release of nutrients following the sequential nutrient requirement of plants, are usually used to improve the nutrient use efficiency (NUE) [8]. SRFs are produced by encapsulating the urea granule using hydrophobic materials, which reduces the nitrogen release to that point where a single application of fertilizer can meet the nitrogen requirement of crops [9]. Diverse range of materials have already been synthesized to coat the urea which includes synthetic (methacrylic acid copolymers [10], inverse vulcanized copolymers [11–13] Polyacrylic acid latex [14], polyurethane [15], waste polystyrene (thermocol) [16], Ca-Mg phosphate polyolefin, polyolefin plus dicyandiamide, poly(lactic acid-co-ethylene terephthalate) [17]) and natural (starch [18], lignin [19], ethylcellulose [20], and bio-based polyurethane) polymers. These materials have shown encouraging lab results in term nutrient release, however, the nonbiodegradability of synthetic polymers contaminates the soil and on the other hand the poor mechanical and physical properties of natural polymers requires modification using plasticizers or crosslinkers increasing the overall cost. These disadvantages arise the need to look for other sustainable and biodegradable materials.

Malaysia being the 3rd largest chicken consumer in the world produces chicken manure in abundance [21]. Chicken manure compared to other manures is a nitrogen, phosphorous and potassium (NPK) enriched organic fertilizer but its direct application to soil leads environmental pollution such as greenhouse gas (GHG) emission, soil acidification and water contamination due to leaching [22]. Therefore, proper treatment process is needed for its application to soil. Our lab recently treated the chicken manure to produce biogas after which chicken manure does not any harmful gases and safe to be use. Herein, we reported the production of slow-release fertilizer by coating chicken manure using paraffin wax as a binder on urea. Then nitrogen release test was conducted on coated urea in distilled to evaluate its slow-release properties.

Materials

Urea was procured from PETRONAS Fertilizer Kedah Sdn. Bhd. Malaysia. Diacetyl monoxime (DAM), thiosemicarbazide (TSC), and O-phosphoric acid (85%) were purchased from Sigma-Aldrich, USA. Paraffin wax was purchased from Kinetic chemical Sdn Bhd Malaysia. All materials were used as received without purification.

Methods

Coating of Urea

First of all, a coating solution was prepared, by melting 5 g of paraffin wax at 80 °C and mixed with 8 g of chicken manure under continuous mixing to obtain homogenous mixture. After this, 10 g of urea was coated using a dip coating method and allowed to cool at room temperature.

Nitrogen Release Test

DAM calorimetry method was used to estimate the nitrogen release from coated urea. This method includes the preparation of color reagent by mixing the acid reagent (460 ml), TSC (15 ml) and DAM (25 ml) solution which were prepared by dissolving 0.25 g Thiosemicarbazide and 2.5 g of Diacetyl monoxime in a 100 ml of distilled water, separately and for acid reagent phosphoric acid (250 ml), sulfuric acid (10 ml) and distilled water (240 ml) were mixed. In order to analyze the nitrogen release properties of the coated urea, 2.0±0.1 g of the developed fertilizer was immersed in 200 ml distilled water in an Erlenmeyer flask sealed with cling wrap. After every 12 h the homogenously mixed 2.5 ml of sample was taken out and mixed with 7.5 ml of color reagent and placed in water bath heated at 85 °C and allowed to react for 30 mins after the sample was allowed cool at room temperature. The resultant solution was then analyzed using UV-vis spectrophotometer.

Kinetics of Release

First-order, Parabolic diffusion, Ritger-Peppas and Elovich models were used to investigate release kinetic data to understand mechanism of release from the newly developed fertilizer. Models used are given in Table 1. R^2 value was used to evaluate these models.

Table 1:Models used for describing nitrogen leaching kinetics data.

Model	Equation	Ref
First-order	$\ln(N_o - N_t) = \ln(N_o) - kt$	[23,24]
Parabolic diffusion	$N_t = b + kt^{0.5}$	[25,26]
Ritger-Peppas	$\ln\left[\frac{M_t}{M_o}\right] = \ln k + n\ln t$	[27,28]
Simple Elovich	$N_t = b + k\ln t$	[29,30]

Where t is the release time, N_o is the highest release rate, N_t is the cumulative release rate of nitrogen in time t, the ratio of M_t to M_o is the rate of Nitrogen release in time t, K is the constant of diffusion (which depends on the type of material and osmotic medium), and n is the diffusional exponent which indicates the transport mechanism.

Results and Discussion

Nitrogen elease in Distilled Water. Figure 1 shows the nitrogen release performance of coated and uncoated urea. The slow-release properties of coated urea were evaluated in distilled water at room temperature. The integrity of the coating film is reflected by the initial nutrient release rate. Thoroughly coated urea releases less nutrient compared to uncoated urea. The uncoated urea released almost 99.9 % of its total nutrient in first 12 h. In case of the coated urea, only 11.8 % of its nutrient in first 12 h. Results are almost similar to the biopolymers [31], which demonstrate that chicken manure can be used as a effective coating material.

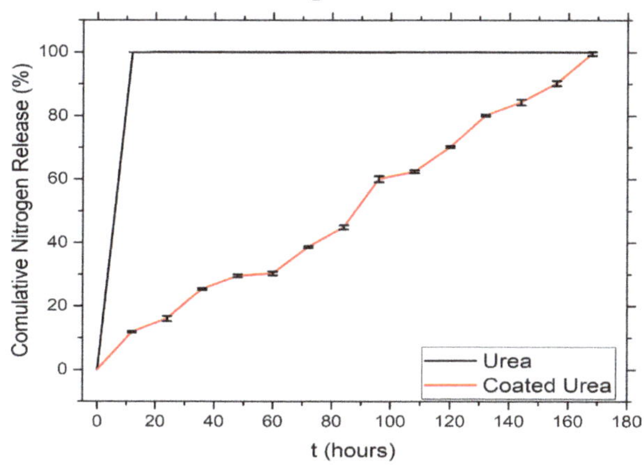

Figure 1: Cumulative nitrogen release percentage as a function of time for coated and uncoated urea.

Release Kinetics

Few models were used to elucidate the functioning of the newly developed coated fertilizer and to get the parameters of release kinetics. Kinetic of release of nutrient from coated urea based on four different models are presented in Figure 2. The kinetic data best fit the Ritger-Peppas law with very high coefficient of determination (R^2) value which is more than0.97. This is due to the fact that the nutrient releases follows both Fickian and Non-Fickian as indicated by the mass transport indicating number n =0.85 which is known as anomalous phenomena [13].

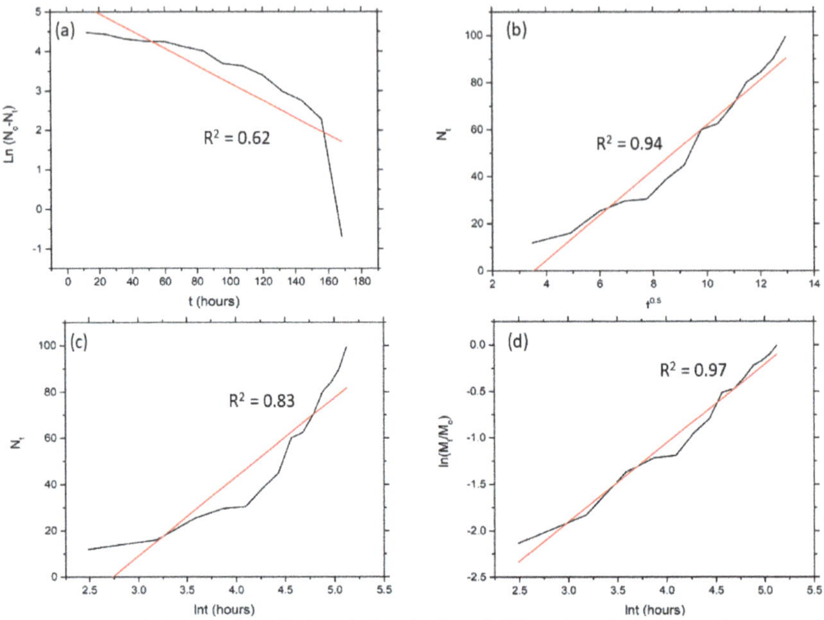

Figure 2: (a) First order, (b) Parabolic, (c) Simple Elovich and (d) Ritger-Peppas models.

Conclusion

Using paraffin wax as a binder urea granule were dip coated with chicken manure to produce novel slow-release coated urea (SRCU). Slow-release properties of the developed fertilizers were examined using nitrogen release test in distilled water. Pristine urea releases more than 99% of its total nutrient in less than 12h, whereas it took almost 680 h in case of SRCU to release ≥99% nutrients. R^2 value of 0.97 for Ritger-Peppas model suggested that the kinetic release data best fir this model.

Acknowledgement

The authors would like to acknowledge the funding supports from Ministry of Higher Education (MOHE) Malaysia through HICoE grant (Cost Centre 015MA0-052/015MA0-104/015MA0-136) to CBBR.

References

[1] M.E. Brown, B. Hintermann, N. Higgins, Markets, climate change, and food security in West Africa, Environ. Sci. Technol. 43 (2009) 8016–8020. https://doi.org/10.1021/es901162d

[2] S.H. Chien, L.I. Prochnow, H. Cantarella, Recent Developments of Fertilizer Production and Use to Improve Nutrient Efficiency and Minimize Environmental Impacts, 1st ed., Elsevier Inc., 2009. https://doi.org/10.1016/S0065-2113(09)01008-6

[3] E.L. Carter, N. Flugga, J.L. Boer, B. Mulrooney, R.P. Hausinger, E.L. Carter, Interplay of metal ions and urease, Metallomics. 1 (2009) 207–221. https://doi.org/10.1039/b903311d

[4] V. Achal, X. Pan, Characterization of Urease and Carbonic Anhydrase Producing Bacteria and Their Role in Calcite Precipitation, Curr Microbiol. 62 (2011) 894–902. https://doi.org/10.1007/s00284-010-9801-4

[5] Suherman, D.D. Anggoro, Producing slow release urea by coating with Starch/Acrylic Acid in fluid bed spraying, Int. J. Eng. Technol. 11 (2011) 77–80.

[6] A. Zarei, V. Ghaffarian, Preparation and Characterization of Biodegradable Cellulose Acetate-Starch Membrane, Polym. - Plast. Technol. Eng. 52 (2013) 387–392. https://doi.org/10.1080/03602559.2012.752831

[7] M.Y. Naz, S.A. Sulaiman, B. Ariwahjoedi, K.Z.K. Shaari, Characterization of modified tapioca starch solutions and their sprays for high temperature coating applications, Sci. World J. 2014 (2014). https://doi.org/10.1155/2014/375206

[8] A. Shaviv, Advances in Controlled Release of Fertilizers 71, before printing., Adv. Agron. 71 (2000) 1–49. https://doi.org/10.1016/S0065-2113(01)71011-5

[9] B. Azeem, K. Kushaari, Z.B. Man, A. Basit, T.H. Thanh, Review on materials & methods to produce controlled release coated urea fertilizer, J. Control. Release. 181 (2014) 11–21. https://doi.org/10.1016/j.jconrel.2014.02.020

[10] M.W. Donida, S.C.S. Rocha, Coating of urea with an aqueous polymeric suspension in a two-dimensional spouted bed, Dry. Technol. 20 (2002) 685–704. https://doi.org/10.1081/DRT-120002824

[11] A.S.M. Ghumman, R. Shamsuddin, M.M. Nasef, C. Maucieri, O.U. Rehman, A.A. Rosman, M.I. Haziq, A. Abbasi, Degradable Slow-Release Fertilizer Composite Prepared by Ex Situ Mixing of Inverse Vulcanized Copolymer with Urea, Agronomy. 12 (2022). https://doi.org/https://doi.org/10.3390/agronomy12010065

[12] A.S.M. Ghumman, R. Shamsuddin, M.M. Nasef, E.G. Krivoborodov, S. Ahmad, A.A. Zanin, Y.O. Mezhuev, A. Abbasi, A Degradable Inverse Vulcanized Copolymer as a Coating Material for Urea Produced under Optimized Conditions, Polymers (Basel). 13 (2021). https://doi.org/https://doi.org/10.3390/polym13224040

[13] A.S.M. Ghumman, R. Shamsuddin, M.M. Nasef, W.Z.N. Yahya, A. Abbasi, H. Almohamadi, Sulfur enriched slow-release coated urea produced from inverse vulcanized copolymer, Sci. Total Environ. 846 (2022) 157417. https://doi.org/10.1016/j.scitotenv.2022.157417

[14] R. Lan, Y. Liu, G. Wang, T. Wang, C. Kan, Y. Jin, Experimental modeling of polymer latex spray coating for producing controlled-release urea, Particuology. 9 (2011) 510–516. https://doi.org/10.1016/j.partic.2011.01.004

[15] Q. Li, S. Wu, T. Ru, L. Wang, G. Xing, J. Wang, Synthesis and performance of polyurethane coated urea as slow/controlled release fertilizer, J. Wuhan Univ. Technol. Mater. Sci. Ed. 27 (2012) 126–129. https://doi.org/10.1007/s11595-012-0421-7

[16] Y.C. Yang, M. Zhang, Y. Li, X.H. Fan, Y.Q. Geng, Improving the quality of polymer-coated urea with recycled plastic, proper additives, and large tablets, J. Agric. Food Chem. 60 (2012) 11229–11237. https://doi.org/10.1021/jf302813g

[17] N. Jintakanon, P. Opaprakasit, A. Petchsuk, M. Opaprakasit, Controlled-release materials for fertilizer based on lactic acid polymers, Adv. Mater. Res. 55–57 (2008) 905–908. https://doi.org/10.4028/www.scientific.net/amr.55-57.905

[18] R. Ito, B. Golman, K. Shinohara, Design of multi-layer coated particles with sigmoidal release pattern, Chem. Eng. Sci. 60 (2005) 5415–5424. https://doi.org/10.1016/j.ces.2005.04.023

[19] W.J. Mulder, R.J.A. Gosselink, M.H. Vingerhoeds, P.F.H. Harmsen, D. Eastham, Lignin based controlled release coatings, Ind. Crops Prod. 34 (2011) 915–920. https://doi.org/10.1016/j.indcrop.2011.02.011

[20] M. Fernández-Pérez, F.J. Garrido-Herrera, E. González-Pradas, M. Villafranca-Sánchez, F. Flores-Céspedes, Lignin and ethylcellulose as polymers in controlled release formulations of

urea, J. Appl. Polym. Sci. 108 (2008) 3796–3803. https://doi.org/10.1002/app.27987

[21] M.D. Manogaran, R. Shamsuddin, M.H. Mohd Yusoff, M. Lay, A.A. Siyal, A review on treatment processes of chicken manure, Clean. Circ. Bioeconomy. 2 (2022) 100013. https://doi.org/10.1016/j.clcb.2022.100013

[22] N.O. Chijioke, M. Uddin Khandaker, K.M. Tikpangi, D.A. Bradley, Metal uptake in chicken giblets and human health implications, J. Food Compos. Anal. 85 (2020) 103332. https://doi.org/10.1016/j.jfca.2019.103332

[23] Y. He, Z. Wu, L. Tu, Y. Han, G. Zhang, C. Li, Encapsulation and characterization of slow-release microbial fertilizer from the composites of bentonite and alginate, Appl. Clay Sci. 109–110 (2015) 68–75. https://doi.org/10.1016/j.clay.2015.02.001

[24] A. Rashidzadeh, A. Olad, M.J. Hejazi, Controlled Release Systems Based on Intercalated Paraquat onto Montmorillonite and Clinoptilolite Clays Encapsulated with Sodium Alginate, Adv. Polym. Technol. 36 (2017) 177–185. https://doi.org/10.1002/adv.21597

[25] H. Shariatmadari, M. Shirvani, A. Jafari, Phosphorus release kinetics and availability in calcareous soils of selected arid and semiarid toposequences, Geoderma. 132 (2006) 261–272. https://doi.org/10.1016/j.geoderma.2005.05.011

[26] Y. Liang, X. Cao, L. Zhao, X. Xu, W. Harris, Phosphorus Release from Dairy Manure, the Manure-Derived Biochar, and Their Amended Soil: Effects of Phosphorus Nature and Soil Property, J. Environ. Qual. 43 (2014) 1504–1509. https://doi.org/10.2134/jeq2014.01.0021

[27] P.L. Ritger, N.A. Peppas, A simple equation for description of solute release II. Fickian and anomalous release from swellable devices, J. Control. Release. 5 (1987) 37–42. https://doi.org/10.1016/0168-3659(87)90035-6

[28] L. Wu, M. Liu, Slow-release potassium silicate fertilizer with the function of superabsorbent and water retention, Ind. Eng. Chem. Res. 46 (2007) 6494–6500. https://doi.org/10.1021/ie070573l

[29] M. Park, J.S. Kim, C.L. Choi, J.E. Kim, N.H. Heo, S. Komarneni, J. Choi, Characteristics of nitrogen release from synthetic zeolite Na-P1 occluding NH4NO3, J. Control. Release. 106 (2005) 44–50. https://doi.org/10.1016/j.jconrel.2005.02.029

[30] N. Abdu, Formulation of a first-order kinetic model and release of added phosphorus in a savanna soil, Arch. Agron. Soil Sci. 59 (2013) 71–81. https://doi.org/10.1080/03650340.2011.600308

[31] B. Azeem, K. KuShaari, M. Naqvi, L.K. Keong, M.K. Almesfer, Z. Al-Qodah, S.R. Naqvi, N. Elboughdiri, Production and characterization of controlled release urea using biopolymer and geopolymer as coating materials, Polymers (Basel). 12 (2020). https://doi.org/10.3390/polym12020400

Kinetics study of nutrients removal from synthetic wastewater using media as submerged in continuous activated sludge system

BAKER NASSER Saleh Al-dhawi[1,a*], SHAMSUL RAHMAN Mohamed Kutty[1,b], LAVANİA Baloo[1,c], AHMED HUSAİNİ Jagaba[1,d], AIBAN ABDULHAKIM Saeed Ghaleb[1,e], NAJİB MOHAMMED Yahya Almahbashi[1,f], VİCKY Kumar[1,g], AZMATULLAH Noor[1,h], ANWAR AMEEN Hezam Saeed[2,i], AL-BARAA Abdulrahman Al-Mekhlafi[3,j], YASER ABDULWAHAB Ali Alsaeedi[4,k]

[1]Department of Civil and Environmental Engineering, Universiti Teknologi PETRONAS, Bandar Seri Iskandar 32610, Perak Darul Ridzuan, Malaysia

[2]Department of Chemical Engineering, Universiti Teknologi PETRONAS, Bandar Seri Iskandar 32610, Perak Darul Ridzuan, Malaysia

[3]Department of Management & Humanities, Universiti Teknologi PETRONAS, Seri Iskandar 32610, Perak, Malaysia

[4]Department of Civil and Environmental Engineering, Universiti Putra Malaysia, 43400 Serdang, Selangor

[a]baker_18002007@utp.edu.my, [b]shamsulrahman@utp.edu.my, [c]lavania.baloo@utp.edu.my, [d]ahmad_19001511@utp.edu.my, [e]aiban_17004546@utp.edu.my, [f]najib_17005375@utp.edu.my, [g]vicky_19000193@utp.edu.my, [h]azmatullah_16005709@utp.edu.my, [i]anwar_17006829@utp.edu.my, [j]albaraa901@gmail.com, [k]yaseralsaeedi50@gmail.com

Keywords: Kinetic, Nutrients, Synthetic wastewater, Palm Oil Clinker, Submerged, Continuous Flow

Abstract. Domestic wastewater effluents are one of the main sources of environmental contaminants such as nutrients. Wastewater has been treated with biological processes for over a century to remove contaminants. Conventional wastewater treatment plants continue to struggle to meet Malaysian discharge limits. Stringent regulation enforced by governing authorities makes it obligatory to comply with discharge guidelines to fulfill ammonia and nitrate levels. To assist the system in meeting these limits, it is recommended that a submerged attached growth Palm Oil Clinker (POC) can be incorporated into the conventional treatment system. The study was conducted in a continuous submerged attach growth conventional activated sludge which was evaluated for the treatment of wastewater (CSAR). A basket was installed in the aeration tank of the reactor to submerge (POC). Two identical reactors were operated for each reactor of study which (A) was referred to as submerged media reactor while (B) was referred to as control. The studies were carried out at various influent flow rates between 5 and 30 L/d, and constant organic load rate OLR. Parameters such as NH_4-N, and NO_3-N, were monitored. Generally, Ammonia and Nitrate were highly removed. At all conditions of flow rate (5-30 L/d), the maximum and minimum NH_4-N removal is 92% and 85%. The experimental data were validated through well-established mathematical bio-kinetic models such as the First order model, and Monod models. The kinetic coefficients R^2 of the first-order model of the substrate removal rate were 0.97 for Ammonia. The steady-state data was fitted to both models obtained at various flowrate. Monod's kinetic model was appropriate for describing experimental results in terms of microbial growth parameters. The kinetic coefficients R^2 (0.984) and Ks 303 for the removal of Ammonia, respectively. While μ_{max} 10 g/L.d for Ammonia removal respectively.

Introduction

Drop water level of quality and Increasing water requirements for human use for necessary enhancement of cost-efficient wastewater treatment technologies. the biological mechanisms have been providing currently much publicity owing to the price, simpler operation, better management, and less adverse impact on the environment [1]. Suspended and attach growth systems are a type of biological process [2, 3]. Activated sludge the most common process frequently used in developing nations, removed biological solids by sedimentation, low settlement of solid contaminants will result in higher costs of treatment of solids, the total concentrations of effluent solids realized, decreased efficiencies in disinfecting, increasing downstream threats environments and health [4, 5]. When discharged without sufficient treatment, these pollutants in wastewater have a harmful influence on aquatic bodies [6, 7]. Organics, in particular, reduce oxygen levels in normal water systems while meeting their biochemical oxygen demand (BOD), causing aquatic life to suffer [8, 9]. Consequently, the anaerobic decomposition of organics additionally leads to the production of foul-smelling gases [10, 11]. Through this, emerging contaminants and heavy metals build in flora and fauna inside the water, eventually affecting the human food chain Moreover, apart from solid media to suspended growth to have biofilm attached to a surface, the growth process is such a successful process, hence improve and increase microbial concentration and rates of pollutant deterioration [12, 13]. Biofilms benefit from an amount of removal of process mechanisms such as biodegradation [14]. Biofilm is shaped on rigid media such as rocks, plastic profiled, or membranes in a fixed bed system. Both biofilm processes with constantly moving media, sustained by high air were moving bed systems [6, 15]. Nutrients and the aqueous bulk in the conditioning film are used by surface-attached bacterial cells to expand and generate further EPS, contributing to the development of microorganisms [16]. In the end, the microorganisms gradually structure a layer covering the surface [1]. It was proposed that the rock particles should be standardized so that 95% of the particles be between 7 and 10 cm in diameter. Due to, the porous surface, the appropriate attachment of biomass is ideal for a specific area and void ratio for Palm Oil Clinker. The size is not too vital, but the media recommended to be uniform to allow sufficient ventilation across the void space, accessible locally, and cost-effective. Due to the economic benefit, the simplicity of availability, and the high potential for linking biofilm, the use of palm oil clinker media seems to be feasible. Its use as a filter media for wastewater treatment in attached growth reactors is not a new concept [12, 17]. Moreover, there are very few literatures available on conventional methods utilized to search biofilm in such bioreactors and on comparative studies on the efficiency of fixed-film reactors and suspended growth reactors in wastewater treatment. The present research was made to use palm oil clinker in the attached growth system as submerged attached media. Biofilms can be developed in most moist environments on surfaces (biotic and abiotic) [18, 19]. Bacteria move through chemotaxis or Brownian motion with mass transport to the substrate during biofilm progression, producing a temporary bacteria-surface association [20]. The nutrients in the conditioning film and the aqueous bulk are used by bacterial cells attached to the surface to grow and develop further EPS, resulting in microcolony forming [21, 22].

The Malaysian Department of Environment (DOE) establishes criteria that must be met to comply with discharge guidelines. New wastewater discharges must be designed to meet the ammonia and nitrate limits [23, 24]. However, the present conventional treatment systems will not be able to meet the limits as conventional treatment systems were not designed to meet the ammonia and nitrate limits. It is proposed that a submerged attached growth system is to be introduced into the conventional treatment system to help the system to meet these limits [25, 26]. Palm (POC) is a waste material that comes from the steam boiler of the palm oil industry. The POC is dumped at dumpsites or landfills. In this research, the POC by-product, the residue from the steam boiler of the palm oil mill, is to be used as an attached growth system [27, 28]. The POC will be processed before being used as an attached growth system. The performance of POC as a media as an attached growth treatment process will be evaluated. The effectiveness and efficiency of using POC as attached growth media in the combined attached growth/conventional system in the removal of ammonia and nitrates will be evaluated. Consequently, the current research was planned to utilize the palm oil clinker as a supporting media biofilm in submerged attached growth reactors to investigate the efficiency of palm oil clinker as media in an attached submerged growth system in traditional treatment system in the aerobic system in removal of organic and ammonia matter. Monitoring of biofilm growth in succession on palm oil clinker media subjected to activated sludge and evaluation of the stability of biofilm growth promoting POC in aerobic conditions to be used for biological treatment of wastewater in attached growth reactors was also considered in-depth for the removal of organic and nutrient from activated sludge under aerobic conditions to be used in attached growth.

Materials Methods

Palm Oil Clinker Preparation

POC was crushed to obtain the nominal size of 20mm, and comply with a classification of aggregates in Particle shape and surface texture following the British Standard BS 812: Part 1: 1975[29, 30]. Palm Oil Clinker with a nominal size of 35 to 25 mm was utilized as sunken media in this project. The load-bearing capacity of Palm Oil Clinker is significantly affected by its porous structure and weight [31]. The highly porous nature of the Palm Oil Clinker aggregate easily induces crack propagation. Raw wastewater taken from Universiti Teknologi Petronas (UTP)'s Sewage Treatment Plant (STP) was used as the influent to both reactors. A packing media enable the continuous usurp of bacteria to inhibit plug issues and introduce the greatest surface area for ventilation, attaching, and linking to wastewater [32].

Synthetic Wastewater Preparation

To resemble domestic wastewater of medium strength, domestic wastewater was synthesized and acclimatized according to Metcalf and Eddy [33]. The wastewater concentration was customized to the medium strength of the domestic wastewater. Purian Alpo High protein puppy dog diet was used to imitate medium-strength domestic wastewater. Due to its compositional similarity to primary sludge, the Alpo dog food was primarily picked [34]. The recipes include grinding the dog food for about 10 minutes in a blender. About 6.5 to 7 grams of the dog food formula was mixed with 1 g of an ammonium salt and mixed with 100 mL of distilled water for approximately 10 minutes.100 mL of the formulated stock solution was weighed once the reactors were to be fed and then mixed with about 10 liters of tap water. The prepared synthetic wastewater recipe was poured into the influent tanks and was mixed well to be kept suspended using pneumatic mixers. The wastewater was continually pumped into both reactors A and B using a Masterflex Precision Pump [35].

Reactors Set-up and Experimental

This study discussed the results obtained from the biological treatment system; continuous flow (complete mixed activated sludge process) was carried out to investigate the viability of utilizing POC as a submerged media in treating medium synthetic wastewater for the removal of nutrients. Two reactors were used (A) referred to as the submerged media reactor while (B) referred to as the control, reactor is rectangular shaped and was made with Plexiglas with dimensions $36.5\times16.0\times24$ cm^3. The reactor's operational capacity was 10L. The air is delivered using an air pump from Hailea® (Model HAP-100), with a 100 L/min maximum output. Both reactors feed from wastewater treatment (STP). The biomass utilized as inoculum in the reactor was taken from return activated sludge (RAS) pipe, at the institutional wastewater treatment plant (IWTP), UTP. The solids retention time (SRT) and the chosen mixed liquid suspended solids (MLSS) of the inoculum were 1500 mg/L and 5 days, respectively. The experiments were then started. In this continuous flow study, synthetic wastewater was fed to both reactors [36]. The design provides influent to flow into both reactors continuously. Change in the flow rate (5-30 L/d). The tests will be conducted according to the Standard Methods for the examination of water and wastewater [29, 37]

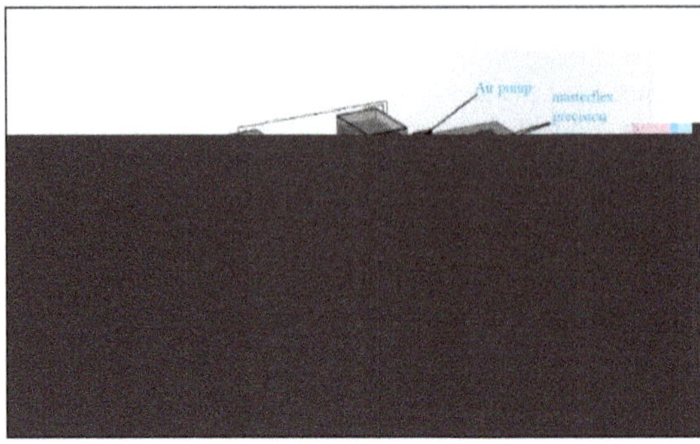

Fig. 1: continuous activated sludge reactor

Kinetic Models

In the study of biological processes at their most fundamental level, Mathematical models have been used to test theories, examine parameter interactions, drive experimental design, and estimate experiment results. These models could also be used to monitor and investigate the performance of the treatment process, as well as to enhance the plant setup and the effect of pilot scale-up experiments [38]. In this research, the steady-state data were suited to two biokinetic models namely, the Monod Kinetic Model and First Order Model.

Results and Discussions

This study discussed the results obtained from two identical reactors that were operated for each reactor of study in which (A) was referred to as submerged media reactor while (B) was referred to as control. The results obtained from both reactors were discussed according to all the monitored parameters discussed including NH_4-N, and NO_3.N.

Experimental Result for Removal of NH_4-N and NO_3-N

For the entire study time, the removal of NH_4.N from the reactors was monitored. The removal of NH_4.N from wastewater typically requires two phases, involving the process of nitrification and denitrification. There was no denitrification in this analysis since the anoxic component was not used in the design of the reactors. However, since the rectors were run in the conventional plug flow process, nitrification was expected. During the decomposition of organic materials by biomass, Ammonia is produced in the reactor, and the influent NH_4 plus the produced ammonia area is degraded in the aeration tank. The concentration of (NH_4-N) and (NO_3-N) from the effluent in all reactors was determined during the research period. However, the oxidation of influent organic matter along with endogenous respiration and cell growth of the microorganism includes concentrations of dissolved oxygen [39]. The dissolved oxygen level in the reactors was held and monitored during the research. This was to provide oxygen and enough mixing for nitrification within the reactors. Fig 2, and 3 indicated the concentration of ammonia and nitrate experiments from the reactors in the study.

Fig.2: a) Influent of Ammonia Concentration vs Sampling Days, b) Effluent of Ammonia Concentration vs Sampling Days

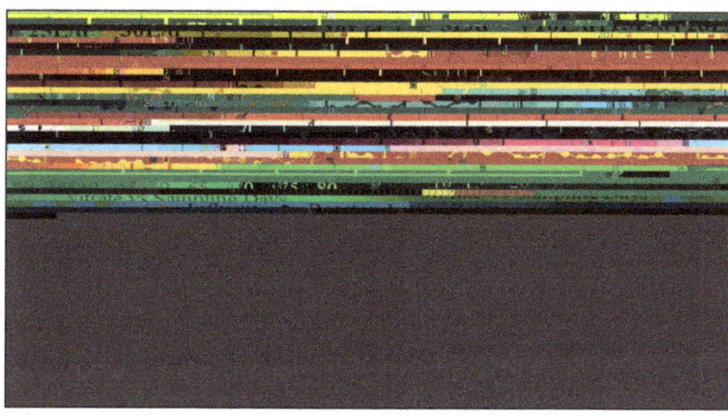

Fig.3: Nitrate Concentration vs Sampling Days

It can be noted from Fig 2, and Fig 3, that successful ammonia oxidization occurred after the startup of the reactors from day 10. The oxidized ammonia was transformed by Nitrosomonas bacteria to nitrate. Although some fluctuation in the NH_4-N and NO_3-N concentrations of the reactors have occurred through the first start days, the ammonia effluent concentration of both reactors has been adequately enough decreased by less than 5 mg/L. This shows that ammonia was used by the biomass as a nutrient to break down the biodegradable organic matter. The NH_4-N in (Reactor B) is almost fully nitrified during the study time. Reactor B effluent nitrate concentration begins to rise steadily and then stable from day 26 to day 30. This suggests that ammonia is used as a nutrient efficiently by the microbes. This was apparent in the growing trend in nitrate concentration from the reactor effluent. A sign of good system efficiency is the presence of high NO_3 with decreased NH_4 in the discharge of the treatment system [40].

Increament in the effluents of ammonia concentrations of the reactors was detected when the influencing flow rate increased the flow rate of both reactors. This suggests that the behaviors of the nitrifying microorganisms were not inhibited due to the increasing flow rate. The concentration of nitrate in the reactor effluent has been found to increase dramatically as the flow rate increases. The increase was due to oxidized ammonia, which was transformed into NO_2 and then NO_3. The mean discharge NH_4 concentration of the reactors was determined to be 2,4,5,6,7, and 9±0.5 mg/L and 4, 6, 7, 8, 9, and 11±0.5 mg/L, respectively, for reactors A and B. The mean concentration of effluent nitrates in the reactors was determined to be 5±1 mg/L and 7±1 mg/L for the A and B reactors, respectively. In comparison, nitrifiers in the mixed liquor are small and have a sluggish growth rate relative to heterotrophic microorganisms. Additionally, Small numbers of nitrifiers are present in mixed liquor, and they develop slowly in comparison to heterotrophic bacteria. The substrate removal rate and the biokinetic coefficients for Ammonia removal were determined by put in the application of the steady-state data to various kinetic models.

Table 1 Steady-state results for Ammonia removal

Influent rate L/d	HRT (days)	Effluent A %	Effluent A (mg/L)	Effluent B (%)	Effluent B (mg/L)
5	2	92	2.5	80	4
10	1	88	4	70	6
15	0.7	80	5	67	7
20	0.5	76.2	6	62	8
25	0.4	65	7.4	55	9.5
30	0.3	57	9.5	48	11

First-Order Kinetic

The first-order kinetic model shows how adjustments in HRT affect system efficacy. Many scholars have used their literature, on the first-order kinetic model to describe bioreactors. The first-order bio-kinetic model coefficients k_1 and R^2 were calculated from the slope of the line at steady-state by plotting (S_0–Se) /HRT versus Se wastewater. The plot of (S_0 –Se) /HRT and Se yielded a straight line as depicted in Fig 4. Ammonia removal constants, k_1 were found to be 5 and 3.9 per day, whereas R^2 was 0.97 [41, 42]. For first-order kinetics, the k_1 value was low, and the higher the k_1 value, the larger the degrading capability of the microorganism. [43]. Although the first-order model parameter in the current study was little, it was bigger than the values reported by other studies. Variations in k_1 levels may be caused by changes in operative circumstances and effluent type [44].

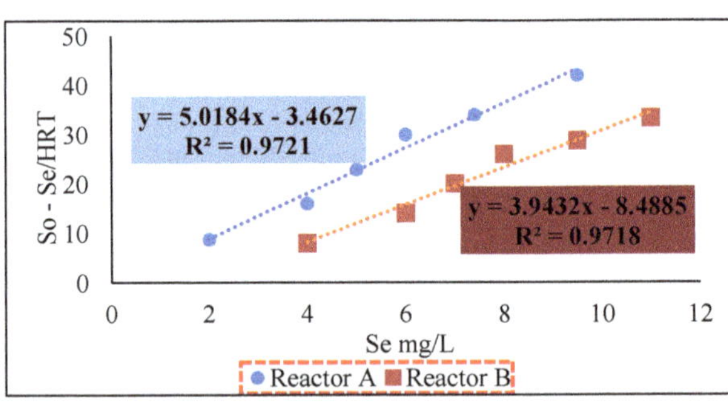

Fig. 4: First-order substrate removal plot for Ammonia

Monod Model

In Fig 5 plot of X*HRT/(So-Se) VS 1/Se, the steady-state data were fitted. The intercept and slope of the plot were then used to calculate the half-saturation constant (Ks) and the maximal specific growth rate (max). For reactors A and B, Ks were determined to be 303 and 600 g/m^3, respectively, whereas (max) 10 and 15 per day were observed with correlation coefficients (R^2) of 0.984 and 0.980, respectively. The biomass yield (Y) and death rate constant (K_d) are kinetic parameters that can be influenced by the DO concentration in the bioreactor. A low Ks, value is widely recognized to imply a stronger affinity between bacteria and substrate [45]. The low KS found in this model suggests that the microbes' affinity for the substrate was great and suggests that (CASR) can effectively remove contaminants at high efficiency under various operating conditions of flow rate.

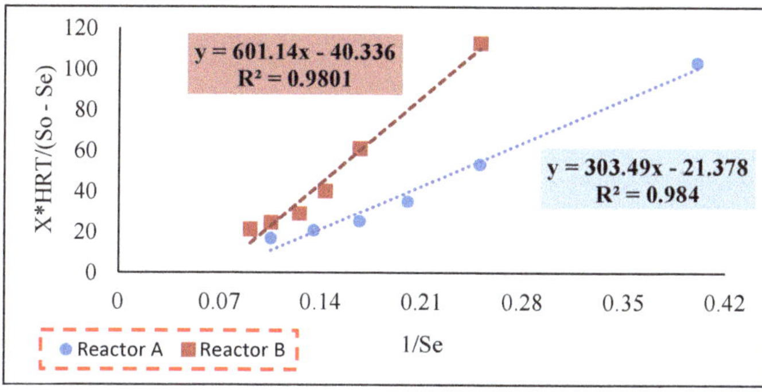

Fig.5: μ_{max} and K_B Plot

The kinetic constants for the substrate removal and growth models are illustrated in Table 2.

Table 2 Kinetic constants for Ammonia removal for all models

kinetic model	Kinetic parameter	Kinetic value		R^2	
		A	B	A	B
First-order	K_1 (1/d)	5	3.9	0.972	0.972
Monod	Ks (g/m^3)	303	600	0.984	0.981
	μ_{max} (1/d)	10	15	0.984	0.981

The kinetic constants obtained in this phase show that the first-order kinetic model appropriately described the experimental data in terms of substrate removal. For the microbial growth parameters, the Monad model appropriately described the experimental data in terms of the maximum specific growth rate (Table 2). This suggests that microbial growth was linked to the concentration of the medium's limiting nutrient. The first-order and Monad kinetic models also have correlation coefficients (R^2) greater than 0.9, implying that these models can be applied to a wide variety of bioreactors. In biological treatment systems, the rate of product formation is influenced by microbial population, environmental factors, temperature, media type, and morphology [46].

Conclusion

Treatment of wastewater (synthetic) was successfully done in a reactor. Ammonia, Nitrate was highly removed at all experimental conditions investigated. In the subsequent stage of the study, the effect of medium synthetic domestic wastewater (DWW) and the hydraulic retention time (HRT) was investigated to examine the performance of conventional activated sludge reactor (CAS) and submerged attached growth activated sludge reactor with similar operating conditions of 1500 mg/L mixed liquor suspended solids (MLSS) as inoculum, in a continuous flow mode. The result revealed that the maximum NH_4^+–N removal by using POC was 92%, higher than using the conventional treatment. Thus, indicating that the biofilm attached to the POC improved the

NO_3, and NH_4^+–N removal in treated wastewater. Thus, it was evident that the POC performance was better than CAS due to the submerged attached growth POC.

Acknowledgment
The authors would like to offer their deep gratitude to UTP and (YUTP) under grant cost center 015LC0-361.

References

[1] I. Naz, S. Seher, I. Perveen, D. P. Saroj, and S. Ahmed, "Physiological activities associated with biofilm growth in attached and suspended growth bioreactors under aerobic and anaerobic conditions," Environ Technol, vol. 36, no. 13-16, pp. 1657-71, Jul-Aug 2015. https://doi.org/10.1080/09593330.2014.1003614

[2] L. Machineni, "Review on biological wastewater treatment and resources recovery: attached and suspended growth systems," Water Science and Technology, vol. 80, no. 11, pp. 2013-2026, 2019. https://doi.org/10.2166/wst.2020.034

[3] A. A. H. Saeed et al., "Pristine and magnetic kenaf fiber biochar for Cd2+ adsorption from aqueous solution," International journal of environmental research and public health, vol. 18, no. 15, p. 7949, 2021. https://doi.org/10.3390/ijerph18157949

[4] F. Sher et al., "Removal of micropollutants from municipal wastewater using different types of activated carbons," Journal of Environmental Management, vol. 278, p. 111302, 2021. https://doi.org/10.1016/j.jenvman.2020.111302

[5] A. Jagaba et al., "Derived hybrid biosorbent for zinc (II) removal from aqueous solution by continuous-flow activated sludge system," Journal of Water Process Engineering, vol. 34, p. 101152, 2020. https://doi.org/10.1016/j.jwpe.2020.101152

[6] B. N. S. Al-dhawi et al., "Treatment of synthetic wastewater by using submerged attached growth media in continuous activated sludge reactor system," 2022.

[7] J. N. Edokpayi, J. O. Odiyo, and O. S. Durowoju, "Impact of wastewater on surface water quality in developing countries: a case study of South Africa," Water quality, vol. 10, p. 66561, 2017. https://doi.org/10.5772/66561

[8] N. K. Singh and A. A. Kazmi, "Environmental performance and microbial investigation of a single stage aerobic integrated fixed-film activated sludge (IFAS) reactor treating municipal wastewater," Journal of environmental chemical engineering, vol. 4, no. 2, pp. 2225-2237, 2016. https://doi.org/10.1016/j.jece.2016.04.001

[9] N. K. Singh, A. A. Kazmi, and M. Starkl, "Treatment performance and microbial diversity under dissolved oxygen stress conditions: Insights from a single stage IFAS reactor treating municipal wastewater," Journal of the Taiwan Institute of Chemical Engineers, vol. 65, pp. 197-203, 2016. https://doi.org/10.1016/j.jtice.2016.05.002

[10] M. Ali, N. K. Singh, A. Bhatia, S. Singh, A. Khursheed, and A. A. Kazmi, "Sulfide production control in UASB reactor by addition of iron salt," Journal of Environmental Engineering, vol. 141, no. 6, p. 06014008, 2015. https://doi.org/10.1061/(ASCE)EE.1943-7870.0000911

[11] R. K. Sinha, S. Herat, G. Bharambe, and A. Brahambhatt, "Vermistabilization of sewage sludge (biosolids) by earthworms: converting a potential biohazard destined for landfill disposal into a pathogen-free, nutritive and safe biofertilizer for farms," Waste Management & Research, vol. 28, no. 10, pp. 872-881, 2010. https://doi.org/10.1177/0734242X09342147

[12] G. Tchobanoglous et al., "Waterwater Engineering: Treatment and Resource Recovery, Metcalf and Eddy Inc," ed: McGraw-Hill, New York, 2014.

[13] A. A. Hezam Saeed, N. Y. Harun, S. Sufian, and M. F. Bin Aznan, "Effect of adsorption parameter on the removal of nickel (II) by low-cost adsorbent extracted from corn cob," International Journal of Advanced Research in Engineering and Technology (IJARET), vol. 11, no. 9, 2020.

[14] S. Pal, U. Sarkar, and D. Dasgupta, "Dynamic simulation of secondary treatment processes using trickling filters in a sewage treatment works in Howrah, west Bengal, India," Desalination, vol. 253, no. 1-3, pp. 135-140, 2010. https://doi.org/10.1016/j.desal.2009.11.019

[15] J. Bassin, I. Dias, S. Cao, E. Senra, Y. Laranjeira, and M. Dezotti, "Effect of increasing organic loading rates on the performance of moving-bed biofilm reactors filled with different support media: Assessing the activity of suspended and attached biomass fractions," Process Safety and Environmental Protection, vol. 100, pp. 131-141, 2016. https://doi.org/10.1016/j.psep.2016.01.007

[16] S. Pandit, S. Sarode, and K. Chandrasekhar, "Fundamentals of bacterial biofilm: Present state of art," in Quorum Sensing and its Biotechnological Applications: Springer, 2018, pp. 43-60. https://doi.org/10.1007/978-981-13-0848-2_3

[17] A. A. H. Saeed et al., "Moisture content impact on properties of briquette produced from rice husk waste," Sustainability, vol. 13, no. 6, p. 3069, 2021. https://doi.org/10.3390/su13063069

[18] J. A. Glaser, "Biological degradation of polymers in the environment," in Plastics in the Environment: IntechOpen, 2019. https://doi.org/10.5772/intechopen.85124

[19] A. A. H. Saeed, N. Y. Harun, and N. Zulfani, "Heavy metals capture from water sludge by kenaf fibre activated carbon in batch adsorption," Journal of Ecological Engineering, vol. 21, no. 6, 2020. https://doi.org/10.12911/22998993/123249

[20] H. J. Busscher, W. Norde, P. K. Sharma, and H. C. Van der Mei, "Interfacial re-arrangement in initial microbial adhesion to surfaces," Current Opinion in Colloid & Interface Science, vol. 15, no. 6, pp. 510-517, 2010. https://doi.org/10.1016/j.cocis.2010.05.014

[21] K. Biswas and S. J. Turner, "Microbial community composition and dynamics of moving bed biofilm reactor systems treating municipal sewage," Applied and environmental microbiology, vol. 78, no. 3, pp. 855-864, 2012. https://doi.org/10.1128/AEM.06570-11

[22] N. Y. Harun, A. A. H. Saeed, A. Vegnesh, and L. A. Ramachandran, "Abundant nipa palm waste as Bio-pellet fuel," Materials Today: Proceedings, vol. 42, pp. 436-443, 2021. https://doi.org/10.1016/j.matpr.2020.10.169

[23] S. Tabassum, Y. Zhang, and Z. Zhang, "An integrated method for palm oil mill effluent (POME) treatment for achieving zero liquid discharge-a pilot study," Journal of Cleaner Production, vol. 95, pp. 148-155, 2015. https://doi.org/10.1016/j.jclepro.2015.02.056

[24] B. Al-dhawi, S. R. Kutty, N. Almahbashi, A. Noor, and A. H. Jagaba, "Organics removal from domestic wastewater utilizing palm oil clinker (POC) media in a submerged attached growth systems," International Journal of Civil Engineering and Technology, vol. 11, no. 6, pp. 1-7, 2020. https://doi.org/10.34218/IJCIET.11.6.2020.001

[25] B.-T. Dang et al., "Non-submerged attached growth process for domestic wastewater treatment: Influence of media types and internal recirculation ratios," Bioresource Technology, vol. 343, p. 126125, 2022. https://doi.org/10.1016/j.biortech.2021.126125

[26] N. Y. Harun, T. J. Han, T. Vijayakumar, A. Saeed, and M. T. Afzal, "Ash deposition characteristics of industrial biomass waste and agricultural residues," Materials Today: Proceedings, vol. 19, pp. 1712-1721, 2019. https://doi.org/10.1016/j.matpr.2019.11.201

[27] J. Kanadasan and H. Abdul Razak, "Utilization of palm oil clinker as cement replacement material," Materials, vol. 8, no. 12, pp. 8817-8838, 2015. https://doi.org/10.3390/ma8125494

[28] A. H. Jagaba et al., Palm oil clinker as a waste by-product: Utilization and circular economy potential. IntechOpen London, 2021.

[29] A. P. H. Association, A. W. W. Association, W. P. C. Federation, and W. E. Federation, Standard methods for the examination of water and wastewater. American Public Health Association., 1915.

[30] M. Z. Jumaat, U. J. Alengaram, R. Ahmmad, S. Bahri, and A. S. Islam, "Characteristics of palm oil clinker as replacement for oil palm shell in lightweight concrete subjected to elevated temperature," Construction and Building Materials, vol. 101, pp. 942-951, 2015. https://doi.org/10.1016/j.conbuildmat.2015.10.104

[31] J. Kanadasan, A. F. A. Fauzi, H. A. Razak, P. Selliah, V. Subramaniam, and S. Yusoff, "Feasibility studies of palm oil mill waste aggregates for the construction industry," Materials, vol. 8, no. 9, pp. 6508-6530, 2015. https://doi.org/10.3390/ma8095319

[32] S. A. Kabir, U. J. Alengaram, M. Z. Jumaat, S. Yusoff, A. Sharmin, and I. I. Bashar, "Performance evaluation and some durability characteristics of environmental friendly palm oil clinker based geopolymer concrete," Journal of cleaner production, vol. 161, pp. 477-492, 2017. https://doi.org/10.1016/j.jclepro.2017.05.002

[33] F. L. Burton and G. Tchobanoglous, Wastewater Engineering: treatment, disposal, and reuse. McGraw-Hill, 2018.

[34] C. W. Teo and P. C. Y. Wong, "Enzyme augmentation of an anaerobic membrane bioreactor treating sewage containing organic particulates," Water research, vol. 48, pp. 335-344, 2014. https://doi.org/10.1016/j.watres.2013.09.041

[35] K. Gajanan, "Nutrients Requirements in Biological Industrial Waste Water Treatment," 2015.

[36] J. T. Novak, S. Banjade, and S. N. Murthy, "Combined anaerobic and aerobic digestion for increased solids reduction and nitrogen removal," Water research, vol. 45, no. 2, pp. 618-624, 2011. https://doi.org/10.1016/j.watres.2010.08.014

[37] E. W. Rice, R. B. Baird, A. D. Eaton, and L. S. Clesceri, Standard methods for the examination of water and wastewater. American public health association Washington, DC, 2012.

[38] D. Manu and A. K. Thalla, "Artificial intelligence models for predicting the performance of biological wastewater treatment plant in the removal of Kjeldahl Nitrogen from wastewater," Applied Water Science, vol. 7, no. 7, pp. 3783-3791, 2017. https://doi.org/10.1007/s13201-017-0526-4

[39] L. Metcalf, H. P. Eddy, and G. Tchobanoglous, Wastewater engineering: treatment, disposal, and reuse. McGraw-Hill New York, 1979.

[40] M. E. Argun, S. Dursun, C. Ozdemir, and M. Karatas, "Heavy metal adsorption by modified oak sawdust: Thermodynamics and kinetics," Journal of hazardous materials, vol. 141, no. 1, pp. 77-85, 2007. https://doi.org/10.1016/j.jhazmat.2006.06.095

[41] M. M. Amin, M. H. Khiadani, A. Fatehizadeh, and E. Taheri, "Validation of linear and non-linear kinetic modeling of saline wastewater treatment by sequencing batch reactor with adapted and non-adapted consortiums," Desalination, vol. 344, pp. 228-235, 2014. https://doi.org/10.1016/j.desal.2014.03.032

[42] P. Mullai and M. Yogeswari, "Substrate removal kinetics of hydrogen production in an anaerobic sludge blanket filter," Separation Science and Technology, vol. 50, no. 7, pp. 1093-1100, 2015. https://doi.org/10.1080/01496395.2014.969806

[43] R. Bhattacharya and D. Mazumder, "Evaluation of nitrification kinetics for treating ammonium nitrogen enriched wastewater in moving bed hybrid bioreactor," Journal of Environmental Chemical Engineering, vol. 9, no. 1, p. 104589, 2021. https://doi.org/10.1016/j.jece.2020.104589

[44] Y. J. Chan, M. F. Chong, and C. L. Law, "Performance and kinetic evaluation of an integrated anaerobic-aerobic bioreactor in the treatment of palm oil mill effluent," Environmental technology, vol. 38, no. 8, pp. 1005-1021, 2017. https://doi.org/10.1080/09593330.2016.1217053

[45] H. A. Hasan, S. R. S. Abdullah, S. K. Kamarudin, N. T. Kofli, and N. Anuar, "Kinetic evaluation of simultaneous COD, ammonia and manganese removal from drinking water using a biological aerated filter system," Separation and Purification Technology, vol. 130, pp. 56-64, 2014. https://doi.org/10.1016/j.seppur.2014.04.016

[46] S. Katoh, F. Yoshida, and J.-i. Horiuchi, Biochemical engineering. Wiley Online Library, 2015. https://doi.org/10.1002/9783527684984

Thermal stability of CO_2 hydrates in porous media with varying grain size in brine solution

AMIRUN NISSA Rehman[1,2,a], CORNELIUS B. Bavoh[1,2,b], BHAJAN Lal[1,2,c*]

[1]Chemical Engineering Department, Universiti Teknologi PETRONAS, Seri Iskandar, 32610 Perak Darul Ridzuan, Malaysia

[2]Research Centre for CO_2 Capture (RCCO_2C), Universiti Teknologi PETRONAS, Seri Iskandar, 32610, Perak, Malaysia

[a]amirun_19001666@utp.edu.my, [b]cornelius_17005250@utp.edu.my, [c]bhajan.lal@utp.edu.my

Keywords: CO_2 Hydrates, Porous Media, Dissociation, Heat Transfer, Particle Size

Abstract. In the present work, the heat transfer behavior of CO_2 hydrate dissociation was studied in three quartz sand particles (QS-1, QS-2 and QS-3) with varying grain sizes. The heat transfer behavior was evaluated by determining the heating rates of the porous media (quartz sand) during the CO_2 hydrate dissociation process in 3.3 wt.% NaCl. The experiment was performed using sandstone hydrate reactor by first forming the CO_2 hydrates at 4 MPa and 274.15 K and then dissociating the hydrates from 274.15 to 277.15 K, respectively. The results indicate that the thermal response of the porous sediment was significantly influenced by the hydrates as well as the porous sediment properties. The heating rate of the porous media increased when the grain size increased. However, the presence of CO_2 hydrates reduced the heat transfer behavior of the porous sediment due to the endothermic nature of hydrate dissociation. The heating behavior of the porous media with hydrates mainly depends on the type and pattern of hydrate formed (pore-filling, load-bearing, and cementation) and the location of the hydrates within the pores of the porous sediment. The pore-filling type of hydrate formation in porous sediments provides high thermal stability for CO_2 hydrate storage due to its less contact with the quartz sand particles. However, the pore-filling hydrate formation type is challenged with low or undesired CO_2 hydrate storage capacity. These findings will provide meaningful insights to select favorable sediment properties/sites for CO_2 storage in the hydrate form in porous sediments.

Introduction

The stability of CO_2 hydrates in porous media is a great concern for CO_2 sequestration. Unstable hydrates could release CO_2 into the ocean leading to ocean acidification and, in severe cases, could affect the marine ecosystem [1], [2]. Generally, hydrate dissociation exhibits an endothermic nature where heat transfer plays a major role. The thermal stability of carbon dioxide hydrates in sediments is determined by the heat transfer rate [3]. Heat transfer is highly controlled by the thermal conductivity of the porous media arising from the temperature variations within the hydrate stable zone [4], [5]. Therefore, understanding the heating rate of the porous media could fundamentally provide suitable insight into the thermal stability of hydrates in different porous media systems.

In lab-scale analysis, the reactor temperature profiles could be used to evaluate the heating behavior or response of hydrates stability in porous media. Yang et al., [6] investigated the dissociation of carbon dioxide hydrates in porous media using the temperature profiles. They claimed that the hydrate dissociation process is affected by the heating rate. Hence, understanding the heating temperature profiles during CO_2 hydrate dissociation will facilitate selecting proper sites for CO_2 storage. There is limited investigated on the thermal effect of CO_2 hydrate dissociation in sediments. However, most studies in the literature focus on methane production by thermal simulation or depressurization in porous media [7], [8], [17]–[22], [9]–[16]. According to Zhao et al., [9] during methane production, water saturation and porous media affects the heat transfer. They reported that the high thermal conductivity of the porous media initially promotes hydrate dissociation but later inhibits it. On the contrary, Song et al.,[23] claimed that hydrate dissociation is inhibited with increasing thermal conductivity at an early stage but subsequently promotes over time. However, one may expect an initial hydrate inhibition owing to the endothermic nature of hydrate dissociation.

The hydrate dissociation rate is influenced by the thermal conductivity of the porous media [14]. Li et al., [8] indicated an increase in the dissociation rate with the increase in the thermal conductivity of the porous media. Thus, understanding the thermal conductivity of the porous media could serve as a guideline to select suitable location for CO_2 storage as hydrate with high thermal stability and less gas release. For such applications, a low thermal conductive porous media is highly preferred. However, the thermal conductivity of porous sediments is affected by different sediment properties including grain size, pore size, surface area, and porosity [24]. Hence more studies related to the properties of the porous media are required to facilitate the understanding of the dissociation of hydrates in the porous media. In this regard, evaluating the thermal effect of the dissociation of CO_2 hydrates is required to fully comprehend the thermal stability of CO_2 hydrates in porous sediments.

In this work, the thermal stability behavior of carbon dioxide hydrates in quartz sand was evaluated via a thermal simulation method by estimating the heating rates of the quartz sand and the hydrate systems during the hydrate dissociation process. In addition, the effect quartz sand average pore size, surface area, porosity and grain size on carbon dioxide hydrate dissociation was also studied. This knowledge will enable selecting a proper location with high thermal stability for CO_2 storage as hydrates. The heating rate of carbon dioxide hydrate dissociation was estimated at 4 MPa and 274.15 K in brine system employing quartz sand to simulate the porous media.

Methodology

Materials, Apparatus and Procedure

CO_2 hydrate dissociation kinetics was studied in 3 different quartz systems (porous media) with varying properties as displayed in Table 2. Quartz sand properties were selected to evaluate the heating rate of the quartz sand on the CO_2 hydrate dissociation kinetics. Pure CO_2 gas with a purity 99.9% was used for all the experimental systems. All the studies were performed in 3.3 wt% NaCl solution as brine to simulate the thermal stability of CO_2 storage as hydrate under real seabed conditions.

Fig. 1 shows the schematic diagram of the complete experimental setup with the internal reactor specifications and sand bed. More details on the apparatus and the experimental procedure followed for carbon dioxide hydrate formation and dissociation in this work using quartz sand can be found in our previous studies [25], [26].

Materials Research Proceedings 29 (2023) 98-108 https://doi.org/10.21741/9781644902516-13

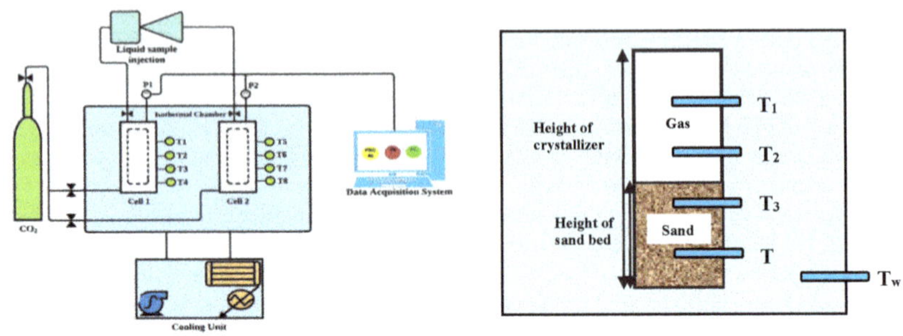

Figure 1. Schematic diagram of the gas hydrate experimental apparatus and internal reactor specification [25].

The recorded changes in the temperature and pressure were extracted to estimate the heating rate of each temperature probe in the gas phase and the porous media, respectively. Fig. 2 illustrates a typical temperature vs time profile during the hydrate dissociation process in this study. The estimated heating rates were used to investigate the carbon dioxide hydrate dissociation behaviour within the quartz sand bed.

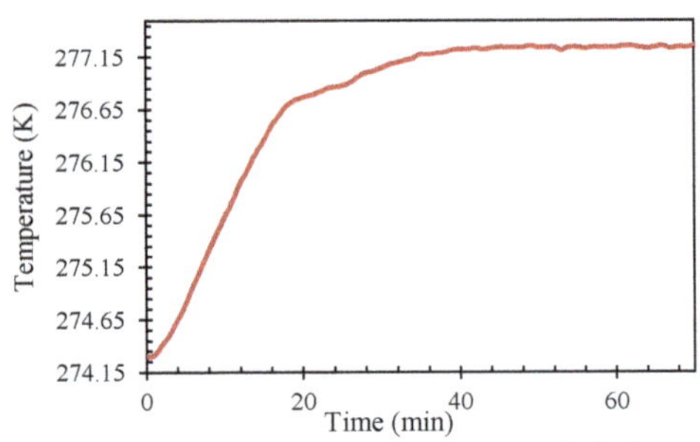

Figure 2. Typical Temperature - time curve of carbon dioxide hydrate dissociation in quartz sand.

Evaluation Parameters

Since the main objective of this study is to measure the response of the hydrates to the thermal changes, the heating rates of CO_2 hydrate dissociation were used to quantify the heat transfer within the hydrate phase and the gas phase. To understand the correlation between the moles of hydrate formed and the heat transfer, the moles of CO_2 consumed, and the hydrate saturation was determined from the hydrate formation experiments.

CO_2 gas consumed (moles) is estimated using the expression below (Eq. 1)

$$\Delta n_H = \left[\frac{PV}{zRT}\right]_0 - \left[\frac{PV}{zRT}\right]_t \quad (1)$$

Hydrate saturation (h_s) is evaluated by the following Eq. 2;

$$h_s\,(\%) = I_s \times 1.1C_{wh} \quad (2)$$

The heating rate (Hr) is calculated from the temperature vs time curve (Fig. 2) for every three minutes for the first one hour of heating (using Eq. 3). The differences in the heating rate within the temperature profiles were estimated statistically using the analysis of variance (ANOVA) at 95% confidence level [27]. The dissociation rate was determined using Eq. 4.

$$Hr = \left(\frac{\Delta T}{dt}\right)_t \tag{3}$$

$$\left(\frac{d\Delta n_H}{dt}\right)_t = \frac{(\Delta n_H)_{t+\Delta t} - (\Delta n_H)_t}{\Delta t} \tag{4}$$

Results and Discussion

Effect of the Heating Rate on Hydrate Dissociation

In this work, the effect of the heating rate on carbon dioxide hydrate dissociation was determined in the porous media (quartz sand). The experiments were conducted by first forming the hydrates at 274.15 K and 4 MPa with brine and then heating/dissociating at temperature 277.15 K to evaluate the heating rate of carbon dioxide hydrate dissociation experiment.

Fig. 3 illustrates the temperature profile of the heating curves in the reactor where, T_w is the water bath temperature and is fixed at 274.15 K, the experimental temperature value. In Fig. 3, T_1, T_2 and T_3, T_4 represent the temperature profiles in the gas phase and the hydrate phase within the quartz sand, respectively. Here, the water bath temperature (T_w) was heated from 274.15 K to 277.15 K initiating the dissociation process to attain equilibrium (Fig. 3). It is seen that the temperature curves in the hydrate phase (T_3 and T_4) deviate from T_w (system temperature) while the gas phase temperature curves (T_1 and T_2) go along with the system temperature (T_w) profile (Fig. 3). The reason for this deviation is the due to CO_2 hydrates formed within the porous sediment (quartz sand). The hydrates delay the heat transfer rate during CO_2 hydrate dissociation because of the endothermic nature of hydrate dissociation [11], [28].

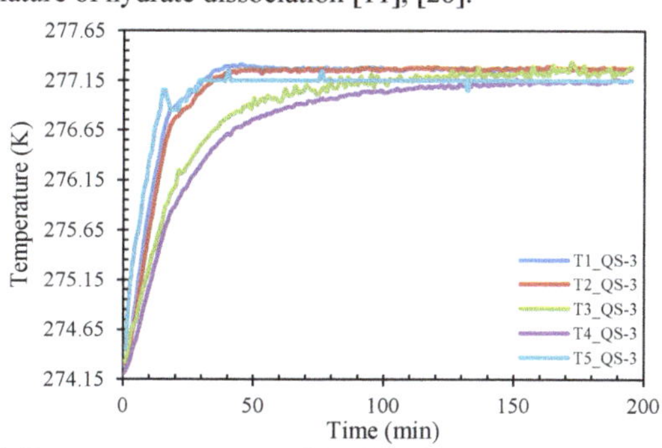

Figure 3. Temperature - time profile of the heating curves in quartz sand.

Effect of the Heating rate

The heating rates of the quartz sand during carbon dioxide hydrate formation and dissociation were calculated to quantify the thermal behavior of the system. The heating rate of the porous media was estimated as a measure to compare the thermal effect in the absence of hydrates. Analysis of variance (ANOVA) was employed to statistically establish the difference in the heating rate of the porous media, respectively. Table 1 represents the calculated heating rates and the obtained P-values from ANOVA analysis in this study.

Table 1. Heating rate of the porous media with and without CO_2 hydrates.

Systems	Mean heating rate					P- value	F- value
	T1	T2	T3	T4	Tw		
Systems with CO_2 hydrates							
QS-1	0.026	0.026	0.026	0.026	0.026	0.999	0.003
QS-2	0.039	0.039	0.036	0.032	0.040	0.875	0.304
QS-3	0.042	0.041	0.036	0.035	0.040	0.971	0.130
System without hydrates							
QS-1	0.104	0.104	0.094	0.090	0.129	0.008	3.641
QS-2	0.099	0.098	0.095	0.089	0.117	0.022	2.983

The P-values in Table 1 suggest a statistical difference in the heating rate of the temperature curves in the gas phase (T_1 and T_2) and the hydrate phase (T_3 and T_4), as shown earlier in Fig. 3. The difference in the heating rate further confirms that the decrease in the heat transfer rate during the dissociation of CO_2 hydrates is caused by the formed hydrates in the quartz sand. The reduction in the heating rate is due to the absorption of heat by the CO_2 hydrates indicating the endothermic nature of hydrate dissociation [15].

However, the heating rates were similar with acceptable P-values below 0.05 (Table 1), in the absence of hydrates indicating that the response of the porous media towards the thermal behavior is the same. This further suggests that the temperature variation observed for the temperature profiles in the quartz sand/porous media (T_3 and T_4), is solely due to the presence of hydrates. These findings are in line with the findings of Linga et al., [6] and Li et al., [29] where they claimed that the delay in the heat transfer was due to the hydrates present in the system and the endothermic dissociation behavior of the hydrates. They further confirmed that the magnitude of reduction in the heat transfer depends on the amount of hydrate formed within the porous media.

Effect of the Porous media properties on the Heating rate of CO_2 hydrates

Heat transfer is a crucial phenomenon in hydrate formation and dissociation which is directly related to the thermal stability of CO_2 hydrates [30]–[32]. Table 2 shows the details of the different quartz sand properties with the results obtained for carbon dioxide hydrate kinetics in this work.

Table 2. Details of the porous media properties and obtained results for hydrate formation and dissociation.

Sand Sample	System	Surface area (m^2/g)	Grain size (mm)	Porosity (%)	Average pore size (nm)	Gas consumed (moles)	Time for 90% CO_2 release (hr)
QS-1	Brine	0.8499	< 0.6	44	9.8721	0.2390	5.23
QS-2	Brine	1.1000	0.6-0.8	38	8.1342	0.3900	1.96
QS-3	Brine	0.6449	0.8-2	42	8.3147	0.2760	2.33

However, when hydrates are present in the porous sediment, the response of the hydrates with respect to the porous sediment is complex in nature and not only depend on the thermal conductivity of the porous sediment [5]. In such instances, any major influence on the heat transfer behavior of the hydrates in the porous media is mainly described based on the exact location of the formed hydrates within the porous sediment. Based on the location, three types of hydrate forming behavior (Fig. 4) could be present, which significantly influences the heat transfer of the hydrates in the porous sediment [33].

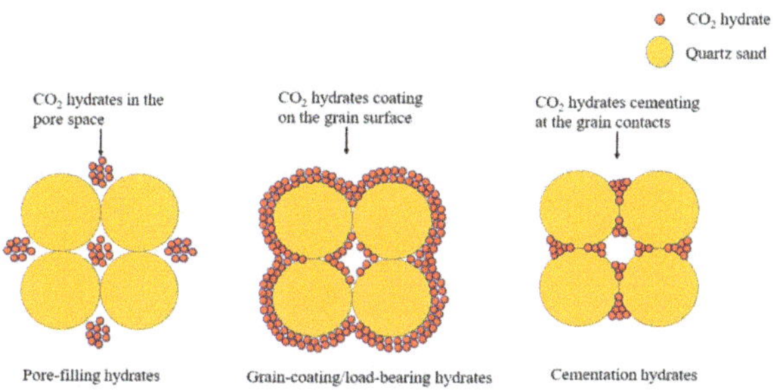

Figure 4. Different types of hydrate forming behavior in porous media.

Carbon dioxide hydrate formation in the porous sediment may occur as pore-filling (formed within the pore space), load-bearing (as particles coating) and cementation [34] type as shown in Fig. 4. For example, in a porous media with the same properties, hydrates formed on the surface of the pores (particles coating/load-bearing) will respond to the heat transfer faster than those formed at the centre of the pore space (pore-filling). It means, hydrate dissociation will be quicker for hydrates formed at the surface of the pores as opposed to the hydrates present within the pore space [33]. This action is because the hydrates formed as load-bearing assumes the same thermal conductivity as the porous media particles due to their direct contact with the surface of the particles. Therefore, CO_2 hydrate thermal stability in this study was further discussed with respect to the porous sediment properties and the hydrate formation behavior. This is necessary to know the heat transfer mechanism of the hydrates with the varying porous sediment properties (Table 2).

Effect of porosity and particle size

Fig. 5a illustrates the thermal response of the three different quartz sand particles (QS1- <0.6mm, QS2- 0.6-0.8mm, and QS3- 0.8-2mm). The thermal conductivity of the porous media particles increases with the particle size [35]–[37]. The heat transfer in the larger particles was 36% faster than the smaller particle sizes. In comparison, the moderate particle size conducts heat about 25% faster compared to the smaller particles (Fig. 5). This finding agrees with the results of Li et al., [29], where they reported that smaller particles respond slowly to heat transfer. This response indicates that the heat transfer behavior for any porous media is constant for specific particle sizes, but dependant on the hydrate formation mechanism/behavior and the petrophysical properties within that particular particle size. Also, the porosity of the porous media correlated with the heating rates showing that high porosity transfers the heat slowly due to the large void spaces between the particles, thus agreeing with the findings of Ahn and Jung [24].

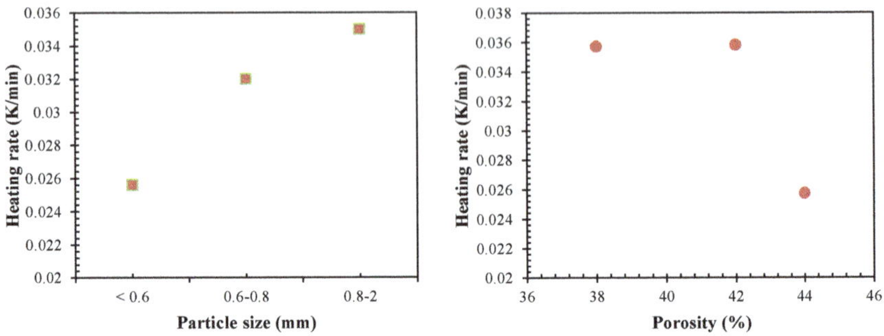

Figure 5. Effect of particle size and porosity on the heating rate of the quartz sand.

Effect of CO_2 Uptake and Hydrate Saturation

To know the thermal response of the carbon dioxide hydrates in the porous sediment, the measured heating rates were related to the amount of hydrate formed and the hydrate saturation in the porous media systems (Figure. 6). The heat absorption rate of the carbon dioxide hydrates formed are not consistent with the heating rates of the porous media particle sizes, as shown in Figure 5, especially for the QS-3 system. This is due to the heat transfer effect or variation in the quartz sand properties arising from the location at which the carbon dioxide hydrates formed within the quartz sand. The system with the largest porosity (QS-1) exhibited the slowest heating rate though it had the least hydrate uptake and saturation. However, it took the maximum time (5.23 hr) to achieve 90% of the carbon dioxide hydrate dissociation amongst the studied sand systems (Table 2). This is because QS-1 likely exhibits a pore filling hydrate formation mechanism owing to its large porosity. Which results in a slow heat transfer from the particle's surface to the hydrate crystals mainly due to the loose contact between the hydrates and the particle surface. This further implies that porous sediments with high porosity and pore-filling hydrate behavior are more suitable to store carbon dioxide hydrates with high thermal stability. However, such sediments might face challenges of high carbon dioxide storage capacity.

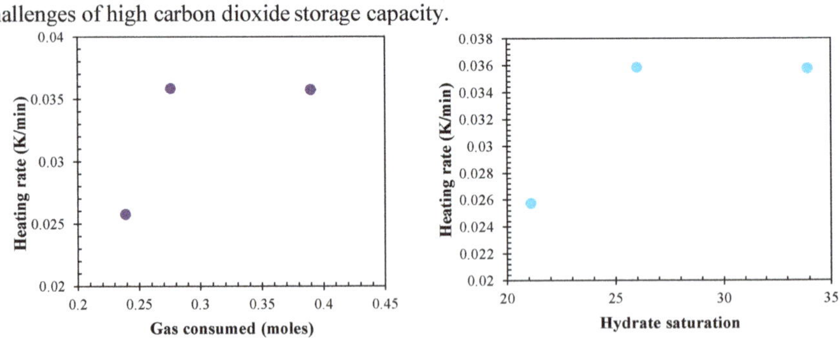

Figure 6. Effect of gas consumption and hydrate saturation on the heating rate.

On the contrary, the system with the lowermost and moderate porosity (QS-2 and QS-3), showed high heating rates and formed more hydrates with high gas uptake and hydrate saturation compared to the QS-1 system, respectively. However, the time elapsed to obtain 90% hydrate dissociation in these systems was 124 -166% faster than the QS-1 system (Table 2). The fast dissociation time could be because both the systems QS-2 and QS3 have larger particle sizes and low porosity resulting in a higher rate of heat transfer between the porous media particles and the hydrates [37]. In the QS-2 system, more hydrates formed, and the dissociation time was less than QS-3 though they have relatively similar heating rates within the hydrate zone (Table 2 and Figure 7). Based on the effect of particle size, one may expect QS-3 (0.8-2 mm) to dissociate the hydrate faster than QS-2 (0.6-0.8 mm). This unusual dissociation trend between QS-2 and QS-3 is related to the heat transfer responses of the type of carbon dioxide hydrate formed within the pores and the location of the hydrate formed in the porous media. In QS-2, where the amount of hydrate formed is much higher (Table 2), it is most likely that the CO_2 hydrate formed within the pores are compact and acts as load bearing framework to the sand particles. This mechanism provides a direct contact of the hydrate to the particles with an enhanced surface area for fast heat conduction from the particles to the hydrates, leading to a faster hydrate dissociation. On the other hand, the hydrate formed in QS-3 is relatively less and probably have much less contact with the particles compared to QS-2. Thus, the hydrate exists in pores, causing a slow response of the hydrate to the heat from the porous media particles. This response suggests that the type of hydrate forming behavior and the location of the hydrate within the porous media are critical phenomenon to consider for the successful storage of carbon dioxide as hydrate in porous sediment.

Conclusion

Herein, the heating rates of CO_2 hydrate dissociation in three quartz sand particles were evaluated to study the response of hydrates to heat transfer during carbon dioxide storage in quartz sand. The endothermic nature of hydrates controls their dissociation behavior in the porous sediment. Hydrate dissociation in porous media is highly controlled by the heat transfer between the porous sediment particles and the hydrates formed within the pores of the porous media. Porous media particles with high hydrate storage capacity most likely dissociate faster due to the high heating rate of the hydrates originating from the close contact of the hydrates formed within the pore network of the porous media. The type of hydrate forming behavior (pore-filling, load-bearing, cementation) influenced the thermal stability of carbon dioxide hydrates significantly. This mechanism suggests that smaller particle size is highly preferable to achieve optimum carbon dioxide hydrates thermal stability in porous media.

Acknowledgment

This work is supported by the Chemical Engineering Department of Universiti Teknologi PETRONAS. GaSTech PETRONAS (MRA Grant No. 015MD0-028) is the financial supporter.

References

[1] R. P. Singh, K. S. Shekhawat, M. K. Das, and K. Muralidhar, "Geological sequestration of CO 2 in a water-bearing reservoir in hydrate-forming conditions," *Oil Gas Sci. Technol. – Rev. IFP Energies Nouv.*, vol. 75, no. 51, pp. 1–19, 2020.

[2] B. K. Sahu, P. Pati, and R. C. Panigrahy, "Impact of climate change on marine plankton with special reference to Indian seas," *Indian J. Geo-Marine Sci.*, vol. 47, no. 2, pp. 259–268, 2018.

[3] B. Lal and O. Nashed, *Chemical Additives for Gas Hydrates*. Cham: Springer International Publishing, 2020.

[4] A. Hosseini Zadeh, I. Kim, and S. Kim, "Experimental study on the characteristics of formation and dissociation of CO2hydrates in porous media," in *E3S Web of Conferences*, 2020, vol. 205, https://doi.org/10.1051/e3sconf/202020502004

[5] D. D. Cortes, A. I. Martin, T. S. Yun, F. M. Francisca, J. C. Santamarina, and C. Ruppel, "Thermal conductivity of hydrate-bearing sediments," *J. Geophys. Res. Solid Earth*, vol. 114, no.

11, pp. 1–10, 2009, https://doi.org/10.1029/2008JB006235

[6] S. H. B. Yang, P. Babu, S. F. S. Chua, and P. Linga, "Carbon dioxide hydrate kinetics in porous media with and without salts," *Appl. Energy*, vol. 162, pp. 1131–1140, Jan. 2016, https://doi.org/10.1016/j.apenergy.2014.11.052

[7] J. Zhao *et al.*, "Heat transfer analysis of methane hydrate sediment dissociation in a closed reactor by a thermal method," *Energies*, vol. 5, no. 5, pp. 1292–1308, 2012, https://doi.org/10.3390/en5051292

[8] X. Y. Li, Y. Wang, X. Sen Li, Y. Zhang, and Z. Y. Chen, "Experimental study of methane hydrate dissociation in porous media with different thermal conductivities," *Int. J. Heat Mass Transf.*, vol. 144, p. 118528, 2019, https://doi.org/10.1016/j.ijheatmasstransfer.2019.118528

[9] J. Zhao, D. Liu, M. Yang, and Y. Song, "Analysis of heat transfer effects on gas production from methane hydrate by depressurization," *Int. J. Heat Mass Transf.*, vol. 77, pp. 529–541, 2014, https://doi.org/10.1016/j.ijheatmasstransfer.2014.05.034

[10] X. Guo *et al.*, "Effect of thermal properties on the production behavior from water-saturated methane hydrate-bearing sediments using depressurization," in *Energy Procedia*, 2019, vol. 158, no. 2018, pp. 5453–5458, https://doi.org/10.1016/j.egypro.2019.01.602

[11] Q. C. Wan, H. Si, B. Li, and G. Li, "Heat transfer analysis of methane hydrate dissociation by depressurization and thermal stimulation," *Int. J. Heat Mass Transf.*, vol. 127, pp. 206–217, 2018, https://doi.org/10.1016/j.ijheatmasstransfer.2018.07.016

[12] Z. Yin, G. Moridis, Z. R. Chong, H. K. Tan, and P. Linga, "Numerical Analysis of Experiments on Thermally Induced Dissociation of Methane Hydrates in Porous Media," *Ind. Eng. Chem. Res.*, vol. 57, no. 17, pp. 5776–5791, 2018, https://doi.org/10.1021/acs.iecr.7b03256

[13] X. Guo *et al.*, "Analytical Investigation of Gas and Water Production from Aqueous-Rich Hydrate-Bearing Sediments by Depressurization," *Energy and Fuels*, 2021, https://doi.org/10.1021/acs.energyfuels.0c04060

[14] M. Roostaie and Y. Leonenko, "Gas production from methane hydrates upon thermal stimulation; an analytical study employing radial coordinates," *Energy*, vol. 194, p. 116815, 2020, https://doi.org/10.1016/j.energy.2019.116815

[15] Q. C. Wan, L. L. Chen, B. Li, K. Peng, and Y. Q. Wu, "Insights into the Control Mechanism of Heat Transfer on Methane Hydrate Dissociation via Depressurization and Wellbore Heating," *Ind. Eng. Chem. Res.*, vol. 59, no. 22, pp. 10651–10663, 2020, https://doi.org/10.1021/acs.iecr.0c00705

[16] E. Chuvilin and B. Bukhanov, "Thermal conductivity of frozen sediments containing self-preserved pore gas hydrates at atmospheric pressure: An experimental study," *Geosci.*, vol. 9, no. 2, 2019, https://doi.org/10.3390/geosciences9020065

[17] P. Wang and Y. Teng, "Numerical and Experimental Research on Gas Production from Methane Hydrate under Water–Excess Conditions," *Energy & Fuels*, vol. 35, no. 6, pp. 4848–4857, 2021, https://doi.org/10.1021/acs.energyfuels.0c04106

[18] C. B. Bavoh, M. S. Khan, B. Lal, N. I. Bt Abdul Ghaniri, and K. M. Sabil, "New methane hydrate phase boundary data in the presence of aqueous amino acids," *Fluid Phase Equilib.*, vol. 478, pp. 129–133, 2018, https://doi.org/10.1016/j.fluid.2018.09.011

[19] C. B. Bavoh, O. Nashed, M. S. Khan, B. Partoon, B. Lal, and A. M. Sharif, "The impact

of amino acids on methane hydrate phase boundary and formation kinetics," *J. Chem. Thermodyn.*, vol. 117, pp. 48–53, 2018, https://doi.org/10.1016/j.jct.2017.09.001

[20] K. M. Sabil, O. Nashed, B. Lal, L. Ismail, and A. Japper-Jaafar, "Experimental investigation on the dissociation conditions of methane hydrate in the presence of imidazolium-based ionic liquids," *J. Chem. Thermodyn.*, vol. 84, pp. 7–13, 2015, https://doi.org/10.1016/j.jct.2014.12.017

[21] M. Idress, M. Jasamai, I. M. Saaid, B. Lal, B. Partoon, and K. M. Sabil, "Effect of Porosity to Methane Hydrate Formation in Quartz Sand," in *ICIPEG 2016*, 2017, pp. 231–240, https://doi.org/10.1007/978-981-10-3650-7_19

[22] C. B. Bavoh, B. Partoon, B. Lal, and L. Kok Keong, "Methane hydrate-liquid-vapour-equilibrium phase condition measurements in the presence of natural amino acids," *J. Nat. Gas Sci. Eng.*, vol. 37, pp. 425–434, 2017, https://doi.org/10.1016/j.jngse.2016.11.061

[23] Y. Song, J. Wang, Y. Liu, and J. Zhao, "Analysis of heat transfer influences on gas production from methane hydrates using a combined method," *Int. J. Heat Mass Transf.*, vol. 92, pp. 766–773, 2016, https://doi.org/10.1016/j.ijheatmasstransfer.2015.08.102.

[24] J. Ahn and J. Jung, "Effects of fine particles on thermal conductivity ofmixed silica sands," *Appl. Sci.*, vol. 7, no. 7, 2017, https://doi.org/10.3390/app7070650

[25] A. N. Rehman *et al.*, "Kinetic insight on CO2 hydrate formation and dissociation in quartz sand in presence of brine," *Int. J. Greenh. Gas Control*, vol. 114, no. February 2021, p. 103582, Feb. 2022, https://doi.org/10.1016/j.ijggc.2022.103582

[26] A. N. Rehman, R. Pendyala, and B. Lal, "Effect of brine on the kinetics of Carbon dioxide hydrate formation and dissociation in porous media," *Mater. Today Proc.*, vol. 47, pp. 1366–1370, 2021, https://doi.org/10.1016/j.matpr.2021.04.024

[27] G. T. Sudha, B. Stalin, and M. Ravichandran, "Optimization of powder metallurgy parameters to obtain low corrosion rate and high compressive strength in Al-MoO3 composites using SN ratio and ANOVA analysis," *Mater. Res. Express*, vol. 6, no. 9, 2019, https://doi.org/10.1088/2053-1591/ab2cef

[28] M. Vasheghani Farahani *et al.*, "Effect of thermal formation/dissociation cycles on the kinetics of formation and pore-scale distribution of methane hydrates in porous media: a magnetic resonance imaging study," *Sustain. Energy Fuels*, vol. 5, no. 5, pp. 1567–1583, 2021, https://doi.org/10.1039/d0se01705a

[29] X.-Y. Li *et al.*, "Influence of Particle Size on the Heat and Mass Transfer Characteristics of Methane Hydrate Formation and Decomposition in Porous Media," *Energy & Fuels*, 2021, https://doi.org/10.1021/acs.energyfuels.0c03812

[30] S. Choi, J. Park, and Y. T. Kang, "Experimental investigation on CO2 hydrate formation/dissociation for cold thermal energy harvest and transportation applications," *Appl. Energy*, vol. 242, no. March, pp. 1358–1368, 2019, https://doi.org/10.1016/j.apenergy.2019.03.141

[31] Y. Wang, J. C. Feng, X. Sen Li, Y. Zhang, and Z. Y. Chen, "Fluid flow mechanisms and heat transfer characteristics of gas recovery from gas-saturated and water-saturated hydrate reservoirs," *Int. J. Heat Mass Transf.*, vol. 118, pp. 1115–1127, 2018, https://doi.org/10.1016/j.ijheatmasstransfer.2017.11.081

[32] S. Liu, Y. Liang, B. Li, Q. Wan, and X. Han, "Interaction relationship analysis between

heat transfer and hydrate decomposition for optimization exploitation," *Fuel*, vol. 256, no. May, 2019, https://doi.org/10.1016/j.fuel.2019.115742

[33] Z. A. Jarrar, K. A. Alshibli, R. I. Al-Raoush, and J. Jung, "3D measurements of hydrate surface area during hydrate dissociation in porous media using dynamic 3D imaging," *Fuel*, vol. 265, no. December 2019, p. 116978, 2020, https://doi.org/10.1016/j.fuel.2019.116978.

[34] W. F. Waite *et al.*, "PHYSICAL PROPERTIES OF HYDRATE-BEARING SEDIMENTS," no. 2008, pp. 1–38, 2009, https://doi.org/10.1029/2008RG000279.Table.

[35] L. Zhan, Y. Wang, and X.-S. Li, "Experimental study on characteristics of methane hydrate formation and dissociation in porous medium with different particle sizes using depressurization," *Fuel*, vol. 230, no. May, pp. 37–44, Oct. 2018, https://doi.org/10.1016/j.fuel.2018.05.008

[36] M. Muraoka, M. Ohtake, N. Susuki, H. Morita, M. Oshima, and Y. Yamamoto, "Thermal properties of highly saturated methane hydrate-bearing sediments recovered from the Krishna–Godavari Basin," *Mar. Pet. Geol.*, vol. 108, no. October 2018, pp. 321–331, 2019, https://doi.org/10.1016/j.marpetgeo.2018.10.037

[37] P. Guo, Y. K. Pan, L. L. Li, and B. Tang, "Molecular dynamics simulation of decomposition and thermal conductivity of methane hydrate in porous media," *Chinese Phys. B*, vol. 26, no. 7, 2017, https://doi.org/10.1088/1674-1056/26/7/073101

Adsorption of hydrogen sulphide (H_2S) using xerogel synthesized from palm kernel shell biochar

DEANA Qarizada[1,2,a], NOR MOHD Razif Noraini[1,3,b], AZIL Bahari Alias[1,c*], HAMASA Kambakhsh[4], NURUL Syafiqah Ahmad Anuar[1,d]

[1]Industrial Process Reliability & Sustainability (INPRES) Research Group, School of Chemical Engineering, College of Engineering, Universiti Teknologi MARA, 40450 Shah Alam, Selangor, Malaysia

[2]Faculty of Chemical Engineering, Jawzjan University, Jawzjan, Afghanistan

[3]National Institute of Occupational Safety & Health (NIOSH)

Lot 1, Jalan 15/1, Section 15, 43650 Bandar Baru Bangi, Selangor, Malaysia

[4]Engineering Faculty, Faryab University, Faryab, Afghanistan

[a]deanaqarizada11@gmail.com, [b]h.kambakhsh1989@gmail.com, [c]azilbahari@uitm.edu.my, [d]syafiqah0anuar@gmail.com

Keywords: Palm Kernel Shell Biochar, Xerogel, Hydrogen sulphide, Adsorption

Abstract: Xerogel is a typical porous material with a large internal surface area, causing them to have significant gas adsorption. Adsorption performance was investigated to determine the potential contribution of xerogel to removing Hydrogen Sulphide (H_2S) in this research. Adsorption is a well-known energy-efficient approach for removing acid gases at low temperatures. H_2S gas harms human health, such as headaches, eye irritation, and loss of smell if exposed to a low concentration. Furthermore, the physical and chemical properties of the raw material and synthesized xerogel were evaluated by various techniques: Fourier Transform Infrared Spectroscopy (FTIR), thermogravimetric Analysis (TGA), and Scan Electron Microscopy (SEM). Results showed that the removal of H_2S increased with increasing adsorbent mass from 3 to 12 g and decreased flow rate from 40 to 26 L/h. The maximum Adsorption capacity of Xerogel for H2S was 27.5 mg/g, and the surface area was 0.2686 m^2/g. This research shows the significant potential of using adsorbent materials obtained from waste to absorb H_2S.

Introduction

Various technologies have been established to decrease and restrict H_2S emission, including wet flue gas desulfurization, Adsorption, catalytic reduction [1], and electron beam technology [2]. Adsorption is the most practical option among these technologies because it is cost-effective, efficient, and simple to handle and run. The adsorbents can be recycled for numerous uses by doing the Adsorption, making them cost-effective. There are several examples of solid adsorbent materials that can be adsorbents: zeolites, calcium oxide, activated carbon, metal oxides, hydrogels, xerogels, polymers, and metal-organic frameworks (MOFs) [4]. These adsorbents can be produced from coal, agricultural residues, and other waste biomass. The correct choice of adsorbent materials is a critical step in constructing adsorption systems since the decision significantly impacts the process' economics and performance [5]. The abundance of palm kernel shells obtained by cracking oil palm nuts in Malaysia demonstrates their use as activated carbon precursors. Furthermore, the amount of solid residue from oil palms will cause the problem of waste disposal at the oil palm factory [6, 7, 8].

Presently, the production of porous materials (xerogel) has attracted worldwide attention in the oven-drying process. The oven drying method is much more straightforward compared to other methods. Xerogel is microporous and can be defined as a low-density biomaterial employed for various promising applications in different fields [9]. Xerogel is produced from gels by replacing a liquid phase, commonly solvent crystals, with a gaseous phase with no ruin of the created porous solid architecture [10, 11, 12]. The significance of Xerogel-based PKS is that it will become an effective and low-cost material widely used globally as an absorbent, not only effective on toxic gases but also in absorbing engine oil and heavy metal removal in wastewater. This research focuses on the synthesis of a new, economical, and versatile product that can serve to adsorb the H2S as an absorbent.

Methodology

Material

This research used palm kernel shell biochar as raw material, H_2S gas, Sodium Alginate, Calcium Carbonate, distilled water, and Glucono Delta Lactone (GDL). Then, the adsorption process used a multi-layered adsorption column.

Preparation of Xerogel Based on Palm Kernel Shell Biochar

A magnetic stirrer was used to stir 0.5g of Sodium Alginate into 50 ml of distilled water until the mixture was homogenized. 0.1g CaCO3 is applied to the homogeneous mixture and mixed until homogeneous, followed by 0.5g grind PKS biochar. Then, 0.15g of GDL was applied to the mixture and mixed until homogenized before transferring to the square mould and put in the refrigerator until it was fully gelled. Following that, the hydrogels were removed from the square mould and placed in a petri dish. Oven drying of gels removes the solvent from the colloidal gel to afford a solid dry gel or xerogel. Many types of xerogel dried by oven can lose 50-70% of the original gel's volume with drying. The hydrogel was dried in an oven at 60 °C for 48 hours.

Characterization of xerogel

The chemical structure of the PKS biochar and synthesized xerogel were measured using a Bruker FTIR spectrophotometer following the ASTM standard test method (E1252- 98). The practice covers a spectral range of 4000–450 cm^{-1} according to the ASTM E1252-98. The chemical structure includes the functional group of the biochar, and xerogel was analyzed [14].

To estimate the element content such as moisture content, volatile matter content, carbon content and maximum temperature profile of PKS biochar and synthesized xerogel, the thermogravimetric (TG) method was used by using the Mettler Toledo Thermogravimetric Analyzer [13]. Thermogravimetric analysis is done to study the stability of the extracted xerogel following ASTM standard method D5142-02a. The xerogel biochar and palm kernel shell (PKS) biochar samples were heated up to 950°C with a heating rate of 20°C/min in the presence of air involving a nitrogen atmosphere at 100ml/min flow rate of gas. Then, the sample atmosphere is switched into the air and heated to 1200 °C [15].

SEM has been used to determine the morphologies of the organisms. The sample has been coated with a sheet before being scanned with electron beams with acceleration forces ranging from 10-15 kV to reach 1,000 – 5,000 times resolution. The technique was accompanied by a standard test defined in detail by the American Society for Testing and Content Specification, ASTM E766 – 14 [15].

Adsorption Test

The adsorption test was carried out in a multilayer's adsorption column, as shown in Figure 1. The column was made of stainless steel as the hazardous gas was used. The fixed parameters used are 25 ppm inlet concentration of H_2S. Then, the adsorption test was started with the different flowrate of H_2S and masses of the adsorbent xerogel to achieve the best Adsorption parameter. The system was run at a temperature of 30°C. Finally, the outlet concentration of H_2S was monitored continuously using a biogas analyzer (MRU OPTIMA) [20]. The results of the adsorption test were interpreted by analyzing breakthrough time curves and the adsorption capacity, qt (mg/g) of hydrogen sulphide, which was calculated by using Eq. (1) [13]:

$$q_t = \frac{(C_O - C_t)V}{W} \quad (1)$$

where: qt is adsorption capacity (mg/g), C_0 is the initial concentration of H_2S supplied (mg/L), Ce is the concentration of H_2S at any time, V is the volume of the adsorbent (L), and W is the mass of the adsorbent used (g).

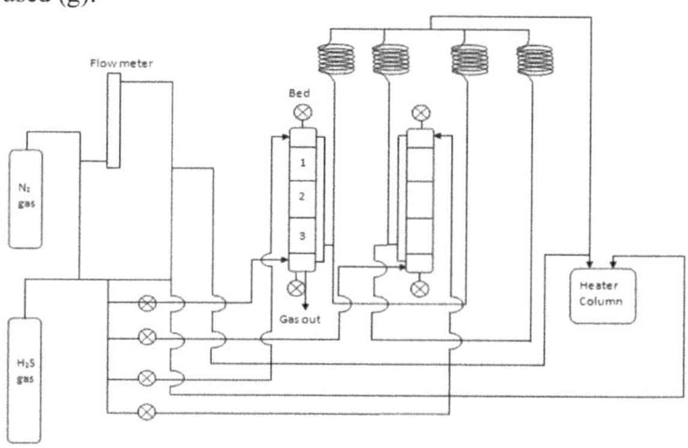

Figure 1: Schematic diagram of Multilayers Adsorption Column.

Results and discussion

Characterization of Raw Material and Xerogel

Figure 2 shows the FTIR spectrum of (a) palm kernel shell (PKS) biochar and (b) xerogel biochar for comparison. Broad characteristic peaks demonstrate the presence of O-H stretching at 3356.82 cm^{-1} for PKS biochar and 3333.76 cm^{-1} for xerogel biochar in the spectra and show hydrogen-bonded hydroxyl group (O-H) in which identical results were reported by [16]. Ma et al. (2015) [17] reported that the C-H stretching vibration with a wavelength of 2942.51 cm^{-1} was observed for PKS biochar and 2917.64-2887.73 cm^{-1} for xerogel, showing the alkane and aldehyde group. Functional group C=C stretching was observed on 1637 cm^{-1} for PKS biochar, and 1616.98 cm^{-1} indicated the presence of an aromatics compound. Aromatic ester is found at bending vibration 1314.05 cm^{-1} in xerogel [18]. The bending vibration around 1466 cm^{-1} represents the existence of the biochar C-C and C-H (alkane) groups. A peak at 1100 cm-1- 1029 cm^{-1} shows the presence of the S= O group in xerogel. Figure 6 illustrates the PKS and xerogel FTIR analysis [17].

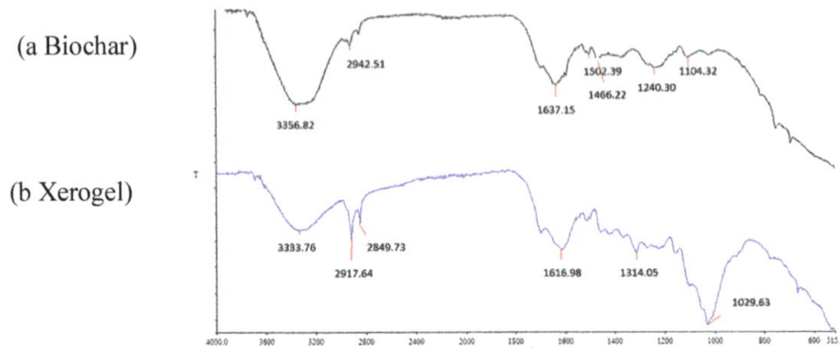

Figure 2: PKS and xerogel FTIR analysis.

Figure 3 illustrates PKS and xerogel TGA analysis. The TGA curve was divided into three sections to evaluate the sample's thermal stability. The first stage is the mass loss of 10.9% (2.2 mg) up to 144.55°C due to moisture removal from the PKS biochar sample. Meanwhile, mass loss of 7.03% (1.4 mg) was found in the first step of the TGA curve of xerogel biochar at around 146.64°. This mass loss is mainly attributed to the physical Adsorption of water in the xerogel at the first step of the TGA curve [18]. The mass loss at the second step is 52.1% (10.42 mg). A significant mass loss in the xerogel sample is higher than PKS biochar in step 2 due to the chemically bound water from the weight reduction that was attributed to chemically bonded water from the sol-gel manufacturing technique [17].

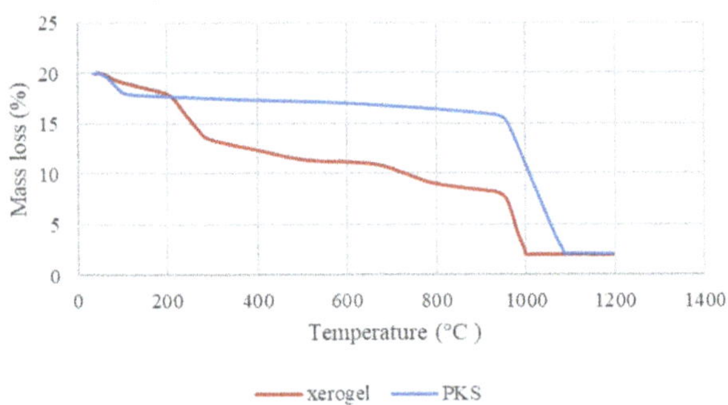

Figure 3: TGA analysis

After switching the atmosphere to oxidizing atmosphere combustion of air at 950°C, a remarkable mass loss of 68.5% (13.7 mg) is shown at 938.16°C, and a mass loss of 30.96% (6.2 mg) is shown at between 950-1000°C for PKS biochar and xerogel biochar, respectively. The final peak at 950°C represents carbon thermal breakdown, and the remaining curve represents ash content [18].

The SEM analysis of PKS biochar and xerogel showed in Figure 4. The surface structure of the xerogel and palm kernel shell (PKS) biochar samples is very irregular and fibrous, with limited regular structuring. In contrast to the surface of PKS biochar, the surface of xerogel had a highly complex network and a rough surface.

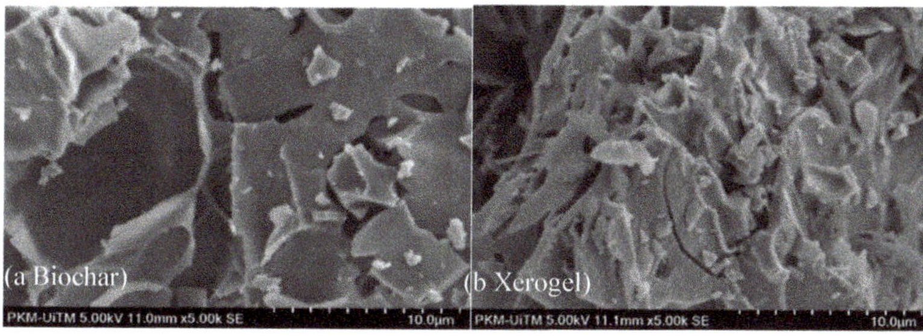

Figure 4: SEM Analysis (a) biochar, (b) xerogel

Based on xerogel SEM, the presence of hydrogel has clogged the pores and created more unstable holes due to polymerization. As an outcome of this situation, we predicted that the hydrogel's ability to swell in the presence of a hydrophilic liquid would assist in removing impurities trapped in the pores and holes of the material [19].

Adsorption Test

Effect of Mass

Figure 5 shows the breakthrough curve for the effect of the mass of adsorbent in the removal of H_2S. The mass of the adsorbent ranging from 3 g to 12 g, was analyzed for the adsorption capacity test. The test was run at a temperature of 30 °C and a flow rate of 30 L/h.

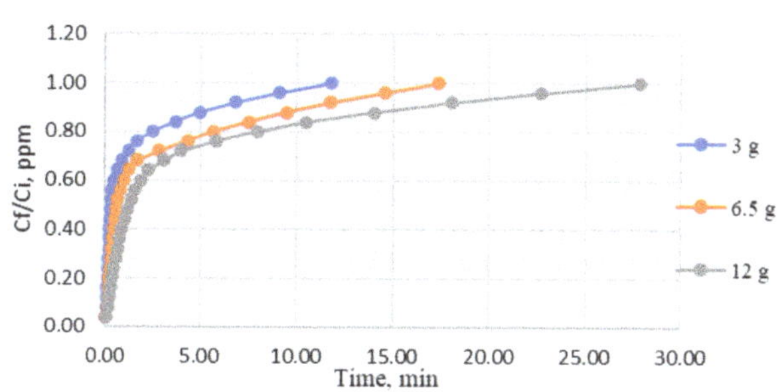

Figure 5: Breakthrough Curve for Different Mass of Adsorbent

Based on Figure 2, the 3 g of adsorbents need around 11 minutes for the H_2S fully adsorb. Next, the 6.5 g of adsorbents need 17 minutes to adsorb H_2S, which is longer than before. Meanwhile, for the 12 g of adsorbents, it takes the longest time to adsorb H2S, which is 28 minutes fully. The same results were reported by [16] for removing H_2S gas using palm kernel shell-activated carbon.

Effect of Flowrate

Figure 6 shows the breakthrough curve for the effect of flow rate in the removal of H_2S, which ranges from 26 L/h to 40 L/h. The test was run at a temperature of 30°C and the mass of the adsorbent at 6.5 g.

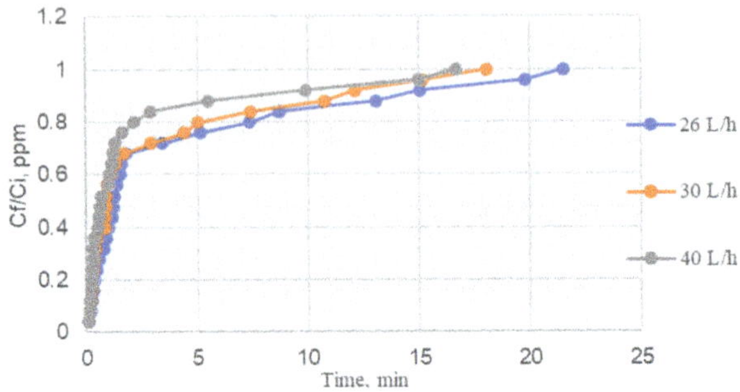

Figure 6: Breakthrough Curve for Different Flowrate

Based on the Figure 6, with the flow rate of 26 L/h, it took 22 min for the H_2S fully adsorb, which is the longest time to adsorb H_2S. The flow rate of 30 L/h needs 18 minutes, which is shorter than before. However, at the flow rate of 40 L/h, it took the fastest time to adsorb H_2S, which was 17 minutes. When the flow rate was increased, the breakthrough curve gradient became steeper, and the breakthrough time became shorter. This phenomenon was explained by [19], that the gas stream had lesser contact time when the flow rate was higher.

Conclusion

This experiment showed the great capability of the PKS biochar xerogel for removing H_2S using a multilayer adsorption column. The breakthrough time for different parameters such as the mass of adsorbent and H_2S flowrate were investigated. It can be concluded that the lower mass has a shorter breakthrough time. However, the higher flowrate of H_2S gave a shorter breakthrough time. The 3 g of adsorbent and the 40 L/h of H_2S are the parameter for the shorter breakthrough time, but 12 g of adsorbent and 26 L/h flow rate are the optimum parameters for the high adsorption capacity. Based on characterizations, the modified xerogel palm kernel shell biochar can be used as a low-cost adsorbent for adsorption in industries.

Acknowledgement

The authors would like to thank the Universiti Teknologi MARA (UiTM) for the financial support. The research is conducted at the School of Chemical Engineering, College of Engineering, UiTM, under the support of KEPU grant (600- RMC/KEPU 5/3 (017/2021).

References

[1] "Hydrogen Sulfide - Overview | Occupational Safety and Health Administration," 2017. https://www.osha.gov/hydrogen-sulfide (accessed Aug. 14, 2022).

[2] S. E. Mousavi, H. Pahlavanzadeh, M. Khani, H. A. Ebrahim, and A. Mozaffari, "Selective catalytic reduction of SO2 with methane for recovery of elemental sulfur over nickel alumina catalysts," Reaction Kinetics, Mechanisms and Catalysis, vol. 124, no. 2, pp. 669–682, Aug. 2018, https://doi.org/10.1007/s11144-018- 1360-x

[3] J. H. Park, J. W. Ahn, K. H. Kim, and Y. S. Son, "Historic and futuristic review of electron beam technology for the treatment of SO2 and NOx in flue gas," Chemical Engineering Journal, vol. 355. Elsevier B.V., pp. 351–366, Jan. 01, 2019. https://doi.org/10.1016/j.cej.2018.08.103

[4] W. S. Chai et al., "A review on conventional and novel materials towards heavy metal adsorption in wastewater treatment application," Journal of Cleaner Production, vol. 296. Elsevier Ltd, May 10, 2021. https://doi.org/10.1016/j.jclepro.2021.126589

[5] N. K. Gupta, J. Bae, S. Baek, and K. S. Kim, "Sulfur dioxide gas adsorption over ZnO/Znbased metal-organic framework nanocomposites," Colloids and Surfaces A: Physicochemical and Engineering Aspects, vol. 634, Feb. 2022, https://doi.org/10.1016/j.colsurfa.2021.128034

[6] M. Zhang et al., "Novel cationic polymer modified magnetic chitosan beads for efficient adsorption of heavy metals and dyes over a wide pH range," International Journal of Biological Macromolecules, vol. 156, pp. 289– 301, Aug. 2020, https://doi.org/10.1016/j.ijbiomac.2020.04.020

[7] Yusuf IT, Babatunde YO, and Abdullah A, "Investigation On The Flexural Strength of Palm Kernel Shell Concrete for Structural Applications," 2018.

[8] M. Yahayu, F. Z. Abas, S. E. Zulkifli, and F. N. Ani, "Utilization of Oil Palm Fiber and Palm Kernel Shell in Various Applications," in Sustainable Technologies for the Management of Agricultural Wastes, Springer Singapore, 2018, pp. 45–56. https://doi.org/10.1007/978-981-10-5062-6_4

[9] U. G. Eziefula, "Developments in utilization of agricultural and aquaculture by-products as aggregate in concrete–a review," Environmental Technology Reviews, vol. 7, no. 1. Taylor and Francis Ltd., pp. 19–45, Jan. 01, 2018. https://doi.org/10.1080/21622515.2017.1423399.

[10] P. Shafigh, S. Salleh, H. Ghafari, and H. bin Mahmud, "Oil palm shell as an agricultural F. A . Idrus et al./MJCET Vol. X (X) (20XX) XX –XX 6 solid wastes in artificial lightweight aggregate concrete," European Journal of Environmental and Civil Engineering, vol. 22, no. 2, pp. 165– 180, Feb. 2018, https://doi.org/10.1080/19648189.2016.1182084

[11] P. E. Imoisili, K. O. Ukoba, and T. C. Jen, "Green technology extraction and characterization of silica nanoparticles from palm kernel shell ash via sol-gel," Journal of Materials Research and Technology, vol. 9, no. 1, pp. 307–313, Jan. 2020, https://doi.org/10.1016/j.jmrt.2019.10.059

[12] Anirban C., "List of Palm Oil Companies in Malaysia: 2018 Overview," 2018. https://blog.bizvibe.com/blog/foodbeverages/list-of-palm-oil-companies-inmalaysia (accessed Aug. 14, 2022).

[13] G. Vijayakumar, R. Tamilarasan, and M. Dharmendirakumar, "Adsorption, Kinetic, Equilibrium and Thermodynamic studies on the removal of basic dye Rhodamine-B from aqueous solution by the use of natural adsorbent perlite," J. Mater. Environ. Sci, vol. 3, no. 1, pp. 157–170, 2011.

[14] Kaya, G. G., & Deveci, H. (2020). Synergistic effects of silica aerogels / xerogels on properties of polymer composites : A review. Journal of Industrial and Engineering Chemistry, 89, 13– 27. https://doi.org/10.1016/j.jiec.2020.05.019

[15] Álvarez, S., Ribeiro, R. S., Gomes, H. T., Sotelo, J. L., & García, J. (2015). Synthesis of carbon xerogels and their application in adsorption studies of caffeine and diclofenac as emerging contaminants. Chemical Engineering Research and Design, 95, 229–238. https://doi.org/10.1016/j.cherd.2014.11.001

[16] Imoisili, P. E., Ukoba, K. O., & Jen, T. C. (2020). Green technology extraction and characterization of silica nanoparticles from palm kernel shell ash via sol-gel. Journal of Materials Research and Technology, 9(1), 307–313. https://doi.org/10.1016/j.jmrt.2019.10.059.

[17] Ma, Z., Chen, D., Gu, J., Bao, B., & Zhang, Q. (2015). Determination of pyrolysis characteristics and kinetics of palm kernel shell using TGA-FTIR and model-free integral methods. Energy Conversion and Management, 89, 251–259. https://doi.org/10.1016/j.enconman.2014.09.074

[18] Meri, N. H., Alias, A. B., Talib, N., Rashid, Z. A., Wan, W. A., & Karim Ghani, A. (2018). Effect of Chemical Washing Pre-treatment of Empty Fruit Bunch (EFB) biochar on Characterization of Hydrogel Biochar composite as Bioadsorbent. IOP Conference Series: Materials Science and Engineering, 358(1). https://doi.org/10.1088/1757- 899X/358/1/012018

[19] H. S. Choo, L. C. Lau, A. R. Mohamed, and K. T. Lee, "Hydrogen sulfide adsorption by alkaline impregnated coconut shell activated carbon," *J. Eng. Sci. Technol.*, vol. 8, no. 6, pp. 741–753, 2013.

[20] A. A. Abd Rahman*, A. B. Alias, N. N. Jaafar, M. A. Amir Hamzah, and W. A. W. A. Karim Ghani, "Adsorption of Hydrogen Sulphide by Commercialized Rice Husk Biochar (RHB) & Hydrogel Biochar Composite (RH-HBC)," *Int. J. Recent Technol. Eng.*, vol. 8, no. 4, pp. 6864–6870, 2019, https://doi.org/10.35940/ijrte.d5207.118419

Sustainable Processes and Clean Energy Transition - ICSuPCET2022 Materials Research Forum LLC
Materials Research Proceedings 29 (2023) 117-127 https://doi.org/10.21741/9781644902516-15

Biodiesel purification by solvent-aided crystallization using n-hexane as solvent

WAN NUR AISYAH Wan Osman[1,2,a], HATEEM Feroze[1,b], SHAFIRAH Samsuri[1,2,c *] and EDUARD Hernandez Yanez[3,d]

[1]Department of Chemical Engineering, Universiti Teknologi PETRONAS, 32610 Seri Iskandar, Perak, Malaysia

[2] HICoE-Centre for Biofuels and Biochemical Research (CBBR), Institute of Self-Sustainable Building (ISB), Universiti Teknologi PETRONAS, 32610 Seri Iskandar, Perak, Malaysia

[3] Universitat Politecnica de Catalunya (UPC) Barcelona TECH,Department of Agrifood Engineering and Biotechnology, Castell-defels 08860, Spain

[a]wan_22000197@utp.edu.my, [b]hateem.feroze_25178@utp.edu.my, [c]shafirah.samsuri@utp.edu.my, [d]eduard.hernandez@upc.edu

Keywords: Cooling Temperature, Fatty Acid Methyl Ester, Environmentally Friendly, Response Surface Methodology, Solvent Concentration

Abstract. Biodiesel is a form of fuel that has a natural origin as it is derived from organic components such as vegetable oil and animal fats. However, biodiesel has to be purified based on the required biodiesel purity standards before it can be used as diesel fuel. This study focuses on a lab-based purification method which is solvent-aided crystallization (SAC) using a typical solvent used for oilseed extraction that is n-hexane. Response surface methodology was used to optimize the process parameters. The purified biodiesel was analyzed via GC-MS to determine its fatty acid methyl ester (FAME) content, reflecting the purity of the biodiesel. The effect of cooling temperature and the concentration of the solvent was studied. The highest purity was obtained at intermediate parameter levels; 12°C and 1.5 wt% of n-hexane. The predicted optimum process parameters within the experimental range were 9.924°C and 1.131 wt%, with FAME purity of 99.789%. The data was validated with an experimental run, and the FAME purity obtained was 99.88%, a 0.1% difference from the predicted value. The FAME purity obtained was above the biodiesel purity standards making this environmentally friendly process viable to be used on a much larger scale in the biodiesel industry.

Introduction

Biodiesel is a form of fuel that is derived from organic components such as vegetable oil, animal fats and waste cooking oil. Biodiesel is produced from a chemical process that converts the oils and fats of natural origins into fatty acid methyl esters (FAMEs) through a process known as transesterification [1]. In Malaysia, biodiesel consumption has seen a sharp increase in biodiesel consumption yearly since 2000, with an average annual rate of 76.06%. In 2019, Malaysia recorded the highest biodiesel consumption with a total of 16.2 thousand barrels per day and showing an increase of 42.42% from 2018 [2]. This shows that biodiesel consumption is certainly growing and has a promising future in the fuel industry, and has emerged as a viable replacement for diesel fuel derived from petroleum and natural gas.

However, biodiesel cannot be used directly without purification as its contaminants might cause fouling of the engine as well as engine failure [3]. The conventional method of biodiesel purification that has been used in the industry is wet washing, either using water, acid, ether or absorbents [4]. However, this method is not environmentally friendly as it uses a large amount of water and produces undesirable by-products that must be treated before they can be discharged. It contains high BOD and COD components and high pH values. Hence, dry washing was introduced as a substitute that uses the application of the adsorption separation process using ion exchange resin and magnesium silicate [3, 5]. However, this method has its limitations, too, due to the lack of knowledge of its chemistry as well as the regeneration of the used adsorbent [6]. Currently, the application of membrane technology for biodiesel purification is being studied to be applied in the industry. However, this technology is considered very costly due to high membrane fouling which leads to higher energy demand and shortens the lifetime of the membrane. Crystallization has been used in various industrial applications as it is a high separation and purification efficiency for solutions with high concentrations or viscosity [7].

For biodiesel purification, solvent-aided crystallization (SAC) can be utilized as it is able to produce biodiesel with high purity or high content of FAMEs [8]. The solvent will assist in reducing the viscosity of the solution and affecting the crystallization kinetics favourably, making the process more efficient. n-Hexane has been used as part of the conventional technology for oilseed solvent extraction; for example, to extract vegetable oils from crops like soybeans [9]. This is because it has various positive attributes such as low latent heat of vaporization, high selectivity, non-polar nature and is easy to be recovered or removed [10]. In a study carried out to compare a few available technologies for oil extraction from leaf feedstock, using n-hexane as the solvent in the solvent extraction technique produced the highest oil yield [11]. Hence, this solvent is expected to have similar attributes when used in the SAC process for biodiesel purification as palm oil is derived from organic components.

The final purity of the biodiesel should meet the purity standards such as the European EN14214 and American D6751 standards before it can be used as a fuel [12]. In Malaysia, the biodiesel standard, MS 2008:2008 conforms to the international standards mentioned earlier. In all these standards, the minimum ester or FAME content is 96.5%. Therefore, this study focuses on the utilization of n-hexane as the solvent for the SAC method. The purity was determined by measuring the fatty acid methyl ester (FAME) content via gas chromatography-mass spectrometry (GC-MS) which was also useful in determining the biodiesel composition. Response surface methodology (RSM) was applied to the design of the experiment (DOE) generated to predict the optimum process parameters.

Methodology

Materials. The palm oil used for the production of biodiesel was the cooking oil of the Saji brand. Methanol with 99.97% purity and potassium hydroxide were obtained from Avantis Laboratory Supply. For the SAC process, n-hexane was used. ethylene glycol solution of 50% (v/v) with water was used as a coolant in a temperature-controlled chiller.

Production of Biodiesel. The experimental setup for the transesterification is illustrated in Fig. 1 below. The necessary apparatus for this experimental procedure was a conical flask, a heating and stirring mantle, a magnetic stirrer and a thermometer. Approximately 300mL of the palm oil was poured into the conical flask, and the temperature was controlled and maintained between 55 to 60°C. 3.8 g of potassium hydroxide (KOH) was dissolved in 67.5mL of methanol. The methoxide solution was added to the heated oil. The mixture was then stirred rapidly at 1000rpm for 10-15 minutes. Once the reaction was completed, the mixture, crude biodiesel was poured into a separatory funnel, as shown in Fig. 2 below. The crude biodiesel was left for 24 hours. Once the crude biodiesel was separated into two visibly distinct layers that are the heavier glycerol at the bottom and the lighter biodiesel at the top, as shown in Fig. 3, 1mL of the biodiesel and glycerol were extracted and inserted into a glass vial separately. The two samples were then sent for differential scanning calorimetry (DSC) analysis to determine their crystallization point. The biodiesel sample was first equilibrated to 30°C and then cooled to -15°C. After one minute, the sample was reheated at 30°C. The temperature range for the glycerol sample was between 0 to 30°C. The ramping rate for both samples was set at 5°C min^{-1}.

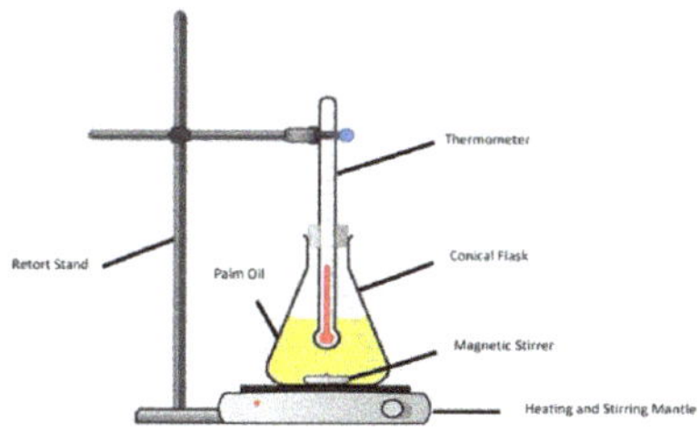

Fig. 1 Experimental Setup for Production of Biodiesel

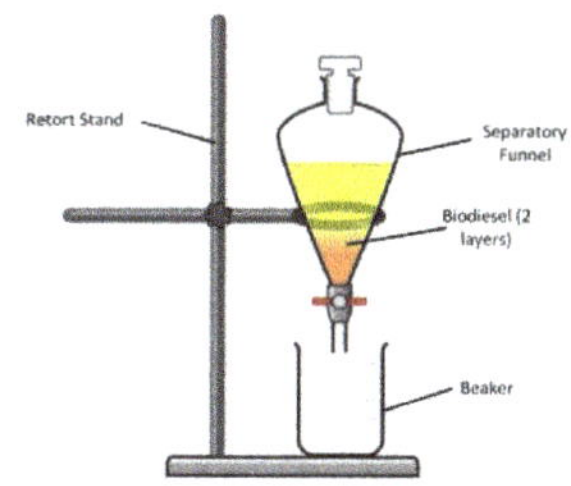

Fig. 2 Biodiesel Gravitational Settling Setup

Solvent-Aided Crystallization (SAC). The experimental setup for the solvent-aided crystallization (SAC) is illustrated in Fig. 3 below. The necessary apparatus for this experimental procedure was a chiller, a stainless-steel cylindrical vessel (crystallizer) and a stirrer. The temperature of the chiller was pre-eminently set to the desired temperature. The crude biodiesel was fed into the vessel. For the SAC process, the crude biodiesel did not undergo any gravity settling process as the purpose of the SAC is to separate and purify the biodiesel. n-Hexane was then added to the vessel as well. The amount of n-hexane required was calculated by weight percentage. Once the coolant temperature in the chiller reached the desired temperature, the vessel was placed in the chiller with the stirrer speed set at 220rpm. The SAC process was left for 40 minutes (cooling time). Two phases were formed in the vessel with the solidified glycerol melt at the bottom of the vessel. The purified biodiesel in the liquid phase was drained into a beaker, and a sample was collected. The solidified glycerol melt was quickly extracted, and a sample was collected. The biodiesel sample was then sent for analysis via GC-MS to determine its purity, while the glycerol sample was stored for future analysis. The entire experiment, the transesterification reaction followed directly by the SAC process, was repeated with different n-hexane concentrations (wt%) and cooling temperatures. The experiment was carried out based on the number of runs that were set by the design of the experiment (DOE) determined prior to the experiment.

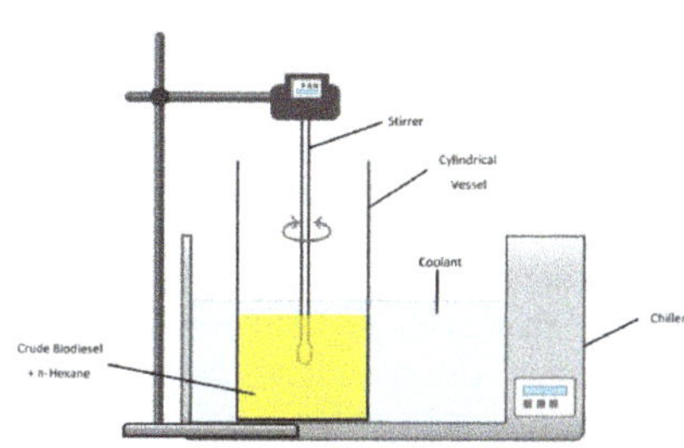

Fig. 3 Experimental Setup for Solvent-Aided Crystallization

Biodiesel Purity Assessment. A portion of the sample of the purified biodiesel was extracted and tested using gas chromatography-mass spectrometry (GC-MS). The initial oven temperature was set at 210°C with a holding time of 13 mins. The temperature was then raised to 230°C and was maintained for 15 mins. The ramping rate used was 5°C min^{-1}.

Process Optimization. The process optimization was carried out by applying response surface methodology (RSM) using Design Expert Software version 13.0.5. The software was utilized from the beginning of the experiment where the design of the experiment (DOE) was generated. Once the experimental results were obtained, the data was input into the software to generate quadratic models and to determine the optimum process parameters. For the DOE, a five-level-two factor central composite design (CCD) was used for this study. A minimum of 10 runs was required to construct the response surface model. The two parameters that were input were n-hexane concentration (A) and cooling temperature (B). There was only one response which was biodiesel purity (K). The biodiesel purity value was reflected by the FAME composition obtained from the GC-MS analysis carried out earlier. The values for the five levels of parameters are shown in Table 1. The SAC for each run was carried out based on the process parameters generated by the statistical DOE. Once all the runs were completed and the biodiesel purity for each run was obtained, the data was inserted into the software. The most suitable regression model was selected that fit the experimental data. Finally, the optimum process parameters were obtained by minimizing the parameters and maximizing the response. A final run of the experiment was carried out to validate the optimization results obtained through the RSM.

Table 1 Levels of Parameters used for CCD

Parameter	Type	Level				
		Min	Low	Central	High	Max
n-Hexane Concentration, A (wt%)	Numeric	0.79	1.0	1.5	2.0	2.21
Cooling Temperature, B (°C)	Numeric	4.48	8.0	12.0	16.0	20.0

Results and discussion

Production of Biodiesel. Based on the observation, the production of biodiesel via transesterification reaction was successfully carried out. The crude biodiesel that was left unattended for 24 hours had visibly separated into two distinct layers, as shown in Figure 3. The top layer is pure biodiesel comprising various types of fatty acid methyl esters (FAME) whereas the bottom darker layer is the contaminants, mainly glycerol. The visibly distinct formation of the two layers was caused by the difference in densities and polarities of the biodiesel and glycerol. From the crude biodiesel of approximately 370 mL, the glycerol obtained after the gravitational settling was approximately 35 mL. This value corresponds with the value reported by [13] stated that about 10 wt% of glycerol is produced as the major by-product in biodiesel production. Other studies also corroborate this data, proving that from every 100 pounds of biodiesel produced, approximately 10% of glycerol is produced [14, 15].

Differential scanning calorimetry (DSC) is a widely used thermal analysis technique for various purposes providing valuable test data for samples. For this study, DSC was carried out to determine the crystallization point of the biodiesel and the glycerol. The thermogram generated from the DSC analysis for the biodiesel sample is shown in Fig. 4, which illustrates the graph of heat flow (mW) against temperature (°C). A single clear peak can be observed, which shows that an exothermic process occurred during crystallization. The onset temperature of the nucleation process which is the point at which the peak begins, was approximately 7.5°C, whereas the crystallization point of the biodiesel at the top of the peak was 6.9°C. The maximum temperature of crystallization of the biodiesel is the end point of the peak at 3.8°C. Fig. 5 shows the thermogram generated from the DSC analysis of the glycerol sample. Unlike the graph in Fig. 4, the graph in Fig. 5 clearly shows no exothermic or endothermic peaks developed throughout the process. The result obtained can be corroborated by a recent study showing that no peak was found when DSC analysis was carried out for glycerol due to supercooling [16]. A similar study also stated that the melting point of glycerol is approximately 17.8°C. Since the crystallization temperature of the biodiesel sample

Materials Research Proceedings 29 (2023) 117-127 https://doi.org/10.21741/9781644902516-15

obtained from the DSC thermogram is approximately 7.5°C, the range of temperature selected for this study was from 8 to 16°C, which is below the melting point of glycerol, 17.8°C obtained from the study.

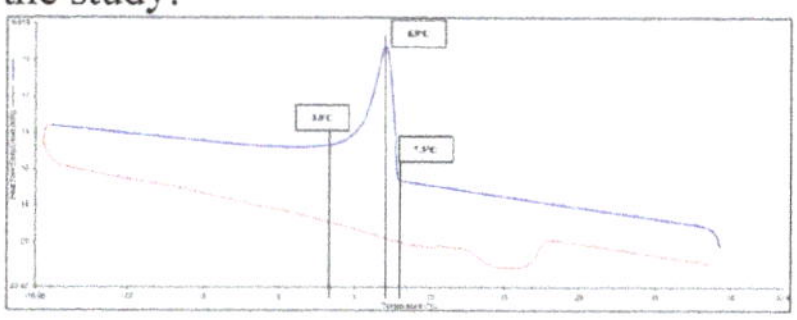

Fig. 4 Graph of Heat Flow vs Temperature (Biodiesel)

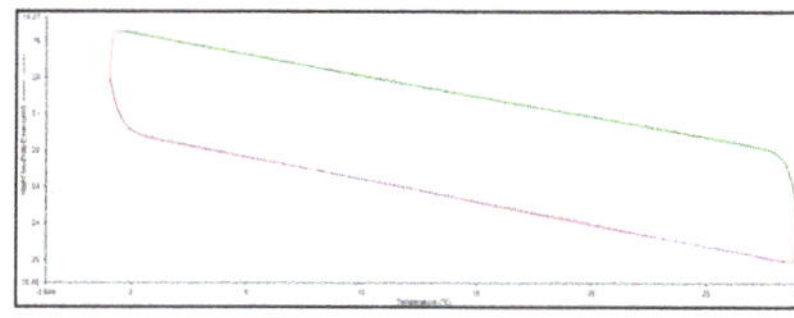

Fig. 5 Graph of Heat Flow vs Temperature (Glycerol)

Solvent-Aided Crystallization (SAC). Fig. 6 shows the pictures taken inside the vessel during and after the SAC process was completed. It can be clearly seen that after the SAC process was completed, two distinct phases were formed. The pure biodiesel was obtained in the liquid phase, and the glycerol crystallized at the bottom and sides of the vessel. n-Hexane is a non-polar solvent, and according to Bogataj and Bagajewicz [17], non-polar solvents can influence the solid crystals to be in an elongated form, as can be seen in Fig. 7. The solidified glycerol is in layer form. This observation concludes that the SAC process was successfully carried out.

Fig. 6 Left: During SAC; Right: After SAC

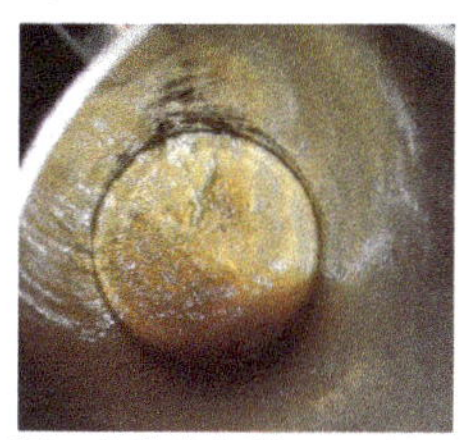

Fig. 7 Solid Glycerol in Layer Form

Biodiesel Purity Assessment. The purified biodiesel obtained from the SAC process was sent for analysis using GC-MS to determine its FAME content. Fig. 8 below shows the chromatogram obtained for the analysis of the first run of the experiment. The chromatogram shows the graph of signal intensity (abundance) against the retention time of the biodiesel sample. The peak data obtained from the GC-MS result is shown in Fig. 9. The percentage composition of the FAME is calculated using Eq. 1.

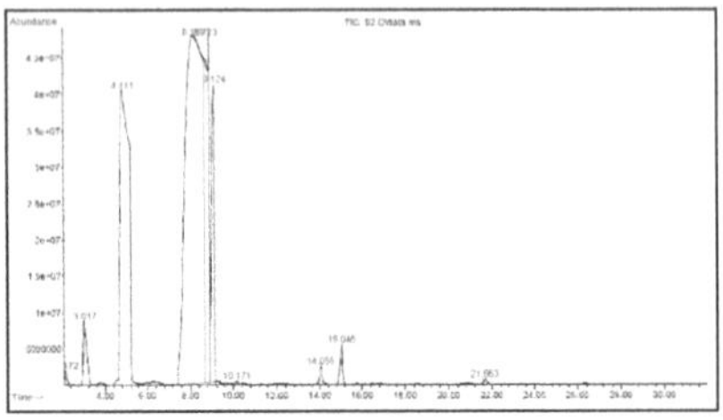

Fig. 8 Graph of Abundance vs Time (Sample from First Run)

peak #	R.T. min	first scan	max scan	last scan	PK TY	peak height	corr. area	corr. % max.	% of total
1	2.172	5	15	54	BB	1773283	113459091	0.42%	0.220%
2	3.017	115	163	240	BB 4	8768916	961693857	3.59%	1.862%
3	4.811	384	476	632	BB 10	40449119	12482249118	46.54%	24.165%
4	8.159	902	1061	1151	BV 10	47995837	26822359763	100.00%	51.928%
5	8.723	1151	1160	1198	VV 9	44663845	6789445313	25.31%	13.144%
6	9.124	1198	1230	1243	VV 2	40694375	3705424598	13.81%	7.174%
7	10.171	1384	1413	1444	BB 3	420628	21613862	0.08%	0.042%
8	14.055	2032	2092	2174	BB 2	2401407	224314990	0.84%	0.434%
9	15.046	2192	2265	2307	BV	5370705	452507347	1.69%	0.876%
10	21.663	3369	3421	3492	BB 5	736501	80360171	0.30%	0.156%

Sum of corrected areas: 51653428112

Fig. 9 Peak Data from GC-MS Results (Sample from First Run)

$$\% \, Composition \, of \, FAME = \frac{Correction \, Area \, of \, Individual \, Component}{Sum \, of \, Correction \, Area} \times 100\% \tag{1}$$

For example, the percentage composition of the FAME in the first peak is calculated as follows:

$$\% \, Composition \, of \, FAME = \frac{113459091}{51653428112} \times 100\% = 0.22\%$$

The FAME present in the sample was identified based on the specific residence time, and the type of fatty acid was determined. The total FAME present was calculated by adding the percentage composition of each FAME present in the sample. The analyzed results are tabulated in Table 2. The total percentage of FAME for the purified biodiesel obtained from the first run of the experiment was 99.959% which is way above the biodiesel purity standards. To clarify, the composition of FAME obtained from this method may be different from the composition of raw palm oil as the FAME sample obtained in this study undergoes purification via SAC.

Table 2 GC-MS Analyzed Results for Sample from First Run

Peak Number	Retention Time (min)	Systematic Name (Library ID)	Trivial Name	Type	Composition of FAME (%)
1	2.172	Dodecanoic Acid	Lauric	Saturated	0.22
2	3.017	Methyl Tetradecanoate	Myristic	Saturated	1.862
3	4.811	Hexadecanoic Acid	Palmitic	Saturated	24.165
4	8.159	11-Octadecenoic Acid	Oleic	Unsaturated	51.928
5	8.723	9-Octadecenoic Acid	Oleic	Unsaturated	13.144
6	9.124	Methyl Stearate	Stearic	Saturated	7.174
8	14.055	cis-Methyl 11-Eicosenoate	Gadoleic	Unsaturated	0.434
9	15.046	Eicosanoic Acid	Arachidic	Saturated	0.876
10	21.663	Docosanoic Acid	Behenic	Saturated	0.156
				TOTAL	**99.959**

Response Surface Methodology – Process Optimization. Various types of models could be used to fit the experimental design and the. Design Expert software suggested the most suitable polynomial model by using the Whitcomb score, which is a heuristic scoring system. The proposed model by the software was the quadratic model as it was not aliased and was adequately significant in representing the FAME purity that was the response for this study. The summary report comparing each of the parameters is shown in Table 3.

Table 3 Fit Summary of Each Model

Model	Sum of Squares	DF	Mean Square	F-value	p-value	Remarks
Mean	99482.87	1	99482.87	-	-	-
Linear	0.0533	2	0.0267	0.6586	0.5469	-
2FI	0.0130	1	0.0130	0.2875	0.6111	-
Quadratic	0.1302	2	0.0651	1.85	0.2692	Suggested
Cubic	0.1404	4	0.0351	-	-	-
Quartic	0.0	0	-	-	-	Aliased
Residual	0.0	0	-	-	-	-
TOTAL	99483.21	10	9948.32	-	-	-

Based on the selected model which was the quadratic model, the final empirical model in terms of the coded factor was generated as shown in Eq. 2.

$$K = 99.91 - 0.0071A - 0.0019B - 0.1197AB - 0.0906A^2 - 0.1251B^2 \quad (2)$$

The experimental and predicted response for each run is tabulated in Table 4. From Table 4, the data obtained from experimental and predicted were in the range of 99.46 to 99.96% since this study used cooking oil (refined oil) as the feed to produce biodiesel. For better clarification, biodiesel produced from refined oil is expected to have a high purity of FAME compared to unrefined oil such as crude palm oil processed from a palm mill. Another study using unrefined oil will be used to support this claim in the near future. On the other hand, the adequacy of the model was further justified by determining the absolute average deviation (AAD), as shown in Eq. 3 [18].

Table 4 Parameters and Response for Each Run

Run	Parameters		Response, K [FAME Purity (%)]		
	n-Hexane Concentration, A (wt%)	Cooling Temperature, B (°C)	Experimental	Predicted	Residual
1	1.5	12.0	99.96	99.91	0.0471
2	1.5	16.0	99.92	99.78	0.1331
3	1.0	8.0	99.51	99.59	-0.0755
4	1.0	20.0	99.46	99.56	-0.1024
5	0.79	16.0	99.87	99.78	0.0859
6	2.0	10.0	99.57	99.84	-0.2698
7	1.5	10.0	99.82	99.88	-0.0616
8	1.5	4.48	99.55	99.47	0.0716
9	2.0	8.0	99.88	99.81	0.0703
10	2.21	10.0	99.88	99.77	0.1012

$$AAD = \frac{1}{n}\sum_{i=1}^{n}\left|\frac{K_{i,exp} - K_{i,pred}}{K_{i,exp}}\right| \times 100\% \quad (3)$$

where n is the number of experiments, K_{exp} is the experimental value of the response and K_{pred} is the predicted value of the response. Hence, from the calculation, the AAD value obtained was 0.102%. Based on this significantly low AAD value, it can be interpreted that the model fitted precisely to the experimental data as a very low AAD value was obtained [19].

Effect of Parameters – Cooling Temperature. The first parameter that was studied was the cooling temperature. The cooling temperature was controlled by manipulating the temperature of the chiller to the desired temperature. Using Design Expert, a graph showing the relationship between the temperature and FAME purity was generated as shown in Fig. 10 below. The value of the concentration of n-hexane kept constant at the central level is 1.5 wt%. The red dots are the expected design points, the black line indicates the trend of the findings, and the dotted blue lines are the bands of the confidence interval of 95%. The highest FAME purity can be achieved at the intermediate temperature of 12°C. Based on this finding, it can be concluded that the most desirable cooling temperature is at 12°C as the FAME is not solidifying while the glycerol and other by-products are well crystallized. According to a recent study [20], the solid grows in a more ordered pattern at a high temperature. This prevents the FAME from being trapped together with the glycerol during the SAC process, subsequently increasing the FAME purity. However, the FAME purity decreases as the temperature increases beyond 12°C. This is because the temperature is now approaching the melting point of the glycerol which is approximately 17.8°C. This reduces the crystallization efficiency of the glycerol, causing some of the glycerol to remain in its liquid form and combine with the FAME in one phase. Thus, for this experiment, the most desirable cooling temperature is in the intermediate region.

Effect of Parameters – Solvent Concentration. The next parameter studied was the concentration of the solvent, n-hexane. Using Design Expert, a graph showing the relationship between the n-hexane concentration and FAME purity was generated, as shown in Fig. 11. The value of the cooling temperature was kept constant at the central level which is 12°C. The most desirable concentration is at 1.5 wt% at which the efficiency of the SAC is at the optimum level. Almost all the glycerol is crystallized leaving only the FAME in the liquid phase. As a result, a high FAME purity can be observed. However, the FAME purity declines as the concentration of solvent is increased beyond 1.5 wt%. This shows that although the addition of n-hexane as the solvent in the SAC process helps to increase the solubility of the glycerol and reduce the density difference between the glycerol and the FAME, adding in too much solvent leads to a poor separation and purification process [21]. This finding is similar to a study carried out in 2015 [22] in which a higher separation of glycerol from the water was achieved with small quantities of the solvent, 1-butanol. The main point from this finding is that an increase in solvent concentration does not necessarily increase the FAME purity as an excessive amount of n-hexane creates a reverse effect by increasing the viscosity of the solution and subsequently, retarding the SAC process.

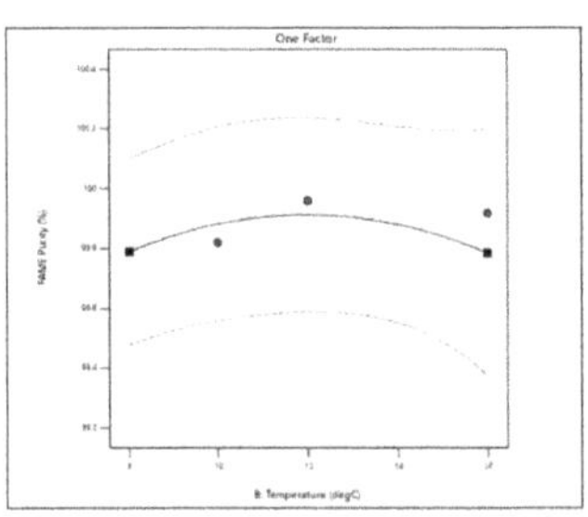

Fig. 10 Graph of Temperature vs FAME Purity

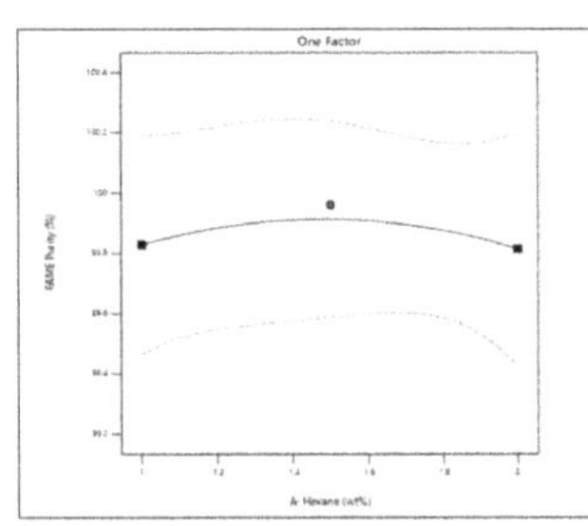

Fig. 11 Graph of n-Hexane Concentration vs FAME Purity

Optimum Conditions. Since there were only two parameters involved, the goal was to maximize the FAME purity while minimizing both parameters within the experimental range. The optimum operating conditions that were generated by the software based on the goals mentioned earlier are shown in Table 5. The desirability of the optimal solution selected was 0.758. The goal in determining the optimum conditions is to minimize the desirability value as a desirability of 1.00

translates that the goals were easy to reach and there might be better results available. Once the optimum conditions were obtained, the experiment was carried out again with the parameters generated by the software as shown in Table 5. The FAME purity obtained from the experiment was 99.88%. The reliability of the predicted biodiesel purity was evaluated by calculating the percentage error between the predicted and experimental value. The percentage error obtained was 0.1%. Since the value of percentage error is way lower than the maximum allowable percentage error of 10%, it can be concluded that the optimum conditions generated by the Design Expert as well as the predicted biodiesel purity are accurate as the model's prediction is almost equal to the actual biodiesel purity which was obtained through experiment. This value also justifies the regression model that was generated, and it confirms that this model was acceptable to be used for this project.

Table 5 Optimum Parameters with Predicted and Experimental Response Values

Parameters	Goal	Response Type	Optimum Conditions
A: Concentration of n-Hexane (wt%)	Minimize	-	1.131
B: Cooling Temperature (°C)	Minimize	-	9.924
K: FAME Purity (%)	Maximize	Predicted	99.789
		Experimental	99.880

Conclusion

This study focused on the utilization of SAC as a biodiesel purification method by studying the effect of two parameters, namely the concentration of the solvent, n-hexane and the cooling temperature. The purity of the biodiesel was reflected by its FAME content which was measured via GC-MS. The highest purity was obtained at intermediate parameter levels; 12°C and 1.5 wt% of n-hexane. Response surface methodology (RSM) was applied via Design Expert software. The adequacy of the selected quadratic model was justified with an absolute average deviation (AAD) value of only 0.102%, showing that the experimental data fitted well with the model. The predicted optimum process parameters within the experimental range were 9.924°C and 1.131 wt%, with FAME purity of 99.789%. The data was validated with an experimental run, and the FAME purity obtained was 99.88%. In conclusion, the biodiesel was well separated and purified as the FAME content is above the conformed purity standards.

Due to time and accessibility constraints caused by the pandemic, only two parameters could be studied. The effect of other parameters such as stirring speed and cooling time can be studied in the future, and the regression model generated can be further improved to provide a more comprehensive outcome. Besides that, other assessment methods can be carried out in future studies to completely characterize biodiesel. Also, different types of feed or oil sources can be used to produce biodiesel to compare the finding with this study. Currently, the use of green solvent has received a lot of attention in various industries as it is environmentally friendly and more cost-efficient. Hence, in future studies related to SAC purification of biodiesel, the effect of using green solvents can be studied, and the outcome can be compared to the results obtained from this study.

Acknowledgment

The authors would like to acknowledge the funding supports from Ministry of Higher Education (MOHE) Malaysia through HICoE Grant (cost centre 015MA0-052/015MA0-104/015MA0-136) to Centre of Biofuels and Biochemical Research (CBBR).

References

[1] Pacific Biodiesel, "What is Biodiesel," 3 March 2020. [Online]. Available: https://www.biodiesel.com/what-is-biodiesel/. [Accessed 5 June 2021].

[2] Knoema, "Malaysia - Biodiesel consumption," knoema.com, 2019. [Online]. Available: https://knoema.com/atlas/Malaysia/topics/Energy/Renewables/Biodiesel-consumption. [Accessed 5 June 2021].

[3] I. M. Atadashi, M. K. Aroua, A. R. Abdul Aziz and N. M. N. Sulaiman, "Refining technologies for the purification of crude biodiesel," Applied Energy, vol. 88, no. 12, pp. 4239-4249, 2011. https://doi.org/10.1016/j.apenergy.2011.05.029

[4] H. Bateni, A. Saraeian and C. Able, "A comprehensive review on biodiesel purification and upgrading," Biofuel Research Journal, vol. 4, no. 3, pp. 668-690, 2017. https://doi.org/10.18331/BRJ2017.4.3.5

[5] I. M. Atadashi, M. K. Aroua, A. R. Abdul Aziz and N. M. N. Sulaiman, "Membrane biodiesel production and refining technology: A critical review," Renewable and Sustainable Energy Reviews, vol. 15, no. 9, pp. 5051-5062, 2011. https://doi.org/10.1016/j.rser.2011.07.051

[6] I. M. Atadashi, "Purification of crude biodiesel using dry washing and membrane technologies," Alexandria Engineering Journal, pp. 1265-1272, 2015. https://doi.org/10.1016/j.aej.2015.08.005

[7] Z. Gao, S. Rohani, J. Gong and J. Wang, "Recent developments in the crystallization process: toward the pharmaceutical industry," Engineering, vol. 3, no. 3, pp. 343-353, 2017. https://doi.org/10.1016/J.ENG.2017.03.022

[8] S. Samsuri, N. L. Jian, F. W. Jusoh, E. Hernández Yáñez and N. Y. Yahya, "Solvent-Aided Crystallization for Biodiesel Purification," Chemical Engineering & Technology, vol. 43, no. 3, pp. 447-456, 2020. https://doi.org/10.1002/ceat.201900433

[9] C. P. Davis, "Medical Definition of n-Hexane," MedicineNet, 29 March 2021. [Online]. Available: https://www.medicinenet.com/n-hexane/definition.htm. [Accessed 10 July 2021].

[10] S. J. Kumar, S. R. Prasad, R. Banerjee, D. K. Agarwal, K. S. Kulkarni and K. V. Ramesh, "Green solvents and technologies for oil extraction from oilseeds," Chemistry Central Journal, vol. 11, no. 1, pp. 1-7, 2017. https://doi.org/10.1186/s13065-017-0238-8

[11] M. M. K. Bhuiya, M. Rasul, M. Khan, N. Ashwath and M. Mofijur, "Comparison of oil extraction between screw press and solvent (n-hexane) extraction technique from beauty leaf (Calophyllum inophyllum L.) feedstock," Industrial Crops and Products, vol. 144, p. 112024, 2020. https://doi.org/10.1016/j.indcrop.2019.112024

[12] M. Hayyan, F. S. Mjalli, M. A. Hashim and I. M. AlNashef, "A novel technique for separating glycerine from palm oil-based biodiesel using ionic liquids," Fuel Processing Technology, vol. 91, no. 1, pp. 116-120, 2010. https://doi.org/10.1016/j.fuproc.2009.09.002

[13] F. Yang, M. A. Hanna and R. Sun, "Value-added uses for crude glycerol--a byproduct of biodiesel production," Biotechnology for biofuels, p. 13, 2012. https://doi.org/10.1186/1754-6834-5-13

[14] N. Binhayeeding, S. Klomklao and K. Sangkharak, "Utilization of waste glycerol from biodiesel process as a substrate for mono-, di-, and triacylglycerol production," Energy Procedia, vol. 138, pp. 895-900, 2017. https://doi.org/10.1016/j.egypro.2017.10.130

[15] A. Rodrigues, J. C. Bordado and R. G. D. Santos, "Upgrading the glycerol from biodiesel production as a source of energy carriers and chemicals-A technological review for three chemical pathways," Energies, vol. 10, no. 11, p. 1817, 2017. https://doi.org/10.3390/en10111817

[16] W. Gao, M. Zhang and H. Wu, "Differential Scanning Calorimetry Studies on the Cold Flow Properties of Fuel Mixtures from Bio-Oil, Crude Glycerol, Methanol, and/or Biochar," Energy & Fuels, vol. 31, no. 8, pp. 8348-8355, 2017. https://doi.org/10.1021/acs.energyfuels.7b01475

[17] M. Bogataj and M. J. Bagajewicz, "Synthesis of non-isothermal heat integrated water networks in chemical processes," Computers & Chemical Engineering, vol. 32, no. 12, pp. 3130-3142, 2008. https://doi.org/10.1016/j.compchemeng.2008.05.006

[18] L. Ma, E. Lv, L. Du, J. Lu and J. Ding, "Statistical modeling/optimization and process intensification of microwave-assisted acidified oil esterification," Energy Conversion and Management, vol. 122, pp. 411-418, 2016. https://doi.org/10.1016/j.enconman.2016.06.001

[19] P. Ghorbannezhad, A. Bay, M. Yolmeh, R. Yadollahi and J. Y. Moghadam, "Optimization of coagulation-flocculation process for medium density fiberboard (MDF) wastewater through response surface methodology," Desalination and Water Treatment, vol. 57, no. 56, pp. 26916-26931, 2016. https://doi.org/10.1080/19443994.2016.1170636

[20] S. Samsuri, N. A. Amran, L. J. Zheng and M. M. M. Bakri, "Effect of coolant temperature and cooling time on fractional crystallization of biodiesel and glycerol," Malays. J. Fundam. Appl. Sci, vol. 13, pp. 676-679, 2017. https://doi.org/10.11113/mjfas.v13n4.925

[21] J. Van Gerpen and G. Knothe, "Bioenergy and biofuels from soybeans," in Soybeans, New York, AOCS Press, 2008, pp. 499-538. https://doi.org/10.1016/B978-1-893997-64-6.50019-6

[22] F. J. Eisenbart and J. Ulrich, "Solvent-aided layer crystallization-Case study glycerol-water," Chemical Engineering Science, vol. 133, pp. 24-29, 2015. https://doi.org/10.1016/j.ces.2014.12.060

Materials Research Proceedings 29 (2023) 128-134 https://doi.org/10.21741/9781644902516-16

Comparison of crystallinity index computational methods based on lignocellulose X-ray diffractogram

CHEAH Yong Sing[1], MASAHARU Komiyama[1,2,3,a,b*]

[1]Chemical Engineering Department, Universiti Teknologi PETRONAS, 32610, Seri Iskandar, Perak, Malaysia

[2]HICoE−Centre for Biofuel and Biochemical Research, Institute of Sustainable Building, Universiti Teknologi PETRONAS, 32610, Seri Iskandar, Perak, Malaysia

[3]Clean Energy Research Centre, University of Yamanashi, Takeda, Kofu, 400-8511, Japan

[a] masaharu.komiyama@utp.edu.my, [b] masaharu@yamanashi.ac.jp

Keywords: Lignocellulosic Biomass, Crystallinity Index, Computational Methods, Comparison, Torrefaction, Artificial Lignocellulose

Abstract. Crystallinity index (CrI) obtained from X-ray diffraction (XRD) technique is often utilized as a characterization parameter of lignocellulosic biomass. There exist a few methodologies to calculate CrI but the respective merit as lignocellulose characterization parameter is not very clear. Here four commonly employed CrI computational methods were applied to raw and torrefied biomasses (palm kernel shell and sugarcane bagasse), cellulose- and lignin-added raw biomasses and artificial mixtures of cellulose, hemicellulose and lignin in order to compare the effect of the composition of lignocellulosic biomass toward CrI calculated from X-ray diffractogram. Calculated CrI systematically showed larger value than the weight percentage of cellulose contained in the samples. Among the four computational methods compared, Segal (single peak height ratio) method and Ruland-Vonk (two-peak area ratio) method appeared to give reasonable CrI numbers although they are still overestimating the cellulose weight ratio. The Ruland-Vonk method consistently gave the lowest CrI values among the methods examined.

Introduction

Crystallinity is one of the primary parameters in describing the nature of lignocellulose. The degree of crystallinity could be made an indicator to predict the rate of cellulose hydrolysis by cellulase in bioethanol production [1] or the strength of biocomposite material [2]. Furthermore, crystallinity is often utilized to monitor the effect of various treatments on lignocellulosic biomass [3-6]. Due to its importance, various methodologies have been proposed to yield a measure of crystallinity in lignocellulosic biomass [7], including X-ray diffraction, Fourier-transform infrared spectroscopy, Raman spectroscopy and solid state ^{13}C nuclear magnetic resonance. X-ray diffraction (XRD) is one of the most popular among those proposed methodologies for the quantification of crystallinity of lignocellulosic biomass, partly due to its wide availability and straight-forward data interpretation.

Crystallinity of lignocellulosic biomass is commonly expressed in the form of crystallinity index (CrI), which is intended to reflect the fraction (could be weight, volume or molar) of crystalline cellulose within the lignocellulosic biomass sample. When XRD is employed to derive CrI of a lignocellulose sample, first powder X-ray diffractogram of the sample is obtained, which constitutes a few diffraction peaks from the crystalline part of its cellulose and a single broad peak (called halo) from the amorphous parts of the sample. There exist a few methodologies to calculate CrI from the powder X-ray diffractogram: Segal peak height method [8], Jayme-Knolle method [9], Ruland-Vonk method [10,11] and peak deconvolution method [12]. The first two methods utilize peak heights to calculate CrI, while the third method employ peak area rather than peak height. The more recent entry to the field, peak deconvolution method, deconvolutes the diffractogram peaks using standard distribution function to obtain respective peak contributions.

Comparison of these methodologies has been attempted in the past [13,14]. However, those comparisons were done on a fixed sample like powder cellulose (Avicel PH-101) along with literature data analysis on the same sample. Park et al. [13] reports that in their literature compilation of 18 measurements, CrI of Avicel PH-101 reported varied from ca. 50 % to 91.7 %. The present paper took a different approach for CrI computational methodology comparison by preparing samples with systematic change of lignocellulose compositions: (1) through torrefaction of raw palm kernel shell and sugarcane bagasse, (2) through addition of pure cellulose and lignin to those raw biomasses and (3) through synthesis of a series of artificial lignocellulose samples by physically mixing known lignocellulosic components.

Experimental

Lignocellulosic Sample Preparation. Palm kernel shell was obtained from an oil palm plantation of FELCRA Processing & Engineering Sdn. Bhd. at Teluk Intan, and sugarcane bagasse from Lian Hup Hin Provision Sdn. Bhd., Kamunting, both in Perak, Malaysia. These raw biomass samples were washed with distilled water to remove dirt and dust, cut into small pieces (ca. 1cm in length) with a mechanical knife and left in a laboratory open atmosphere for two days before proceeding for drying in an air oven at 105 °C for 24 hours. The moisture content of the samples was measured following ASTM E871-82 protocol and was less than 10 wt%. The dried biomass was then grinded and sieved to obtain fractions of uniform particle size. The sugarcane bagasse was then undergone palletization process to increase its density due to its inherently characteristic low density.

In order to manipulate chemical composition of these raw biomasses, torrefaction was performed at 200 °C for 30 min under flowing nitrogen. Furthermore, a modified lignocellulosic biomass was prepared by mixing microcrystalline cellulose (Avicel PH-101, Sigma-Aldrich) and pure lignin (Sigma-Aldrich) to the raw biomasses. In addition, several artificial lignocellulosic biomass samples were produced by mixing microcrystalline cellulose, xylan (representative of hemicellulose) and pure lignin to simulate the changes in crystallinity in the artificial lignocellulosic sample. The amount of cellulose, hemicellulose and lignin in the raw and torrefied lignocellulosic biomass was determined following the method described by Yang et al. [15] and listed in Table 1.

Crystallinity Index Measurement. X-ray diffractogram of the above samples were obtained by using Malvern Panalytical's HighScore Plus X-ray diffractometer. The scanning ranged from 5° to 40°, with an exposure time of 0.4 s per step in steps of 0.01415° using Cu-$K_{\alpha 1}$ radiation at 1.541 Å. The extracted data were computed and analyzed using Origin Pro 8.5 software.

Table 1. Chemical Compositions of the Employed Biomass

	Cellulose [wt%]	Hemicellulose [wt%]	Lignin [wt%]	Extractives [wt%]
Sugarcane Bagasse	43	26	24	7
Torrefied Sugarcane Bagasse	36	15	44	5
Palm Kernel Shell	24	28	46	4
Torrefied Palm Kernel Shell	19	17	61	3

Results and Discussion

Crystallinity Index of Raw and Torrefied Biomass. The CrIs of raw and torrefied biomass as well as microcrystalline cellulose (Avicel PH-101) were computed from respective diffractogram using the four computational methods, viz., Segal peak height method, Jayme-Knolle method, Ruland-Vonk method and peak deconvolution method, and the results are shown in Fig. 1.

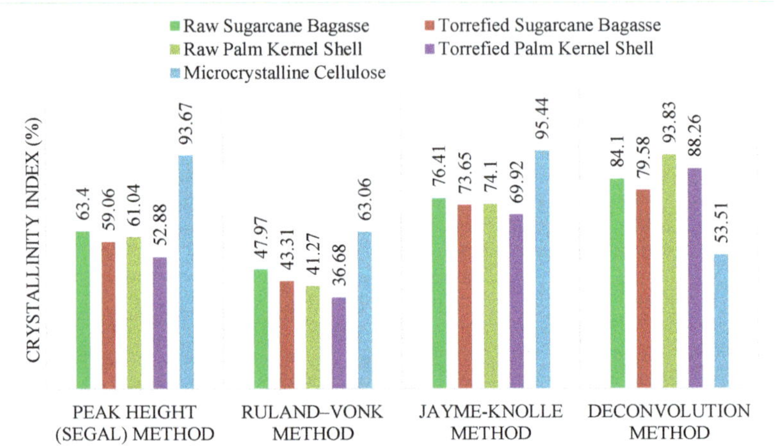

Fig. 1. Crystallinity index obtained using different computational methods

Overall comparison of the obtained CrI shown in Fig. 1 indicates that the first three computational methods listed in the figure, viz., Segal peak height method, Ruland-Vonk method and Jayme-Knolle method seems to show similar trend among the measured samples: (1) Avicel PH-101 (blue bars) shows the highest CrI within the same computational method, (2) raw sugarcane bagasse (green bars) always show higher CrI than raw palm kernel shell (olive green bars), (3) torrefaction always decrease CrI for both sugarcane bagasse and palm kernel shell. On the other hand, the last computational method, deconvolution, seems to contradict in many of these trends indicated by the first three methods.

Among the first three computational methodologies, Jayme-Knolle method gives highest overall CrI values for all the five samples, followed by Segal peak height method and then the Ruland-Vonk method. It is noted that the lowest absolute CrI values provided by the Ruland-Vonk method still seems to be higher than actual cellulose crystal contents when compared to the cellulose content of the respective samples shown in Table 1, particularly for the lower cellulose content samples such as palm kernel shell and torrefied palm kernel shell. Attentions should be paid, however, that the cellulose content in Table 1 is in the unit of wt%, while CrIs calculated and shown in Fig. 1 are the X-ray diffraction intensity ratios.

Modified Lignocellulosic Biomass Sample Analysis. In order to examine the effect of the amount of lignin and cellulose towards the measured crystallinity of lignocellulosic biomass, 10 wt%, 20 wt% and 40 wt% of microcrystalline cellulose or lignin were added into the two raw biomass samples, respectively. This translates to the actual cellulose content of 43 wt%, 48 wt%, 53 wt% and 59 wt% for Avicel-added sugarcane samples, 43 wt%, 39 wt%, 36 wt% and 31 wt% for lignin-added sugarcane samples, 24 wt%, 31 wt%, 37 wt% and 46 wt% for Avicel-added palm kernel shell samples and 24 wt%, 22 wt%, 20 wt% and 17 wt% for lignin-added palm kernel shell samples, respectively. Obtained CrIs shown in Figs. 2 and 3 indicate that addition of cellulose monotonously increases, and addition of lignin monotonously decreases, the calculated CrIs for the three computational methods, viz., Segal peak height, Ruland-Vonk and Jayme-Knolle. Deconvolution method again fails to follow these trends, and its calculated CrIs show irregular trend with the increase of added cellulose or lignin.

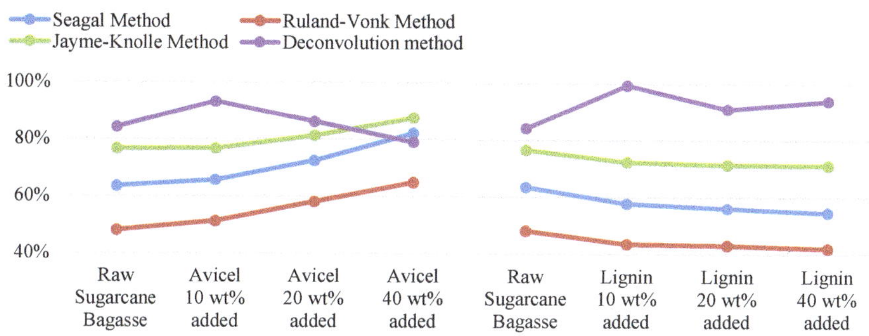

Fig. 2. Crystallinity index of raw and modified Sugarcane Bagasse

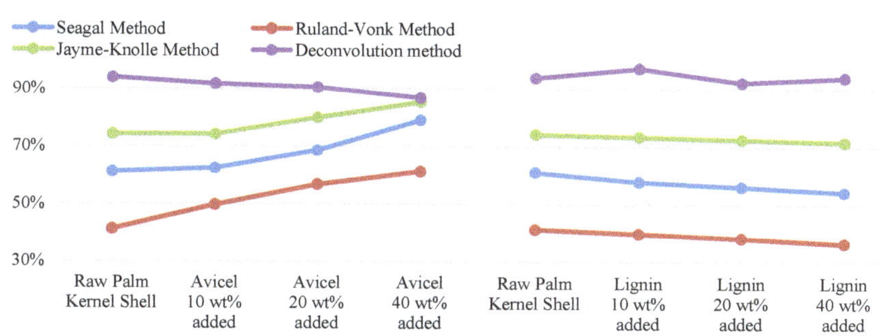

Fig. 3. Crystallinity index of raw and modified Palm Kernel Shell

Among the three computational methodologies that appear to give reasonable trend with the systematic increase (by addition of cellulose) and decrease (by addition of lignin) of cellulose amount, the absolute values of CrIs obtained are in the decreasing order of Jayme-Knolle, Segal peak height and Ruland-Vonk, in parallel with the results found in Fig. 1 for raw and torrefied biomasses. It is further noted that the two methods, Segal peak height and Ruland-Vonk, gives almost the same CrI increment when compared with raw biomass and 40 wt% Avicel (17-19 % CrI increase for sugarcane bagasse and 18-20 % for palm kernel shell). This is in good agreement with the actual increment of cellulose in those samples, 16 % and 22 %, respectively.

On the other hand, lignin addition gave much less impact on measured CrI: for the actual decrease of 12 % in cellulose content CrI decreased only 6-9 % in sugarcane bagasse, and for palm kernel shell 7 % decrease caused 5-7 % CrI decrease. Apparently, depending on the samples, non-crystalline part of the sample contributes less for the calculation of CrI for those two calculation methods.

Artificial Lignocellulosic Biomass Crystallinity Analysis. A series of artificial lignocellulosic biomass is prepared by physically mixing microcrystalline cellulose (Avicel PH-101), xylan (representative of hemicellulose) and pure lignin. Six sets of artificial lignocellulosic biomass were produced with the compositions listed in Table 2. The compositions chosen are more or less arbitrary, except that Sample No. 1, 2, 4 and 6 mimic those of the raw and the torrefied samples found in Table 1.

Figure 4 shows the CrIs of those artificial lignocellulosic biomasses listed in the increasing order of cellulose content. As has been observed in Figs. 2 and 3, the CrIs of these samples monotonously increased with increasing cellulose content, except for the deconvolution method which is not stable and did not show the similar trend that are apparent in other three computational methods. Among the three computational methods that gave similar monotonous trend with the systematic change of the amount of cellulose, Jayme-Knolle method gave the highest absolute CrI values, followed by Segal peak height method and Ruland-Vonk method, again in line with the observation made in Figs. 1 through 3.

The two methods that gave the low absolute CrI values, Seagal and Ruland-Vonk methods, gave almost the same CrIs, although they are still higher than actual cellulose contents (in wt%) of the respective samples (abscissa). Note that in the present samples, the increment of CrI (18.7 for Ruland-Vonk and 18.0 for Seagal) between sample No. 6 and No. 1 is lesser than the increment of cellulose content (26.6 wt%).

Table 2. Chemical Compositions of the Artificial Biomass

Sample No.	Cellulose [wt%]	Hemicellulose [wt%]	Lignin [wt%]
1	46.23	27.96	25.81
2	37.89	15.79	46.31
3	35.00	25.00	40.00
4	30.00	30.00	40.00
5	25.00	28.10	46.90
6	19.59	17.53	62.89

Comparison of Computational Methods. The foregoing data obtained on three types of lignocellulose biomasses, viz., raw/torrefied, cellulose/lignin added and artificial, all points to the following conclusions: (1) Jayme-Knolle method, Segal peak height method and Ruland-Vonk method give similar tendencies of monotonous increase of CrI with increase of cellulose content, (2) the absolute value of CrI decreases in the order of Jayme-Knolle method, Segal peak height method and Ruland-Vonk method, (3) the CrI values are systematically higher than actual cellulose content expressed in wt% and (4) the deconvolution method does not seem to reflect the biomass crystallinity very accurately.

The present work indicated that the deconvolution method seems to give fluctuated results in terms of sample cellulose content. The challenge for the deconvolution method may be to identify the location of the peak, which might vary according to sample [14]. Sometimes certain diffraction peaks such as (10−1) and (021) may not be obvious in the obtained Xray diffractogram, which may cause the results of computed CrI to be rather inaccurate and unreliable.

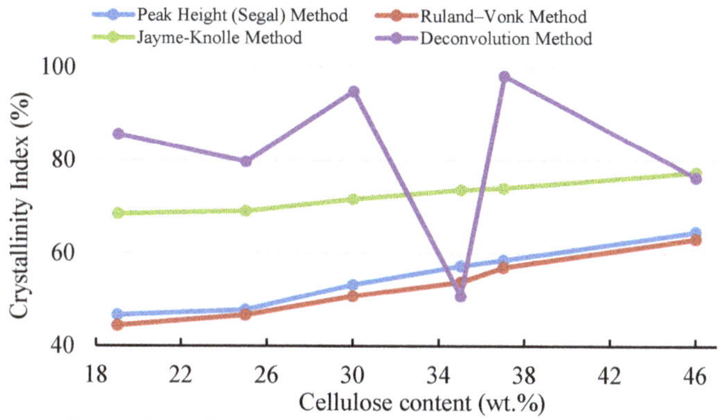

Fig. 4. Crystallinity index of artificial lignocellulosic biomass

While both Jayme-Knolle method and Seagal peak height method utilizes diffraction peak heights as calculation parameters, the former gives higher CrI than the latter due to the fact that the former incorporates (101) peak height in addition to (002) peak height while the latter uses only (002) peak. The Ruland-Vonk method, which measures peak areas rather than peak heights, led to the lowest CrI values, may be due to the fact that it considers all the amorphous region of the X-ray diffractogram without excluding any data except for the background noise.

It is interesting to note that the Segal method which utilizes simple peak heights gives CrI values closer (Figs. 1 through 3) or almost the same (Fig. 4) with Ruland-Vonk method which uses peak areas. This is also notable in view of the possibility that XRD peak height would be influenced by the cellulose crystallite sizes when they are small enough to give peak broadening [13].

Summary

Three lignocellulosic samples with systematic change of lignocellulose compositions were prepared and submitted for XRD CrI measurement. Comparison of four calculation methods showed that Jayme-Knolle method, Segal peak height method and Ruland-Vonk method give similar tendency of monotonous increase of CrI with increase of cellulose content while deconvolution method does not seem to reflect the biomass crystallinity very accurately. Among the three methods which gave a reasonable CrI trend, Segal peak height method and Ruland-Vonk method seem to be more reliable and suitable in examining the sample crystallinity. In any of the examined calculation methodologies, the amorphose part of lignocellulose samples appears to be less represented in X-ray diffractgram compared to crystalline part, giving systematically larger CrI values than actual cellulose content expressed in wt%. There also was a strong indication that CrI depends on sample composition, most likely due to its elemental composition difference which may give different X-ray scattaring characteristics among the samples.

Acknowledgement

Support from the Ministry of Education Malaysia through HICoE award to Centre for Biofuel and Biochemical Research is duly acknowledged.

References

[1] M. Hall, P. Bansal, J.H. Lee, M.J. Realff, A.S. Bommarius, Cellulose crystallinity - a key predictor of the enzymatic hydrolysis rate, FEBS J. 277 (2010) 1571-1582. https://doi.org/10.1111/j.1742-4658.2010.07585.x

[2] I. Siró, D. Plackett, Microfibrillated cellulose and new nanocomposite materials: a review, Cellulose, 17 (2010) 459-494. https://doi.org/10.1007/s10570-010-9405-y

[3] U. Rofiqah, A. Safitri, Fadhilah, Study of delignification process and crystallinity index on lignocellulose components of corn cob in different pretreatments: a combination of pretreatment (ionic cholineacetate and NaOH) and NaOH pretreatment, IOP Conf. Ser.: Mater. Sci. Eng. 625 (2019) 012029. https://doi.org/10.1088/1757-899X/625/1/012029

[4] R. Abu Darim, A. Azizan, J. Salihon, Study of crystallinity index (CrI) of oil palm frond pretreatment using aqueous [EMIM][OAc] in a closed system, IOP Conf. Ser.: Mater. Sci. Eng. 358 (2018) 012007. https://doi.org/10.1088/1757-899X/358/1/012007

[5] A.A.N. Gunny, D. Arbain, P. Jamal, Effect of structural changes of lignocelluloses material upon pre-treatment using green solvents, AIP Conf. Proc. 1835 (2017) 020022. https://doi.org/10.1063/1.4981844

[6] Y. Pu, F. Hu, F. Huang, B.H. Davison, A.J. Ragauskas, Assessing the molecular structure basis for biomass recalcitrance during dilute acid and hydrothermal pretreatments, Biotechnol. Biofuels, 6 (2013) 15. https://doi.org/10.1186/1754-6834-6-15

[7] K. Karimi, M.J. Taherzadeh, A critical review of analytical methods in pretreatment of lignocelluloses: composition, imaging, and crystallinity, Bioresour. Technol. 200 (2016) 1008-1018. https://doi.org/10.1016/j.biortech.2015.11.022

[8] L. Segal, J.J. Creely, A.E. Martin Jr., C.M. Conrad, An empirical method for estimating the degree of crystallinity of native cellulose using the x-ray diffractometer, Text. Res. J. 29 (1959) 786-794. https://doi.org/10.1177/004051755902901003

[9] G. Jayme, H. Knolle, Introduction into empirical X-ray determination of crystallinity of cellulose materials, Das Papier, 18 (1964) 249-255.

[10] W. Ruland, X-ray determination of crystallinity and diffuse disorder scattering, Acta Crystallogr. 14 (1961) 1180-1185. https://doi.org/10.1107/S0365110X61003429

[11] C.G. Vonk, Computerization of Ruland's X-ray method for determination of the crystallinity in polymers, J. Appl. Crystallogr. 6 (1973) 148-152. https://doi.org/10.1107/S0021889873008332

[12] R. Ibbett, D. Domvoglou, D.A.S. Phillips, The hydrolysis and recrystallisation of lyocell and comparative cellulosic fibres in solutions of mineral acid, Cellulose, 15 (2008) 241-254. https://doi.org/10.1007/s10570-007-9157-5

[13] S. Park, J.O. Baker, M.E. Himmel, P.A. Parilla, D.K. Johnson, Cellulose crystallinity index: measurement techniques and their impact on interpreting cellulase performance, Biotechnol. Biofuels, 3 (2010) 10. https://doi.org/10.1186/1754-6834-3-10

[14] N. Terinte, R. Ibbett, K.C. Schuster, Overview on native cellulose and microcrystalline cellulose I structure studied by X-ray diffraction (WAXD): comparison between measurement techniques, Lenzinger Ber. 89 (2011) 118-131.

[15] H. Yang, R. Yan, H. Chen, C. Zheng, D.H. Lee, D.T. Liang, In-depth investigation of biomass pyrolysis based on three major components: hemicellulose, cellulose and lignin, Ener. Fuels, 20 (2006) 388-393. https://doi.org/10.1021/ef0580117

Human reliability assessment for tank overfilling incident utilizing minimized human performance shaping factors

ASHER AHMED Malik[1,a], RISZA Rusli[1,b*], FATIN AFIFAH Binti Mohd Nasir[1,b], SALMAN Nazir[2,c] and RIZAL HARRIS Wong[3,d]

[1]Center of Advanced Process Safety (CAPS), Chemical Engineering Department, Universiti Teknologi PETRONAS, Seri Iskandar, Perak Malaysia

[2]Training and Assessment Research Group, Department of Maritime Technology and Innovation,University of South-Eastern Norway, Postboks 4, 3199 Borre, Norway

[3]Principal Engineer, Process Safety Management (PSM) of Group Technical Solutions (GTS), PETRONAS Malaysia

[a]asher_19000349@utp.edu.my, [b]risza.rusli@utp.edu.my, [c]fatin_17004913@utp.edu.my, [d]salman.nazir@usn.no, [e]rizalharris@petronas.com

Keywords: Human Reliability Assessment, Quantitative Risk Analysis, Bayesian Networks, Reliability, Safety

Abstract. Human errors are identified as significant contributors to process industry accidents. Human reliability analysis (HRA) has been conducted in previous studies to improve human performance in several industrial operations. However, human error predictions are greatly influenced by various performance-shaping factors (PSFs). Research also demonstrates that PSFs are interdependent, which thereby complicates the modeling and analysis. Therefore, this study performs HRA, for a tank overfilling accident scenario that resulted due to human failure.Fewer independent PSFs through careful classification were used to estimate tank overfilling probability resulting from different human-triggered factors. For HRA, this study uses a combination of the Standardized Plant Analysis Risk Human Reliability Analysis (SPAR-H) andBayesian Belief Network (BBN). The failure probability distributions of individual interconnecting tasks were calculated using SPAR-H, and the probabilistic interdependence of each task to the final tank overfilling scenario was modeled using a BBN. From the current analysis, divergent stream identification is determined as the key to lead tank overfilling with 40% probability. This study concludes that BBN can be reliably employed in the Quantitative Risk Analysis (QRA) framework to examine human factors in industrial failure probability estimation for various other human-related industrial accident scenarios.

Introduction

The industrial sector has grown to meet global demand. Chemical process industries (CPIs) must use sophisticated techniques that incorporate human-machine interactions (HMIs) [1]. Despite regulatory action, there has been a continuous increase in accidents at CPIs. Moreover, more than 80% of such accidents in the process industries directly or indirectly correlate to human factors [2]. Legislative bodies are concerned about risk minimization in CPIs and avoid enormous financial and environmental losses [3].

Due to the importance of human factors in safety management, many industries have been conducting human reliability analysis (HRA) [4]. Several HRA approaches have been developed for chemical process industries, including structured probabilistic risk assessment and cognitive theory of control. Fault Tree Analysis (FTA) and Fuzzy Bayesian Network (FBN) are utilized to investigate human factor contribution in flammable storage tanks [5,6]. Standardized Plant Analysis Risk-HRA (SPAR-H) was used to study probabilistic human error in storage and transportation where SPAR-H proved as a simplified method to represent human behavior [6,7]. Human error in Malaysian oil and gas sector has recorded considerable accident rate with 35% as of 2009 (Ministry of Human Resource) [8]. Due to increased oil and gas operations, the risk of accident occurring could further increase [9]. Nonetheless, with the exception of Oktarinandaand Norazahar[10], and Alaw et al.[11], only a few studies have attempted to estimate the probability of human error in the oil and gas sector while taking Malaysia's demographics into account. Hence, there is still a need to accurately identify the human factors, work culture and their interrelationships associated with a chemical plant operation in Malaysia.

QRA is a standardized tool for industrial plant accident analysis and risk mitigation [12]. The current QRA framework also makes use of advanced Computational Fluid Dynamics (CFD) for consequential analysis together with probabilistic modelling [13,14]. Even though QRA is not an exact reflection of reality, it is currently the most widely utilized tool for analyzing the hazardsof complex processes and storage facilities [15]. Although QRA is a consolidated tool, the inclusion of the human factor in its framework is still trivial [16,17].

Humans have contributed to past chemical process industry accidents [18]. The QRA studies in industries could merge in the state-of-the-art human reliability analysis methodologies to quantitatively assess process industries accidents more comprehensively [19]. In this regard, Bayesian Networks (BNs) are effective for analyzing random variable connections, displaying a domain's probabilistic model, and deducing causal probabilities for scenarios given data.Bayesian modeling in QRA framework has been applied in marine research [20–25] and fire- induced domino effect probabilistic modeling [26–31]. Nevertheless, HRA in onshore process industries using Bayesian networking has received less attention. SPAR-H is another second- generation HRA approach that, unlike earlier human error probability (HEP) methods, uses generalizations to represent the spectrum of human behavior. Furthermore, human performanceis influenced by PSF, and their inter-relationship study has only been conducted in regard to the nuclear power plants [32]. Hence, there is also a need to establish an approach to identify the keycorrelating PSFs in oil and gas applications.

This research investigates human error probability for storage tank overfilling in Malaysian process facility demographics and operator culture. Chemical mishaps are common at storage and distribution facilities and the fuel level monitoring is performed by operators and thus necessitates the human reliability analysis [9]. Previous studies on HRA reckon that integratingSPAR-H with BN could deliver enough accuracy with a limited dataset. Hence, in this study, the SPAR-H technique is used to quantify prior human error probability by incorporating modified PSFs (time, stress/stressors, and complexity) and then the Bayesian Belief Network (BBN) model is built to integrate human factors in QRA. The model is validated against a published overfilling storage tank case study.

Methodology

This research aims to improve the QRA framework by including human factors. Figure 1 shows the study's process flow.

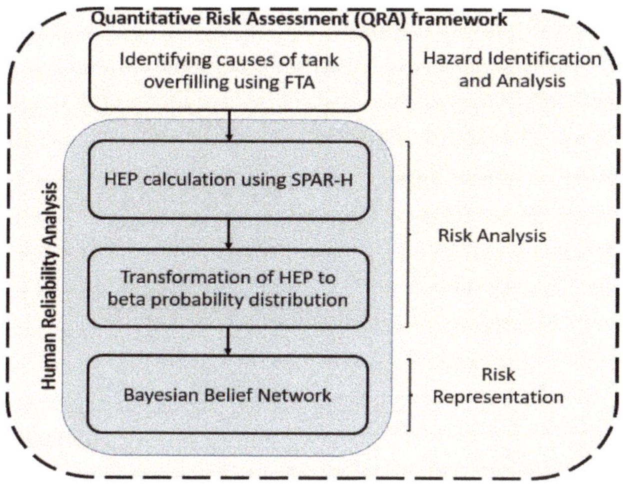

Fig. 1. Process flow of HRA inclusion in QRA framework developed in this study.

Hazard identification:
QRA and risk management frameworks include hazard identification to recognize the causesof any incident. Most of the common hazard identification techniques used in process industriessuch as Hazard and Operability Study (HAZOP), What-If Analysis, Preliminary Hazard Analysis, etc. are qualitative based as summarized by [31]. However, FTA provides a qualitativeand quantitative estimate of hazard and a logical, quantified description of unwanted events, including basic events from human factors as one of the causes of failure to the top event. Therefore, this study adopts the fault tree analysis from Steijn et al. [7] for the estimation of modified human error probability as initial basis or input to BBN model.

Human error probability calculation using SPAR-H:
Studies show that integrating SPAR-H method with BBN delivers enough accuracy with a limited dataset. Based on earlier research, BBN-SPAR-H has the simplest structure and is the most likely method to be included in this extended QRA framework. SPAR-H classifies PSFs that affect human reliability in process industry operations. As illustrated in figure 2, the existingSPAR-H has eight PSFs and can't effectively manage dependencies [32,33]. Liu and Li [34] suggested to deduce a SPAR-H model that comprises of independent set of well-defined PSFs toreduce the overall model complexity. In this respect and relating to the tank overfilling scenario,the present study carefully considers three key independent PSFs, namely, available time, processcomplexity, and mental stress/stressors as suggested by Park et al. [32]. The consensus was reached through careful evaluation by an experienced team of process safety experts and operators that have experienced in handling the same tasks as in this case study.

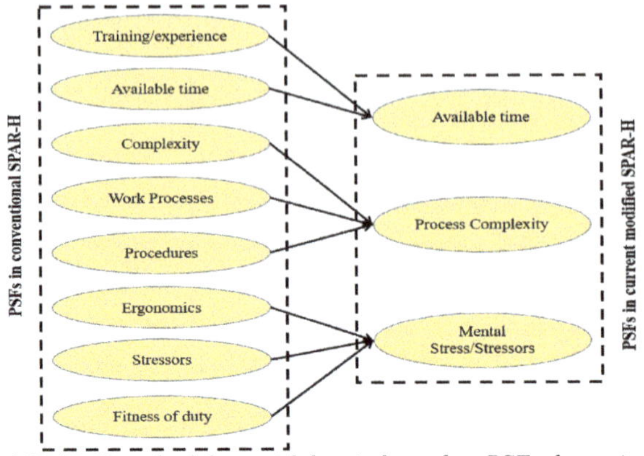

Fig. 2. Modified SPAR-H methodology with less independent PSFs determined in this study.

Expert judgments are needed to identify the appropriate multipliers for each PSF depending on three conditions: worst-case scenario, modal-case scenario, and best-case scenario. Further details on the PSF multiplier definition can be found elsewhere [35]. Multiplier ranges for currentstudy are listed in Table 1.

Table 1. Multiplier Ranges for PSFs considered in current research [35,36].

PSF	Multiplier range
Available time	0.01-10 (less available time corresponds to greater multiplier)
Mental Stress and stressors	1-5 (high stress corresponds to greater multiplier)
Process Complexity	1-5 (high complexity corresponds to greater multiplier)

The corresponding multipliers of each independent PSF are used to estimate the overall Human Error Probability (HEP) given by equation 1[36,37].

$$HEP = HEP_{nominal} \prod_{1}^{3} PSF_i i = 1,2,3, \ldots \tag{1}$$

'$HEP_{nominal}$' for diagnosis task is 0.01 while for action task is 0.001 [38]. Diagnosis relates to cognition, which includes everything from analyzing information to making decisions whereas pressing a button etc. is referred to as action task. If the calculated product of multipliers and the nominal human error probability yield an illogical value of higher than 1, the adjusted HEP needsto be used as shown in Equation 2 [36,37].

$$HEP_{adjusted} = \frac{HEP_{nominal} \Pi_1^3 PSF_i}{HEP_{nominal} (\Pi_1^3 PSF_i - 1) + 1} \tag{2}$$

Single-point HEP transformation to Beta distribution. Uncertainty may develop in the conventional method of calculating human error due to varied thinking styles. The multipliers are necessary because the numerous point inputs must be translated to the Beta distribution (probability distribution), which can decrease the gap by allowing the analyst to express their uncertainty across a large range. This research calculates the beta probability distribution's standard deviation and mean using the three-point estimate approach (equations 3 and 4) [39,40].

$$\text{Mean } (\mu) = \frac{\min + (4\text{mod}) + \max}{6} \tag{3}$$

$$\text{Std, dev } (\sigma) = \frac{(\max - \max)}{6} \tag{4}$$

The lowest realistic case is called the min (best case scenario), normal operation is called the mod (modal case scenario), and the highest realistic case is called the max (worst case scenario). The min number indicates the quickest time to complete a task and vice versa. In this work, an operator with 20 years of experience at one of Malaysia's multinational oil and gas firms determined failure probabilities through expert judgment. The formula to obtain the beta probability distribution is given in equation 5 [7].

$$\text{P(x)} = \frac{(\text{x}-\text{p})^{\alpha-1}(\text{q}-\text{x})^{\beta-1}}{\text{B}(\alpha,\beta)(\text{q}-\text{p})^{\alpha+\beta+1}}; \alpha = \left(\frac{1-\mu}{\sigma^2}-\frac{1}{\mu}\right)\mu^2, \beta = \alpha\left(\frac{1}{\mu}-1\right) \tag{5}$$

The α and β represent the number of success (α) and failures (β) that have been observed.

Bayesian Belief Network (BBN):
Computing resources have progressed sufficiently to allow probabilistic modeling using Bayesian networks a viable option. Furthermore, BBN modeling is gaining popularity in severalareas of industrial risk management [41] using GeNIe Modeler [42]. Beta Distribution providesnetwork prior probability distributions. The continuous beta distribution model is converted to discrete data. GeNIe Modeler [42] uses Monte-Carlo sampling to generate random variables. This stochastic model will generate random inputs inside the defined boundary. Each run buildsconditional probability. The model has 1,000 samples and discretization. Current scenario statesfail or not fail for conditional probability. The model's maximum and lowest boundaries are basedon each node's starting value. To limit node sensitivity, BBN should have fewer than five layers,but this relies on the topology since asymmetrical intermediary nodes may be deliberate [43]. BBN outputs failure probabilities for each Failure Tree gate. The software employs Bayes Theorem propagation, which incorporates prior node beliefs. The failure probability of storage tank overfilling is predicted with this knowledge.

Modelled Scenario:
Pumping activities from a ship to two land-based atmospheric storage tanks are included in the scenario. Offloaded fuel exceeded the initial (empty) target Tank X's capacity. To compensate, flow is transferred to Tank Y while the ship's pumps remain operating. This resultedin tank overfilling caused by human error.

Results and discussions
The nodes representation and the interdependency between nodes follow the already established fault tree network from the study of Steijn et al. [7]. The well-established fault tree for tank overfilling scenario is shown in figure 3 is then translated into BBN for probabilistic assessmentas shown in figure 4.

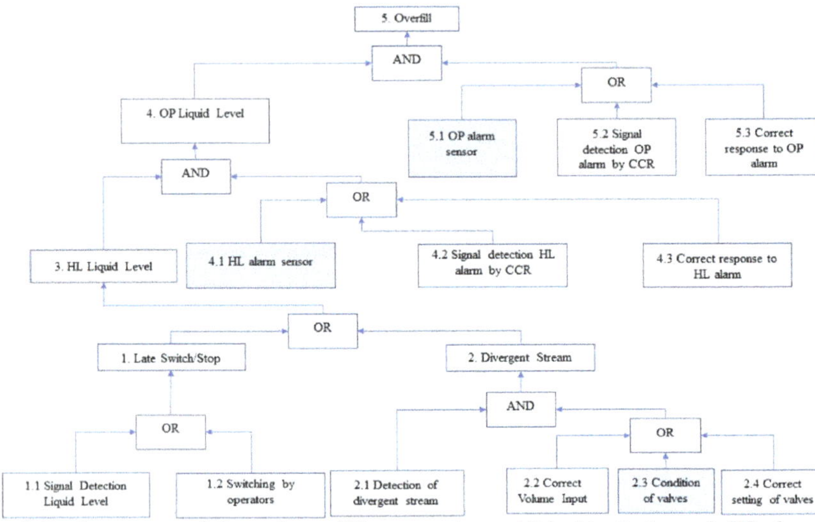

Fig. 3. Fault tree network for tank overfilling scenario established by Steijn et al. [7] where, HL =High Level, OP = Overfill Protection, CCR= Central Control Room.

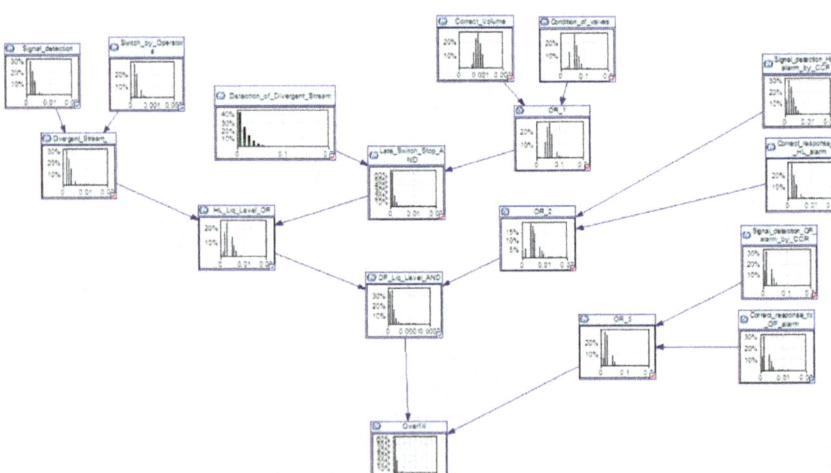

Fig. 4. Probabilistic BBN model for tank overfilling scenario developed based on current PSFs.

Figure 4 of the BBN model shows a 60% possibility of tank overfilling, with Overfill Protection (OP) liquid level failure being more likely (40%) than OP alarm detection and reactionfailure. This is because the high-level liquid valve requires operator intervention to stop the pumpand thus possibly be mishandled. OP liquid level is highly influenced by the failure ofhigh liquidlevel, which is caused by late switch operation (60%) and divergent stream detection failure (40%chance). The reason in this scenario could be that both processes (switch operation and divergent stream detection) are being performed by a human operator in control room, and any delay in valve operation or tank liquid detection leads to undesired overfilling scenario. The operators may not provide continuous monitoring in such operations due to eyestrain and the task is to justswitch the tank back or maintain routing to the tank if it is empty after long durations. The use of autonomous switches could provide on-time response to switching and divergent detection totimely avoid the overfilling without operator straining. The BBN model identified the major factors (nodes) likely to contribute to tank overfilling. This study used fewer separate PSFs thanearlier investigations, which improved failure probability prediction. This discovery lowered thecomplexity of estimating human error probability. The BBN model quantifies the interdependence of several contributing components.

Table 2 shows comparative analysis where Steijn et al. [7] random value distribution resultedin more scattered data than the existing questionnaire datasets. This may be because Steijn et al. [7] had a greater standard deviation. However, the difference in the mean failure probability is only 6% and a lower standard deviation of this study is more reliable because the data is closer to the mean. Moreover, the probability data skewness in present work is less which corresponds to greater accuracy and thus the present work appears to be of greater significance in terms of simplicity and accuracy. Overall, the tank overfilling probability is low because failure probability didn't surpass 1×10^{-3} [44].

Table 2: Comparison of the statistical parameters in failure probability distribution against published work.

Data Source	Mean	Standard Deviation	Skewness
Steijin et al. [7]	4.9e-07	8.8e-04	3.86
Current work	4.6e-07	6.2e-04	3.63

Since the method involves elicitation of prior probability from the expert judgment, the data collection from virtual simulation can also further improve the reliability of the data as it is closely imitating the real digital HMI interface environment. The scarcity of empirical data alongwith uncertainty and variability creates further challenges in the human failures' quantification. Therefore, assessing and identifying suitable prediction tools with limited dataset is essentially required for future research to overcome the current limitation of data availability for human error probability estimation in process industry.

Conclusions

This study reduces model complexity by limiting the number of PSFs from eight to three factors. Instead of a subjective point probability assessment, the SPAR-H point probability is transformed into a beta probability distribution to represent the spectrum of human error probability. This reduces the uncertainty element in the prediction of human error probability. It was discovered that the current model with fewer independent PSFs conforms to the one with all eight PSFs involved. A BBN was developed using a previous study's tank overfilling scenario. The causal components (nodes) of BBN were determined using fault-tree analysis from past research. The risk of tank overfilling was found to be substantially dependent on the fundamental cause of failure in divergent stream identification by operator. According to the study, BN may be utilized effectively in QRA to examine the dependence of human-related factors on the probability of failure associated to the outcome. There is still potential in training of non-technical staff and development of novel assessment methods to further exploit the performance shaping factors as necessitated in high-risk systems. Furthermore, diverse research analyses are plausible in terms of cognitive load of operators in complex dynamic work environments using advancing techniques such as machine learning and Artificial Intelligence (AI) and CFD.

Acknowledgment
The authors greatly acknowledge inputs from various process safety professionals that were involvedin the data collections of the case study. The authors also acknowledge the funding and support provided by the Centre of Advanced Process Safety (CAPS), Universiti Teknologi PETRONAS,Malaysia and the European Union's Horizon 2020 research and innovation program under the Marie Skłodowska-Curie grant agreement No 823904 (Project: ENHANCing Human Performance in Complex Socio-Technical SystEms, ENHANCE).

References

[1] C. Sharma, P. Bhavsar, B. Srinivasan, R. Srinivasan, Eye gaze movement studies of control room operators: A novel approach to improve process safety, Comput. Chem. Eng. 85 (2016) 43–57. https://doi.org/10.1016/j.compchemeng.2015.09.012

[2] E. Zarei, F. Khan, R. Abbassi, Importance of human reliability in process operation: A critical analysis, Reliab.Eng. Syst. Saf. 211 (2021) 107607. https://doi.org/10.1016/j.ress.2021.107607

[3] Directive 2012/18/EU. European Parliament and Council Directive 2012/18/EU of July 2012 on Control of Major-Accident Hazards Involving Dangerous Substances, Amending and Subsequently Repealing Council Directive 96/82/EC., Off. J. Eur. Communities L 197/1, Brussels, 24.7.2012. (n.d.).

[4] M. Musharraf, J. Smith, F. Khan, B. Veitch, S. MacKinnon, Incorporating individual differences in human reliability analysis: An extension to the virtual experimental technique, Saf. Sci. 107 (2018) 216–223.https://doi.org/10.1016/j.ssci.2017.07.010

[5] S.T. Ung, Evaluation of human error contribution to oil tanker collision using fault tree analysis and modified fuzzy Bayesian Network based CREAM, Ocean Eng. 179 (2019) 159–172.https://doi.org/10.1016/j.oceaneng.2019.03.031

[6] F. Laal, M. Pouyakian, M.J. Jafari, F. Nourai, A.A. Hosseini, A.R. Khanteymoori, Technical, human, and organizational factors affecting failures of firefighting systems (FSs) of atmospheric storage tanks: Providing a risk assessment approach using Fuzzy Bayesian Network (FBN) and content validity indicators, J. Loss Prev.Process Ind. (2020).

[7] W.M.P. Steijn, J.N. Van Kampen, D. Van der Beek, J. Groeneweg, P.H.A.J.M. Van Gelder, An integration of human factors into quantitative risk analysis using Bayesian Belief Networks towards developing a 'QRA+,' Saf. Sci. 122 (2020) 104514. https://doi.org/10.1016/j.ssci.2019.104514

[8] U. Alkhaldi, M, Pathirage, C and Kulatunga, The role of human error in accidents within oil and gas industry in Bahrain, Usir. / (2017) 821–834.

[9] M.H. Wood, L. Fabbri, Challenges and opportunities for assessing global progress in reducing chemical accident risks, Prog. Disaster Sci. 4 (2019) 100044. https://doi.org/10.1016/j.pdisas.2019.100044

[10] R. Oktarinanda, N. Norazahar, Bayesian Analysis for Assessing Risks of Rotating Equipment and Its Financial Loss, IOP Conf. Ser. Mater. Sci. Eng. 808 (2020). https://doi.org/10.1088/1757-899X/808/1/012015

[11] F.A. Alaw, N. Sulaiman, H. Tan, Incorporation of human factors in risk analysis of oil and gas pipeline using Bayesian Network, (2018) 1–9.

[12] P. Swuste, W. Zwaard, J. Groeneweg, F. Guldenmund, Safety professionals in the Netherlands, Saf. Sci. 114 (2019) 79–88. https://doi.org/10.1016/j.ssci.2018.12.015

[13] R. Yang, F. Khan, E.T. Neto, R. Rusli, J. Ji, Could pool fire alone cause a domino effect?, Reliab. Eng. Syst. Saf. 202 (2020) 106976. https://doi.org/10.1016/j.ress.2020.106976
[14] A.A. Malik, M.S. Nasif, A.A. Mokhtar, M.Z. Mohd Tohir, Numerical investigation of the effect of weather conditions on the escalation and propagation of fire-induced domino effect, Process Saf. Prog. (2021) 1–13. https://doi.org/10.1002/prs.12251
[15] S. Mannan, Lee's Loss Prevention in the Process Industries, 2005.
[16] L. Torres, O.P. Yadav, E. Khan, A review on risk assessment techniques for hydraulic fracturing water and produced water management implemented in onshore unconventional oil and gas production, Sci. Total Environ. 539 (2016) 478–493. https://doi.org/10.1016/j.scitotenv.2015.09.030
[17] X. Zhen, J.E. Vinnem, C. Peng, Y. Huang, Quantitative risk modelling of maintenance work on major offshore process equipment, J. Loss Prev. Process Ind. 56 (2018) 430–443. https://doi.org/10.1016/j.jlp.2018.10.004
[18] R.D. Calvo Olivares, S.S. Rivera, J.E. Núñez Mc Leod, A novel qualitative prospective methodology to assess human error during accident sequences, Saf. Sci. 103 (2018) 137–152. https://doi.org/10.1016/j.ssci.2017.10.023
[19] W.M.P. Steijn, J. Groeneweg, F.A. van der Beek, J. van Kampen, P. van Gelder, An integration of human factors into quantitative risk analysis: A proof of principle, Saf. Reliab. - Theory Appl. - Proc. 27th Eur. Saf. Reliab. Conf. ESREL 2017. (2017) 321–328. https://doi.org/10.1201/9781315210469-43
[20] H. Rozuhan, M. Muhammad, U. Muhammad, Probabilistic risk assessment of o ff shore installation hydrocarbon releases leading to fire and explosion, incorporating system and human reliability analysis, Appl. Ocean Res. 101 (2020) 102282. https://doi.org/10.1016/j.apor.2020.102282
[21] J.E. Vinnem, R. Bye, B.A. Gran, T. Kongsvik, O.M. Nyheim, E.H. Okstad, J. Seljelid, J. Vatn, Journal of Loss Prevention in the Process Industries Risk modelling of maintenance work on major process equipment on offshore petroleum installations, J. Loss Prev. Process Ind. 25 (2012) 274–292. https://doi.org/10.1016/j.jlp.2011.11.001
[22] X. Zhen, J. Erik, C. Peng, Y. Huang, Quantitative risk modelling of maintenance work on major offshore process equipment, 4794207768 (1990) 1–27.
[23] B.A. Gran, R. Bye, O.M. Nyheim, E.H. Okstad, J. Seljelid, S. Sklet, J. Vatn, J.E. Vinnem, Journal of Loss Prevention in the Process Industries Evaluation of the Risk OMT model for maintenance work on major offshore process equipment, J. Loss Prev. Process Ind. 25 (2012) 582–593. https://doi.org/10.1016/j.jlp.2012.01.001
[24] M. Hänninen, O.A. Valdez, P. Kujala, Expert Systems with Applications Bayesian network model of maritime safety management, 41 (2014) 7837–7846. https://doi.org/10.1016/j.eswa.2014.06.029
[25] N. Khakzad, F. Khan, P. Amyotte, Safety Scien ce Quantitative risk analysis of offshore drilling operations : A Bayesian approach, Saf. Sci. 57 (2013) 108–117. https://doi.org/10.1016/j.ssci.2013.01.022
[26] N. Khakzad, G. Reniers, Cost-effective allocation of safety measures in chemical plants w.r.t land-use planning, Saf. Sci. 97 (2017) 2–9. https://doi.org/10.1016/j.ssci.2015.10.010
[27] N. Khakzad, G. Reniers, Risk-based design of process plants with regard to domino effects and land use planning, J. Hazard. Mater. 299 (2015). https://doi.org/10.1016/j.jhazmat.2015.06.020

[28] N. Khakzad, P. Amyotre, F. Khan, Valerio Cozzani, Domino Effect Analysis Using Bayesian Networks, Risk Anal. 33 (2013) 292–306. https://doi.org/10.1111/j.1539-6924.2012.01854.x
[29] N. Khakzad, F. Khan, P. Amyotte, V. Cozzani, Risk Management of Domino Effects Considering Dynamic Consequence Analysis, Risk Anal. 34 (2014) 1128–1138. https://doi.org/10.1111/risa.12158
[30] N. Khakzad, Modeling wildfire spread in wildland-industrial interfaces using dynamic Bayesian network, Reliab. Eng. Syst. Saf. 189 (2019) 165–176. https://doi.org/10.1016/j.ress.2019.04.006
[31] N. Khakzad, A Tutorial on Fire Domino Effect Modeling Using Bayesian Networks, Modelling. 2 (2021) 240–258. https://doi.org/10.3390/modelling2020013
[32] J. Park, W. Jung, J. Kim, Inter-relationships between performance shaping factors for human reliability analysis of nuclear power plants, Nucl. Eng. Technol. 52 (2020) 87–100. https://doi.org/10.1016/j.net.2019.07.004
[33] R.L. Boring, SPAR-H Step-by-Step Guidance, (2015).
[34] P. Liu, Z. Li, Human Error Data Collection and Comparison with Predictions by SPAR-H, Risk Anal. 34 (2014)1706–1719. https://doi.org/10.1111/risa.12199
[35] D.I. Gertman, BlackmanH.S., J. Byers, L. Haney, C. Smith, J. Marble, NUREG/CR-6883-The SPAR-H method-US Nuclear regulatory commision, Washington D.C., (2005).
[36] K.M. Groth, L.P. Swiler, Bridging the gap between HRA research and HRA practice: A Bayesian network version of SPAR-H, Reliab. Eng. Syst. Saf. 115 (2013) 33–42. https://doi.org/10.1016/j.ress.2013.02.015
[37] B. Hallbert, A. Whaley, R. Boring, P. McCabe, Y. Chang, O. of N.R.R. U.S. Nuclear Regulatory Commission, Human Event Repository and Analysis (HERA): The HERA Coding Manual and Quality Assurance (NUREG/CR-6903, Vol. 2), U.S. Nucl. Regul. Comm. Off. Nucl. Regul. Res. Washington, DC 20555-0001. 2 (2007) 133.
[38] M. Mirzaei Aliabadi, R. Esmaeili, I. Mohammadfam, M. Ashrafi, Human Reliability Analysis (HRA) Using Standardized Plant Analysis Risk-Human (SPAR-H) and Bayesian Network (BN) for Pipeline Inspection Gauges (PIG) Operation: A Case Study in a Gas Transmission Plant, Heal. Scope. 8 (2019). https://doi.org/10.5812/jhealthscope.87148
[39] J.R. Van Dorp, S. Kotz, A novel extension of the triangular distribution and its parameter estimation, J. R. Stat. Soc. Ser. D Stat. 51 (2002) 63–79. https://doi.org/10.1111/1467-9884.00299
[40] W.E. Stein, M.F. Keblis, A new method to simulate the triangular distribution, Math. Comput. Model. 49 (2009) 1143–1147. https://doi.org/10.1016/j.mcm.2008.06.013
[41] P. Weber, G. Medina-Oliva, C. Simon, B. Iung, Overview on Bayesian networks applications for dependability, risk analysis and maintenance areas, Eng. Appl. Artif. Intell. 25 (2012) 671–682. https://doi.org/10.1016/j.engappai.2010.06.002
[42] GeNie Modeler- BayesFusion, (2020). https://www.bayesfusion.com/2020/04/24/genie-2-5-released/
[43] S.H. Chen, C.A. Pollino, Environmental Modelling & Software Good practice in Bayesian network modelling, Environ. Model. Softw. 37 (2012) 134–145. https://doi.org/10.1016/j.envsoft.2012.03.012
[44] G. Song, F. Khan, M. Yang, Integrated risk management of hazardous processing facilities, Process Saf. Prog. 38 (2019) 42–51. https://doi.org/10.1002/prs.11978

Materials Research Proceedings 29 (2023) 145-151 https://doi.org/10.21741/9781644902516-18

Study of damage and repair of flax/Elium biocomposites under dynamic loading

RACHA Manaii[1,2], LAURENT Guillaumat[1*], SVETLANA Terekhina[1], DAVY Duriatti[2]

[1]LAMPA, ENSAM, esplanade des Arts et Metiers, 49100 Angers, France

[2]Depestele, 5 rue de l'église, Bourguébus 14540, France

*laurent.guillaumat@ensam.eu

Keywords: Biocomposites, Flax Fibres, Thermoplastic Matrix, Repair, Dynamic Behavior, Damage

Abstract. Laminated composites based on the new thermoplastic Elium 188 from Arkema and woven flax fibers has been manufactured by the infusion process at room temperature, still yet reserved to thermoset-based composites. The low velocity impact behaviour has been investigated between energy 7 to 22J for two stacking sequences $[0/90]_6$ and $[\pm 45]_6$.Repair after impact has been investigated too. The impact resistance was established by measuring load, deflexion, absorbed energy, contact duration and damage. Visual, high-speed images and microscopic observations were performed on impacted samples to show the effect of incident energy on the damage extension in the composites, by revealing the fiber/matrix debonding as principal damage mode. 3 points bending tests were carried out to assess the residual impact performance. In addition, a simple thermo-compression damage repair process was applied to carry out the multiple impact/repair cycles on the impacted plates. A significant recovery of stiffness and maximum impact force up to 4th cycle at 4J of $[0/90]_6$ plate has been revealed, by highlighting the interesting potential of flax/Elium® bio-composite to its repair aptitude.

Introduction

A large number of composites using synthetic and non-synthetic fibres with different thermoset and thermoplastic matrices have been already investigated. But, thanks to their eco-friendly property and low cost production with interesting physical and mechanical properties natural fibres are gathering a growing scientific and economic interest [1].

Thus, some studies have been conducted over recent years to investigate the mechanical behaviour of natural fibre-based composites. Moreover, composite structures are loaded under static, cyclic and dynamic loadings during their life cycle. However, low Velocity Impact (LVI) is one of the most important loadings and it is difficult to set up these parameters because of its random nature. Understanding the behaviour of composite when subjected to LVI loading is necessary for designing structures as it is a common issue and damages are usually not apparent to the naked eye.

Finally, this study focuses on the LVI damage and the reparability of thermoplastic biocomposite plates. For this purpose, impact tests were carried out and post-impact bending tests as well. Moreover, to demonstrate repairability, two batches of impact tests were conducted with and without fibre breaks.

Material, manufacturing and samples

Material

The biocomposite used in this study is made of thermoplastic matrix with flax fibres. Fibres are available as woven twill flax fabric from the French company Depestele and with an area

Materials Research Proceedings 29 (2023) 145-151 https://doi.org/10.21741/9781644902516-18

density of 360 g/cm^2. The matrix is a liquid thermoplastic matrix named Elium®188, recently formulated by Arkema company.

Plates, of dimensions 1000×750 mm^2 with a thickness between 4.5 and 4.8 mm, were manufactured by CMP Company using Vacuum Resin Infusion (VRI) (figure 1). This latter is usable although it used thermoplastic matrix because of the very low resin viscosity.

Two stacking sequences were considered for this study: $[0/90]_6$ and $[\pm 45]_6$.

Fig. 1. biocomposite plate

Impact samples

Samples with dimensions of 100 x 150 mm^2 were cut from the previous plates according to Airbus standard AITM 1-0010 [2]. However, a main difference in our study is in the boundary conditions as discussed below.

Two batches have been prepared according to both stacking sequences.

Repair method

An in-house thermocompression device (Figure 2) was designed in our laboratory to manufactured thermoplastic-based composites. But, in our case, it has been used to repair the impacted plates. For that we put on the impacted sample a coupon of a comingled fabric with flax and polypropylene fibres. The size of this latter was approximately the damaged area. Using a temperature of 200 °C and a very low pressure to avoid changing the dimensions and more especially the thickness the plates were repaired.

Samples were put in between the two aluminium trays and we just added an weight of 5 kg on the upper plate. 5 minutes was enough to make flat again the impacted samples.

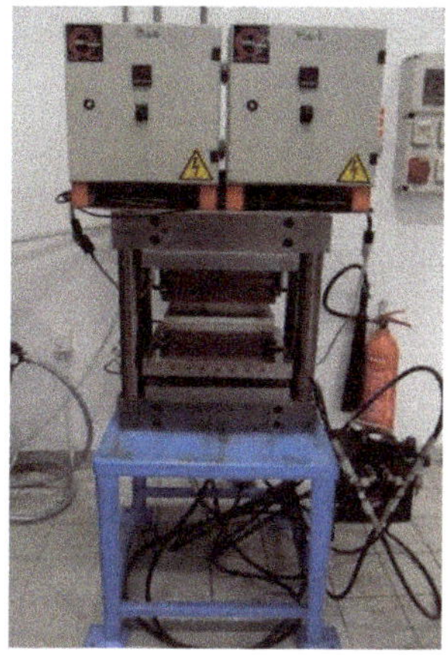

Fig. 2. Conventional hot press.

Devices

Drop tower

Low velocity impact (LVI) means that the velocity ranges from approximately 1 m/s to 10 m/s [3]. The falling weight impact setup, (so called drop tower), is the most powerfull device to reproduce low energy impacts.

Our drop tower (Figure 3) consists of two parts. The first one is the dropping mass and the second one is the boundary conditions. As mentioned previously, the boundary conditions used were not those recommended by the standard. We clamped the sample on two metallic supports closed to 3 points bending configuration.

Displacements (centre of the impacted plate and position of the striker) were measured using laser sensors and the force by a piezoelectric sensor.

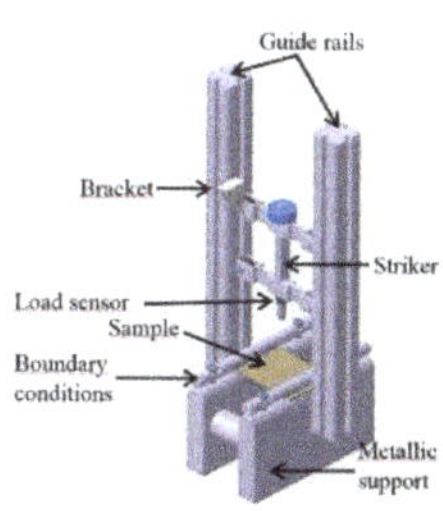

Fig. 3. Scheme of the Drop Tower.

The mass of the striker was around 1 kg and we used 3 heights: 1, 2 and 3 m. Moreover, one batch of samples with the stacking sequence $[0/90]_6$ was aged at 75%HR at room temperature.

Quasi-static tests

Quasi-static tests have been done to compare results to dynamical tests. To ensure the same experimental conditions in both cases, the plates were clamped with a similar support and the same hemispherical tip used for impact applied the quasi-static loading (Figure 4).

Fig. 4 Quasi-static device.

Finally, conventional 3 points bending tests were conducted to estimate the residual resistance. This test was chosen over the compression after impact (CAI) because in the thermoplastic- based composites delamination is not the main damage mode during an impact due to the ductility of the matrix.

Results and discussion

Post-impact visual inspections revealed an indentation on the impacted face which increase with the energy of impact for all impacted specimens even at 1 meter height.

The back face, as expected, exhibits a damage shape according to the stacking sequence (Figure 5).

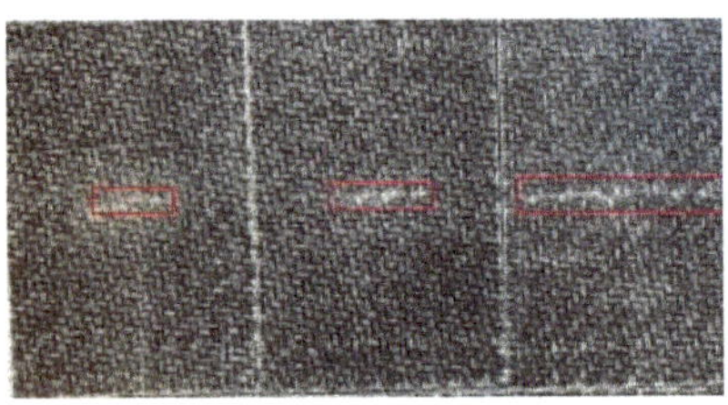

1 m 2 m 3 m
(a)

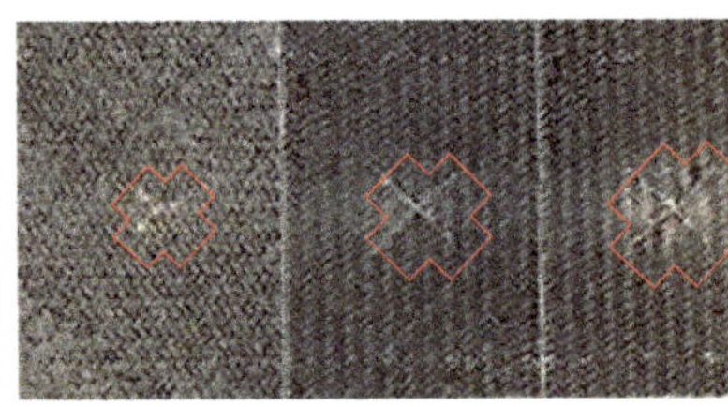

1 m 2 m 3 m
(b)

Fig. 5. Damage evolution from 1 to 3 m (a) $[0/90]_6$, (b) $[\pm 45]_6$.

Optical microscopy observations confirm that no large delamination occurred during the impact loading. It is why CAI is probably not the best test to estimate the residual properties. But we observed cracks inside the yarns or often at the border with the surrounding matrix (Figure 6).

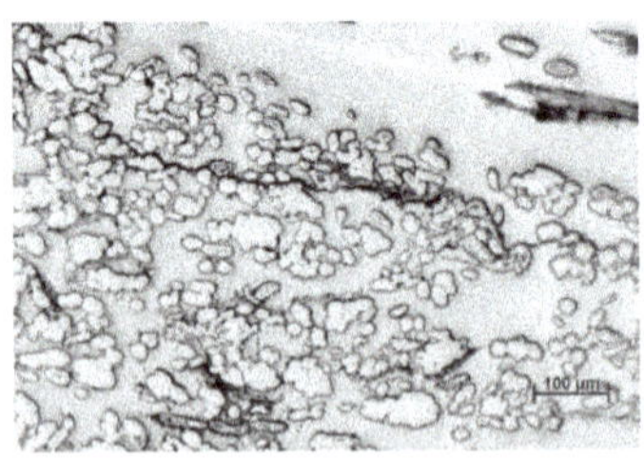

Fig 6 Crack in a yarn.

The force versus time curves for the 3 different energy levels for the 3 batches ($[0/90]_6$, $[\pm 45]_6$ and ([0/90]6 aged) are shown at Figure 7. Curves exhibit significant oscillations which can be mainly due to the vibration of the specimen according to its eigen modes [4].

It can be observed that, for both stacking sequences, the contact duration between the sample and the striker during impact increases slowly with respect to time. This phenomenon is due to the decrease of the stiffness because of the growth of the macro-crack located at the back face as illustrated in the figure 5. Thus, it is well known that the contact duration depends strongly on the striker mass and on the stiffness of the impacted structure. It can be noted that the maximum force increases slightly with the incoming energy too. The maximum force increases slightly too with the incoming energy as expected.

Post-impact bending tests were preferred to compression as discussed above to evaluate the residual properties. The results (Figure 8) for both orientations. Indicate that the performance of the stacking $[0/90]_6$ is much more sensitive to specimens $[\pm45]_6$ compared to the refence one. It is well known that he fibres angle has a strong influence on the mechanical responses and of course a angle of 0° increases strongly the young modulus but then these fibres break this latter decreases strongly too. This explanation is confirmed by the damage induced by the impact (figure 5). Effectively, we can observe that the crack for the $[0/90]_6$ is transverse to the span creating a strong effect on the sample stiffness contrary to the $[\pm45]_6$ for which the damage is more located at the centre of the specimen.

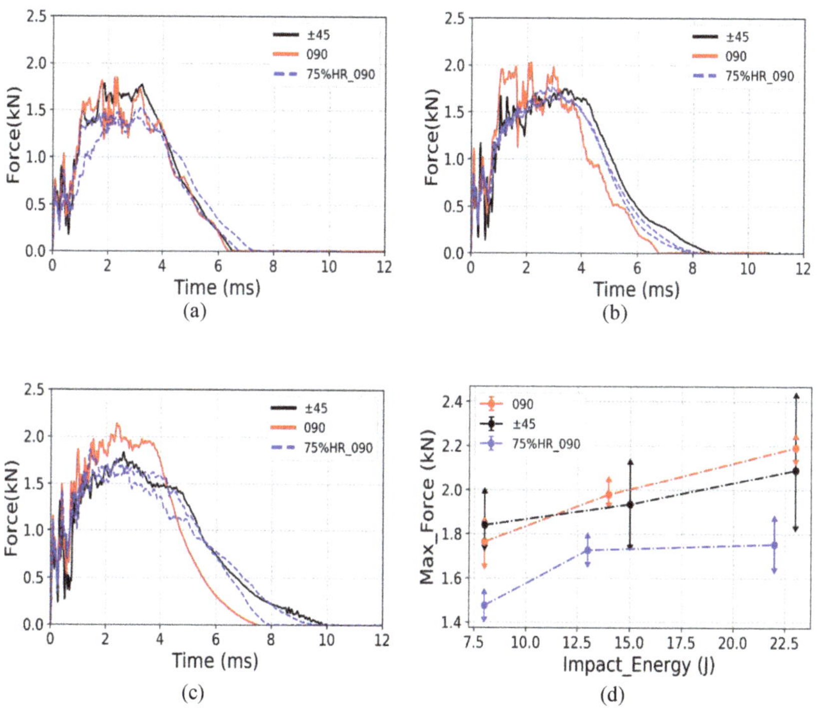

Figure 7. Load versus time during impact for around (a) 8 J, (b) 15 J, (c) 22J, (d) max force evolution versus impact energy.

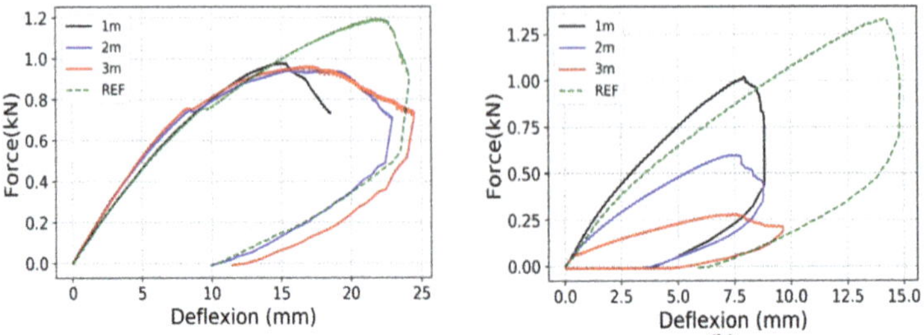

(a)
[±45]6

(b)
[0/90]6

Figure 8. Residual performance after impact compared with the quasi-static test (called REF).

Finally, several cycles impact – repair have been done on the [0/90]6 composite with two levels of the incoming energy (0,5 and 1 m). The main objective was to prepare samples with and without broken fibres on the back face but matrix cracks can exist.

The curves figure 9 show the evolution of the force versus time according to the number of cycle. It can be seen that in the case of the sample with broken fibres (a) the stiffness of the sample decrease with the number of cycles: the maximum force decreases and the contact duration increase. On the contrary, if the sample doesn't contain broken fibres (b) it can maintain the same stiffness even after several cycles impact-repairs (figure 9 – b). Moreover, we observed that the macro-cracks increase slightly with the number of cycles in both cases but the repair was sufficient to ensure constant mechanical properties in the case (b).

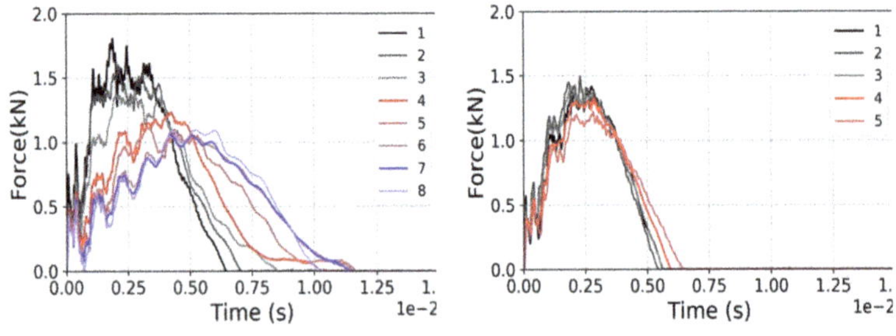

(a) 1m

(b) 0.5m

Fig. 9. Force vs Time curves for the multiple impact - repair cycle on [0/90]6 plate conducted at 1m (a) and 0.5m (b).

Conclusion

This paper dealed with behaviour of flax/Elium composites LVI loading. The experiments have to main objective to determine the behaviour of plates, to identify damages and the reparability of thermoplastic bio-composite plates. Visual observations allowed to determine the damage mechanisms which was mainly microcracks and plastic deformation. Quasi-static tests have been conducted and had showed that specimens exhibit, barely the same response excepted of oscillations representing the vibratory response in case of the impacted specimens. Bending tests have been performed to evaluate the residual strength after impact in place of compression because specimens did not develop large delamination. The repair of the plates was successful for the impacted plates without broken fibres. Repetitive impact and repair cycles have shown that the force and time of impact are a function of the stiffness.

References

[1] O. Faruk, A. Bledzki, H.-P. Fink, and M. Sain, "Biocomposites reinforced with natural fibres: 2000-2010," *Prog. Polym. Sci.*, vol. 37, pp. 1552–1596, 2012, doi: 10.1016/j.progpolymsci.2012.04.003

[2] AITM1-0010, "Airbus Test Method: Determination of Compression Strength after Impact," Blagnac, 2005.

[3] G. A. Schoeppner and S. Abrate, "Delamination threshold loads for low velocity impact on composite laminates," Compos. Part A Appl. Sci. Manuf., vol. 31, no. 9, pp. 903–915, 2000, doi: 10.1016/S1359-835X(00)00061-0

[4] S. Liang, L. Guillaumat, and P. B. Gning, "Impact behaviour of flax/epoxy composite plates," Int. J. Impact Eng., vol. 80, pp. 56 –64, 2015, doi: 10.1016/j.ijimpeng.2015.01.006

Effect of particle size of fly ash and solid to liquid ratio on microstructure and mechanical properties of geopolymer

AHMER Ali Siyal[1,a], RASHIDAH Mohamed Hamidi[1,b], RASHID Shamsuddin[1,c*], MUHAMMAD Moniruzzaman[2,b]

[1]HICoE, Centre for Biofuel and Biochemical Research (CBBR), Institute of Self-sustainable Building, Department of Chemical Engineering, Universiti Teknologi PETRONAS, 32610 Bandar Seri Iskandar, Perak Darul Ridzuan, Malaysia

[2]Center of Research in Ionic Liquids (CORIL), Institute of Contaminant Management,

Department of Chemical Engineering, Universiti Teknologi PETRONAS, 32610 Bandar Seri Iskandar, Perak Darul Ridzuan, Malaysia

[a]ahmer.ali_g03690@utp.edu.my, [b]rashidah_g02348@utp.edu.my, [c]mrashids@utp.edu.my, [d] m.moniruzzaman@utp.edu.my

Keywords: Fly Ash, Geopolymer, Microstructure, Mechanical Activation, Mechanical Strength

Abstract. This paper describes the effect of particle size of fly ash and solid to liquid ratio on the microstructure and flexural strength of fly ash geopolymer for coating applications. Sodium hydroxide concentration of 12 M was used for the preparation of geopolymers. The mechanically activated fly ash was divided into two parts based on the particle size (MFA A (11.81 μm) & MFA B (8.59 μm)). The decrease of particle size of fly ash and increase of S:L ratio produced dense and compact geopolymer which is due to the increase of reactivity of fly ash with the decrease of particle size and increase of water content. The decrease of particle size increased the flexural strength of geopolymer from 13.39 MPa (OFA-1) to 23.84 MPa (MFA B-III). The decrease of S:L ratio showed irregular trend with optimum flexural strengths of 13.39 MPa, 16.74 MPa, and 23.84 MPa obtained for OFA-1, MFA A-II, and MFA B-III respectively. The optimum flexural strength of 23.84 MPa was obtained. The mechanical activation is a useful technique to increase the reactivity of fly ash and produce dense and compact geopolymer with higher flexural strength.

Introduction

Geopolymers are inorganic polymer materials introduced by J. Davidovits in 1979 [1,2]. In geopolymer preparation, aluminosilicate materials provide silica and alumina, the major building blocks of geopolymer network which can be obtained naturally (kaolin, metakaolin) and from industrial byproducts (fly ash, slag and etc.) [3]. Alkaline activation by sodium hydroxide (NaOH) and sodium silicate (Na_2SiO_3) or mixture of the two dissolves the reactive species of silicate and alumina ions from aluminosilicate source materials via dissolution. Subsequently, polycondensation forms the aluminosilicate gel where the last stage involves rearrangement and reorganization of the network resulting in 3D geopolymer network comprising of tetrahedral SiO_4 and AlO_4 [1, 2].

Geopolymers are considered as green and sustainable materials due to low CO_2 production during their synthesis and they can be produced from various waste or byproduct materials such as coal fly ash, slag, and others. The excellent properties of geopolymers such as high early strength [12], resistance towards aggressive environment [13,14], stable at high temperature and long-term durability [15,16] makes it a prominent material in structural (geopolymer brick, cement, concreter etc.) and nonstructural applications (coating, soil stabilizer, radioactive and heavy metals waste treatment etc.). Geopolymer as a coating material has been used for metal plates and urea fertilizer [3, 4]. Investigation on the mechanical properties is crucial to evaluate the capability and performance of geopolymer in handling sudden impact and prolonged exposure to load. Flexural strength plays an important role in the coating application of geopolymers.

Many studies have been conducted on the increase of flexural strength of geopolymers using various techniques such as increase of NaOH concentration, reinforcement by reduced graphene oxide (rGO), wollastonite ($CaOSiO_2$), additively manufactured metallic rebars (Ti6Al4V), steel fibres, steel and polymer fibre, self-assembled nanofiber networks, and graphene oxide/geopolymer nanocomposite [1, 5-13]. However, no study has focused on the improvement of flexural strength of geopolymer by particle size reduction of fly ash through mechanical activation. Therefore, this study determines the effect of particle size of fly ash and different solid to liquid (S: L) ratios.

This paper describes the effect of particle size reduction of fly ash and variation of S:L ratios on flexural strength of fly ash geopolymer and microstructure. S:L ratio in the range of 3:1-2.2:1 was used to synthesize geopolymer.

Materials and Methods:

Materials: Class F fly ash obtained from a local coal thermal power plant located in the northern region of Malaysia was utilised for the synthesis of geopolymer. Chemical composition of fly ash is shown in Table 1. Sodium hydroxide (NaOH) and sodium silicate (Na_2SiO_3) were purchased from Merck in pellet form with 99% purity.

Table 1: Chemical Composition of fly ash [14]

Oxide	SiO_2	Al_2O_3	Fe_2O_3	CaO	K_2O	Na_2O	P_2O_5
Composition (wt%)	40.8	35.4	1.86	3.71	0.69	2.96	1.36

Methods

Mechanical activation: Original fly ash (OFA) was subjected to pre-treatment via mechanical activation which produced mechanically activated fly ash (MFA). High energy planetary ball mill (FITSCH, Pulverisette 5 classic line) was used for mechanical activation which consists of two milling slots of stainless-steel jar (250 ml capacity) and 5 mm diameter of stainless-steel balls as the grinding chamber and media respectively. The mill was run at 1300 W full rated power of the milling motor engine capacity and operating voltage of 200-220 V. Dry mechanical milling method was implemented where 150 g of grinding media was first loaded into each of the grinding chambers followed by the fly ash and subsequently closed with the lid that comes together with a Teflon seal then tightened and clamped with the safe locks. The hood of high planetary ball mill was closed and automatically locked as the process began. Milling operation was performed in accordance with the parameters; Ball-to-powder (BPR) weight ratio, residence time of mechanical activation, and revolution speed as summarized in Table 2. Throughout the process, impact and collision occurred in the chamber causing the heat to build up. Cooling interval was allowed to take place with 3 min pause for every 15 min of continuous milling to prevent overheating. When the grinding process completed, the ground fly ash was separated from the grinding media. Fine sand grinding at slow operational speed carried out at moderate time also helped in removing any remaining samples attached to both chambers and media. Later they were cleaned with water, dried, and kept in a dry place. Finally, mechanical activated fly ash (MFA) with particle size of interest was characterized and used in the preparation of geopolymer.

Table 2: Operational parameters of mechanical activation

Revolution speed (rpm)	Ball to powder ratio (BPR)	Residence time (min)
300	2:1, 3:1, 4:1, 5:1, 6:1	60

Synthesis of geopolymer

The 12 M alkaline solution was prepared by dissolving NaOH pellets in distilled water, the desired quantity of sodium silicate was also added into the alkaline solution. Fly ash and alkaline solution mixed in a plastic container using overhead mechanical mixer (IKA RW20 Digital) for 10 minutes. Geopolymer samples were prepared at varying S:L ratios as described in Table 3. Geopolymer slurry was put into the mould and cured at 80 °C for 24 hours in an electric oven.

Table 3: Composition of fly ash geopolymer samples

S: L ratio (by weight)	Code	Sample Codes		
		OFA (15.33 µm)	MFA A (11.81 µm)	MFA B (8.6 µm)
3:1	I	OFA-I	MFA A-I	MFA B-I
2.8:1	II	OFA-II	MFA A-II	MFA B-II
2.6:1	III	OFA-III	MFA A-III	MFA B-III
2.4:1	IV	OFA-IV	MFA A-IV	MFA B-IV
2.2:1	V	OFA-V	MFA A-V	MFA B-V

Microstructural characterization

Microstructural images of fly ash geopolymer samples were captured by scanning electron microscope (Phenom Pro X). The specimen was directly mounted onto the clamping device on the sample holder by adjusting the gap of 4 mm for the sample prior to analysis. Deposition of any conductive layer of metallic coating was not required as the sample holder was specifically designated as metallurgical charge reduction. Microscopic examination of the samples was done using an operating set up of 2.00 kV EHT acceleration voltage and magnification from 300 X to 5000 X.

Flexural strength of geopolymer

Flexural strength of prepared geopolymer samples was investigated using Universal Testing Machine (UTM), model series of Al-7000S, GOTECH with 5 kN load capacity. The flexural test was performed as per specified in ASTM D790, adopting a three-point bending fixture method with a centre loading. Rectangular bar shape specimens with dimensions of 127 mm length, 12.7 mm height and 3 mm wide were produced by casting the geopolymer paste onto the acrylic mould as depicted in Fig.1. The cured and hardened geopolymer bar specimen was loaded and horizontally aligned on the 48 mm length of the support span. The test was run with the crosshead speed of 1.28 mm/min and the results were recorded and analysed with software (U60) equipped with the UTM. Replicates of 5 geopolymer specimens were tested and the average was taken as the result.

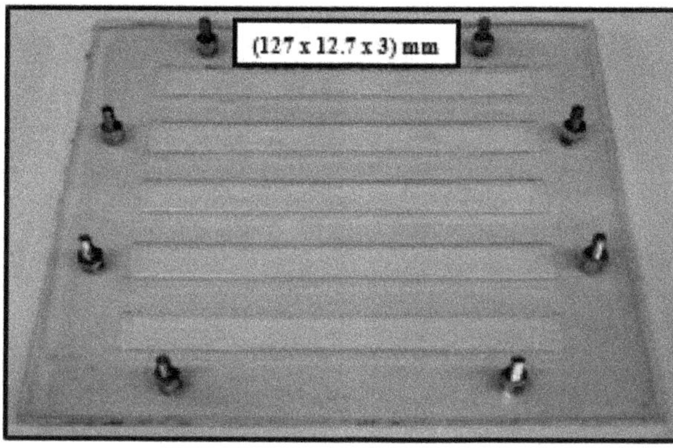

Figure 1: Acrylic mould for flexural strength sample

Results and Discussion

Microstructure of mechanically activated fly ash geopolymer

Morphological and microstructural characteristics of OFA, MFA A, and MFA B based geopolymer samples prepared at varying S: L ratios are shown in Fig. 2 (a-c). Fly ash geopolymer samples are only shown at highest S:L ratio of 3:1. Generally, dense, and compact microstructure was obtained at higher S: L due to higher degree of reaction while incomplete or moderate degree of geopolymerisation resulted in the formation of loose microstructure with the emergence of voids or cavities as well as the presence of considerable large portion of unreacted and partially reacted fly ash particles at lower S:L ratios.

Geopolymer samples show heterogeneous microstructure consisting of geopolymer gel with traces of unreacted and partially reacted fly ash particles as shown in Fig. 2 (a-c). The appearance of voids and cavities in geopolymer microstructure could be due to the air bubbles that were appeared during mixing or from the water molecules entrapped in the gel. As the specimens were cured, the water or moisture eventually evaporated leaving a hollow shaped space (voids).

Samples prepared at higher S:L ratio (MFA A-I) displayed more unreacted and partially reacted particles embedded within the geopolymer matrix. For MFA B based geopolymer, it is worth noting that, though geopolymer matrix still coexists with unreacted and partially reacted fly ash particles, the amount is considerably less than that of OFA and MFA A based geopolymers. Moreover, it is evident that high formation of geopolymer with better microstructure where more uniform geopolymeric gel is produced as shown in Fig.2(c). It can be inferred that geopolymerisation occurred at higher degree and way more efficiently with fine fly ash particles.

According to Cheng et al. high S:L ratio caused the increase of oligomer size which affected the exchange kinetics of the silicate unit during geopolymerisation [15]. This lead to increase of viscosity and subsequently reduced the workability of geopolymer mixture [16]. Low workability cause difficulties in moulding and compacting the mixture during sample preparation, consequently, results in porous and non-uniform microstructure. Addition of water would enhance the workability allowing the mixture to flow nicely in the mould and compact easily. Thereby, avoiding the presence of air bubbles that may cause the formation of voids. However, excessive liquid is undesirable as it also could result in highly porous geopolymer structure due to inefficient geopolymerisation caused by the gap created by the water molecules that hinder the formation of geopolymer network. These results are consistent with literature findings [17, 18].

Figure 2: Microstructure of fly ash geopolymer (a) OFA-I, (b) MFA A-I, (c) MFA B-I.

Mechanical properties of mechanically activated fly ash geopolymer

Flexural strength of geopolymer samples is shown in Fig.3. The low S: L ratio negatively affected the quality of geopolymer as shown in OFA based geopolymer where flexural strength dropped from 13.39 MPa to 4.65 MPa. The low S: L ratio indicates high water content. Water does not take part in the chemical reactions involved during geopolymerization, thus excessive amount of water results in the dilution of NaOH activator and consequently lowers the concentration. It has been noticed that strong alkalis are required to completely or partially dissolve aluminosilicate source materials [19]. Dilution of alkali solution by addition of water hinders the interaction of fly ash particles with Na^+ and OH^- and hampers the dissolution process of Si and Al active species. Furthermore, water also serves as the medium in transporting the dissolved active species (Si and Al) to form geopolymer network with other active species. However, high amount of water molecules reduces the particle-to-particle reaction and the active species are unable to form the geopolymer network [20]. Delayed dissolution process and inefficient formation of geopolymer network, affect the quality of geopolymer. It was evidently confirmed by microstructural analysis where porous geopolymer was obtained at low S: L ratio. The presence of pores and voids produces a less compact geopolymer and eventually results in lower flexural strength.

Conversely, for MFA A and MFA B based geopolymer, decrease of S: L ratio improved the flexural strength up to a certain point before it dropped. Similar trend was also exhibited in microstructural results. The MFA A-I and MFA A-II displayed an increase in flexural strength from 15.39 MPa to 16.74 MPa, respectively. As the specimen prepared from highly reactive fine particle of MFA A, the hardening process occurred rapidly especially when using high S: L ratio. Shorter setting time had caused the geopolymer slurry to lose its workability. Lower tendency in achieving homogenous mixture as well as difficult moulding process resulted in imperfect structure of MFA A-I, consequently lower flexural strength.

Water added to the system managed to improve the workability which allowed better formation of geopolymer structure and further enhanced the mechanical strength of geopolymer as shown by MFA A-II. Nonetheless, geopolymer prepared from lower S: L ratio of 2.6:1 to 2.2:1 showed gradual decrease in flexural strength from 10.14 MPa to 5.76 MPa, respectively owing to a highly porous geopolymer microstructure. Since MFA B possessed smaller particle size, lower S:L ratio is much desirable in producing geopolymer with better quality. This can be observed by the increasing flexural strength from 18.94 MPa (MFA B-I) and 20.32 MPa (MFA B-II) to 23.84 MPa (MFA B-III). When fine fly ash was mixed with the activator solution at high S:L ratio, thick geopolymer slurry was formed instantaneously due to its high reactivity. High viscosity of geopolymer slurry is less workable and led to the formation of low quality geopolymer specimen which causes the low strength of MFA B-I. More liquid was required for MFA B based geopolymer to prolong the setting time to extend the dissolution process of Si and Al species. Therefore, higher reaction degree may occur with intense formation of geopolymer network which resulted in improved geopolymer formation and subsequently enhanced the strength development as shown by MFA B-II and FA B-III. Nevertheless, the decrease of flexural strength was shown for sample MFA B-IV and MFA B-V with 19.47 MPa and 10.42 MPa, respectively when prepared at too low S: L ratio.

The influence of fly ash particle size was clearly visible where major strength was attained when geopolymer produced from the finest fly ash in this research which is MFA B. The highest strength achieved in OFA based geopolymer batch was OFA-I with 13.39 MPa, which was prepared at the highest S: L ratio of 3:1. Meanwhile, further increase in flexural strength was observed by geopolymer prepared from MFA A as shown by MFA A-II with 16.74 MPa, the highest in the batch. It is worth to note that, though produced from slightly lower S:L ratio (2.8:1), the strength of MFA A-II is better than OFA-I which was synthesised from 3:1 ratio. Finally, the ultimate flexural strength in MFA B batch was demonstrated by MFA B-III with 23.84 MPa which was prepared at much lower S: L ratio (2.6:1). Larger fly ash particle was deemed to be inert with low surface area. When subjected to mechanical activation, smaller fly ash particle size distribution was obtained. This allows higher reaction degree of geopolymerisation and more network formation [21]. It enables the production of geopolymer with denser microstructure and higher mechanical strength. These findings agree with those reported in literature [22, 23].

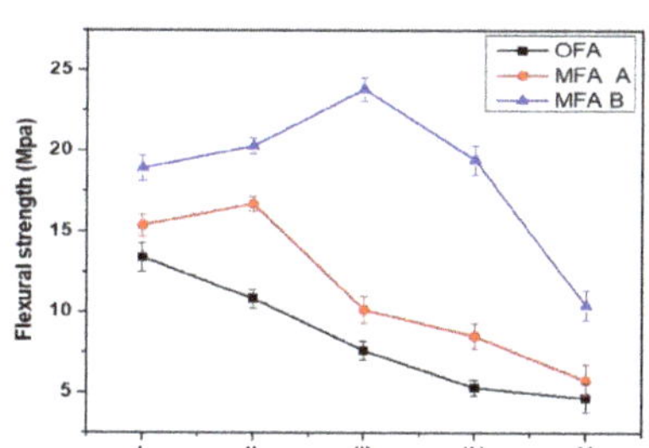

Figure 3: Flexural strength of OFA, MFA A and MFA B based geopolymer

Conclusions

The mechanical activation of fly ash decreased the particle size and resulted in the formation of dense and compact geopolymer with higher flexural strength. The increase of S:L ratio (2.2:1-3:1) resulted in the increase of geopolymer formation with dense and compact geopolymer microstructure. The highest flexural strength of 23.84 MPa was obtained from MFA B-III. The future work can be conducted on the use of different nanomaterials to see their effect on microstructure and mechanical strength of geopolymer.

Acknowledgement

The authors would like to acknowledge the funding supports from Ministry of Higher Education (MOHE) Malaysia through HICoE Grant (cost centre 015MA0-052/015MA0-104/015MA0-136) to CBBR.

References

[1] H. Wang, H. Li, and F. Yan, Synthesis and mechanical properties of metakaolinite-based geopolymer, Colloids and Surfaces A: Physicochemical and Engineering Aspects, 268 (2005) 1-6. https://doi.org/10.1016/j.colsurfa.2005.01.016

[2] W.-J. Long, T.-H. Ye, Q.-L. Luo, Y. Wang, and L. Mei, Reinforcing Mechanism of Reduced Graphene Oxide on Flexural Strength of Geopolymers: A Synergetic Analysis of Hydration and Chemical Composition, 9 (2019) 1723. https://doi.org/10.3390/nano9121723

[3] M. Irfan Khan, K. Azizli, S. Sufian, and Z. Man, Sodium silicate-free geopolymers as coating materials: Effects of Na/Al and water/solid ratios on adhesion strength, Ceramics International, 41 (2015) 2794-2805. https://doi.org/10.1016/j.ceramint.2014.10.099

[4] B. Azeem, K. Kushaari, M. Naqvi, L. Kok Keong, M. K. Almesfer, Z. Al-Qodah, S. R. Naqvi, and N. Elboughdiri, Production and Characterization of Controlled Release Urea Using Biopolymer and Geopolymer as Coating Materials, 12 (2020) 400. https://doi.org/10.3390/polym12020400

[5] D. M. Nurjaya, S. Astutiningsih, and A. J. I. J. O. T. Zulfia, Thermal effect on flexural strength of geopolymer matrix composite with alumina and wollastonite as fillers, 6 (2015) 462-470. https://doi.org/10.14716/ijtech.v6i3.1441

[6] W.-J. Long, T.-H. Ye, Q.-L. Luo, Y. Wang, and L. Mei, Reinforcing Mechanism of Reduced Graphene Oxide on Flexural Strength of Geopolymers: A Synergetic Analysis of Hydration and Chemical Composition, Nanomaterials, 9 (2019) 1723. https://doi.org/10.3390/nano9121723

[7] D. M. Samuel, A. Sutrisno, and W. M. Kriven, Relative importance of Al(V) and reinforcement to the flexural strength of geopolymer composites, Journal of the American Ceramic Society, 104 (2021) 3452-3460. https://doi.org/10.1111/jace.17656

[8] I. Farina, M. Modano, G. Zuccaro, R. Goodall, and F. Colangelo, Improving flexural strength and toughness of geopolymer mortars through additively manufactured metallic rebars, Composites Part B: Engineering, 145 (2018) 155-161. https://doi.org/10.1016/j.compositesb.2018.03.017

[9] A. Nazari, A. Maghsoudpour, and J. G. Sanjayan, Flexural strength of plain and fibre-reinforced boroaluminosilicate geopolymer, Construction and Building Materials, 76 (2015) 207-213. https://doi.org/10.1016/j.conbuildmat.2014.12.002

[10] M. A. Yazdi, M. Liebscher, S. Hempel, J. Yang, and V. Mechtcherine, Correlation of microstructural and mechanical properties of geopolymers produced from fly ash and slag at room temperature, Construction and Building Materials, 191 (2018) 330-341. https://doi.org/10.1016/j.conbuildmat.2018.10.037

[11] A. Bagheri, A. Nazari, and J. G. Sanjayan, Fibre-reinforced boroaluminosilicate geopolymers: A comparative study, Ceramics International, 44 (2018) 16599-16605. https://doi.org/10.1016/j.ceramint.2018.06.085

[12] S. Zhang, X. Ji, W. Zhou, X. Liu, Q. Wang, X. Chang, J. Tang, C. Huang, and Y. Lu, High-flexural-strength of geopolymer composites with self-assembled nanofiber networks, Ceramics International, 47 (2021) 31389-31398. https://doi.org/10.1016/j.ceramint.2021.08.014

[13] S. Yan, P. He, D. Jia, X. Duan, Z. Yang, S. Wang, and Y. Zhou, In Situ Processing of Graphene/Leucite Nanocomposite Through Graphene Oxide/Geopolymer, Journal of the American Ceramic Society, 99 (2016) 1164-1173. https://doi.org/10.1111/jace.14089

[14] A. A. Siyal, K. A. Azizli, L. Ismail, Z. Man, and M. I. Khan, Suitability of Malaysian Fly Ash for Geopolymer Synthesis, Advanced Materials Research, 1133 (2016) 201-205. https://doi.org/10.4028/www.scientific.net/AMR.1133.201

[15] H. Cheng, K.-L. Lin, R. Cui, C.-L. Hwang, T.-W. Cheng, and Y.-M. Chang, Effect of solid-to-liquid ratios on the properties of waste catalyst-metakaolin based geopolymers, Construction and Building Materials, 88 (2015) 74-83. https://doi.org/10.1016/j.conbuildmat.2015.01.005

[16] C. Y. Heah, H. Kamarudin, A. M. Mustafa Al Bakri, M. Bnhussain, M. Luqman, I. Khairul Nizar, C. M. Ruzaidi, and Y. M. Liew, Study on solids-to-liquid and alkaline activator ratios on kaolin-based geopolymers, Construction and Building Materials, 35 (2012) 912-922. https://doi.org/10.1016/j.conbuildmat.2012.04.102

[17] J. Xie, J. Wang, B. Zhang, C. Fang, and L. Li, Physicochemical properties of alkali activated GGBS and fly ash geopolymeric recycled concrete, Construction and Building Materials, 204 (2019) 384-398. https://doi.org/10.1016/j.conbuildmat.2019.01.191

[18] S. Onutai, S. Jiemsirilers, P. Thavorniti, and T. Kobayashi, Fast microwave syntheses of fly ash based porous geopolymers in the presence of high alkali concentration, Ceramics International, 42 (2016) 9866-9874. https://doi.org/10.1016/j.ceramint.2016.03.086

[19] C.-L. Hwang and T.-P. Huynh, Effect of alkali-activator and rice husk ash content on strength development of fly ash and residual rice husk ash-based geopolymers, Construction and Building Materials, 101 (2015) 1-9. https://doi.org/10.1016/j.conbuildmat.2015.10.025

[20] Y.-M. Liew, C.-Y. Heah, A. B. Mohd Mustafa, and H. Kamarudin, Structure and properties of clay-based geopolymer cements: A review, Progress in Materials Science, 83 (2016) 595-629. https://doi.org/10.1016/j.pmatsci.2016.08.002

[21] S. Xu, M. A. Malik, Z. Qi, B. Huang, Q. Li, and M. Sarkar, Influence of the PVA fibers and SiO2 NPs on the structural properties of fly ash based sustainable geopolymer, Construction and Building Materials, 164 (2018) 238-245. https://doi.org/10.1016/j.conbuildmat.2017.12.227

[22] P. Nuaklong, V. Sata, and P. Chindaprasirt, Properties of metakaolin-high calcium fly ash geopolymer concrete containing recycled aggregate from crushed concrete specimens, Construction and Building Materials, 161 (2018) 365-373. https://doi.org/10.1016/j.conbuildmat.2017.11.152

[23] M. Sumesh, U. J. Alengaram, M. Z. Jumaat, K. H. Mo, and M. F. Alnahhal, Incorporation of nano-materials in cement composite and geopolymer based paste and mortar - A review, Construction and Building Materials, 148 (2017) 62-84. https://doi.org/10.1016/j.conbuildmat.2017.04.206

Life cycle analysis on energy generation via anaerobic digestion of chicken manure

M. DEVENDRAN Manogaran[1,2,a], MOHD FIRDAUS Rosly[2,b], RASHID Shamsuddin[1,2,c*], LIM Jun Wei[1,3,d] and HAMAD Almohamadi[4,e]

[1]HICoE-Centre for Biofuel and Biochemical Research (CBBR), Institute for Self-Sustainable Building (ISB), Universiti Teknologi PETRONAS (UTP), 32610 Seri Iskandar, Perak, Malaysia

[2]Chemical Engineering Department, Universiti Teknologi PETRONAS (UTP), 32610 Seri Iskandar, Perak, Malaysia

[3]Fundamental and Applied Sciences Department, Universiti Teknologi PETRONAS (UTP), 32610 Seri Iskandar, Perak, Malaysia

[4]Department of Chemical Engineering, Faculty of Engineering, Islamic University of Madinah, Madinah, Saudi Arabia

[a]devendran.m@utp.edu.my, [b]mohd_17004687@utp.edu.my, [c]mrashids@utp.edu.my, [d]junwei.lim@utp.edu.my, [e]hha@iu.edu.sa

Keywords: Life Cycle Analysis, Global Warming Potential, Eutrophication Potential, Acidification Potential, Anaerobic Digestion

Abstract. Fossil fuels are the primary reservoir of Malaysia's energy supply hence, the government and major stakeholders are looking into renewable means for energy generation. Chicken manure on the other hand is an abundant biomass waste as chicken is the second most staple food item in the country after rice. Thus, green energy generation via anaerobic digestion of chicken manure is an option. The motivation behind this research is to dismiss claims that green energy generation is more detrimental to the environment compared to the conventional fossil fuel driven approach. A Life Cycle Analysis comprising of Life Cycle Inventory and Life Cycle Impact Assessment was conducted using Microsoft Excel to evaluate the environmental constraints of energy generation via anaerobic digestion specifically in the global warming, eutrophication, and acidification potential impact categories. The findings revealed that the global warming potential is up to 832248.183 kg CO_2 equivalents (eq), outweighing the concern of the other two impact categories. This is because carbon dioxide is the main greenhouse gas emitted during the process primarily due to poor management of chicken manure at the broiler house. Thus, recommendations were put forth in terms of introducing other practices parallel to anaerobic digestion such as composting and gasification which also yield value added products.

Introduction

The steady growth in the nation's population has resulted in enhanced energy demand. The main challenge put forward by such circumstances is the high dependency on fossil fuels derived energy, up to 92.31 % in the year 2020 [1]. Accordingly, there is a relevant demand for more renewable energy reservoirs which prevents any rising issues associated to energy security. One approach to better invest in renewable energy reservoirs is by adopting circular economy practices instead of the linear economy norm [2]. This means that any form of waste or by-product from the process is recycled to some extend in consecutive cycles of the similar process or is incorporated elsewhere for other applications.

Chicken manure which is been produced in abundance as of lately due to the popularity of chicken meat as a source of protein [3], is oftentimes misused as organic fertilizer without any form of pre-treatment [4]. This normalized practice observes adverse effects to the environment particularly in the form of enhanced greenhouse gas (GHG) emissions which contradicts with the Sustainable Development Goals (SDG) established by the United Nations in 2015, especially SGD 13 which aspires to solve the issue of climate change. On the contrary, chicken manure exhibits potential as feedstock for multiple waste-to-wealth treatment methods such as composting, pyrolysis and gasification [5] however, anaerobic digestion (AD) specifically has been receiving substantial attention. This is because such efforts are in line with the 12th Malaysia plan which observes the nation making a progressive shift to be carbon-neutral by the year 2030. Consequently, renewable energy derived from solar, biomass and biogas specifically will be put at the forefront [6].

Life Cycle Assessment (LCA) is a holistic, standard means of evaluating the environmental burdens of new products or technology designs by distinguishing the energy, consumables and emissions released to the surrounding [7]. Additionally, LCA is also a method of assessing the GHG emissions of different technologies associated to the desired outcome hence, aiding the identification of environmentally preferred technologies [8]. This paper evaluates the environmental constraints for biogas generation via AD of chicken manure primarily on global warming potential (GWP).

Methodology

The research is executed by firstly defining the goals and scope, followed by the Life Cycle Inventory (LCI), Life Cycle Impact Assessment (LCIA), and finally the results' interpretation as depicted in Figure 1.

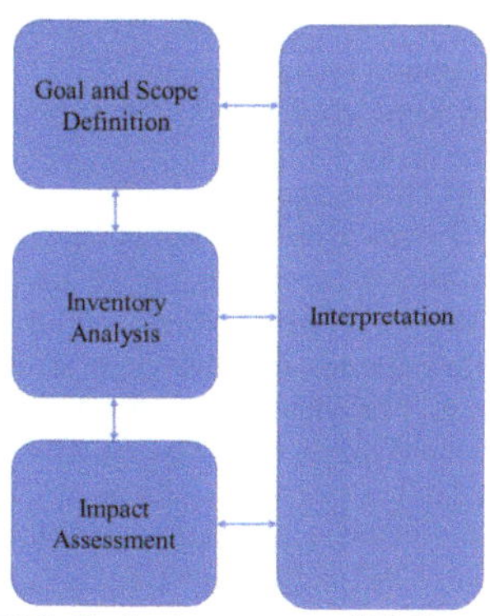

Figure 1: Life Cycle Assessment (LCA) workflow [9].

Define Goal and Scope. The items that need to be defined are the functional unit and the system boundary. The functional unit is defined as 30000 units of chickens in a broiler house with the duration of one cycle defined as 30 days. The system boundary is as shown in Figure 2.

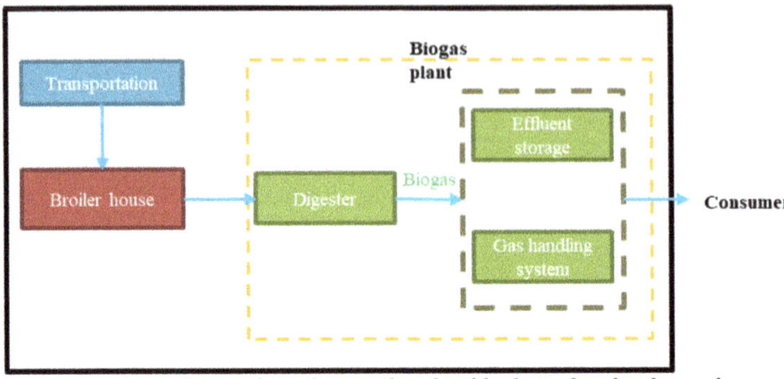

Figure 2: System boundary outlined in black marker for the study.

Life Cycle Inventory (LCI). Essentially, the LCI stage involves an inventory of input and output flows for a biogas production which revolves around data associated to the operations of the broiler house and biogas plant. Use of water, energy and raw materials are a few examples of the inputs whereas emission released to the air or leachates in the soil and water bodies are considered as outputs with respect to the system boundary [10]. The inventory data can be acquired from literature study primarily from journal articles. Additionally, statistical data associated to the operations of the poultry house and chicken production were obtained via personal contact from Dindings Poultry Processing Sdn. Bhd. (DPP). The data comprises of the total number of chickens, generation of chicken manure as well as energy consumption for heating and electricity supply. Journal articles on the other hand were used as reference to estimate the hazardous emissions within the system boundary [11].

Life Cycle Impact Assessment (LCIA). The consecutive step is to perform LCIA utilising the inventory database acquired in LCI with respect to the system boundary. First and foremost, the impact categories to be evaluated within the system boundary need to be defined. Impact categories aid in screening the different emissions to observe which exhibits the most prominent effect on the environment. Accordingly, the different emissions are integrated into one representative unit with respect to a characterization factor. For instance, in the GWP impact category, the different emissions are expressed as kg CO_2 equivalents (kg CO_2-eq). The kg CO_2-eq for different emissions were calculated using in Equation 1:

$$IC_i = \sum(E_j) \times CF_{ij} \quad (1)$$

where IC_i (impact category indicator) is the indicator value per functional unit for the impact category i, E_j is the release of emission j or consumption of resource j per functional unit, CF_{ij} is characterization factor for emission j or resource j contributing to impact category i [12].

Interpretation and LCA Tool. Life cycle interpretation is a systematic approach to identify, quantify and evaluate information and data from the LCI and LCIA stages with respect to the ISO 14040:2006 standards. These standards entail identifying apparent concerns and constraints based on the LCI and LCIA stages as well as drawing conclusions and providing recommendations [13]. The findings aid in the identification of key challenges with respect to the scope definition for the study. Microsoft Excel is the primary tool used in this research to tabulate the inventory data during the LCI phase and generate graphical data during the LCIA stage as executed in past research by Feiz, et al. [14].

Results and Discussion

Data associated to the broiler house operations such as the daily feed intake and utility requirements are attained through personal contact from Dindings Poultry Processing Sdn Bhd., a primary chicken supplier in Malaysia, accounting to 10 % of the national chicken supply. Cumulatively, 3.5 kg of feed is required per chicken during each cycle. For chicken bedding, it is estimated 2.76 kg of bedding needed per chicken. Inventory data associated to the biogas plant is extracted from a study by Morero, et al. [15] which revolves around anaerobic digestion as well. The data from the LCI stage for broiler house and biogas plant operations are depicted in Table 1 and 2 respectively.

Table 1: Inventory data for a broiler house.

Type of Input	Measurement/Type	Type of Output	Measurement
Feed intake		Solid waste	
Pre starter	325.00 g	Municipal solid waste	3.30 kg
Starter	807.00 g		
Grower	2.40 kg	Air emission	
Average	3.50 kg/chicken	NH_3	11.40 kg
Transportation related		SO_2	0.38 g
Distance	213.20 km	CO	4.89 kg
Diesel	81.02 litre	NO_X	47.10 kg
Type of transport	Lorry	CO_2	336234.81 kg
Physical parameters		N_2O	3677.08 g
Start weight	100 g	CH_4	27724.93 g
Finish weight	2.50-3.00 kg	Water emission	
Mortality rate	3.60 %	PO_4^{-3}	3.60 kg
Bedding	2.76 kg/chicken	NO_3^-	159.30 kg
Utilities			
Amount of water for drinking	0.10 m^3/chicken		
Electricity	15580.00 W		

Table 2: Inventory data for biogas plant.

Parameter	unit	Value
Feedstock (poultry litter)	kg/year	680400.00
Digester capacity	m^3	500.00
Hydraulic retention time (HRT)	days	34.00
Working temperature	°C	37.00
Biogas production	m^3/year	107231.04
CH_4 content of biogas	%	56.00
CH_4 leakage	%	2.40
Installed electrical power	MW	1.50
Electrical efficiency of biogas conversion	%	30.00
Electricity generated	kWh/year	214462.08
Electricity used	kWh/year	1404000.00
Heat production	kWth/year	268077.60

Digestate production	kg/year	595350.00
Water	m^3	15.00
Output in terms of emissions (30 days)		
CO_2	kg	355140.00
CH_4	g	19305.90
N_2O	g	3352.50
O_2	kg	2976.75
N_2	kg	58968.00

In terms of transportation consideration for the broiler house operations, the distance defined is that between the broiler house and feed mill with 13 trips assumed to be taken in the period of one cycle. As there are large amounts of broiler feed to be transported, it is assumed that the mode of transportation is a lorry, hence, is estimated that 38 L of diesel is consumed per 100 km.

The data gathered is accordingly used in the manual calculations which employ the gas emission factors as provided by the guidelines in the Environmental Protection Agency guideline. The results for the calculations are depicted in Table 3.

Table 3: Total of GHG emission for the whole system boundary.

Type of material/activity	**Value**	**Carbon dioxide emission (kg CO_2 eq)**	**Methane emission (g CH_4 eq)**	**Nitrous oxide emission (g N_2O eq)**
Transportation: Diesel	81.02 liter	214.70	4.41	1.78
Distance	213.20 km			
Feed material	105000.00 kg	336000.00	27720.00	3675.00
Electricity	117015.60 kW	117015.60	1275.47	971.23
Manure	56700.00 kg	238140.00	18030.60	2381.40
Bedding	82800.00 kg	135792.00	10432.80	5216.40
Total		827162.30	57463.28	12245.80

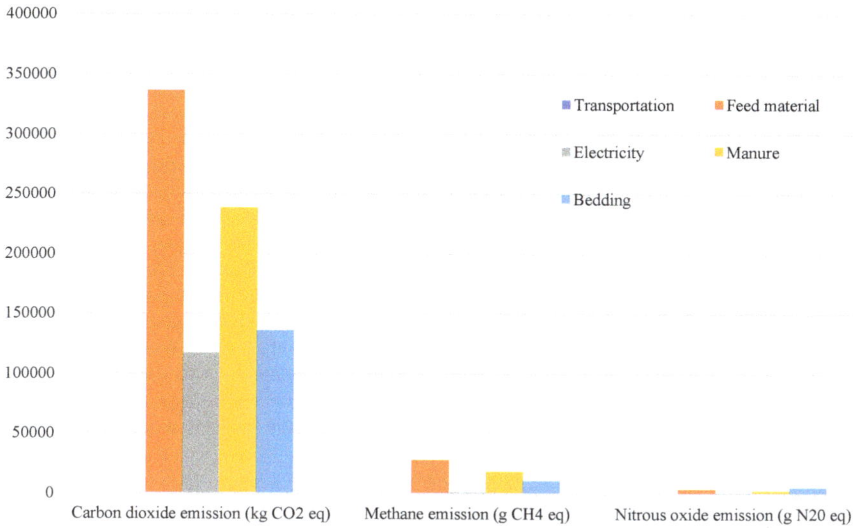

Figure 3: Total GHG emission by material/activity.

From Figure 1, it is apparent that the CO_2 emission is significantly higher in comparison to CH_4 and N_2O emissions. Additionally, it can be observed that the feed for the chickens which is primarily from agricultural biomass waste and the chicken manure which acts as the feedstock for the biogas plant contribute significantly to GHG emissions in comparison to the other factors. In terms of N_2O emission, the attribute that contributed dominantly is the use of bedding materials as it constitutes of nitrogenous rich elements such as wood and tree barks.

The impact category observed is the Global Warming Potential (GWP). The different emissions are standardized to a single attribute which in the case of GWP is kg CO_2-eq using the characterization factor, otherwise known as the equivalent factor as reported by Fallahpour, et al. [16]. GWP is arguably one of more significant impact categories to be taken into consideration especially when it revolves around discussions on sustainable energy generation. As CO_2 is the standard for GWP, it can be described as the heat absorbed by any GHG in the atmosphere, as a multiple of the heat that would be absorbed by the same mass of CO_2. Each GHG species significantly contributes to GWP as they persist for different durations in the atmosphere. Accordingly, other constituents that contribute to GWP are converted to kg CO_2-eq as well. In the case of CH_4 and N_2O emissions, it is identified as equivalent to 25-year and 298-year kg CO_2-eq per kg GHG respectively. The GWP calculation is as summarized in Table 4.

Table 4: GWP calculation

Emission	Values (in kg)	CO_2 equivalent factor	GWP (kg CO_2-eq)
CO_2	827162.30	1.00	827162.30
CH_4	57.46	25.00	1436.58
N_2O	12.25	298.00	3649.31
Total			832248.18

Based on the results depicted in Table 4, it can be observed that CO_2 is the primary contributor to GWP in comparison to CH_4 and N_2O. As such, variables significantly contributing to CO_2 emission is identified such that practical steps are taken to control CO_2 emission. In a study by Montemayor, et al. [17], extremely low GWP values were observed in the case of digested manure slurry application, to which the GHG offset due to bioenergy generation from digested manure slurry production was the primary contribution. Krexner, et al. [18] on the other hand deduced that biogas generation attributing to manure management only contributed to 10.07 % of GWP with respect to the cradle-to-gate system boundary of his study inclusive of manure acquisition up to synthesis of product. Nonetheless, chicken manure can also be treated through other waste-to-wealth approaches such as composting, pyrolysis, and gasification [19].

Eutrophication Potential (EP) on the other hand entails the impact on environments caused by over-fertilisation or excess supply of nutrients. The surplus of nutrients leads to enhanced growth of plants, especially plankton algae. In the ecosystem of a water body, excessive development of microorganisms deters the supply of oxygen and sunlight, adversely effecting the plants. The characteristics of the water body is also bound to change. For example, a formerly clean lake with drinking water quality can evolve into water with an anoxic (free of oxygen) depth layer [20]. Similar to GWP, the different constituents are expressed in a single attribute which is kg NO_3-eq. The EP calculation is as depicted in Table 5.

Table 5: EP calculation

Emission	Values (in kg)	NO_3^- equivalent factor	EP (kg NO_3^--eq)
NO_3	159.30	1.00	159.30
NH_3	11.40	3.64	41.50
PO_4^3	3.60	10.45	37.62
Total			238.42

Based on Table 5, the EP is largely contributed by nitrate, NO_3 emission from the broiler house stage. This is because chicken manure is rich in nitrogen hence the practice of soil conditioning without any form of pre-treatment is not a sustainable management approach as major accumulation of NO_3 in the soil profile increases the NO_3 concentration in groundwater, contributing to EP [21]. This correlation has also been observed in a study conducted by Huang, et al. [22] which observed rise in EP when nitrogen rich fertilizers are used to enhance crop yield. The phosphate, PO_4^3 and ammonia, NH_3 emissions are relatively low compared to the NO_3 emission. However, in comparison to GWP, EP is still manageable.

Acidification Potential (AP) is associated to acid deposition of acidifying contaminants on soil, groundwater, surface waters, biological organisms, ecosystems, and substances [23]. There are a few inventory items relevant to the AP including sulphur dioxide, SO_2 which is the standard equivalent for AP. The AP calculation is as depicted in Table 6.

Table 6: AP calculation

Emission	Values (in kg)	SO_2 equivalent factor	AP (kg SO_2-eq)
SO_2	0.00038	1.00	0.00038
NH_3	11.40	1.88	21.43
NO_X	47.10	0.28	13.19
Total			34.61

With respect to Table 6, it can be deduced that NH_3 is the main contributor for emissions followed by nitrogen oxides, NO_x resulting in acidification despite SO_2 being the standard in this impact category. The broiler house exhibits notable NH_3 emission due to poor manure management. This is coherent with a study by Kacprzak, et al. [24] which observed that NH_3 emission from poultry manure is higher in comparison to cattle and cow manure. The emission of NO_x is justified in a similar manner due to the volatilization of nitrogenous content in the manure [25]. Additionally, the nature of NH_3 which evolves to NO_x in the atmosphere is also used to infer this observation [26]. Hence, apart from considering circular economy-based treatment approaches for the management of chicken manure, reinforcing legislations entailing proper livestock manure management practices as executed in China is a pragmatic measure to alleviate adverse effects to the environment [27].

Conclusion

In a LCA framework, GWP, EP and AP are oftentimes regarded as the more prominent impact categories that need to be evaluated as it is directly correlates to GHG emissions which is rising concern. The findings revealed that EP and AP are not much of a concern however control measures need to be put in place to reduce the impact of GWP with respect to GHG emissions. Of the few GHG emissions, CO_2 seems to be the most apparent primarily due to broiler house activities associated to feedstock acquisition for the chickens and manure management. Consequently, more sustainable practices should be introduced to further reduce the GWP.

Acknowledgement

The authors would like to acknowledge the funding support from Ministry of Higher Education (MOHE) Malaysia through HICoE Grant (cost centre 015MA0-052/015MA0-104/015MA0-136) to Centre for Biofuel and Biochemical Research (CBBR). Prototype Research Grant Scheme, PRGS/1/2022/TK09/UTP/02/2 by MOHE Malaysia is also recognized.

References

[1] D. Jones, Global electricity review 2021. (2021).

[2] P. Lacy, J. Rutqvist, Waste to wealth: The circular economy advantage. Springer, 2016. https://doi.org/10.1057/9781137530707

[3] N. O. Chijioke, M. Uddin Khandaker, K. M. Tikpangi, D. A. Bradley, Metal uptake in chicken giblets and human health implications, Journal of Food Composition and Analysis. 85 (2020) 103332. https://doi.org/10.1016/j.jfca.2019.103332

[4] L. Zhang L. Li, X. Pan, Z. Shi, X. Feng, B. Gong, J. Li, L. Wang, Enhanced Growth and Activities of the Dominant Functional Microbiota of Chicken Manure Composts in the Presence of Maize Straw, Front. Microbiol. 9 (2018). https://doi.org/10.3389/fmicb.2018.01131

[5] M. D. Manogaran, R. Shamsuddin, M. H. Mohd Yusoff, M. Lay, A. A. Siyal, A review on treatment processes of chicken manure, Cleaner and Circular Bioeconomy. 2 (2022) 100013. https://doi.org/10.1016/j.clcb.2022.100013

[6] M. Vaka, R. Walvekar, A. K. Rasheed, M. Khalid, A review on Malaysia's solar energy pathway towards carbon-neutral Malaysia beyond Covid'19 pandemic, Journal of Cleaner Production. 273 (2020) 122834. https://doi.org/10.1016/j.jclepro.2020.122834

[7] Y. H. Chan, R. R. Tan, S. Yusup, H. L. Lam, A. T. Quitain, Comparative life cycle assessment (LCA) of bio-oil production from fast pyrolysis and hydrothermal liquefaction of oil palm empty fruit bunch (EFB), Clean Technologies and Environmental Policy. 18 (2016) 1759-1768. https://doi.org/10.1007/s10098-016-1172-5

[8] M. A. Nasution, D. S. Wibawa, T. Ahamed, R. Noguchi, Comparative environmental impact evaluation of palm oil mill effluent treatment using a life cycle assessment approach: A case study based on composting and a combination for biogas technologies in North Sumatera of Indonesia, Journal of Cleaner Production. 184 (2018) 1028-1040. https://doi.org/10.1016/j.jclepro.2018.02.299

[9] D. Cahyani, A. Haryanto, G. A. Putra, R. Fil'aini, D. S. S. Marpaung, Life cycle assessment of biogas digester in small scale tapioca industry, IOP Conference Series: Earth and Environmental Science. 258 (2019) 012017. https://doi.org/10.1088/1755-1315/258/1/012017

[10] S. González-García, Z. Gomez-Fernández, A. C. Dias, G. Feijoo, M. T. Moreira, L. Arroja, Life Cycle Assessment of broiler chicken production: a Portuguese case study, Journal of Cleaner Production. 74 (2014) 125-134. https://doi.org/10.1016/j.jclepro.2014.03.067

[11] E. M. M. Esteves, A. M. N. Herrera, V. P. P. Esteves, C. d. R. V. Morgado, Life cycle assessment of manure biogas production: A review, Journal of Cleaner Production. 219 (2019) 411-423. https://doi.org/10.1016/j.jclepro.2019.02.091

[12] J. T. E. Lee, M. U. Khan, H. Tian, A. W. L. Ee, E. Y. Lim, Y. Dai, Y. W. Tong, B. K. Ahring, Improving methane yield of oil palm empty fruit bunches by wet oxidation pretreatment: Mesophilic and thermophilic anaerobic digestion conditions and the associated global warming potential effects, Energy Conversion and Management. 225 (2020) 113438. https://doi.org/10.1016/j.enconman.2020.113438

[13] C. Cao, Sustainability and life assessment of high strength natural fibre composites in construction, in: M. Fan, F. Fu (Eds.), Advanced High Strength Natural Fibre Composites in Construction, Woodhead Publishing, 2017, pp. 529-544. https://doi.org/10.1016/B978-0-08-100411-1.00021-2

[14] R. Feiz, M. Johansson, E. Lindkvist, J. Moestedt, S. N. Påledal, N. Svensson, Key performance indicators for biogas production-methodological insights on the life-cycle analysis of biogas production from source-separated food waste, Energy. 200 (2020) 117462. https://doi.org/10.1016/j.energy.2020.117462

[15] B. Morero, E. Groppelli, E. A. Campanella, Life cycle assessment of biomethane use in Argentina, Bioresource Technology. 182 (2015) 208-216. https://doi.org/10.1016/j.biortech.2015.01.077

[16] F. Fallahpour, A. Aminghafouri, A. Ghalegolab Behbahani, M. Bannayan, The environmental impact assessment of wheat and barley production by using life cycle assessment (LCA) methodology, Environment, Development and Sustainability. 14 (2012) 979-992. https://doi.org/10.1007/s10668-012-9367-3

[17] E. Montemayor, A. Bonmatí, M. Torrellas, F. Camps, C. Ortiz, F. Domingo, V. Riau, A. Antón, Environmental accounting of closed-loop maize production scenarios: Manure as fertilizer and inclusion of catch crops, Resources, Conservation and Recycling. 146 (2019) 395-404. https://doi.org/10.1016/j.resconrec.2019.03.013

[18] T. Krexner, A. Bauer, W. Zollitsch, K. Weiland, A. Bismarck, A. Mautner, F. Medel-Jimenez, A. Gronauer, I. Kra, Environmental life cycle assessment of nano-cellulose and biogas production from manure, Journal of Environmental Management. 314 (2022) 115093. https://doi.org/10.1016/j.jenvman.2022.115093

[19] M. D. Manogaran, M. R. Shamsuddin, M. H. M. Yusoff, M. Lay, An overview on available treatment processes of poultry manure in Malaysia, AIP Conference Proceedings. 2610 (2022) 040005. https://doi.org/10.1063/5.0099555

[20] A. Arias, C. R. Behera, G. Feijoo, G. Sin, M. T. Moreira, Unravelling the environmental and economic impacts of innovative technologies for the enhancement of biogas production and sludge management in wastewater systems, Journal of Environmental Management. 270 (2020) 110965. https://doi.org/10.1016/j.jenvman.2020.110965

[21] S. Zahedi, C. Martín, R. Solera, M. Pérez, Evaluating the Effectiveness of Adding Chicken Manure in the Anaerobic Mesophilic Codigestion of Sewage Sludge and Wine Distillery Wastewater: Kinetic Modeling and Economic Approach, Energy & Fuels. 34 (2020) 12626-12633. https://doi.org/10.1021/acs.energyfuels.0c01852

[22] J. Huang, C. Xu, B. G. Ridoutt, X. Wang, P. Ren, Nitrogen and phosphorus losses and eutrophication potential associated with fertilizer application to cropland in China, Journal of Cleaner Production. 159 (2017) 171-179. https://doi.org/10.1016/j.jclepro.2017.05.008

[23] J. Zhang, B. Ling, Y. He, Y. Zhu, Z. Wang, Life cycle assessment of three types of hydrogen production methods using solar energy, International Journal of Hydrogen Energy. 47 (2022) 14158-14168. https://doi.org/10.1016/j.ijhydene.2022.02.150

[24] M. Kacprzak, M. Krystyna, G. Anna, S. Jolanta, W. Katarzyna, D. Danuta, J. Anna, M. Erik, Cycles of carbon, nitrogen and phosphorus in poultry manure management technologies - environmental aspects, Critical Reviews in Environmental Science and Technology, (2022) 1-25.

[25] T. J. van der Weerden, A. Noble, C. A. de Klein, N. Hutchings, R. E. Thorman, M. A. Alfaro, B. Amon, I. Beltran, P. Grace, M. Hassouna, D. J. Krol, Ammonia and nitrous oxide emission factors for excreta deposited by livestock and land-applied manure, Journal of Environmental Quality. 50 (2021) 1005-1023. https://doi.org/10.1002/jeq2.20259

[26] T. Sepperer, G. Tondi, A. Petutschnigg, T. M. Young, K. Steiner, Mitigation of Ammonia Emissions from Cattle Manure Slurry by Tannins and Tannin-Based Polymers, Biomolecules. 10 (2020) 581. https://doi.org/10.3390/biom10040581

[27] S. Wei, Z. Zhu, J. Zhao, D. R. Chadwick, H. Dong, Policies and regulations for promoting manure management for sustainable livestock production in China: A review, Frontiers of Agricultural Science Engineering. 8 (2021) 45-57. https://doi.org/10.15302/J-FASE-2020369

Utilizing membrane technologies in advancing the recycling of spent lithium-ion batteries using green electrochemical method – A review

MUHAMMAD Faiz Aizamddin[1,a], SITI Nur Amira Shaffee[2,b], MUHAMMAD Zharfan Mohd Halizan[1,c], SAIFUL Arifin Shafiee[3,d], AWIS Sukarni Mohmad Sabere[4,e], ZARIF Mohamed Sofian[5,f], NURAINI Daud[6,g], FADZLI Irwan Bahrudin[7,h], IRINA Harun[8,i], NORAZAH Abd Rahman[9,j] and MOHD Muzamir Mahat[1,k*]

[1]Faculty of Applied Sciences, Universiti Teknologi MARA, Shah Alam, 40450 Selangor, Malaysia

[2]Group Research and Technology, PETRONAS Research Sdn. Bhd., Bandar Baru Bangi 43000 Selangor, Malaysia

[3]Department of Chemistry, Kulliyyah of Science, International Islamic University Malaysia, Bandar Indera Mahkota, 25200 Kuantan, Pahang, Malaysia

[4]Department of Pharmaceutical Chemistry, Kulliyyah of Pharmacy, International Islamic University Malaysia, Bandar Indera Mahkota, 25200 Kuantan, Pahang, Malaysia

[5]Department of Pharmaceutical Technology, Faculty of Pharmacy, Universiti Malaya, 50603 Kuala Lumpur, Malaysia

[6]School of Computing & Creative Media, University of Technology Sarawak, 96000 Sibu, Sarawak, Malaysia

[7]Kulliyyah of Architecture and Environmental Design, International Islamic University Malaysia, Jalan Gombak, 50728 Kuala Lumpur, Malaysia

[8]Department of Environment, Faculty of Forestry and Environment, Universiti Putra Malaysia, 43400 Serdang, Selangor, Malaysia

[9]School of Chemical Engineering, College of Engineering, Universiti Teknologi MARA, Shah Alam, 40450 Selangor, Malaysia

[a]faizaizamddin@gmail.com, [b]nuramira_shaffee@petronas.com, [c]halizanzharfan@yahoo.com, [d]sabs@iium.edu.my, [e]awissabere@iium.edu.my, [f]ms_zarif@um.edu.my, [g]nuraini@uts.edu.my, [h]fadzliirwan@iium.edu.my, [i]irina@upm.edu.my, [j]noraz695@uitm.edu.my and [k]mmuzamir@uitm.edu.my*

Keywords: Spent Lithium-Ion Batteries, Membrane Technology, Lithium Green Recycling, Electrochemical Method, Solid Polymer Electrolytes, Interpenetrating Network Polymer

Abstract. The demand for lithium (Li) resources is soaring due to the widespread production (or consumption) of electronic products such as mobile and electronic devices, (or laptops, tablets, and home-appliances goods in accommodating current global lifestyle). The other lithium-ion batteries (LiBs) applications include electric vehicles, solar panels, wind turbines, and electric toys. The increasing demand for Li, while driving the economic progress of the industry, is putting a strain on the resource reserves. Therefore, the production industry is promptly searching for an efficient spent LiBs recycling process to counterbalance the highly sought-after element. Current Li recycling systems, in which extraction and recovery are typically accomplished by hydrometallurgical processes, have a significant impact on the environment, are energy-intensive, and necessitate vast operational capacities. Recently, electrochemical methods are seen as sustainable and green approaches to Li production. The use of membrane materials for Li recovery together with electrochemical processes provides a means to reduce energy consumption and scale up the spent LiBs recycling. In this overview, Li recovery technologies through pyrometallurgy, hydrometallurgy and green electrochemical extraction are explored along with their benefits and drawbacks. Recent advances in membrane materials selection that lead to significant improvement in Li production are also discussed.

Introduction to Lithium-ion Battery Recycling

Lithium-ion batteries (LiBs), as the most significant candidate for energy storage devices, have quickly occupied the global electrical consumer market due to their relatively high energy density, advanced operating voltage, and lack of a memory effect [1]. Statistics by GlobeNewswire reported that the global market for LiBs is forecasted to increase from USD 41.1 billion in 2021 to USD 116.1 billion by 2030, where the compound annual growth rate (CAGR) is projected to grow by 12.3% from 2021 to 2030. Factors such as declining LiBs prices and the high demand for batteries from the automotive industry are expected to drive the battery market during the forecast period. The increasing need for a 'battery material' supply has led to the continuous search for Li global reserves. Hence, it is expected that the amount of Li in the world is shrinking rapidly and will not be able to meet future demand [2]. Therefore, there is a need to develop or improve the current Li production and recovery system to ensure supply chain resiliency.

There are three (3) main methods that have been developed to recover Li from waste LiBs. They are electrochemical extraction, pyrometallurgy, and hydrometallurgy [3]. Fig. 1 shows the comparison of the three Li recycling methods based on energy efficiency, chemical cost, recycling efficiency, simplicity, environmental friendliness, and scale-up state. As shown in the figure, by and large, the electrochemical extraction method is a more environmentally benign option than pyrometallurgy and hydrothermal processes [3–6]. The electrochemical extraction can be operated at room temperature, produce less chemical waste, and is more energy efficient. Furthermore, it can yield Li with >99% purity. Meanwhile, the other two methods need to be run at the expense of a higher temperature setup and have the potential to cause acid, volatile compounds, and organic waste pollution. Nonetheless, the recycling efficiency of Li for the electrochemical extraction method is only 75% to 98% when the conductive ceramic solid electrolyte is utilized as the membrane separator. In addition, the electrochemical extraction process is currently at the proof-of-concept stage and is yet to be commercialized despite its advantages. Complex electrochemical setup and the fabrication of Li separators remain a major concern. Such limitations bring more value to Li mining, underscoring the significance of Li recycling.

The purpose of this paper is to serve as a reference for researchers, scientists, engineers, and technologists working to improve spent LiB recycling and recovery processes in the future. The review also details the processes, benefits, and downsides of the membrane materials for Li recycling that are already in use. The direction for future Li recycling technology research and development is suggested.

Materials Research Proceedings 29 (2023) 170-191 https://doi.org/10.21741/9781644902516-21

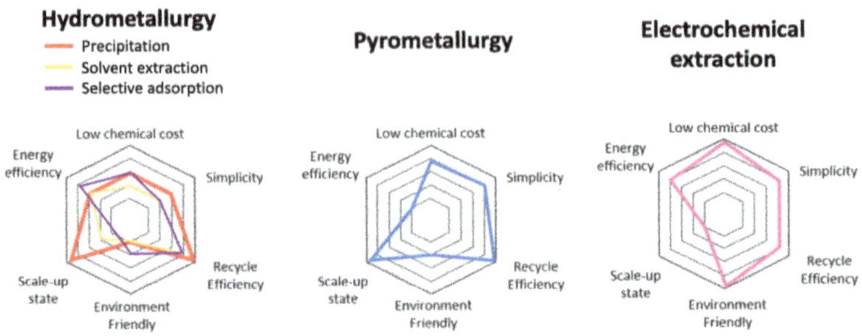

Fig. 1. Comparison of lithium recycling methods. Adapted from Bae et al. [3].

Lithium Recovery Technologies

The efficiency of Li extraction is a major consideration in selecting recovery technologies. Technically, the metal-extraction process is employed to transform the solid metals found in used LiBs into their alloy form or solution state. Techniques like pyrometallurgy, hydrometallurgy, and electrochemical are commonly used during the extraction of Li in active materials (see Fig. 2). However, towards the sustainable environment and life, the most preferred Li ion recycling method globally is the method which could support and stand for the green technology concept.

The 'green technology' concept which is emphasized for LiBs recycling methods in this article can consist of several temperaments. The most important ones are the characteristics which are included within recycling process such as waste prevention especially into the earth, energy usage reduction which can be translated into application of ambient temperature and pressure, inclusion of synthetic methods or chemical products which are designed to significantly reducing toxicity, inclusion of synthetic methods that able to ensure all materials incorporated within the process to be fully utilized and the minimization of auxiliary substances. More on that, the recycling process should prevent the application of stoichiometric reagents and replace it with catalytic reagents after careful consideration. All substances included in the recycling process should be able to minimize the chemical or physical accidents. Besides, the characteristics which should be included in the raw materials selections is that the materials should exist in renewable form [6]. The following sections discuss each extraction method and its efficiency as reported in the literature.

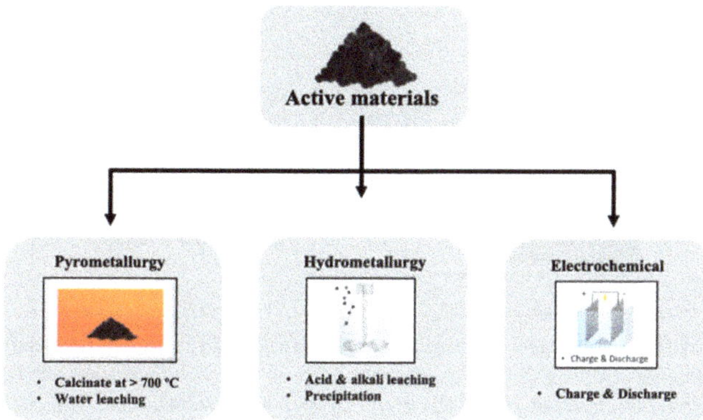

Fig. 2. Schematic diagram of three (3) main recycling stages and methods from active materials.

Pyrometallurgy

Much research highlights that pyrometallurgical is a prevalent and practical method for metal extraction, including for LiBs recycling. To recycle old batteries, for instance, a Umicore process could be used [8]. During the process, the active materials are melted down in a furnace in a smelting process. Compounds like polymers, organic solvents, and graphite in the batteries are decomposed while metal components are reduced and converted to alloys. The resulting alloys are further treated using sulfuric acid (H_2SO_4) as the leaching agent and solvent extraction to obtain cobalt oxides (CoO) and nickel hydroxide ($Ni(OH)_2$) [9]. Although this process does not require pre-treatment, it could significantly impact the recovery of Li compositions [10]. Due to this issue, researchers have proposed a unique approach that combines pyrometallurgy and hydrometallurgy to recycle valuable metals from wasted LiBs, intending to minimize Li loss during recovery.

In a study, Georgi et al. [11] were able to salvage precious metals from used LiBs by employing a reduction smelting technique. Important metals, including Fe, Co, Ni, and Mn, were transformed into alloys. However, Li went into slag or dust throughout the process. H_2SO_4 leaching was used to further isolate Li. Further, Werner et al. [12] proposed a method to evaporate Li from used LiBs using a combination of vacuum evaporation and selective carrier gas evaporation at high temperatures. But, the process temperature was higher than 1400 °C, which requires substantial energy usage. Another study by Peng et al. [13] reported a novel approach to recycling wasted LiBs, roasting them at a low temperature (25 °C to 70 °C) in an argon environment (see mechanism illustrated in Fig. 3), then extracting the Li_2CO_3 using water leaching. After the crystallization and leaching processes, the Li concentration in the leaching solution reached 4.36 g/L. Generally, the pyrometallurgy process to recover important metals from spent LiBs is straightforward, but it is not eco-friendly because of the high energy consumption and the potential secondary pollution [14,15] which disobeyed the green technology concept that should aimed to reduce energy consumption and prevent substance or products wastes. In addition, Li loss during the recycling process is also significant and needs to be addressed in future studies. Table 1 summarised the past reviews on Li extraction using pyrometallurgy.

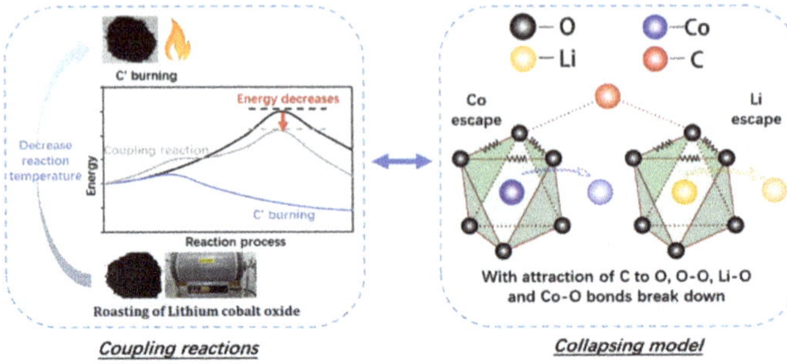

Fig. 3. Illustration mechanism of coupling reaction and collapsing model in carbothermic reduction roasting. During this process, carbothermic reaction facilitates the reduction by destroying oxygen octahedrons (O) in the metal oxide crystal formations. The collapse of the O octahedrons allows Li and Co to escape at ease, driving the reaction forward. Adapted from Peng et al. [13].

Table 1. List of past reviews and their research summary on Li extraction using pyrometallurgy.

Method	Steps	Summary	Ref.
Pyrometallurgy	Umicore process	The active materials were melted down in a furnace and converted into alloys. Further recovery of Li was achieved using H_2SO_4 as the leaching agent	[8]
	Reduction smelting	Li went into slag or dust throughout the process and the use of H_2SO_4 as the leaching agent promotes Li's isolation	[11]
	Vacuum evaporation and selective gas carrier	Li extraction was recovered through evaporation process from the used LiBs at high temperatures up to 1400 °C	[12]
	Roasting and water leaching	Roasting the used LiBs at low temperatures of 25 °C to 70 °C in an argon environment, thus extracting Li_2CO_3 using water leaching	[13]

Hydrometallurgy

Hydrometallurgy is the most widely used laboratory metal recovery technique and it is on the brink of commercialization. The method secures high recovery rates with few supplementary equipment [16,17]. First, acids and bases are used to ionize Li in pre-treated active materials. Following that, Li is recovered from the resulting Li^+ solutions by leaching of acids. Sulfuric acid, hydrochloric acid, citric acid, and nitric acid are examples of acids that were employed [18,19]. The leaching efficiency was improved by applying redox reactions with hydrogen peroxide (H_2O_2). Due to its high reducibility and low toxicity, the redox agent could speed up the leaching procedure. Nevertheless, a low-pH acid has the potential to release Cl_2 and NO_2, which have negative impacts on the environment (soil and air pollution) [20]. Therefore, it should be avoided if possible. In a past study, Zhang et al. [21] utilized biodegradable tritrifluoroacetic acid (TFA) and trichloroacetic acid (TCA) as the leaching agent, while H_2O_2 as the reducing agent (see Fig. 4 (a)). As a result, the Li leaching efficiencies reached 99.7%, under optimal conditions.

Another study by Chen et al. [22] found that a 96% leaching rate of Li extraction could be achieved by employing multiple processes including acid leaching, alkali leaching, solvent extraction, and chemical precipitation. An H_2SO_4 was used as a leaching agent and H_2O_2 as a reducing agent. Li et al. [23] devised a three-step procedure that includes ultrasonic cleaning, roasting, and organic acid leaching. A 1.25 mol/L of ascorbic acid was used as both the leaching agent and the reducing agent (25 g/L ratios), resulting in 98.5% of Li leaching rates. Chen et al. [24] invented a cost-effective method for recycling precious metals from used LiBs (see Fig. 4 (b)). In the extraction process, citric acid was selected as the leaching agent, and D-glucose was employed as the reducing agent to dissolve the spent cathode material. The dissolution reaction resulted in approximately 99% leaching efficiency of the Li. Furthermore, Chen et al. [25] reported that $LiCoO_2$ could be extracted into a precipitate and a Li solution using tartaric acid, which demonstrated a 97% recovery performance of Li. Overall, the recycling of Li via hydrometallurgy had an utmost efficiency of up to ~98%. However, this process necessitates a huge amount of acids and bases as leaching agents which does not come cheap. This factor alone already diverted from the green technology concept which should reduce toxicity and substances should have less risk towards safety and health. Moreover, there may be substantial secondary costs to dispose the used solutions and disobey the green technology concept which should prevent waste substances or products. Not to forget that environmental pollution still emerged as an issue in hydrometallurgy. Table 2 summarised the past reviews on Li extraction using hydrometallurgy.

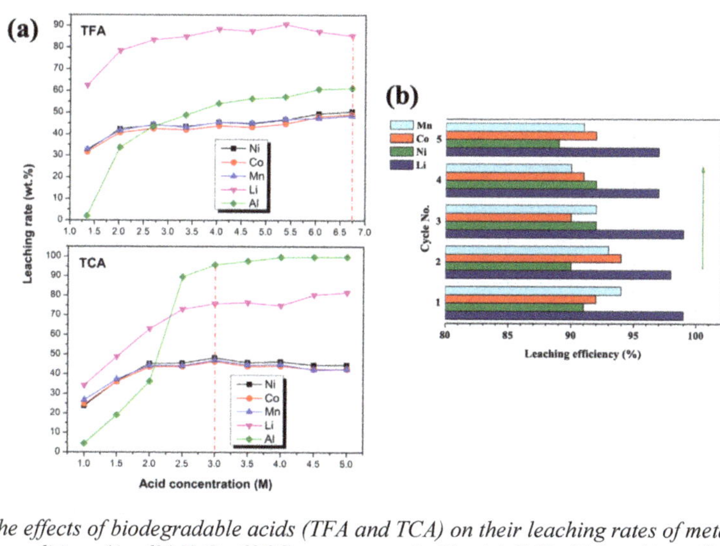

Fig. 4. The effects of biodegradable acids (TFA and TCA) on their leaching rates of metals under optimal conditions (4 vol% H_2O_2, 60 °C temperature in 30 min) [19]. (b) Leaching efficiency of citric acid to precious metals; Mn, Co, Ni, and Li under conditions of 1.5 mol/L citric acid concentration, 80 °C temperature, and 120 min reaction time [22].

Table 2. List of past reviews and their research summary on Li extraction using hydrometallurgy.

Method	Summary	Ref.
Hydrometallurgy	The uses of acid and bases *e.g.* HCl as leaching agents could be used to ionize Li^+. Moreover, the additional redox reaction with H_2O_2 could significantly improve the leaching efficiency. However, low pH acid could impact the release of Cl_2 and NO_2which promotes soil and water pollution.	[16,17]
	95% leaching rate of Li extraction could be achieved by employing multi-steps processes including acid and alkali leaching, solvent extraction and chemical precipitation.	[22]
	98.5% leaching rate of Li extraction was achieved in three steps procedure including ultrasonic cleaning, roasting and organic acid leaching.	[23]
	A cost-effective utilising citric acid and D-glucose as the leaching and reducing agent resulted in approximately 99% leaching efficiency of Li.	[24]
	$LiCoO_2$ could be extracted into precipitates using tartaric acid, which demonstrated 97% leaching efficiency.	[25]

Electrochemical extraction

Mainly, based on multiple dimensions, the electrochemical extraction method is considered more environmentally friendly than the pyrometallurgy and hydrothermal processes [26, 27]. It can be run at room temperature, does not produce a lot of chemical waste, and does not require much energy to operate. These characteristics made this electrochemical extraction method into the green technology list and tick all the necessary green characteristics. Furthermore, it can yield Li up to 99% purity [6]. The schematic diagram of the electrochemical setup is illustrated in Fig. 5 (a). However, the Li recycling effectiveness for the electrochemical extraction approach using an LTAP ceramic solid electrolyte as the membrane to recover Li from pre-treated active materials is only 75%. Also, despite its benefits, the electrochemical extraction study has only been conducted on a lab scale and has not yet been commercialized.

The electrochemical Li extraction system is capable of extracting Li via the charging and discharging mechanism [28]. Kanoh et al. [29] in their early study performed the first electrochemical capture of Li cations from a solution in battery material. Li intercalation, the most common way for Li-ion batteries to recharge, formed the basis of the capture process. They generated oxygen or hydrogen at a manganese oxide (MnO_2) working electrode and a platinum (Pt) wire counter electrode. However, this technique has some drawbacks due to the expensive process of water splitting reaction undergoing simultaneously with the Li recovery [28]. Following this, Pasta et al. [30] proposed an alternative method to capture and release Li ions and anions. The phrase "*electrochemical ion pumping*" was recently used to describe this innovation [30–32]. Since electrochemical ion pumping only involves breaking and production of chemical bonds, it requires (OR reasonably/rationally consume) less energy.

Electrodialysis is another method used in electrochemistry to extract Li from various materials (brines and seawater) [2,32]. This method relies on a selective membrane that allows only Li^+ ions to pass through. The primary benefit this (OR of the) technique is the voltage which provides the driving force for capturing Li. Therefore, no other chemical species is required to replenish the active material. To date, approximately 95% to 97% purity extraction was reported by using this method. Other works of literature reported that this method requires less time consumption [34, 35]. The increased capture rate of Li is another direct result of applying current to the electrode in electrodialysis [36]. The charging and discharging process (see Fig. 5 (b)) are the basis needed for the electrochemical method with no further chemicals or heat involved. It is worth noticing that this method is orchestrated as a green Li recycling.

Moreover, Li purity can be extracted through a designed solid electrolyte that enables only Li to flow through. However, this technique has only been validated on a lab scale, whilst the properties of the membrane extractor are important to facilitate the movement of Li-ions and recover Li with high purity. Several studies have been conducted to upgrade the properties of the membrane. Membrane extractors such as solid polymer electrolytes (SPE), gel polymer electrolytes (GPE), and composite polymer electrolytes (CPE) were designed to increase the efficiency of recycling Li purity [37–39]. These membrane strategies are explained in the following sections which include their functionalities, and an ongoing process in this area. Although this method could exhibit the highest efficiency (97.20% to 99.01%) of Li recovery compared to hydrometallurgy and pyrometallurgy processes, the drawbacks such as complex electrochemical setup and the selection of the membrane separator are crucial. Table 3 summarised the past reviews on Li extraction using electrochemical extraction.

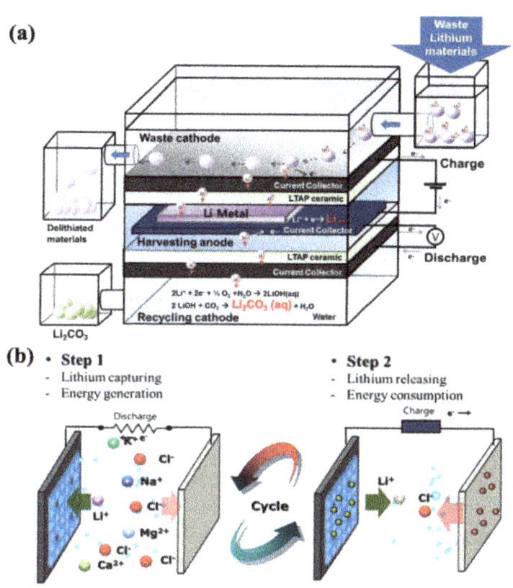

Fig. 5. (a) A schematic diagram of electrochemical setup in recycling Li. During charging, Li materials in the waste cathode compartment migrate to the harvesting anode compartment, where they combine to generate Li metal. Following this, by reacting with H_2O, O_2, and CO_2, the recycling cathode compartment generates Li_2CO_3 or LiOH during system discharge [5]. (b) In step 1, the discharging process involves capturing Li ions in the source water, and in step 2, the releasing Li process occurs in the reservoir solution [36].

Table 3. List of past reviews and their research summary on Li extraction using the electrochemical method.

Method	Summary	Ref.
Electrochemical extraction	A 75% Li extraction could be achieved via LTAP ceramic solid electrolyte as the membrane separator	[6]
	Li extraction through the charging and discharging mechanism in electrochemical setup. The separator was used to capture Li, while MnO2 and Pt as the working electrode and counter electrode, respectively	[28]
	Electrochemical ion pumping was discovered to capture and release Li ions. This method requires less energy consumption	[30]
	A 95% to 97% of Li extraction was considered in the electrodialysis step utilising brines and seawater separators	[2,32]
	A 97.20% to 99.01% of Li purity could be extracted through a designed SPE, GPE, and CPE as membrane separator	[37–40]

Membrane Polymeric Materials for Lithium-ion Recycling in Green Electrochemical Method

In the electrochemical method, the ionic conductivity of membrane materials is the prime factor in recycling Li as it will ensure an efficient migration and high purity of Li ions to pass through the membrane. Previously, high Li ion conductivity was provided by the traditional liquid electrolytes, but safety concerns are raised due to the flammability of organic solvents, like ethylene carbonate and dimethyl carbonate in the electrochemical medium [41,42]. As a result, the development of polymer-based, solid polymer electrolytes (SPEs) has been the focus of study for quite some time. The rate of electrode/electrolyte interface deformation is reduced, leading to improved thermal, mechanical, and electrochemical stability [43,44]. While these polymer electrolytes meet all of the necessary safety standards, their ionic conductivity is typically lower than that of liquid electrolytes. In a past study, Johari et al. [45] reported polyethylene oxide (PEO) based Li/SPE membrane possessed low ionic conductivity, σ, at approximately $\sim 10^{-7}$ S/cm at room temperature. This phenomenon could be explained mainly due to the crystalline phase of polymers, which restricted their wide utilization [46,47]. Therefore, a lot of materials and methods were explored to reduce the glass transition temperature (T_g) and improve the conductive network of SPEs.

SPEs consist of a primarily polymeric matrix with one or more fillers, which may be ceramic, Li salts, or ionic liquids. Generally, the ionic conductivity of an SPE is a key factor, therefore fillers are often added to the material to enhance its conductivity. The introduction of those fillers could decrease the polymer's crystallinity, which significantly improves ion migration and ionic conductivity. Poly(ethylene oxide) (PEO), poly(vinylidene fluoride) (PVDF) and its copolymers, poly(ethylene glycol) (PEG), poly(acrylonitrile) (PAN), and poly(ethylene carbonate) ($PECO_3$) are the most used (OR utilized/applied) polymer matrices for SPEs [48–50]. Li salts, ionic liquids, composite polymers, and interpenetrating network polymers are all examples of modified membranes. The following section discussed the current findings on the abovementioned membrane materials, their potentials, and their drawbacks in recycling Li in the green electrochemical method.

SPEs-based Li salts and ionic liquid

In recent studies, Li salts such as lithium bis(trifluoromethanesulfonyl)imide (LiTFSI) and lithium perchlorate ($LiClO_4$) have been shown to be the most effective active fillers for SPEs' membrane [51,52]. The ionic conductivity and the number of charge carriers both rise (OR grow/increase/raise) directly (OR instantly/promptly) as a result of (OR resulted from or due to) the presence of these salts. Moreover, ionic liquids like 1-ethyl-3- methylimidazolium bis(trifluoromethylsulfonyl)imide (EMIM)(TFSI), 1-butyl-3-methylimidazolium chloride (BMIM)(Cl) and lithium tetrafluoroborate ($LiBF_4$), lithium hexafluorophosphate ($LiPF_6$) could also improve the ionic conductivity of SPEs [53,55]. According to Yoon et al. [53], the ionic liquid could influence the ion-dipole interactions between ions and the polymer matrix. Therefore, these interactions promote Li salt dissociation and boost the number and mobility of Li-ion transfers. In the end, it could produce a high yield of Li recovery.

Mendes et al. [56] designed polyurethane acrylate (PUA) and bis(trifluoromethanesulfonyl) imide (LiTFSI) salts as SPE membranes. The introduction of LiTFSI significantly modified the surface of SPEs. The ionic conductivity of UV photocurable in 30 wt% LiTFSI content reached a maximum of 0.0032 mS/cm at ambient temperature. In addition, the limited oxidative stability and poor interfacial charge transport of existing SPEs can be ameliorated by employing ether-based electrolytes with the presence of aluminum fluoride (AlF_3) [57]. Incorporating PEO into a matrix containing lithium trifluoromethyl silane (LiTFSI) resulted in the creation of an SPE with impressive flame retardant capacity, enhanced mechanical stability, and a wide electrochemical stability window of up to 4.6 V. In addition, their study also embarks on the addition of sodium alginate (SA) to improve the mechanical properties of SPE and the safety of the membrane through the nanofiber structure [58]. Figs. 6 (a and b) illustrate the physical appearance and the morphological images of the membrane.

In a previous study by Kale et al. [59], a more complicated composite of poly(ethylene oxide) monomethacrylate, PEGMA, and cellulose filled with LiTFSI and 1-butyl-1-methylpyrrolidinium bis(trifluoromethanesulfonyl)imide (Pyr14TFSI) (see schematic diagram in Fig. 6 (c)) also (OR have) achieved a good cycling stability at room temperature with an average transference number of 0.43 and electrochemical stability window up to 5 V. Further, the membrane does offer/present/provide flexible attribute/characteristic as it does not succumb to bending and even twisting (Figs. 6 (d and e)). The crystallinity of the SPE can be lowered by including several Li salts into a PMMA matrix grafted with natural rubber, which in turn increases the Li transference number and, by extension, the ionic conductivity of the material. However, the use of this type of ionic liquid and Li salts-based SPEs has a drawback, which is the toxicity impact due to the possibility of release into the soil or water courses, where they could become persistent pollutants and pose environmental risks. Table 4 depicts the natural polymers and their fillers that have been the basis for recent developments in SPEs membranes.

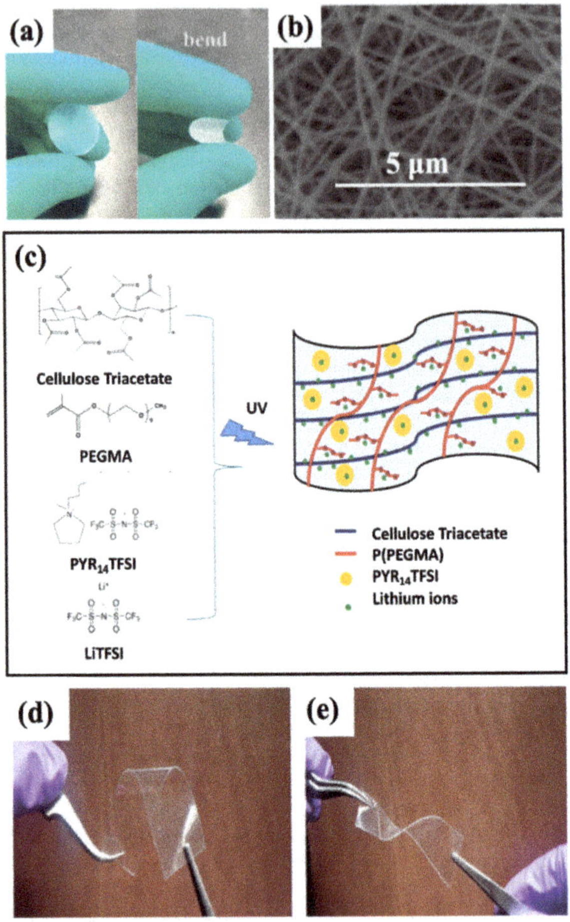

Fig. 6. (a) A free-standing film of PEO/SA-SPE membrane, (b) which is randomly pilled at ~170 nm observed by SEM images [56]. This is believed as an ideal 3D framework for structural support of membranes. (c) Li coordination site accessibility and polymer synthesis reaction scheme. (d and e) SPE membranes that can stand on their own, are transparent, and flexible in bending and twisting [59].

Table 4. Natural polymers and their fillers have been the basis for recent developments in SPEs membranes.

Polymer Matrix	Fillers	Ionic conductivity (S/cm)	Reference
PMMA/natural rubber	LiBF	1.89×10^{-6}	[60]
Cellulose	$LiPF_6$	2.71×10^{-2}	[61]
Triacetate/PEGMA	LiTFSI	5.24×10^{-3}	[59]
PEO/SA	LiTFSI	$\sim 10^{-4}$	[58]

Composite polymer electrolytes (CPEs) membrane

In a recent study, Kim et al. [62] utilized the UV radiation approach to increase the mechanical strength of the PEO membrane. It is clear that the polymer can be solidified following UV radiation in corresponds with trimethylolpropane ethoxylate triacrylate (ETPTA) ceramic filler included in the precursor solution. This resulted in the production of a freestanding membrane. At room temperature, the ionic conductivity can reach as high as 3.3×10^3 S/cm with a 0.76 Li^+ transference number. Din et al. [63] created the CPEs with varied structures, such as mixes of PEO with ceramic particles, ceramic networks, and ceramic layers. The 50 to 70 μm thick membrane was created when the nanowire of $Li_7La_3Zr_2O_{12}$ (LLZO) was combined with PEO containing $LiClO_4$ salt. The maximum ionic conductivity of 4.42×10^4 S/cm was achieved with the addition of 52.5% weight of LLZO. Fig. 7 (a) shows the filler particles of LLZO are packed densely throughout the polymer matrix. By introducing the filler, the PEO undergoes a change in crystallinity that is conducive to high chain mobility, and, by extension, high Li-ion conductivities were produced inside the polymer matrix.

High ionic conductivities could be observed in CPEs including well-aligned conducting nanowires by varying their orientations [64]. Figs. 7 (b-e) illustrates a comparison of the Li conductive pathway in a different alignment. Well-aligned nanowires exhibit 6.05×10^5 S/cm at 30 °C conductivity which is 10 times higher than that of their randomly-dispersed counterparts. The calculated surface conductivity of the nanowires at 30 °C is 1.26×10^2 S/cm, which is quite close to the ion conduction in liquid electrolytes. A rapid ion conduction channel without surface-to-surface connections is responsible for the notable conductivity enhancement observed in nanowires. The addition of the nanowires also enhances the polymer electrolytes' long-term conductivity stability. Solid electrolytes with in-plane orientation could also pave the way for more efficient Li migration in the membrane. Li et al. [64] developed free-standing nanofibers based on a 3D $Li_{0.35}La_{0.55}TiO_3$ (LLTO) framework. In order to prevent the Li-ion route from being blocked by clumps of ceramic nanoparticles, a 3D framework may be necessary. Their study concluded that the LLTO 3D framework is superior to one made with LLTO nanoparticles. At room temperature, the conductivity reaches its maximum of 8.8×10^5 S/cm. According to Feng et al. [65], issues like brittleness (poor mechanical properties) and less compatibility with ceramic-polymer adhesion could diminish the Li recovery process. More previous studies following the types of CPEs and their respective conductivity are summarised in Table 5.

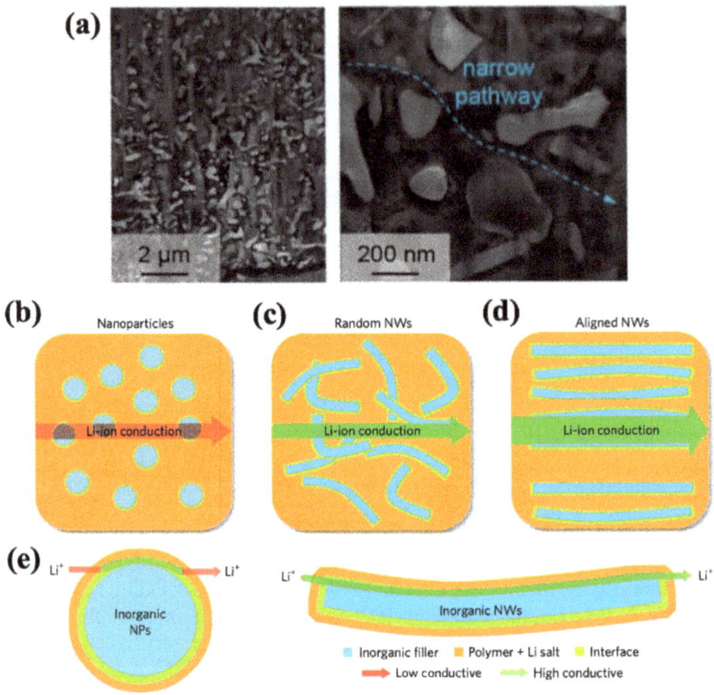

Fig. 7. (a) SEM images of the cross-sectional area of PEO consisting of 20 wt% nanowires LLZO [61]. The Li-ion conduction routes in CPEs contain (b) nanoparticles, (c) random nanowires, and (d) aligned nanowires. Unpredictable nanowires may provide a more consistent rapid conduction pathway for Li ions than isolated nanoparticles. The crossing junctions do not exist in aligned nanowires, which is a major advantage over random nanowires. (e) Li-ion conduction occurs along the surface of inorganic nanoparticles and nanowires [64].

Table 5. Types of ceramic fillers in their electrochemical performances in CPEs.

Structural type	**Ceramic filler**	**Polymer host**	**Ionic conductivity (S/cm)**	**Reference**
Nanoparticle	$Li_{6.75}La_3Zr_{1.75}Nb_{0.25}O_{12}$	$PVDF/LiClO_4$	9.20×10^{-5}	[67]
	LAGP	PET/LiTFSI	7.78×10^{-5}	[68]
	LATP	$PEG/PEO/LiClO_4$	5.20×10^{-5}	[69]
Nanowire	$Li_{6.4}La_3Zr_2Al_{0.2}O_{12}$	PEO/LiTFSI	2.50×10^{-4}	[70]
	$Li_{6.75}La_3Zr_{1.75}Nb_{0.25}O_{12}$	$LiClO_4$	2.20×10^{-5}	[71]
	LLZO	PVDF/LiTFSI	9.50×10^{-4}	[72]
3D framework	$Li_{6.28}La_3Zr_2Al_{0.24}O_{12}$	LiTFSI	8.50×10^{-5}	[73]
	$Li_{6.4}La_3Zr_2A_{10.2}O_{12}$	LiTFSI	1.80×10^{-5}	[74]
	LLZO	PEO/LiTFSI	1.14×10^{-4}	[75]

Interpenetrating network polymer (IPNs) membrane

The combination of CPEs/SPEs with IPNs is an innovative way to improve the efficiency of Li's recovery performance. Generally, IPNs are made up of two or more cross-linked polymers that are not bonded to one another [76]. Reviews have found that IPNs can improve ion conductivities and mechanical properties by reducing their crystallinity [77]. A past study by Zeng et al. [78] developed the IPN-based poly(ether-acrylate) that possessed 2.2×10^4 S/cm at room temperature with a low glass transition temperature at 64.2 °C. The newly ventured electronically conductive polymers (CPs) namely poly(3,4-ethylene dioxythiophene) (PEDOT) have taken place in designing the basic electrode of supercapacitors [79]. In this semi-IPN framework, it is possible to create new interconnecting materials or membrane separators that are based on CPs [80].

This strategy was designed to replace the older method, which uses organic liquid electrolytes. These CPs are chosen due to their unique IPNs that are capable of decreasing the crystallinity and therefore enhancing ionic conductivities and mechanical properties. IPN structure in SPE-CPs could be compromised into multiple networks which are at least partially interlaced on a polymer scale but not covalently bonded to each other. Other characteristics such as (1) good thermal stability, (2) high electrolyte wettability, (3) redox-active separator, (4) electron-rich groups, (5) fibrous structure, and (6) capacity-enhancing could drive the ion exchange through the SPE-CP membrane. When applied as the membrane separator for Li extraction, the electronic nature of CPs should be eliminated to avoid the risk of a short circuit between positive and negative electrodes. To solve the issue, Wang et al. [81] embarked on the electrochemical overoxidation of CPs, prior to their incorporation into the SPE membrane. It offers a loss of electronic conductivity, the formation of carbonyl groups that can attract cations, and hinders the diffusion of anions through the membrane. Based on this, the electron-rich groups are introduced into the conjugated CP's structure and converted into a rapid ion-exchange membrane. This newly emerged technique aims to enhance and broaden the useful features of original SPE compounds.

The CP-based SPEs could offer (OR provide) ionic conductivity that exceeds 10^{-5} S/cm, facilitating the migration of Li ions through the membrane. Besides, Li^+ is typically less dynamic than its anionic counterpart since its motion is coupled with the lewis basic sites of the polymer matrix. The ratio of migrating Li^+ to all migrating ions, including anions in the electrolyte is defined as the Lithium Ion Transference Number (LITN) [82]. During charge/discharge cycling, only migrating Li^+ contributes to the performance of the battery. Overall, the ideal LiTN is equal to 1, which is achievable by CP. In terms of physical properties, CPs can offer great fiber orientation, as their porosity is tuneable and can be designed accordingly. This allows the acceleration mechanism of Li^+ during the extraction process. Fong et al. [79] developed the semi-IPNs-based PEDOT. The material's shape enables these enhancements in specific capacitance, as depicted schematically in Fig. 8 (a). The ionically PEDOT in post-overoxidation provides an ion reservoir throughout the electrode, making pseudocapacitive polymer available for charge storage processes even inside the bulk of the material. This technology improves the material use efficiency of an electrically CP without requiring sophisticated production procedures or compromising mechanical stability. In addition, the flexible framework structure characteristic of the ionically conductive matrix offer multiple advantages including; (1) able to accommodate the volumetric variations associated with ion intercalation/deintercalation in the pseudocapacitive polymer, (2) decrease mechanical stress in the electrode and (3) produce high cycling stability.

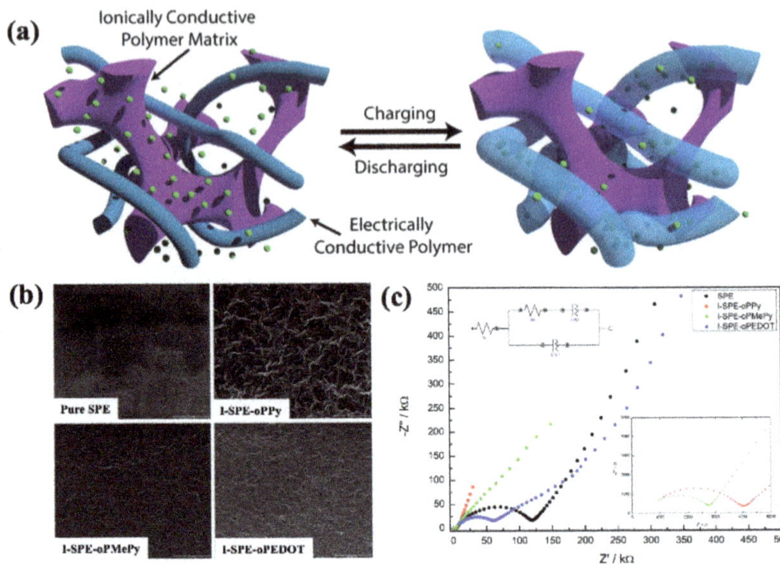

Fig. 8. (a) Schematic illustration of semi-IPN PEDOT-PEO matrix. The cross-linked PEO matrix acts as a reservoir of electrolyte ions and locally constraints swelling of the PEDOT during charging-discharging cycling [77]. (b) The morphological images of pure SPE, I-SPE-oPPy, I-SPE-oPMePy, and I-SPE-oPEDOT. The pyrrole-derivatives I-SPE-oPPy (b) and I-SPE-oPMePy (c) display a uniform network without any depositions. Consequently, there is a link between the materials' structural analysis and their ionic conductivity [81].

Leš et al. [83] in their work reported SPEs with the encapsulation of the CPs; poly(1H-pyrrole) (I-SPE-oPPy), poly(N-methyl pyrrole) (I-SPE-oPMePy), and poly(3,4-ethylene dioxythiophene) (I-SPE-oPEDOT). Each CPs was overoxidized subsequently by an electrochemical procedure. The morphology of the fabricated membranes is shown in Fig. 8 (b). The surface of the SPE polymer matrix pure SPE is quite dense with few inhomogeneous regions with pores (areas with brighter contrast). As a result, the LiTFSI is not uniformly distributed throughout the matrix. The polymer composites, on the other hand, construct a 3D structure surface despite having significantly thicker polymer chains. More space was observed between the individual polymer pieces. The I-SPE-PEDOT membrane exhibits some crystals attached to the polymer chains. A link between the dissociation of the Li salt in the polymer matrix and the ionic conductivity is dominant since the dissolution leads to a bigger number of free ions for Li transport [84]. Meanwhile, in Fig. 8 (c), the I-SPEs-oPMePy exhibited higher ionic conductivity at 7.00×10^{-6} S/cm compared to pure SPE at 1.3×10^{-7} S/cm. The calculated activation energies for Li-ion conduction display the effect of the generated IPNs, which lower the energy barrier for Li^{+} conduction. Such materials could lead to a novel group of interpenetrated solid polymer electrolytes as separator materials based on mixed conductors. Moreover, the combination of a polyacrylate-matrix with cation-selective properties of overoxidized conducting polymers leads to 3D materials with higher ionic conductivity than SPEs and selective ion-exchange membrane properties with good stability by facile fabrication. These kinds of synthesis methods appear to provide an alternative way of varying the properties and applications of CPs.

Conclusions and Future Outlooks

In the last decade, Li has emerged as a highly sought after resource. The rapid increase in demands for Li in lithium batteries (LiBs) for various applications such as electric vehicle batteries, energy storage grid systems, and personal electronics during the past few years have prompted concerns about the supply among the international community. Despite the resource reserve, the environmental sustainability of Li extraction, production and recovery is also critical. However, the current methods of Li extraction such as pyrometallurgy and hydrometallurgy are time-consuming and energy-intensive. In addition, these procedures are harmful to the environment because of the large amounts of chemicals used. Therefore, there is a lot of interest all around the world in refining and improving existing Li extraction in green methods. The electrochemical extraction method is seen as more environmentally benign due to manifold benefits/advantages, such as (1) produced less chemical waste production, (2) operation (OR could be operated) at room temperature, (3) does not require much energy to operate and, (4) it could yield Li up to 99% purity. This approach is capable of extracting Li from LIBs using a separator membrane, as explained in this article. The use of SPEs and CPEs-based membranes for Li harvesting provides many benefits, including (1) a low-cost operation, (2) high separation efficiency, (3) high selectivity, (4) high permeability. In addition, the separation process is less harmful to the environment. Sadly, membrane technologies have fallen victim to their own advantages. Membrane fouling, membrane longevity, and difficulties in scaling up operations are all drawbacks of membrane technologies. Poor physical and ionic conductivity of membranes are also some of the significant obstacles, necessitating more development in the future. Therefore, modification of the membranes and the addition of fillers are often used to scale up their ionic conductivity. The introduction of those fillers could decrease the polymer's crystallinity, which significantly improves ion migration and ionic conductivity. Li salts, ionic liquids, composite polymers, and interpenetrating network polymers are all examples of modified membranes. This includes focus on developing novel membranes with enhanced selectivity and stability, as well as optimizing existing procedures. For instance, using CPs integrated into the SPE membrane could tune a specific framework architecture and present promising avenues for developing continuous Li extraction in green recycling.

Conflicts of Interest

There are no conflicts of interest to declare.

Acknowledgments

Authors would like to thank Universiti Teknologi MARA (UiTM), Shah Alam and PETRONAS Research Sdn. Bhd. for the support to conduct this research.

References

[1] W. Liu, T. Placke, K.T. Chau, Overview of batteries and battery management for electric vehicles, Energy Reports. 8 (2022) 4058-4084. https://doi.org/10.1016/j.egyr.2022.03.016

[2] V. Lundaev, A.A. Solomon, U. Caldera, C. Breyer, Material extraction potential of desalination brines: A technical and economic evaluation of brines as a possible new material source, Miner Eng. 185 (2022) 107652. https://doi.org/10.1016/j.mineng.2022.107652

[3] Petersen, H. A., Myren, T. H., O'Sullivan, S. J., & Luca, O. R. Electrochemical methods for materials recycling. Materials Advances, 2(4), (2021)1113-1138. https://doi.org/10.1039/D0MA00689K

[4] H. Bae, Y. Kim, Technologies of lithium recycling from waste lithium-ion batteries: A review, Mater Adv. 2 (2021) 3234-3250. https://doi.org/10.1039/D1MA00216C

[5] W. Jin, Y. Zhang, Sustainable Electrochemical Extraction of Metal Resources from Waste Streams: From Removal to Recovery, ACS Sustain Chem Eng. 8 (2020) 4693-4707. https://doi.org/10.1021/acssuschemeng.9b07007

[6] H. Bae, S.M. Hwang, I. Seo, Y. Kim, Electrochemical Lithium Recycling System toward Renewable and Sustainable Energy Technologies, J Electrochem Soc. 163 (2016) 199-205. https://doi.org/10.1149/2.0691607jes

[7] Anastas, P. T., & Warner, J. C. Principles of green chemistry. Green chemistry: Theory and practice, 29 (1998).

[8] S. Windisch-Kern, A. Holzer, C. Ponak, H. Raupenstrauch, Pyrometallurgical lithium-ion-battery recycling: Approach to limiting lithium slagging with the indured reactor concept, Processes. 9 (2021) 1-15. https://doi.org/10.3390/pr9010084

[9] X. Zheng, W. Gao, X. Zhang, M. He, X. Lin, H. Cao, Y. Zhang, Z. Sun, Spent lithium-ion battery recycling - Reductive ammonia leaching of metals from cathode scrap by sodium sulphite, Waste Management. 60 (2017) 680-688. https://doi.org/10.1016/j.wasman.2016.12.007

[10] A. Holzer, S. Windisch-Kern, C. Ponak, H. Raupenstrauch, A novel pyrometallurgical recycling process for lithium-ion batteries and its application to the recycling of lco and lfp, Metals (Basel). 11 (2021) 1-22. https://doi.org/10.3390/met11010149

[11] T. Georgi-Maschler, B. Friedrich, R. Weyhe, H. Heegn, M. Rutz, Development of a recycling process for Li-ion batteries, J Power Sources. 207 (2012) 173-182. https://doi.org/10.1016/j.jpowsour.2012.01.152

[12] D.M. Werner, T. Mütze, U.A. Peuker, Influence of Cell Opening Methods on Electrolyte Removal during Processing in Lithium-Ion Battery Recycling, Metals (Basel). 12 (2022) 663. https://doi.org/10.3390/met12040663

[13] C. Peng, F. Liu, Z. Wang, B.P. Wilson, M. Lundström, Selective extraction of lithium (Li) and preparation of battery grade lithium carbonate (Li2CO3) from spent Li-ion batteries in nitrate system, J Power Sources. 415 (2019) 179-188. https://doi.org/10.1016/j.jpowsour.2019.01.072

[14] R. Golmohammadzadeh, F. Rashchi, E. Vahidi, Recovery of lithium and cobalt from spent lithium-ion batteries using organic acids: Process optimization and kinetic aspects, Waste Management. 64 (2017) 244-254. https://doi.org/10.1016/j.wasman.2017.03.037

[15] S. Doose, J.K. Mayer, P. Michalowski, A. Kwade, Challenges in eco-friendly battery recycling and closed material cycles: A perspective on future lithium battery generations, Metals (Basel). 11 (2021) 1-17. https://doi.org/10.3390/met11020291

[16] J. Neumann, M. Petranikova, M. Meeus, J.D. Gamarra, R. Younesi, M. Winter, S. Nowak, Recycling of Lithium-Ion Batteries-Current State of the Art, Circular Economy, and Next Generation Recycling, Adv Energy Mater. 12 (2022) 2102917. https://doi.org/10.1002/aenm.202102917

[17] L.F. Zhou, D. Yang, T. Du, H. Gong, W. bin Luo, The Current Process for the Recycling of Spent Lithium Ion Batteries, Front Chem. 8 (2020) 578044. https://doi.org/10.3389/fchem.2020.578044

[18] L. ye Sun, B. Rui Liu, T. Wu, G. de Wang, Q. Huang, Y. Feng Su, F. Wu, Hydrometallurgical recycling of valuable metals from spent lithium-ion batteries by reductive

leaching with stannous chloride, Int Journal of Minerals, Metallurgy and Materials. 28 (2021) 991-1000. https://doi.org/10.1007/s12613-020-2115-z

[19] N. Vieceli, R. Casasola, G. Lombardo, B. Ebin, M. Petranikova, Hydrometallurgical recycling of EV lithium-ion batteries: Effects of incineration on the leaching efficiency of metals using sulfuric acid, Waste Management. 125 (2021) 192-203. https://doi.org/10.1016/j.wasman.2021.02.039

[20] W. Mrozik, M.A. Rajaeifar, O. Heidrich, P. Christensen, Environmental impacts, pollution sources and pathways of spent lithium-ion batteries, Energy Environ Sci. 14 (2021) 6099-6121. https://doi.org/10.1039/D1EE00691F

[21] X. Zhang, H. Cao, Y. Xie, P. Ning, H. An, H. You, F. Nawaz, A closed-loop process for recycling LiNi1/3Co1/3Mn1/3O2 from the cathode scraps of lithium-ion batteries: Process optimization and kinetics analysis, Sep Purif Technol. 150 (2015) 186-195. https://doi.org/10.1016/j.seppur.2015.07.003

[22] X. Chen, B. Fan, L. Xu, T. Zhou, J. Kong, An atom-economic process for the recovery of high value-added metals from spent lithium-ion batteries, J Clean Prod. 112 (2016) 3562-3570. https://doi.org/10.1016/j.jclepro.2015.10.132

[23] L. Li, J. Lu, Y. Ren, X.X. Zhang, R.J. Chen, F. Wu, K. Amine, Ascorbic-acid-assisted recovery of cobalt and lithium from spent Li-ion batteries, J Power Sources. 218 (2012) 21-27. https://doi.org/10.1016/j.jpowsour.2012.06.068

[24] X. Chen, B. Fan, L. Xu, T. Zhou, J. Kong, An atom-economic process for the recovery of high value-added metals from spent lithium-ion batteries, J Clean Prod. 112 (2016) 3562-3570. https://doi.org/10.1016/j.jclepro.2015.10.132

[25] X. Chen, D. Kang, L. Cao, J. Li, T. Zhou, H. Ma, Separation and recovery of valuable metals from spent lithium-ion batteries: Simultaneous recovery of Li and Co in a single step, Sep Purif Technol. 210 (2019) 690-697. https://doi.org/10.1016/j.seppur.2018.08.072

[26] Y. Zhang, W. Sun, R. Xu, L. Wang, H. Tang, Lithium extraction from water lithium resources through green electrochemical-battery approaches: A comprehensive review, J Clean Prod. 285 (2021) 124905. https://doi.org/10.1016/j.jclepro.2020.124905

[27] X. Zhao, H. Yang, Y. Wang, Z. Sha, Review on the electrochemical extraction of lithium from seawater/brine, Journal of Electroanalytical Chemistry. 850 (2019) 113389. https://doi.org/10.1016/j.jelechem.2019.113389

[28] Y. Bai, N. Muralidharan, J. Li, R. Essehli, I. Belharouak, Sustainable direct recycling of lithium-ion batteries via solvent recovery of electrode materials, Chem Sus Chem. 13 (2020) 5664-5670. https://doi.org/10.1002/cssc.202001479

[29] H. Kanoh, K. Ooi, Y. Miyai, S. Katoh, Electrochemical recovery of lithium ions in the aqueous phase, Sep Sci Technol. 28 (1993) 643-651. https://doi.org/10.1080/01496399308019512

[30] M. Pasta, C.D. Wessells, Y. Cui, F. la Mantia, A desalination battery, Nano Lett. 12 (2012) 839-843. https://doi.org/10.1021/nl203889e

[31] A. Battistel, M.S. Palagonia, D. Brogioli, F. la Mantia, R. Trócoli, Electrochemical Methods for Lithium Recovery: A Comprehensive and Critical Review, Advanced Materials. 32 (2020) 1905440. https://doi.org/10.1002/adma.201905440

[32] N. Kim, X. Su, C. Kim, Electrochemical lithium recovery system through the simultaneous lithium enrichment via sustainable redox reaction, Chemical Engineering Journal. 420 (2021) 127715. https://doi.org/10.1016/j.cej.2020.127715

[33] G. Zhou, L. Chen, Y. Chao, X. Li, G. Luo, W. Zhu, Progress in electrochemical lithium ion pumping for lithium recovery, Journal of Energy Chemistry. 59 (2021) 431-445. https://doi.org/10.1016/j.jechem.2020.11.012

[34] V. Flexer, C.F. Baspineiro, C.I. Galli, Lithium recovery from brines: A vital raw material for green energies with a potential environmental impact in its mining and processing, Science of the Total Environment. 639 (2018) 1188-1204. https://doi.org/10.1016/j.scitotenv.2018.05.223

[35] J.F. Song, L.D. Nghiem, X.M. Li, T. He, Lithium extraction from Chinese salt-lake brines: Opportunities, challenges, and future outlook, Environ Sci (Camb). 3 (2017) 593-597. https://doi.org/10.1039/C7EW00020K

[36] J. Lee, S.H. Yu, C. Kim, Y.E. Sung, J. Yoon, Highly selective lithium recovery from brine using a λ-MnO 2-Ag battery, Physical Chemistry Chemical Physics. 15 (2013) 7690-7695. https://doi.org/10.1039/c3cp50919b

[37] T. Or, S.W.D. Gourley, K. Kaliyappan, A. Yu, Z. Chen, Recycling of mixed cathode lithium-ion batteries for electric vehicles: Current status and future outlook, Carbon Energy. 2 (2020) 6-43. https://doi.org/10.1002/cey2.29

[38] W. Gu, F. Li, T. Liu, S. Gong, Q. Gao, J. Li, Z. Fang, Recyclable, Self-Healing Solid Polymer Electrolytes by Soy Protein-Based Dynamic Network, Advanced Science. (2022) 2103623. https://doi.org/10.1002/advs.202103623

[39] F. Baskoro, H.Q. Wong, H.J. Yen, Strategic Structural Design of a Gel Polymer Electrolyte toward a High-Efficiency Lithium-Ion Battery, ACS Appl Energy Mater. 2 (2019) 3937-3971. https://doi.org/10.1021/acsaem.9b00295

[40] S.K. Kim, Y. Yoon, J.H. Ryu, J.H. Kim, S. Ji, W. Song, S. Myung, J. Lim, H.K. Jung, S.S. Lee, J. Lee, K.S. An, Recyclable High-Performance Polymer Electrolyte Based on a Modified Methyl Cellulose-Lithium Trifluoromethanesulfonate Salt Composite for Sustainable Energy Systems, Chem Sus Chem. 13 (2020) 376-384. https://doi.org/10.1002/cssc.201902756

[41] L. Fan, S. Wei, S. Li, Q. Li, Y. Lu, Recent Progress of the Solid-State Electrolytes for High-Energy Metal-Based Batteries, Adv Energy Mater. 8 (2018) 1702657. https://doi.org/10.1002/aenm.201702657

[42] K. Liu, Y. Liu, D. Lin, A. Pei, Y. Cui, Materials for lithium-ion battery safety, Science advances. 4 (2018) 9820. https://doi.org/10.1126/sciadv.aas9820

[43] Z. Sun, Y. Li, S. Zhang, L. Shi, H. Wu, H. Bu, S. Ding, G-C3N4 nanosheets enhanced solid polymer electrolytes with excellent electrochemical performance, mechanical properties, and thermal stability, J Mater Chem A Mater. 7 (2019) 11069-11076. https://doi.org/10.1039/C9TA00634F

[44] W. Tang, S. Tang, C. Zhang, Q. Ma, Q. Xiang, Y.W. Yang, J. Luo, Simultaneously Enhancing the Thermal Stability, Mechanical Modulus, and Electrochemical Performance of Solid Polymer Electrolytes by Incorporating 2D Sheets, Adv Energy Mater. 8 (2018) 1800866. https://doi.org/10.1002/aenm.201800866

[45] S.N. 'Afini M. Johari, N.A. Tajuddin, H. Hanibah, S.K. Deraman, A Review: Ionic Conductivity of Solid Polymer Electrolyte Based Polyethylene Oxide, Int J Electrochem Sci. 16 (2021) 1-15. https://doi.org/10.20964/2021.10.53

[46] Y.L. Ni'Mah, Z.H. Muhaiminah, S. Suprapto, Increase of solid polymer electrolyte ionic conductivity using nano-SiO2 synthesized from sugarcane bagasse as filler, Polymers (Basel). 13 (2021) 4240. https://doi.org/10.3390/polym13234240

[47] Z. Wei, Y. Ren, M. Wang, J. He, W. Huo, H. Tang, Improving the Conductivity of Solid Polymer Electrolyte by Grain Reforming, Nanoscale Res Lett. 15 (2020) 1-8. https://doi.org/10.1186/s11671-020-03355-4

[48] N. Verdier, D. Lepage, R. Zidani, A. Prébé, D. Aymé-Perrot, C. Pellerin, M. Dollé, D. Rochefort, Cross-Linked Polyacrylonitrile-Based Elastomer Used as Gel Polymer Electrolyte in Li-Ion Battery, ACS Appl Energy Mater. 3 (2020) 1099-1110. https://doi.org/10.1021/acsaem.9b02129

[49] G. Homann, L. Stolz, J. Nair, I.C. Laskovic, M. Winter, J. Kasnatscheew, Poly(Ethylene Oxide)-based Electrolyte for Solid-State-Lithium-Batteries with High Voltage Positive Electrodes: Evaluating the Role of Electrolyte Oxidation in Rapid Cell Failure, Sci Rep. 10 (2020) 1-9. https://doi.org/10.1038/s41598-020-61373-9

[50] C.Y. Chiang, Y.J. Shen, M.J. Reddy, P.P. Chu, Complexation of poly(vinylidene fluoride): LiPF6 solid polymer electrolyte with enhanced ion conduction in "wet" form, J Power Sources. 123 (2003) 222-229. https://doi.org/10.1016/S0378-7753(03)00514-7

[51] M. Serhan, M. Sprowls, D. Jackemeyer, M. Long, I.D. Perez, W. Maret, N. Tao, E. Forzani, Total iron measurement in human serum with a smartphone, in AIChE Annual Meeting, Conference Proceedings, American Institute of Chemical Engineers, 2019. https://doi.org/10.1109/JTEHM.2020.3005308

[52] E. Nasybulin, W. Xu, M.H. Engelhard, Z. Nie, S.D. Burton, L. Cosimbescu, M.E. Gross, J.G. Zhang, Effects of electrolyte salts on the performance of Li-O2 batteries, Journal of Physical Chemistry C. 117 (2013) 2635-2645. https://doi.org/10.1021/jp311114u

[53] I.J. Shamsudin, A. Ahmad, N.H. Hassan, H. Kaddami, Biopolymer electrolytes based on carboxymethyl κ-carrageenan and imidazolium ionic liquid, Ionics (Kiel). 22 (2016) 841-851. https://doi.org/10.1007/s11581-015-1598-5

[54] M.D. Widstrom, K.B. Ludwig, J.E. Matthews, A. Jarry, M. Erdi, A. v. Cresce, G. Rubloff, P. Kofinas, Enabling high-performance all-solid-state lithium metal batteries using solid polymer electrolytes plasticized with ionic liquid, Electrochim Acta. 345 (2020) 136156. https://doi.org/10.1016/j.electacta.2020.136156

[55] H.K. Yoon, W.S. Chung, N.J. Jo, Study on ionic transport mechanism and interactions between salt and polymer chain in PAN based solid polymer electrolytes containing LiCF3SO3, Electrochim Acta. 50 (2004) 289-293. https://doi.org/10.1016/j.electacta.2004.01.095

[56] C. Mendes-Felipe, J.C. Barbosa, R. Gonçalves, D. Miranda, C.M. Costa, J.L. Vilas-Vilela, S. Lanceros-Mendez, Lithium bis(trifluoromethanesulfonyl)imide blended in polyurethane acrylate photocurable solid polymer electrolytes for lithium-ion batteries, Journal of Energy Chemistry. 62 (2021) 485-496. https://doi.org/10.1016/j.jechem.2021.01.030

[57] C.Z. Zhao, Q. Zhao, X. Liu, J. Zheng, S. Stalin, Q. Zhang, L.A. Archer, Rechargeable Lithium Metal Batteries with an In-Built Solid-State Polymer Electrolyte and a High Voltage/Loading Ni-Rich Layered Cathode, Advanced Materials. 32 (2020) 1905629. https://doi.org/10.1002/adma.201905629

[58] Y. Chen, Y. Wang, Z. Li, D. Wang, H. Yuan, H. Zhang, Y. Tan, A flame retarded polymer-based composite solid electrolyte improved by natural polysaccharides, Composites Communications. 26 (2021) 100774. https://doi.org/10.1016/j.coco.2021.100774

[59] S.B. Kale, T.C. Nirmale, N.D. Khupse, B.B. Kale, M. v. Kulkarni, S. Pavitran, S.W. Gosavi, Cellulose-Derived Flame-Retardant Solid Polymer Electrolyte for Lithium-Ion Batteries, ACS Sustain Chem Eng. 9 (2021) 1559-1567. https://doi.org/10.1021/acssuschemeng.0c06309

[60] R.A.G. Whba, L. TianKhoon, M.S. Su'ait, M.Y.A. Rahman, A. Ahmad, Influence of binary lithium salts on 49% poly(methyl methacrylate) grafted natural rubber based solid polymer electrolytes, Arabian Journal of Chemistry. 13 (2020) 3351-3361. https://doi.org/10.1016/j.arabjc.2018.11.009

[61] Q. Sabrina, C.R. Ratri, A. Hardiansyah, T. Lestariningsih, A. Subhan, A. Rifai, R. Yudianti, H. Uyama, Preparation and characterization of nanofibrous cellulose as solid polymer electrolyte for lithium-ion battery applications, RSC Adv. 11 (2021) 22929-22936. https://doi.org/10.1039/D1RA03480D

[62] J. il Kim, Y.G. Choi, Y. Ahn, D. Kim, J.H. Park, Optimized ion-conductive pathway in UV-cured solid polymer electrolytes for all-solid lithium/sodium ion batteries, J Memb Sci. 619 (2021) 118771. https://doi.org/10.1016/j.memsci.2020.118771

[63] M.M.U. Din, M. Häusler, S.M. Fischer, K. Ratzenböck, F.F. Chamasemani, I. Hanghofer, V. Henninge, R. Brunner, C. Slugovc, D. Rettenwander, Role of Filler Content and Morphology in LLZO/PEO Membranes, Front Energy Res. 9 (2021) 532. https://doi.org/10.3389/fenrg.2021.711610

[64] W. Liu, S.W. Lee, D. Lin, F. Shi, S. Wang, A.D. Sendek, Y. Cui, Enhancing ionic conductivity in composite polymer electrolytes with well-aligned ceramic nanowires, Nat Energy. 2 (2017) 1-7. https://doi.org/10.1038/nenergy.2017.35

[65] B. Li, Q. Su, L. Yu, D. Wang, S. Ding, M. Zhang, G. Du, B. Xu, Li0.35La0.55TiO3 Nanofibers Enhanced Poly(vinylidene fluoride)-Based Composite Polymer Electrolytes for All-Solid-State Batteries, ACS Appl Mater Interfaces. 11 (2019) 42206-42213. https://doi.org/10.1021/acsami.9b14824

[66] J. Feng, L. Wang, Y. Chen, P. Wang, H. Zhang, X. He, PEO based polymer-ceramic hybrid solid electrolytes: a review, Nano Converg. 8 (2021) 1-12. https://doi.org/10.1186/s40580-020-00251-6

[67] R. Li, D. Wu, L. Yu, Y. Mei, L. Wang, H. Li, X. Hu, Unitized Configuration Design of Thermally Stable Composite Polymer Electrolyte for Lithium Batteries Capable of Working Over a Wide Range of Temperatures, Adv Eng Mater. 21 (2019) 1900055. https://doi.org/10.1002/adem.201900055

68] F. Ma, Z. Zhang, W. Yan, X. Ma, D. Sun, Y. Jin, X. Chen, K. He, Solid Polymer Electrolyte Based on Polymerized Ionic Liquid for High Performance All-Solid-State Lithium-Ion Batteries, ACS Sustain Chem Eng. 7 (2019) 4675-4683. https://doi.org/10.1021/acssuschemeng.8b04076

[69] H. Zhai, P. Xu, M. Ning, Q. Cheng, J. Mandal, Y. Yang, A Flexible Solid Composite Electrolyte with Vertically Aligned and Connected Ion-Conducting Nanoparticles for Lithium Batteries, Nano Lett. 17 (2017) 3182-3187. https://doi.org/10.1021/acs.nanolett.7b00715

[70] K. Fu, Y. Gong, J. Dai, A. Gong, X. Han, Y. Yao, C. Wang, Y. Wang, Y. Chen, C. Yan, Y. Li, E.D. Wachsman, L. Hu, Flexible, solid-state, ion-conducting membrane with 3D garnet nanofiber networks for lithium batteries, Proc Natl Acad Sci U S A. 113 (2016) 7094-7099. https://doi.org/10.1073/pnas.1600422113

[71] J. Sun, Y. Li, Q. Zhang, C. Hou, Q. Shi, H. Wang, A highly ionic conductive poly(methyl methacrylate) composite electrolyte with garnet-typed Li6.75La3Zr1.75Nb0.25O12 nanowires, Chemical Engineering Journal. 375 (2019) 121922. https://doi.org/10.1016/j.cej.2019.121922

[72] Y. Li, W. Zhang, Q. Dou, K.W. Wong, K.M. Ng, Li7La3Zr2O12 ceramic nanofiber-incorporated composite polymer electrolytes for lithium metal batteries, J Mater Chem A Mater. 7 (2019) 3391-3398. https://doi.org/10.1039/C8TA11449H

[73] Q. Han, X. Chi, S. Zhang, Y. Liu, B. Zhou, J. Yang, Y. Liu, Durable, flexible self-standing hydrogel electrolytes enabling high-safety rechargeable solid-state zinc metal batteries, J Mater Chem A Mater. 6 (2018) 23046-23054. https://doi.org/10.1039/C8TA08314B

[74] J. Dai, K. Fu, Y. Gong, J. Song, C. Chen, Y. Yao, G. Pastel, L. Zhang, E. Wachsman, L. Hu, Flexible Solid-State Electrolyte with Aligned Nanostructures Derived from Wood, ACS Mater Lett. 1 (2019) 354-361. https://doi.org/10.1021/acsmaterialslett.9b00189

[75] P. Yao, H. Yu, Z. Ding, Y. Liu, J. Lu, M. Lavorgna, J. Wu, X. Liu, Review on Polymer-Based Composite Electrolytes for Lithium Batteries, Front Chem. 7 (2019). https://doi.org/10.3389/fchem.2019.00522

[76] Sperling LH, Hu R. Interpenetrating polymer networks. InPolymer blends handbook Springer, Dordrecht (2014) 677-724. https://doi.org/10.1007/978-94-007-6064-6_8

[77] H. Xin Xie, Q. gang Fu, Z. Li, S. Chen, J. min Wu, L. Wei, X. Guo, Ultraviolet-Cured Semi-Interpenetrating Network Polymer Electrolytes for High-Performance Quasi-Solid-State Lithium Metal Batteries, Chemistry - A European Journal. 27 (2021) 7773-7780. https://doi.org/10.1002/chem.202100380

[78] X.X. Zeng, Y.X. Yin, N.W. Li, W.C. Du, Y.G. Guo, L.J. Wan, Reshaping Lithium Plating/Stripping Behavior via Bifunctional Polymer Electrolyte for Room-Temperature Solid Li Metal Batteries, J Am Chem Soc. 138 (2016) 15825-15828. https://doi.org/10.1021/jacs.6b10088

[79] K.D. Fong, T. Wang, H.K. Kim, R.V. Kumar, S.K. Smoukov, Semi-Interpenetrating Polymer Networks for Enhanced Supercapacitor Electrodes, ACS Energy Lett. 2 (2017) 2014-2020. https://doi.org/10.1021/acsenergylett.7b00466

[80] W. Hua, B. Schwarz, R. Azmi, M. Müller, M.S. Dewi Darma, M. Knapp, A. Senyshyn, M. Heere, A. Missyul, L. Simonelli, J.R. Binder, S. Indris, H. Ehrenberg, Lithium-ion (de)intercalation mechanism in core-shell layered Li(Ni,Co,Mn)O2 cathode materials, Nano Energy. 78 (2020) 105231. https://doi.org/10.1016/j.nanoen.2020.105231

[81] Z. Wang, R. Pan, C. Ruan, K. Edström, M. Strømme, L. Nyholm, Redox-Active Separators for Lithium-Ion Batteries, Advanced Science. 5 (2018) 1700663. https://doi.org/10.1002/advs.201700663

[82] Y. Zhao, L. Wang, Y. Zhou, Z. Liang, N. Tavajohi, B. Li, T. Li, Solid Polymer Electrolytes with High Conductivity and Transference Number of Li Ions for Li-Based Rechargeable Batteries, Advanced Science. 8 (2021) 2003675. https://doi.org/10.1002/advs.202003675

[83] K. Leš, C.S. Jordan, Ionic conductivity enhancement in solid polymer electrolytes by electrochemical: In situ formation of an interpenetrating network, RSC Adv. 10 (2020) 41296-41304. https://doi.org/10.1039/D0RA07966A

[84] Z. Xue, D. He, X. Xie, Poly(ethylene oxide)-based electrolytes for lithium-ion batteries, J Mater Chem A Mater. 3 (2015) 19218-19253. https://doi.org/10.1039/C5TA03471J

Advanced thermo-chemical treatment of waste Bambusa Vulgaris for sustainable resource recovery

MUKESH Bhatt[1,a], KALPIT Shah[2,b], JYESHTHARAJ Bhalchandra Joshi[3,c], MADHURA Yadav[4,d], ABHISHEK Sharma[1,2,d*]

[1]Department of Chemical Engineering, Manipal University Jaipur, Jaipur 303007, India

[2]Department of Chemical and Environmental Engineering, School of Engineering, RMIT University, Melbourne, Victoria, 3000 Australia

[3]Department of Chemical Engineering, Institute of Chemical Technology, Mumbai-400019, Maharashtra, India

[4]School of Architecture & Design, Manipal University Jaipur, Jaipur 303007, India

[a]bhattz1989@gmail.com, [b]kalpit.shah@rmit.edu.au, [c]jbjoshi@gmail.com, [d]madhura.yadav@jaipur.manipal.edu, [e]abhishek.sharma@jaipur.manipal.edu

Keywords: Biomass, Thermochemical Conversion, Bio-Oil, Waste Management, Biochar

Abstract. Bambusa Vulgaris or Bamboo is one of the rapidly growing woody plants among different species and is not being utilized properly around the world. Hence, the utilization of bamboo for sustainable resource recovery is required to avoid deforestation. This study is focused on thermo-chemical treatment of bamboo waste to investigate the physico-chemical properties of generated products. Experiments were carried out in semi batch, four zone electrically heated, rotary kiln reactor, with reaction time of 90 minutes and heating rate of 25 °C/min, for yield optimization. Products were analyzed through TGA, XRF, SEM/EDS, GCMS and FTIR. From the GCMS study, it was found that the bio-oil obtained through thermo-chemical treatment, contains high amount of alcohols, aldehydes, ketone and phenolic compounds. XRF of biochar showed 11 % inorganic components, among them K and Ca were found to be the major elements. XRF and EDS results were in good agreement and confirmed high carbon content. Biochar had the calorific value of 30.97 MJ/kg, which makes it suitable as energy source for commercial applications. Biochar contained carbonyl (C=O), hydroxyl (HO-), and carboxyl (HOOC) groups, which are responsible for ion exchange reactions during soil amendment. Low ash and high carbon content along with high porosity promotes its usage as high value material in different non-agricultural applications. Additionally, high amount of carbon is beneficial for maximizing the carbon storage and making the process environmentally sustainable.

Introduction

India is in the second position in bamboo cultivation and produces approximately 3.23 milliontons annually. Bamboo is fast growing biomass and it can usually be harvested in 1-4 years. More than 100 species of bamboo are found in Asia [1–3]. Planned harvesting of bamboo cangenerate huge raw material resources for producing value added materials. Other thantraditional applications of bamboo such as furniture manufacturing, building material or medicinal uses, bamboo is being used to produce high value chemicals and biofuels.

Bamboo is made up of hemicellulose, cellulose, and lignin, which are chemically different and have different characteristics. All these components of bamboo plants decompose in different temperature range. Pyrolysis is the process of organic material decomposition under oxygen deprived condition with the application of heat [4,5]. Earlier studies in literature have tried to enhance product yield and optimize the pyrolysis conditions. Depending on the slow or fast pyrolysis and the final use of biochar, the required parameters may be different. The most common use has been as fuel for energy production, heating, and cooking [6].

Higher yield of biochar is obtained through slow pyrolysis in comparison with syngas and bio-oil. Therefore, slow pyrolysis is the adopted technology to produce biochar [7]. Biochar produced from biomass is generally in the range of 20-40 wt % of dry lignocellulose. Pyrolysisconditions such as feedstock, particle size, pressure, holding time, heating rate, and temperature strongly influenced product yield and chemical characteristics of the pyrolysis products [8]. Pyrolysis oil produced from bamboo contain mixture of different chemical compounds such asalcohols, phenols, and acids [9]. During the pyrolysis, micro pores are created which results inlarger microscopic surface area of biochar. Biochar is used for the pollutant removal and nutrients retention in soil, due to its high ion exchange capacity. It contains different functionalgroups which facilitates the ion exchange reactions on the surface of biochar. The transformation of native bamboo by pyrolysis to produce char can be an alternative and potential option for wastewater treatment in the industries, carbon sequestration, oil amelioration and steel making [10].

Zhao et al. showed that during pyrolysis, both operating parameters and feedstock affects the properties of biochar [11]. High lignin content (20 – 40 wt %) makes it an interesting feedstockfor the recovery of aromatic chemicals. [12]. Hernandez-Mena et al. performed slow pyrolysisof bamboo with particle size of 0.7 mm, in fixed bed reactor having 230 mm height and 115 mm of inner diameter. Experiments were performed in the temperature range of 300-600 °C. Oil yield was 35 wt% along with 25 wt% of biochar and 40 wt% of gases. Oil yield increasedwith increasing temperature but no significant difference in the oil yield was observed beyond500 °C. Biochar produced from the pyrolysis contained high amount of carbon (up to 80 wt%)and low amount of ash with heating value of approximately 30 MJ/kg [13]. Another bamboo pyrolysis study was carried out by Wang et al. in fixed bed reactor (oven type) in the temperature range of 400-600 °C with the bamboo particle size of 100 mm. Study revealed thatyield of char decreased with increase in temperature from 46 wt% at 400 °C to 31 wt% at 600 °C, and high amount of gases were produced at higher temperature due to secondary cracking.Oil contained mainly levoglucosan, furfurals and aromatics with small quantity of tar. Char produced was highly porous with high carbon content [14].

Most of the previous studies have been carried out using lab scale experimental set-ups [9,15–17]. In this study, pyrolysis of bamboo (Bambusa Vulgaris) was conducted in semi-pilot rotarykiln reactor. Detailed analysis of the bio-oil and char has been carried out for the improved understanding of pyrolysis of bamboo and the quality of the products. Results provided in this study will add necessary information for bamboo pyrolysis for large-scale system developmentand parameter optimization.

Material and Methods

Experimental Procedure:

Bamboo chips were received from local market and the size of bamboo chips were in the rangeof 50 to 120 mm. The thermal behavior of the bamboo was investigated using TGA by heatingin presence of pure nitrogen (100 ml/min) at 20 °C/min up to 900 °C. Pure nitrogen was used to remove gaseous condensable products and air in pyrolysis zone. Each experiment in rotary kiln reactor was carried out with 2 kg bamboo chips, that was initially preheated for 1 hr at 150°C for moisture removal. The experimental set up has been shown in Fig. 1 and details explained in our previous study [18].

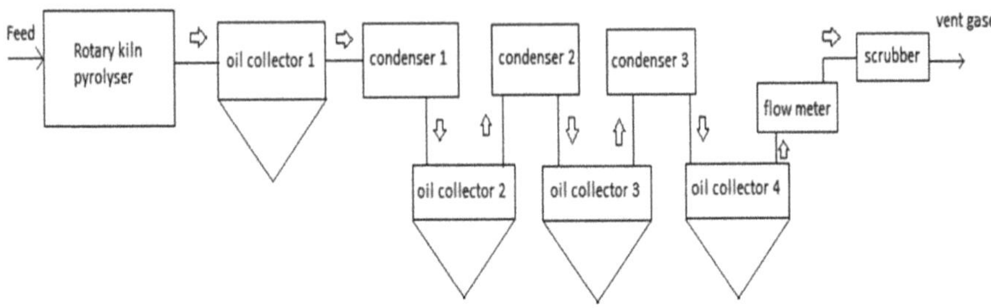

Figure 1: Rotary kiln reactor set-up flow diagram

Characterization Techniques:
The collected oil samples from different biomass feed, at various temperatures were analysed with a Shimadzu-GCMS QP2020 to determine their chemical composition. The GCMS was equipped with a 30 m capillary column. The oven temperature was raised from 50 to 270 °C inthe presence ofhelium with the flow rate of 2 ml/min. For the GCMS analysis of the oil, the method was adopted from the earlier study carried out by Lyu et al. [2015][19].

The feed and product samples were also characterized using various instrumentation techniques such as Scanning Electron Microscope (SEM JEOL make JSM-7610 F-Plus), X-ray fluorescence (XRF) (Epsilon 1 Benchtop Panalytical Germany), Fourier Transform Infrared Spectroscopy (FTIR, Bruker alpha), for detailed analysis. The proximate and ultimate analysiswas carried out using ASTM D7582 and ASTM D3176 standard, respectively. The elemental analysis was carried out using LECO-Truspec CHNS analyser. FTIR was carried out to determine the functional group present in the char. XRF was conducted to determine the elemental composition of solid material and SEM/EDX was carried out for surface topographyand the composition of the pyrolysis char. Products produced at 500 °C were analysed in this study.

Results and Discussion
Thermal decomposition profile of bamboo was investigated through the thermogravimetric analyser and it was observed that bamboo decomposed in the temperature range of 240 °C to 700 °C. At 240-300 °C, the decomposition was mainly due to xylan. Lignin undergoes slow decomposition over a wide temperature range from 300 to 700 °C (Fig. 2). Once the temperature reached close to 400 °C or above, hemicellulose and cellulose decomposed to charand volatiles. Approximately 70 % weight loss occurred between 240 to 600 °C. Hence, pyrolysis experiments were carried out in the range of 400 to 600 °C in rotary kiln reactor.

Materials Research Proceedings 29 (2023) 192-200 https://doi.org/10.21741/9781644902516-22

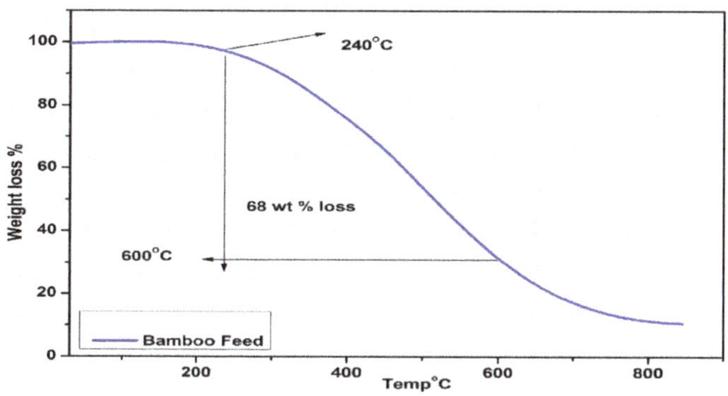

Figure 2: TGA of bamboo feed

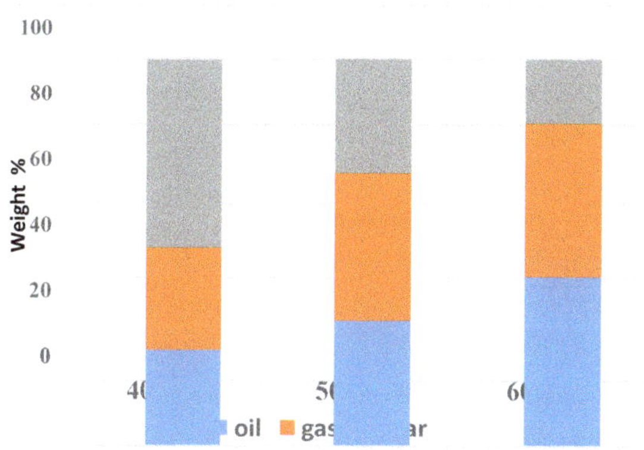

Figure 3: Pyrolysis products yield

As shown in Fig. 3, at temperature around 400 °C, amount of oil produced was around 25 wt% due to partial decomposition of volatile matters and increased with increase in temperature. As the temperature increased from 500 to 600 °C, amount of gases increased withno significant change in the oil yield. Higher amount of volatiles were released and the yield of biochar decreased from around 50% to 25% with increase in temperature from 400 to 600 °C. Results were in well agreement with the previous studies [13-14]. Yield of bio-oil and charcould be increased with optimization of the operating parameters.

Oil Analysis

From the GCMS study, it was found that the bio-oil obtained from bamboo pyrolysis contains alcohols, aldehydes, and ketones with phenolics and aromatic hydrocarbons. Slow pyrolysis also produced high yields of acetic acid. These compounds accounted for 50 area % in total and these compounds are shown in Table 1. Overall, the GCMS data indicated different class of compounds which can be separated from the bio-oil for the recovery of valuable platform chemicals.

Table 1: Bio-oil GCMS analysis

Peak#	R.Time	Area%	Compound Name
1	3.105	2.28	Ethane, 1,1-diethoxy-
2	3.200	10.44	Acetic acid
3	3.753	1.94	2-Propanone, 1-hydroxy-
4	4.348	2.15	2,4-Dimethyl-1-heptene
5	5.572	2.01	Cyclopentanone
6	7.201	2.18	3,5-Dimethylpyrazole-1-methanol
7	8.370	2.92	3-Furanmethanol
8	10.053	1.98	Benzene, 1,2-dichloro-
9	11.379	3.74	Tricyclohex-3-ene-3-carbonitr
10	13.621	4.74	Phenol
11	15.618	3.36	4,4-Dimethyl-cyclohex-2-en-1-ol
12	17.478	2.34	1-Tetradecene
13	21.034	2.48	1-Nonadecene
14	22.405	2.29	,4-Di-tert-butylphenol
15	24.066	2.53	1-Nonadecene
16	26.763	2.38	1-Nonadecene

Char Analysis

X-ray fluorescence spectroscopy

Mineralogical composition of the bamboo char was analysed through XRF. K and Ca were found to be the major micronutrients present in the biochar (Table 2). Overall XRF showed that bamboo char contains low amount of inorganic impurities.

Table 2: XRF analysis of Bamboo char

Compound (wt%)	Sample 1
MgO	0.16
Al O	0.12
SiO_2	1.09
P O	0.173
SO	0.2
K O	6
CaO	2.11
TiO	0.14
Fe O	0.314
Cl	0.65
Loss on ignition	**89**

Ultimate and Proximate Analysis

Char obtained from bamboo contains low amount (4.32 wt %) of moisture. At 500 °C, the sample contained 77.8 wt % of volatile matter, with high total carbon. It is high in calorific value which makes it suitable as the alternate fuel source (Table 3).

Table 3: Ultimate and Proximate analysis of Bamboo char

Sample ID	Bamboo Char
LOD @ 110 ^{0}C	4.32
Ash @ 800 ^{0}C	13.85
Volatile Matter	77.8
Fix Carbon	4.03
Carbon (%)	78.60
Hydrogen (%)	1.92
Nitrogen (%)	1.823
Sulphur (%)	0.028
Calorific Value MJ/kg	30.97

Scanning electron microscopy-energy dispersive X-ray

Bamboo biochar developed longitudinal pores and high porosity (Fig. 4). Thermal breakdown and large release of volatile matters from bamboo was the reason for the formationof channel like structure. Images showed the surface contains pore size in the range of 2 to 6 μm. It was distinct from the SEM images that biochar prepared at 500 °C from bamboo also showed honeycomb-like pore structures with defined pores

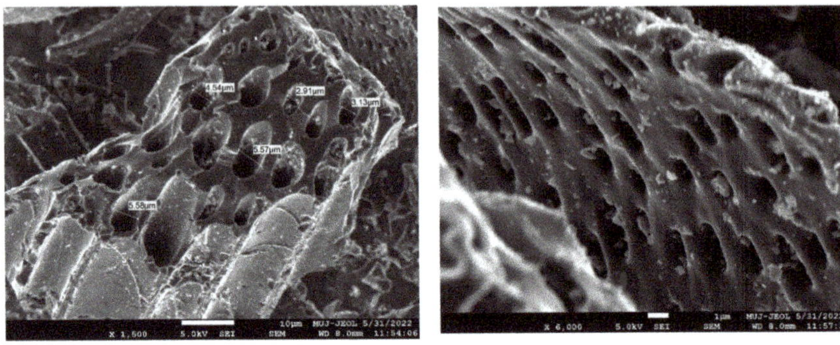

Figure 4: SEM images of Bamboo char

Table 4: Bamboo char EDX analysisFigure 5: EDX graph of bamboo char

Element	Weight %	Atomic %
C	77	85.2
N	0.4	0.4
O	15.7	13
Mg	0.7	0.4
Si	0.4	0.2
K	1.2	0.4

As
per

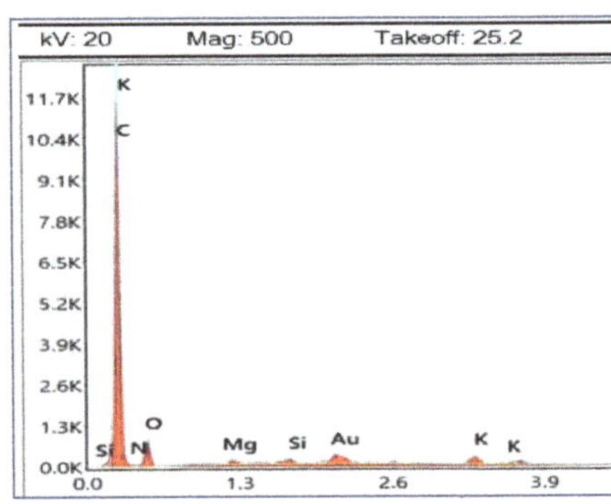

As per Table 4, and Fig. 5, approximately 77 % of char constituted of carbon. The remaining percentage consisted of N, Mg, K and O. These are the essential elements required for soil applications and adsorption process.

Fig. 6 shows that major bands appeared at 875 cm^{-1}, 1517 cm^{-1}, 1697 cm^{-1} and broad range of (3000-3750) cm^{-1}. These peaks correspond to aromatic CH and C-O from carbonates, and C-O vibrations from carbohydrates present in the bamboo feedstock. The peak at around 3750cm^{-1} of O-H groups present in the biochar. Polar functional groups that can enhance bond ability include carbonyl (C=O), carboxyl (HOOC), hydroperoxide (HOO-), and hydroxyl (HO-) groups, were also present.

Fourier transform infrared

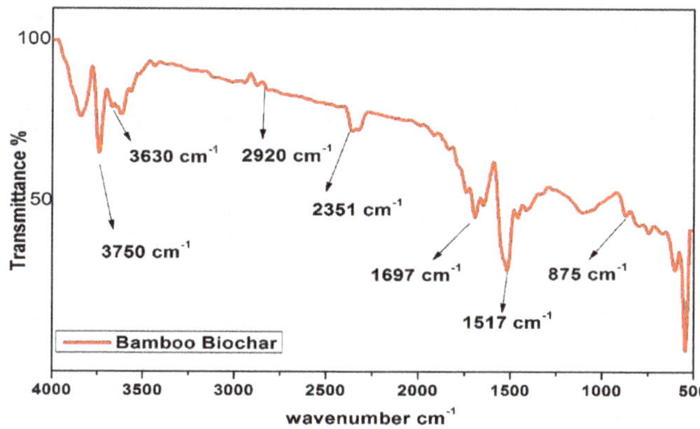

Figure 6: FTIR of Bamboo char

Conclusions

In this research work, thermochemical conversion of bamboo was carried out in temperature range of 400-600°C using semi-pilot reactor system. Results were in well agreement with the literature studies and revealed information about biochar and bio-oil quality. Study revealed that oil obtained from bamboo contained higher concentrations of acetic acid and phenols, which can be extracted through suitable chemical extraction techniques. Bamboo char has low ash content and high porosity, which render it attractive in many applications. High amount ofcarbon present in biochar is beneficial for maximizing the carbon storage and could be used asenergy source and for soil applications. This also indicates that bamboo has the potential to bea promising precursor to produce activated carbon for different commercial applications.

References

[1] M.D. Burger, Evaluating different value adding processing systems for bamboo developments, (2018).

[2] A. Kumar, S. Shabnam, M.S. Malik, Doubling Farmers Income through Agroforestry, Scientific Publishers, 2021.

[3] S. Ross, Bamboo construction as a sustainable building technology from a structural and materials engineering perspective, (2021).

[4] X. Hu, M. Gholizadeh, Biomass pyrolysis: A review of the process development and challenges from initial researches up to the commercialisation stage, Journal of Energy Chemistry. 39 (2019) 109-143. https://doi.org/10.1016/j.jechem.2019.01.024

[5] A.J. Ridout, M. Carrier, F.-X. Collard, J. Görgens, Energy conversion assessment of vacuum, slow and fast pyrolysis processes for low and high ash paper waste sludge, Energy Conversion and Management. 111 (2016) 103-114. https://doi.org/10.1016/j.enconman.2015.12.043

[6] S. Fawzy, A.I. Osman, H. Yang, J. Doran, D.W. Rooney, Industrial bio-char systems for atmospheric carbon removal: a review, Environmental Chemistry Letters. 19 (2021) 3023-3055. https://doi.org/10.1007/s10311-021-01210-1

[7] A.N. Amenaghawon, C.L. Anyalewechi, C.O. Okieimen, H.S. Kusuma, Biomass pyrolysis technologies for value-added products: a state-of-the-art review, Environment, Development and Sustainability. 23 (2021) 14324-14378. https://doi.org/10.1007/s10668-021-01276-5

[8] T. Kan, V. Strezov, T.J. Evans, Lignocellulosic biomass pyrolysis: A review of product properties and effects of pyrolysis parameters, Renewable and Sustainable Energy Reviews. 57 (2016). https://doi.org/10.1016/j.rser.2015.12.185. https://doi.org/10.1016/j.rser.2015.12.185

[9] I.P. Woyinbrakemi, A.J. Gunorubon, A.O. Michael, Production and Characterization of Bio Oil from Bamboo, American Journal of Chemical Engineering. 9 (2021) 134. https://doi.org/10.11648/j.ajche.20210906.11

[10] J. Zhang, J. Li, D. Ye, X. Zhu, Q. Liao, B. Zhang, Tubular bamboo charcoal for anode in microbial fuel cells, Journal of Power Sources. 272 (2014) 277-282. https://doi.org/10.1016/j.jpowsour.2014.08.115

[11] L. Zhao, X. Cao, O. Mašek, A. Zimmerman, Heterogeneity of bio-char properties as a function of feedstock sources and production temperatures, Journal of Hazardous Materials. 256 (2013) 1-9. https://doi.org/10.1016/j.jhazmat.2013.04.015

[12] T. Stedile, L. Ender, H.F. Meier, E.L. Simionatto, V.R. Wiggers, Comparison between physical properties and chemical composition of bio-oils derived from lignocellulose and triglyceride sources, Renewable and Sustainable Energy Reviews. 50 (2015) 92-108. https://doi.org/10.1016/j.rser.2015.04.080

[13] L.E Hernandez-Mena, A.A. Pécoraa, and A.L., Beraldob,. Slow pyrolysis of bamboo biomass: analysis of bio-char properties. Chemical Engineering transactions, 37(2014), 115-120.

[14] H. Wang, X. Wang, Y. Cui, Z. Xue, Y. Ba, Slow pyrolysis polygeneration of bamboo (Phyllostachys pubescens): Product yield prediction and bio-char formation mechanism, Bioresource Technology. 263 (2018) 444-449. https://doi.org/10.1016/j.biortech.2018.05.040

[15] H.C. Lin, Preliminary Safety Evaluation of Bamboo Pyrolysis Products: Charcoal and Vinegar, in: Bamboo-Current and Future Prospects, IntechOpen, 2017. https://doi.org/10.5772/intechopen.68542

[16] Y. Hou, Z. Feng, Y. He, Q. Gao, L. Ni, M. Su, H. Ren, Z. Liu, W. Hu, Co-pyrolysis characteristics and synergistic interaction of bamboo residues and disposable face mask, Renewable Energy. (2022). https://doi.org/10.1016/j.renene.2022.05.111

[17] D. Zhang, J.B. Ji, C.F. Song, S.D. Shan, Catalytic Pyrolysis of Bamboo Residues for Composite Bio- char and Bamboo Oil, in: Applied Mechanics and Materials, Trans Tech Publ, 2014: pp. 921-925. https://doi.org/10.4028/www.scientific.net/AMM.472.921

[18] M. Bhatt, S. Wagh, A.G. Chakinala, K.K. Pant, T. Sharma, J.B. Joshi, K. Shah, A. Sharma, Conversion of refuse derived fuel from municipal solid waste into valuable chemicals using advanced thermo- chemical process, Journal of Cleaner Production. 329 (2021) 129653. https://doi.org/10.1016/j.jclepro.2021.129653

[19] G. Lyu, S. Wu, H. Zhang, Estimation and comparison of bio-oil components from different pyrolysis conditions, Frontiers in Energy Research. 3 (2015) 28. https://doi.org/10.3389/fenrg.2015.00028

Sustainable Processes and Clean Energy Transition - ICSuPCET2022 Materials Research Forum LLC
Materials Research Proceedings 29 (2023) 201-208 https://doi.org/10.21741/9781644902516-23

Molecular screening of metal oxides for arsenic removal from water

NOOR E Hira[1,a*], SERENE Sow Mun Lock[1,b*], KHADIJA Asif [1,c], FARMAN Ullah[2,d] and ABID Salam Farooqi[3,e]

[1]CO2 Research Center (CO2RES), Universiti Teknologi PETRONAS, 32610 Seri Iskandar, Perak, Malaysia

[2]Department of Chemical Engineering, Universiti Teknologi PETRONAS, 32610 Seri Iskandar, Perak, Malaysia

[3]Centre of Innovative Nanostructures & Nanodevices, Universiti Teknologi PETRONAS, 32610 Seri Iskandar, Perak, Malaysia

[a]noor_22000794@utp.edu.my, [b]sowmun.lock@utp.edu.my, [c]khadija_20000006@utp.edu.my, [d]farman_19001014@utp.edu.my, [e]abid_20001276@utp.edu.my

Keywords: Water Treatment, Adsorption, Arsenic, Monte Carlo Simulation, Metal Oxides

Abstract. Toxic metal arsenic in the ground water is poisonous and harmful that should be treated to ensure human health and safety. For many years, different technologies have been developed for the treatment of contaminated water and adsorption is an economical method in which a large number of adsorbents are being used including metal oxides. The selection of these metal oxides needs to be done systematically to choose the best metal oxide with good potential for arsenic removal from water. Previous work has been mostly focused on experimental study, which is time-consuming and expensive. Only a limited number of simulation study has been conducted, which is confined to only several specific adsorbents, such as oxides of iron. There is a need to do research for other metal oxides to evaluate which one is more capable of removing arsenic from water. In this research work, screening of metal oxides was done using Molecular Dynamics and Monte Carlo Simulation. The molecular structures were optimized and sorption calculations were performed at fixed pressure of 100kPa and temperature of 298K to observe the adsorption capability of metal oxides. Al_2O_3 and SnO_2 were found to be good adsorbent for arsenic removal from water with adsorption capacity of 1681.80 g/g and 975.03 g/g respectively. Previously used Al_2O_3 was used as a benchmark for this research and adsorption capacity results also proved it. It was observed that SnO_2 has potential to remove arsenic from water with adsorption capacity 975.03 g/g. The results displayed that SnO_2 can be one of the best adsorbents for application of arsenic removal from water. It is concluded that apart from using conventional metal oxides for arsenic removal, other metal oxides should be studied and can also be used as an adsorbent as they can provide great adsorption capacity for arsenic removal from water.

Introduction

Water is an essential element for living things to thrive. Humans can't survive without water on this planet. However, with the increasing population and industrialization, access to safe drinking water is becoming more difficult. Water contamination has become a global problem. Developing countries are facing even more severe problems in this regard. According to World Health Organization (WHO), 80% of diseases are due to unsafe and polluted water [1]. In particular due to domestic, agricultural, medical, technological applications heavy metals have been widely distributed in our ecosystem, thus, raising concerns for human health and the environment.

Heavy metal ions, such as Cadmium (Cd), Lead (Pb), Mercury (Hg), Arsenic (As), are severe threat to living organisms and metals are harmful for health and their prolonged intake may result in worst consequences [2]. Water treatment is necessary if these metals are present in water.

Among all the heavy metals, arsenic accumulation in groundwater is increasing with time due to human activities and natural phenomenon [3]. Approximately 6.12×10^{10} and 2.380×10^{11} g of arsenic is added into our oceans per annum through soil erosion and leaching respectively [4].

Different processes have been employed for water treatment such as coagulation, adsorption and membrane separation. Adsorption is a very popular and effective method for water treatment. Many adsorbents are used for adsorption process like metal oxides, zeolites etc. Excellent adsorbents have a large area-to-volume ratio to provide maximum adsorption sites for metallic ions [5]. Adsorption is the best method for water treatment and specifically for arsenic removal from water as it is easy in operation and economical with greater efficiency. The removal of arsenic depends on the tendency of the adsorbent to sorb arsenic on its surface. There are a lot of adsorbents that can be used for arsenic removal from water and metal oxides tend to be a better candidate due to their surface area and adsorption capacity [6].

Every metal oxide has a different capacity to sorb a compound to be removed. Some metal oxides like iron oxides and aluminum oxide have been studied by experiments for arsenic removal from water and the experimental studies showed that Fe_2O_3 and Al_2O_3 are good adsorbents [7, 8, 9]. For arsenic removal from water, molecular simulation has evolved as an emerging technology. It has been performed to circumvent limitation of instruments and materials from experimental perspective [10]. Moreover, it provides a platform to study the molecular level interpretation for parameters like energy, enthalphy or entropy [11].

From the review, it is found that although a myriad of experimental work has been available for study of heavy metal removal from water, the investigations using molecular simulation are scarcely available typically for arsenic separation. Among the limited number of studies, they are only confined to specific adsorbents of iron oxide.[12, 13]. There should be research work related to molecular simulation and screening study for arsenic removal from water to select the best adsorbent for arsenic removal from water with good and effective adsorption capacity, it is important to be conducted but has received less scrutiny to date.

In this research work, 11 metal oxides will be studied and adsorption capacity of these adsorbents will be unveiled by adopting Monte Carlo Simulation on Material Studio Software. The metal oxides selected to study for arsenic adsorption are Fe_2O_3, Al_2O_3, TiO_2, ZrO_2, Ag_2O, CaO, CeO_2, La_2O_3, MgO and SnO_2 because of their applications in water treatment [14, 15, 16, 17, 18, 19, 20]. The metal oxides Fe_2O_3 and Al_2O_3 are used as commercial adsorbents for arsenic removal from water and they are included as a benchmark [9] for sorption calculations whereas the other eight metal oxides are studied for sorption loading to remove arsenic from water.

Methodology

The simulation study was performed using Material Studio 8.0 Software [21]. Adsorption phenomenon was studied using molecular dynamics simulation on MS [21]. The geometry optimizations of all the metal oxide structures given in figure 1 was done using the Focite module with a Universal forcefield. Universal force field provides full coverage of the periodic table. It is good for predicting geometry and energy differences of organic molecules, inorganics and metal complexes. This force field corrects the angles and optimize the bond distance of molecules [22]. The structures of metal oxides adsorbents are given in figure 1.

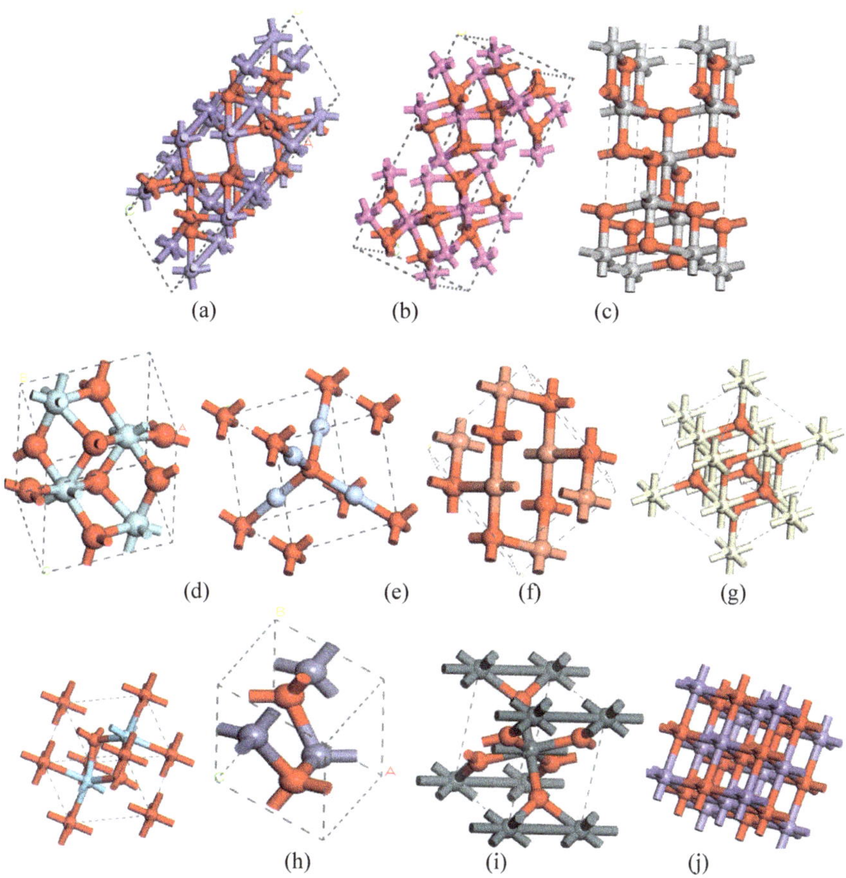

Figure 1: Illustration of adsorbents in Material Studio (a) Fe_2O_3 (b) Al_2O_3 (c) TiO_2 (d) ZrO_2 (e) Ag_2O (f) CuO (g) CeO_2 (h) La_2O_3 (i) ZnO (j) SnO_2 (k) FeO

First of all, selected adsorbents structure was imported from Material Studio library and surface was built for each metal oxide. Then using Forcite module, Geometry Optimization calculations were performed. Geometry optimization was done to achieve stability of structure by energy minimization [23]. After that, sorption calculations were performed using Universal forcefield at temperature 298.0 K. From practical observation, it was seen that most water treatments plants are operated at room temperature and 100kPa, previous simulation studies were also performed at 298.0K [24, 25, 26, 27] and 100kPa [28]. The computations were performed using Monte Carlo Simulation at fixed pressure of 100kPa. This type of simulation is called as Grand Monte Carlo Simulation where temperature is kept constant and pressure is fixed and potential of adsorbent is estimated as sorption loading value [21].

Sorption at fixed pressure task resembles experimental conditions and provides average loading of sorbate component at a given temperature. The Metropolis method was employed using the

Universal forcefield in Materials Studio. The Metropolis method is a conventional Monte Carlo simulation approach that handles the sorbent as rigid structure and solely incorporates sorbate translations and reorientations [21]. Under the Metropolis method, the fixed pressure sorption function, also known as Grand canonical Monte Carlo (GCMC) simulations was used to predict the amount of sorbate at the designated temperature and pressure. It was performed using "Sorption" module in material software with medium quality. The sorption loading value of arsenic within the metal oxide showed its adsorption capacity.

Result and Discussion

The geometery optimization parameters of initial structures (before optimization) and final structure (after performing geometery optimization) for selected metal oxides are given in Table 1. The calculations showed that the total enthalpy of the structure was lowered. This reduction in enthalpy value stabilizes the structure and then it was further used for sorption calculations. The optimization of van der waals energy and non-bond energy are major contributor for optimizing the total energy of the molecular structure [29]. The total molecular structure energy is reduced in this way to acheive stability.

Table 1: Structure parameters of metal oxides before and after Geometry Optimization on Material Studio

Sr. No.	Metal Oxide	Before geometry optimization of stucture parameters	After geometry optimization of structure parameters
		Total Enthalpy (kcal/mol)	**Total Enthalpy (kcal/mol)**
1	Fe_2O_3	179525.1	158163.3
2	Al_2O_3	501334.3	444405.8
3	TiO_2	9245.8	7144.9
4	ZrO_2	291102.3	80875.7
5	Ag_2O	407.6	217.5
6	CuO	7439.88	5757.75
7	CeO_2	9768.1	9726.7
8	La_2O_3	31471.5	27291.1
9	ZnO	17667.9	17326.5
10	SnO_2	70863.4	67335.6
11	FeO	19802.6	17971.7

Arsenic adsorption result from water using selected metal oxides are shown in table 2. The sorption loading was obtained from sorption module that is further used to calculate adsorption capacity.

Table 2: Sorption Loading and average total energy of metal oxides after loading, result for all selected metal oxides.

Metal Oxide	Average Sorption Loading	Maximum Sorption Loading	Adsorption Capacity (g/g)
Fe_2O_3	0.973	7	13.51
Al_2O_3	81.80	91	1681.80
TiO_2	0.045	2	1.58
ZrO_2	8.82×10^{-3}	2	0.436
Ag_2O	0.083	3	4.05
CuO	0.086	3	5.619
CeO_2	4.4×10^{-4}	1	0.647
La_2O_3	8.04×10^{-3}	3	3.908
ZnO	0.239	3	17.59
SnO_2	14.21	18	975.03
FeO	3.3×10^{-4}	1	1.126

Fe_2O_3, Al_2O_3 were used as a benchmark to compare the sorption loading since they are conventionally used as adsorbent for arsenic removal from water with good adsorption capacity [9, 10]. The adsorption capacity obtained for Fe_2O_3 and Al_2O_3 was 13.51 g/g and 1681.80 g/g respectively, while for SnO_2 the adsorption capacity obtained was 975.03 g/g. It was the second adsorption capacity among all these metal oxides. The arsenic adsorption on SnO_2 is shown in the Figure 2, in which the red dots above the SnO_2 indicates arsenic.

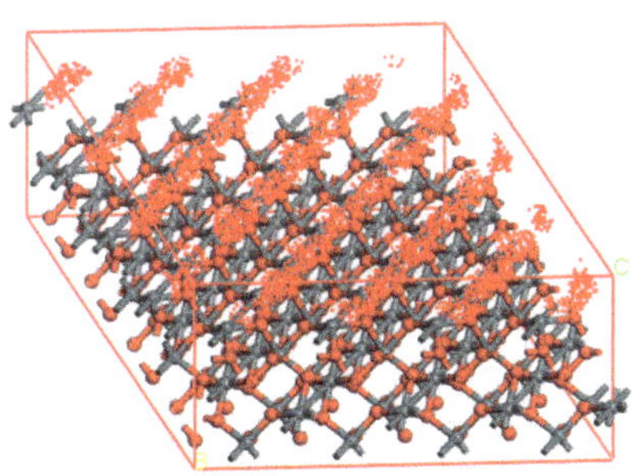

Figure 2: Adsorption of Arsenic on the surface of SnO_2 at fixed pressure

This finding implied that it has a greater tendency to sorb arsenic and has potential to be a good adsorbent to remove arsenic from water. It should be studied as adsorbent for arsenic removal from water. Review from literature suggests that SnO_2 is an effective adsorbent to remove other toxic metals like lead and cadmium [20] since it has the excellent capability and adsorption capacity. It is due to its interesting semiconducting property with band gap of 3.6 eV between O_2 and Sn that give great capability of heavy metal ions removal [20]. Moreover, according to literature SnO_2

nano-particles have high surface area of 128 m^2/g [30] which makes it good adosrbent with great efficiency while microporous Fe_2O_3 has 111m^2/g [31] that is less than SnO_2surface area.

Conclusion

The screening of different metal oxides for arsenic removal from water was done using Material studio software and sorption calculations were performed. The adsorbents according to sorption capacity value were in the order of: $Al_2O_3 > SnO_2 > ZnO_2 > Fe_2O_3 > CuO > Ag_2O > La_2O_3 > TiO_2$ $FeO > CeO_2 > ZrO_2$. The results obtained from this present research work showed that among above 11 chosen metal oxides, SnO_2 has second adsorption capacity of 975.03 g/g. Al_2O_3 has the maximum adsorption capacity of 1681.80 g/g among all and has been used previously for arsenic removal while Fe_2O_3 had adsorption capacity of 13.51g/g that is less than SnO_2. SnO_2 is a metal oxide that had not been used for arsenic removal from water but is used to remove other toxic metals like cadmium and lead from water. This study suggests that SnO_2 has good adsorption capacity and potential to remove arsenic from water and should be explored as an adsorbent to remove arsenic from water. Moreover, that other metal oxides like CuO, MnO, ZnO and FeO should also be studied for arsenic removal form water. The effect of operating variables towards adsorption capacity should also be studied in the future to optimize the separation performance.

Acknowledgment

The authors acknowledge financial support from Universiti Teknologi PETRONAS, Malaysia. This work is done with the financial support from Yayasan Universiti Teknologi PETRONAS (Grant No: 015LC0-322).

References

[1] C. Hewitson, "Water," *Water*, vol. 5, no. July, pp. 1–68, 2013, https://doi.org/10.4324/9780203769287
[2] and Á. I. A. Arruti, I. Fernández-Olmo, "Evaluation of the contribution of local sources to trace metals levels in urban PM2.5 and PM10 in the Cantabria region (Northern Spain)," *J. Environ. Monit., vol. 12, no. 7, pp. 1451–1458.*
[3] P. Kavcar, A. Sofuoglu, and S. C. Sofuoglu, "A health risk assessment for exposure to trace metals via drinking water ingestion pathway," *Int. J. Hyg. Environ. Health*, vol. 212, no. 2, pp. 216–227, 2009, https://doi.org/10.1016/j.ijheh.2008.05.002
[4] F. T. Mackenzie, R. J. Lantzy, and V. Paterson, "Global trace metal cycles and predictions," J. Int. Assoc. Math. Geol., vol. 11, no. 2, pp. 99–142, Apr. 1979.
[5] S. De Gisi, G. Lofrano, M. Grassi, and M. Notarnicola, "Characteristics and adsorption capacities of low-cost sorbents for wastewater treatment: A review," *Sustain. Mater. Technol.*, vol. 9, pp. 10–40, 2016, https://doi.org/10.1016/j.susmat.2016.06.002
[6] L. Chen *et al.*, "Application of metal oxide heterostructures in arsenic removal from contaminated water," *J. Nanomater.*, vol. 2014, 2014, https://doi.org/10.1155/2014/793610
[7] J. H. Gulledge and J. T. O'Connor, "Removal of arsenic (V) from water by adsorption on aluminum and ferric hydroxides," J. / Am. Water Work. Assoc., vol. 65, no. 8, pp. 548–552, 1973, https://doi.org/10.1002/j.1551-8833.1973.tb01893.x
[8] M. Gallegos-Garcia, K. Ramírez-Muñiz, and S. Song, "Arsenic removal from water by adsorption using iron oxide minerals as adsorbents: A review," Mineral Processing and Extractive Metallurgy Review, vol. 33, no. 5, pp. 301–315, 2012, https://doi.org/10.1080/08827508.2011.584219
[9] Youngran Jeong, Maohong Fan, Shilpi Singh, Chia-Line Chuang, Basudeb Saha, J. Hans van Leeuwen, Evaluation of iron oxide and aluminum oxide as potential arsenic(V) adsorbents, Chemical Engineering and Processing - Process Intensification, Volume 46, Issue 10,2007

[10] S.S.M. Lock, K.K. Lau, A.M. Shariff, Y.F. Yeong, M.A. Bustam, Norwahyu Jusoh, Faizan Ahmad, An atomistic simulation towards elucidation of operating temperature effect in CO2 swelling of polysulfone polymeric membranes,Journal of Natural Gas Science and Engineering,

[11] Wanyonyi, F. S., Fidelis, T. T., Mutua, G. K., Orata, F., & Pembere, A. M. (2021). Role of pore chemistry and topology in the heavy metal sorption by zeolites: From molecular simulation to machine learning. Computational Materials Science, 195, 110519.

[12] J. Farrell and B. K. Chaudhary, "Understanding arsenate reaction kinetics with ferric hydroxides," *Environ. Sci. Technol.*, vol. 47, no. 15, pp. 8342–8347, 2013, https://doi.org/10.1021/es4013382

[13] X. Zhang *et al.*, "Stability study of the As(V)-Fe(III) oxyhydroxide coprecipitate over a broad pH range: Characteristics and mechanism," *Sci. Total Environ.*, vol. 806, p. 150794, 2022, https://doi.org/10.1016/j.scitotenv.2021.150794

[14] Sandoval, R., Cooper, A. M., Aymar, K., Jain, A., & Hristovski, K. (2011). Removal of arsenic and methylene blue from water by granular activated carbon media impregnated with zirconium dioxide nanoparticles. Journal of hazardous materials, 193, 296-303.

[15] Rahmanifar, B., & Moradi Dehaghi, S. (2014). Removal of organochlorine pesticides by chitosan loaded with silver oxide nanoparticles from water. Clean Technologies and Environmental Policy, 16(8), 1781-1786.

[16] Shimabuku, Q. L., Arakawa, F. S., Fernandes Silva, M., Ferri Coldebella, P., Ueda-Nakamura, T., Fagundes-Klen, M. R., & Bergamasco, R. (2017). Water treatment with exceptional virus inactivation using activated carbon modified with silver (Ag) and copper oxide (CuO) nanoparticles. Environmental technology, 38(16), 2058-2069.

[17] Kurian, M. (2020). Cerium oxide based materials for water treatment–A review. Journal of Environmental Chemical Engineering, 8(5), 104439.

[18] Xie, J., Lin, Y., Li, C., Wu, D., & Kong, H. (2015). Removal and recovery of phosphate from water by activated aluminum oxide and lanthanum oxide. Powder Technology, 269, 351-357.

[19] Baruah, S., K Pal, S., & Dutta, J. (2012). Nanostructured zinc oxide for water treatment. Nanoscience & Nanotechnology-Asia, 2(2), 90-102.

[20] Kumar, K. Yogesh; Raj, T.N. Vinuth; Archana, S.; Prasad, S.B. Benaka; Olivera, Sharon; Muralidhara, H.B. (2016). SnO2 nanoparticles as effective adsorbents for the removal of cadmium and lead from aqueous solution: Adsorption mechanism and kinetic studies. Journal of Water Process Engineering, 13, 44–52.

[21] L. Li *et al.*, "Adsorption of xanthate from aqueous solution by multilayer graphene oxide: an experimental and molecular dynamics simulation study," *Adv Compos Hybrid Mater*, vol. 4, no. 3, pp. 725–732, 2021, https://doi.org/10.1007/s42114-021-00310-4

[22] A.K. Rappé, C.J. Casewit, K. Colwell, W.A. Goddard III, W.M. Skiff, UFF, a full periodic table force field for molecular mechanics and molecular dynamics simulations, J.Am. Chem. Soc. 114 (25) (1992) 10024–10035.

[23] Friesecke, G., Theil, F. (2015). Molecular Geometry Optimization, Models. In: Engquist, B. (eds) Encyclopedia of Applied and Computational Mathematics. Springer, Berlin, Heidelberg. https://doi.org/10.1007/978-3-540-70529-1_239

[24] Abdelhameed, R. M., Taha, M., Abdel-Gawad, H., Mahdy, F., & Hegazi, B. (2019). Zeolitic imidazolate frameworks: Experimental and molecular simulation studies for efficient capture of pesticides from wastewater. Journal of Environmental Chemical Engineering, 7(6), 103499.

[25] Bayantong, A. R. B., Shih, Y. J., Ong, D. C., Abarca, R. R. M., Dong, C. D., & de Luna, M. D. G. (2021). Adsorptive removal of dye in wastewater by metal ferrite-enabled

graphene oxide nanocomposites. Chemosphere, 274, 129518.

[26] Bigdeli, A., Khorasheh, F., Tourani, S., Khoshgard, A., & Bidaroni, H. H. (2020). Molecular simulation study of the adsorption and diffusion properties of terephthalic acid in various metal organic frameworks. Journal of Inorganic and Organometallic Polymers and Materials, 30(5), 1643-1652.

[27] Düren, T., Bae, Y.-S., & Snurr, R. Q. (2009). Using molecular simulation to characterise metal–organic frameworks for adsorption applications. Chemical Society Reviews, 38(5), 1237. https://doi.org/10.1039/b803498

[28] Q. Zhang, Y. Han, and L. Wu, "Influence of electrostatic field on the adsorption of phenol on single-walled carbon nanotubes: A study by molecular dynamics simulation," Chem. Eng. J., vol. 363, no. January, pp. 278–284, 2019, https://doi.org/10.1016/j.cej.2019.01.146

[29] Hongwei Yan, Baisheng Nie, Chao Peng, Peijun Liu, Xiaotong Wang, Feifei Yin, Jie Gong, Yueying Wei, and Shuangshuang Lin Energy & Fuels 2021 35 (21), 17602-17616 https://doi.org/10.1021/acs.energyfuels.1c02658

[30] Haq, S., Rehman, W., Waseem, M., Rehman, M. ur, & Khan, B. (2019). Adsorption of Cd2+ Ions onto SnO2 nanoparticles synthesized via sol-gel method: Physiochemical study. Materials Research Express. doi:10.1088/2053-1591/ab38c8

[31] Yu, C., Dong, X., Guo, L., Li, J., Qin, F., Zhang, L., … Yan, D. (2008). Template-Free Preparation of Mesoporous Fe2O3 and Its Application as Absorbents. The Journal of Physical Chemistry C, 112(35), 13378–13382. https://doi.org/10.1021/jp8044466

Sustainable Processes and Clean Energy Transition - ICSuPCET2022 Materials Research Forum LLC
Materials Research Proceedings 29 (2023) 209-217 https://doi.org/10.21741/9781644902516-24

Development of web-based interface for dry-low emission gas turbine using raspberry Pi

MOCHAMMAD Faqih[1,a*], JEREMY Peter James[1,b], MADIAH Binti Omar[1,c], and ROSDIAZLI Bin Ibrahim[2,d]

[1] Chemical Engineering, Universiti Teknologi PETRONAS, Perak, Malaysia

[2] Electrical and Electronics Engineering, Universiti Teknologi PETRONAS, Perak, Malaysia

[a]mochammad_22000035@utp.edu.my, [b]jeremy_17005433@utp.edu.my, [c]madiah.omar@utp.edu.my, [d]rosdiazli@utp.edu.my

Keywords: Dry-Low Emission Gas Turbine, Rowen's Model, Scilab/XCos, Raspberry Pi, Python, HTML

Abstract. In recent years, Dry-Low Emission (DLE) mode has been popular in gas turbine applications which can reduce emission production while maintaining high performance by operating the combustion at a low firing temperature. However, this type of combustor is prone to experience trips due to lean blowout occurrences and combustion instability. Dynamic simulation is needed to investigate the system's behavior in mitigating the tripping problem. Therefore, this paper proposed a simulator of the DLE gas turbine using Raspberry Pi, which can be accessed through the local server. Rowen's model is modified to represent the DLE gas turbine system, and its physical model is developed using open-source software, namely Scilab/XCos. The model is then attached to a web-based interface developed using Python and HTML programming language. The deployment of the web-based interface enables the gas turbine model to be accessed remotely without caring about the device's location by connecting the mobile to the same Wi-Fi router. This simulator will help improve the technical know-how of DLE gas turbine operation and can be embedded in the plant control system as a predictive maintenance tool.

Introduction

A compressor, combustor, and turbine are the three main sections of a conventional gas turbine [1]. Gas turbine operates on the thermodynamic cycle referred to as Brayton cycle. Brayton cycle is a thermodynamic cycle in which compressed air is mixed with fuel and burned under constant pressure conditions, followed by mechanical power being generated by the hot gas that expands via a turbine to accomplish work [2]. However, the downside of a conventional gas turbine is that because it operates at a high temperature at the primary zone, it produces high emission of CO_x and NO_x. As a result of the high emission of CO_x and NO_x gases, environmental factors become a worry such as the increase of greenhouse gases in the environment which contributes to respiratory diseases, and a challenge is provided to reduce emissions without endangering the turbine's power efficiency. Because of that, Dry-Low Emission (DLE) gas turbines are introduced to solve this problem.

Even though the benefits of DLE gas turbine are apparent, DLE gas turbines also has its flaws. Typically, in DLE combustion, fuel is burned at double the amount of air with a low flame temperature [3,4]. While operating at a low flame temperature reduces NO_x emissions, it also raises the risk of incomplete combustion. Incomplete combustion results in higher contents of CO. As a result, the flame temperature in the combustor must be kept within the temperature range depicted in Fig. 1 to keep NO_x and CO_x gas emissions to a minimum.

Operating at low flame temperature also makes the DLE gas turbine prone to tripping. This is because at low flame temperature, the flame is very weak and unstable which then becomes very easy to flameouts hence causing trips to occur. Therefore, a dynamic simulation for DLE gas turbine model is required to analyse the system's behaviour to ensure stable operation of the DLE gas turbine.

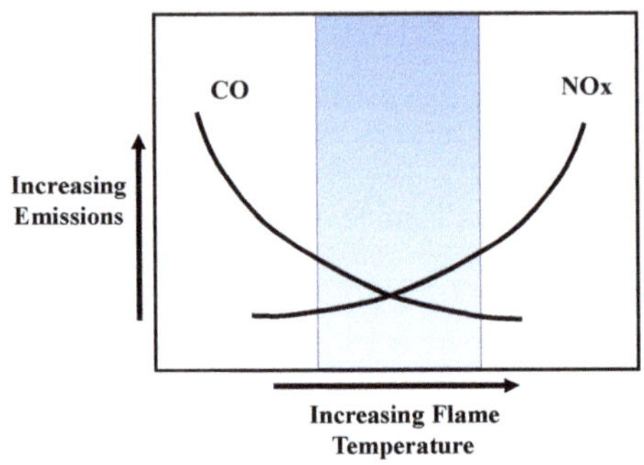

Fig.1 Behaviour of NOx and CO versus flame temperature.

In this paper, Rowen's model is selected to represent the behavior of the DLE gas turbine as shown in Fig.2. Preceding model was developed in MATLAB/Simulink which required high amount of license cost. This becomes a limitation for those who does not have license subscription. Therefore, in this paper, an alternative open source-based software namely Scilab/XCos is utilized to develop the Rowen's model. Scilab/XCos is selected due to the ability of performing simulation of physical model which is as powerful as MATLAB/Simulink with a free access and easy installation.

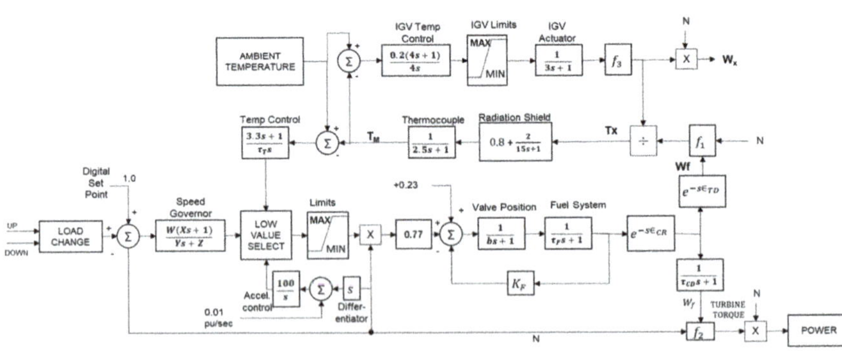

Fig.2 Rowen's model for dynamic studies [5].

Deployment of Dry-Low Emission Gas Turbine Model in Raspberry pi

Rowen's Model

Models that are derived directly from dynamic physical thermodynamic properties and principles are known as physical models. It involves utilizing laws governing thermodynamic behaviour,

Materials Research Proceedings 29 (2023) 209-217 https://doi.org/10.21741/9781644902516-24

such as conservation of mass/power/energy, in the Brayton cycle with several assumptions to obtain the differential equations to represent the dynamic gas turbine behaviour. As stated in [6], the ideal process of Brayton cycte is implemented into differential equations of total mass balance conservation as shown in Equation 1 and Equation 2:

$$\frac{dm}{dt} = \dot{m}_{in} - \dot{m}_{out} \tag{1}$$

$$\frac{dE}{dt} = \dot{m}_{in}i^{*}{}_{in} - \dot{m}_{out}i^{*}{}_{out} + Q + W \tag{2}$$

Where, $\dot{m}$ is mass flow rate, E is total energy, Q is heat input, and W is work Produced.

Rowen's Model is the first physical model of a gas turbine system, which depicts a heavy-duty gas turbine mathematically. The load-frequency and temperature controls are two of the most important controls illustrated in the diagram [7,8]. The assumptions that were made for this model are as followed: 1) Ambient Pressure is 1 atm; 2) Ambient Temperature is 15 ° C; 3) Relative Humidity is 60%; 4) Model developed is for a simple cycle and a single-shaft gas turbine.

With this developed model, calculation of output values, such as turbine temperature, can be calculated from the process inputs such as load. However, this method is very difficult to implement especially in a larger system due to numerous derivations that must be done to achieve the outputs [9,10].

Raspberry Pi

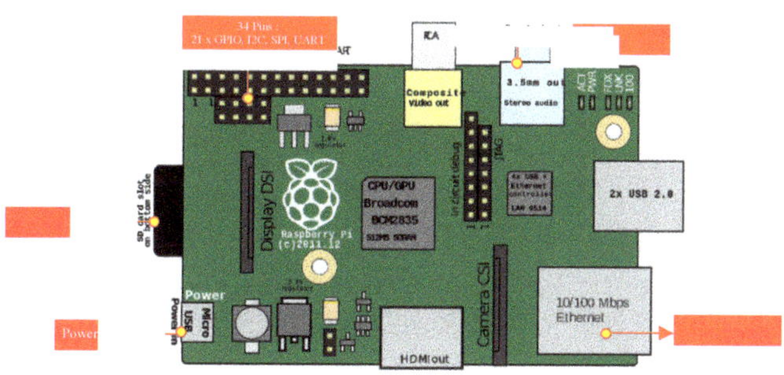

Fig.3 Layout of Raspberry Pi

Raspberry Pi and other single-board computers are low-cost development devices for testing and education [11]. Raspberry Pi operates on the Raspbian Operating System (OS) that is based on Debian (Linux) [12]. Operating on the Raspbian OS enables Raspberry Pi to have access to multiple packages and programs such as the Python programming language that is widely used in machine learning and deployment development. Layout of Raspberry Pi is displayed on Fig. 3. Assuming that the referred model hardware is the Raspberry Pi 4, this hardware includes a system-on-chip (SoC) Broadcom BCM2711, a Quad core Cortex-A72 CPU running at 1.5GHz, a graphical processor unit (GPU) VideoCore IV, and 4GB of RAM memory. The device also does not include a hard disk or solid-state disk by default, instead opting for the use of an SD card for the operating system and for nonvolatile storage. Power supply required by the device is a 5V 3A and the power port on the PCB of Raspberry Pi is a USB type-C connection. General Purpose Input Output (GPIO) ports on the Raspberry Pi function is to facilitate connecting a variety of peripheral devices to Raspberry Pi. USB/Lan ports are also available to connect devices onto the device and with both the USB/Lan ports and GPIO pins, it is made possible to connect devices to the Raspberry Pi in a variety of ways that can be used for data collection or data dumping. Finally, the HDMI out port located at the device enable the use of a monitor on the Raspberry Pi and through connecting a keyboard and mouse on the USB ports in the device, Raspberry Pi can essentially function as a standalone device. Raspberry Pi can also be access without needing a screen, a mouse or a keyboard connected to it. This can be done through secure shell (SSH) from other devices such as laptops, personal computers, or mobile phones.

Methodology

The development method of the interface is depicted in Fig.4 including the Rowen's Model Integration and Configuration, Development of Web Hosting Service through Python, and Deployment on Raspberry Pi.

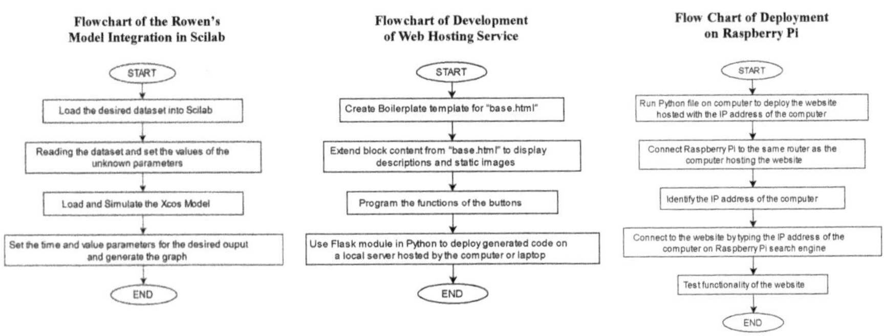

Fig. 4 Flowchart of web-based interface for DLE gas turbine.

The Rowen's model developed using Xcos in the previous section is integrated with Scilab to generate the graphs for the outputs of exhaust flow, temperature after turbine, turbine power, speed, valve position, and IGV.

The flow of development of web hosting service is starting by configuring boilerplate template. Boilerplate template is a template with the necessities needed to first start creating a HTML website. "Base.HTML" is the name of the file that the boilerplate template will be placed it and, in the template, "<%block content%>" are included to extend the "Base.HTML" into another HTML file where the necessary code for the other functions of the website is placed. The extended block content for the "Base.HTML" is written in a new file called "Home.HTML". In the new HTML file, it houses the code for the descriptions and static images placed on the website. Buttons functions for the exhaust flow, temperature after turbine, IGV, speed, turbine power, and valve position of the website is also located in "Home.HTML".To program the functions of the buttons located on the website, the programming language used will be Python and the modules used will be PyTesseract and PyAutoGUI. Finally, to deploy the created website, Flask module on Python is used.

By ensuring that the Python file on the computer is running and the website is hosted with the IP address of the computer to act as a local server, the Raspberry Pi can now be switched on to connect to the deployed website. For the Raspberry Pi to have access to the website hosted locally by the computer, the Raspberry Pi must be connected to the same Wi-Fi router as the hosting computer. After ensuring the connections to the same router is completed, the IP address of the hosting computer can be identified, and the Raspberry Pi can connect to the hosted website by typing in the IP address of the computer on its own dedicated search engine which is "Chromium". If connection is successful, the Raspberry Pi will now access to the functions available on the website to generate the simulation of the gas turbine model and also retrieve the graphically data of the outputs of the gas turbine model which are exhaust flow, valve position, speed, IGV, temperature after turbine, and turbine power.

Result and discussion

From the generated Xcos model simulation, output parameters such as the exhaust flow, valve position, temperature after turbine, speed, IGV, and turbine power of the simulated gas turbine can be retrieved. To retrieve the output parameters of the simulated model, the gas turbine load dataset was inputted, and the variables used in the Xcos model simulation of the output values were identified as it is case-sensitive. Scilab was used to generate the Xcos simulation for a period of 10 seconds, hence, data simulated will be for the stated time period. With the variables identified, the average values and graphical representation of the output parameters was able to be retrieve using Scilab as described in Table 1 and the simulation result is shown in Fig.5.

Table 1. Parameters of simulation.

Parameters	Average Value
Exhaust Flow	0.845 kg/hr
Valve Position	0.858 %
Speed	1.0 %
IGV	0.382 %
Temperature After Turbine	857.71 °C
Turbine Power	0.869kW

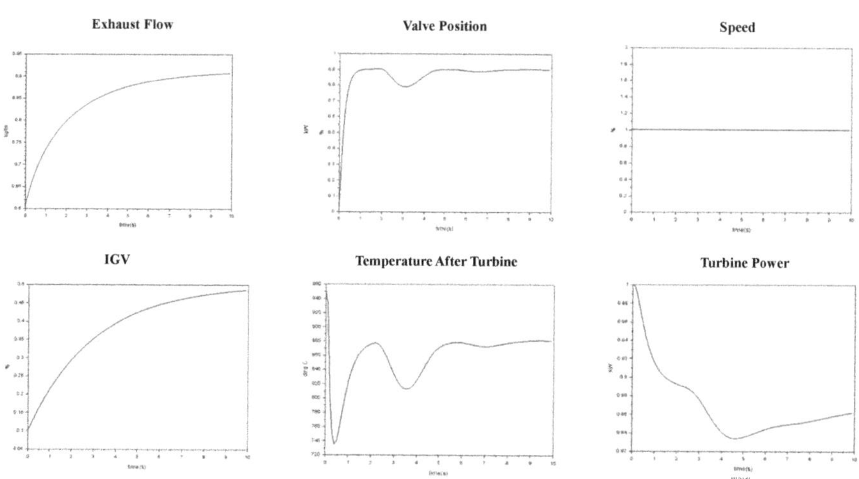

Fig. 5 Simulation result.

For the deployment of web-based interface, "Base.html" file was created using a boilerplate template that contains multiple scripts from "jQuery" and "bootstrap" which was used to indicate the appearance of the website generated. In the "base.html", extending blocks are also available to further improve the website on a separate html file saved as "home.html".

In "home.html" file, the general description of the developed gas turbine model and all push able buttons for the output parameters of the gas turbine model are included. Static images such as the UTP logo and gas turbine image is also included in "home.html" to further aid in the visual representation of the gas turbine model interface.

Two (2) main Python modules were used to give functions to the buttons available on the website which are PyTesseract and PyAutoGUI. As stated previously, PyTesseract is an optical character recognition (OCR) tool for Python that is used to recognize and read the text embedded in images while PyAutoGUI is a module that lets Python scripts control the mouse and keyboard to automate interactions with other applications. Extraction of the graphs and value generated from the previous subsection is done using the two (2) stated Python modules. As an example, to extract the graph and mean value of exhaust flow from the Xcos simulation, PyAutoGUI is used to automate the mouse and keyboard of the local server (computer used to host the website) to input the dataset submitted by the client or user through the "Choose File" s on Scilab to start the Xcos simulation of the gas turbine model and screenshot the graph and mean values of the exhaust flow simulated. From the screenshotted mean value of the exhaust flow, PyTesseract is then used to convert the image into text to display the value on the website. Other clickable buttons such as the buttons for "Valve Position", "IGV", "Temperature After Turbine", "Turbine Power", "Speed", and "Run" also perform the similar functions as explained for the exhaust flow button function. However, for the "Choose File" button, the function of this button is to open the file explorer browser of the client's or user's computer to input their desired dataset and after the file has been submitted, the file will then be saved locally on a folder on the computer hosting the website.

As all the clickable buttons functions are completely programmed, the website was then tested. Firstly, to submit the desired dataset of the load of the gas turbine, the "Choose File" button is clicked, and the file explorer browser will be prompted open to select the file that is to be submitted.

After locating the load dataset from the file explorer browser, the file is then submitted through clicking the "Submit" button. Moving on, the "Run" button is then clicked which will then run a Python script to retrieve the data generated from the simulated Xcos model of the gas turbine model. After the website has been updated with the data retrieved from the Xcos model, the exhaust flow, speed, valve position, temperature after turbine, and turbine power buttons can be clicked to display the graphs of each of the parameters.

As the website html is generated and all button functions are programmed, the website can then be deployed using the Flask module in Python. The computer hosting the deployed website is acting as the local server for the website and any devices connected to the same Wi-Fi router with the local server will have access to the website hosted.

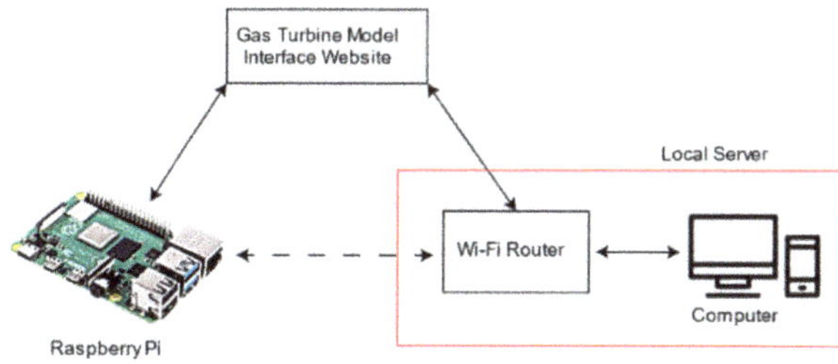

Fig.6 Connections for Deployment into Raspberry Pi

Fig.6 displays the connections that were made to make the Raspberry Pi have access to the gas turbine model interface website that was created in the previous subsection. The computer hosting the server and the Raspberry Pi must be connected to the same Wi-Fi router which is done wirelessly without the need of cables to have access to the website as the website is run locally on the computer instead on a cloud computing web server. Besides connecting to the same Wi-Fi router, the Raspberry Pi access the website for the gas turbine model interface by typing the URL of the hosted website in the web browser available in Raspberry Pi which is Chromium Browser. The URL of the hosted website is the IP address of the local server at port 5000.

After successfully connecting, the Raspberry Pi has the ability to conduct the simulation of the gas turbine model remotely regardless of the distance it is from the local server as long as it is still connected to the same Wi-Fi router. Loading the new datasets and generating the average values and graphs of the outputs of the gas turbine simulation is also possible to be done from the Raspberry Pi itself and this shows that the deployment of the website is successful as the gas turbine model interface website is able to be access remotely. Fig.7 presents the gas turbine model interface on Raspberry Pi accessed through Chromium browser.

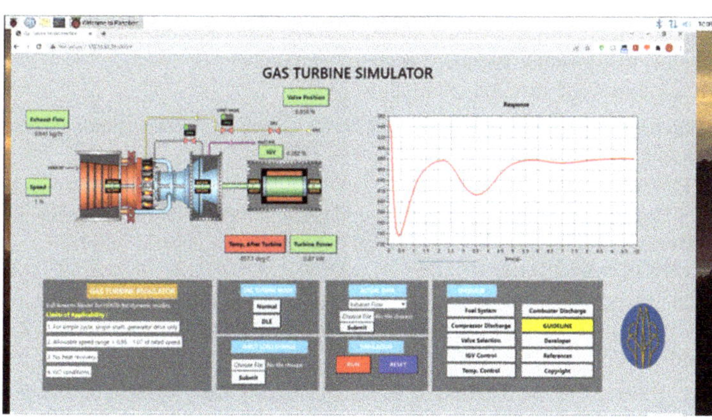

Fig.7 Web Interface Display of Temperature After Turbine Graph

Conclusion

In conclusion, the development and deployment of the gas turbine model interface was successfully conducted. The development of the gas turbine model, that was based on a previously done gas turbine model using MATLAB/Simulink, was converted into Scilab/Xcos model to as Scilab/Xcos is an open-source language compared to MATLAB/Simulink. From the generated Scilab/Xcos model, the acquiring of data was conducted using Scilab to select the needed dataset for the model and the graphs and mean values were acquired. With the Scilab/Xcos model built, the creation of the gas turbine model interface and deployment was done using Python and HTML. HTML was used to create the buttons and overall frontend of the website to host the gas turbine model interface, however, for the backend of the website that focusses on the function of the buttons, Python programming language was used instead. The functions of the button used mainly two (2) python modules which are PyTesseract and PyAutoGUI. PyTesseract is an optical character recigbutuib (OCR) tool for python while PyAutoGUI enables Python scripts to control the mouse and keyboard to automate interactions with other applications. Through using both modules, the functioning of the buttons is done to simulate the gas turbine model developed in Scilab/Xcos according to the dataset file submitted by the user. Upon simulating of the gas turbine model, the mean values and graphs generated for each output parameters of the gas turbine model, such as exhaust flow and speed, can be acquired. As for the deployment of the website, the Python module known as Flask was used. Flask is a micro web framework written in Python and using Flask, the website created was able to be deployed on a local server (which is the hosting computer) and any devices connected to the same Wi-Fi router as the local server will have access to the hosted website. Hence, the Raspberry Pi is able to establish a connection to the deployed website and through that, the remote access of the gas turbine model interface is able to be conducted as well as data collection and storage.

Acknowledgement

The authors are thankful to Ministry of Higher Education Malaysia (MOHE) and Universiti Teknologi PETRONAS for the support in carrying this research through grant PRGS (PRGS/1/2020/TK09/UTP/02/2) and YUTP (015LC0-382).

References

[1] Anheden, M. (2000). Analysis of gas turbine systems for sustainable energy conversion (PhD dissertation, Kemiteknik). Retrieved from http://urn.kb.se/resolve?urn=urn:nbn:se:kth:diva-2914

[2] Olumayegun, O., Wang, M., & Kelsall, G. (2016). Closed-cycle gas turbine for power generation: A state-of-the-art review. Fuel, 180, 694-717. https://doi.org/10.1016/j.fuel.2016.04.074

[3] Hazel, T., Peck, G., & Mattsson, H. (2014). Industrial Power Systems Using Dry Low Emission Turbines. IEEE Transactions On Industry Applications, 50(6), 4369-4378. https://doi.org/10.1109/tia.2014.2346697

[4] Nemitallah, M., Rashwan, S., Mansir, I., Abdelhafez, A., & Habib, M. (2018). Review of Novel Combustion Techniques for Clean Power Production in Gas Turbines. Energy & Fuels, 32(2), 979-1004.

[5] Faqih, M., Omar, M. B., & Ibrahim, R. B. (2022, August). Development of Rowen's Model for Dry-Low Emission Gas Turbine Dynamic Simulation using Scilab. In *2022 IEEE 5th International Symposium in Robotics and Manufacturing Automation (ROMA)* (pp. 1-5). IEEE.

[5] Asgari, H., Chen, X., Morini, M., Pinelli, M., Sainudiin, R., Spina, P., & Venturini, M. (2016). NARX models for simulation of the start-up operation of a single-shaft gas turbine.

Applied Thermal Engineering, 93, 368-376. https://doi.org/10.1016/j.applthermaleng.2015.09.074

[6] Omar, M., Ibrahim, R., Abdullah, M., & Tarik, M. (2018). Modelling and System Identification of Gas Fuel Valves in Rowen's Model for Dry Low Emission Gas Turbine. 2018 IEEE Conference On Big Data And Analytics (ICBDA). https://doi.org/10.1109/icbdaa.2018.8629705

[7] Faqih, M., Omar, M. B., Ibrahim, R., & Omar, B. A. (2022). Dry-Low Emission Gas Turbine Technology: Recent Trends and Challenges. *Applied Sciences*, *12*(21), 10922.

[8] Bahashwan, A. A., Ibrahim, R. B., Omar, M. B., & Faqih, M. (2022). The Lean Blowout Prediction Techniques in Lean Premixed Gas Turbine: An Overview. *Energies*, *15*(22), 8343.

[9] Omar, M., Tarik, M. H. M., Ibrahim, R., & Abdullah, M. F. (2017, November). Suitability study on using rowen's model for dry-low emission gas turbine operational performance. In *TENCON 2017-2017 IEEE Region 10 Conference* (pp. 1925-1930). IEEE.

[10] Tarik, M. H. M., Omar, M., Abdullah, M. F., & Ibrahim, R. (2017, November). Modelling of dry low emission gas turbine using black-box approach. In *TENCON 2017-2017 IEEE Region 10 Conference* (pp. 1902-1906). IEEE.

[11] Shalan, H. E., Hassan, M. M., and Bahgat, A. B. G. (2011). Parameter estimation and dynamic simulation of gas turbine model in combined cycle power plants based on actual operational data. J. Am. Sci. 7,303–310.

[12] Lima, Z., García-Vázquez, H., Rodríguez, R., Khemchandani, S., Dualibe, F., & del Pino, J. (2018). A System for Controlling and Monitoring IoT Applications. Applied System Innovation, 1(3), 26. https://doi.org/10.3390/asi1030026

Materials Research Proceedings 29 (2023) 218-225 https://doi.org/10.21741/9781644902516-25

Gas turbine combustion profile modelling for predictive maintenance using am artificial neural network

FARAH Adlina Md Jafrry[1,a,*], MADIAH Binti Omar[1,b], MOCHAMMAD Faqih[1,c], and ROSDIAZLI Bin Ibrahim[2,d]

[1]Department of Chemical Engineering, Universiti Teknologi PETRONAS, Bandar Seri Iskandar, Perak, Malaysia

[2]Department of Electrical and Electronics Engineering, Universiti Teknologi PETRONAS, Perak, Malaysia

[a]farahadlina2000@gmail.com, [b]madiah.omar@utp.edu.my, [c]mochammad_22000035@utp.edu.my, [d]rosdiazli@utp.edu.my[2]

Keywords: DLE Gas Turbine, Predictive Maintenance, Multi-Layer Perceptron (MLP), Artificial Neural Network (ANN)

Abstract. Dry Low Emission (DLE) gas turbine has been developed as a solution to encounter the harmful high NOx emission from conventional gas turbine. However, it is prone to create a Lean Blowout (LBO) error that causes frequent shutdown due to its stringent condition that needs to be operate inside its desired operating condition that can be monitored through the temperature, NOx and CO emission concentration. This paper develops an Artificial Neural Network – Multilayer Perceptron (ANN-MLP) predictive maintenance model using actual DLE gas turbine data that predict trips from the gas exhaust emission and classification of warning stages on the LBO error. 94.12% of R^2 for the regression model and 100% accuracy of the classification model using Python is obtained using four months period data. This proposed ANN-MLP model manage to predict the suitable maintenance time of DLE gas turbine using real time data which can help reduce cost lost from unscheduled shutdown.

Introduction

Gas turbines are internal combustion engine that converts mechanical to electrical energy from the air-fuel combustion that spins the turbine. However, this process emits high Nitrogen Oxide (NOx) emissions due to the high temperature at stoichiometric conditions, which is a poisonous and dangerous pollutant to humans and the environment. To encounter this problem, Dry Low Emission (DLE) gas turbines is introduced to lower the NOx emission by applying lean premixed combustion that lowers its flame temperature [1-4].

This lean premixed condition, however, causes the DLE gas turbine to frequently trip due to its flame and combustion instability, which leads to frequent unscheduled shutdown of gas turbine affecting the maintenance cost and energy production [5]. This instability is observed to occur when the flame fuel/air ratio (equivalence ratio) is too low outside of the desired operating range that causes Lean Blowout (LBO) condition, as shown in Fig. 1. The desired operating range and LBO area can be monitored through the CO and NOx emission as well as the fuel/air ratio or temperature.

This paper proposes that DLE gas turbine [1,2] trips can be prevented using Predictive Maintenance (PdM) approach by early prediction of the LBO error from the gas emission concentration (CO and NOx). Predictive Maintenance is a method that determines the necessary time for maintenance by early failure prediction through continuous monitoring of the system [6]. Based on the literature reviewed, data-driven methods using artificial intelligence (AI) such as

Artificial Neural Network (ANN) and Multilayer Perceptron (MLP) are some of the preferable methods for PdM due to a large amount of data of the system that can be obtained easily.

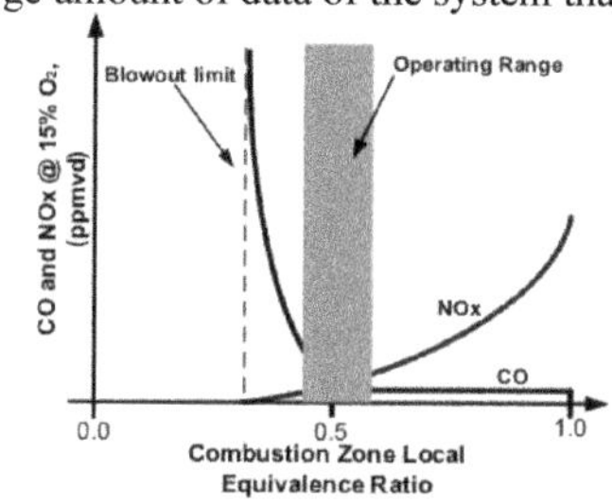

Fig. 1. Effect of Emission Concentration to Lean Blow-Out Condition

Literature review has been done on DLE Gas Turbine based on seven journals where the model used, predicted value and mode of study is observed and compared. From the literature, [7] uses NARX model using actual gas turbine data as an artificial intelligent simulation method to predict the gas turbine pressure, temperature, power and fuel flow rate. [8] uses Computational Fluid Dynamic Method using experimental data from a combustor to predict the LBO error of a DLE Gas Turbine through its temperature and emission concentration respectively. [9] perform emission identification study through predicting emission concentration on experimental data using CFD and ANN model respectively. [10] and [11] perform fault identification of the gas turbine system where uses ANN model to predict the gas exhaust temperature, gas turbine speed and angle while uses numerical ROWEN model [12,13,14] to predict the gas turbine efficiency.

Various work on PdM for conventional gas turbine were also compared. Among those that use ANN architecture, [15,16] were simulated using actual gas turbine data. uses ANN to predict the emission concentration, fuel flow, pressure and temperature, while [15,16] uses MLP to predict emission concentration, mass flowrate, temperature, power, pressure ratio, efficiency and flow capacity. [4] performed ANN or MLP modelling on an experimental mode that predicts the temperature, power, pressure ratio, temperature, flowrate, instability frequency and amplitude. [17] develop modelling in experimental mode to predict the LBO error using numerical modelling. Numerical modelling however takes much time to be developed compared to data-driven methods of modelling such as ANN and MLP due to the abundance of data collected from the plant.

Based on the literature review done, ANN is a favourable model to use for PdM of gas turbine systems. ANN is an artificial intelligence method that requires many data to get accurate result for regression and classification analysis. ANN is suitable for modelling complex and uncertain data, and is applicable for real and live operating data. There are many types of ANN model, and the most basic model is MLP [18,19]. MLP is a feedforward neural network that consists of an input layer, one or more hidden layers and an output layer. The nodes in the hidden and output layer contain their respective transfer function that allows the model to predict and develop a trendline for complex and non-linear system. MLP as a universal approximator that can predict most of the system with a few lines of codes without developing any specific mathematical model makes it a time and cost-efficient method [20] to be used for this project.

It is summarized that less study on prediction model of LBO error and gas emission concentration is observed where mostly were done on experimental mode for DLE gas turbine. PdM on actual gas turbine has been studied for conventional gas turbine where mostly utilizes ANN model. Therefore, this project will study further on developing a PdM model through the prediction of the LBO error and emission concentration using actual DLE gas turbine data from the plant while implementing similar approach that have been done on conventional gas turbine studies.

Methodology
In this chapter, the methodology of the project is elaborated starting from data pre-processing, regression model, classification model and lastly on the model performance.

A. Data Pre-processing and Analysis
39,324 clean data from three months of DLE gas turbine operation from a plant that consists of fuel gas flow, fuel gas pressure interstage, stop ratio valve (SRV), opening, gas turbine control valve (GCV) opening, splitter opening, compressor discharge, ambient air temperature, speed, exhaust press, inlet guide vanes (IGV) opening, temperature, load, nitrogen oxide (NOx) concentration and carbon monoxide (CO) concentration is used to develop the machine learning model. The data are visualized and screened to remove outliers consisting of its transients during any start-up and shut-down operation, as well as corrupted data that gives error signals. Other outliers are further identified through boxplot and regression-based method [21]. Next, the tripping period of the DLE gas turbine is identified based on its load and speed, and the dataset is labelled to be either in 90, 75, 60, 45, 30 or 15 minutes before tripping as the unhealthy condition dataset. The remaining is labelled as normal condition. The datasets are randomly separated into three types which are training, testing and validation datasets with a ratio of 60:15:25, respectively [22].

B. Regression Models Development
Regression model using machine learning models from python *sklearn* library such as artificial neural network (ANN), multilayer perceptron (MLP), k-nearest neighbors (kNN), random forest (RF), decision tree (DT) and logistic regression (LR) is developed to predict NOx, CO and temperature from the remaining parameter. The models' accuracy is evaluated and compared in terms of their root mean square error (RMSE), mean absolute error (MAE) and regression coefficient (R^2) to clarify if ANN-MLP is a suitable model. The ANN-MLP regression model is further optimized through a trial-and-error method by changing its hyperparameters in order of activation function, the number of neurons followed by the number of hidden layers [3,17]. The hyperparameters tuned are as in Table 1. The model's R^2 is tabulated and compared to identify the highest result as the best model.

Table 1. ANN-MLP Regression Model Hyperparameters

Hyperparameter	Value
Activation Function	Identity, relu, logistic, tanh
Number of Hidden Layers	1 to 5
Number of Neurons	10 to 1000
Number of iterations	10000

b. Classification Model Development
The predicted value from the regression model is then plotted to be NOx and CO concentration against temperature. The trend of emission concentration during the DLE Gas Turbine tripping and the normal condition is analysed and classified into two healthy and unhealthy state regions based on the gas emission concentration and temperature range as in Fig. 2. The yellow region is classified as a healthy condition represented as '0', while the rest is identified as unhealthy defined by '1'. After determining the classification type, a classification model is developed using temperature, CO and NOx emission concentration as the input data and DLE gas turbine performance of healthy and unhealthy as the predicted output data. The model will be tuned by adjusting the hyperparameters of the ANN-MLP model, such as activation function, number of hidden layers, and number of nodes until the highest accuracy model is identified. The classification model is evaluated by its accuracy percentage and confusion matrix.

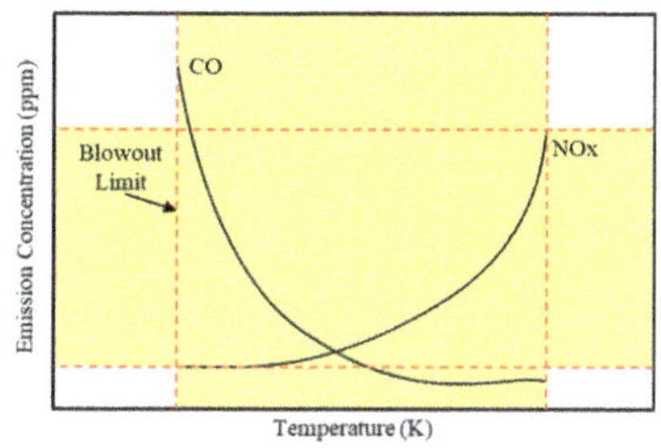

Fig. 2. Healthy and Unhealthy Data Classification from Regression Result

D. Model Performance and Evaluation

To validate the model's performance, the validation dataset set aside is run through both ANN-MLP regression and classification model simultaneously. The R^2, MAE and RMSE values of the training and validation data are compared for the regression model, while the accuracy and confusion matrix are compared for the classification model. Using validation dataset can validate that the ANN-MLP can predict NOx, CO, temperature and DLE gas turbine healthiness of different datasets that the model has never trained. Next, the significance of NOx and CO in predicting the healthiness of DLE gas is further investigated by developing another classification model using ANN-MLP with three different sets of inputs namely MLP-1, MLP-2 and MLP-3. MLP-1 is set to have only NOx, CO and temperature as the input. MLP-2 uses all inputs except NOx and CO, while MLP-3 uses all the inputs including NOx and CO. The accuracy and confusion matrix of the models are compared and studied to investigate if having NOx and CO as the input parameters give significant results in predicting the healthiness of the DLE Gas Turbine.

Result And Discussion

In this chapter, results from the regression model development, classification model development and model performance are discussed.

A. Regression Models Development

The accuracy of five machine learning models developed to predict DLE gas turbine temperature, NOx and CO concentration is compared and tabulated in Table 2. All models have higher accuracy than Linear Regression model for NOx and CO emission prediction, indicating that NOx and CO have a non-linear behaviour. The average correlation coefficient, R^2, of NOx, CO and temperature (T) prediction is calculated to ease the comparison between models. The highest average R^2 of 94.12% from the ANN-MLP model indicates its suitability for this system. Based on the comparison, MLP has the highest accuracy in predicting the overall model with an average R^2 of 94%, followed by RF. However, RF is slightly more accurate than MLP when predicting temperature. Since both models can predict temperature up to 99% accuracy, MLP is considered an excellent model to predict NOx, CO and temperature compared to the other ML models.

Through tuning of the hyperparameters of the ANN-MLP regression model, 94.12% accuracy is obtained using ANN-MLP using logistic activation function with 200 neurons in 4 hidden layers. The first hyperparameter tuned is the activation function where logistic gives the highest accuracy. Using logistic as the activation function, the number of neurons is varied from 1 until 500 and its R^2 is plotted. R^2 increases significantly as the number of neurons increase from 1 to 50 until it reaches 87% R^2 and starts to become constant from 200 to 500 neuron. Using the range of 50 to 300 neurons, the number of hidden layers is varied from 1 to 4, and its R^2 is presented as in Fig. 3. Four hidden layers provide the highest R^2, where five layers drop the model's performance.

Therefore, the optimized ANN-MLP model is selected to use a logistic activation function with two hundred neurons in four hidden layers.

Table 2. Machine Learning Regression Model Comparison

		MLP	RF	kNN	DT	LR
NOx	R^2	89.04	88.52	80.34	75.88	62.74
	MAE	1.836	1.859	2.471	2.587	3.501
	RMSE	2.470	2.529	3.309	3.665	4.556
CO	R^2	93.98	92.91	82.77	84.97	57.07
	MAE	5.114	5.734	8.979	7.486	16.29
	RMSE	8.981	9.747	15.19	14.18	23.99
T	R^2	99.32	99.61	97.91	98.9	98.64
	MAE	1.006	0.765	1.918	1.306	1.71
	RMSE	1.512	1.144	2.651	1.92	2.139
Average R^2		94.12	93.63	87.00	86.58	72.82

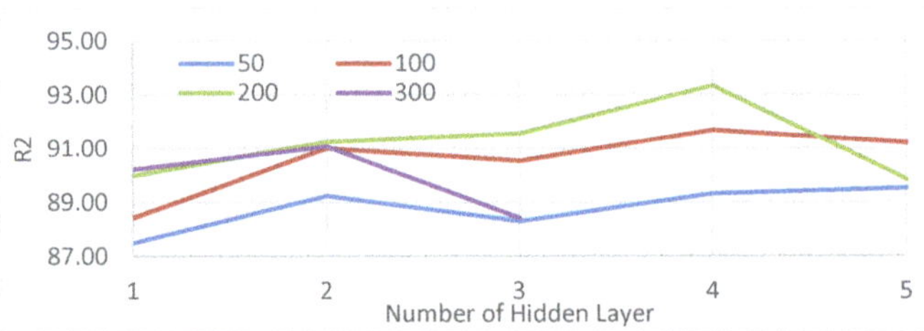

Fig. 3. R^2 of MLP-ANN by Number of Neuron and Hidden Layer

B. Classification Model Development

The predicted NOx, CO and Temperature from the ANN-MLP regression model is then plotted as in Fig. 4 to identify the baseline for the desired operating condition of the DLE gas turbine. The desired operating condition is identified to be between temperature 775 K to 915 K and NOx concentration between 74 ppm to 140 ppm. Too low temperature causes the flames to become unstable leading to LBO condition and tripping [1,2]. From the data plot, the results show similar trends of temperature, NOx and CO concentration according to literature where NOx increases while CO decreases with temperature [23]. The healthiness of the gas turbine can be reflected on the NOx and CO emission, where higher NOx emission indicates that the DLE gas turbine operates at higher temperature similar to conventional gas turbines, that will emit higher amount of the harmful NOx. Lower NOx concentration indicates that the combustion occurs at very low temperature near the lean extinction point, resulting to the possibility of flame extinction and tripping [1]. This result is similar to the analysis on actual data, where datapoints out of this range occur when approaching tripping, indicating that temperature and emission concentration can predicts the healthiness of gas turbine before tripping.

The predicted output of the ANN-MLP regression model [24,25] is used as the input to train the ANN-MLP Classification model based on the operating range, resulting in an accuracy of 100%. The model can accurately predict all healthy conditions that are represented as 0 and unhealthy conditions that are represented as 1.

Materials Research Proceedings 29 (2023) 218-225 https://doi.org/10.21741/9781644902516-25

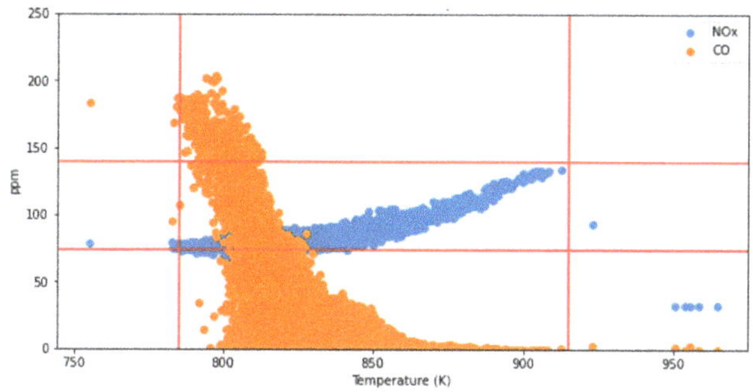

Fig. 4. Desired Operating Range of DLE Gas Turbine between 775K to 915K and 74ppm to 140ppm

C. Model Performance and Evaluation

The regression model performance of the validation dataset presents similar R^2, RMSE and MAE as the training and testing value tabulated in Table 3. Similar results in R^2, RMSE and MAE indicates that the ANN-MLP regression model is not overfitting and manage to predict a new set of data that it has never trained before and can predict future set of data from the plant. The predicted NOx, CO and temperature validation data is then input into the ANN-MLP classification model, where 99.996% accuracy is achieved. The performance of the validation data is described in the confusion matrix where all 22,2261 healthy datasets are accurately predicted. 12 out of 13 unhealthy datasets manage to be accurately predicted for the unhealthy dataset, indicating that this model can predict another untrained dataset with great precision.

Table 3. Regression Performance of ANN-MLP Validation Data

	NOX	CO	TEMPERATURE
R^2	88.92	94.39	99.49
RMSE	2.469	8.577	1.302
MAE	1.818	4.899	0.988

To further investigate if NOx and CO contribute to the prediction of DLE gas turbine healthiness, another classification model is developed using ANN-MLP with three different sets of inputs. MLP-1 is set to have the same input as the classification model developed. MLP-2 uses all inputs apart from NOx and CO, while MLP-3 uses all the inputs available from the dataset. The comparison is made to observe the impact of NOx and CO on predicting the healthiness of gas turbines. Comparing the accuracy, MLP-1 (99.71%) and MLP-2 (99.72%) have very little difference in their accuracy in predicting the healthiness of DLE gas turbines. However, when both inputs from MLP-1 and MLP-2 are combined, creating MLP-3, the model's accuracy increases to 99.77%. The increment in accuracy indicates that adding NOx and CO as the input and other parameters helps increase the model's prediction accuracy.

Conclusion

In conclusion, this project manages to develop a regression and classification MLP-ANN model to predict the healthiness of DLE gas turbines by predicting temperature, NOx and CO emission. ANN-MLP is the best model for this study by achieving the best prediction performance of 94.12% R^2 accuracy of prediction from the regression model and 100% accuracy from the classification model compared to RF, DT, kNN and LR. The optimized ANN-MLP model identified uses a logistic activation function with 4 hidden layers with 200 neurons for the regression model and a *relu* activation function with 1 hidden layer of 100 neurons for the classification model. The desired operating condition of the DLE gas turbine that is used to classify the healthiness of the model is identified to be in the range of between temperature 775K to 915K and NOx concentration between 74 ppm to 140 ppm. The classification model developed shows that incorporating CO and NOx as input could better predict the healthiness of the DLE gas turbine.

Acknowledgement
The authors are thankful to Universiti Teknologi PETRONAS for the support in carrying this research through grant YUTP (015LC0-382).

Reference

[1] M. Faqih, M.B. Omar, R. Ibrahim, B.A.A., "Dry-Low Emission Gas Turbine Technology: Recent Trends and Challenges", Applied Sciences 12 (21), 10922, 2022. https://doi.org/10.3390/app122110922
[2] A.A. Bahashwan, R.B. Ibrahim, M.B. Omar, M. Faqih, "The Lean Blowout Prediction Techniques in Lean Premixed Gas Turbine: An Overview", Energies 15 (22), 8343, 2022 https://doi.org/10.3390/en15228343
[3] S. Kwak, J. Choi, M. C. Lee, and Y. Yoon, "Predicting instability frequency and amplitude using artificial neural network in a partially premixed combustor," Energy (Oxf.), vol. 230, no. 120854, p. 120854, 2021. https://doi.org/10.1016/j.energy.2021.120854
[4] I. Parrella, F. Bardi, G. Salerno, D. Gronchi, M. Cannavò, and E. Sparacino, "Using analytics to assess health status of DLE combustion Gas turbines," in Day 3 Wed, November 13, 2019, 2019. https://doi.org/10.2118/197679-MS
[5] A. Kaluri, P. Malte, and I. Novosselov, "Real-time prediction of lean blowout using chemical reactor network," Fuel (Lond.), vol. 234, pp. 797-808, 2018. https://doi.org/10.1016/j.fuel.2018.07.065
[6] T. P. Carvalho, F. A. A. M. N. Soares, R. Vita, R. da P. Francisco, J. P. Basto, and S. G. S. Alcalá, "A systematic literature review of machine learning methods applied to predictive maintenance," Comput. Ind. Eng., vol. 137, no. 106024, p. 106024, 2019. https://doi.org/10.1016/j.cie.2019.106024
[7] M. H. M. Tarik, M. Omar, M. F. Abdullah, and R. Ibrahim, "Modelling of dry low emission gas turbine using black-box approach," in TENCON 2017 - 2017 IEEE Region 10 Conference, 2017. https://doi.org/10.1109/TENCON.2017.8228169
[8] P. C. Nassini, D. Pampaloni, R. Meloni, and A. Andreini, "Lean blow-out prediction in an industrial gas turbine combustor through a LES-based CFD analysis," Combust. Flame, vol. 229, no. 111391, p. 111391, 2021. https://doi.org/10.1016/j.combustflame.2021.02.037
[9] Z. Chen, T. Yang, S. Zhang, S. Li, and Z. Ren, "Efficient emission modelling in lean premixed flames with pre-tabulated formation characteristics," Fuel (Lond.), vol. 301, no. 121043, p. 121043, 2021. https://doi.org/10.1016/j.fuel.2021.121043
[10] Y. Park, M. Choi, X. Li, C. Jung, S. Ahn, and G. Choi, "Fault detection in fuel distribution characteristics of industrial gas turbine combustor using artificial neural network method," SSRN Electron. J., 2021. https://doi.org/10.2139/ssrn.3874483
[11] M. B. Omar, R. Ibrahim, M. F. Abdullah, and M. H. M. Tarik, "Modelling of Dry-Low Emission Gas Turbine Fuel System using First Principle Data-Driven Method," Journal of Power Technologies, vol. 100, no. 1, p. 1, 2020.
[12] M. Omar, R. Ibrahim, M.F. Abdullah, M.H.M. Tarik, "Modelling and System Identification of Gas Fuel Valves in Rowen's Model for Dry Low Emission Gas Turbine, in 2018 IEEE

Conference on Big Data and Analytics (ICBDA), pp. 33-37, 2018.
https://doi.org/10.1109/ICBDAA.2018.8629705
[13] M. Omar, M.H.M. Tarik, R. Ibrahim, M.F. Abdullah, "Suitability study on using rowen's model for dry-low emission gas turbine operational performance", in TENCON 2017-2017 IEEE Region 10 Conference, pp. 1925-1930, 2017. https://doi.org/10.1109/TENCON.2017.8228173
[14] M. Faqih, M.B. Omar, R.B. Ibrahim, "Development of Rowen's Model for Dry-Low Emission Gas Turbine Dynamic Simulation using Scilab", in 2022 IEEE 5th International Symposium in Robotics and Manufacturing Automation (ROMA), IEEE, pp. 1-5, 2022.
https://doi.org/10.1109/ROMA55875.2022.9915695
[15] D. Huang, S. Tang, and D. Zhou, "A Nitrogen Oxides emission prediction model for gas turbines based on interpretable multilayer perceptron neural networks," in Volume 9: Oil and Gas Applications; Organic Rankine Cycle Power Systems; Steam Turbine, 2020.
https://doi.org/10.1115/GT2020-15478
[16] Z. Liu and I. A. Karimi, "Gas turbine performance prediction via machine learning," Energy (Oxf.), vol. 192, no. 116627, p. 116627, 2020. https://doi.org/10.1016/j.energy.2019.116627
[17] Kirubakaran and N. Shankar, "Prediction of lean blowout performance on variation of combustor inlet area ratio for micro gas turbine combustor," Aircr. Eng. Aerosp. Technol., vol. 93, no. 5, pp. 915-924, 2021. https://doi.org/10.1108/AEAT-02-2021-0042
[18] M. Salehi, H. Eivazi, M. Tahani, and M. Masdari, "Analysis and prediction of gas turbine performance with evaporative cooling processes by developing a stage stacking algorithm," J. Clean. Prod., vol. 277, no. 122666, p. 122666, 2020.
https://doi.org/10.1016/j.jclepro.2020.122666
[19] O. I. Abiodun et al., "Comprehensive review of artificial neural network applications to pattern recognition," IEEE Access, vol. 7, pp. 158820-158846, 2019.
https://doi.org/10.1109/ACCESS.2019.2945545
[20] B. Akkaya and N. Çolakoğlu, "Comparison of Multi-class Classification Algorithms on Early Diagnosis of Heart Diseases."
[21] A. Smiti, "A critical overview of outlier detection methods," Comput. Sci. Rev., vol. 38, no. 100306, p. 100306, 2020. https://doi.org/10.1016/j.cosrev.2020.100306
[22] A. I. Abubakar, H. Chiroma, and S. Abdulkareem, "Comparing performances of neural network models built through transformed and original data," in 2015 International Conference on Computer, Communications, and Control Technology (I4CT), 2015.
https://doi.org/10.1109/I4CT.2015.7219599
[23] M. Chmielewski and M. Gieras, "Small gas turbine gtm-120 bench testing with emission measurements," J. KONES Powertrain Transp., vol. 22, no. 1, pp. 47-54, 2015.
https://doi.org/10.5604/12314005.1161610
[24] M.H.M. Tarik, M. Omar, M.F. Abdullah, R. Ibrahim, "Optimization of neural network hyperparameters for gas turbine modelling using Bayesian optimization", IET Digital Library, pp. 10-15, 2018. https://doi.org/10.1049/cp.2018.1308
[25] A.A. Bahashwan, R.B. Ibrahim, M.B. Omar, M. Faqih, "Data-driven lean blowout prediction based on industrial dry low emission gas turbine dataset Using decision tree", 2022 IEEE Industrial Electronics and Applications Conference (IEACon), 90-93, 2022.
https://doi.org/10.1109/IEACon55029.2022.9951749

Impact of parameters on gas ratios obtained from air gasification of date palm waste

MUHAMMAD Shahbaz[1,2,a*], MUDDASSER Inayat[2,3,c], TAREQ Al Ansari[1,b], GORDON Mckay[1,d]

[1]College of Science and Engineering, Hamad Bin Khalifa University, Qatar Foundation, Education City, P.O. Box 5825, Doha, Qatar

[2]Universiti Teknologi PETRONAS, 32610 Seri Iskandar, Perak Darul Ridzuan, Malaysia

[3]Research Group of Energy Conversion, Department of Mechanical Engineering, School of Engineering, Aalto University, 02150 Espoo, Finland

[a]mshahbaz@hbku.edu.qa, [b]talansari@hbku.edu.qa, [c]muddasser.inayat@aalto.fi, [d]gmckay@hbku.edu.qa

Keywords: Date Palm Waste, Gas Ratio, Particle Size, Temperature, Air Flow Rate

Abstract. Utilizing date palm waste as an energy source is a potential low-carbon energy source, especially in areas such as the Middle East. This study investigates date palm waste air gasification performed in a downdraft gasifier. The study aims to analyze the impact of various parameter ranges such as temperature of 600-900 °C, air flow rate of 1.0 to 4.0 l/min, and particle size of <2-6> mm on the gas ratios of producer gas. H_2/CO ratio is an important parameter which is an increase from 0.53 to 0.71 with a rise in temperature from 600 to 900 °C. A similar profiling of other gas ratios was noticed with the temperature increase. The H_2/CO ratio and H_2/CO_2 are found maximum at the air flow rate of 2.5 l/min and 3.0 l/min.H_2/CH_4 showed a very steady trend with an increase of air flow rate up to 2.5 l/min, but a sharp hike was noticed by increasing the air flow rate of 2.5 to 4.0 l/min. Larger particle size shows a lower value of H_2/CO, H_2/CH_4, and H_2/CO_2 due to the lower heat and mass transfer diffusion compared to smaller particle size.

Introduction

The extraction of energy from fossil fuels to satisfy energy demand is challenged by finite sources, oil price volatility, greenhouse gas emissions, and climate change [1]. As such, there are global efforts to identify alternative, renewable, sustainable, and carbon-neutral sources to satisfy the growing energy demand, which is expected to increase by 53% in 2030 [2]. Incidentally, biomass waste has emerged as a reliable, effective, and dependable source due to its extensive availability across the globe up to 70 EJ, no carbon footprint, and energy density have the potential to replace the crude-based fuel refinery with biorefinery [3]. The extraction of energy from biomass-based waste is based on the biological and thermochemical conversion process. Among all thermal treatment processes, Gasification is a well-known and progressive technology that converts organic biomass and wastes into gaseous fuel at higher temperatures using the oxidizing agent (steam, air, and oxygen) [4].

The Middle East, especially Gulf Cooperation Council (GCC), is known for its petroleum reserves and export of petroleum products. Date fruit is the main agricultural product and is well integrated with their culture, which contributes to 67% of global date exports and 75% of date production worldwide [5]. The date fruit industry including its cultivation and processing produces an ample quantity of waste including leaves, fronds, and date pits. According to an estimation, a date palm tree produces 20 kg of leaves and fronds waste per year [6]. The conversion of date palm waste can provide a valuable clean energy source, which has been significantly explored in recent

Materials Research Proceedings 29 (2023) 226-233 https://doi.org/10.21741/9781644902516-26

years. The thermal kinetic study and pyrolysis of date palm wastes have been investigated to understand the thermal degradation as well as char and oil production [7, 8].

The production of syngas from the conversion of date palm waste is also investigated both in simulation and experimental studies, in which there are numerous simulation models that have been developed for the gasification of date palm wastes to predict the syngas by varying the temperature and steam flow rate [9, 10]. The quality and utilization of produced gas depend on the H_2/CO ratio and other gas ratios. These ratios have immense importance for converting into value-added products such as methanol, methane, dimethyl ether (DME), and F-T diesel. A few studies have been reported to investigate the impact of parameters on gas ratios. Shahbaz et al. [2] have reported the H_2/CO ratio optimization for the steam gasification of date palm wastes. Most of the reported literature is focused on the impact of temperature, steam flow rate, and catalysts on the H_2/CO ratio [11]. However, very few studies have focused on the other gas ratio such as H_2/CO, H_2/CH_4, H_2/CO_2, CO/CH_4, and CO/CO_2 and there is no study available for date palm waste gasification. This study aims to perform the air gasification of date palm waste in the downdraft gasifier. The impact of temperature (600-900 °C), air flow rate (1.0-4.0 l/min), and particle size of (<2-6> mm) on H_2/CO, CO/CO_2, H_2/CO_2, H_2/CH_4, and CO/CH_4 ratio. This study is very helpful to identify the ratios for the further utilization of producer gas into fuels and chemical conversion.

Methodology

For this study, the dried palm leaves from the Middle East were collected from one of the date fields. The received dried palm waste was characterized to determine the waste's heating values, and chemical compositions in terms of the proximate and ultimate analysis as presented in Table 1 was also reported in one of our previous studies[12]. In the next stage, the feedstock was sun-dried, crushed in the jaw crusher, and sieved using the sieve shaker into the desired particle sizes of <2, 4, and 6> mm, respectively. The prepared feedstock was stored in airtight jars to avoid moisture.

Table 1. Chemical composition of properties of date palm waste [12].

Feedstock	Analysis (weight% dry basis)								Higher heating value (MJ/kg)
	Ultimate					Proximate			
	C	H	N	S	O*	Volatile matter	Carbon	Ash	
DPF	37.12	5.78	3.12	0.20	53.78	80.60	12.69	6.71	16.61

*By difference method

The air gasification of date palm waste is instituted in the setup presented in Figure 1. The gasifier reactor is made of high-quality steel to perform the high-temperature process. The gasifier is insulated with high-quality ceramic material to obtain the isothermal condition. The mode of the gasification system was downdraft. The heating system consists of electric heaters that were externally mounted and jacketed to heat up the system. A system was mechanically cleaned and purged with N_2 to ensure the removal of dust and unburned material before each experiment. The gasifier was heated up to the targeted temperature by the electric heater. The temperature was measured and controlled through the thermocouple and microcontroller. The produced gas was passed through the pipes and gas analyzer system to measure the gas composition at intervals of 1.0 second. The 100 grams of feedstock was used for each batch. The air flow rate varied from 1.0-4.0 l/min and the temperature varied from 600-900 °C. The particle size of <2, 2, 2-6, and 6> mm was investigated. These experimental conditions ranges were identified with the series of test runs.

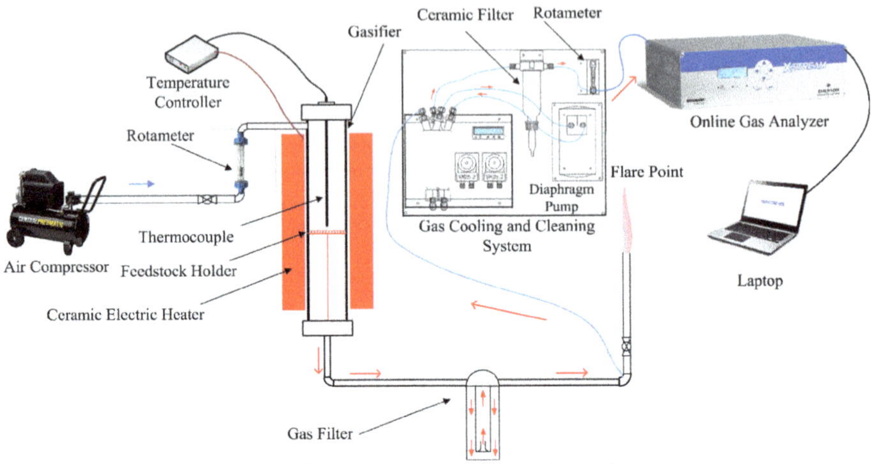

Figure 1. Air gasification system for date palm waste.

Results and Discussion

Impact of Gasification Temperature

The temperature has ample impact on the gas composition and the gas ratio is very important for the conversion of syngas into value-added products. H_2/CO is increased from 0.53 to 0.71 with the increase of temperature, as shown in Figure 2. This is due to the activation of the water shift reaction which enhanced the H_2 content [12]. Similarly, the H_2/CH_4 ratio increase from 0.77 to 2.05 shows the activation of methane reforming reaction at higher temperature [13]. The increasing trend of CO/CO_2 and H_2/CO_2 ratio shows that the CO_2 is decreased with the activation of endothermic reactions such as bourdourd reaction. This trend was also reported in the literature with the temperature increase [2]. The CO/CH_4 ratio also follows a similar profile due to the reduction of methane and the activation of methane reforming reaction at a higher temperature [13].

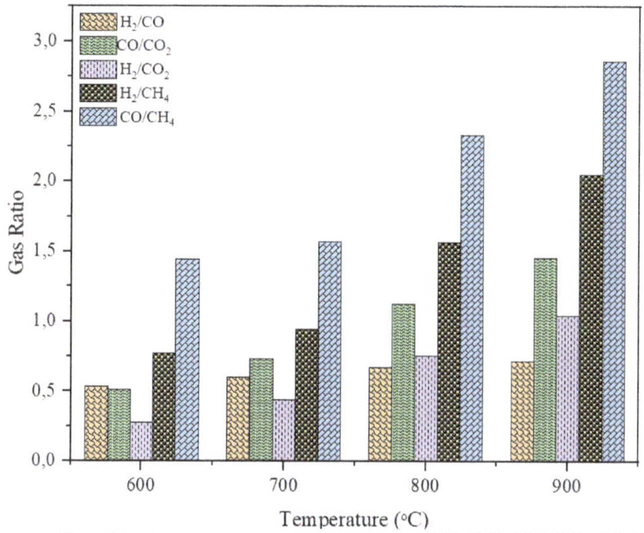

Figure 2. Impact of gasification temperature on gas ratio of H_2/CO, CO/CO_2, H_2/CO_2, H_2/CH_4, and CO/CH_4 at 3.0 l/min and 2-6 mm particle size.

Impact of Air Flow Rate

The air flow rate is very important because it works as an oxidizing agent to instigate the gasification reaction and defines the concentration of gases in the product gas. H_2/CO ratio is increased and reached 1.03 at the air flow rate of 2.5 l/min but has dropped to 0.5 at a higher air flow rate of 4.0 l/min as presented in Figure 3. A similar pattern was also noticed for H_2/CO_2 ratio air flow rate. This trend shows that 2.5 l/min provides enough oxygen to activate the water gas shift reaction and methane reforming reactions [14]. Furthermore, it was noticed that H_2/CH_4 and CO/CO_2, and CO/CH_4 also increased with the increase of air flow rate up to 3.5 l/min and higher values 1.52, 1.27, and 2.4 1respeitevly. These also follow the decreased trend at a higher flow rate. The increasing trend shows that a higher flow rate enables the higher content of CO and CO_2 gases due to the higher oxygen availability that enhances the combustion reaction and forms the CO and CO_2 [15].

Materials Research Proceedings 29 (2023) 226-233 https://doi.org/10.21741/9781644902516-26

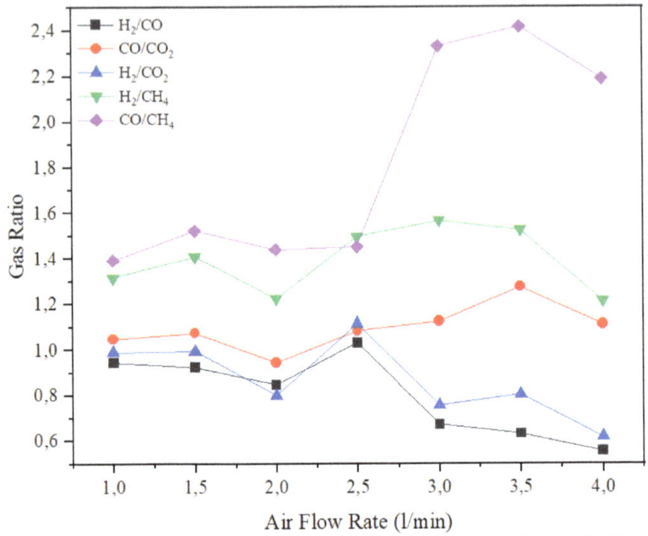

Figure 3. Impact of air flow rate on gas ratio of H_2/CO, CO/CO_2, H_2/CO_2, H_2/CH_4, and CO/CH_4 at 800 ℃ and 2-6 mm particle size.

Impact of Particle Size

Particle size also plays a very important role in the gasification process. It is directly related to the heat and mass transfer with the particle and from the oxidizing agent as well as the heating medium which directs the reactions placed in the gasification process [16]. Particle size is directly related to heat and mass transfer diffusion. H_2/CO is found to be higher 0.85 at smaller particle size due to better heat and mass transfer at higher temperature [12, 17]. A similar profile was noticed for H_2/CO_2 ratio and CO/CO_2 ratio, showing that smaller particle size helps the water gas shift reactions. But in the case of CO/CH_4 and H_2/CH_4, it shows an increasing or no effect trend with the smaller particle size. The reason for this is related to the CH_4 content. The gasification is performed at a higher temperature of 800 °C such that the particle size has less influence on methane formation or reforming [18].

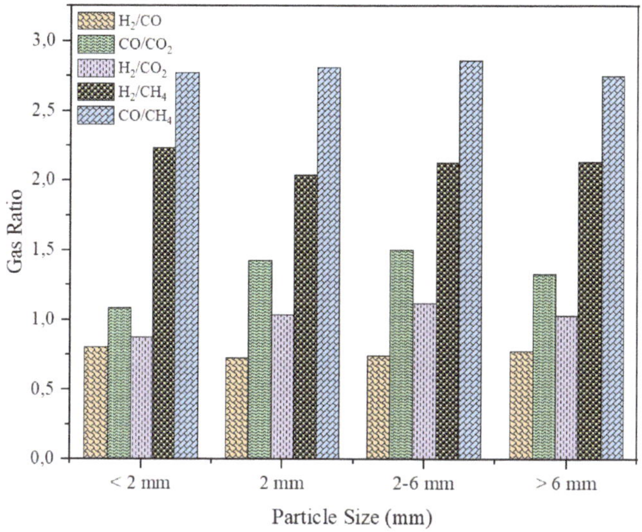

Figure 3. Impact of particle size on gas ratio of H_2/CO, CO/CO_2, H_2/CO_2, H_2/CH_4, and CO/CH_4 at 800 °C and 3.0 l/min air flow rate.

Conclusion

This study investigated the conversion of date palm wastes into gas using a downdraft gasifier. In the investigation, the impact of parameters temperature (600-900 °C), particle size <2, 2, 2-6, and 6> mm), and air flow rate (1.0-4.0 l/min) on the gas ratio produced from date palm waste. Generally, the gas ratio has an increasing trend with the increase in temperature. It was noticed that H_2/CO and H_2/CH_4 increased from 0.53 to 0.71 and 0.77 and 2.05 by increasing the temperature due to the activation of endothermic reactions. The air flow rate of 2.5 l/min was the best in which H_2/CO ratio and H_2/CO_2 were maximum. However, H_2/CH_4 was 1.5 at air flow rate of 3.0 l/min. The CO/CH_4 shows a very sharp increase at a higher air flow rate, which may be ineffective for methane reforming reactions. The smaller particle size is more effective in obtaining the higher H_2/CO and H_2/CH_4 ratio. This study supports the understanding of the gas ratios, which is important for the conversion of producer gas into valuable products such as methanol, DME, CH_4, etc.

Acknowledgment

The authors are thankful to the Hamad Bin Khalifa University, Doha, Qatar, and Universiti Teknologi PETRONAS, Perak Darul Ridzuan, Malaysia for financial and technical support.

References

[1] A. Al-Rumaihi, M. Shahbaz, G. McKay, H. Mackey, T. Al-Ansari, A review of pyrolysis technologies and feedstock: A blending approach for plastic and biomass towards optimum biochar yield, Ren. Sustain. Energ. Rev. 167 (2022) 112715. https://doi.org/10.1016/j.rser.2022.112715

[2] M. Shahbaz., T. Al-Ansaria, G. McKay, S. Yusup, M. Inayat, Optimization Study of H2/CO Ratio in the Steam Gasification of PKS using Coal Bottom ash for fuel Production through Response Surface Methodology, in: S. Pierucci, F. Manenti, G.L. Bozzano, D. Manca (Eds.),

Computer Aided Chemical Engineering, Elsevier 2020, pp. 1021-1026. https://doi.org/10.1016/B978-0-12-823377-1.50171-3

[3] M. Shahbaz, S. Yusup, M. Ammar, A. Inayat, D.O. Patrick, Development of process flow sheet for syngas production from sorption enhanced steam gasification of palm kernel shell, Chem. Eng. Trans. 61 (2017) 1675-1680.

[4] A. AlNouss, G. McKay, T. Al-Ansari, Superstructure Optimization for the Production of Fuels, Fertilizers and Power using Biomass Gasification, Computer Aided Chemical Engineering 46 (2019) 301-306. https://doi.org/10.1016/B978-0-12-818634-3.50051-5

[5] B. Dhehibi, M.B. Salah, A. Frija, Date palm value chain analysis and marketing opportunities for the Gulf Cooperation Council (GCC) countries, Agricultural Economics-Current Issues, IntechOpen 2018. https://doi.org/10.5772/intechopen.82450

[6] M.R. Kabli, A.M. Ali, M. Inayat, A.A. Zahrani, K. Shahzad, M. Shahbaz, S.A. Sulaiman, H2-rich syngas production from air gasification of date palm waste: an experimental and modeling investigation, Bio. Conver. Bioref. (2022) 1-13. https://doi.org/10.1007/s13399-022-02375-7

[7] H.H. Sait, A. Hussain, A.A. Salema, F.N. Ani, Pyrolysis and combustion kinetics of date palm biomass using thermogravimetric analysis, Biores. Technol. 118 (2012) 382-389. https://doi.org/10.1016/j.biortech.2012.04.081

[8] G. Bensidhom, A. Ben Hassen-Trabelsi, K. Alper, M. Sghairoun, K. Zaafouri, I. Trabelsi, Pyrolysis of Date palm waste in a fixed-bed reactor: Characterization of pyrolytic products, Biores. Technol. 247 (2018) 363-369. https://doi.org/10.1016/j.biortech.2017.09.066

[9] A. AlNouss., G. McKay, T. Al-Ansari, Optimum Utilization of Biomass for the Production of Power and Fuels Using Gasification, in: J.J.K.S.R.P.S.V.T.W. Anton Friedl (Ed.) Proceedings of the 28th European Symposium on Computre Aided Process Engineering, Elsevier B.V., Graz, Austria, 2018, pp. 1481-1486. https://doi.org/10.1016/B978-0-444-64235-6.50258-8

[10] M. Bassyouni, S.W. ul Hasan, M.H. Abdel-Aziz, S.M.S. Abdel-hamid, S. Naveed, A. Hussain, F.N. Ani, Date palm waste gasification in downdraft gasifier and simulation using ASPEN HYSYS, Energ. Conver. Manag. 88 (2014) 693-699. https://doi.org/10.1016/j.enconman.2014.08.061

[11] M. Inayat, S.A. Sulaiman, J.C. Kurnia, M.Y. Naz, Catalytic and noncatalytic gasification of wood-coconut shell blend under different operating conditions, Env. Prog. Sustain. Energy 38(2) (2019) 688-698. https://doi.org/10.1002/ep.13003

[12] A.M. Ali, M. Inayat, A.A. Zahrani, K. Shahzad, M. Shahbaz, S.A. Sulaiman, H. Sadig, Process optimization and economic evaluation of air gasification of Saudi Arabian date palm fronds for H2-rich syngas using response surface methodology, Fuel 316 (2022) 123359. https://doi.org/10.1016/j.fuel.2022.123359

[13] M. Shahbaz, Y. Suzana, A. Inayat, M. Ammar, D. Patrick, A. Pratama, S. Raza Naqvi, Syngas production from steam gasification of Palm kernel shell with subsequent CO2 Capturing using CaO sorbent: An Aspen plus modelling, Energ. Fuels 31(11) (2017) 12350-12357. https://doi.org/10.1021/acs.energyfuels.7b02670

[14] F.M. Guangul, S.A. Sulaiman, A. Ramli, Study of the effects of operating factors on the resulting producer gas of oil palm fronds gasification with a single throat downdraft gasifier, Ren. Energ. 72 (2014) 271-283. https://doi.org/10.1016/j.renene.2014.07.022

[15] M. Shahbaz, S.A.A. Taqvi, M. Inayat, A. Inayat, S.A. Sulaiman, G. McKay, T. Al-Ansari, Air catalytic biomass (PKS) gasification in a fixed-bed downdraft gasifier using waste bottom ash as catalyst with NARX neural network modelling, Comp. Chem. Eng. 142 (2020) 107048. https://doi.org/10.1016/j.compchemeng.2020.107048

[16] M. Shahbaz, S. Yusup, A. Inayat, D.O. Patrick, A. Pratama, Application of response surface methodology to investigate the effect of different variables on conversion of palm kernel shell in steam gasification using coal bottom ash, App. Energ. 184 (2016) 1306-1315. https://doi.org/10.1016/j.apenergy.2016.05.045

[17] M. Inayat, S.A. Sulaiman, J.C. Kurnia, Investigation on the effect of blending ratio and airflow rate on syngas profile produced from co-gasification of blended feedstock, EDP Sciences, p. 03015. https://doi.org/10.1051/matecconf/201713103015

[18] C. Guizani, F.J. Escudero Sanz, S. Salvador, Influence of temperature and particle size on the single and mixed atmosphere gasification of biomass char with H2O and CO2, Fuel Process. Technol. 134 (2015) 175-188. https://doi.org/10.1016/j.fuproc.2015.01.031

Application of machine learning models to predict ecotoxicity of ionic liquids (Vibrio fischeri) using VolSurf principal properties

GRACE Amabel Tabaaza[1,a], BENNET Nii Tackie-Otoo[2,b], DZULKARNAIN B Zaini[1,c] and BHAJAN Lal[1,d *]

[1]Chemical Engineering Department, Universiti Teknologi PETRONAS, 32610, Bandar Seri Iskandar, Perak Darul Ridzuan, Malaysia

[2]Petroleum Engineering Department, Universiti Teknologi PETRONAS, 32610, Bandar Seri Iskandar, Perak Darul Ridzuan, Malaysia

[a]grace_20000207@utp.edu.my, [b]bennet_17006974@utp.edu.my, [c]dzulkarnain.zaini@utp.edu.my, [d]bhajan.lal@utp.edu.my

Keywords: Ionic liquids, Vibrio Fischeri, Toxicity, Machine learning, QSPR/QSAR and Principal Properties

Abstract. Owing to the rapid growth in IL synthesis due to feasible cation–anion combinations, knowledge of their toxicity is pertinent for their successful application. Toxicity information measurement of various ILs on a broad spectrum of conditions through experimental techniques is way demanding on time, resources, and is at times impractical. Various research works have been performed in Quantitative Structure Activity/Property Relationship (QSAR/QSPR) for IL toxicity prediction. In this study, ML models have been trained and tested on Vibrio fischeri toxicity data set using in silico principal properties (PPs) as descriptors. Deploying this properties aid in considering both the effect of cations and anions on *Vibrio fischeri* toxicity prediction. Among the models trained, the Random Forest model proved to be the most precise nevertheless, decision tree model was the most accurate and consistent. Considering the importance of the descriptors to *Vibrio fischeri* toxicity selection techniques and model optimization.

Introduction

Ionic liquids have unique characteristics viz low melting point, low vapor pressure and cation and anion composition, which allow them to serve as solvents, catalysts [1], [2], electrolytes and separating agents for numerous industrial solutions such as gas hydrate inhibition in Oil and gas pipeline[3]–[6] and promotion in carbon sequestration [7]–[9]. Ionic Liquids have low vapor pressure and may be regarded as "green" solvents. They are, therefore, not expected to have high toxicity in comparison with conventional volatile solvents in the environment. However, due to their high solubility and stability in water, they are likely to persist in wastewater. Therefore, it is essential to determine the level of risk to the aquatic life to successfully use these ILs.

Toxicity assessments of ILs are usually carried out on a marine luminescent bacterium, *Vibrio fischeri*. This is a photobacterium fischeri which emits light through normal metabolic process. When exposed to contaminants or pollutants, their metabolic process is affected which reflects as reduction in the amount of light emitted. This is a measure of ecotoxicity expressed as EC50 [10]–[16]. The international standard ecotoxicological bioassay (DINEN ISO11348) [17] recommends this bioassay with the bioluminescent bacterium *Vibrio fischeri* and is often considered as the reagent mostly used for toxicity determination [18]. Studies on environmental toxicity have adopted this assessment for testing chemicals [19]. This is attributed to the fact that bioassays with *Vibrio fischeri* are simple and fast in comparison to other procedures. Furthermore, this toxicity evaluation is applicable for many ILs because it shows optimum hydrophilic/lipophilic balance which is attributed to the fact that *Vibrio fischeri* is a gram-negative bacterium [16], [20]–[23].

Experimental toxicity measurement is the most effective and direct way of finding ILs with desired low toxicity. In addition, synthesis of IL structures that are novel is rapidly on the rise as a result of many feasible cation–anion combinations. Due to this rapid growth, toxicity measurement of various ILs under extensive conditions through experimental techniques requires lots of time and resource and are impractical. Therefore, various studies have been performed in Quantitative Structure Activity/Property Relationship (QSAR/QSPR) for IL toxicity prediction [24]. The models are developed using data on their properties acquired through experiments and this data can produce predictions on the toxicities of new ILs that are satisfactory. These studies use the "univariate" approach. In this approach, the cation effect is considered when the anion is fixed, and the anion effect is considered when the cation is fixed; assuming there is no toxicity variation with changing cationic or anionic IL counterparts. The other assumption is that there is no effect from the interaction between the cation and anion. Some studies also used electrostatic and topological structural descriptors or heuristic descriptors [25], [26].

To improve upon this approach and overcome the difficulty in the acquisition of experimental descriptors, Games approach in VolSurf+ to derived the in silico cation and anion physicochemical descriptors. The VolSurf+ descriptors calculate interaction energy moments and capacity factors, assess hydrophilic and hydrophobic regions, molecular size and shape, compute amphiphilic moments, hydrophobic-lipophilic balance, as well as partition coefficient in different solvents, diffusion in water solvent, molecular flexibility in different solvents and pH dependent water solubility. A QSPR model was first used to validate the VolSurf+ descriptors using aquatic toxicity scores as responses. A good correlation was achieved yet there was difficulty in handling large number of descriptors especially for big data set [28]. Therefore, they developed nine principal properties (PPs) as physicochemical descriptors for 38 anions (4 PPs-) and 218 cations (5 PPs+) [27]. These PPs were used to come up with QSPR models for assessing and predicting IPC-81 rat cell line cytotoxicity, acetylcholinesterase inhibition [27] and *Vibrio fischeri* toxicity [16].

In this current study, the QSPR approach is extended by deploying other machine learning (ML) to predict *Vibrio fischeri* toxicity using the PPs developed by Games et al. [27] on the same IL data set used. In this study the cationic PPs are denoted as PP(n)C and the anionic ones are denoted as PP(n)A where the "n" is the numbering of the PPs. Five different ML models are trained and tested in this study. Their performances on the test data are compared to choose the best model.

Models used for supervised machine learning

Numerous industries have adapted supervised machine learning methods to solve complicated problems. In this current study QSPR studies has been extended by training and testing five ML models on *Vibrio fischeri* toxicity data using in silico descriptors developed by Paternò et al. [16], [27]. The five supervised ML models deployed in this study are briefly described below. Detailed description could be found in [29].

Decision tree models are used in machine learning to unite a series of the basic test efficiently where a numeric feature is compared to a threshold value in each test. Decision trees are an easy to use and well-known approach for statistical learning; such trees aim to identify the splitting criteria which describes the relationship between a set of input combinations, and regions of the output space. It is used to solve both regression and classification problems [30]–[32].

Random Forest (RF) model is usually used on datasets that are randomly sampled and functions by training decision trees on the datasets to have their predictions averaged. Features are generated from the predictions made by trees, and prediction made by one tree automatically becomes a feature for the prediction of the final model. It achieves good model results because it is easy to interpret and is robust against overfitting. RF is biased against features which are highly cardinal.

Extra tree regression: Geurts et al. [33] proposed that the random forest regression model has extra trees as extensions. These extra trees belong to the class of decision tree-based ensemble learning methods. Multiple decision trees are used to perform classification and regression tasks in decision tree-based ensemble methods [34].

Gradient boosting (GB) regressor combines simple parameterized functions with "poor" performance (high prediction error) using an iterative algorithm in order to come up with a prediction rule that is highly accurate. Contrary to other methods of statistical learning which usually provide comparable accuracy (such as support vector machines and neural networks), GB produces results that are interpretable, and this is done without requiring a lot of data preprocessing and parameter tuning. GB is a technique used in machine learning for both regression and classification problems and produces an ensemble of weak prediction models simply called a prediction model [35].

Extreme Gradient Boosting (XGBoost) regression is carried out by means of the stochastic gradient boosting algorithm. It is a portable, flexible and efficient model for machine learning [36]. The stochastic gradient boosting algorithm has a parallel boosting technique that allows efficient, accurate and fast machine learning. As reported by Chen and Guestrin [36], XGBoost has one desired attribute of maximization of loss function. This helps prevent overfitting when the final weights are smoothened out and an extra regularization term is added to it. The successful performance of this model is due to its scalability in all setups. This model prevents over-fitting and shows great robustness against multicollinearity. It penalizes the irrelevant input variables coefficients, drawing these closer to zero and eliminating these zeroed inputs to minimize standard errors. This process allows for optimization of the algorithms.

Criteria for model evaluation

Models were evaluated using error matrices which indicate the deviation from true probabilities as well as the time taken to evaluate the models. These are Mean Absolute Error (MAE), Root Mean Squared Error (RMSE) and Akaike Information Criteria.

Mean Absolute Error (MAE): This indicates the deviation from true probability and is the mean of the absolute value of the errors. Mathematically expressed as,

$$MAE = \frac{\sum_{i=1}^{n} |y_i - x_i|}{n}, \tag{1}$$

Root mean Squares Error (RMSE): This is interpretable as the standard deviation of the prediction errors, a popular performance evaluation metric for models [37].

$$RMSE = \sqrt{\frac{\sum_{i=1}^{n} (x_i - y_i)^2}{n}}, \tag{2}$$

Akaike Information criteria (AIC): To determine which of the many models is the best for dataset at hand, AIC is the single number score to use. The criteria estimates the model quality for each model relative to other models [38]. It functions by checking fitness of the model's on data training and for the complexity of the model, it adds a penalty. The result that is desirable is the lowest possible AIC, which depicts model fitness. The model is evaluated by AIC on the basis of maximum likelihood estimation. This is expressed as:

$$AIC = 2K - 2(log - likelihood) \tag{3}$$

AIC uses the maximum likelihood estimation of the model (log-likelihood) to measure fitness. Log-likelihood is a measure of how likely one is to see their observed data, given a model.

Maximum likelihood indicates the best fitness of the model to the dataset. AIC uses computational convenience instead of natural log of the likelihood.

Methodology

Data Collection and Description. The most widely used methods to determine toxicological risk in an aqueous medium are inhibition assays which use Vibrio fischeri (formerly known as Photobacterium Phosphoreum), a marine gram-negative bacterium[39]. Many different luminescence inhibition tests involving Vibrio fischeri have been developed for the analysis of aqueous samples [10]. The toxicity data used in this study are log (EC50) where EC50 is in $\mu molL^{-1}$. The EC50 is the concentration of a compound at which 50% of its maximal effect is observed [16]. These totoxicity values which span over five log units are reported for 74 ILs including thirty-five heterocyclic cations and eighteen organic and inorganic anions available in literature [27]. The descriptors used in this study were developed by Paternò et al. [27] using principal component analysis (PCA) to compact VolSurf+ derived in silico molecular properties into nine principal properties. The variables used are 128 cationic and 38 anioinc variables available in VolSurf+. Five principal components which described 77.5% of variance are considered significant to the PCA model and chosen as cationic PPs. On the other hand, four principal components explained 73.5% of the variance were chosen as anionic PPs [27].

The first cationic PP decribes the solubility, size, flexibility and molecular weight of the cation. The second describes cation's interaction with water and the hydrophobic volume of the cation. The third cationic PP evidences the difference between more amphiphilic ILs from those with a higher hydrophobic character. The fourth cationic PP includes descriptors such as skin permeability, hydophilic/hydrophobic ratio, permeability into CACO2 cells blood-brain barrier permeation and ability to form hydrogen bond as donor. The fifth cationic PP is mainly required to discriminate hydrogen bond donor descriptors exhibiting high PPC5 values from hydrogen bond acceptors exhibiting very high negative PPC5 values. On the other hand, first and second anionic PPs encompasses descriptors related to hydrophobicity/hydrophilicity balance, and critical packing. The first aninic PP distinguishes anions based on their hydrophobicity/hydrophilicity balance. The third and fourth anionic PPs descriptors related to the size/shape of the anions and to the anion's ability to form hydrogen bond as donor or acceptor. The third anionic PP differentiate anions with hydrophilicity due to polarizability from those with hydrophilicity due to the anion ability to form hydrogen bond as donor or acceptor. Detailed explanation of the various PPs could be found in [27].

Data Analysis and Visualization. For the entire workflow in this study, python 3.7 programming language of Jupyter notebook was used. The data was subjected to a series of techniques to process, clean and select variables that are relevant to the machine learning models. The relationship between the input (principal properties) and output (Vibrio fischeri toxicities) data was explored as a summary statistic of the input and output data are given in Table 1. A pair plot gave the pictorial relationship between the descriptors and target data as indicated in Fig. 1. Linear regression will not be an appropriate model for this study as indicated by the nonlinear distribution of the data. There is the need to remove colinear features which could be achieved by using advanced feature selection methods. This is because the distribution shows presence of multicollinearity that is likely to cause uncertainty in any model for machine learning. To quantify the degree of correlation among the descriptors spearman rho correlation covariance matrix was calculated as shown in Fig. 2. This analysis useful in the determination of the distribution of data and helps in telling their expected behaviour.

Table 1. Summary statistics of input and output data

	Vibrio_f	PP1C	PP2C	PP3C	PP4C	PP5C	PP1A	PP2A	PP3A	PP4A
count	74	520	520	520	520	520	520	520	520	520
mean	3.18	0.84	-1.67	0.20	0.03	-0.05	0.01	-2.13	-0.65	-0.08
std	1.36	5.50	4.99	3.59	3.01	1.87	2.40	3.27	2.49	2.31
min	-0.18	-21.63	-8.24	-9.65	-8.83	-9.43	-13.16	-5.73	-5.87	-3.83
25%	2.50	-2.70	-5.91	-2.39	-1.31	-1.17	-1.19	-5.52	-2.31	-1.25
50%	3.24	0.58	-3.31	1.33	0.51	-0.19	-0.57	-2.15	-1.63	-1.13
75%	4.14	4.71	2.25	2.76	1.70	0.86	1.35	-0.06	1.04	2.10
max	6.10	14.74	15.45	6.77	5.90	5.63	4.81	7.57	6.72	4.62

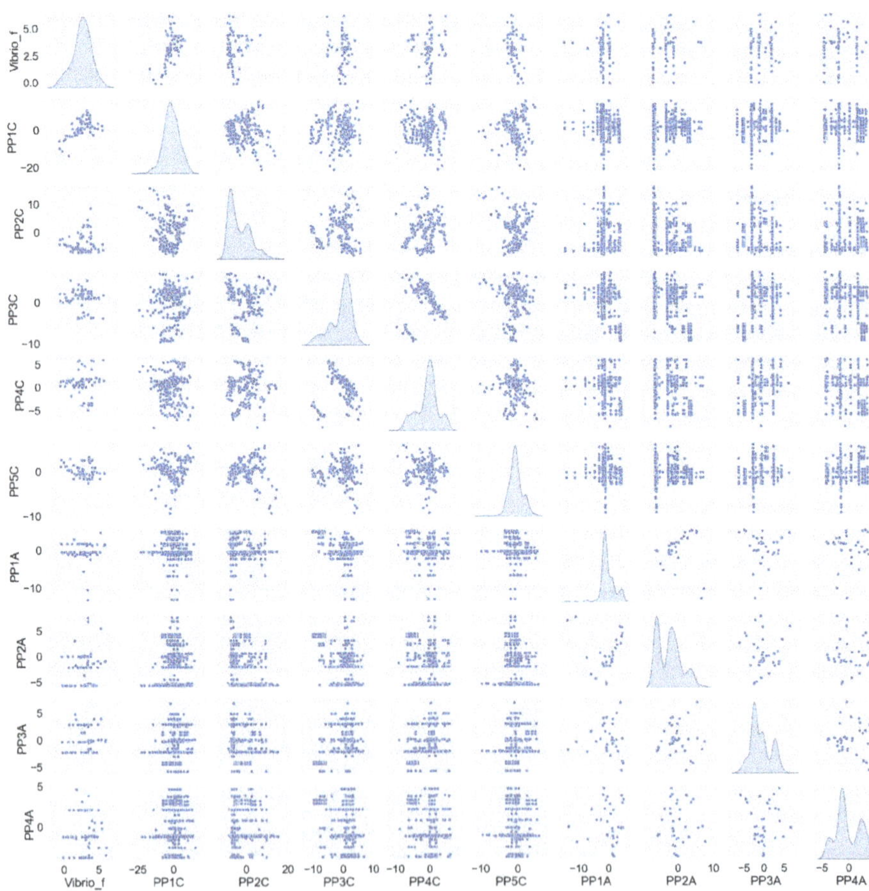

Fig. 1. Pair plot distribution of PPs and Vibrio fischeri toxicity

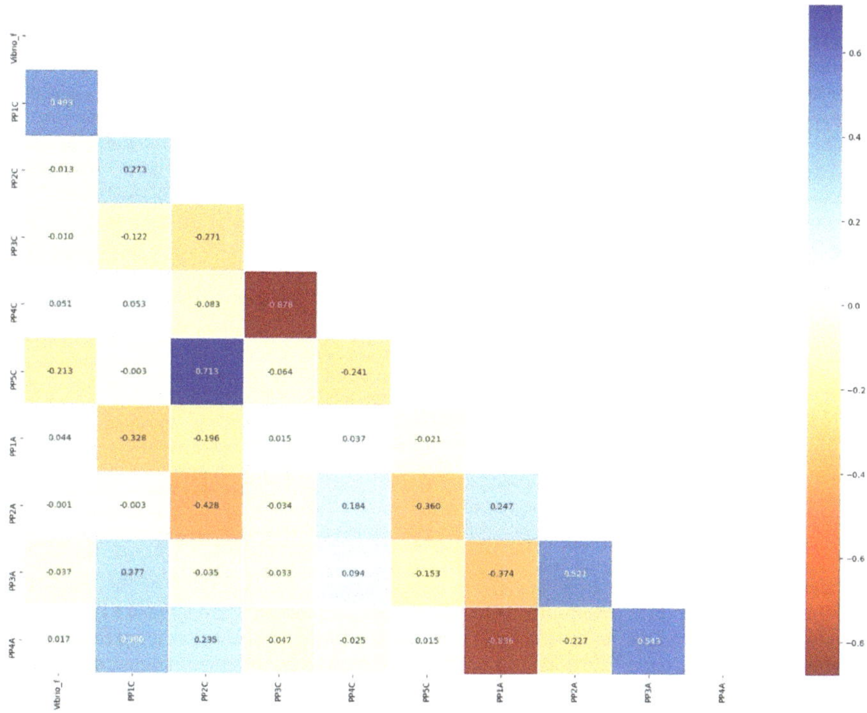

Fig. 2. Plot of the Spearman rho correlation covariance matrix for all the data ranging from -1 to 1.

The various ML models were used to estimate *Vibrio fischeri* toxicities from all the PPs. A section of data that was not used in the learning process (secondary section) was used to cross validate the model. Cross validation was to check model accuracy in predicting the whole dataset. The technique used here for cross validation was holdout which works by using a new subset of data (i.e., data not used in model training) on the model. This technique finds ways to prevent occurrence of over or under-fitting, offer an insight into bias-variance trade-offs and minimize estimation errors on unobserved data [40]. The held-out data was later utilized in the validation of the accuracy of the prediction model. The data was split into two, i.e., 85% (62 data points) was used as training data and 15% (12 data points) as test data. This is to ensure that validation of the prediction accuracy of the model is done with another set of data set aside from model training. It is imperative to note that the focus is not the model accuracy but the prediction error. The variance and bias of the model is better understood by analyzing the errors. The bias is the error rate, and the selection of input data is the most influencing parameter on a model's bias. On the other hand, the variance is the performance deterioration of the of the model in terms of accuracy when used on the test data compared to when used on the training data.

Results and Discussion

Model Performances Evaluation. Model performance for various models is done using the test data which was held out in training these models. The training and test scores for the various models are presented in Table 2. As seen from Table 2, the training accuracy for each model is very high as expected. This is because the model tries to replicate data that has already been seen. Nevertheless, a high training accuracy does not necessary depict that the model is a good one as it could be a non-generalized model produced due to overfitting where inherent noise is captured by the model. Another scenario could be an out-of-sample accuracy which is the correct prediction percentage that the model makes on data not used in training it. It is imperative for models to have an out-of-sample accuracy that is high as they are meant to make correct predictions on unknown data. All the models had low out-of-sample accuracy due to small data set with high dimension. However, decision tree proved to have the most accurate out-of-sample prediction among the models trained.

Table 2. Correlation coefficient score for train and test data

Models	Train Score	Test Score
Decision Tree	1	0.613051016
Random Forest	0.930593458	0.571406118
Extra tree Regression	1	0.474419201
Gradient Boosting Regression	0.981733285	0.571829785
XGBoost	0.999996143	0.444828751

In terms of probabilistic ranking of models' performances, the likelihood of the models to reduce loss of information is determined by AIC. A more parsimonious model usually has lower AIC values compared to other models [41]. This makes the chosen model better for prediction. As shown in Fig. 3 the decision tree is best in predicting *Vibrio fischeri* toxicity among the models trained in this study. The AIC assesses the convergence of the models fit to actual data by computing its relative quality. Nevertheless, the model's tendency to overfit or underfit still exists. Therefore, the precision, consistency, and accuracy of the models' performance were also evaluated.

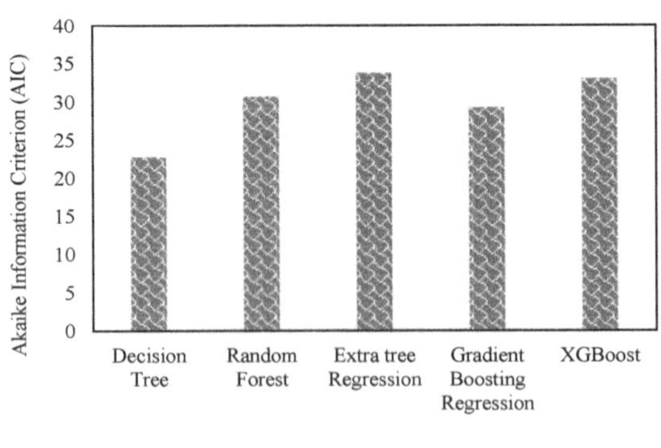

Fig. 3. Comparison of AIC for all models on test data

The models are further assessed on their improvement in accuracy (i.e., reduction in error). The models deployed in this study have varying performances that are very much due to their statistical and theoretical framework. Cross-validation analysis is performed to examine the model performance in terms of precision, accuracy, and consistency in predicting unknown toxicities. This analysis is performed using the RMSE and MAE as shown in Fig. 4. The RMSE of all the models indicated that the decision tree model is the most consistent in predicting *Vibrio fischeri* toxicity. Nevertheless, the MAE results showed that the most precise model for *Vibrio fischeri* toxicity prediction is RF model. Therefore, as regards fitness, consistency and accuracy of the model, decision tree model outperforms all the other models deployed in this study. Comparing the performance of the models based on these error matrices and the AIC, the decision tree proves to be the most robust and accurate model trained on this data set.

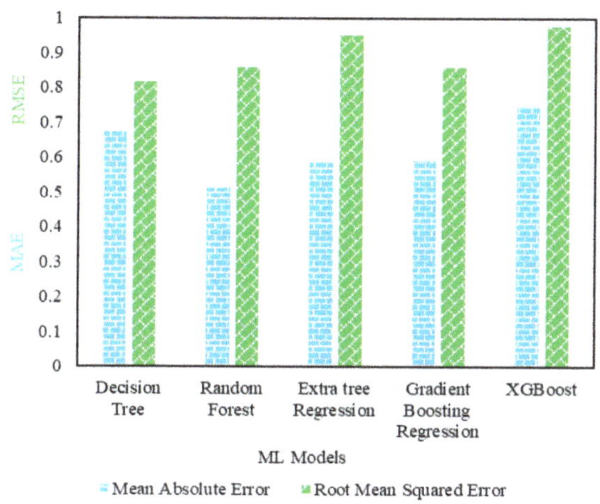

Fig. 4. Comparison of prediction errors for all models

It is imperative to point out that variation in model performance could also be attributed to the data set given. Some models work better on large data set while others work better on smaller data set. Generally, XGBoost model tend to perform best among the models deployed in this study however, the decision tree proved robust owing to the small data set used in this study. Another inherent property with XGBoost model is the tendency to convey the significance of the descriptors to the output based on F-score as portrayed in Fig. 5. From Fig. 5, PP1C is the most important followed by PP1A then PP3C. Paternò et al. [16] in their prediction of *Vibrio fischeri* toxicity also used a plot of the Variable Importance for Projection (VIP) to show the significance of the various PPs to *Vibrio fischeri* toxicity. PP1C was the most pertinent descriptor followed by PP2C then PP5C (which had high error) then PP3A. The other five descriptors were less important. Though different variations of the descriptor importance have been obtained, the PP1C prove to be the most important descriptor in both cases. This shows how the cationic part of IL liquids are important to their toxicity. The size, flexibility, solubility and molecular weight of the cation is information represented by PP1C. In this study, the XGBoost also shows how the anionic part of IL is also important to their toxicity and the need to consider both counterparts in toxicity prediction.

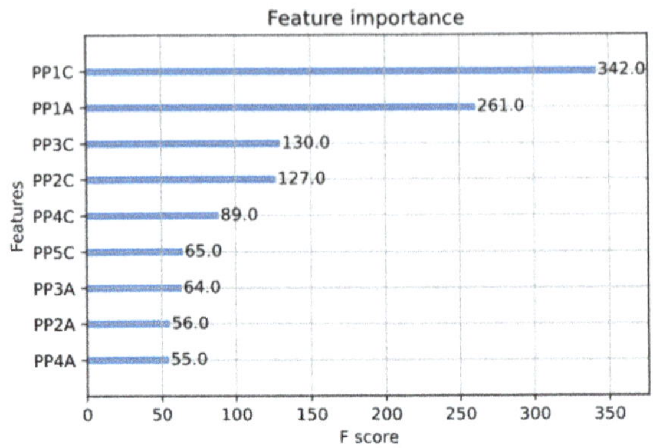

Fig. 5. Feature importance of all descriptors to Vibrio fischeri toxicity prediction

Further evaluation of model performances is conducted by comparing the predicted toxicities using descriptors of test data to the actual test toxicities. Figs. 6 and 7 show joint plots of correlation plot of predicted toxicities versus actual toxicities together with the distributions of the toxicities for decision tree and XGBoost. The shaded area around the correlation line shows the deviation of the predicted values. The wider the shaded area, the more deviated the predicted toxicities are from the actual toxicity values. By comparing Fig. 6 and 7, the decision tree proved to have better predictive power over the XGBoost in predicting *Vibrio fischeri* toxicity. Nevertheless, the predicted values for both models had different distribution compared to the actual values. This shows how less consistent even the decision tree is in predicting *Vibrio fischeri* toxicity.

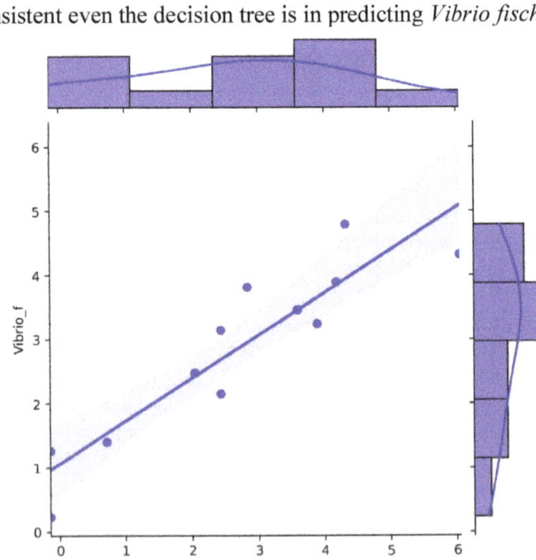

Fig. 6. Joint plot of Actual against predicted toxicities by decision tree

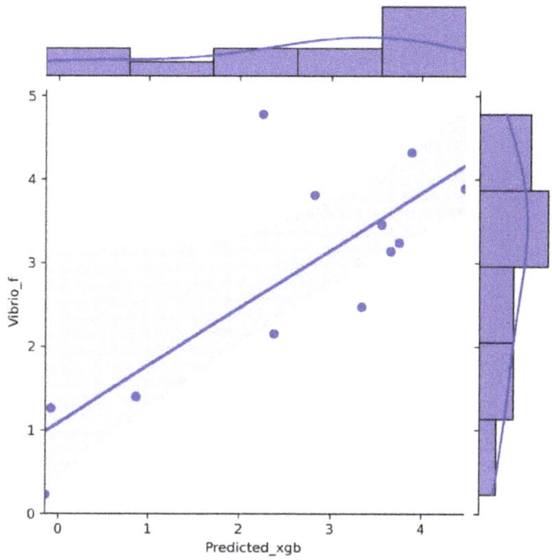

Fig. 7. Joint plot of Actual against predicted toxicities by XGBoost

Sensitivity Analysis. The kernel density estimates of *Vibrio fischeri* toxicity by decision tree and XGBoost are presented in Fig. 8 and 9 respectively. The red curve represents actual values while the blue one is predicted values. In this case, the XGBoost had its predicted values closer to the actual values than the decision tree. The XGBoost showed better predictive power in higher values than lower values while the values predicted by the decision tree showed wider distribution with lower mode. The toxicity of *Vibrio fischeri* is expressed as EC50 which signifies the compound's concentration at which 50% of its maximal effect is observed, hence the lower this value is, the more toxic the IL. Therefore, the XGBoost was biased toward predicting lower toxicities more accurately than higher toxicities. The decision tree showed consistency in its prediction irrespective of the error.

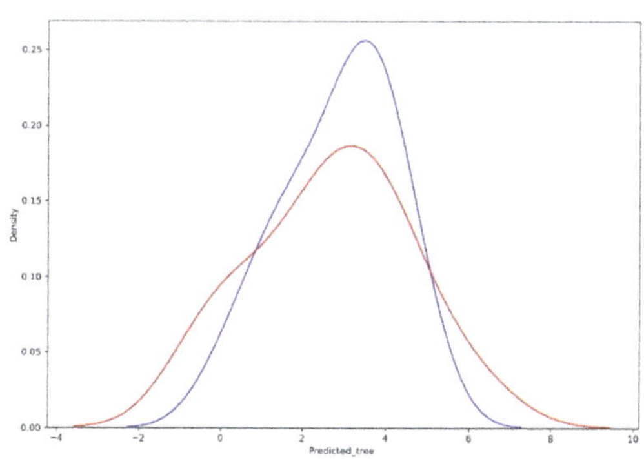

Fig. 8. Kernel density estimation: The closeness of predicted to actual values for Decision tree Vibrio fischeri toxicity prediction

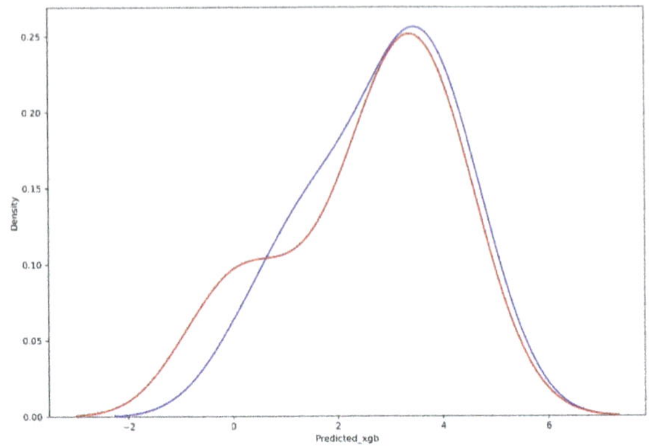

Fig. 9. Kernel density estimation: The proximity of predicted values to actual values for the prediction of Vibrio fischeri toxicity using the XGBoost model.

Conclusion

In this study, ML models have been trained and tested for prediction of *Vibrio fischeri* toxicities of IL using nine in silico descriptors developed by Paternò et al. [16], [27]. Comparing their training and test scores, most of the model showed evidence of overfitting. Nevertheless, the decision tree model proved to be the most consistent and accurate model among the models deployed in this study. Model performance based on RMSE showed that the RF model was the most precise. Other models are generally deemed more accurate than the decision tree, therefore, the observed performance could be attributed to the small data set size used. The inherent attribute of the XGBoost showed that both PP1C and PP1A are very important to *Vibrio fischeri* toxicity prediction. This proved that both the cationic and anionic counterparts of IL are important in determining their level of toxicities. Further ML techniques could be deployed to make the models more robust including feature selection and optimization. This study proves that QSPR studies on toxicity prediction of new ILs could be improved via ML application.

Acknowledgement

The authors would like to acknowledge the FRGS Grant (015MA0-126), Universiti Teknologi PETRONAS for the financial assistance.

Declaration of conflict of interest

On behalf of all the co-authors, the corresponding author states that there is no conflict of interest.

References

[1] F. D'Anna, S. Marullo, P. Vitale, and R. Noto, "Synthesis of aryl azides: A probe reaction to study the synergetic action of ultrasounds and ionic liquids," *Ultrason. Sonochem.*, vol. 19, no. 1, pp. 136–142, 2012, https://doi.org/10.1016/j.ultsonch.2011.06.010.

[2] N. Taccardi *et al.*, "Gallium-rich Pd-Ga phases as supported liquid metal catalysts," *Nat. Chem.*, vol. 9, no. 9, pp. 862–867, 2017, https://doi.org/10.1038/NCHEM.2822.

[3] M. Tariq, D. Rooney, E. Othman, S. Aparicio, M. Atilhan, and M. Khraisheh, "Gas hydrate inhibition: A review of the role of ionic liquids," *Ind. Eng. Chem. Res.*, vol. 53, no. 46, pp. 17855–17868, 2014, https://doi.org/10.1021/ie503559k.

[4] M. El-Harbawi, "Toxicity Measurement of Imidazolium Ionic Liquids Using Acute

Toxicity Test," *Procedia Chem.*, vol. 9, no. December 2014, pp. 40–52, 2014, https://doi.org/10.1016/j.proche.2014.05.006.

[5] C. B. Bavoh, O. Nashed, A. N. Rehman, N. A. A. B. Othaman, B. Lal, and K. M. Sabil, "Ionic Liquids as Gas Hydrate Thermodynamic Inhibitors," *Ind. Eng. Chem. Res.*, vol. 60, no. 44, pp. 15835–15873, 2021, https://doi.org/10.1021/acs.iecr.1c01401.

[6] C. B. Bavoh, T. N. Ofei, and B. Lal, "Investigating the Potential Cuttings Transport Behavior of Ionic Liquids in Drilling Mud in the Presence of sII Hydrates," *Energy and Fuels*, vol. 34, no. 3, pp. 2903–2915, 2020, https://doi.org/10.1021/acs.energyfuels.9b04088.

[7] E. E. L. Tanner, R. R. Hawker, H. M. Yau, A. K. Croft, and J. B. Harper, "Probing the importance of ionic liquid structure: A general ionic liquid effect on an SNAr process," *Org. Biomol. Chem.*, vol. 11, no. 43, pp. 7516–7521, 2013, https://doi.org/10.1039/c3ob41634h.

[8] J. Scholz *et al.*, "Ethylene to 2-butene in a continuous gas phase reaction using silp-type cationic nickel catalysts," *ChemCatChem*, vol. 6, no. 1, pp. 162–169, 2014, https://doi.org/10.1002/cctc.201300636.

[9] R. Mancuso, C. S. Pomelli, C. Chiappe, R. C. Larock, and B. Gabriele, "A recyclable and base-free method for the synthesis of 3-iodothiophenes by the iodoheterocyclisation of 1-mercapto-3-alkyn-2-ols in ionic liquids," *Org. Biomol. Chem.*, vol. 12, no. 4, pp. 651–659, 2014, https://doi.org/10.1039/c3ob41928b.

[10] P. M. Ventura, C. S. Marques, A. A. Rosatella, C. A. M. Afonso, and F. Gonc, "Ecotoxicology and Environmental Safety Toxicity assessment of various ionic liquid families towards Vibrio fischeri marine bacteria," vol. 76, pp. 162–168, 2012, https://doi.org/10.1016/j.ecoenv.2011.10.006.

[11] S. P. F. Costa, P. C. A. G. Pinto, R. A. S. Lapa, and M. L. M. F. S. Saraiva, "Toxicity assessment of ionic liquids with Vibrio fischeri: An alternative fully automated methodology," *J. Hazard. Mater.*, vol. 284, pp. 136–142, 2014, https://doi.org/10.1016/j.jhazmat.2014.10.049.

[12] R. N. Das, T. E. Sintra, J. A. P. Coutinho, S. P. M. Ventura, K. Roy, and P. L. A. Popelier, "Development of predictive QSAR models for: Vibrio fischeri toxicity of ionic liquids and their true external and experimental validation tests," *Toxicol. Res. (Camb).*, vol. 5, no. 5, pp. 1388–1399, 2016, https://doi.org/10.1039/c6tx00180g.

[13] F. G. Doherty, "A review of the Microtox® toxicity test system for assessing the toxicity of sediments and soils," *Water Qual. Res. J.*, vol. 36, no. 3, pp. 475–518, 2001.

[14] M. Ishaq, D. Zaini, A. Mohd, and M. Moniruzzaman, "Framework for Ecotoxicological Risk Assessment of Ionic Liquids," *Procedia Eng.*, vol. 148, pp. 1141–1148, 2016, https://doi.org/10.1016/j.proeng.2016.06.567.

[15] M. I. Khan, D. Zaini, and A. M. Shariff, "Estimation of safe environmental concentrations of ionic liquids towards bacteria by chemical toxicity distribution method Estimation of safe environmental concentrations of ionic liquids towards bacteria by chemical toxicity distribution method," 2018, https://doi.org/10.1088/1757-899X/458/1/012038.

[16] A. Paterno', S. Scire, and G. Musumarra, "A QSPR approach to the ecotoxicity of ionic liquids (Vibrio fischeri) using VolSurf principal properties," *Toxicol. Res. (Camb).*, vol. 5, no. 4, pp. 1090–1096, 2016.

[17] A. Romero, A. Santos, J. Tojo, and A. Rodr, "Toxicity and biodegradability of imidazolium ionic liquids," vol. 151, pp. 268–273, 2008,

https://doi.org/10.1016/j.jhazmat.2007.10.079.

[18] L. Maltby, "Small-Scale Freshwater Toxicity Investigations: Volume 1 ? Toxicity Test Methods," *Freshw. Biol.*, vol. 52, no. 1, pp. 198–198, 2007, https://doi.org/10.1111/j.1365-2427.2006.01662.x.

[19] S. M. Steinberg, E. J. Poziomek, W. H. Engelmann, and K. R. Rogers, "A review of environmental applications of bioluminescence measurements," *Chemosphere*, vol. 30, no. 11, pp. 2155–2197, 1995, https://doi.org/10.1016/0045-6535(95)00087-O.

[20] S. P. F. Costa, V. D. Justina, K. Bica, M. Vasiloiu, P. C. A. G. Pinto, and M. L. M. F. S. Saraiva, "Automated evaluation of pharmaceutically active ionic liquids'(eco) toxicity through the inhibition of human carboxylesterase and Vibrio fischeri," *J. Hazard. Mater.*, vol. 265, pp. 133–141, 2014.

[21] M. Jafari, M. H. Keshavarz, and H. Salek, "A simple method for assessing chemical toxicity of ionic liquids on Vibrio fischeri through the structure of cations with specific anions," *Ecotoxicol. Environ. Saf.*, vol. 182, p. 109429, 2019.

[22] M. G. Montalbán, J. M. Hidalgo, M. Collado-González, F. G. D. Baños, and G. Víllora, "Assessing chemical toxicity of ionic liquids on Vibrio fischeri: Correlation with structure and composition," *Chemosphere*, vol. 155, pp. 405–414, 2016.

[23] K. M. Docherty and C. F. Kulpa, "Toxicity and antimicrobial activity of imidazolium and pyridinium ionic liquids," pp. 185–189, 2005, https://doi.org/10.1039/b419172b.

[24] B. Peric, J. Sierra, E. Martí, R. Cruañas, and M. A. Garau, "Quantitative structure–activity relationship (QSAR) prediction of (eco) toxicity of short aliphatic protic ionic liquids," *Ecotoxicol. Environ. Saf.*, vol. 115, pp. 257–262, 2015.

[25] F. Yan, Q. Shang, S. Xia, Q. Wang, and P. Ma, "Topological study on the toxicity of ionic liquids on Vibrio fischeri by the quantitative structure-activity relationship method," *J. Hazard. Mater.*, vol. 286, pp. 410–415, 2015, https://doi.org/10.1016/j.jhazmat.2015.01.016.

[26] D. J. Couling, R. J. Bernot, K. M. Docherty, J. N. K. Dixon, and E. J. Maginn, "Assessing the factors responsible for ionic liquid toxicity to aquatic organisms via quantitative structure–property relationship modeling," *Green Chem.*, vol. 8, no. 1, pp. 82–90, 2006, https://doi.org/10.1039/b511333d.

[27] A.釗Paternò *et al.*, "Cyto- and enzyme toxicities of ionic liquids modelled on the basis of VolSurf+ descriptors and their principal properties," *SAR QSAR Environ. Res.*, vol. 27, no. 3, pp. 221–244, 2016, https://doi.org/10.1080/1062936X.2016.1156571.

[28] A. Paternò, G. Bocci, L. Goracci, G. Musumarra, and S. Scirè, "Modelling the aquatic toxicity of ionic liquids by means of VolSurf in silico descriptors," *SAR QSAR Environ. Res.*, vol. 27, no. 1, pp. 1–15, 2016.

[29] T. Hastie, R. Tibshirani, J. H. Friedman, and J. H. Friedman, *The elements of statistical learning: data mining, inference, and prediction*, vol. 2. Springer, 2009.

[30] Priyanka and D. Kumar, "Decision tree classifier: A detailed survey," *Int. J. Inf. Decis. Sci.*, vol. 12, no. 3, pp. 246–269, 2020.

[31] K. Mittal, D. Khanduja, and P. C. Tewari, "An insight into 'Decision Tree Analysis'"," *World Wide J. Multidiscip. Res. Dev.*, vol. 3, no. 12, pp. 111–115, 2017.

[32] Y. Zhao and Y. Zhang, "Comparison of decision tree methods for finding active objects," *Adv. Sp. Res.*, vol. 41, no. 12, pp. 1955–1959, 2008.

[33] P. Geurts, D. Ernst, and L. Wehenkel, "Extremely randomized trees," *Mach. Learn.*, vol. 63, no. 1, pp. 3–42, 2006.

[34] J. Gall, A. Yao, N. Razavi, L. Van Gool, and V. Lempitsky, "Hough forests for object detection, tracking, and action recognition," *IEEE Trans. Pattern Anal. Mach. Intell.*, vol. 33, no. 11, pp. 2188–2202, 2011.

[35] J. H. Friedman, "Greedy function approximation: a gradient boosting machine," *Ann. Stat.*, pp. 1189–1232, 2001.

[36] T. Chen and C. Guestrin, "Xgboost: A scalable tree boosting system," in *Proceedings of the 22nd acm sigkdd international conference on knowledge discovery and data mining*, 2016, pp. 785–794.

[37] D. A. Otchere, T. O. A. Ganat, J. O. Ojero, M. Y. Taki, and B. N. Tackie-Otoo, "Application of gradient boosting regression model for the evaluation of feature selection techniques in improving reservoir characterisation predictions," *J. Pet. Sci. Eng.*, p. 109244, 2021.

[38] H. Bozdogan, "Model selection and Akaike's information criterion (AIC): The general theory and its analytical extensions," *Psychometrika*, vol. 52, no. 3, pp. 345–370, 1987.

[39] S. Parvez, C. Venkataraman, and S. Mukherji, "A review on advantages of implementing luminescence inhibition test (Vibrio fischeri) for acute toxicity prediction of chemicals," *Environ. Int.*, vol. 32, no. 2, pp. 265–268, 2006, https://doi.org/10.1016/j.envint.2005.08.022.

[40] W. Ertel, *Introduction to artificial intelligence*. Springer, 2018.

[41] D. A. Otchere, T. O. A. Ganat, R. Gholami, and M. Lawal, "A novel custom ensemble learning model for an improved reservoir permeability and water saturation prediction," *J. Nat. Gas Sci. Eng.*, vol. 91, no. April, p. 103962, 2021, https://doi.org/10.1016/j.jngse.2021.103962.

Sustainable Processes and Clean Energy Transition - ICSuPCET2022 Materials Research Forum LLC
Materials Research Proceedings 29 (2023) 248-254 https://doi.org/10.21741/9781644902516-28

Preliminary study of carbon dioxide (CO_2) adsorption using chicken eggshell

NOR ADILLA Rashidi[1,2,a*], SHARINA NATASHA Sheikh Saghir[2,b]

[1]HICoE-Centre for Biofuels and Biochemical Research (CBBR), Institute of Self-Sustainable Building (ISB), Universiti Teknologi PETRONAS, 32610 Seri Iskandar, Perak, Malaysia

[2]Department of Chemical Engineering, Universiti Teknologi PETRONAS, 32610 Seri Iskandar, Perak, Malaysia

[a]adilla.rashidi@utp.edu.my, [b]Sharinanatasha29@gmail.com

Keywords: Bio-Calcium, Calcination, Carbonation, Carbon Dioxide, Chicken Eggshell

Abstract. Carbon dioxide (CO_2) emission is a major concern nowadays due to its contribution to global warming scenarios. Therefore, valorization of inexpensive and green material to adsorb the CO_2 prior to emission has been emphasized. In this research study, calcined chicken eggshell waste had been tested as a potential CO_2 adsorbent. Specifically, calcined eggshell at 700°C, 800°C, and 900°C has been used in subsequent carbonation reaction at 650°C for 1 hr. The result shows that increasing the calcination temperature improves the eggshell decomposition (weight loss) and CO_2 adsorption capacity (weight gain). Specifically, the weight gain by the calcined eggshell at 900°C is roughly 61 wt.%. Both pristine and calcined eggshells are characterized by the Fourier-Transform Infrared Spectroscopy (FTIR), thermogravimetric analysis (TGA), Scanning Electron Microscopy (SEM), and N_2 physisorption. Overall, utilization of the chicken eggshell waste is indeed promising as this can contribute to both environmental and economic advantages.

Introduction

The climate change specifically due to anthropogenic carbon dioxide (CO_2) emission has become a major issue amongst the researchers, policy makers, and other related professionals, due to these arising problems: global warming, ecological imbalance, socio-economic issue, and other climate-related disasters. Due to these, carbon reduction has become a prominent global focus these days. It is worth noting that calcium looping (CaL) technology is one of the promising short-term method for CO_2 capture [1, 2]. Fundamentally, two reactions are involved in this process, i.e., calcination and carbonation, as described in Eq. (1) and (2), respectively.

$$Endothermic\ calcination: CaCO_3 \leftrightarrow CaO + CO_2, \Delta H = 178\ kJ/mole \quad (1)$$

$$Exothermic\ carbonation: CaO + CO_2 \leftrightarrow CaCO_3, \Delta H = -178\ kJ/mole \quad (2)$$

To date, limestone has been extensively used as a source of calcium oxide (CaO), due to its low prices and wide availability. Nevertheless, there has been a great concern on the mineral limestone extraction and mining process, which will jeopardize the environment [3, 4]. Due to ecological destruction, it is worthy to shift to the greener calcium-based feedstock i.e., chicken eggshell waste. Her et al. [5] reported that the eggshell has high resemblance to the limestone, in which it possesses >95% calcium carbonate ($CaCO_3$) as well. Being used as a main commodity in Malaysia, it is expected that approximately 70,000 tons of eggshells will be produced annually.

Since the eggshell waste is traditionally useless, majority of this calcium-rich wastes are dumped in the landfill without any pretreatment, thus, it will result in serious environmental problems [6]. Furthermore, Jayasankar et al. [7] reported that the accumulation of eggshell waste will attract worms and rats, in which leading to health problems. Therefore, transformation of the discarded eggshell wastes to value-added material (i.e., CO_2 sorbent) is promising as it could be a novel and valuable approach in the solid waste management, and at the same time, to reduce CO_2 emission.

Materials and Methods

Materials. The waste chicken eggshells were obtained from burger stalls and cafeteria in Universiti Teknologi PETRONAS. Upon washing with tap water several times, these eggshells were oven-dried at 105°C for 24 hr. The dried wastes were then crushed and pulverized by using agate mortar and grinder, respectively, before being screened to < 80 µm. The samples were kept in an air-tight container to prevent any reaction with the surroundings, prior to the further experimentation. In addition, high purity CO_2 (99.8% purity) and nitrogen (N_2, 99.9% purity) tank were supplied from Linde Malaysia Sdn. Bhd.

Calcination-carbonation process. Following López-Periago et al. [8], the calcination-carbonation testing of the chicken eggshell had been carried out by using a tubular furnace. In calcination process, ~25 g of eggshell was weighted inside a ceramic crucible and then placed in the central part of the tube furnace. In the process, the samples were heated to the desired calcination temperature (700-900°C) with fixed heating rate of 10°C/min under N_2 flow rate of 200 mL/min. The weight loss in the calcination process was calculated, following Eq. (3) [9]. Meanwhile, carbonation process was carried out by heating the calcined samples to 650°C, in presence of CO_2 flow. The carbonation process was maintained for 1 hr upon reaching the desired temperature. Further, weight gain that corresponds to CO_2 adsorption was measured by following Eq. (4) [10].

$$\text{Weight loss (calcination)} = \frac{m_o - m_f}{m_o} x100\% \qquad (3)$$

$$\text{Weight gain (carbonation)} = \frac{m_f - m_o}{m_o} x100\% \qquad (4)$$

Referring to Eq. (3) and (4), m_o and m_f represents the initial and final weights, respectively.

Characterizations. The decomposition profile of the pristine eggshells was investigated by using a Perkin Elmer thermal analyzer under purified N_2 atmosphere, at heating rate of 10°C/min. Further, the functional groups of the bio-calcium were analyzed using the Perkin Elmer Fourier Transform Infrared (FTIR) spectrometer within the range of 4000-400 cm^{-1}. The surface area and porosity analysis of both raw and calcined eggshell were measured by the Micromeritics ASAP 2020. Prior to analysis, the samples were outgassed at 250°C. Lastly, the exterior morphology of these samples were observed by using the Scanning Electron Microscope (SEM), coupled with the Energy Dispersive X-ray (EDX) spectroscopy.

Results and Discussion

Calcination-carbonation studies. The suitable temperature for the calcination is deduced from the thermal decomposition analysis from ambient temperature to 800°C. Referring to Fig. 1, a minor weight loss (~5 wt.%) at temperature of < 300°C is attributed to the loss of moisture and volatile organic substance, while a significant decomposition (~40 wt.%) at the temperature of >650°C is attributed to the $CaCO_3$ conversion to CaO [11-13]. Salauddeen et al. [14] reported that the CO_2 release is taken from the point where a significant weight loss is observed till the end of analysis.

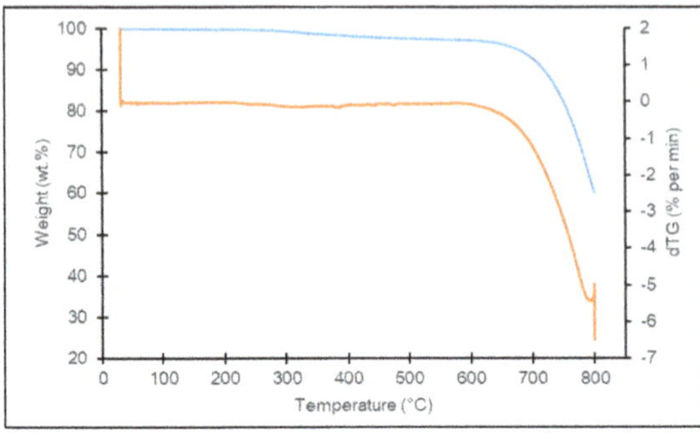

Fig. 1. TG-dTG curve of raw chicken eggshell

Similarly, Rohim et al. [11] observed that the decomposition of eggshell has two distinct stages, which is at < 640°C and 640-840°C. The decomposition behavior of eggshell is further validated by using the furnace, where the sample is heated from ambient to 700-900°C. In this study, 700°C is taken as the minimum point, as Gao and Xu [15] previously agreed that temperature of <700 °C do not contribute to the CaO formation. Further, calcination temperature should not be excessively high as it may result in particle sintering and attrition [14], hence, explains why 900°C is taken as the maximum point. Referring to Fig. 2, the calcination temperature is directly proportional to weight loss, where weight loss of 5.96% and 45.00% has been encountered at 700°C and 900°C, respectively. This behavior is expected owing to the endothermic reaction of calcination, as described in Eq. (1). Besides, accelerating decomposition at higher temperatures is due to increment in kinetic energy of the particles, which will enhance the calcination rate. On top of that, higher calcination temperatures are found to positively correlate with weight increase of CaO (Fig. 2), which is plausibly due to the changes in textural and chemical compositions. Moreover, it shows that at the calcination temperature of 900°C, color of eggshell that has transformed to white color (Fig. 3) indicates more formation of metal oxide.

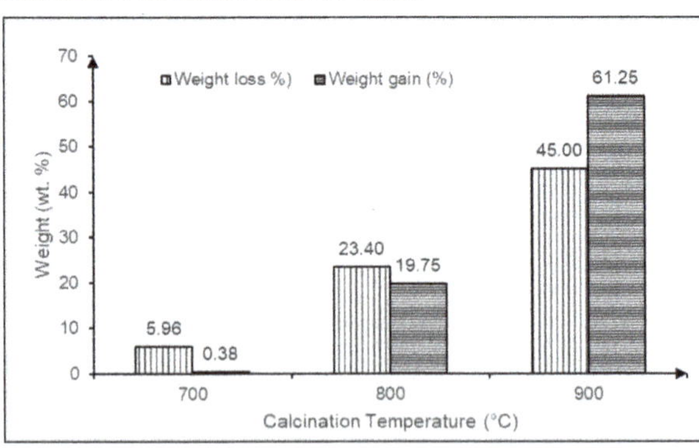

Fig. 2. Weight loss and weight gain at different calcination temperatures. (Note: The weight loss measures the degree of calcination (decomposition) while weight gain corresponds to carbonation conversion)

Fig. 3. Chicken eggshell powder before calcination (a), and after calcination at (b) 700°C and (c) 900°C

Likewise, similar experimental findings have been obtained by Mohadi et al. [16] and Commey and Mensah [17]. Besides, it has been reported that in an atmospheric condition, the calcination temperature of ~950°C is considered as an optimal condition as it does not contribute to sorbent decay even at prolong calcination time [18]. Future work may incorporate the surface area and porosity analysis as well as elemental compositions at each calcination temperatures. In addition, a comparative work with the commercial CaO has been conducted, at which the weight gain (CO_2 adsorption capacity) is approximately 12 wt.%.

Functional group analysis: The FTIR spectra of calcined and carbonated eggshell are illustrated in Fig. 4. Based on Fig. 4, it demonstrates that the calcinated eggshell consists of the hydroxyl O-H stretching, where there is a large peak at ~3600 cm^{-1}. Meanwhile, for the carbonated eggshell, sharp peaks are observed at 1427 and 873 cm^{-1}, which indicative the presence of C-O stretching and bending of $CaCO_3$, respectively [6, 11]. Other noticeable peaks for the carbonated eggshell that are at 2515, 1798, and 712 cm^{-1} indicates the presence of $CaCO_3$ [19, 20].

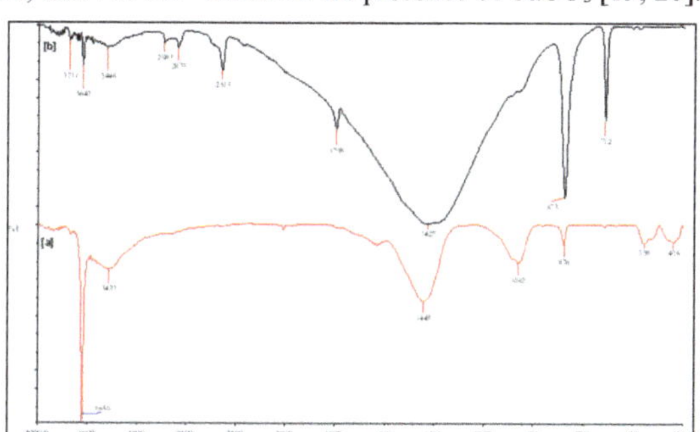

Fig. 4: FTIR spectra of (a) calcined eggshell at 900°C and (b) carbonated eggshell waste

SEM-EDX analysis: The exterior surface of raw and calcined eggshell at 900°C is shown in Fig. 5. In reference to Ghouti and Salih [21], raw eggshell is composed of irregular-shaped particles and in agglomerated form. In contrast, calcined eggshell has a porous structure, which is plausibly due to release of gases (i.e., CO_2) that occupy the pores and complete transformation to metal oxide (CaO), as supported by the EDX analysis. From analysis, it shows that this calcined eggshell is merely composed of calcium (55 wt.%) and oxygen (45 wt.%) content. Likewise, Sree et al. [22] also reported that the calcined eggshell predominantly contains 65 wt.% calcium and balanced by oxygen compositions.

Surface area and porosity analysis: From the N_2 physisorption analysis, BET surface area and total pore volume of calcined eggshell is found to be 1.18 m^2/g and 0.003 cm^3/g, respectively. Herein, these textural characteristics are similar to Viriya-empikul et al. [23], who previously reported that the calcined eggshell has surface area of 1.1 m^2/g and total pore volume of 0.005 cm^3/g. Further, the calcined eggshell at 900°C is mesoporous in nature, where the average pore diameter is 10.13 nm. It is noteworthy to mention that the surface properties are not appealing for the CO_2 adsorption, hence, this sorption process is most likely due to intrinsic nature of calcined eggshell.

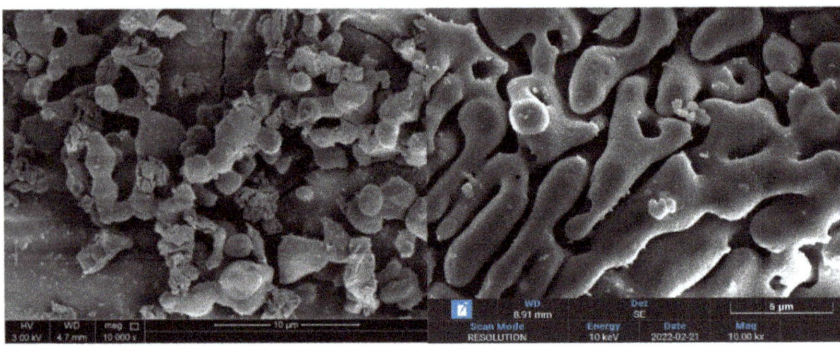

Fig. 5. SEM micrography of raw eggshell waste (left) and calcined eggshell at 900°C (right)

Conclusions

The results prove the suitability of eggshell wastes as potential substitute for the limestone in the CaL process, wherein the maximum CO_2 uptake of around 61 wt.% has been attained by calcined eggshell at 900°C. Further, CO_2 adsorption capacity by eggshell derived CaO that is higher than commercial CaO (~60% vs. 12%) is attractive due to its inexpensiveness and environmental-friendly properties. Upon calcination, eggshell is found to be primarily composed of calcium and oxygen, thus verifies the CaO formation. Further, based on structural analysis, the calcined eggshell follows the Type II isotherm while the surface morphology indicates that it has a porous and irregular structure. In terms of the carbonation process, the calcite formation can be noticeably observed from the functional groups.

Acknowledgement

The authors would like to acknowledge the funding supports from Ministry of Higher Education (MOHE) Malaysia through HICoE Grant (cost center 015MA0-052/015MA0-104/015MA0-136) to CBBR. Besides, the authors would like to acknowledge the Chemical Engineering department and Centralized Analytical Laboratory, Universiti Teknologi PETRONAS (UTP) for the facilities and financial support given in completing the research study.

References

[1] M. C. Romano et al., The calcium looping process for low CO_2 emission cement and power, Energy Procedia 37 (2013) 7091-7099. https://doi.org/10.1016/j.egypro.2013.06.645

[2] J. Chen, L. Duan, and Z. Sun, Review on the development of sorbents for calcium looping, Energy Fuels 34 (2020) 7806-7836. https://doi.org/10.1021/acs.energyfuels.0c00682

[3] S. Kittipongvises, Assessment of environmental impact of limestone quarrying operations in Thailand, Environ. Clim. Technol. 20 (2017) 67-83. https://doi.org/10.1515/rtuect-2017-0011

[4] H. Ganapathi and M. Phukan, Environmental hazards of limestone mining and adaptive practices for environment management plan, in: R. Singh, P. Shukla, P.Singh (Eds.), Environmental Processes and Management, Springer, Cham, 2020, pp. 121-134. https://doi.org/10.1007/978-3-030-38152-3_8

[5] S. Her, J. Park, P. Li, and S. Bae, Feasibility study on utilization of pulverized eggshell waste as an alternative to limestone in raw materials for Portland cement clinker production, Constr. Build. Mater. 24 (2022) 126589. https://doi.org/10.1016/j.conbuildmat.2022.126589

[6] S. Sethupathi, Y. C. Kai, L. L. Kong, Y. Munusamy, M. J. K. Bashir, and N. Iberahim, "Preliminary study of sulfur dioxide removal using calcined egg shell, Malays. J. Anal. Sci. 21 (2017) 719-725. https://doi.org/10.17576/mjas-2017-2103-21

[7] R. Jayasankar, N. Mahindran, and R. Ilangovan, Studies on concrete using fly ash, rice husk ash and egg shell powder," Int. J. Civ. Struct. Eng. 1 (2010) 362-372.

[8] A. López-Periago, J. Fraile, P. López-Aranguren, L. Vega, and C. Domingo, CO_2 capture efficiency and carbonation/calcination kinetics of micro and nanosized particles of supercritically precipitated calcium carbonate, Chem. Eng. J. 226 (2013) 357-366. https://doi.org/10.1016/j.cej.2013.04.071

[9] A. F. Santos, A. L. Arim, D. V. Lopes, L. M. Gando-Ferreira, and M. J. Quina, Recovery of phosphate from aqueous solutions using calcined eggshell as an eco-friendly adsorbent," J. Environ. Manag. 238 (2019) 451-459. https://doi.org/10.1016/j.jenvman.2019.03.015

[10] M. Mohamed, S. Yusup, and M. A. Bustam, Synthesis of CaO-based sorbent from biomass for CO2 capture in series of calcination-carbonation cycle, Procedia Eng. vol. 148, pp. 78-85, 2016. https://doi.org/10.1016/j.proeng.2016.06.438

[11] R. Rohim, R. Ahmad, N. Ibrahim, N. Hamidin, and C. Z. A. Abidin, Characterization of calcium oxide catalyst from eggshell waste, Adv Environ Biol. 8 (2014) 35-38.

[12] M.-M. Pedavoah, M. Badu, N. O. Boadi, and J. A. Awudza, Green bio-based CaO from guinea fowl eggshells, Green Sustain. Chem. 8 (2018) 208-219. https://doi.org/10.4236/gsc.2018.82015

[13] Y. Kim, Y. Yoo, M. G. Kim, J.-H. Choi, and K. S. Ryoo, Evaluation of Raw and calcined eggshell for removal of Cd2+ from aqueous solution, J. Korean Chem. Soc. 64 (2020) 249-258.

[14] S. A. Salaudeen, S. H. Tasnim, M. Heidari, B. Acharya, and A. Dutta, Eggshell as a potential CO_2 sorbent in the calcium looping gasification of biomass, Waste Manag. 80 (2018) 274-284. https://doi.org/10.1016/j.wasman.2018.09.027

[15] Y. Gao and C. Xu, Synthesis of dimethyl carbonate over waste eggshell catalyst, Catal. Today 190 (2012) 107-111. https://doi.org/10.1016/j.cattod.2011.12.004

[16] R. Mohadi, K. Anggraini, F. Riyanti, and A. Lesbani, Preparation calcium oxide from chicken eggshells, Sri. J. Env. 1 (2016) 32-35. https://doi.org/10.22135/sje.2016.1.2.32-35

[17] A. Commey and M. Mensah, An experimental study on the use of eggshell powder as a pH modifier: Production of lime from eggshells, Int. J. Innov. Sci. Res. Technol. 4 (2019) 766-768.

[18] G. S. Grasa and J. C. Abanades, CO_2 capture capacity of CaO in long series of carbonation/calcination cycles, Ind. Eng. Chem. Res. 45 (2006) 8846-8851. https://doi.org/10.1021/ie0606946

[19] A. Borhade and A. Kale, Calcined eggshell as a cost effective material for removal of dyes from aqueous solution, Appl. Water Sci. 7 (2017) 4255-4268. https://doi.org/10.1007/s13201-017-0558-9

[20] T. Witoon, Characterization of calcium oxide derived from waste eggshell and its application as CO_2 sorbent, Ceram. Int. 37 (2011) 3291-3298. https://doi.org/10.1016/j.ceramint.2011.05.125

[21] M. A. Al-Ghouti and N. R. Salih, Application of eggshell wastes for boron remediation from water, J. Mol. Liq., 256 (2018) 599-610. https://doi.org/10.1016/j.molliq.2018.02.074

[22] G. V. Sree, P. Nagaraaj, K. Kalanidhi, C. Aswathy, and P. Rajasekaran, Calcium oxide a sustainable photocatalyst derived from eggshell for efficient photo-degradation of organic pollutants, J. Clean. Prod. 70 (2020) 122294. https://doi.org/10.1016/j.jclepro.2020.122294

[23] N. Viriya-empikul, P. Krasae, B. Puttasawat, B. Yoosuk, N. Chollacoop, and K. Faungnawakij, Waste shells of mollusk and egg as biodiesel production catalysts, Bioresour. Technol. 101 (2010) 3765-3767. https://doi.org/10.1016/j.biortech.2009.12.079

Comparison of cellulose properties from different agriculture biomass

SYAZMI ZUL Arif Hakimi Saadon[1,2,a], NORIDAH B. Osman[1,2,b*], NURALIFAH ILYANA Hamzah[1,c], SURRYA SHOVANESH Ganesh[1,d], MUHAMMAD FAIZ FAIZUL Zainif[1,e], NOR ADILLA Rashidi[1,2,f], MYLENE Uy[3,g]

[1]Department of Chemical Engineering, Universiti Teknologi PETRONAS, Bandar Seri Iskandar, 32610 Perak, Malaysia

[2]HICOE – Center for Biofuel and Biochemical Research, Institute of Self-Sustainable Building, Department of Chemical Engineering, Universiti Teknologi PETRONAS, 32610, Seri Iskandar, Perak, Malaysia

[3]Premier Research Institute of Science and Mathematics Chemistry Department, College of Sciences and Mathematics, MSU-Iligan Institute of Technology, Andres Bonifacio Avenue, 9200 Iligan City, Philippines

[a]syazmi_19001700@utp.edu.my, [b]noridah.osman@utp.edu.my, [c]nuralifah_23885@utp.edu.my, [d]surrya_24923@utp.edu.my, [e]muhammad_16005497@utp.edu.my, [f]adilla.rashidi@utp.edu.my, [g]mylene.uy@g.msuiit.edu.ph

Keywords: Cellulose, Bamboo, Kenaf, Alkaline Extraction, Napier Grass

Abstract. Cellulose is one of the most abundant molecules on the planet that presence in all plant cell walls on Napier grass and bamboo. Cellulose has been extensively used in industries like textile, food, composites, and pharmaceuticals, due to its outstanding properties. This research project aims to obtain a yield of cellulose from Napier grass, bamboo and Kenaf fiber by manipulating the operating condition of alkaline treatment. The extracted cellulose fiber was then being characterized to identify its yield, chemical composition, and crystallinity of the fiber by applying moisture content, chemical analysis, FTIR and X-ray Diffraction. As the temperature of the alkali reaction increases, the cellulose yield decreases due to the formation of purified cellulose after chemical process. Cellulose content increased after the fibers were water retted. The results were then validated by using FTIR analysis, which the amount of hemicellulose and lignin content decreased as the concentration, temperature, and time increases after treatment and the peak intensity band for lignin and hemicellulose reduces which indicate the removal of these two components. XRD analysis showed a declining trend for its crystallinity index as the temperature elevates. The crystallinity of the treated fibers did not affect the crystalline structure of the cellulose and that water retted fibers have higher crystallinity index.

Introduction

The development of cellulosic biomass materials has received a great deal of attention in researching their utilization towards relevant applications. According to Sonali Mohapatra et al., biomass is a renewable organic substance and form of sustainable energy that originates from various resources that give multiple benefits economically and environmentally [1]. In the last three decades, biomass energy has been the most widely used renewable energy source after hydropower (Annual Energy Review, 2016). Biomass is a cheap, safe, and easily accessible resource that minimizes the world's major pollution, emissions of greenhouse gases. Thus, cellulosic biomass-based materials caught the attention industries and policy makers to further research and transform them into value-added products. Lignocellulosic biomass is a type of polymer that is renewable in which is derived from various sources such as agricultural residues, forestry residues, and energy crops. Cellulose, lignin, and hemicellulose are complexly bundled in the structure of the cell wall of the lignocellulosic biomass. One of the compositions of the lignocellulosic biomass, which is cellulose, specifically has been by far the most abundant organic polymer

Cellulose is a polymer made up of glucose units joined together by condensation polymerization. Plant-based cellulose fiber has many unique benefits such as high heat resistance, being renewable, and inexpensive [2]. Hence, cellulose is used in a wide range of products such as synthetic products, thickening agents in shampoo, cosmetics, plastics, and even fabrics. However, various sources of cellulose are utilized in different applications. Generally, the cellulose that is isolated from hardwood fibers such as bamboo is commonly used as a raw material for the pulp and paper industry to make paper, fiberboard, sanitary pads, and other related products. Furthermore, cellulose fibers can be extracted from a hardwood plant in different ways to achieve high purity of cellulosic fiber. Barbash and Yashchenko states that parameters and conditions in the extraction process immensely affect the physical and mechanical properties of cellulose [3]. The statistic of the global cellulose market by end-use industry displayed that paper and pulp industry encompassed the largest share of the global cellulose market while closely followed by textiles, chemical synthesis and construction [4]. Due to the concern of sustainability and waste issue created by the synthetic materials, natural resources have become an attractive alternative instead of the synthetic which led to further demand for cellulose-based materials.

In this study, cellulose was extracted using the alkaline extraction method with necessary pretreatments to isolate unbleached cellulose fiber from bamboo, Napier grass and Kenaf with varying alkali concentration, temperature, and time of treatment to observer the properties of the extracted cellulose.

Materials and Methods

Materials

Napier grass Pakchong (NG) were obtained from Bota, Perak. Napier grass leaves and roots were chopped off and their stems were cleaned using damp cloth to remove dirt. They were then chopped into pieces having the length of approximately 15 cm. Next, the chopped stems were sun-dried for 8 hours before oven-dried for 12 hours at temperature of 105°C. The dried stems were then granulated using granulator and afterwards grinded into 500 microns using a power mill. The grinded Napier grass were then sieved into sizes of 300 μm and 500 μm.

2 types of bamboo were used in this project. Bamboo 1 (*Schizostachyum brachycladum*) (BL) was collected from Kampung Gajah, Perak and Bamboo 2 (*Bambusa Vulgaris*) (BM) used was obtained from Kampung Pintu Pandang, Gopeng, Perak. Bamboo culms were collected in poles of approximately 30cm and they were cleaned with a damp cloth to remove dirt. The bamboo poles were stored in the cold storage room at 10°C before being chopped into small pieces at an approximate length of 5 cm. The bamboo pieces were let to sun-dried for 8 hours and then oven-dried using biomass ovens for 48 hours at 105 °C. The dried bamboos were then granulated using a granulator to an approximate size of 3-5mm, grinded and sieved into 250μm using biomass power mill.

Fresh Kenaf was obtained from Lambor, Perak, Malaysia and the stems were separated from the leaves. The stems from the Kenaf were chopped into small pieces to be around 5cm with a machete for the extraction of fibers from the grass stem internodes. The fibers were then separated into non-water retting (NRK) and water retting (WRK) method. The fibers from Kenaf were water retted by soaking in a bucket of water for 7 days. It was then washed with distilled water followed by drying in the oven at temperature 105°C for 24 hours. The dried samples were then granulated with a granulator followed by grinding by a power mill and sieved into a particle size of 250 microns using a sieve shaker. The samples were then stored separately in polyethylene bags based on the type of feedstocks and retting method for 3 days.

Acetic acid, ethanol, sodium bisulfite, sodium hydroxide pellets, nitric acid, sodium chlorite and toluene used were of analytical grade and purchased from Avantis Laboratory Supply.

Extraction of cellulose process
Similar processes were done to extract cellulose, but some methods were modified according to each sample. For Napier grass, the methodology was adopted from Reddy et al [5] with a few modifications. NG and BL were dewaxed using Soxhlet extraction apparatus using ethanol: toluene (2:1, v/v) for 6 hours. The samples were then dried for 24 hours at 105 °C. The samples were delignified and bleached using 0.7 % sodium chlorite, acidified using acetic acid at 100 °C for 2 hours. The samples were then filtered and washed with 2 % sodium bisulfite and water. The filtrate was treated with 17.5 % sodium hydroxide solution at 25 °C for 45 minutes. After alkali treatment, it was filtered, washed with 10 % acetic acid and distilled water to neutralize the solution. It was then acid hydrolyzed using 80 % acetic acid and 70 % nitric acid (10:1 ratio) at 120 °C for 20 minutes. The samples were then cooled, washed with 95 % ethanol and water and dried at 105 °C until constant weight. The experiment was repeated using different alkali treatment temperature at 35 °C.

For BM, Soxhlet extraction was used onto 5.0 g of grinded sample using 95% ethanol for 4 hours to remove extractives. The sample was treated according to Fitriana et al. [6] with some modifications. The extractive- free bamboo was then mixed with 50ml aqueous sodium hydroxide (NaOH) solution at 5wt% and 10wt%. The solution obtained was incubated in a shaker flask at 35ºC and 45 °C. This incubation process was done in an incubator for 16 hours at 100 rpm. The solid residue of the incubated sample which was the cellulose fibers were filtered using the gravitational filtration method. The solid residue will be washed thoroughly with 100ml deionized water and dried at 45 °C in an oven to remove water until constant weight.

For Kenaf sample, Soxhlet extraction was also used to remove extractive similar to NG and BL. The extractive-free kenaf was then dried at 105 °C and let cooled in a desiccator. It then underwent digestion with 400 mL of hot water at 100 °C for 3 hours and washed with 100 mL of hot water followed by 50 mL of ethanol. The subsequent alkali treatment procedures were adopted from Kommula et al. [7] with some modifications. The extractives-free samples were treated with 5 % (w/v) NaOH solution for 4 hours at temperature of 25°C. A liquor ratio of 30:1 was maintained for the alkali hydrolysis to remove the hemicellulose, lignin, and surface impurities. The extractives-free samples were treated with 5 % and 15 % (w/v) sodium hydroxide, NaOH solution for 4 hours and 6 hours, and at temperature of 25 °C and 35°C. Next, the fibers were neutralized using 1% (w/v) acetic acid solution. The solid residue was filtered, washed with distilled water, and dried for 24 hours at 100 °C. The dry solid residue was soak in a 4% (w/v) NaOH solution at 50°C for bleaching purpose with constant stirring at 200 rpm. Then 30% hydrogen peroxide (H_2O_2) to the same amount of NaOH was added by drops and stirred continuously for 1 hour. The samples were then filtered, washed, and dried until constant weight.

Moisture Content Analysis
The procedure for conducting the experiment was based on ASTM D4442 with a few changes. The following Eq. 1 used to calculate the moisture content:

$$Moisture\ content(\%) = \frac{W_2 - W_3}{W_3 - W_1} \times 100\% \quad (1)$$

where W_1 is the weight of crucible and cover, W_2 is the weight of the sample with crucible and cover and W_3 is the weight of dried sample with crucible and cover.

Yield analysis

Yield analysis was done to observe each sample at different treatment conditions which indicate the fibre quality from the extracted biomass. The yield is calculated based on the ratio of extracted cellulose weight over the initial weight of sample used. Eq. 2 is used to calculate the yield percentage.

$$Yield\ (\%) = \frac{W_c}{W_b} \times 100\% \tag{2}$$

where Wc is dry weight of extracted cellulose after the treatment process and Wb is dry weight of bamboo sample after Soxhlet extraction.

Fourier Transformed Infrared (FTIR) Spectroscopy Analysis

FTIR with Attenuated total reflectance (ATR) was used to evaluate the pretreated sample. Changes in the intensity of functional group of the pretreated sample are determined and analyzed based on the FTIR spectrum. The spectra were recorded in the frequency range of 4000-400 cm^{-1} with a resolution of 1 cm^{-1}. The expected functional groups that shall present were C-H, C-O, and O-H.

X-ray Diffraction (XRD) Analysis

The pretreated samples are scanned in XRD at $10°\leq 2\theta \leq 40°$ at an exposure time of 1 seconds with a step size of $0.02°/2\theta$. The crystallinity index for cellulose is calculated using the empirical method in Eq. 3 below

$$Crytallinity\ index, CI\ (\%) = \frac{I_{002} - I_{AM}}{I_{002}} \times 100\% \tag{3}$$

where I_{002} is main crystalline plane in which the intensity of the crystalline peak at the maximum 2θ at 22.5° and I_{AM} refers to amorphous fraction in which the intensity at the maximum 2θ at 16°.

Results and discussion

Moisture content

The average moisture content of each sample was shown in Fig. 1. It is important that the moisture content is kept under 5 %. Higher moisture content contributes to less lignocellulose content per weight of sample since much of the weight is contributed by the weight of water molecules.

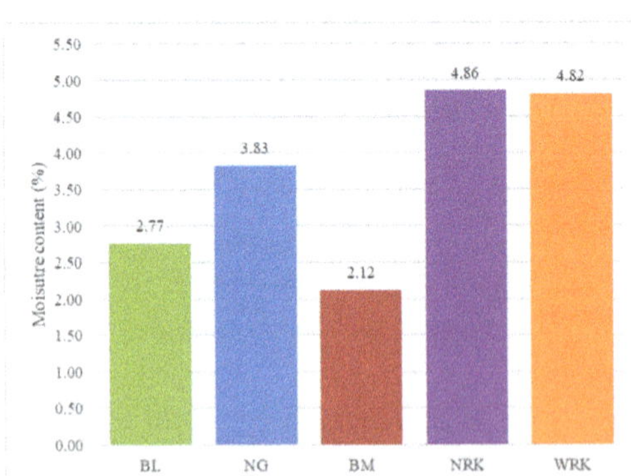

Figure 1: Moisture Content % against the type of sample

Yield Analysis
The yield of each set is shown in Fig. 2.

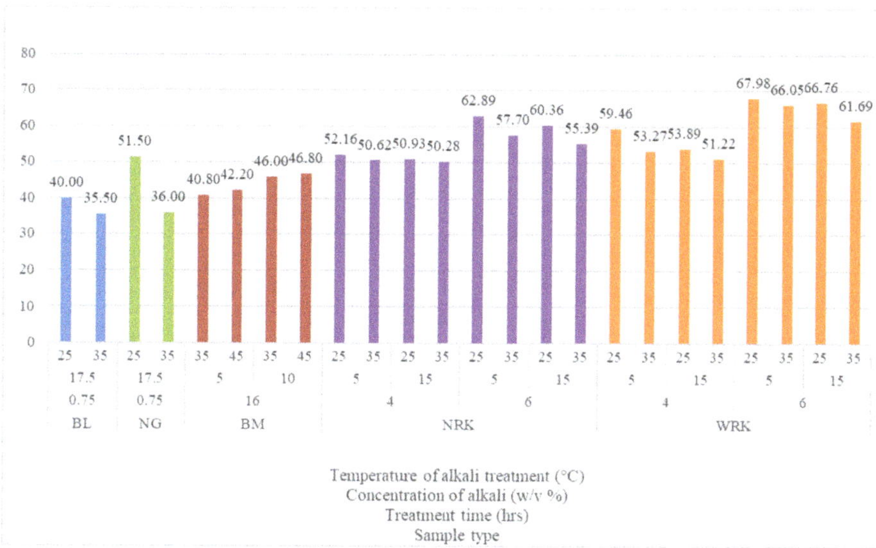

Figure 2: Yield of Cellulose (%) against Temperature, Concentration of alkali, Treatment Time, and Sample type

Bamboo BL and Napier grass
When comparing between BL and NG, NG seems to have higher extracted cellulose. It is seen that the yield decreases as the temperature of alkali treatment increases. The decline in NG is much more significant than BL. This is due to the enhanced corrosive nature of sodium hydroxide at higher temperature [8]. This also causes the extracted samples to be finer and easily distinguishable. Using high temperatures for alkaline conditioning, problems such as the formation of toxic compounds and loss of sugars have been reported [9].

Bamboo BM
For BM, it is found that the yield of cellulose increases with both increment of concentration and temperature. It is significant that treatment conditions using 10 wt.% of NaOH at 45 °C have the highest yield percentage. This shows a similar trend of increases in cellulose yield along with the temperature. However, the increase of incubation temperature beyond 45 °C no longer shows significance in the yield of cellulose. Therefore, it can be concluded that the optimum temperature for BM using alkaline treatment is at 45 °C.

Kenaf
It is found that water retted samples of Kenaf produces higher extracted lignin. The objective of water retting in this experiment was to allow water to penetrate the portion of central stalk and burst the outer layer through swelling the inner cells. This process enhances the absorption of moisture as well as bacteria that decomposition. This allows the substances surrounding the fibers such as cellular tissues and gummy substance to rot away while also separating the fibers from the

stem. Water retting also removed substances surrounding the fibers such as pectin and other impurities thus increasing the proportion of cellulose within the fibers.

Increase in temperature and concentration causes the yield to decrease. Similar to NG and BL, this is due to the more intense disruption from the alkali. The time effect improved the yield where 6 hours produces more cellulose than 4 hours. Longer reaction time is found to improve component penetration and better in removing the hemicellulose and impurities resulting in higher yield.

FTIR analysis

Characterization of functional group and chemical composition were analyzed using FTIR analysis. For cellulose, there are several functional groups that should be observed such as hydroxyl, carbonyl, methoxyl, carboxyl, aromatic and phenolic. Assignment of functional group to FTIR spectrum wavenumber are listed in Table 1 below.

Table 1: Assignment of functional group to FTIR spectrum wavenumber [10-12]

Wavenumber (Cm^{-1})	Assignments	Band assignment
3600-3100	O-H	Stretching vibration of alcoholic and phenolic OH groups involve in hydrogen bonds
2960-2820	C-H	$-CH_2$, $-CH_3$
1770-1685 **1680-1650**	C=O	Conjugated p-substituent carbonyl and carboxyl
1600-1500, **1430-1420**	Aromatic skeletal	Aromatic ring vibrations
1515-1511	C=C	Aromatic skeletal breathing with C-O stretching
1470-1450, **1370-1360**	C-H	C–H deformations methyl and methylene
1427-1423	C-H	Aromatic skeletal vibrations combined with C–H in-plane deformation
1375-1397	O-H C-H	Phenolic OH and aliphatic C-H in methyl groups
1170-1150	C-H	Aromatic C-H in-plane deformation in the guaiacyl ring
1145-1140	C-H	Aromatic C-H in-plane deformation in the Syringyl ring
1035-1025	C-O, C-H	Aromatic ring and primary alcohol

Bamboo BL and Napier grass

The spectra of raw BL, treated BL and pure commercial cellulose are shown in Fig. 3. For BL, it is found that the FTIR spectra after alkali treatment at 25 °C is almost similar to the untreated sample in which the only slight decrease in absorbance intensity is seen. This shows that the treatment is not as effective to produce cellulose. At this temperature, the alkali might not penetrate the lignocellulosic matrix in which only the surface is degraded and causes the yield to be only at 40.0 %. When the temperature was increased, a prominent impact is observed where the absorbance spectra is much lower. This means that the increase in temperature breaks much of the chemical bonds. The intensity at 1730 cm^{-1} decreased for after alkali treatment which clearly showed hemicellulose removal after the chemical treatment. The band near 1514 cm^{-1} and 1250 cm^{-1} which indicated the aromatic skeletal vibration and C-O stretch of ester group was significantly reduced as compared to the raw sample. Also, it is worth noting that the peak near 897 cm^{-1} was presented as the C-H group in cellulose is noticeable after treatment which shows that cellulose is obtained.

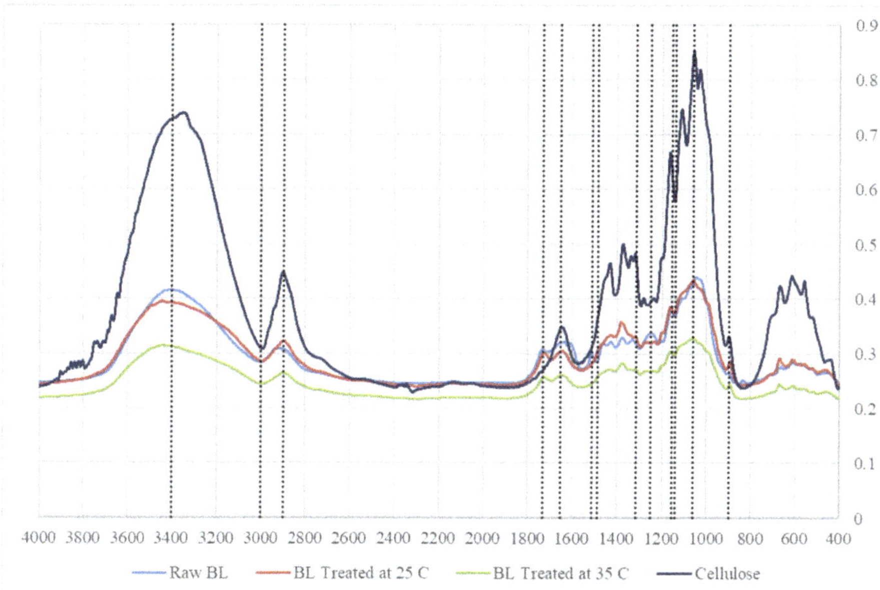

Figure 3: FTIR Spectrum for Bamboo in terms of its absorbance

The spectra of raw NG, treated NG and pure commercial cellulose are shown in Fig. 4. The spectra for NG after alkali treatment is much distant when compared to raw NG. Similar to BL, increase in temperature causes the spectra to be shifted downwards indicating much of the chemical bonds are broken at higher temperature. Looking at the peaks at 1250 cm-1, 1514 cm-1 and 1730 cm-1, the peaks seem to diminish suggesting delignification is successful. The peaks at 897 cm-1 for both treated samples are more prominent as compared to the untreated sample and it is similar to the peak at pure cellulose. This shows that cellulose is well extracted from the NG samples.

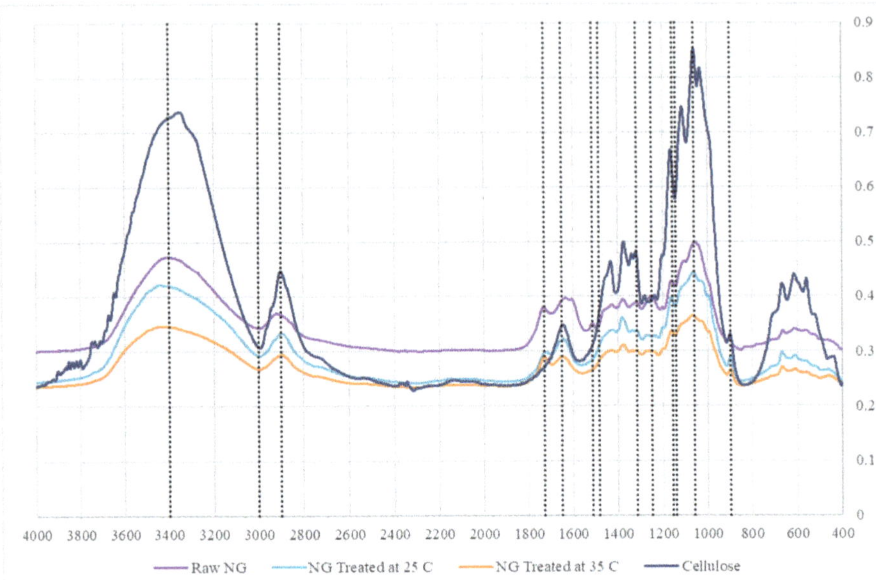

Figure 4: FTIR Spectrum for Napier Grass in terms of its absorbance

Bamboo BM

The spectra for untreated BM, treated BM and pure commercial cellulose is shown in Fig. 5. It is noticeable that all 4 conditions of treatments show a similar trend of the spectrum. The C=C double bond stretching at 1513 cm^{-1} which attributes to the aromatic ring group in lignin had totally disappeared in the treated samples. This corresponds to the removal of lignin from the alkaline treatment process. The O-H bending group at 1650 cm^{-1} still shows dominance in the peak after the treatment. This is because, the O-H bending is assigned as the interaction with water on the molecular surface [13]. Thus, hydrophilic cellulose strongly interacts with water and remains the O-H bending. Cellulose structure does not dissolve in water, but it can absorb a large quantity of water which causes it to swell. Similar to BL and NG, the peaks at 897 cm^{-1} which indicates the presence of β-glycosidic linkages which associates with the C-H group are visible after alkali treatment [14]. The bamboo samples that were treated with alkaline treatment method using NaOH indicates that the extracted cellulose samples were mainly compromised of cellulose structure.

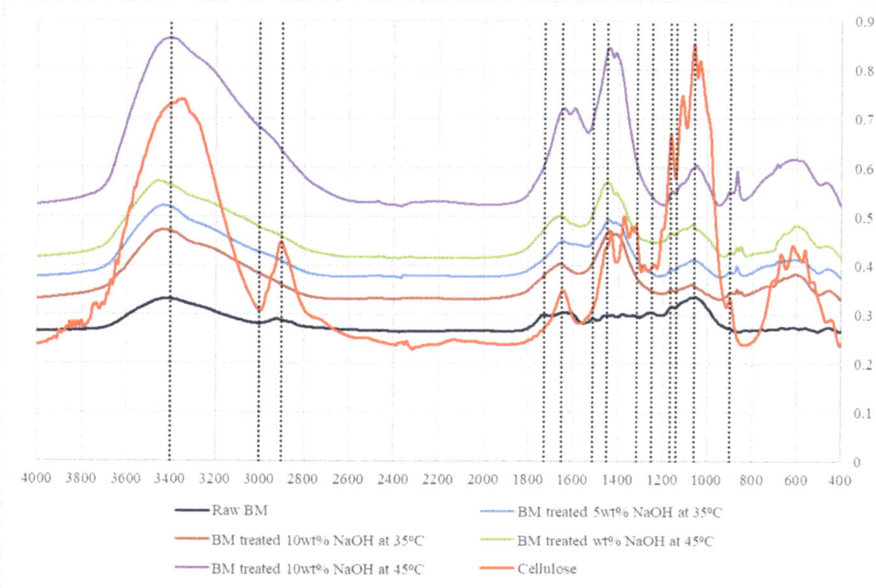

Figure 5: FTIR Spectrum for Bamboo BM in terms of its absorbance

Kenaf

The spectra of NRK, WRK and Kenaf samples after alkali treatment are shown in Fig. 6, 7, 8 and 9. When comparing the NRK and WRK before alkali treatment, it is seen that the peaks are almost similar but differing in absorbance values. The biggest difference between NRK and WRK is at 3339 cm^{-1} where WRK produces much higher absorbance. This is due to the swelling effect after the sample are soaked in water and accumulation of water molecules.

After alkali treatment, the FTIR spectra appear to be lower when comparing with before the treatment. It can be seen that the treated WR-Kenaf have higher absorbance than treated Kenaf following the same trend of untreated WR-Kenaf and Kenaf. This is due to the water retting process removing the surrounding substance around the fibers such as pectin and gummy substances thus increasing the quality of the fibers resulting in increasing amplitude intensity of treated WR-Kenaf fibers than treated Kenaf fibers. similar absorption bands and the band at 1240 and 1735 cm^{-1} can be seen to decreases as compared to the untreated fibers. This shows the C-O stretching of acetyl of hemicellulose and C=O stretching vibration in ester groups of hemicellulose to decreases. As for the absorption band at 1505 cm^{-1} shows the C=C stretching vibrations in aromatic rings of lignin still having a peak. This shows that the lignin component was not fully removed during the treatment. Although traces of lignin component were present, the absorption band at 1505 cm^{-1} of treated fibers were lower than the untreated fibers. can be seen that for concentration of 15% NaOH have higher absorbance than concentration 5% NaOH that have lower absorbance. As for the 5% NaOH, the soaking time of 6h have lower absorbance as compared to soaking time of 4h. As for K 15% 4h, the absorption bands do not follow the trend of the other treated fibers though it shows the decreasing in peak at 1505 cm^{-1} indicating full removal of lignin.

Materials Research Proceedings 29 (2023) 255-271 https://doi.org/10.21741/9781644902516-29

This can be considered pure cellulose although a replication of FTIR analysis could uncover the reason on different trend.

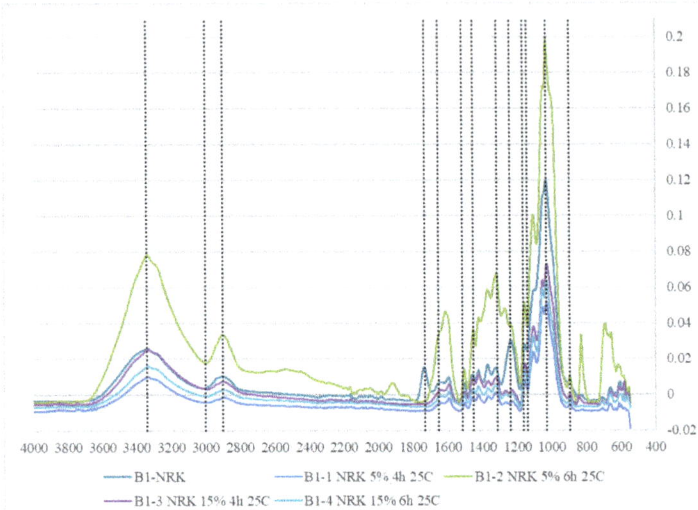

Figure 6: FTIR Spectrum for NR-Kenaf at 25 °C in terms of its absorbance

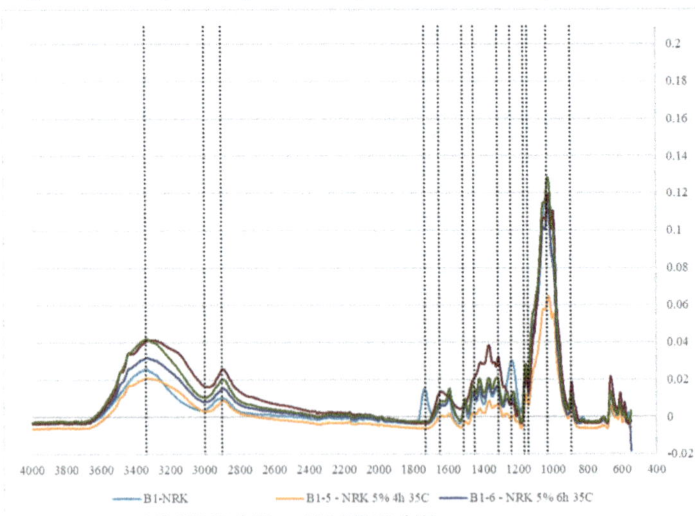

Figure 7: FTIR Spectrum for NR-Kenaf at 35 °C in terms of its absorbance

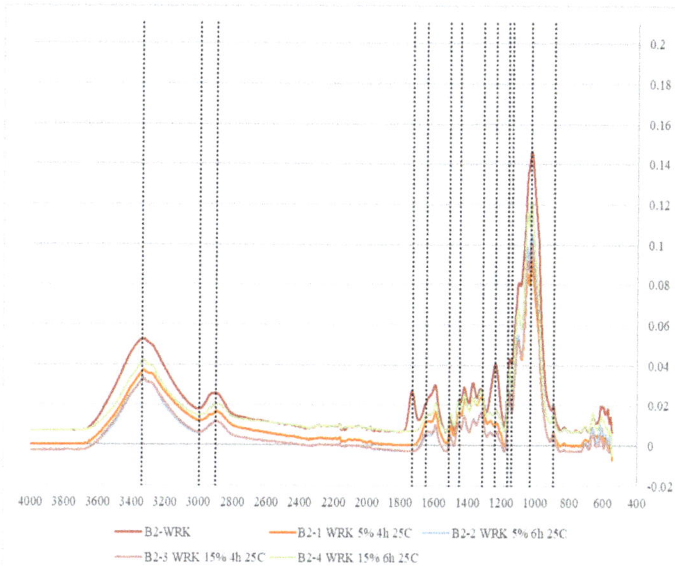

Figure 8: FTIR Spectrum for WR-Kenaf at 25 °C in terms of its absorbance

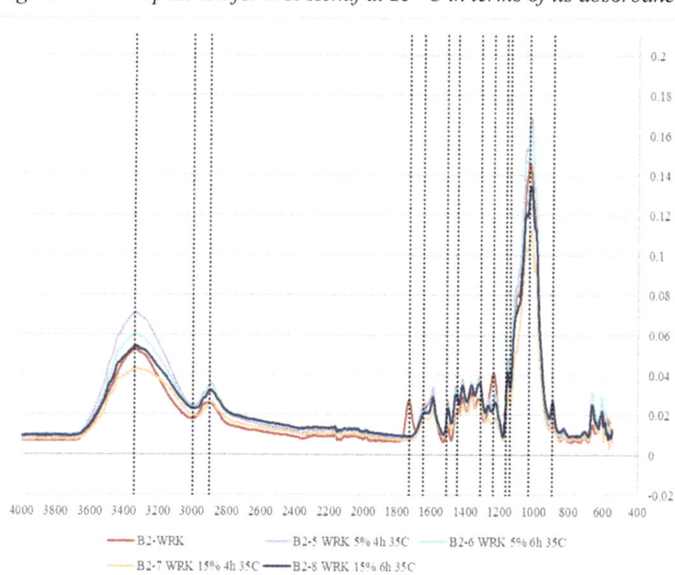

Figure 9: FTIR Spectrum for WR-Kenaf at 35 °C in terms of its absorbance

Materials Research Proceedings 29 (2023) 255-271 https://doi.org/10.21741/9781644902516-29

XRD Analysis
The calculated crystallinity indices for the samples are shown in Fig. 10.

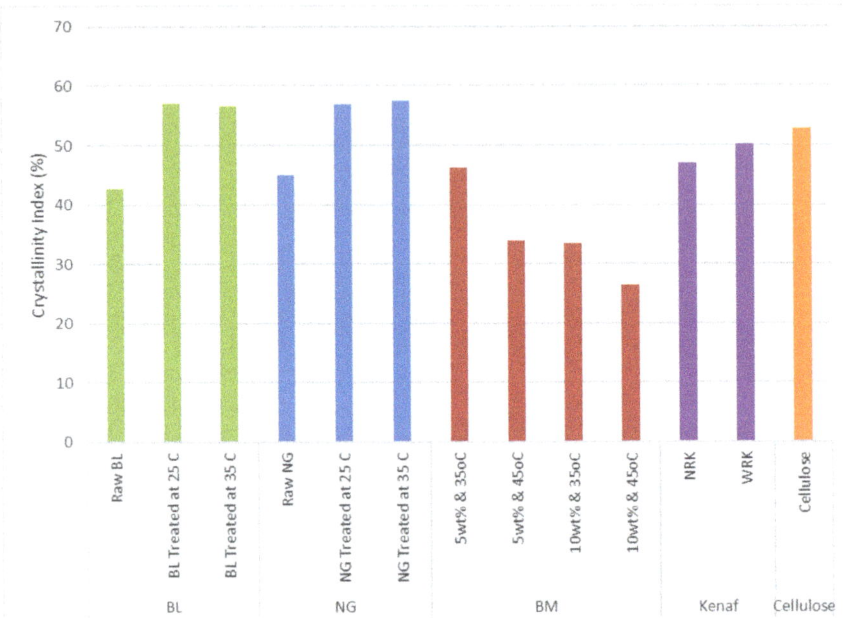

Figure 10: Crystallinity Index of Samples

Bamboo BL and Napier grass
For BL, the XRD spectra is shown in Fig. 11. The crystallinity index (CrI) shows that the alkali treatment increases the CrI. When comparing between the samples treated with different temperature, the change is not significant with regards to CrI. Fig. 12 shows the XRD spectra for NG. Based on the crystallinity results for NG, the CrI seems to also increase with increase in alkali treatment temperature. The increasing in crystallinity index was due to the removal of lignin and hemicellulose during the chemical process as compared to the raw samples where lignin and hemicellulose were still present in the samples. The treated fibers appeared to be more crystalline than the untreated ones since the loss of their amorphous constituent happened after treatment [12].

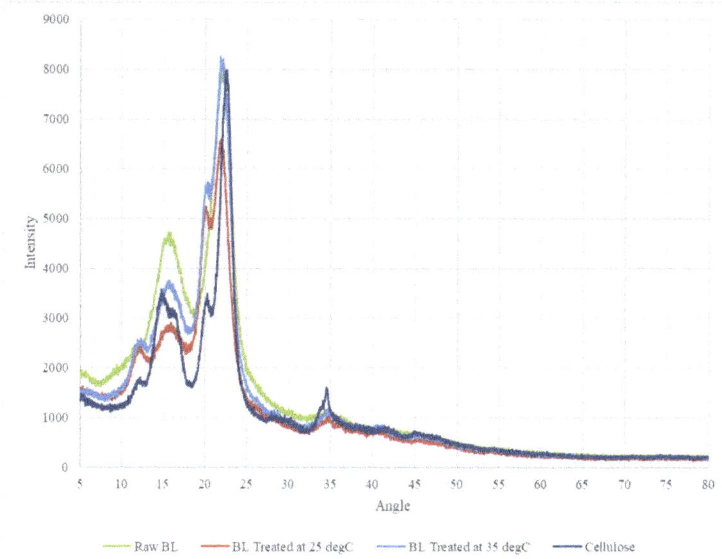

Figure 11: XRD Spectra for BL

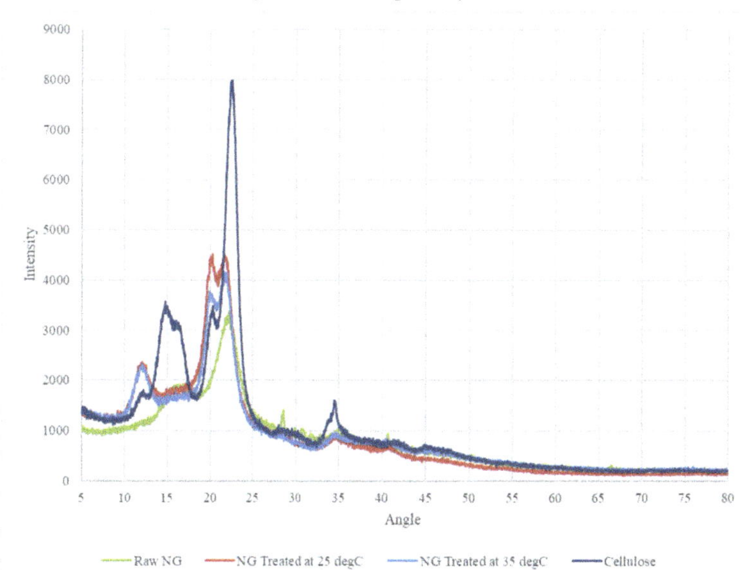

Figure 12: XRD Spectra for NG

Bamboo BM

From Fig. 13, the cellulose crystallinity diffraction pattern for all 4 different conditions shows a similar trend. It is significant that cellulose sample treated with 10 wt.% of NaOH at 45 °C have the highest crystallinity index at 61% and 5 wt.% of NaOH at 35 °C have the lowest crystallinity at 52%. Besides, an untreated sample of bamboo is observed to have a crystallinity index of 45.57%. The crystallinity index of bamboo cellulose pulp (Bambusa Vulgaris) is reported to be 51.48% which was produced by the mechanical defibrillation process of 0.3% anthraquinone ($C_{14}H_8O_2$) per unit solid mass. Thus, this crystallinity index obtained was within the range of bamboo cellulose fiber. The crystallinity index obtained also clearly reflects accordingly to the absorbance (%) in the FTIR analysis in Fig. 5. Therefore, the treated cellulose sample for 10 wt.% of NaOH at 45 °C is the closest to the analytical grade of pure cellulose which associates with the highest crystallinity index and absorbance (%). This result explains that the concentration of 10 wt.% of NaOH as well as the incubation temperature of 45 °C is preferred for a higher crystallinity index of cellulose from bamboo.

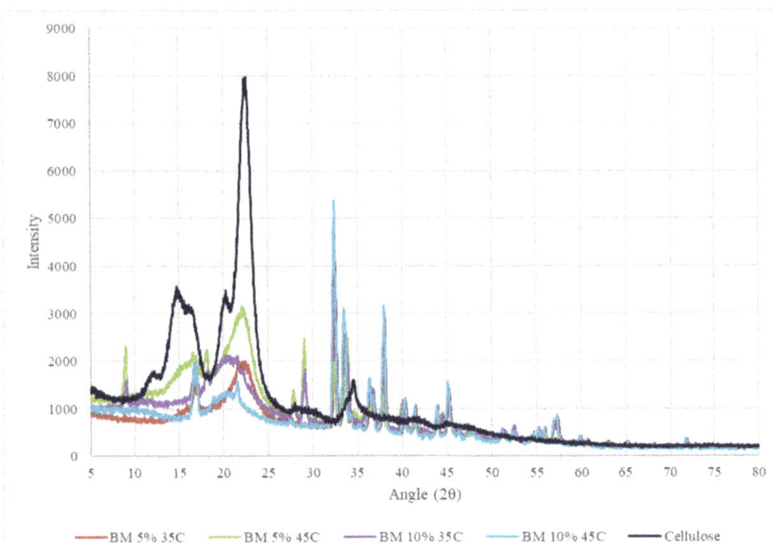

Figure 13:XRD Spectra for BM

Kenaf

Fig. 14 shows the XRD spectra of alkali-treated NR-Kenaf (ATNR-Kenaf) fibers and alkali-treated WR-Kenaf (ATWR-Kenaf) fibers at constant parameters during the alkali treatment and bleaching process. A similar diffraction pattern can be seen for both the alkali-treated NR-Kenaf and alkali-treated WR-Kenaf. The diffraction for WR-Kenaf have a higher peak at 22.53° showing higher intensity as compared to a lower peak intensity of NR-Kenaf. The peak at 22.53° was presumed to be cellulose while the peak at 16.23° was presumed to be the presence of hemicellulose and lignin results in an amorphous phase. It can also be seen that the retting process does not affect the reformation of cellulose crystalline structure. The removal of amorphous components, pectin, hemicellulose, and some lignin were aided by the retting, alkali treatment, and bleaching process.

Crystallinity index has been calculated and presented into Fig. 10. As shown in the table, the CrI of ATNR-Kenaf and ATWR-Kenaf were 47.01% and 50.12% respectively. It shows that the treated WR-Kenaf have a higher CrI than the treated Kenaf at a difference of 3.11%. The water retting process removes the pectin allowing the alkali treatment and bleaching process to easily remove amorphous components, hemicellulose, and lignin. The treated Kenaf does not utilize water retting resulting in the alkali treatment and bleaching process to remove pectin, amorphous components, hemicellulose, and lignin by the process itself which results in lower CrI for non-water retting.

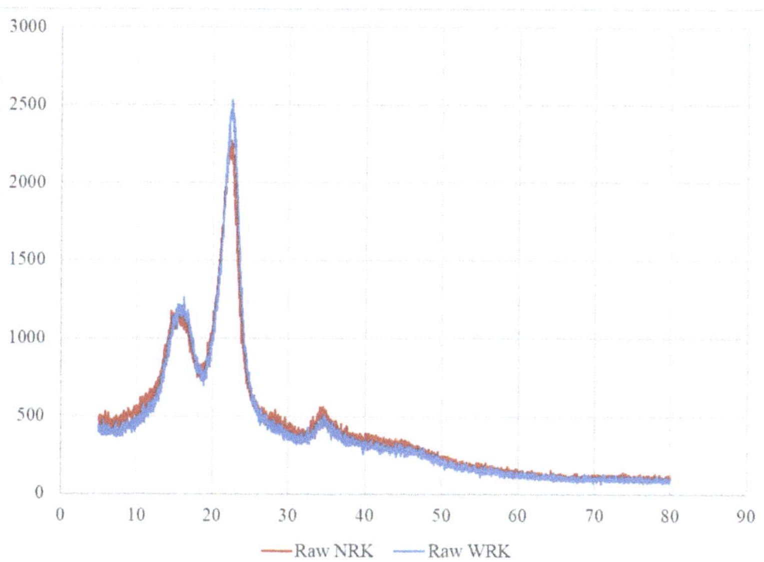

Figure 14: XRD Spectra for Kenaf

Conclusion

The variation of parameter of cellulose extraction is to quantify and evaluate the extraction via alkali treatment. The feedstocks were varied, and different concentration, temperature and treatment time were applied to optimize the extraction factor and the extracts were analyzed for its properties. The moisture content before alkali treatment was checked so that it is well-below 5 % in order to not compromise the chemical process. The yield of cellulose decreases with increasing alkali concentration and temperature while increasing the time would improve the yield. Of all the feedstocks, cellulose was able to be extracted best from Kenaf. Based on the FTIR analysis, it is evident that extractive removal and delignification was successful in removing lignin, hemicellulose, and extractives. After analyzing the crystallinity indices of the samples, it was found that increasing the temperature and concentration also decrease the crystallinity of cellulose, but the changes of temperature are less significant towards the crystallinity. The extracts were assessed with analytical pure cellulose for comparison and much of the aspects of cellulose were exhibited in the samples which shows that the process done is effective, but more optimizations are needed. This would unlock the potential of utilizing and commercialize agricultural biomass feedstock for production of cellulose and also towards other component extraction.

Acknowledgements
The authors would like to acknowledge the funding support from the Ministry of Higher Education (MOHE) Malaysia through the Higher Education Center of Excellence (HICoE) grant to Center for Biofuel and Biochemical Research (cost centre: 015MA0-052) and supported by the Department of Chemical Engineering (ChE), Universiti Teknologi PETRONAS.

References

[1] S. Mohapatra, R.C. Ray, S. Ramachandran, Chapter 1 - Bioethanol From Biorenewable Feedstocks: Technology, Economics, and Challenges, in: R.C. Ray, S. Ramachandran (Eds.), Bioethanol Production from Food Crops, Academic Press2019, pp. 3-27. https://doi.org/10.1016/B978-0-12-813766-6.00001-1

[2] A. Kafy, H.C. Kim, L. Zhai, J.W. Kim, L.V. Hai, T.J. Kang, J. kim, Cellulose long fibers fabricated from cellulose nanofibers and its strong and tough characteristics, Scientific Reports 7(1) (2017) 17683. https://doi.org/10.1038/s41598-017-17713-3

[3] V. Barbash, O. Yashchenko, Preparation, Properties and Use of Nanocellulose from Non-Wood Plant Materials, 2020. https://doi.org/10.5772/intechopen.94272

[4] Cellulose Market Size, Share and Industry Analysis, By Derivatives Type (Commodity Cellulose Pulp, Cellulose Fibers, Cellulose Ethers, Cellulose Esters, Microcrystalline Cellulose, Nanocellulose and Others), By End-Use Industry (Textile, Food, Chemical Synthesis, Pharmaceuticals, Construction, Paper & Pulp, Paints & Coatings, and Others), and Regional Forecast, 2019-2026, 2020. https://www.fortunebusinessinsights.com/cellulose-market-102062. 2020).

[5] K.O. Reddy, C.U. Maheswari, M.S. Dhlamini, B.M. Mothudi, V.P. Kommula, J. Zhang, J. Zhang, A.V. Rajulu, Extraction and characterization of cellulose single fibers from native african napier grass, Carbohydr Polym 188 (2018) 85-91. https://doi.org/10.1016/j.carbpol.2018.01.110

[6] N. Fitriana, A. Suwanto, T. Jatmiko, S. Mursiti, D.J. Prasetyo, Cellulose extraction from sugar palm (Arenga pinnata) fibre by alkaline and peroxide treatments, IOP Conference Series: Earth and Environmental Science 462 (2020) 012053. https://doi.org/10.1088/1755-1315/462/1/012053

[7] V.P. Kommula, K.O. Reddy, M. Shukla, T. Marwala, A.V. Rajulu, Physico-chemical, Tensile, and Thermal Characterization of Napier Grass (Native African) Fiber Strands, International Journal of Polymer Analysis and Characterization 18(4) (2013) 303-314. https://doi.org/10.1080/1023666X.2013.784935

[8] V.G. Hambate, S. Ngouadjio, L. Daï-Yang, A.E. Ofudje, M.J. Ketcha, B. Loura, CHEMICAL TREATMENT OF SUGARCANE BAGASSE FOR THE PRODUCTION OF CELLULOSIC FIBERS, Acta Technica Corviniensis - Bulletin of Engineering 10(3) (2017) 33-38.

[9] G. Antonopoulou, H.N. Gavala, I.V. Skiadas, G. Lyberatos, The Effect of Aqueous Ammonia Soaking Pretreatment on Methane Generation Using Different Lignocellulosic Biomasses, Waste and Biomass Valorization 6(3) (2015) 281-291. https://doi.org/10.1007/s12649-015-9352-9

[10] T. Rashid, C.F. Kait, T. Murugesan, A "Fourier Transformed Infrared" Compound Study of Lignin Recovered from a Formic Acid Process, Procedia Engineering 148 (2016) 1312-1319. https://doi.org/10.1016/j.proeng.2016.06.547

[11] J.S. Lupoi, S. Singh, R. Parthasarathi, B.A. Simmons, R.J. Henry, Recent innovations in analytical methods for the qualitative and quantitative assessment of lignin, Renewable and Sustainable Energy Reviews 49 (2015) 871-906. https://doi.org/10.1016/j.rser.2015.04.091

[12] J. Shi, D. Xing, J. Lia, FTIR Studies of the Changes in Wood Chemistry from Wood Forming Tissue under Inclined Treatment, Energy Procedia 16 (2012) 758-762. https://doi.org/10.1016/j.egypro.2012.01.122

[13] E. Galiwango, N.S. Abdel Rahman, A.H. Al-Marzouqi, M.M. Abu-Omar, A.A. Khaleel, Isolation and characterization of cellulose and α-cellulose from date palm biomass waste, Heliyon 5(12) (2019) e02937. https://doi.org/10.1016/j.heliyon.2019.e02937

[14] K.O. Reddy, K.R.N. Reddy, J. Zhang, J. Zhang, A. Varada Rajulu, Effect of Alkali Treatment on the Properties of Century Fiber, Journal of Natural Fibers 10(3) (2013) 282-296. https://doi.org/10.1080/15440478.2013.800812

Materials Research Proceedings 29 (2023) 272-280 https://doi.org/10.21741/9781644902516-30

Activated carbon/graphene composite electrodes with green polymeric binder for carbon dioxide reduction: Preliminary study

MUHAMAD FARHAN HAQEEM Bin Othman[1,a], INTAN SYAFIQAH Ismail[1], MUHAMAD HAFIZHAN HAFIY Mohammad Azmi[2], NOR ADILLA Rashidi[1,b*], SUZANA Yusup[3] and NORIDAH Osman[1,c]

[1]Chemical Engineering Department, Higher Institution of Center of Excellence: Centre for Biofuel and Biochemical Research, Institute of Self-Sustainable Building, Universiti Teknologi PETRONAS, 32610 Bandar Seri Iskandar, Perak, Malaysia

[2]Process & Process Safety, Exen International Sdn Bhd, Suite 10-23, Tower 1, Wangsa 118, Jalan Wangsa Delima, Wangsa Maju, 53300, Kuala Lumpur, Malaysia

[3]Generation Unit (Fuel & Combustion Section), Tenaga Nasional Berhad Research (TNBR), No1, Kawasan Institusi Penyelidikan, Jln Ayer Hitam, 43000 Kajang, Selangor, Malaysia

[a]farhanhaqeem123@gmail.com, [b]adilla.rashidi@utp.edu.my, [c]noridah.osman@utp.edu.my

Keywords: Carbon Dioxide Reduction, Activated Carbon-Graphene Oxide Composite, Green Polymeric Binder, Biopolymer, Electrocatalytic Reaction

Abstract: In the present study, a few types of green polymer binder such as Xanthan gum (XG), Carboxyl methylcellulose (CMC) and Sodium alginate (SA) have been prepared with activated carbon-graphene oxide composite electrode (AC/GO). Throughout the experiment, CMC as an individual polymeric binder produce the least amount of the weight loss (8.28%). To enhance the adhesion, the binder was mixed with synthetic binder (PEI), which results in mass loss of 1.86%. Furthermore, the electrochemical analysis depicts that AC/GO-SA produces the highest current rate (-7.8 mA/cm^{-2}) throughout the 30 min of reduction process, with 16.9% Faradaic efficiency. The experimental results show that a facile, low-cost, eco-friendly design of green polymer binder is promising for the carbon dioxide reduction, thus, promote a greener technology.

Introduction

Ever since the Industrial Revolution, anthropogenic carbon dioxide (CO_2) level in the atmosphere has continuously increasing from 280 ppm to over 400 ppm due to extensive fossil fuels burning [1]. Excessive consumption of fossil fuels has cautiously led to the depletion of these finite natural resources while rising the atmospheric CO_2 concentration. The increasing CO_2 concentration in the Earth's atmosphere has raised numerous of environmental concerns, such as global warming, rising sea level as well as species extinction, which all cause a significant disruption to the ecosystem [2]. The release of CO_2 into the environment is uncontrollable as a consequence of various current industrial processes. In contrast with the natural carbon cycle of having a closed loop process, the open loop of these anthropogenic activities which releases CO_2 into the environment is unsustainable. Minimizing the CO_2 emissions assist in solving the environmental issues such as switching to low carbon-fuel (hydrogen or syngas) or upgrading the energy efficiency and enhance the energy conservation [3].

Electrochemical reduction of CO_2 has attracted much attention in recent years as it represents a clean technique at the expense of facilitating sustainable supply for energy storage and conversion. This technique has the advantage where water can be used as the proton source and procedure can be carried out a room temperature. However, some issues need to be resolved before the electrochemical reduction of CO_2 becomes appealing for technological applications, related to the high efficiency and product selectivity [4]. The electrochemical approach mainly relates with measurement and control of a redox reaction in electrochemical cells. In electrochemical cells, electrodes are set into an electrolyte for chemical reaction to occur. The composite electrode is considered as one of the important configurations in electrochemical performance testing [1].

Typically, fabrication of composite electrode includes active material, polymer binder material and conductive material. As an improvement in the electrode fabrication, activated carbon electrode particularly the functionalization of active material and extensive performance of conductive material has gained many research interests, with minimal attention to improve the binder material. [5]. Binder plays an important role as the surfactant that binds the material altogether and control the functionality of the electrode. The binder forms a matrix for the active material and allows build-up contact with the current collector, which then provides mechanical durability. The binder has to be electrochemically stable, insoluble in the electrolyte solution, and must be evenly distributed in the slurry during the electrode fabrication. Common polymeric binder in electrochemical capacitor is based on fluorinated polymer like polyvinylidene fluoride (PVDF), polyvinyl alcohol (PVA) and Nafion. However, these polymer binders suffer atrocious drawbacks such as high toxicity and poor adhesion [6]. These fluorinated polymers often require wetting agents or organic solvents during the fabrication process and release fluorinated monomers which has severe environmental effects [7]. Thus, green polymeric binder is said to be a great alternative as it maintains the environmental concerns of reducing toxicity [8]. Derived from nature, Sodium Alginate (SA), Xanthan Gum (XG), and Carboxyl Methylcellulose (CMC) are the examples of green binders as substitute for electrode fabrication and the performance of each binder has been carried out and described in this study.

Experimental and methods

Chemicals and reagents

Green polymer such as carboxymethyl cellulose (CMC) (CAS No: 9000-11-7; Food grade), xanthan gum (XG) (CAS No: 11138-66-2; Food grade), and sodium alginate (SA) (CAS No: 9005-38-3; Food grade) were supplied by Take It Global Sdn Bhd, Penang, Malaysia. Activated carbon (AC) powder (from coconut shell) was acquired from Kekwa Indah Sdn Bhd, Negeri Sembilan, Malaysia. Graphene oxide (GO) solution (4 mg/mL, in H_2O), polyethylenimine (PEI) solution (50% in H_2O) and corn starch (CAS No: 99005-25-8) were supplied by Sigma Aldrich. Potassium hydroxide (KOH) and potassium bicarbonate ($KHCO_3$) (Chemically Pure Grade) were procured from the R&M Chemical. Copper (II) sulfate pentahydrate powder (CAS No: 7758-99-8), Sodium hydroxide pellets (CAS NO: 1310-73-2) and L (+) ascorbic acid powder (CAS No: 50-81-7) were purchased from Merck Millipore. All chemicals were used as supplied.

Catalyst synthesis and electrode fabrication

The synthesis of catalyst or material for the working electrode starts with the preparation of copper nanoparticle (CuNP) as a conductive material for the composite electrode. The chemical reduction process by using corn starch as capping agent was inspired from [9]. Next, the process continued with mixing all the material for composite electrode such as AC, GO (as active material), CuNP (as conductive material) and SA, XG, PEI and CMC (as polymer binder) in a one-pot synthesis with a composition ratio of 8:1:1 (active material: conductive material: binder).

Dipping layer-by layer technique was selected to fabricate the electrode; the detailed process was already been reported elsewhere [10], [11]. Four electrodes having the similar active and conductive material, but different polymer binder have been fabricated, and was labelled according to the type of binder, as: AC/GO-x where x represent the polymer binder (x = SA, XG, CMC and CMC/PEI). The CMC-PEI electrode used a mixed binder of green polymer with synthetic polymer at a ratio of 1:1 (CMC:PEI) as the ionic interaction of anionic polymer (CMC) and cationic polymer (PEI) during crosslinking process could improve the binding energy [12].

Physical and electrochemical measurement

Excess slurry of the coating material was dried at 50°C overnight to obtain the catalyst in powder form for further characterization. The powder's structure was examined using Fourier Transmission Infrared Spectroscopy (FT-IR) with a wavelength ranged from 4000 to 400 cm^{-1}. The IR spectrum were detected by using a KBr disc to study effect of adding green polymer binder towards the AC-GO solution. Next, zeta potential for the green polymer was determined by using the Dynamic light scattering (DLS) with the Standard Operating Procedure of the DLS instrument.

The electrochemical measurement was carried out by using the H-cell, three electrodes setup with dual compartments, separated by Nafion 117 membrane and single-channel potentiostat (AUTOLABPGSTAT3002, Metrohm, Malaysia). The prepared electrodes act as a working electrode, with platinum coil as counter electrode and silver chloride served as reference electrode. Before and after each run, the working electrode mass was recorded to measure the percentage mass loss. The electrochemistry activity towards CO_2 reduction was determined using cyclic voltammetry (CV) and chronoamperometry (CA) analysis. The CV analysis was performed from 0 to -1.4V *vs.* Ag/AgCl applied potential to determine the onset potential for the CO_2RR. The CV was in both CO_2 and inert argon atmosphere with gas flow rate of 20 m^3/s. Later, the optimum potential will be selected and applied in CA for 30 min. Finally, the liquid product of the reaction was sent for FTIR and DLS analysis to study the surface properties of the materials.

Result and discussions

Surface properties

The FT-IR spectra of AC/GO-raw (without binder), AC/GO-CMC/PEI, AC/GO-SA, AC/GO-XG and AC/GO-CMC are illustrated in Figure 1. All of the composite shared a distinct peaks at 1075 and 3437 cm^{-1} which correspond to C-O and C-OH bonds in the carbon skeleton network [13]. For raw composite, the presence of oxygen-containing functional groups at peaks 1436 and 1719 cm^{-1} refers to the O-H bending of and C=O stretching of carboxyl group, respectively. According to [14], the absorption bands 1635 and 1436 cm^{-1} are assigned to the pyrrole ring vibrations which can only be clearly seen on the raw composite. The addition of polymer binder onto the composite clearly deforms the peaks of the pyrrole ring which attributed to C=C that indicates a good interaction between the AC/GO and the polymer binder. Moreover, peak at 2346 cm^{-1} is attributed to the O=C=O stretching which only clearly visible for AC/GO-CMC, AC/GO-CMC/PEI and AC/GO-XG. The broad peak at 3430 to 3420 cm^{-1} correspond to the hydroxyl (C-OH stretching) group. However, AC/GO-CMC/PEI shows a broader peak around 3434 cm^{-1} as compared to others due to the N-H stretching vibration from PEI solution [15]. All the functional groups detected promote intermolecular bonding such as covalent or hydrogen bond, which could possibly improve the active material adhesion.

Furthermore, the behavior of the green polymers is studied through the zeta-potential as it reflects the interpretation of the polymers and the charged of the polymers [16]. The value of zeta potential is important when dispersing the composite in aqueous solution to ensure the stability of the colloid in solution. The zeta potential for the green polymer composite (AC/GO-SA, AC/GO-XG and AC/GO-CMC) is shown in Figure 1 (b). Based on the analysis, AC/GO-CMC has the highest charge, followed by AC/GO-SA and AC/GO-XG with zeta-potential of -84.6, -75.2 and -27.5 mV, respectively.

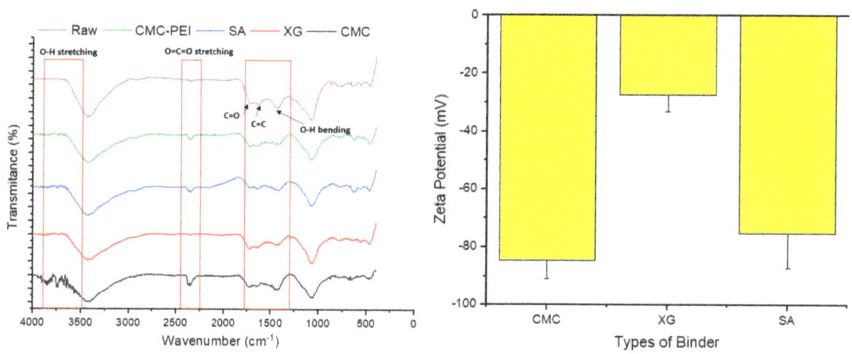

Figure 1: FTIR analysis was done to examine the chemical bonding of the composites (left), Zeta-potential of AC/GO-x (x= CMC; XG; and SA) solution (right).

According to [16], the charge of the green polymer can be increased by mixing different type of polymers could initiate mutual masking of surface charges among multiple polymers. Zeta-potential value higher than 60 mV is already considered as excellent stability except for AC/GO-XG which have zeta-potential value less than 30 mV which considered as incipient instability [17]. The higher the value of zeta potential, the more stable the dispersion is, as these particles repel each other and will prevent agglomeration during cross-linking process. Nevertheless, overall efficiency of the binder is still depending on the adhesiveness of the coating material between the current collector and composite materials.

To study the reusability of the fabricated electrodes, adhesiveness of the coating material is measured and recorded in Table 1. The mass of the fabricated electrode before and after the reduction reaction is recorded to calculate the percentage mass loss. Based on the results, AC/GO-XG has the highest mass loss with 19.89% followed by AC/GO-SA with 19.78% of mass loss. Further, AC/GO-CMC has the least mass loss with 8.28%; hence this supports the zeta potential findings shown in Figure 1 (b). To further improve the adhesiveness of the composite catalyst, mixed polymeric binder is being synthesized, as cationic and anionic interaction of CMC and PEI is expected to produce a positive impact to the adhesiveness. The concept of cross-linked binder is to enhance the characteristics of the binder, increase chemical stability and maximize acid-base interactions between material [18]. PEI solution is chosen as synthetic binder to be mixed with CMC (mixing ratio 1:1) as it has the ability to tune the carbon backbone in the AC-GO, and to promote the CO_2RR reaction [19]. Table 1 shows that AC/GO-CMC/PEI has the highest stability and recyclability electrode among others as it leached out the least (1.86%). This might due to the non-covalent dispersion interactions based on Van der Waals forces and hydrogen bonding which eliminates the hydrophobic interface of PEI with other materials [20]. Though, according to [21], determining optimum amount of polymeric binder is important as the surplus amount contributes to pore clogging and destruction of catalysts' active sites.

Table 1: Percentage mass loss of electrode in adhesiveness test.

Composites	**Mass (g)**		**Percentage mass loss**
	Before	**After**	**(%)**
AC/GO-SA	0.172	0.138	19.78
AC/GO-XG	0.191	0.153	19.89
AC/GO-CMC	0.145	0.133	8.28
AC/GO-CMC/PEI	0.161	0.158	1.86

Electrochemical measurement of CO_2 reduction

The CV analysis is performed at potential ranges of 0 to -1.4 V *vs.* Ag/AgCl at a scan rate of 50 mV/s, using 0.1M $KHCO_3$ as catholyte. The CV result is compared between CO_2 (red) and argon atmosphere (black) in order to observe the anodic shifting, and the findings are illustrated in Figure 2. Among all the electrodes, AC/GO-SA produces the highest current in the CO_2 atmosphere (-0.08 A) at potential of -1.4 V and followed by AC/GO-XG (0.045 A), AC/GO-CMC (0.04 A) and AC/GO-CMC/PEI (0.03 A). However, only the AC/GO-SA and AC/GO-CMC/PEI shows a clear anodic shifting whereas the other electrode produce the same current as in the inert atmosphere. Fundamentally, if the current produced is similar under the inert and CO_2 atmosphere, there is high possibility of dominant hydrogen (H_2) evolution, instead of the CO_2 evolution [22].

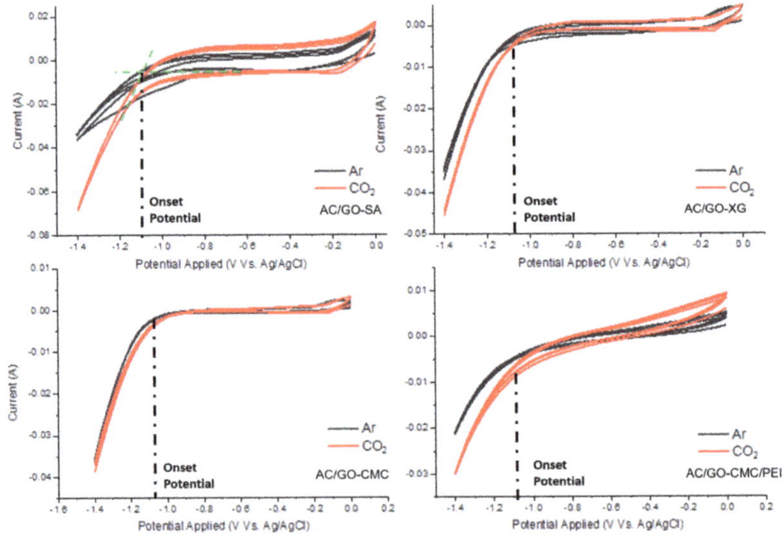

Figure 2: Cyclic voltammetry of AC/GO-x (where x = SA, XG, CMC and CMC/PEI) at the potential ranging between 0 to -1.4 V vs. Ag/AgCl

From Figure 2, the onset potential can be determined as well. This onset potential refers to the starting potential where all the thermodynamics and kinetic barriers are overcome, and an occurrence of the reduction (H_2 adsorption) process. The fast-rising current section of the peak is linearly extrapolated to the baseline and the intercept potential to obtain the onset potential as drawn in Figure 2 [23], [24]. The onset potential is selected to be applied for CA, since it is the lowest voltage for the reaction to begin, and to prevent a rapid reaction that will shift the reaction to the HER evolution. Referring to [25], the reduction of CO_2 to ethanol will go through multiple steps of kinetic barriers and if the reaction is too hasty, the reaction will promotes H_2 evolution instead of CO_2RR. Upon the CV analysis, the research proceeds with CA analysis with applied potential of -1.1V *vs.* Ag/AgCl as all the electrodes showing an equivalent onset potential value. Figure 3 depicts the result of CA analysis for these electrodes at the selected onset potential. Herein, the CA analysis aims to study the stability of the reduction reaction such as kinetics of chemical reactions [26]. At *t*=0, the current is stepped up to reduce all the adsorbed species on the working electrode. Then, the current flow is controlled by rate of diffusion/adsorption process and the reaction that occur on the surface of working electrode [27].

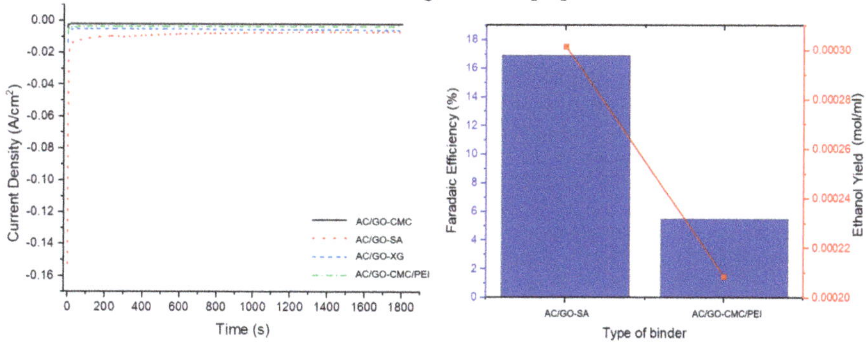

Figure 3: Chronoamperometry analysis (CA) at the potential of -1.1 V vs. Ag/AgCl for 30 mins (left). Faradaic efficiency and total ethanol yield for AC/GO-SA and AC/GO-CMC/PEI electrodes (right).

Based on Figure 3, AC/GO-SA shows the highest current density (cathodic) throughout the experiment, followed by AC/GO-XG, AC/GO-CMC/PEI and AC/GO-CMC with the corresponding value of 7.8, 5.1, 3.3 and 1.8 mA, respectively. The addition of PEI into CMC proves that mixed binder could increase the electrochemical performance as compared to the individual binders. Nevertheless, the produced current density is still lower when compared to AC/GO-SA, and AC/GO-XG. The poor performance of AC/GO-CMC/PEI is plausible due to high mixing ratio of the mixed polymeric binder as the amount of the binder affected the surface resistance of the electrodes [28]. If excess amount of binder is applied, it will slow down the ion transfer and if the opposite, the electrode cannot sustain its shape which both will led to higher current resistance.

Upon the completion of CA analysis, the liquid products are collected and sent for further analysis to determine the amount of ethanol formed. Figure 3 shows the Faradaic efficiency produced by AC/GO-SA and AC/GO-CMC/PEI, while ethanol is not detected by another two electrodes. However, due to poor understanding and lack of surface characterization, the precise explanation in term of the surface functionalities is still unknown. The formula of Faradaic efficiency is shown in Eq (2), where *n* is the amount of product detected, *Q* is the total charge passed during the electrolysis, *F* is the Faraday constant and *Z* is the number of electrons required to obtain 1 molecule of products which is 12 for ethanol [31]. According to literature study, [32] has managed to prepare zirconium oxide (ZrO_2) nanoparticles dispersed on N-doped carbon sheets by using SA and producing a promising FE value of 64% toward carbon monoxide (CO) production at low potential (-0.4 V *vs.* RHE). In 2022, [29] has successfully fabricated a single atom catalyst of nickel supported with carbon hydrogel by using CMC as binder. The working electrode work very well towards CO production at potential applied of -0.8V *vs.* RHE and resulting high FE (99.7%) with 226 mA/cm^2 current density. Besides, chitosan may also be used as polymer binder to replace the synthetic binder. Research made by [30], shows that copper nanoparticle electrode combine with chitosan as polymer could produce high CO (FE = 93.7%) at potential -1.1V *vs.* RHE with 10 mA/cm^2 current density. Referring to Figure 3, the AC/GO-SA has exerted comparable current density (7.8 mA/cm^2) with the literature and sufficiently high Faradaic efficiency (16.9%) reported in literature for green binder towards ethanol production compared to AC/GO-CMC/PEI (5.4%)

$$\text{Faradaic efficiency (FE)} = \frac{n}{Q/ZF} = \frac{nZF}{Q} \quad (2)$$

Conclusion and way forward

In conclusion, the preliminary electrochemistry performance of an AC/GO composite electrode with green polymeric binder such as XG, SA, and CMC was tested in this work. According to physical analysis, CMC is the most adhesive polymeric binder among the others, with the lowest percentage mass loss (8.28%) during the investigation. This work also introduces a mixed polymer binder that is composed of PEI and CMC polymers. The combined polymer improves the coating material's adhesiveness. The electrochemical measurement of the fabricated electrode shows that AC/GO-SA caused a considerable anodic shift in CV while retaining a high current density (-7.8 mA/cm^2) during CA. Nevertheless, ethanol is only found for AC/GO-SA and AC/GO-CMC/PEI electrodes which has Faradaic efficiency of 16.9% and 5.4%, respectively. The preliminary studies show that CMC is a promising binder for coating due to the surface charges and intermolecular forces, while SA is able to produce better reaction toward the electrochemical reduction. Overall, the reaction towards electrochemical process is affected by the functional groups and crystal facet of the active material which will be explored further in future work.

Acknowledgement

The authors would like to acknowledge funding from the Ministry of Higher Education, Malaysia, under the HICoE grant (Grant no: 015MA0-052) and Universiti Teknologi PETRONAS for supporting the research.

References

[1] K. B. Karnauskas, S. L. Miller, and A. C. Schapiro, "Fossil Fuel Combustion Is Driving Indoor CO2 Toward Levels Harmful to Human Cognition," *GeoHealth*, vol. 4, no. 5, 2020, https://doi.org/10.1029/2019GH000237

[2] K. O. Yoro and M. O. Daramola, CO2 emission sources, greenhouse gases, and the global warming effect. Elsevier Inc., 2020.

[3] J. Rissman *et al.*, "Technologies and policies to decarbonize global industry: Review and assessment of mitigation drivers through 2070," *Appl. Energy*, vol. 266, no. March, p. 114848, 2020, https://doi.org/10.1016/j.apenergy.2020.114848

[4] S. Jin, Z. Hao, K. Zhang, Z. Yan, and J. Chen, " Advances and Challenges for the Electrochemical Reduction of CO 2 to CO: From Fundamentals to Industrialization ," *Angew. Chemie*, vol. 133, no. 38, pp. 20795–20816, 2021, https://doi.org/10.1002/ange.202101818

[5] J. H. Lee *et al.*, "Understanding the Role of Functional Groups in Polymeric Binder for Electrochemical Carbon Dioxide Reduction on Gold Nanoparticles," *Adv. Funct. Mater.*, vol. 28, no. 45, pp. 1–6, 2018, https://doi.org/10.1002/adfm.201804762

[6] A. Pizzi, A. N. Papadopoulos, and F. Policardi, "Wood composites and their polymer binders," *Polymers (Basel).*, vol. 12, no. 5, 2020, https://doi.org/10.3390/POLYM12051115

[7] C. Bauer, A. Bilican, S. Braxmeier, G. Reichenauer, and A. Krueger, "Sustainable supercapacitor electrodes based on preagglomerated carbon onions and a green binder," *Carbon N. Y.*, vol. 197, no. March, pp. 555–562, 2022, https://doi.org/10.1016/j.carbon.2022.06.041

[8] V. K. Thakur, M. K. Thakur, P. Raghavan, and M. R. Kessler, "Progress in green polymer composites from lignin for multifunctional applications: A review," *ACS Sustain. Chem. Eng.*, vol. 2, no. 5, pp. 1072–1092, 2014, https://doi.org/10.1021/sc500087z

[9] A. Khan, A. Rashid, R. Younas, and R. Chong, "A chemical reduction approach to the synthesis of copper nanoparticles," *Int. Nano Lett.*, vol. 6, no. 1, pp. 21–26, 2016, https://doi.org/10.1007/s40089-015-0163-6

[10] M. Farhan, H. Othman, H. Hafiy, M. Azmi, and N. Adilla, "Materials Today : Proceedings " One-pot " synthesis of activated carbon – Graphene oxide polyethyleneimine-functionalized supported copper nanoparticles electrode for carbon dioxide reduction reaction," *Mater. Today Proc.*, no. xxxx, 2022, https://doi.org/10.1016/j.matpr.2022.03.048

[11] Z. Ge and Z. He, "An effective dipping method for coating activated carbon catalyst on the cathode electrodes of microbial fuel cells," *RSC Adv.*, vol. 5, no. 46, pp. 36933–36937, 2015, https://doi.org/10.1039/c5ra05543a

[12] H. N. Yang, J. S. Park, S. Y. Jeon, and K. H. Park, "Carboxymethylcellulose (CMC) formed nanogels with branched poly(ethyleneimine) (bPEI) for inhibition of cytotoxicity in human MSCs as a gene delivery vehicles," *Carbohydr. Polym.*, vol. 122, pp. 265–275, 2015, https://doi.org/10.1016/j.carbpol.2014.12.073

[13] G. Folaranmi, M. Bechelany, P. Sistat, M. Cretin, and F. Zaviska, "Comparative investigation of activated carbon electrode and a novel activated carbon/graphene oxide composite electrode for an enhanced capacitive deionization," *Materials (Basel).*, vol. 13, no. 22, pp. 1–14, 2020, https://doi.org/10.3390/ma13225185

[14] S. Dhibar and C. K. Das, "Silver nanoparticles decorated polypyrrole/graphene nanocomposite: A potential candidate for next-generation supercapacitor electrode material," *J. Appl. Polym. Sci.*, vol. 134, no. 16, pp. 1–14, 2017, https://doi.org/10.1002/app.44724

[15] Z. Y. Sui, Y. Cui, J. H. Zhu, and B. H. Han, "Preparation of Three-dimensional graphene oxide-polyethylenimine porous materials as dye and gas adsorbents," *ACS Appl. Mater. Interfaces*, vol. 5, no. 18, pp. 9172–9179, 2013, https://doi.org/10.1021/am402661t

[16] W. Li *et al.*, "Preparation and Characterization of Beads of Sodium Alginate/Carboxymethyl Chitosan/Cellulose Nanofiber Containing Porous Starch Embedded with Gallic Acid: An In Vitro Simulation Delivery Study," *Foods*, vol. 11, no. 10, 2022, https://doi.org/10.3390/foods11101394

[17] J. M. A. Alsharef, M. R. Taha, and T. A. Khan, "Physical dispersion of nanocarbons in composites – A review," *J. Teknol.*, vol. 79, no. 5, pp. 69–81, 2017, https://doi.org/10.11113/jt.v79.7646

[18] J. Zhang *et al.*, "An overview of the characteristics of advanced binders for high-performance Li–S batteries," *Nano Mater. Sci.*, vol. 3, no. 2, pp. 124–139, 2021, https://doi.org/10.1016/j.nanoms.2020.10.006

[19] S. Zhang *et al.*, "Polyethylenimine-enhanced electrocatalytic reduction of CO2 to formate at nitrogen-doped carbon nanomaterials," *J. Am. Chem. Soc.*, vol. 136, no. 22, pp. 7845–7848, 2014, https://doi.org/10.1021/ja5031529

[20] X. Li *et al.*, "Efficient CO2 capture by functionalized graphene oxide nanosheets as fillers to fabricate multi-permselective mixed matrix membranes," *ACS Appl. Mater. Interfaces*, vol. 7, no. 9, pp. 5528–5537, 2015, https://doi.org/10.1021/acsami.5b00106

[21] S. Das *et al.*, "Core-shell structured catalysts for thermocatalytic, photocatalytic, and electrocatalytic conversion of CO2," *Chem. Soc. Rev.*, vol. 49, no. 10, pp. 2937–3004, 2020, https://doi.org/10.1039/c9cs00713j

[22] Z. Kou *et al.*, "Fundamentals, On-Going Advances and Challenges of Electrochemical Carbon Dioxide Reduction," *Electrochem. Energy Rev.*, vol. 5, no. 1, pp. 82–111, 2022, https://doi.org/10.1007/s41918-021-00096-5

[23] A. J. R. Botz, M. Nebel, R. A. Rincón, E. Ventosa, and W. Schuhmann, "Onset potential determination at gas-evolving catalysts by means of constant-distance mode positioning of nanoelectrodes," *Electrochim. Acta*, vol. 179, pp. 38–44, 2015, https://doi.org/10.1016/j.electacta.2015.04.145

[24] R. Sanchis-Gual, A. Seijas-Da Silva, M. Coronado-Puchau, T. F. Otero, G. Abellán, and E. Coronado, "Improving the onset potential and Tafel slope determination of earth-abundant water oxidation electrocatalysts," *Electrochim. Acta*, vol. 388, p. 138613, 2021, https://doi.org/10.1016/j.electacta.2021.138613

[25] L. Li, Y. Huang, and Y. Li, "Carbonaceous materials for electrochemical CO2 reduction," *EnergyChem*, vol. 2, no. 1, p. 100024, 2020, https://doi.org/10.1016/j.enchem.2019.100024

[26] R. K. Franklin, S. M. Martin, T. D. Strong, and R. B. Brown, *Chemical and Biological Systems: Chemical Sensing Systems for Liquids*, no. May 2015. Elsevier Ltd., 2016.

[27] F. Scholz, Electroanalytical methods: Guide to experiments and applications. 2010.

[28] Y. K. Lee, "The effect of active material, conductive additives, and binder in a cathode composite electrode on battery performance," *Energies*, vol. 12, no. 4, 2019, https://doi.org/10.3390/en12040658

[29] H. Guo, D.-H. Si, H.-J. Zhu, Q.-X. Li, Y.-B. Huang, and R. Cao, "Ni single-atom sites supported on carbon aerogel for highly efficient electroreduction of carbon dioxide with industrial current densities," *eScience*, vol. 2, no. 3, pp. 295–303, 2022, https://doi.org/10.1016/j.esci.2022.03.007

[30] A. Marcos-Madrazo, C. Casado-Coterillo, J. Iniesta, and A. Irabien, "Use of Chitosan as Copper Binder in the Continuous Electrochemical Reduction of CO2 to Ethylene in Alkaline Medium," *Membranes (Basel).*, vol. 12, no. 8, p. 783, 2022, https://doi.org/10.3390/membranes12080783

[31] E. O. Eren and S. Özkar, "Recent advances in heterogeneous catalysts for the effective electroreduction of carbon dioxide to carbon monoxide," *J. Power Sources*, vol. 506, no. June, 2021, https://doi.org/10.1016/j.jpowsour.2021.230215

[32] Z. Miao, P. Hu, C. Nie, H. Xie, W. Fu, and Q. Li, "ZrO2 nanoparticles anchored on nitrogen-doped carbon nanosheets as efficient catalyst for electrochemical CO2 reduction," *J. Energy Chem.*, vol. 38, pp. 114–118, 2019, https://doi.org/10.1016/j.jechem.2019.01.010

Sustainable Processes and Clean Energy Transition - ICSuPCET2022 Materials Research Forum LLC
Materials Research Proceedings 29 (2023) 281-288 https://doi.org/10.21741/9781644902516-31

Automated determination of optimal component design for a binary solvent for absorption-based acid gas removal

Yogesh Patil[1,a], Yash Shankar Chikorde[1,b], Madiah Binti Omar[2,c*], Mohd Hakimi Bin Rosli[2,d] and ROSDIAZLI Bin Ibrahim[3,e]

[1]Department of Chemical Engineering KLE Technological University Belgavi, India

[2]Department of Chemical Engineering Universiti Teknologi Petronas Bandar Seri Iskandar, Malaysia

[3]Department of Electrical and Electronics Engineering, Universiti Teknologi PETRONAS, Perak, Malaysia

[a]yogeshpatil2001@gmail.com, [b]yashchikorde6633@gmail.com, [c]madiah.omar@utp.edu.my, [d]hakimi_19001502@utp.edu.my, [e]rosdiazli@utp.edu.my

Keywords: Natural gas, Acid gas, Machine learning, Regression

Abstract. Natural gases containing impurities, namely carbon dioxide (CO_2), heavy hydrocarbons, hydrogen sulfide (H_2S), and water vapour, need treatment for removing acidic gases (CO_2 and H_2S) to reduce corrosion and enhance the heat capacity of the gas. This gas is commercially known as "sour", and typically, sour gas is any gas that contains significant levels of hydrogen sulfide. The presence of carbon dioxide can affect natural gas quality, which can also lead to CO_2 freezing issues; hence reliable techniques for reducing CO_2 and H_2S from natural gases is necessary. New blends of amines show CO_2 and H_2S uptake capacity comparable to traditional MEA benchmark solutions. This work aimed to create different regression models using open-source software and estimate the best fit model for a given amine solvent. For this purpose, data were obtained from simulation using Aspen HYSYS V12.1 for MDEA (40-45 wt.%), MDEA +PZ (42-50wt.% + 0-2.5wt.%), DEA (21-26wt.%). Regression models for different amine solvent blends were developed and validated. The study showed that the XGB Regression model was best suited for the MDEA solution, while MDEA + PZ and DEA were best suited for multiple linear regression. The data is generated using simulation from ASPEN HYSYS and models were created in python correlating the simulation-generated values with the model results. These models showed low MSE, RMSE and high R^2 values for the tried solvents.

Introduction

Many chemical processes involve the absorption of gases along with chemical reactions (reactive absorption). The reasons for reactive absorption could be purifying flue and tail gases in refineries to meet pollution legislations, removing acidic gases from the feed in fertilizer industries plants, and purification of feed gases to polymerization units in petrochemical industries to prevent poisoning of catalysts. Aqueous amine solutions are solvents traditionally used in the treatment of refinery and synthesis gas streams [1].

Fig. 1 depicts a flow diagram of the amine scrubbing unit for removing CO_2 and H_2S [2]. The purified gas stream contacts the amine-based solvent in a counter-current manner to facilitate acid gas absorption. The concentrated solvent gets separated as acid gas (top product) and regenerated solvent (bottom product) in the distillation column. The absorber regenerates the solvent. The acid gas treatment by amine absorption is energy intensive due to high energy requirements at reboilers in the regenerating distillation columns. The regeneration step captures about 3.3 GJ / t of CO_2 under optimal conditions [3].

The rapid increasing energy prices have led to research for new solvents and solvent blends, as energy consumption in a process is directly related to the chosen solvent. [4]. The focus on air pollution abatement and the new rules and regulations worldwide mandates the implementation of newer CO_2 and H_2S reactive absorption [5]. Energy generation from fossil fuels in thermal power plants and exhaust gases are the primary supply of CO_2 and account for almost half of greenhouse gases leading to global warming [6]. Acid gases affect the gas–liquid mass switch, which reacts with alkanolamine. Hence, improving gas–liquid mass transfer enhances the acid gas absorption performance [7].

Since there are many factors involved in the energy consumption of a gas treating plant, leveraging statistical modelling to characterize and predict this energy consumption level can be an essential tool for managing and optimizing energy systems. Multivariate regression modelling is one of the most common methodologies for this purpose and can be used to improve energy efficiency (EE) [8]. Statistical modelling helps understand the most significant factors for optimizing energy systems which can be seen in previous studies [9–18].

The ongoing research focuses on finding innovative solvents that can effectively remove CO_2 and H_2S from different gas streams. This paper aims to analyze and compare regression models such as linear, multiple, Ridge and XGB regression in predicting the concentration of H_2S and CO_2.

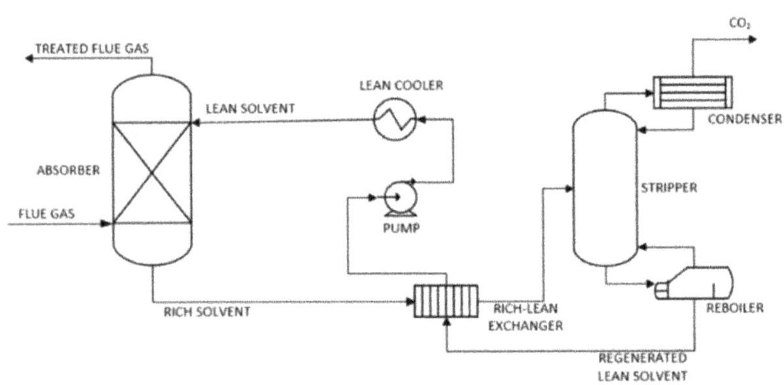

Fig. 1 Typical gas sweetening process by chemical absorption [2]

Methodology

Process flow sheeting

The simulations were conducted for conventional solvents such as MDEA (40-45 wt.%), MDEA +PZ (42-50wt.% + 0-2.5wt.%) and DEA (21-26wt.%) for model and simulation validation. Fig. 2 shows the flowsheet that is developed through Aspen HYSYS V12.1. The case explores the model of a plant that removes H_2S and CO_2 from gas coming from upstream operations. It is a critical operation in the midstream and refining industries to ensure the product gasses meet product quality and emissions standards while operating at optimal conditions.

The flow diagram depicts a packed column (absorber) for removing CO_2 and H_2S from the gas stream. A regenerator (distillation column) regenerates lean amine, and an absorber recycles the

Materials Research Proceedings 29 (2023) 281-288 https://doi.org/10.21741/9781644902516-31

same. The amine intensity and recirculation are maintained by introducing a replenishment block to estimate the make-up for water.

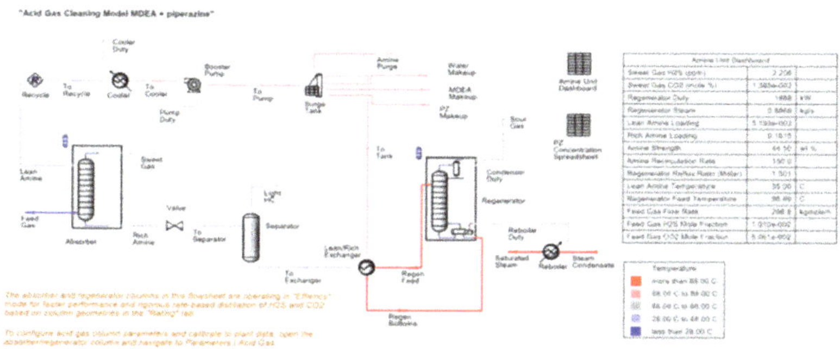

Fig. 2 Flowsheet developed from Aspen HYSYS V12.1

Development of models

Regression models. The choice of the regression analysis technique depends on the data's nature. This paper employs simple, multiple linear, Ridge and XGB regression models for predicting the concentration of H_2S and CO_2. The model of dependent variable y and the independent variables xxi (i=1,2,3...) predicts the trend.

The development of the regression models is a three-step process. First, the data will be collected from various sources and identify the dependent and independent variables in optimization. The multiple parameters are run through python for further regression analysis. Further, RMSE and R^2 quantify are compared to see how well a regression model fits a dataset. Finally, in the simulation, the obtained data is cross verified for model checking.

A design view. For developing regression analysis, the input needs to be made available in the workspace for python to recognize the input and load it into the model. The input can be made available in the base workspace of python by reading from a .xlsx file (excel file) or a .csv data file which is already present in the workspace and accessed through the local device. After the input is loaded, the regression model needs to be run from the coded program.

The input data from excel, as shown in Fig. 3, is used to import any excel data present in the user's pc/laptop and load it into the base workspace of python. The output parameters should be displayed for a specific model for the given data set. For example, if the dependent variable is the absorption efficiency of CO_2, then the R^2, MSE and RMSE values must display the output attributes and their values corresponding to when the dataset is loaded. The plots are also added to the code. When a program is run, its output parameter will be displayed as a graph. The graphical view of the model is shown in Fig. 4.

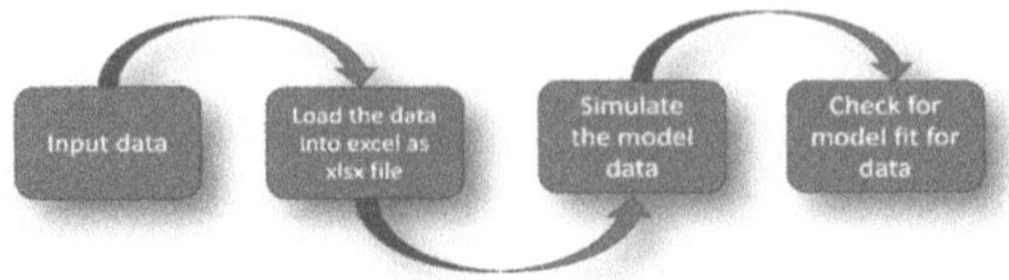

Fig. 3 Block Diagram for developing a Regression model

Fig. 4 Design view of code with markings

Optimization and code view. The coding part for the model is given below. The algorithm block diagram for the code is shown in Fig. 5. The library imported for regression analysis was sklearn, from which the regression model was imported. The program's data was split into two variables, X and Y. X, assigned the parameter to train and tested with the Y variable. The X variables include the parameter, i.e., no of plates, overflow weir height, inlet gas flow rate, pressure, inlet gas-liquid ratio, inlet gas loading, solvent concentration and removal efficiency, and the output variable Y includes the efficiency of a particular solvent.

The test and train were split into 20-80%, and the model was fitted into the X train and Y train by importing the sklearn library. First, the Y test was predicted using the X test, and corresponding values were noted. Then, the full data of X and Y were fitted, model predictions for X were made, and the corresponding values were printed.

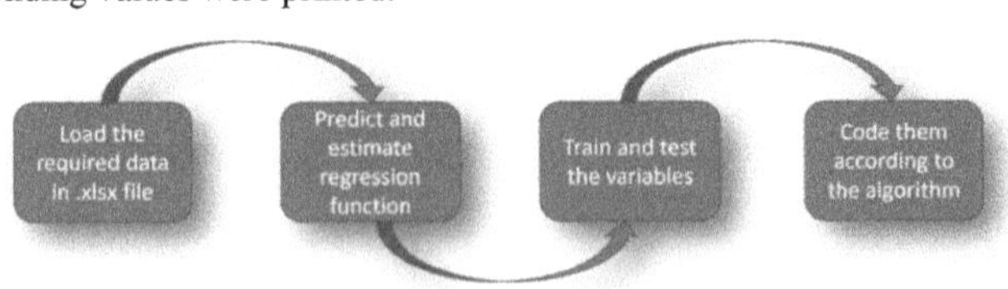

Fig. 5 Block diagram for the code of the application

Materials Research Proceedings 29 (2023) 281-288 https://doi.org/10.21741/9781644902516-31

Results and Discussion

Model simulation results

The CO_2 and H_2S absorption performance of all solvents was ultimately analyzed in terms of the regression analysis to allow practical comparison and future data utilization in process design. Table 1 shows the performance comparison between the models in predicting H_2S and CO_2 in sweet gas while Fig. 6, 7, 8 and 9 shows different graphical representations for other regression models that were coded to get a better understanding of model fit data. This comparison allowed the models to cross-verify with simulated data from ASPEN HYSYS and correlate the model-verified data with simulated generated data .

For the prediction of concentration of H_2S and CO_2 from using MDEA + PZ, it can be observed that high ranges of R^2 values were obtained from linear and multiple regression (0.95 to 0.99) indicating that higher percentage of data is fitted with these proposed models [11]. As for Ridge and XGB regression, based on the coefficient of determination, which largely ranged from 0.53 to 0.93 for Ridge, and 0.07 to 0.1 for XGB regression demonstrates that the H_2S and CO_2 prediction in sweet gas by using MDEA and PZ through Ridge and XGB regression were less accurate.

As for using DEA in predicting the concentration of H_2S and CO_2 in sweet gas, it can be observed that high ranges of R^2 values were obtained from multiple, Ridge and XGB regression, ranging between 0.88 to 0.99. In contrast, low R^2 value is observed through linear regression, ranging from -0.56 to 0.78. As for MSE and RMSE, XGB regression shows promising results for H_2S and CO_2, obtaining MSE values of 0.02 and 0.01, and RMSE value of 0.01 and 0.02. MSE and RMSE can be an indicator on how accurately the model predicts the response and lower values of both MSE and RMSE shows a better fit. Thus, XGB regression suits well for predicting H_2S and CO_2 concentration in sweet gas by using DEA.

It can be observed that high range of R^2 values is obtained from linear, multiple and ridge regression (0.91 to 0.99) in predicting the concentration of H_2S and CO_2 in sweet gas by using MDEA while XGB regression produced a slightly lower range, which is between 0.77 to 0.98 for the R^2. All models showed low range values of MSE (0.001-0.02) and RMSE (0.01-0.086) indicating that all models are able to predict well the values of the response variables.

Table 1: Performance comparison of models

			Linear	Multiple	Ridge	XGB
MDEA + PZ	CO_2	MSE	0.01	0.01	0.01	0.02
		RMSE	0.04	0.02	0.08	0.01
		R^2	0.97	0.99	0.93	0.07
	H_2S	MSE	0.17	0.10	2.51	0.01
		RMSE	0.41	0.31	1.58	0.01
		R^2	0.95	0.98	0.53	0.1
DEA	CO_2	MSE	21.43	1.38	1.43	0.01
		RMSE	4.63	1.17	1.19	0.02
		R^2	-0.56	0.94	0.94	0.99
	H_2S	MSE	5.1	5.0	5.0	0.02
		RMSE	0.45	0.4	2.0	0.01
		R^2	0.78	0.88	0.88	0.99
MDEA	CO_2	MSE	0.001	0.02	0.02	0.01
		RMSE	0.05	0.008	0.01	0.08
		R^2	0.99	0.97	0.99	0.77
	H_2S	MSE	0.03	0.007	0.01	0.01
		RMSE	0.055	0.086	0.09	0.01
		R^2	0.94	0.93	0.91	0.98

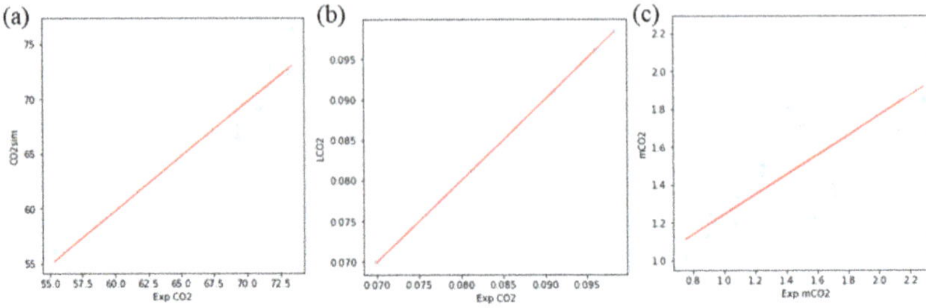

Fig. 1: Linear regression for (a) MDEA+PZ (b) DEA (c) MDEA

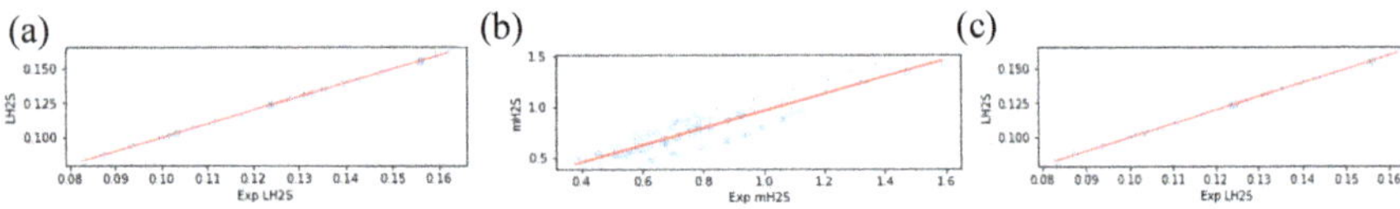

Fig. 2: multi-Linear regression for (a) MDEA+PZ (b) DEA (c) MDEA

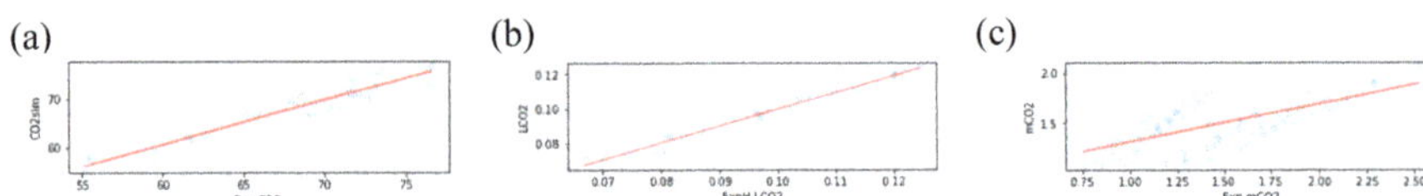

Fig. 3: Ridge regression for (a) MDEA+PZ (b) DEA (c) MDEA

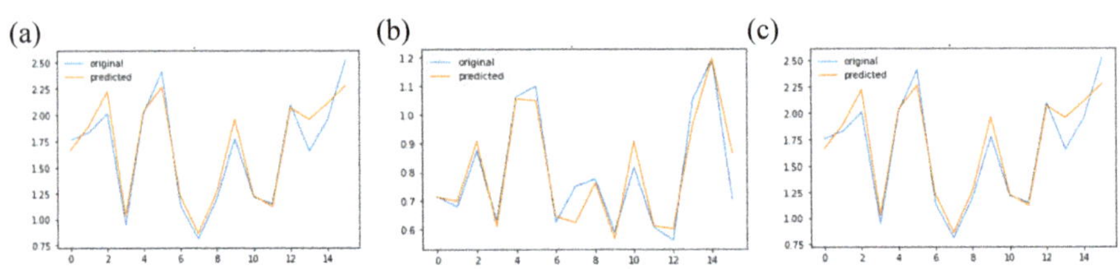

Fig. 4: XGB regression for (a) MDEA+PZ (b) DEA (c) MDEA

Conclusion

This paper aims to analyze and compare regression models such as linear, multiple, Ridge and XGB regression in predicting the concentration of H_2S and CO_2. Acid gas treatment by using solvents such as MDEA (40-45 wt.%), MDEA +PZ (42-50wt.% + 0-2.5wt.%) and DEA (21-26wt.%) were simulated by using Aspen HYSYS 12.1. This study showed that linear and multiple linear regression models were found to be the best fit for MDEA + Piperazine. DEA showed the slightest error for XGB and Ridge regression model for both CO_2 and H2S, and MDEA showed good results from linear, multiple and Ridge regression.

Acknowledgement

The authors are thankful to Universiti Teknologi PETRONAS for the support in carrying this research through grant YUTP (015LC0-362).

References

[1] T. Chakravarty, U.K. Phukan, R.H. Weilund, Reaction of acid gases with mixtures of amines, in: 1985.

[2] S.-Y. Oh, M. Binns, H. Cho, J.-K. Kim, Energy minimization of MEA-based CO2 capture process, Appl. Energy. 169 (2016) 353–362. https://doi.org/https://doi.org/10.1016/j.apenergy.2016.02.046

[3] M.R.M. Abu-Zahra, L.H.J. Schneiders, J.P.M. Niederer, P.H.M. Feron, G.F. Versteeg, CO2 capture from power plants: Part I. A parametric study of the technical performance based on monoethanolamine, Int. J. Greenh. Gas Control. 1 (2007) 37–46. https://doi.org/https://doi.org/10.1016/S1750-5836(06)00007-7

[4] D.D.D. Pinto, H. Knuutila, G. Fytianos, G. Haugen, T. Mejdell, H.F. Svendsen, CO2 post combustion capture with a phase change solvent. Pilot plant campaign, Int. J. Greenh. Gas Control. 31 (2014) 153–164. https://doi.org/https://doi.org/10.1016/j.ijggc.2014.10.007

[5] F. Bashipour, S.N. Khorasani, A. Rahimi, H2S Reactive Absorption from Off-Gas in a Spray Column: Insights from Experiments and Modeling, Chem. Eng. Technol. 38 (2015) 2137–2145. https://doi.org/https://doi.org/10.1002/ceat.201500233

[6] J. Ye, C. Jiang, H. Chen, Y. Shen, S. Zhang, L. Wang, J. Chen, Novel Biphasic Solvent with Tunable Phase Separation for CO2 Capture: Role of Water Content in Mechanism, Kinetics, and Energy Penalty, Environ. Sci. Technol. 53 (2019) 4470–4479. https://doi.org/10.1021/acs.est.9b00040

[7] J. Zhan, B. Wang, L. Zhang, B.-C. Sun, J. Fu, G. Chu, H. Zou, Simultaneous Absorption of H2S and CO2 into the MDEA + PZ Aqueous Solution in a Rotating Packed Bed, Ind. Eng. Chem. Res. 59 (2020) 8295–8303. https://doi.org/10.1021/acs.iecr.9b06437

[8] G. Guido, M. Compagnoni, L. Pellegrini, I. Rossetti, Mature versus emerging technologies for CO2 capture in power plants: Key open issues in post-combustion amine scrubbing and in chemical looping combustion, Front. Chem. Sci. Eng. 12 (2018). https://doi.org/10.1007/s11705-017-1698-z

[9] M. Koolivand Salooki, R. Abedini, H. Adib, H. Koolivand, Design of neural network for manipulating gas refinery sweetening regenerator column outputs, Sep. Purif. Technol. 82 (2011) 1–9. https://doi.org/10.1016/j.seppur.2011.07.015

[10] P. Nimmanterdwong, R. Changpun, P. Janthboon, S. Nakrak, H. Gao, Z. Liang, P. Tontiwachwuthikul, T. Sema, Applied Artificial Neural Network for Hydrogen Sulfide Solubility in Natural Gas Purification, ACS Omega. 6 (2021) 31321–31329. https://doi.org/10.1021/acsomega.1c05169

[11] N.F. Salehuddin, M.B. Omar, R. Ibrahim, K. Bingi, A Neural Network-Based Model for Predicting Saybolt Color of Petroleum Products, Sensors. 22 (2022). https://doi.org/10.3390/s22072796

[12] M.H.M Tarik, M.B. Omar, M.F. Abdullah, R. Ibrahim, Optimization of neural network hyperparameters for gas turbine modelling using Bayesian optimization, IET, 2018. https://doi.org/10.1049/cp.2018.1308

[13] Tarik, M. H. M., Omar, M., Abdullah, M. F., & Ibrahim, R. (2017, November). Modelling of dry low emission gas turbine using black-box approach. In TENCON 2017-2017 IEEE Region 10 Conference (pp. 1902-1906). IEEE.

[14] Omar, M. B., Ibrahim, R., Abdullah, M. F., & Tarik, M. H. M. (2020). Modelling of Dry-Low Emission Gas Turbine Fuel System using First Principle Data-Driven Method. Journal of Power Technologies, 100(1).

[15] Omar, M. B., Ibrahim, R., Mantri, R., Chaudhary, J., Ram Selvaraj, K., & Bingi, K. (2022). Smart Grid Stability Prediction Model Using Neural Networks to Handle Missing Inputs. Sensors, 22(12), 4342.

[16] Hakimi, M., Omar, M. B., & Ibrahim, R. (2023). Application of Neural Network in Predicting H2S from an Acid Gas Removal Unit (AGRU) with Different Compositions of Solvents. Sensors, 23(2), 1020.

[17] Rosli, N. S., Ibrahim, R., Ismail, I., & Omar, M. (2022). Modeling of high voltage induction motor cooling system using linear regression mathematical models. Plos one, 17(11), e0276142.

[18] Mahesh, A., Aadhavan, B. A., Meenaa, V. V., Omar, M. B., Ibrahim, R. B., Salehuddin, N. F., & Sujatha, R. (2022, August). Employment of ANN for Predictive Motor Maintenance and Bearing Fault Detection Using Park's Vector Analysis. In 2022 IEEE 5th International Symposium in Robotics and Manufacturing Automation (ROMA) (pp. 1-6). IEEE.

The effect of the mixing temperature on the viscosity of blood mimicking fluid for application in medical simulator under angiography imaging

REASHINI Sritharan[1,a], ANIS SUHAILA Shuib[1,b], HUAN XIN Wong[1,c*], SIEW WEI Phang[1,d], HAFISOH Ahmad[1,e], VOON LOONG Wong[2,f], AHMAD SOBRI Muda[3,g]

[1]School of Engineering, Taylor's University Malaysia, 47500, Selangor, Malaysia

[2]School of Engineering and Physical Sciences, Heriot-Watt University, 62200, Wilayah, Persekutuan Putrajaya, Malaysia

[3]Radiology Department, Hospital Pengajar University Putra Malaysia, Faculty of Medicine and Health Sciences, 43400 Serdang, Selangor Malaysia

[a]reashinisritharan@sd.taylors.edu.my, [b]anissuhaila.shuib@taylors.edu.my, [c]huanxin.wong@sd.taylors.edu.my, [d]SiewWei.Phang@taylors.edu.my, [e]Hafisoh.Ahmad@taylors.edu.my, [f]V.Wong@hw.ac.uk, [g]asobri@upm.edu.my

Keywords: Shear Thinning, Mixing Temperature, Viscosity, Polysaccharides

Abstract. Bench-top medical simulator is used for rehearsing medical treatment procedure to treat abnormal blood flow in blood vessels. The blood vessels are 3D-printed, and flow is simulated using blood mimicking fluid. It is crucial that the fluid can replicate the tactile feedback of actual blood during simulation which contributed by blood viscosity. There are limited studies regarding the effect of mixing temperature during blood mimicking fluid preparation towards the viscosity of blood mimicking fluid. This experiment aims to investigate the effect of mixing temperature variation of 25°C, 37°C and 88°C towards blood mimicking fluid that made up of xanthan gum, corn starch, water, and glycerol. The viscosities of samples are measured against a range of shear rates, from 0.1 s^{-1} to 1000 s^{-1} using a rheometer. Generally, shear thinning viscosity was observed and the viscosity data was fitted to power law viscosity model. The viscosity decreased as the mixing temperature increased. Power law fluid consistency index, *K* and non-Newtonian fluid behaviour index, *n* showed deviation of more than 8% from the actual blood.

Introduction

Medical surgical simulation is known to be the future for healthcare education. Surgical simulator includes human cadavers, bench top models and virtual reality systems to recreate real life surgical situations for both novice and expert practitioners to sharpen their skills [1]. Among others, bench top surgical simulator provides an economical medical simulation with 3D printing technology and supported by advancement in biomimetic material [23]. Blood mimicking fluids also known as blood analogue fluids used to mimic the physical properties of actual blood for the surgical simulation application to improve the tactile feedback.

The challenge of blood mimicking fluid formulation is to match the actual blood rheological and mechanical properties. Blood is defined as a non-Newtonian and shear thinning fluid. Shear thinning occurs as the apparent viscosity of fluid decreases with increasing shear rate. The blood viscosity model can be described as a power law behavior as shown in Equation 1 [2]

$$\mu = K\dot{\gamma}^{n-1} \qquad (1)$$

where, μ is the apparent viscosity, γ represents the shear rate, K signifies the fluid consistency index whereas n is the non-Newtonian fluid behaviour index. The value of coefficient K and n for actual blood are 18.215 mPa.s and 0.727 respectively [2]. The non- Newtonian characteristics are affected by the size and shape of flow conduits. Cardiovascular circulation system comprises various subsystems such as arteries and veins, blood capillaries and porous tissues where shear thinning properties of blood is significant [3].

Polysaccharides such as xanthan gum and starch has been used in formulating blood mimicking fluid to provide non-Newtonian characteristics. There are limited studies regarding the effect of mixing temperature during the blood mimicking fluid preparation towards the viscosity properties through the addition of different polysaccharides in the water-glycerol base fluid [4,6]. In this work, the blood mimicking fluid formulation is focused to investigate the effect of mixing temperature during the addition of polysaccharides which are xanthan gum and corn starch into water-glycerol base fluid. Addition of iodine-containing contrast media or contrast agent to the formulation is necessary since it enable the visualization of blood vessel under angiography imaging in the medical simulator.

Methodology

The blood mimicking fluid sample formulation contains 99.5% purity of glycerol (Sigma Aldrich, US), distilled water, xanthan gum from Xanthomonas campestris (Sigma Aldrich, US), corn starch (Sigma Aldrich, US) and contrast agent of Iohexol 350 (Omnipaque) from GE Healthcare Shanghai, China. The blood mimicking fluid samples comprises three different temperatures with three different ratios respectively with glycerol-water base, xanthan gum (XG), corn starch (CS) and contrast agent (CA). Blood mimicking fluid composition is tabulated in Table 1.

Table 1: Blood mimicking fluid compositions

Material	**Composition (%w/w)**
Distilled water	59.93
Glycerol	40.00
Xanthan Gum (XG)	0.01
Corn Starch (CS)	0.01
Contrast Agent (CA)	0.05

The samples were prepared by measuring the glycerol, water and CA using a measuring cylinder whereas the XG and CS were measured using an analytical balance. The experimental set up is shown in Figure 1.

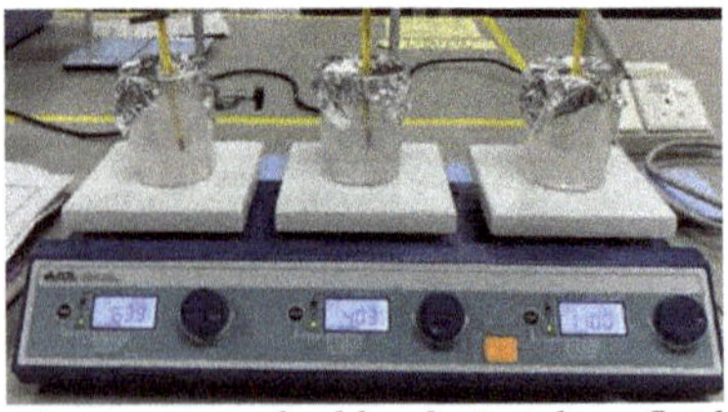

Fig. 1. Experiment set-up for blood mimicking fluid formulation

The fluid was prepared by mixing all of the materials using the hot plate with magnetic stirrer until is homogenously mixed according to the published method [15]. The XG, CS and glycerol were stirred on the hot plate with a temperature of 50 °C until there were no lumps present. Once the formation of lumps diminished, distilled water was gradually added to the side of the beaker.

The speed of the magnetic stirrer was gradually increased to 600 rpm and the temperature of the plate was set. Three variation of mixing temperature was investigated at 25°C, 37°C and 88°C. 25°C is referring to room temperature and 37°C is according to human average body temperature. The temperatures of 88°C is selected since the aqueous xanthan gum solution viscosity decreased when the temperature above 80°C [5]. This implied the weakening of intermolecular interactions that may assist good mixing.

The rapid freeze quench technique is applied to ensure to property at high mixing temperature is retained [7]. Once the samples were rapidly mixed after 30 minutes, it is transferred into a labelled storage bottle and stored in the freezer for quenching at -15 °C. The samples were quenched for 24 hours before taken out to ambient room surrounding conditions. CA was added to the samples, it was stirred at 600 rpm for 30 minutes. Once samples were mixed thoroughly, it was stored in the chiller at 5°C to avoid bacterial growth.

The HAAKE™ MARS™ Modular Advanced Rheometer was utilised to measure the viscosity of each blood mimicking fluid sample. The rheometer functions are based on the torque applied at a certain rotational speed or shear rate. The rheometer studies the relationship between the stress and strain of samples in order to analyse the deformation properties. The blood mimicking fluid viscosity was measured at different shear rates ranging from 0-1000 s^{-1}. During the viscosity measurement, the sample of 10 ml is required. The samples must be homogenized and in room temperature before conducting the test. Every sample testing was conducted at constant, controlled temperature of 25°C to match the ambient conditions during surgical simulation environment. The rheometer was fitted with the coaxial cylinder which are Becher and Cup CCB26 DIN with Rotor size CC26 Ti. Once the geometries were set up, the rheometer was set to measure its zero-level controlled by the RheoWin software prior the blood mimicking fluid sample measurements. The software and its data manager comprise an integrated server for the data plotting of viscosity against shear rate. The time taken for each blood mimicking fluid sample measurement is 10 minutes. The results were tabulated and graphs of viscosity against shear rate were plotted for the data comparison with actual human blood.

The non-Newtonian fluid behaviour is explained through power-law model in Equation 1 and can be described as a logarithmic form as shown in Equation 2.

$$log\,\mu = (n-1)\,log\,\dot{\gamma} + \log\,K \quad (2)$$

Equation 2 is described as a linear fitting graph. The x-axis and y-axis are $log\,\gamma$ and $\log\mu$ respectively. The gradient of graph is (n-1), hence the fluid behaviour index can be taken from the gradient. The fluid consistency index K can be obtained from the y-intercept of linear graph. The results of blood mimicking fluid samples were compared with actual human blood data reported by Jung et al [2]. The average actual blood data was derived from 297 healthy individuals consisting of 123 men and 174 women. The viscosity values from this investigation were taken as a reference basis to compare the measured viscosity of blood mimicking fluid with actual human blood.

Result and Discussion

Figure 2 depicted the relationship between viscosity and shear rate with variation of mixing temperature of 25°C, 37°C and 88°C respectively. A total of one hundred readings were taken to develop the relationship between viscosity and shear rate.

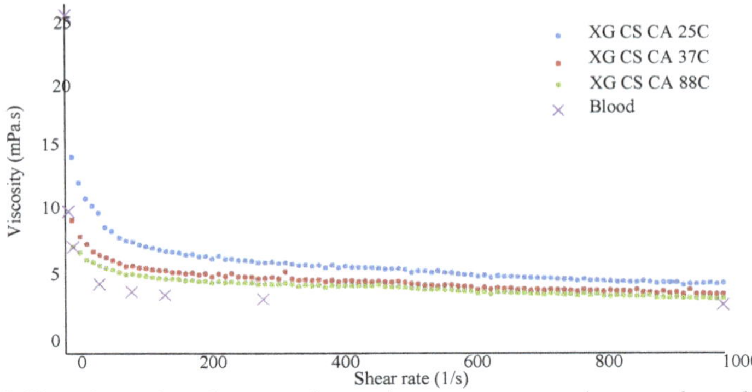

Fig. 2. Viscosity against shear rate for xanthan gum-corn starch-water glycerol base and contrast agent with variation of mixing temperature.

It was observed that the viscosity profile decreases as the mixing temperature increases. Higher temperature may have altered the intermolecular bonding between xanthan gum, corn starch, contrast agent, water and glycerol. The bonds may have weakened at higher temperature that leads to reduce in viscosity. There will be conformational changes in XG chains at mixing temperatures above 36°C [19]. XG has a chemical structure of trisaccharide side chain which is closely aligned with the polymer backbone leading a single, double, or triple helix stiff chain [20]. At higher temperatures, the XG secondary structure will convert from an ordered helical structure to flexible and disordered, where the side chains were extended away from backbone. When XG is mixed in distilled water at 25°C, weak structure is formed since the molecules are more associated to hydrogen bonding [19]. According to Rochefort et al. the XG will convert to coil-like structure due to the dissociation of molecules and major changes in rheological properties [20].

The viscosity of XG based samples are dependent on the dissolution temperature. When the mixing temperature is above 60°C, the viscosity of XG decreases with increasing temperature. This is due to the order-disorder transition of the XG molecule. The mixing temperature of 88°C gradient is almost closer to actual blood because the helix-coil transition of XG will only occur for temperatures exceeding 70°C.

Typically, suspensions of CS and water exhibits shear thickening behaviour [21]. The shear thickening behaviour occurs when particles hydro clustering occurs which are due to hydrodynamic lubrication. It was stated that CS suspensions will show two different viscosity regimes when increasing shear rate, where, at low shear rates ($0.01\text{-}1s^{-1}$) exhibits shear thinning behaviour whereas at higher shear rates ($>1s^{-1}$) shear thickening can be observed [22]. Based on the CS samples prepared, the concentration of CS added was minimal. Shear thinning behaviour may be observed due to the lower starch concentrations ranging 10-30 wt% because the starch suspensions exhibits weak thickening behaviour and increasing in critical thickening shear stress was observed [22]. Due to low concentrations of starch present, shear thickening properties are less visible.

The combination of XG and CS polysaccharides will create a synergism on their respective viscosity and the overall viscosity of the substance. Two polysaccharides will create a stronger effect on viscosity compared to individual contribution of each polysaccharide [18]. This is because XG consists of a distinctive rigid, rod like structure which is highly responsive to shear than a random coil structure of CS. This is due to the close structural alignment of the trisaccharide side chains with (1-4) linked cellulose backbone. Also, the XG is capable to enwrap the starch molecules which slightly increases the viscosity [18]. The XG and CS combined sample with the mixing temperature of 88°C displays a gradient much closer to actual blood data. The significant high mixing temperature was able to break the rigid structure of XG molecules which were bind to the hydrogen bonding to interact with dissolved amylose molecules present in CS.

The data from Figure 2 were fitted into a linear logarithmic plot according to Equation 2 and shown in Figure 3. The graphs were plotted based on the power law model. The data of actual blood is plotted together for comparison [2].

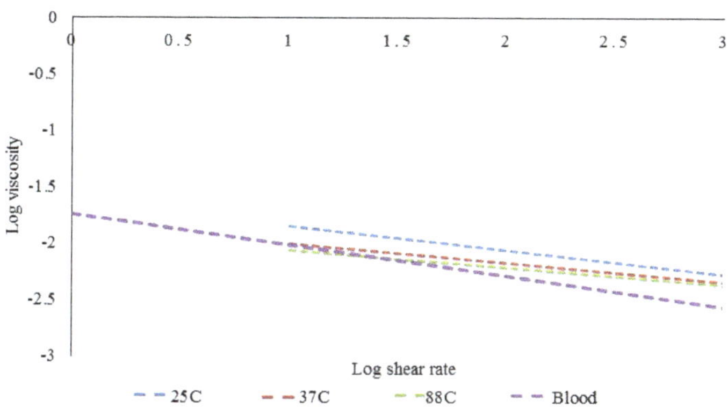

Fig. 3. Logarithmic plot of Figure 2 according to Equation 2.

The *K* and *n* coefficients of each plot were tabulated in Table 2. It is evident that the graph displays a non-Newtonian pseudoplastic behaviour.

Table 2. Linear fittings parameter from Equation 2

Samples	Log *K*	*K*	n-1	*n*	R^2
XG CS CA 25°C	-1.6317	0.023	-0.2123	0.7877	0.9907
XG CS CA 37°C	-1.8323	0.015	-0.1689	0.8311	0.9731
XG CS CA 88°C	-1.9167	0.012	-0.1506	0.8944	0.9601
Blood [4]	-1.7396	0.018	-0.2730	0.7270	0.9096

The data in Table 2 is calculated by using Equation 2. Based on Table 2, it was noticed that the fluid consistency index, *K* decreases with increasing mixing temperature whereas the non-Newtonian fluid behaviour index, *n* increases with increasing mixing temperature. The percentage deviation among the coefficients of blood mimicking fluid samples and actual blood were calculated and summarized as Table 3.

Table 3. Percentage deviation from actual blood data and each blood mimicking fluid sample

Blood mimicking fluid samples	*K*	% deviation (*K*)	*n*	% deviation
XG CS CA 25°C	0.023	27.78	0.7877	8.35
XG CS CA 37°C	0.015	-16.67	0.8311	14.32
XG CS CA 88°C	0.012	-33.33	0.8494	16.84

As shown in Table 3, sample with mixing temperature 25°C had the lowest non- Newtonian index of 8.35% higher than actual blood. However, the consistency index *K* was significantly large, with a differences of 27.78%. The difference mixing temperature used in this study resulted in different *K* and *n* values. As the temperature increased from 25°C to 88°C, the *K* value decreases, and the n value increases. Blood mimicking fluid Samples with mixing temperature 88 has the deviation of *K* with 33.33% lower than actual blood data whereas the deviation of *n* coefficient was 16.84% lower than the actual blood.

Table 4 compares the *K* and *n* values obtained in this investigation with other published results. In general, the shear thinning viscosity is sensitive towards the material and composition added where it will affect the *K* and *n* values. Present study shows that blood mimicking fluid of the same composition of XG and CS mixed at different temperature causes the changes to shear thinning characteristics demonstrated by *K* and *n* value. Increasing the mixing temperature resulted in higher *n* values but lowering the *K* values. Lee et. al [14] conducted the same experiment at a constant mixing temperature of 25°C but without the addition of CS. As the concentration of XG reduced, *K* values increased while *n* values decreased. Elblbesy and Hereba measured viscosity of blood mimicking fluid without CS and CA [17]. It was discovered that decreasing the concentration of glycerol decreases the *K* values while increasing the *n* values. Increasing the concentration of xanthan gum increases the *K* values while decreasing the *n* values.

Table 4 K and n values of the fluids in the current study and published data

Fluid composition				***Mixing temperature***	***K* (Pa.s^n)**	***n***	***Reference***
Gly (v/v%)	**XG (w/v%)**	**CS (w/v%)**	**CA (v/v%)**				
40	0.01	0.01	0.05	88°C	0.012	0.849	Present study
40	0.01	0.01	0.05	37°C	0.015	0.831	Present study
40	0.01	0.01	0.05	25°C	0.023	0.788	Present study
40	0.08	Not studied	0.05	25°C	0.034	0.598	[14]
40	0.07	Not studied	0.05	25°C	0.028	0.621	[14]
40	0.06	Not studied	0.05	25°C	0.025	0.622	[14]
40	0.02	Not studied	Not studied	25°C.	0.016	0.816	[17]
35	0.02	Not studied	Not studied	25°C.	0.013	0.820	[17]
35	0.05	Not studied	Not studied	25°C.	0.046	0.681	[17]

Conclusion

The increasing of mixing temperature was found to affect the viscosity of the fluid. As the mixing temperature increases the measured viscosity decreases. At an increasing temperature, degradation of intermolecular forces and thermal stability explained the decrease in viscosity. Based on the findings from Power Law viscosity analysis, the *K* and *n* values measured in the experiment having deviation from actual blood property for more than 8%. Further improvement is recommended to focus on the viscosity measurement at shear less than $10s^{-1}$. Other non-linear blood viscosity models are recommended to be considered to account for more rheological factors.

Acknowledgements
The authors would like to acknowledge the financial support provided Taylor's University under Taylor's Research Excellence Scholarship awarded to Wong Huan Xin. The research is mainly supported by Ministry of Higher Education Malaysia under Fundamental Research Grant Scheme (FRGS/1/2020/TK0/TAYLOR/02/5).

References

[1] I. Badash, K. Burtt, C. A. Solorzano, and J. N. Carey, "Innovations in surgery simulation: a review of past, current and future techniques," *Ann. Transl. Med.*, vol. 4, no. 23, pp. 2–2, 2016, https://doi.org/10.21037/ATM.2016.12.24

[2] J. M. Jung et al., "Reference intervals for whole blood viscosity using the analytical performance-evaluated scanning capillary tube viscometer," *Clin. Biochem.*, vol. 47, no. 6, pp. 489–493, 2014, https://doi.org/10.1016/j.clinbiochem.2014.01.021

[3] A. M. R. G.P. Galidi R. Rannacher and S. Turek, *Hemodynamical Flows: Modeling, Analysis and Simulation*, vol. 37. 2008.

[4] L. Webb, "Mimicking Blood Rheology for More Accurate Modeling in Benchtop Research," *Pegasus Rev. UCF Undergrad. Res. J.*, vol. 12, 2020.

[5] E. M. Nsengiyumva and P. Alexandridis, "Xanthan gum in aqueous solutions: Fundamentals and applications," International Journal of Biological Macromolecules, vol. 216, pp. 583–604, Sep. 2022, https://doi.org/10.1016/j.ijbiomac.2022.06.189

[6] A. D. Anastasiou, A. S. Spyrogianni, K. C. Koskinas, G. D. Giannoglou, and S. V. Paras, "Experimental investigation of the flow of a blood analogue fluid in a replica of a bifurcated small artery," *Med. Eng. Phys.*, vol. 34, no. 2, pp. 211–218, 2012, https://doi.org/10.1016/j.medengphy.2011.07.012

[7] F. Nami, P. Gast, and E. J. J. Groenen, "Rapid Freeze-Quench EPR Spectroscopy: Improved Collection of Frozen Particles," *Appl. Magn. Reson.*, vol. 47, no. 6, pp. 643–653, 2016, https://doi.org/10.1007/s00723-016-0783-7

[8] C. E. Brunchi, M. Bercea, S. Morariu, and M. Dascalu, "Some properties of xanthan gum in aqueous solutions: effect of temperature and pH," *J. Polym. Res.*, vol. 23, no. 7, Jul. 2016, https://doi.org/10.1007/S10965-016-1015-4

[9] H. A. Khouryieh, T. J. Herald, F. Aramouni, and S. Alavi, "Influence of mixing temperature on xanthan conformation and interaction of xanthan-guar gum in dilute aqueous solutions," *Food Res. Int.*, vol. 39, no. 9, pp. 964–973, 2006, https://doi.org/10.1016/j.foodres.2006.06.001

[10] N. A. Shaari, R. Sulaiman, and C. Y. Cheok, "Rheological properties of native and modified corn starches in the presence of hydrocolloids," *Int. Food Res. J.*, vol. 24, no. 5, pp. 2082–2089, 2017.

[11] N. C. Crawford, L. B. Popp, K. E. Johns, L. M. Caire, B. N. Peterson, and M. W. Liberatore, "Shear thickening of corn starch suspensions: Does concentration matter?," *J. Colloid Interface Sci.*, vol. 396, pp. 83–89, Apr. 2013, https://doi.org/10.1016/J.JCIS.2013.01.024

[12] N. H. Ibrahim and C. H. Weng, “Physical properties of dressing-type oil in water emulsions as affected by Okra gum-xanthan gum-corn starch interactions,” *Malaysian Appl. Biol.*, vol. 46, no. 1, pp. 131–137, 2017.

[13] P. Aspelin, “Effect of ionic and non-ionic contrast media on whole blood viscosity, plasma viscosity and hematocrit in vitro,” *Acta radiol.*, vol. 19, no. 6, pp. 977–989, 1978, doi: 10.1177/028418517801900613

[14] S. H. Lee, A. S. Shuib, D. G. Hassell, M. I. Ahmad Sabri, and A. S. Muda, “Impact of the additional of radio-contrast agent on the behavior of used in surgical simulation,” *4th International Conference on Chemical Innovation*, pp. 65–71, 2019.

[15] N. Perrira, A. S. Shuib, S. W. Phang, and A. S. Muda, “Experimental Investigation of Viscosity for Application in 3D-Printed Medical Simulator,” *Journal of Physics: Conference Series*, vol. 2222, no. 1, p. 012016, May 2022, https://doi.org/10.1088/1742-6596/2222/1/012016

[16] C. Completo, V. Geraldes, and V. Semiao, “Rheological and dynamical characterization of blood analogue flows in a slit,” *International Journal of Heat and Fluid Flow*, vol. 46, pp. 17–28, Apr. 2014, https://doi.org/10.1016/j.ijheatfluidflow.2013.12.008

[17] M. A. Elblbesy and A. T. Hereba, “Computation of the coefficients of the power law model for whole blood and their correlation with blood parameters,” *Applied Physics Research*, vol. 8, no. 2, p. 1, 2016.

[18] N. H. Ibrahim and C. H. Weng, “Physical properties of dressing-type oil in water emulsions as affected by Okra gum-xanthan gum-corn starch interactions,” *Malaysian Appl. Biol.*, vol. 46, no. 1, pp. 131–137, 2017.

[19] C.E. Brunchi, M. Bercea, S. Morariu, and M. Dascalu, “Some properties of xanthan gum in aqueous solutions: effect of temperature and pH,” *Journal of Polymer Research*, vol. 23, no. 7, Jun. 2016, https://doi.org/10.1007/s10965-016-1015-4

[20] H. A. Khouryieh, T. J. Herald, F. Aramouni, and S. Alavi, “Influence of mixing temperature on xanthan conformation and interaction of xanthan-guar gum in dilute aqueous solutions,” *Food Res. Int.*, vol. 39, no. 9, pp. 964–973, 2006, https://doi.org/10.1016/j.foodres.2006.06.001.

[21] N. A. Shaari, R. Sulaiman, and C. Y. Cheok, “Rheological properties of native and modified corn starches in the presence of hydrocolloids,” *Int. Food Res. J.*, vol. 24, no. 5, pp. 2082–2089, 2017.

[22] N. C. Crawford, L. B. Popp, K. E. Johns, L. M. Caire, B. N. Peterson, and M. W. Liberatore, “Shear thickening of corn starch suspensions: Does concentration matter?,” *J. Colloid Interface Sci.*, vol. 396, pp. 83–89, Apr. 2013, https://doi.org/10.1016/J.JCIS.2013.01.024

[23] B. C. dos Santos, P. Y. Noritomi, J. V. L. da Silva, I. A. Maia, and B. M. Manzini, “Biological multiscale computational modeling: A promising tool for 3D bioprinting and tissue engineering,” Bioprinting, vol. 28, p. e00234, Dec. 2022, https://doi.org/10.1016/j.bprint.2022.e00234

Computational study of CO_2 solubility in amino acid-based ionic liquids using COSMO-RS

SULAFA Abdalmageed Saadaldeen Mohammed[1,a], WAN ZAIREEN NISA Yahya[1,2,b*], MOHAMAD AZMI Bustam[1,2,c], AMIN Abbasi[1,d] and MD GOLAM Kibria[3,e]

[1]Chemical Engineering Department, Universiti Teknologi PETRONAS, 32610 Seri Iskandar, Perak, Malaysia

[2]Centre of Research in Ionic Liquids, Universiti Teknologi PETRONAS, 32610 Seri Iskandar, Perak, Malaysia

[3]Chemical and Petroleum Engineering, University of Calgary, Alberta, Canada, 2500 University Drive, NW, Calgary, Alberta, T2N 1N4, Canada

[a]sulafa_19001261@utp.edu.my, [b]zaireen.yahya@utp.edu.my, [c]azmibustam@utp.edu.my, [d]amin_18000407@utp.edu.my, [e]md.kibria@ucalgary.ca

Keywords: Ionic Liquid, Amino Acid, CO_2 Solubility, COSMO-RS

Abstract. The carbon capture, use, and sequestration (CCUS) techniques are proven to be efficient at lowering the atmospheric concentration of carbon dioxide. Notwithstanding the advances in this area, there are still significant restrictions in carbon dioxide (CO_2) capture techniques in industry such as high capital costs, solvent evaporation losses, and low absorption and desorption rates. Ionic liquids (ILs) have received much interest as green solvent due to the benefits of their distinctive properties such as low vapor pressure and their capacity to capture CO_2 making them a suitable replacement for present solvents, such as amines. Amino acid based ILs having close similarity with the alkanolamines may potentially have high affinity for CO_2 absorption. Nevertheless, available database on these ILs is still limited and only focus on the common types of amino acids. Therefore, this paper aims to predict the CO_2 absorption of different amino acid-based ionic liquids as cation/anion using quantum chemical calculation tools namely Conductor like Screening Model for Real Solvents (COSMO-RS) and TURBOMOLE. We evaluated 84 different ILs of different cations and anions based on their CO_2 capacity, activity coefficient at infinite dilution (γ^∞), and Henry's constant (H). The results showed that amino acid as anions significantly enhanced the CO_2 solubility compared to amino acid as cations. However, glycinium tetrafluoroborate [Gly^+][BF_4] showed high affinity for CO_2 absorption compared to other amino acid-cations based with activity coefficient at infinite dilution (γ^∞) = 0.117 and (H) = 8.07. We showed that the selection of anions/cations can significantly change the CO_2 capacity in ILs.

Introduction

The concentration of CO_2 in the atmosphere increases over time because of carbon emissions from the burning of fossil fuels [1]. One successful strategy to reduce CO_2 emissions in the atmosphere is by carbon dioxide capture, utilization, and storage (CCUS) [2]. Many industries currently use the amine scrubbing technique as their primary CO_2 capture technology [3, 4]. This process however, have numerous weaknesses such as equipment corrosion, high energy need for regeneration, solvent loss, and oxidative degradation [5]. Ionic liquids (ILs) are promising for CO_2 capture because of their strong affinity for acid gases [6], high physical and chemical stability [7], low vapor pressure [8], and strong polarity [9]. Moreover, the structures of ILs (cation and/or anion) can be finely tuned to enhance their physical and chemical characteristics.

CO_2 absorption in non-functionalized ILs mainly takes place by physical absorption through weak Lewis acid–base interactions [10]. Subsequently, the amine-functionalization of the cation and/or anion of the ILs have been investigated since amine solvents showed strong reactivity with CO_2 molecules [11, 12]. Similarly, amino acid can also be utilized as cation or anion to enhance the efficiency of CO_2 absorption in ILs [13]. Amino acids contain a primary amine (–NH_2) and carboxyl group (–COOH), thus, the reactivity of CO_2 with amino acids is expected to be similar as that of alkanolamines [14, 15]. However, there is still a conflict on the actual mechanism of the CO_2 reaction with amino acids [16].

Shahrom et al. [13] studied eight ionic liquids that consist of amino acids-based anions namely lysine [Lys], proline [Pro], histidine [His], serine [Ser], taurine [Tau], alanine [Ala], arginine [Arg], and glycine [Gly]) with vinylbenzyltrimethylammonium [VBTMA] as cation. They reported that [Arg] and [Lys] showed the highest CO_2 sorption values followed by [Hist], [Ser], [Tau], [Pro], [Gly], and [Ala]. Another study by Sistla and Khanna [10] evaluated different amino acid based ILs and observed that aminate-anion ILs perform better than amino-cation and non-functionalized ILs. In their study, 1-butyl-3-methylimidazolium [BMIM] was used as the cation and found that the order of CO_2 capacity in the ILs starting from the highest is [Arg]>[Lys]>[His]> Methionine [Met]> Leucine [Leu] >[Gly]> Valine [Val]>[Ala]>[Pro]. They explained it by the availability of more accessible amine (N) groups. Noorani and Mehrdad [1] also studied the CO_2 solubility in 1-butyl-3-methylimidazolium glycine [BMIM][Gly], [BMIM][Ala], and [BMIM][Val], where they reported that the order of CO_2 capacity starting from the highest is [BMIM][Gly]>[BMIM][Ala]>[BMIM][Val].

The solubility of CO_2 in amino acid-based ILs are therefore a subject of interest for further investigation. However, the current database on these ILs is still limited, mainly experimental based, and only focused on the common types of amino acids. This aim of this study is to predict the CO_2 solubility of different amino acid-based ionic liquids as cation/anion and to provide better understanding of their respective roles using Conductor like Screening Model for Real Solvents (COSMO-RS). COSMO-RS is a quantum chemical calculation tool to estimate the chemical potential differences of molecules inside liquids designed by Klamt [17, 18]. The software has been used in various application for prediction of thermodynamic properties and behavior of ionic liquids [19]. The screening of ILs via COSMO-RS will aid in the design for the desired ILs for CO_2 capture and avoid uncertain trails.

Methods

Initially, to generate COSMO files, the molecular structures of ILs (cations/anions) were built based on density functional theory using Turbomole software (TmoleX). For TmoleX settings, BP functional with triple-zeta valence polarized set (BP_TZVP_C30_1301) was used as parameterization for COSMO-RS calculations. The sigma (σ) profile and sigma (σ) surface were generated. The σ-profile of any molecule is calculated based on the weighted sum of the profiles of all segments [17] as per Equation (1);

$$p_i(\sigma) = \sum_i^N x_i \times p_i \tag{1}$$

where x_i is the mole fraction of component (i) in the mixture and p_i is the σ profile of any molecule X. The Henry constant (H) is a significant indication of gas solubility, which is directly proportional to the activity coefficient. A low value of H is an indication of good CO_2 solubility. The Henry's constant values for selected ILs were calculated using Equation (2);

$$H_S^i = \gamma_S^{S,\infty} \times P_S^i \tag{2}$$

where, $\gamma_S^{S,\infty}$ is the activity coefficient at infinite dilution and P_S^i is the vapor pressure.

To evaluate the effect of anions and cations functionalization, the activity coefficient at infinite dilution (γ^∞), CO_2 capacity (mole CO_2/mole of ILs), and Henry's constant (H) of 84 ILs were

Materials Research Proceedings 29 (2023) 297-306 https://doi.org/10.21741/9781644902516-33

estimated using COSMO-RS for the solubility prediction. For this study, the ILs structures of the cations are: 1-butyl-3-methylimidazolium (BMIM), tetrabutylammonium (N4444), dibutylpyrrolidinium (BBPyrro), glycinium (Gly^+), L-lysinium(+1) (Lys^+), and (MGlyB) as depicted in Figure 1. The ILs structures of the anions are: tetrafluoroborate (BF_4), bromide (Br), bis(trifluoromethane)sulfonimide (TFSI), nitrate (NO_3), Gly, Ala, Val, Leu, isoleucine (Ile), phenylalanine (Phe), Pro, serine (Ser), threonine (Thr), tyrosine (Tyr), cysteine (Cys), Met, Lys, Arg, His, tryptophan (Trp), aspartic acid (Asp), glutamic acid (Glu), asparagine (Asn), and glutamine (Gln) as shown in Figure 2.

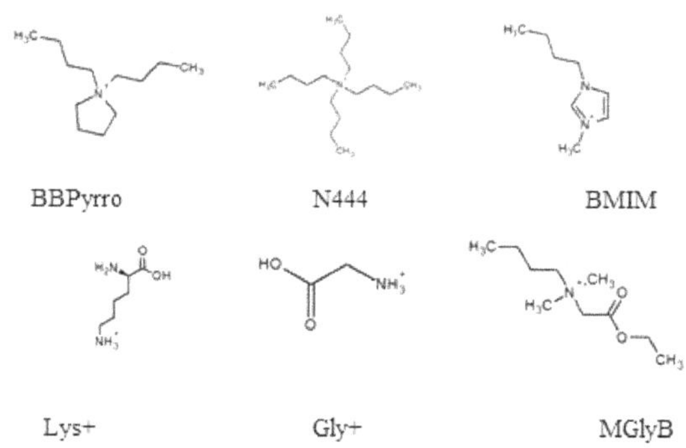

Fig. 1. Different cations structures of the ILs used in this study.

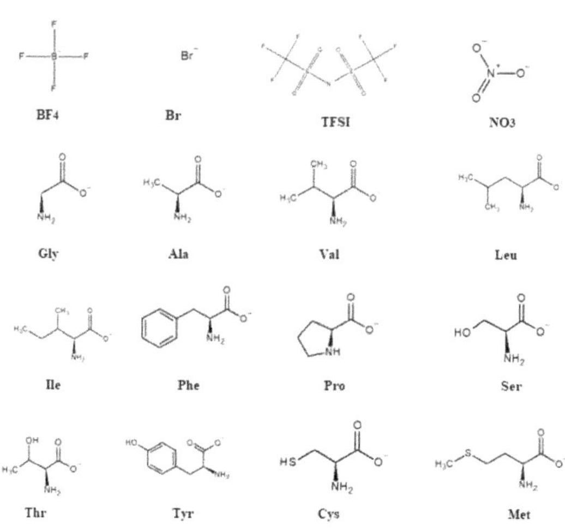

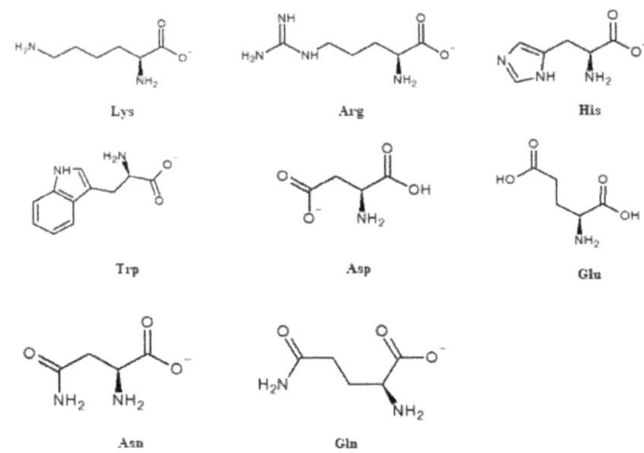

Fig. 2. Different anions structures of the ILs used in this study.

Results and Discussions

To study the effect of amine functional group (–NH_2) and carboxyl group (–COOH) on the capacity of CO_2 in ILs, different aminate-anion ILs, non-functionalized ILs and, amino-cation ILs were evaluated. First, 20 aminate-anion ILs with different combination of cations namely BMIM, BBPyrro, and N4444 were evaluated to study the effect of amino acid based-anion ILs on CO_2 capacity.

For BMIM-cation based ILs, the values of CO_2 capacity are shown in Figure 3. It can be observed that the order of CO_2 capacity starting from the highest is [Arg]>[Lys]>[Met]>[Phe]>[His]>[Trp]>[Leu]>[Val]>[Asn]>[Gly]>[Ala]>[Pro]>[Ile]>[Cys]> [Ser>[Gln]>[Tyr]>[Thr]>[Asp]>[Glu]. Figure 4 shows the Henry's constant (H) and activity coefficient at infinite dilution. It can be noted that the order of Henry's constant (H) values starting from the lowest is [Arg]<[Lys]<[Met]<Phe]<[His]<[Trp]<[Leu]<[Val]<[Asn]<[Gly]<[Ala]<[Pro] <[Ile]<[Cys]<[Ser]<[Gln]<[Tyr]<[Thr]<[Asp]<[Glu]. The trend observed is also generally consistent with the order obtained for γ^{∞} as shown in Figure 4. By comparing CO_2 capacity prediction with the experimental results obtained by Sistla and Khanna [10], almost similar trend with a slight difference was observed where [Arg] and [Lys] showed the highest CO_2 capacity values while [Ala] and [Pro] showed the lowest values. The results of the overall 60 ILs screening, showed that [Arg] and [Lys] have the highest CO_2 capacity, while [Thr], [Asp], and [Glu] showed the lowest affinity for CO_2 absorption.

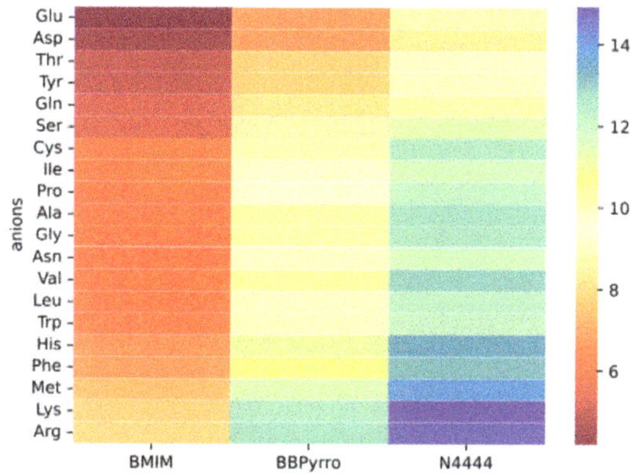

Fig. 3. CO_2 capacity values for different ILs, the red color represents the highest value followed by orange, yellow, green, and blue.

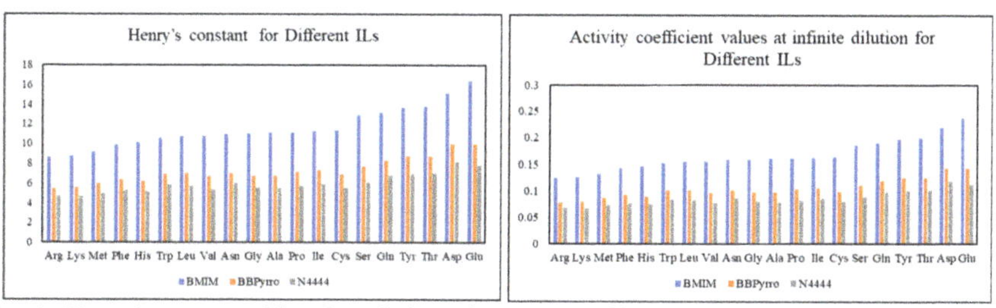

Fig. 4. Henry's constant (H) values (left) and activity coefficient values (γ^{∞}) (right) for different ILs.

The σ-profile and σ-surface for CO_2, [BMIM], [Arg], [Lys], [Thr], [Asp], and [Glu] are presented in Figure 5 to understand the effect of molecular structures on the activity performance. The σ-profile represents the charge density profiles for the structures, and it can be divided into three major parts: hydrogen bond donor for $\sigma < -0.01\ e/Å^2$, hydrogen bond acceptor for $\sigma > +0.01\ e/Å^2$, and nonpolar for $-0.01\ e/Å^2 < \sigma < +0.01\ e/Å^2$. Furthermore, σ-surface is used to estimate the charge distribution of the structure using color coded representation. The neutral characteristic is represented by green color, yellow color represents a partial negative charge while the strong electronegativity is represented by red color, and the blue color represents a strong positive charge. From Figure 5 it can be observed that [Arg] and [Lys] anions are highly nonpolar compared to [Thr], [Asp] and [Glu] anions, which increase the London dispersion interaction with the nonpolar CO_2. Moreover, it can be noted that [Arg] and [Lys] anions have high hydrogen bond acceptance properties which increase the interaction with BMIM that has hydrogen bond donation property. On the other hand [Thr], [Asp], and [Glu] anions, have hydrogen bond donation properties compared to [Arg] and [Lys] anions, which reduce their attraction forces with BMIM as well as the attraction forces with CO_2 since CO_2 is relatively considered as acid.

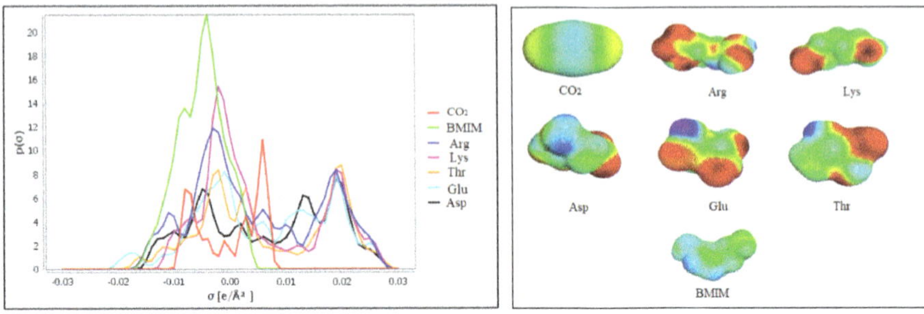

Fig.5 Sigma profile (left) and sigma surface (right) for CO_2, [BMIM], [Arg], [Lys], [Thr], [Asp], and [Glu].

To understand the effect of different cations on CO_2 capacity, [N4444], and [BBPyrro] cations with the same 20 anions were further evaluated. The values of CO_2 capacity for N4444-based ILs and BBPyrro-based ILs are shown in Figure 3. For N4444-based ILs, the order of CO_2 capacity starting from the highest is [Lys]>[Arg]>[Met]>[His]>[Phe]>[Val]>[Ala]>[Cys]>[Gly]>[Leu]>[Pro]>[Trp]>[Ile]>[Asn]>[Ser]>[Gln]>[Tyr]>[Thr]>[Glu]>[Asp].

The result is consistent for the Henry's constant (H) values for N4444 as shown in Figure 4, from the lowest is [Lys]<[Arg]<[Met]<[His]<[Phe]<[Val]<[Ala]<[Cys]<[Gly]<[Leu]<[Pro]<[Trp]<[Ile]<[Asn]<[Ser]<[Gln]<[Tyr]<[Thr]<[Glu]<[Asp]. The same sequence was found for γ^{∞} values as shown in Figure 4. By comparing the predicted CO_2 capacity with the experimental results obtained by Recker et al. [20], a similar trend was observed where [Val] showed higher CO_2 capacity than [Leu], followed by [Ile]. For BBpyrro-based ILs, the predicted CO_2 capacity sequence starting from the highest is [Arg]>[Lys]>[Met]>[His]>[Phe]>[Val]>[Ala]>[Gly]>[Cys]>[Trp]>[Asn]>[Leu]>[Pro]>[Ile]>[Ser]> [Gln]>[Tyr]>[Thr]>[Glu]>[Asp]. While for the order of H values starting from the lowest is [Arg]<[Lys]<[Met]<[His]<[Phe]<[Val]<[Ala]<[Gly]<[Cys]<[Trp]<[Asn]<[Leu]<[Pro]<[Ile]<[Asn]< [Ser]<[Gln]<[Tyr]<[Thr]<[Glu]<[Asp]. Similar trend was then observed for the for γ^{∞} values as shown in Figure 4.

It can be clearly seen that the order of CO_2 capacity slightly varies by changing the cations. It can be noted that when comparing the three cations, ammonium [N4444]-based ILs recorded an overall higher CO_2 capacity than [BBPyrro] and [BMIM]. To investigate the cause of variation of CO_2 capacity order with different cations, the σ-profile for CO_2, [BMIM], [N4444], and [BBpyrro] were generated as shown in Figure 6. From Figure 6, it can be observed that [N4444] is strongly nonpolar, followed by [BBPyrro] then [BMIM] which explains the high CO_2 affinity of [N4444] compared to [BBPyrro] and [BMIM].

Materials Research Proceedings 29 (2023) 297-306 https://doi.org/10.21741/9781644902516-33

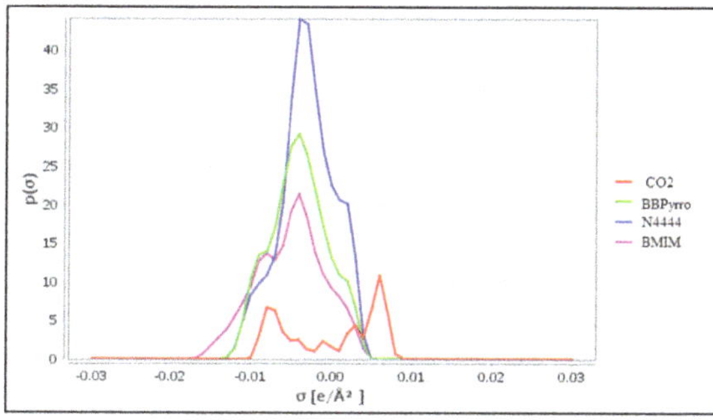

Fig. 6. Sigma profile for CO_2, BBpyrro, N4444, and BMIM

To understand the effect of aminate anions on the cations it is important to look at the nature of the amino acids. Amino acids (AAs) can be classified into five major groups based on the properties of the "R" group in each amino acid. First is AAs with polar uncharged side chain such as ([Ser], [Thr], [Asp], [Glu]), second is AAs with hydrophobic side like ([Ala],[Val], [Ile], [Leu], [Met], [Phe], [Tyr], [Trp]), third is AAs with basic side like ([His], [Arg], [Lys]), forth is AA with acidic side like ([Glu], [Asp]), and fifth is AA with special cases ([Cys], [Gly], [Pro]).
For BBPyrro, N4444, and BMIM -based cations, the nonpolar anions with high hydrogen bond acceptance properties and less hydrogen bond donation properties (with base side chain) like [Arg] and [Lys] shows higher CO_2 capacity, while the anion with acidic properties like [Glu] and [Asp] showed the lowest values. This is because CO_2 is considered as nonpolar and relatively acid. For nonpolar cation [N4444], the nonpolar amino acid-based anion like [Lys] showed high affinity for CO_2 compared to the relatively polar [Arg]. On the other hand, the relatively polar cation like [BMIM] showed higher CO_2 capacity with the relatively polar [Arg] compared to the highly nonpolar anion [Lys].

To study the effect of cation-based amino acids on CO_2 capacity, glycinium [Gly^+], L-lysinium(+1) [Lys^+], and [MGlyB] cations were evaluated along with tetrabutylammonium [N4444], [BMIM] and [BBPyrro] with different anions. From Figure 7, it can be observed that for bromide based ILs, the CO_2 capacity order starting from the highest is [N4444]>[BBPyrro]>[BMIM]>[MGlyB]>[Lys^+]>[Gly^+]. From Figure 8, similar trend was observed for γ^{∞} and H values starting from the lowest [N4444]<[BBPyrro]<[BMIM]<[MGlyB]<[Lys^+]<[Gly^+]. It can be noted that the cation-based amino acids generally have low CO_2 solubility. For further validation of the results, other common anions namely TFSI, BF_4 and NO_3 anion-based were evaluated with the same combinations of cations. From Figure 7 and Figure 8, it can be observed that TFSI and NO_3 anion-based showed the same trend as Br-based ILs. Interestingly, for BF_4 the order of CO_2 capacity starting from the highest is [N4444]>[BBPyrro]>[Gly^+]>[BMIM]>[MGlyB]>[Lys^+]. One striking trend is observed for the case of [Gly^+] cations. According to the combination of anion, the CO_2 absorption vary significantly. It can be observed that [Gly^+] cations showed highest affinity for CO_2 absorption when coupled with BF_4.

Materials Research Proceedings 29 (2023) 297-306 https://doi.org/10.21741/9781644902516-33

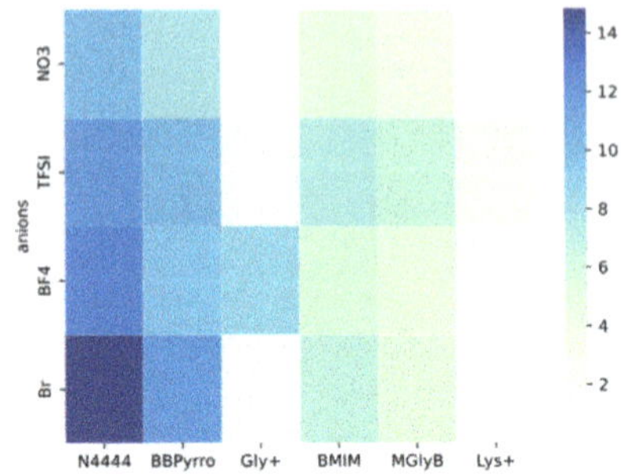

Fig. 7. CO_2 capacity values for different ILs, the red color represent the highest value followed by orange, yellow, and green.

Cations \ Anions	BF4	TFSI	NO3	Br
N4444	0.08012275	0.08593158	0.1003414	0.06747542
BBPyrro	0.10701314	0.09972247	0.13052135	0.08603673
Gly+	0.11682421	0.89772805	13.19474802	1263.34407176
BMIM	0.19396429	0.13791883	0.2356154	0.15027662
MGlyB	0.24145579	0.15690728	0.29848708	0.21102923
Lys+	2.44863768	0.46763114	3.74863892	3.10644848

Cations \ Anions	BF4	TFSI	NO3	Br
N4444	5.53	5.94	6.93	4.66
BBPyrro	7.39	6.89	9.02	5.94
Gly+	8.07	62.01	911.5	8726.89
BMIM	13.4	9.53	16.28	10.38
MGlyB	16.7	10.84	20.62	14.58
Lys+	169.15	32.3	258.5	214.59

Fig. 8. Henry's constant (H) values and infinite dilution activity coefficient (γ^{∞}) values for different ILs

To investigate the activity of [Gly^+] cation with different anions, sigma profiles and sigma potential for [Gly^+], [N4444], [BBPyrro], [BMIM], [MGlyB], [Lys^+], [TFSI], [Br], [BF_4] and [NO_3] were generated as shown in Figure 9. From the sigma profile and sigma potential, it can be observed that [Gly^+]has strong hydrogen bond donation property while BF_4 has strong hydrogen bond acceptance property which increases the molecular interaction between the molecules. It can also be noted that although Br anion is highly electronegative, [BF_4] showed higher CO_2 capacity and this is due to branched chain of B-F in [BF_4] that provides higher interaction surface especially with [Gly^+] since glycine is classified among the special group because it contains a hydrogen at its side chain N^+H_3 (rather than carbon as in all other amino acid), which makes the glycine the most flexible amino acid. It can be observed that the interaction between the CO_2 and big structures like [Lys^+] and other cations is mainly by London dispersion interaction (nonpolar-nonpolar) interaction due to the distance from the charged atoms unlike [Gly^+]. It can be therefore deduced that the cations/anions combination play important roles on CO_2 capacity in ILs through the molecular interaction.

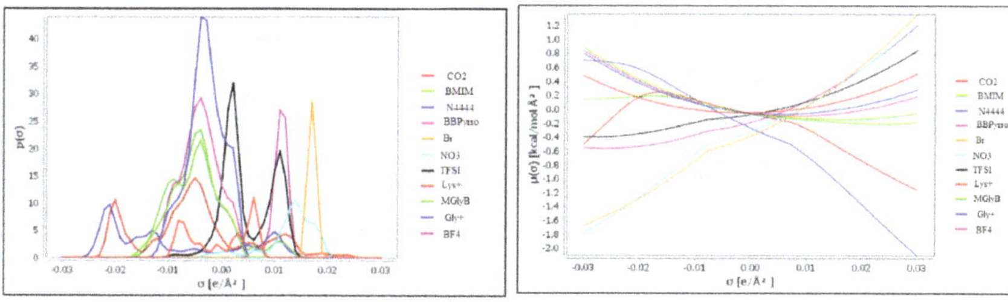

Fig. 9. Sigma profile (left) and sigma potential (right) for different anions and cations

Conclusion

The effect of amine functional group (–NH_2) and carboxyl group (–COOH) on CO_2 absorption of 84 different amino acid based ILs and non-functionalized ILs were elucidated using COSMO-RS and TurboMole by predicting the values of CO_2 capacity, activity coefficient and Henry's constant. The results showed that amino acid-based ionic liquids as anions significantly enhanced the CO_2 solubility in ILs compared to amino acid-based ionic liquids as cations. However, [Gly^+][BF_4] showed interesting CO_2 capacity compared to other amino acid cations-based due to increase molecular interaction. The results showed that the selection of cations/anions combination can significantly vary the CO_2 affinity of CO_2 in ILs. The affinity can be governed by acid/base interaction, hydrogen bond strength, and the nonpolar-nonpolar interaction.

Acknowledgement

This research was funded by Yayasan Universiti Teknologi PETRONAS (YUTP-FRG-015LC0-455). The authors also acknowledge technical support and facilities from the Chemical Engineering Department and Centre of Research in Ionic Liquids of Universiti Teknologi PETRONAS.

References

[1] N. Noorani, A. Mehrdad, CO_2 solubility in some amino acid-based ionic liquids: Measurement, correlation and DFT studies, Fluid Ph. Equilibria. 517 (2020)112591. https://doi.org/10.1016/j.fluid.2020.112591

[2] T. M. Gür, Carbon dioxide emissions, capture, storage and utilization: Review of materials, processes and technologies, Prog. Energy Combust. Sci. 89 (2022)100965. https://doi.org/10.1016/j.pecs.2021.100965

[3] Z. Tan, S. Zhang, X. Yue, F. Zhao, F. Xi, D. Yan, H. Ling, R. Zhang, F. Tang, K. You, Attapulgite as a cost-effective catalyst for low-energy consumption amine-based CO_2 capture, Sep. Purif. Technol. 298 (2022)121577. https://doi.org/10.1016/j.seppur.2022.121577

[4] D. Danaci, M. Bui, C. Petit, N. Mac Dowell, En route to zero emissions for power and industry with amine-based post-combustion capture, Environ. Sci. Technol. 55 (2021) 10619-10632. https://doi.org/10.1021/acs.est.0c07261

[5] G. T. Rochelle, Amine scrubbing for CO_2 capture, Science .359 (2009) 1652-1654. https://doi.org/10.1126/science.1176731

[6] S. Zheng, S. Zeng, Y. Li, L. Bai, Y. Bai, X. Zhang, X. Liang, S. Zhang, State of the art of ionic liquid-modified adsorbents for CO_2 capture and separation, AIChE Journal.68 (2022) 17500. https://doi.org/10.1002/aic.17500

[7] C. Xu, G. Yang, D. Wu, M. Yao, C. Xing, J. Zhang, H. Zhang, F. Li, Y. Feng, S. Qi, Roadmap on ionic liquid electrolytes for energy storage devices, Chem. Asian J.16 (2021) 549-562. https://doi.org/10.1002/asia.202001414

[8] I. I. Sam, S. Gayathri, G. Santhosh, J. Cyriac, S. Reshmi, Exploring the possibilities of energetic ionic liquids as non-toxic hypergolic bipropellants in liquid rocket engines, J. Mol. Liq. 350 (2021) 118217. https://doi.org/10.1016/j.molliq.2021.118217

[9] Z. Lin, Y. Su, R. Dai, G. Liu, J. Yang, W. Sheng, Y. Zhong, L. Tan, Y. Chen, Ionic liquid-induced Ostwald ripening effect for efficient and stable tin-based perovskite solar cells, ACS Appl. Mater. Interfaces. 13 (2021) 15420-15428. https://doi.org/10.1021/acsami.1c01408

[10] Y. S. Sistla, A. Khanna, CO_2 absorption studies in amino acid-anion based ionic liquids, Chem. Eng. J. 273 (2015) 268-276. https://doi.org/10.1016/j.cej.2014.09.043

[11] Y. S. Sistla, A. Khanna, Carbon dioxide absorption studies using amine-functionalized ionic liquids, J. Ind. Eng. Chem. 20 (2014) 2497-2509. https://doi.org/10.1016/j.jiec.2013.10.032

[12] X. Zhu, Z. Chen, and H. Ai, Amine-functionalized ionic liquids for CO_2 capture, J. Mol. Model. 26 (2020) 1-12. https://doi.org/10.1007/s00894-019-4247-5

[13] M. S. R. Shahrom, C. D. Wilfred, D. R. MacFarlane, R. Vijayraghavan, F. K. Chong, Amino acid based poly (ionic liquid) materials for CO_2 capture: effect of anion, J. Mol. Liq. 276 (2019) 644-652. https://doi.org/10.1016/j.molliq.2018.12.044

[14] R. J. Hook, An investigation of some sterically hindered amines as potential carbon dioxide scrubbing compounds, Ind. Eng. Chem. Res.36 (1997) 1779-1790. https://doi.org/10.1021/ie9605589

[15] P. Kumar, J. Hogendoorn, G. Versteeg, P. Feron, Kinetics of the reaction of CO_2 with aqueous potassium salt of taurine and glycine, AIChE Journal.49 (2003) 203-213. https://doi.org/10.1002/aic.690490118

[16] P. Kumar, J. Hogendoorn, P. Feron, G. Versteeg, New absorption liquids for the removal of CO2 from dilute gas streams using membrane contactors, Chem. Eng. Sci.57 (2002) 1639-1651. https://doi.org/10.1016/S0009-2509(02)00041-6

[17] A. Klamt, Conductor-like screening model for real solvents: a new approach to the quantitative calculation of solvation phenomena, J. Phys. Chem. 99 (1995) 2224-2235. https://doi.org/10.1021/j100007a062

[18] A. Klamt, V. Jonas, T. Bürger, J. C. Lohrenz, Refinement and parametrization of COSMO-RS, J. Phys. Chem A.102 (1998) 5074-5085. https://doi.org/10.1021/jp980017s

[19] H. W. Khan, A. V. B. Reddy, M. M. E. Nasef, M. A. Bustam, M. Goto, M. Moniruzzaman, Screening of ionic liquids for the extraction of biologically active compounds using emulsion liquid membrane: COSMO-RS prediction and experiments, J. Mol. Liq.309 (2020) 113122. https://doi.org/10.1016/j.molliq.2020.113122

[20] E. A. Recker, M.Green, M. Soltani, D.H. Paull, G.J. McManus, J.H. Davis Jr, A.Mirjafari. Direct Air Capture of CO_2 via Ionic Liquids Derived from "Waste" Amino Acids," ACS Sustainable Chemistry Engineering, 10(2022) 11885-11890. https://doi.org/10.1021/acssuschemeng.2c02883

Materials Research Proceedings 29 (2023) 307-314 https://doi.org/10.21741/9781644902516-34

Kinetic study for biodiesel production from dairy waste via catalytic microwave heating

MUHAMMAD Ayoub[1,a*], SITI AMINAH Mohd Johari[1,b], MUHAMMAD RASHID Shamsuddin[1,c], BAWADI Abdullah[2,d], SARAH Farukkh[3,e], SALMAN RAZA Naqvi[4,f], MUHAMMED Danish[5,g]

[1] Universiti Teknologi PETRONAS, HiCoE, Centre for Biofuel and Biochemical Research (CBBR), Institute of Sustainable Living (ISB), 32610 Seri Iskandar, Perak, Malaysia

[2] Biomass Processing Laboratory, Centre for Biofuel and Biochemical Research, Chemical Engineering Department, Universiti Teknologi PETRONAS, Bandar Seri Iskandar, 32610 Perak, Malaysia

[3] Department of Petroleum Engineering, Universiti Teknologi PETRONAS, 32610, Seri Iskandar, Perak, Malaysia Department of Petroleum Engineering, Universiti Teknologi PETRONAS 32610, Seri Iskandar, Perak, Malaysia

[4] National University of Sciences & Technology, 54000, Pakistan

[5] Universiti Kuala Lumpur Malaysian Institute of Chemical and Bio-Engineering and Technology, Lot 1988, Taboh Naning, Alor Gajah 78000, Melaka, Malaysia

[a] muhammad.ayoub@utp.edu.my, [b] aminahjohari1402@gmail.com, [c] mrashids@utp.edu.my, [d] bawadi_abdullah@utp.edu.my, [e] maqsood.ahmad@utp.edu.my, [f] salman.raza@scme.nust.edu.pk, [g] danish@usm.my

Keywords: Activation Energy, Frequency Factor, Biodiesel, Dairy Waste Scum Oil, Microwave

Abstract. Biodiesel production from waste such as wastewater sludge, palm oil mill effluent and dairy waste provide more sustainable approach and greener energy. This paper analyzes the kinetic modelling for biodiesel production from dairy waste scum oil via microwave heating transesterification. The results from this study show that transesterification of dairy waste scum oil is a first order kinetic which defines that the biodiesel production is dependent on the concentration of one reactant linearly. The activation energy for the reaction was validated to be 18.12 J/mol and frequency factor was 3.163 s^{-1}. It is validated that the activation energy of transesterification process in this study is lower than the past works have been done. The comparison of the value is as tabulated in the kinetic analysis section.

Introduction

Nowadays, diesel engines are renowned for their high torque and thermal efficiency. However, due to the increased demand for the energy together with the rapid depletion of fossil fuels and the requirement to consider emission regulations have led researchers to develop alternative fuels from renewable resources without any engine modifications. In research from Ashnani et al. [1], the transport sectionproduces almost a quarter of global energy-rated greenhouse gas emissions where theroad transportation covers the largest portion (more than 70%), followed by marine (15%), and aviation (10%).

Biodiesel has been found as a compatible alternative due to its advantages which are renewability, lower greenhouse gas emissions, higher flash points, and biodegradable with low carbon footprint. Biodiesel is an alternative for conventional diesel which is formulated based on plants product, animal fats, and recycled restaurant or factory grease. The conversion of the sources to the biodiesel is commonly undergo transesterification process. Transesterification of triglycerides such as vegetable oils, animal fats, or industrial fatty waste with alcohols (methanol, ethanol, propanol) in the presence of a homogeneous catalyst such as inorganic solid bases like sodium hydroxide (NaOH), potassium hydroxide (KOH) and sodium methoxide (CH_3ONa) or organic acids such as sulphuric acid (H_2SO_4) and hydrochloric acid (HCl), have been the most widely used method to synthesize biodiesel for the past two decades.

The major drawback of using homogeneous catalyst is non-recyclability of catalyst and deactivation by high moisture and higher saponification [2]. As a result, a thorough investigation into the manufacture of biodiesel with a heterogeneous catalyst has been conducted to overcome the homogeneous process economic and environmental limitations. There are many advantages of a heterogeneous catalytic reaction include quick separation from the reaction mixture by mechanical filtration, the ability to reuse the catalyst for further transesterification, ease of re-generation with less corrosion, lower environmental hazard, and lower cost.

Dairy waste scum is produced by dairy industries that handle raw and chilled milk as well as milk products such as butter, yoghurt, and ice-cream [3]. Transformation of dairy waste scum into biodiesel can be done via pyrolysis, dilution or blending, micro-emulsion and transesterification techniques. Among these four methods mentioned above, transesterification is the most common and convenient technology in production of biodiesel [4]. The transesterification reaction occurs between the lipid and an alcohol to form esters and a glycerol by product. Commercially, biodiesel production occurred frequently via alkali-catalysed (e.g., sodium hydroxide, potassium hydroxide) transesterification with excess methanol, which results in formation of fatty acid methyl ester (FAME) and glycerol [5]. Moreover, Marx [6] also added that base catalyzed transesterification is the most preferred method for biodiesel production due to high conversion obtained with shorter reaction time at relatively low cost.

In any study, kinetics modelling is important to determine the activation energy (Ea) and frequency factor (A) of the process. Activation energy acts as a barrier that the reactant in a chemical reaction has to overcome to form a product. The higher theactivation energy, the more energy that a process consumed to achieve the desired product. Thus, it is preferable to have a reaction process with lower activation energyin order to save the energy consumed to form a desired product. Normally, the activation energy can be reduced by using catalyst by altering the reaction's transitionstate. After that, the reaction follows a different path/mechanism than the uncatalyzedreaction. The net energy difference between the reactant and the product is unaffected by the catalyst. On the other hand, the frequency factor represents the frequency of collisions between reactant molecules at a standard concentration. In this paper, the focus was on kinetic modelling of biodiesel production as there has been limited knowledge regarding kinetic modelling for dairy waste scum oil feedstock.

Experimental

Materials and reagents

Methanol (99.9%, Merck 106009), potassium hydroxide, sulfuric acid and isopropyl alcohol. Eggshell from kitchen waste for CaO catalyst generation.

Preparation of catalyst

The discarded eggshells were obtained from a burger stall at university campus.First and foremost, the waste eggshells were cleaned up by using tap water to removethe unwanted contaminants or impurities. Then, the membrane of the eggshells was removed thoroughly. Consequently, the eggshells were left in the drying oven for 24 hours at 100°C to be dried. After 24 hours, the dried eggshells were collected from the oven and crushed into powder by using blender. The powder formed of eggshells were placed in cubicles and calcined for 3.5 hours in a furnace at 900°C to maximize the biodiesel yield. After calcination process completed, a cold solid is obtained and stored in a desiccator for 24 hours.

Extraction of dairy waste scum oil.
The dairy waste scum was collected at a milk factory waste effluent. The raw dairy waste scum needs to undergo extraction process to obtain the scum oil. Firstly, the collected waste scum was put in a 100 ml beaker and heated to 100°C on the hot plate for 30 minutes to evaporate the moisture in the mixture. Then, the liquid scum was allowed to settle for a few minutes until sediments and solid contaminants float to the top surface. By using fine steel coarse mesh, the undesired solid waste and suspended particles were removed. The liquid scum was centrifuged for a few times in a centrifuge machine for several minutes with 5000 rpm.The steps are repeated for a few times until a pure anhydrousscum oil is achieved.

Pre-treatment and Acid Esterification
For the synthesis of biodiesel, esterified scum oil was trans esterified using a calcium oxide catalyst via microwave irradiation method to heat 100 mL of esterified samples. Normally, the free fatty acid (FFA) contents in the extracted scum oil are more than 2%. Therefore, it must undergo pre-treatment and acid esterification process to determine the FFA content value. In case the FFA content is more than 2%, the acid esterification process must be taken. The scum oil was undergone acid esterification process for high FFA (>2%). A 100 mL of the sample's preheated oil is poured in a separate flask and placed on a mechanical stirrer set to 600 rpm. The reaction mixture is stirred for approximately 60minutes at a temperature of 60°C with a known amount of 1% v/v $H_2SO_4$and 20 mL methanol. After cooling the reaction mixture, it was transferred to the separating funnel for phase separation. The top layer is a mixture of excess unreactedmethanol and water was removed, while the bottom layer is an esterified oil sample that will be rinsed with distilled water to remove any residue and heated to evaporate any remaining water. The FFA level of esterified scum oil was determined forsubsequent transesterification reactions.

Transesterification
For the synthesis of biodiesel, esterified scum oil was transesterified using a calcium oxide catalyst (calcined eggshells) where microwave irradiation method was used to heat 100 mL of the esterified samples. The methanol and oil were heated in a microwave oven with 6:1 molar ratio at 45°C, 50°C, 55°C, 60°C by using a magnetic stirrer and an infrared temperature sensor while the oven is able to maintain constant stirring and temperature control without the need for direct contact. 750 W was set upto irradiate the mixture. The microwave reactor was utilized in this process. The reaction time was taken in range of 10, 15, 20, 25 and 30 minutes. Once the reaction mixture has been allowed to settle for 5 minutes at room temperature, the topester layer was poured into a new flask. Afterwards, a vacuum was used to remove thesolid catalyst from the reaction, which was then transferred to the separating funnel and allowed to stand overnight to complete the phase separation process. Layers one,two, and three all contain FAME, while the bottom layer has traces of catalyst. Excessmethanol and calcium oxide particles are removed from the filtered biodiesel by distillation and decalcification with citric acid.

Biodiesel purification
In order to purify the biodiesel, the water washing process was utilized, and thebiodiesel mixture was gently swirled to avoid foam formation. This process was donethree times to guarantee that the biodiesel fraction was clear of most impurities. The washed biodiesel mixture was then maintained overnightin a separating funnel to separate into two phases: water impurities and biodiesel. To eliminate any leftover moisture, the washed biodiesel portion was heated to 100 °C for1 hour. The yield of biodiesel was calculated by comparing the weight of biodiesel tothe original weight of lipid. Subsequently, the biodiesel yield will be evaluated by calculating the amount of

biodiesel obtained after purification process compared to the amount of oil used in transesterification process in percentage. The formula to calculate the percentage of biodiesel yield is shown in Eq. 1.

$$Biodiesel\ yield\ (\%) = \frac{\text{weight of fatty acid methyl ester produced (g)}}{\text{weight of oil used (g)}}\ x\ 100 \tag{1}$$

Kinetic modelling evaluation for biodiesel production

It is necessary to conduct batch trials to determine how well the transesterification process works at various reaction times and temperatures (318 K, 328 K, 338 K), where the lowest reaction temperature is limited to the mesophilic range (338 K) due to the slower glycerol production rates at lower temperatures. The highest reaction temperature is set at 338 K due to the range of 318 K, 328 K, and 338K (Pugazhendi,2020). In assumption, the transesterification process should follow first order kinetics as the function of biodiesel yield. The reaction temperature has thegreatest influence on the time-dependent reaction rate constant. At various temperatures, the transesterification reaction's rate constant is computed. Biodiesel yield t and response time k are the rate constants in Eq. 2, which displays the differential form of a first-order rate equation.

$$Reaction\ rate = -\frac{d(Y)}{dt} \tag{2}$$

$$Reaction\ Rate = k(Y)$$

$$\frac{d(Y)}{dt} = k(Y) \tag{3}$$

Assuming the initial biodiesel yield as Y_0 at time t=0, and the yield at time, t on integration, the Eq. 3 is expressed in Eq. 4.

$$k = \frac{\ln(Y_t) - \ln(Y_0)}{t} \tag{4}$$

It can be shown from Eq. 4 that the slope of the plot (ln (d(Y)/dt) vs ln (Y) supplied the reaction rate constant for the reaction.

Taking the natural logarithm and expressing the Arrhenius equation in Eq. 5 results in the formation of Eq. 6. The activation energy and frequency factor may be derived from the slope and intercept of ln k against 1/T in Eq. 6, which can be obtained here:

$$k = Ae^{\frac{-Ea}{RT}} \quad 5)$$

$$\ln k = \frac{-Ea}{RT} + \ln A \tag{6}$$

Results and discussion

Kinetic modeling analysis

In any study, kinetics modelling is important to determine the activation energy (Ea) and frequency factor (A) of the process. Activation energy acts as a barrier that the reactant in a chemical reaction has to overcome to form a product. The higher the activation energy, the more energy that a process consumed to achieve the desired product. Thus, it is preferable to have a reaction process with lower activation energy in order to save the energy consumed to form a desired product. Normally, the activation energy can be reduced by using catalyst by altering the reaction's transition state. After that, the reaction follows a different path/mechanism than the uncatalyzed reaction. The net energy difference between the reactant and the product is unaffected by the catalyst. On the other hand, the frequency factor represents the frequency of collisions between reactant molecules at a standard concentration

The correlation between activation energy and frequency factor is shown in Arrhenius equation as formulated in Eq. 7.

$$k = Ae^{\frac{-Ea}{RT}} \tag{7}$$

By taking the natural logarithm on both sides, the Eq. 7 is expressed as Eq. 8

$$\ln k = \frac{-Ea}{RT} + \ln A \tag{8}$$

Where Ea is representing activation energy (J/mol) and A is indicating the frequency factor (min^{-1}). The value of the R, universal gas constant is taken as 8.314 J/mol.K andT is the absolute temperature.

In the Eq. 8, it is noted that the linear correlation exists between ln k and 1/T. Therefore, the activation energy and frequency factor can be calculated from the slope of the and intercept of the correlation, respectively.

On top of that, the first-order kinetics equation has to be derived to obtain the reaction rate constant (k) value. The reaction is assumed to be first order kinetics as research reported that in a process that molar ratio of methanol higher than triglyceride, the forward reaction is believed to be at pseudo first-order reaction. Thus, the rate law of equation is formulated as in Eq. 9.

$$Rate = -r = \frac{-d[TG]}{dt} = k[TG] \tag{9}$$

By integrating the Eq. 9, the rate law of equation becomes as follow:

$$ln \frac{[TG]o}{[TG]} = k_{TG}.t \tag{10}$$

Where,

$$X_{ME} = 1 - \frac{[TG]}{[TG]o} \tag{11}$$

Thus,

$$\ln (X_{ME} - 1) = -kt \tag{12}$$

Therefore, a linear correlation between ln (X_{ME} − 1) and times will exhibit slope valuethat indicates the value of rate constant at particular temperature and reaction time. Fig. 1 shows the graph plotted to obtain the reaction rate constant of each temperature.

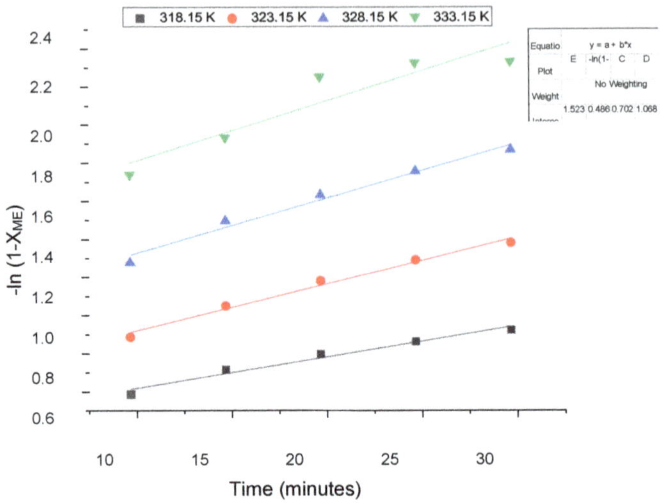

Figure 1: Graph -ln (1-X_{ME}) against time to find the rate constants, k

From graph depicted in Fig. 1, the slope of each temperature is taken as the value of the rate constants, k. The value of the rate constants for each temperature ispresented in Table 1.

Table 1: Value of Rate Constants, k at respective temperature

Temperature (K)	k value	ln k
318.15	0.01633	-4.11475
323.15	0.02467	-3.70217
328.15	0.02887	-3.54495
333.15	0.03164	-3.45333

Consequently, a graph ln k against 1/T is plotted as shown in Fig. 2 to obtain the value of the activation energy and frequency factor. The linear graph exhibits an equation which the slope is equal to -1.29. By substituting in the Eq. 11, the activation energy is defined at 10.73 J/mol.

$$\frac{-Ea}{R} = -2.18$$

$$Ea = 1.29(8.314\frac{J}{mol}) \quad (11)$$

$$= 18.12\frac{J}{mol}$$

On the other hand, the frequency factor of the reaction is determined at 3.163 which isthe intercept of the graph.

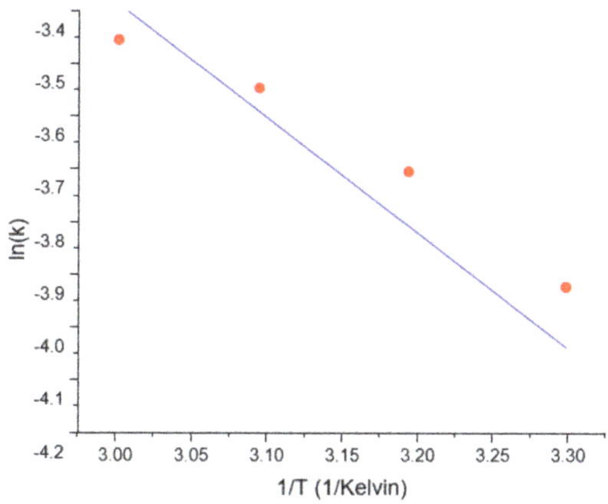

Figure 2: Graph of ln k against 1/T to obtain activation energy and frequency factor

Graph depicted in Fig. 2 shows the linear graph of ln k against 1/T to determine the slope which indicates the activation energy and also the intercept as frequency factor. From the calculation, the activation energy is obtained at 18.12 J/mol and 3.163frequency factor. Therefore, in this study the minimum amount of energy that is required to activate atoms or molecules to a condition in which they can undergo chemical transformation is 18.12 J/mol of energy to form biodiesel. It is a good indication since previous studies with same feedstocks that using different type of catalyst has higher activation energy as described in literature. Table 2 in the data validation section shows the comparison data of activation energy between current study and previous study.

Data validation

The results of the kinetic studies showed that the use of derived calcined eggshells heterogeneous solid catalyst for the production of biodiesel from dairy wastescum oil under the influence of microwave heating has proved that it produces lower activation energy compared to other studies and the reactants achieved the equilibriumin a short interval by giving the maximum yield of biodiesel and conversion of triglycerides. The activation energy stated in the comparison Table 2 has shown thatthe use of dairy waste scum oil combined with calcined eggshells catalyst has minimizethe activation energy of the transesterification process as compared to the other studies previously described in the literature which is the lowest activation energy is the mostpreferable reaction. The activation energy of the previous work is from to the study by Yatish and Krishnamurty.

Table 2: The comparison of activation energy between present work and literatures.

Feedstock	Catalyst	Activation Energy	References
Dairy ScumOil	Calcined Eggshells	18.12	Present Work
Dairy ScumOil	Nano $CaTiO_3$	35.56	[7]
Dairy ScumOil	Snail Shell CaO	67.21	[2]

Conclusions

The investigation of principle of kinetics reaction contributes well knowledge into biodiesel generation from dairy waste scum oil in presence of CaO from eggshell. The microwave assisted transesterification of dairy waste scum oil has produced lower activation energy compared to other studies which is significant to save energy especially when the production of biodiesel is in large scale industry which required to consume a lot of energy. The results also showed that microwave heating only need small amount of activation energy which fasten the biodiesel production with high yield. A lower activation energy process is preferable since it can produce the same production amount with low cost and energy.

Acknowledgement

We thank and appreciate the Center for Biofuel and Biochemical Research (CBBR), Universiti Teknologi PETRONAS, Higher Institution Centre of Excellence (HiCoE) grant (015MA0-104), Malaysia and International Collaborative Research Fund with Kyushu Institute of Technology grant (015ME0-271).

References

[1] M. H. M. Ashnani, T. Miremadi, A. Johari, and A. Danekar, "Environmental Impact of Alternative Fuels and Vehicle Technologies: A Life Cycle Assessment Perspective," *Procedia Environ. Sci.*, vol. 30, pp. 205–210, 2015, https://doi.org/10.1016/j.proenv.2015.10.037

[2] K. N. Krishnamurthy, S. N. Sridhara, and C. S. Ananda Kumar, "Optimization and kinetic study of biodiesel production from Hydnocarpus wightiana oil and dairy waste scum using snail shell CaO nano catalyst," *Renew. Energy*, vol. 146, pp. 280–296, 2020, https://doi.org/10.1016/j.renene.2019.06.161

[3] V. Kavitha, V. Geetha, and P. J. Jacqueline, "Production of biodiesel from dairy waste scum using eggshell waste," *Process Saf. Environ. Prot.*, vol. 125, pp. 279–287, 2019, https://doi.org/10.1016/j.psep.2019.03.021

[4] F. Ma and M. A. Hanna, "Biodiesel production: a review," *Bioresour. Technol.*, vol. 70, no. 1, pp. 1–15, 1999, https://doi.org/10.1016/s0960-8524(99)00025-5

[5] J. Saleh, A. Y. Tremblay, and M. A. Dubé, "Glycerol removal from biodiesel using membrane separation technology," *Fuel*, vol. 89, no. 9, pp. 2260–2266, 2010, https://doi.org/10.1016/j.fuel.2010.04.025

[6] S. Marx, "Glycerol-free biodiesel production through transesterification: A review," *Fuel Process. Technol.*, vol. 151, pp. 139–147, 2016, https://doi.org/10.1016/j.fuproc.2016.05.033

[7] K. V. Yatish, H. S. Lalithamba, R. Suresh, and H. K. E. Latha, "Ochrocarpus longifolius assisted green synthesis of CaTiO3 nanoparticle for biodiesel production and its kinetic study," *Renew. Energy*, vol. 147, pp. 310–321, 2020, https://doi.org/10.1016/j.renene.2019.08.139

The effect of glass fiber length on compressive and flexural strength of reinforced geopolymer

SARINI Mat Yaakob[1,a], SURIATI Sufian[1,b*] and NURUL EKMI Rabat[1,c]

[1]Department of Chemical Engineering, Universiti Teknologi PETRONAS, 32610, Seri Iskandar, Perak, Malaysia

[a]sarinimatyaakob@yahoo.com, [b]suriati@utp.edu.my, [c]nurulekmi.rabat@utp.edu.my

Keywords: Fly Ash, Geopolymer, Glass Fiber, Adhesion, Flexural, Compressive

Abstract. The use of fly ash in the development of geopolymers is considerably practical in minimizing landfills. In addition, to preserve the environment, this approach may also help converting waste material to profitable returns. Currently, geopolymers are used in numerous applications owing to their properties that are comparable as the conventional material of Ordinary Portland Cement (OPC). A few advantageous properties of the geopolymers included resistant to acid, fast setting, high compressive strength and produce low carbon dioxide gas to the atmosphere. However, chemical interaction of raw aluminosilicates reactive precursor with an aqueous alkaline solution during synthesis process has commonly produced a porous geopolymer, which gives the limitation to the geopolymer performance. This weakness may be recovered by implementing fiber reinforcement to improve the properties and reduce the number of internal pores of the geopolymer. This study evaluates the enhancement recorded in geopolymer properties prior to addition of glass fiber into geopolymer matrix. The results of adhesion, flexural and compressive strength of geopolymer significantly increased by addition of glass fiber.

Introduction

Geopolymers are synthetic green polymeric materials formed by chemical process of high aluminosilicates raw materials such as coal fly ash, granulated blast furnace slag, slag and mineral clays (metakaolin) with any alkaline or alkaline silicate solution with produce final product of solid material. The term "geopolymer" was being introduced by J. Davidovits in 1970s. The term was used as to highlight some of the resemblances of the geopolymers with organic thermoset resins [1]. In 1980-1990, a product known as "pyrament" that having superior strength was introduced and it received interest of the research community as a direction towards the future building material [2].

The history of alkali activation technology was first initiated in 1908. Kuhl et.al [3] revealed their first innovation called —slag cement using alkaline component with blast furnace slag (BFS). In 1930-1950, first lab scale research on cement prepared from steel slag and alkalis were published by Prudon et.al in Belgium [4]. Starting from the date, this material was commercially used and applied in building construction. The development of geopolymer by fly ash based began in 1990s by Wastiels et.al [5] with combination of blast furnace slag. Geopolymers are consisted of three main components; raw aluminosilicates materials as precursor, alkaline activator and water. The geopolymer properties are strongly influenced by the type of precursor used, the alkaline activator type and concentration, the particle size of precursor and the water content.

Chemical interaction of raw aluminosilicates reactive precursor with an aqueous alkaline solution during synthesis process has commonly produced a porous geopolymer. This weakness may be recovered by implementing fiber to improve the strength and reduce the number of internal pores of the geopolymer [6], [7]. Fiber in the forms of long unidirectional, whiskers and particles have been utilized as reinforcement in the geopolymer composites to enhance the properties. A few types of fiber including carbon, basalt and natural whose length was approximate with 14 mm, 16 mm and 30 mm, respectively, were used in existing studies [8], [9].

The positive effects of fiber on hardened mechanical properties of reinforced geopolymer have been the subject of numerous studies focusing on the cracking and toughening mechanisms of the fiber according to their shape, volume fraction, and orientation. Despite the superior mechanical properties, fiber reduce the workability of geopolymer paste, which results in the excessive void formation and poor compaction. Thus, a compromise between hardened behavior and fresh properties is essential for applicability.

As seen in Fig. 1 regardless of the fiber type and geometry, increasing the fiber content reduces the flow ability of fiber-reinforced geopolymer. The reduced workability can be explained as a result of an increase in yield stress of the fresh reinforced geopolymer due to the contact network between rigid fiber inside the matrix, proportionate increased with fiber content, equivalent diameter, and aspect ratio. The critical concentration value was suggested to be in the range of 0.2%–2% for the geopolymer and cementitious composites [10]. When the fiber content increases over a critical concentration, the fiber tend to get uneven dispersion and form clumps or balls, and even very valuable matrices might not pass through the congested fiber network properly. Therefore, the fresh matrix requires more vibration to lose its harsh static mode and form the mold [11].

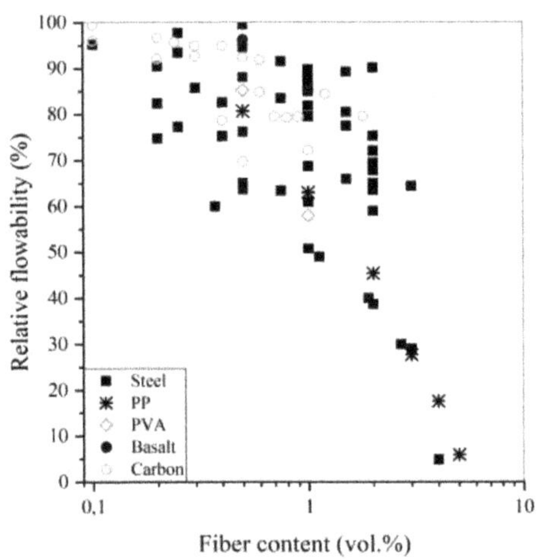

Fig 1: Relative flow ability of fiber reinforced geopolymer composites compared with plain matrices. Adapted from Ref [12]

Drying shrinkage of geopolymers is due to the high capillary pressure produced between wet and dry zones of the micropore network, which leads to specimen deformation and crack initiation. To control the drying shrinkage, two approaches have been often followed: (1) the modification of the pore structure to minimize capillary porosity and to control water loss during curing and (2) the inclusion of inert or reactive fillers and fiber [13]. Several studies have indicated that fiber content, fiber modulus, and fiber-binder interaction are the dominant factors in controlling the shrinkage of reinforced geopolymer composites [14]. Fig. 2 shows the influence of two fiber on controlling the shrinkage of fly ash-based geopolymer. The incorporation of PP and steel fiber even in small amount, 0.5 vol%, reduced the drying shrinkage of the composite specimens significantly. Increasing the steel fiber content to 2 vol% and above resulted in almost no shrinkage. While, the similar reduction was not observed for the same content of PP fiber, and even it showed an adverse effect when fiber content was increased to 4 vol%. This can be ascribed to the poor compaction of the geopolymer lower stiffness of the PP fiber integrated with weak fiber-binder interaction is other reasons for poor performance of the PP fiber compared to the steel fiber. Similar superior shrinkage performance was observed for PVA fiber-reinforced geopolymers as compared with PP fiber due to the hydrophilic nature and higher stiffness of PVA fiber [15]. Besides, it was reported that using a longer fiber can (1) reduce, (2) increase, or not change the shrinkage of geopolymer composites. Furthermore, the shrinkage trend of fiber- reinforced geopolymer composites is affected by the nature of the binder and curing conditions. For example, for the same type and amount of fiber, the shrinkage of slag- based geopolymer composite was reduced by variation of environmental humidity ranging from 50% to 95%, while fly ash-based geopolymer composite was less sensitive to environmental humidity.

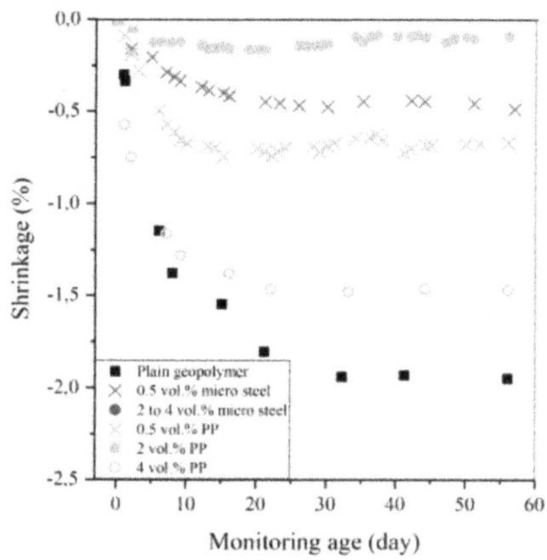

Fig 2: Drying shrinkage of steel and polypropylene fiber reinforced geopolymer. Adapted from Ref [12]

Previous studies by PVA and alumina fiber have shown that as fiber diameter increases, it reduces mechanical properties as well as fiber modulus fiber [16], [17]. This is because fiber imperfections increase when the diameter of the fiber is large compared to the smaller diameter. The fiber aspect ratio denotes the ratio of fiber length to its diameter. The growing interest to add fiber in the fabrication of geopolymer composite make it very important to comprehensively evaluate the properties of that composite via the addition of fiber in the system. Researchers focus on the effect of fiber volume fraction and how this volume does significantly affect the distribution of fiber within the matrix, which indirectly influenced the positioning behavior of fiber and matrix. This part is very crucial to understand as to answer the trend of stress being transferred with the composite. Not only to that aspect, the mechanical properties of geopolymer composite also affected by the fiber aspect ratio.

Shouha et al. [18] observed the effect of fiber aspect ratio in the geopolymer composite reinforced with carbon. The tensile property of geopolymer composite was evaluated. It was found that tensile strength improved until the fiber aspect ratio increases up to 20. Further increasing the ratio, gives nothing enhancement to the composite properties. A similar finding was also supported by Lin et al. [19] in his studies using sheet like short carbon fiber. They found that length of fiber was significantly affected to the tensile strength of composite. However, contracts finding on the effect of fiber aspect ratio on the composite properties were found by several studies. Sudhikumar et al. observed the effect of 25, 38 and 50 fiber aspect ratios on the strength properties of fibrous ferrocement geopolymer composite. Higher compressive strength and flexural strength were recorded at 25 MPa and 38 MPa. But, by increasing the ratio to 50, yield lower strength of composite. They claimed that at lower fiber aspect ratio, more adsorption of stress by composite. The results have been shown with lower aspect ratio, give higher compressive and flexural strength. Study by Jose et al. [20] has successfully developed a reinforced composite by Prosopis juliflora with multiple fiber aspect ratio with fiber length from 46 to 227 mm and fiber loadings from 5 to 32 wt% was evaluated. The study found that the optimum fiber aspect ratio and fiber loading was at 136 and 23.53 wt% for obtaining the composite with high strength. On top of that, there was a limited study explored where the potential of short fibers, especially a glass fiber in the development of reinforced geopolymer. Short glass fiber may offer more advantages over the other kinds of fiber due to their isotropic mechanical behavior, abundantly available in the market and more cost effectively. Thus, the development of the short glass fiber reinforced geopolymer is needed to be explored so that a new outcome from the study may provide beneficial knowledge to community.

Research Methodology

Materials

Raw materials used in the present research including coal fly ash, sodium hydroxide (NaOH), and glass fiber. Coal fly ash was procured from a local Malaysia power plant. To minimize the variation in the particle size distribution, and to separate larger particles, the fly ash was sieved using a 45μm sieve through a vibratory sieve shaker (AS200 Basic RETSCH, Germany). Chemical composition, the presence of different phases and particle size analysis of the fly ash was characterized prior to its use as geopolymer. Sodium hydroxide (99%) was procured from Merck Millipore, Subang, Malaysia. Standard solution of the required concentrations of this base was prepared using distilled water. The synthesis of geopolymer consisted of three steps; (i) preparation of the alkali activator solution, (ii) mixing of alkali activator with fly ash and (iii) curing of geopolymer. Prior to the development of geopolymer the ratio of Na: Al ratio, water: solid ratio and curing days was fixed at 1.0, 0.33 and 3 days, respectively. The glass fiber chopped strands were imported from JN Technologies PvT. Ltd Glass strands were mainly composed of silicon oxide (SiO_2), alumina oxide (Al_2O_3), calcium oxide (CaO), manganese oxide (MnO), sulfur trioxide (SO_3), magnesium oxide (MgO), sodium oxide (Na_2O), ferum oxide (Fe_2O_3), potassium oxide (K_2O), titanium oxide (TiO_2) and phosphorus oxide (P_2O_5). The average diameter of the glass fiber was 14 μm with two different lengths of 6 mm and 12mm. The fiber was incorporated in the geopolymer matrix using an ultrasonic bath sonicator for 10-minute agitation process. This approach was to ensure the glass fiber was well mixed and homogeneous composite was formed. Subsequently, the homogeneity of the composite was reconfirmed by a field emission scanning electron microscope (FESEM). The content of fiber varied from 0.2 -2.0 wt%. Table 1 shows the test matrix used in incorporating the fiber into geopolymer system.

Table 1: Fiber length and fiber weight in incorporating system

Fiber length	wt%									
6 mm	0.2	0.4	0.6	0.8	1	1.2	1.4	1.6	1.8	2
12 mm	0.2	0.4	0.6	0.8	1	1.2	1.4	1.6	1.8	2

Materials Research Proceedings 29 (2023) 315-324 https://doi.org/10.21741/9781644902516-35

Results and Discussions

Diffractometry Analysis of Fly Ash

X-rays diffraction (XRD) analysis was conducted to investigate the qualitative and quantitative analysis of the various phases existing in the ashes. The XRD analysis of the fly ash was shown in Fig. 3. As can be seen, the ash composed of crystalline and amorphous phases. The sharp and visible peaks were representing crystalline phase, whereas the hump from the baseline in the 2-theta range of 18- 40° signified a presence of amorphous materials. The crystalline phase of quartz was detected at 20.91°, 25.45°, 26.52°, 43.26°, 50.29° and 68.37°. The result was found consistent with the pattern of quartzite, ICOD card no. 01-085-0457.

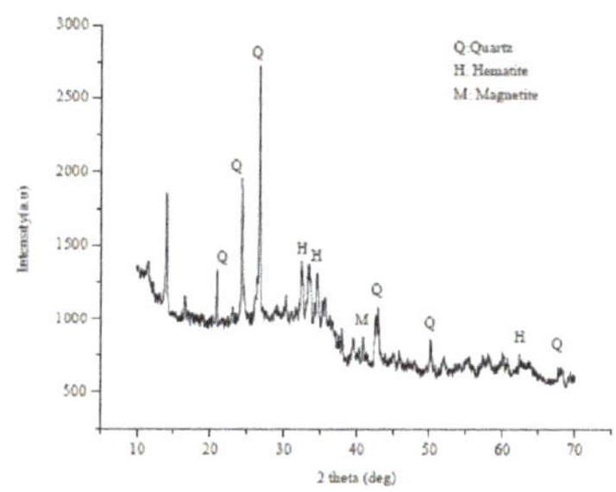

Fig 3: X-rays diffraction analysis of the raw materials of fly ash

Microstructural Image of Geopolymer

Microstructural analysis provides information about the microstructure of materials. Fig. 4 (a) and (b) shows FESEM images of unreinforced and reinforced geopolymers, respectively. Porous internal structure of geopolymer with some unreacted fly ash could be observed in Fig. 4 (a). Apparently, the presence of glass fiber into geopolymer, led to reduce the number of pores inside the system. The fiber significantly closed the pores via bridging effect between fiber surface and geopolymer paste. The fiber was also seemed to dispersed uniformly and well scattered in the paste.

Fig. 4: FESEM image of (a) unreinforced geopolymer; (b) glass fiber reinforced geopolymer

Adhesion strength

The pull-out tests were carried out to analyze the adhesion strength of the reinforced geopolymer to the steel plate. The adhesion test was determined using Elcometer 108, according to ASTM D-

Materials Research Proceedings 29 (2023) 315-324 https://doi.org/10.21741/9781644902516-35

4541 [21]. Fig. 5 shows the adhesion strength of the reinforced geopolymer that was cured isothermally at 60 °C for three days. The adhesion strength showed variations with mixed design.

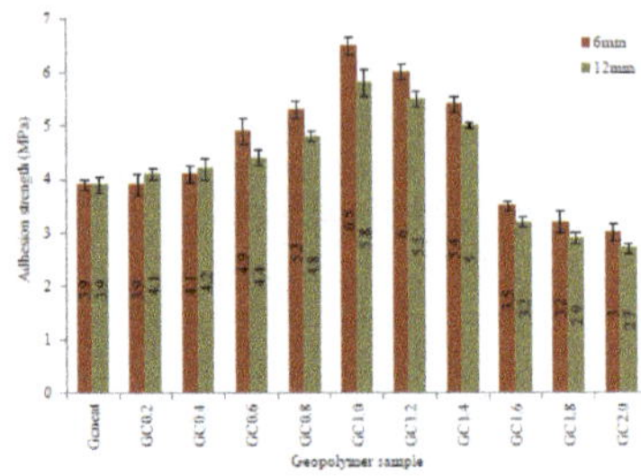

Fig. 5: Adhesion strength of 6mm and 12mm glass fiber reinforced geopolymer at three days curing time

The strength of reinforced geopolymer for both 6 mm and 12 mm was increased with increasing content of fiber up to 1.0 wt%, in with the highest strength of 6.5 MPa for 6 mm and 5.8 MPa for 12 mm fiber. This could be explained by FESEM image shown in Fig 4 (b). The fiber was adhered strongly with the paste, consequently produced geopolymer with compact structure, led to better strengths. However, when the fiber content was further from 1.0 wt% to 2.0 wt%, the strength decreased gradually. The lowest strength of 3.0 MPa and 2.7 MPa was obtained for 6 mm and 12 mm fiber, respectively. This finding was comparable to a study reported by Zhao et al. [22] and Herrera et al. [23].

Flexural strength

Effect of glass fiber content on the flexural strength of the samples was discussed. Fig. 6 presented the flexural strength of the geopolymer sample.

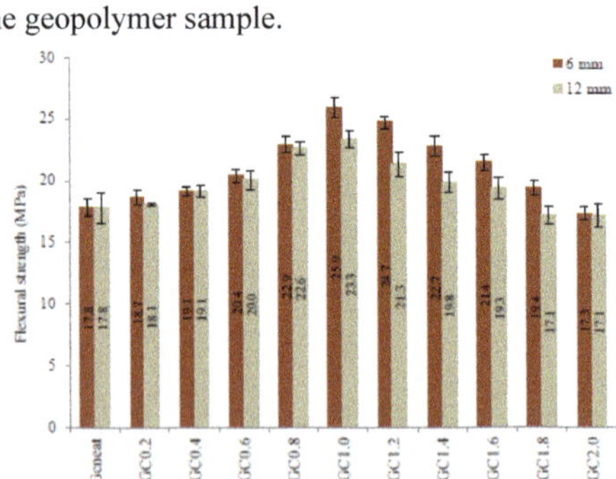

Fig. 6: Flexural strength of 6mm and 12mm glass fiber reinforced geopolymer at three days curing time

Based on the Fig. 6, the reinforced geopolymer confirmed the high-quality advancement in the aspect of flexural strength. The higher the fiber content, the higher the bending strength compared to the neat geopolymer (unreinforced geopolymer). It can be seen, only 17.8 MPa of strength was recorded for unreinforced geopolymer. Meanwhile, the flexural strength of 6 mm reinforced geopolymer glass fiber increased drastically with 0.2 wt% up to 1.0 wt% of glass fiber. Interestingly, the flexural strength decreased with further increase in fiber content up to 2%. A similar trend was also observed by adding 12 mm glass fiber into the matrix. However, the increment of the flexural strength was not as high as the geopolymer that was reinforced with 6 mm length fiber. It was expected with too low and high fiber contents at 0.2 and 2.0 wt%, causes high distribution density of small cracks formed in geopolymer which could consume a lot of energy, consequently lead to low strength in the flexural test [24].

Compressive Strength

Compression test of glass fiber reinforced geopolymer was carried out by Liangdong compression testing machine. Fig. 7 (a) and (b) presents the compressive strength of 6 mm and 12 mm glass fiber reinforced geopolymer, respectively.

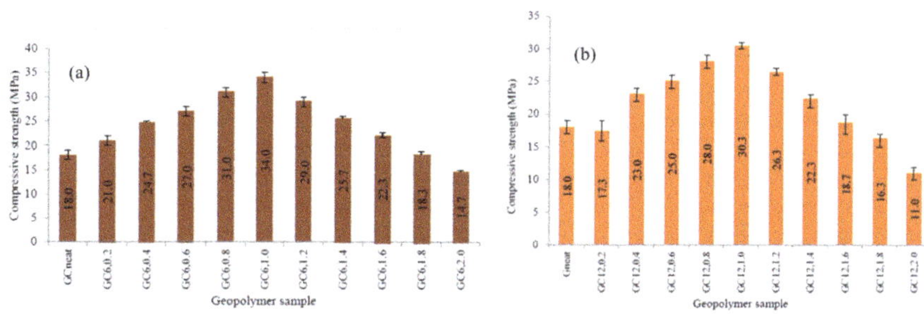

Fig. 7: Compressive strength of (a) 6mm glass fiber reinforced geopolymer; (b) 12mm glass fiber reinforced geopolymer

The addition of different concentrations of glass fiber into geopolymer affected the compressive strength. The addition of the glass fiber led to an improvement of the compressive strength of both 6 mm and 12 mm fiber length systems. The highest compressive strength in the geopolymer was achieved with the addition of 1 wt% of 6 mm length glass fiber at 89% greater neat geopolymer. In 12 mm system, the geopolymer was only managed to record 69% strength increment compared to the neat geopolymer. Noticeably, the highest strength obtained for 6 mm and 12 mm system was 34MPa and 30 MPA, respectively.

Three mechanisms can be related to the enhancement in compressive strength; firstly, the bridging effect of fiber on the pores (shown in Fig. 8), produced a compact and dense geopolymer structure. Secondly, the potential of the fiber to delay a propagation of micro crack, thus indirectly improved the energy absorption and ductility of the composite through a bridging effect, leading to a greater load carrying capacity. Thirdly, the well dispersion of the fiber in geopolymer matrix significantly created a strong bonding at the fiber / matrix interface. The optimal transferring of stress from matrix to fiber might occur by having a good interface interaction, thus significantly improve the compressive strength of composites [14 - 16].

Fig. 8: Bridging effect of fiber into pores

Summary

In present research work, a reinforced geopolymer with short glass fiber was successfully developed and studied for its mechanical properties, i.e. adhesion, flexural and compressive strength. Fly ash was used as a precursor in geopolymer development. Two fiber length of 6 mm and 12 mm was chosen to understand the influence of length to the geopolymer properties. It was noticed the incorporation of fiber into geopolymer managed to enhance the flexural and compressive strength of geopolymer. The present work proven that 6mm fiber length gives better strength to the geopolymer. It was also noticed further increased fiber content 1 wt% onwards, resulted to poor strength of geopolymer.

Acknowledgement

The research work was performed at the Department of Chemical Engineering, Universiti Teknologi PETRONAS. Great appreciations offered to Universiti Teknologi PETRONAS for the facilities and research fund; Yayasan Universiti Teknologi PETRONAS (YUTP), 015LC0-138 and 0153AA-H34.

References

[1] J. Davidovits, "Geopolymers: Inorganic polymeric new materials," Therm. Anal., vol. 37, pp. 1633-1656, 1991. https://doi.org/10.1007/BF01912193

[2] J. Davidovits, "Geopolymer Chemistry and Applications, 3rd ed; Institut Gèopolymère: Saint," 2011.

[3] H. Kuhli, "Slag cement and process of making the same," 1908

[4] A. O. Prudon, "The action of alkalis on blast-furnace slag," J. Soc. Chem. Ind. trnas Commun, vol. 59, pp. 191-202, 1940.

[5] J. Wastiels, X. Wu, S. Faignet, and G. Patfoort, "Mineral polymer based on fly ash," J. Resour. Manag. Technol., vol. 22, pp. 135-141, 1994.

[6] H. Le Chi, P. Louda, A. P. Periyasamy, T. Bakallova, and V. Kovacic, "Flexural Behaviour of Carbon Textile-Reinforced Geopolymer Composite Thin Plate," 2018. https://doi.org/10.3390/fib6040087

[7] M. Alzeer and K. J. D. Mackenzie, "Synthesis and mechanical properties of new fiber reinforced composites of inorganic polymers with natural wool fibers," J. Mater. Sci., vol. 47, pp. 6958-6965, 2012. https://doi.org/10.1007/s10853-012-6644-3

[8] M. Yasir, N. Amir, F. Ahmad, S. Ullah, and M. Jimenez, " Effect of basalt fibers dispersion on steel fi re protection performance of epoxy-based intumescent coatings," Prog. Org. Coatings, vol. 122, no. February, pp. 229-238, 2018. https://doi.org/10.1016/j.porgcoat.2018.05.029

[9] W. Li and J. Xu, "Mechanical properties of basalt fiber reinforced geopolymeric concrete under impact loading," vol. 505, pp. 178-186, 2009. https://doi.org/10.1016/j.msea.2008.11.063

[10] R. N. Swamy and P. S. Mangat, "Influence of fibre aggregate interaction on some properties of steel fiber reinforced concrete," Mater. Constr., vol. 7, pp. 307-314, 2000. https://doi.org/10.1007/BF02473840

[11] N. Ranjbar, S. Talebian, M. Mehrali, C. Kuenzel, H. S. Cornelis Metselaar, and M. Z. Jumaat, "Mechanisms of interfacial bond in steel and polypropylene fiber reinforced geopolymer composites," Compos. Sci. Technol., vol. 122, pp. 73-81, 2016. https://doi.org/10.1016/j.compscitech.2015.11.009

[12] N. Ranjbar and M. Zhang, "Fiber-reinforced geopolymer composites: A review," Cem. Concr. Compos., vol. 107, no. January 2020, pp. 1-43, 2020. https://doi.org/10.1016/j.cemconcomp.2019.103498

[13] N. K. Lee, J. G. Jang, and H. K. Lee, "Shrinkage characteristics of alkali-activated fly ash/slag paste and mortar at early ages," Cem. Concr. Compos, vol. 53, pp. 239-248, 2014. https://doi.org/10.1016/j.cemconcomp.2014.07.007

[14] Y. Ling, K. Wang, and C. Fu, "Shrinkage behavior of fly ash based geopolymer pastes with and without shrinkage reducing admixture," Cem. Concr. Compos., vol. 98, no. October 2018, pp. 74-82, 2019. https://doi.org/10.1016/j.cemconcomp.2019.02.007

[15] G. Masi, W. D. A. Rickard, M. C. Bignozzi, and A. Van Riessen, "The effect of organic and inorganic fibres on the mechanical and thermal properties of aluminate activated geopolymers," Compos. Part B Eng., vol. 76, pp. 218-228, 2015. https://doi.org/10.1016/j.compositesb.2015.02.023

[16] E. Mohseni, M. J. Kazemi, M. Koushkbaghi, B. Zehtab, and B. Behforouz, "Evaluation of mechanical and durability properties of fiber-reinforced lightweight geopolymer composites based on rice husk ash and nano-alumina," Constr. Build. Mater., vol. 209, pp. 532-540, 2019. https://doi.org/10.1016/j.conbuildmat.2019.03.067

[17] A. C. C. Trindade, P. H. R. Borges, and F. de Andrade Silva, "Mechanical behavior of strain-hardening geopolymer composites reinforced with natural and PVA fibers," Mater. Today Proc., vol. 8, pp. 753-759, 2019. https://doi.org/10.1016/j.matpr.2019.02.017

[18] P. Shouha, M. Swain, and A. Ellakwa, "The effect of fiber aspect ratio and volume loading on the flexural properties of flowable dental composite," Dent. Mater., vol. 30, pp. 1234-1244, 2014. https://doi.org/10.1016/j.dental.2014.08.363

[19] T. Lin, D. Jia, P. He, and M. Wang, "In situ crack growth observation and fracture behavior of short carbon fiber reinforced geopolymer matrix composites," Mater. Sci. Eng., vol. 527, pp. 2404-2407, 2010. https://doi.org/10.1016/j.msea.2009.12.004

[20] A. S. Jose and K. G. Sokor, "Effect of aspect ratio and loading on meachinal properties of prosopis juliflora fiber reinforced phenol formaldehyde composites," Fibres Text. East. Eur., vol. 25, pp. 59-64, 2017. https://doi.org/10.5604/01.3001.0010.2664

[21] "Standard test method for pull-off strength of coatings using portable adhesion testers," ASTM D4541-02.

[22] W. Zhao, Y. Wang, X. Wang, and D. Wu, "Fabrication, mechanical performance and tribological behaviors of polyacetal-fiber-reinforced metakaolin-based geopolymeric composites," Ceram. Int., vol. 42, no. 5, pp. 6329-6341, 2016. https://doi.org/10.1016/j.ceramint.2016.01.022

[23] P. J. Herrera-Franco and A. Valadez-González, "A study of the mechanical properties of short natural-fiber reinforced composites," Compos. Part B Eng., vol. 36, no. 8, pp. 597-608, 2005. https://doi.org/10.1016/j.compositesb.2005.04.001

[24] N. Ranjbar, S. Talebian, M. Mehrali, C. Kuenzel, H. S. Cornelis Metselaar, and M. Z. Jumaat, "Mechanisms of interfacial bond in steel and polypropylene fiber reinforced geopolymer composites," Compos. Sci. Technol., vol. 122, pp. 73-81, 2016. https://doi.org/10.1016/j.compscitech.2015.11.009

[25] K. Korniejenko, E. Fr, E. Pytlak, and M. Adamski, "Mechanical properties of geopolymer composites reinforced with natural fibers," vol. 151, pp. 388-393, 2016. https://doi.org/10.1016/j.proeng.2016.07.395

[26] A. T. Akono, S. Koric, and W. M. Kriven, "Influence of pore structure on the strength behavior of particle- and fiber-reinforced metakaolin-based geopolymer composites," Cem. Concr. Compos., vol. 104, no. July, p. 103361, 2019. https://doi.org/10.1016/j.cemconcomp.2019.103361

Deep eutectic solvents as environment-friendly solvents for separation processes in the oil and gas industry

ZAKARIA Hamdi[1,a*], SHABERDI Koshekov[2,b], WENG CHOH Cheong[1,c]

[1]Heriot-Watt University, Malaysia

[2]Politecnico di Torino, Italy

[a]Z.Hamdi@hw.ac.uk, [b]Shaberdi.k99@gmail.com, [c]alex.cheong@hw.ac.uk

Keywords: Deep Eutectic Solvents, Green Solvents, Environmental Friendly Materials

Abstract. Oil and gas are still the primary energy source in the world, despite the advancements in renewable energy. Many researchers showed chemical enhanced oil recovery, particularly surfactant is the most efficient method in many reservoirs. But most of these projects were not only commercially successful (due to high cost), but also involve environmental issues. Recently, Deep Eutectic Solvents (DESs) were discovered as a cheaper and greener alternative to conventional surfactants. DESs are cost-effective, easy to prepare, non-toxic, recyclable, biodegradable and environmental-friendly. Due to these advantages, DESs have found application in different fields including the oil and gas industry. The main applications of DESs in the oil and gas industry are separation processes such as dearomatization, desulfurization, purification of biodiesel and CO_2 capture. In this paper, the performance of DESs in these processes was reviewed as a new environmentally friendly method. The study introduces a way forward for current challenges with environmental sustainability.

Introduction to Deep Eutectic Solvents

Search for green chemical solvents is always in progress. Abbot et al. introduced DESs and defined them as new-generation sustainable solvents and potential alternatives to ionic liquids (ILs) and conventional organic solvents [1]. Fundamental properties of DESs are similar to ILs such as density, viscosity, conductivity, surface tension, high thermal stability and negligible volatility. DESs are cheaper, easy to prepare and more environmentally friendly compared to ILs. They are recyclable, biodegradable, non-corrosive and non-flammable [2].

The eutectic system is defined as "Eutectic system is a mixture of chemical compounds or elements that exhibit a single chemical composition that freezes at a lower temperature than any other composition" [3]. The eutectic point is the intersection of eutectic composition and eutectic temperature. DES is consisting a minimum of two green low-cost components that can form a eutectic mixture, having the melting point lower than its components [1].

Synthesis of DES

DESs are made by a special mixing of some of the hydrogen bond acceptors (HBAs) such as quaternary ammonium salts and a hydrogen-bond donor (HBD) at a specific ratio (molar) [4]. The mechanism by which a DES is formed is not complicated. The HBD interacts with the anion of the HBA (salt) and as a result, the effective size will be increased, at the same time, it reduces interactions of the anion with the involved cation. As a result, the produced DES melting point will be reduced [5, 6]. In general, DESs are characterized by a large depression of the freezing point and by being liquids at temperatures maximum at 150 °C. However, the majority are liquid somewhere between room temperature (20 °C) and as high as 70 °C.

The most studied DES constituents are choline chloride and urea which was first introduced by Abbott and co-workers in 2003. For choline chloride, the melting temperature is 302°C and for urea, the point is known to be 133°C. When the two materials were mixed at a 1:2 molar ratio of choline chloride: urea, a liquid eutectic mixture was formed. Choline chloride is considered green and it doesn't have any harmful effects [7]. It is mostly used as a vitamin for animals. Urea (carbamide) is well-known agricultural fertilizer. Since the constituents are not harmful when mixed resulting DES is also totally green [8].

Heating and grinding methods are two main types of methods that are used for the preparation of DES [9]. The most commonly used one is the heating method which is based on mixing and heating the compounds at 100 °C with stirring constantly until clear liquid is achieved. In the grinding method, compounds are mixed at room temperature and ground with a pestle in a mortar until a homogeneous liquid is formed.

DESs are produced without chemical reactions and need catalysts. While producing ILs involve raw material reacting chemically and by-products that will form impurities and as a result, they require extra purification work [10]. Unlike traditional ILs, the preparation of DESs is simple just by mixing and heating the compounds under stirring without the requirement of a catalyst. Moreover, overall mass efficiency and zero-emission during synthesis are achieved. As a result, all the involved salt and HBD are converted to DES, which as stated earlier, is completely environment-friendly as there are no unwanted by-products, including any vapours [11].

DESs are cheap due to their simple synthesis and cheap components. No purification is required and there is no waste production. DESs have high boiling points and low vapour pressures and most of their components are non-toxic and biodegradable [12, 13]. All these advantages increased interest in DESs as greener alternatives for ILs. Gurkan et al. reported the number of publications and citations for DES by 2019, as shown in figure 1 [14]. We can see that research in this area increases significantly in recent years.

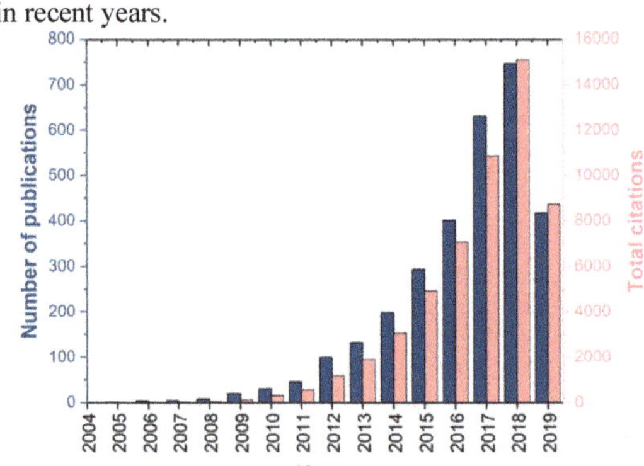

Fig. 1. Total Publications and Citations for DES [14]

Types of DES

DESs are mainly classified into four main types. Smith et al. showed the classification of DESs using the formula generally shown as $Cat^{+}X^{-}zY$, in which the Cat^{+} is a cation, X^{-} is an anion and also Y is Brønsted or Lewis acid. Moreover, z shows the Y molecule numbers that would interact with the anion) [15]. Table 1 shows DESs formed from a general MClx in the first part followed by quaternary ammonium salts.

Table1. A Types of DESs [15]

Types	General formula	Terms	Example
Type I	$Cat^+X^- + zMCl_x$	M = Zn, In, Sn, Al, Fe	$ChCl + ZnCl_2$
Type II	$Cat^+X^- + zMCl_x.yH_2O$	M = Cr, Ni, Cu, Fe, Co	$ChCl + CoCl_2.6H_2O$
Type III	$Cat^+X^- + zRZ$	Z = OH, COOH, $CONH_2$	ChCl + Urea
Type IV	$MCl_x + zRZ$	M = Zn, Al and Z = OH, $CONH_2$	$ZnCl_2$ + Urea

Type I: Composed of organic salts and metal salts.
Type II: Composed of organic salts and metal salt hydrate.
Type III: Composed of organic salts and HBD. This is the most common type.
Type IV: Composed of metal salts and HBD.

Properties of DESs

Freezing Point.

The low freezing point of DESs is one of the important properties which makes them unique. As mentioned earlier Abbot et al. showed the formation of a unique solvent with 12 °C as the freezing point, which is considerably lower than its components when mixed at a 1:2 molar ratio of choline chloride: urea [1]. Moreover, Kareem et al. presented a solvent formation with a melting point of – 66 °C from a 1:2 molar ratio of choline chloride: ethylene glycol [16].

Many factors can contribute towards the freezing point of the DESs such as the nature of the organic salt, the choice of HBD, the organic salt/HBD molar ratio and the anion of the organic salt. No correlation was concluded between the melting points of the individual components and the freezing point of the synthesized eutectic mixture [4].

Viscosity.

DESs have high viscosity due to the hydrogen bonding between components which reduces the mobility of compounds [5]. Van der Waals or electrostatic forces might cause high viscosity as well.

The DES viscosity is mainly affected by the individual component properties such as temperature, water content, and salt/HBD molar ratio [4]. Florindo et al. showed that by adding water to the system, the viscosity decreases drastically around 10-30 times [9]. At 25 °C, by hydrating ChCl: urea alongside 6 wt% of water, viscosity was reduced 13 times compared to the viscosity of dry ChCl: urea [17].

Du et al. studied the effect of water presence (2.5–20 wt%) on the three most common choline chloride-based DESs, namely (1) Reline (ChCl: urea), (2) Ethaline (ChCl: ethylene glycol) and (3) Glycerine (ChCl: glycerol) as shown in figure 2 [17].

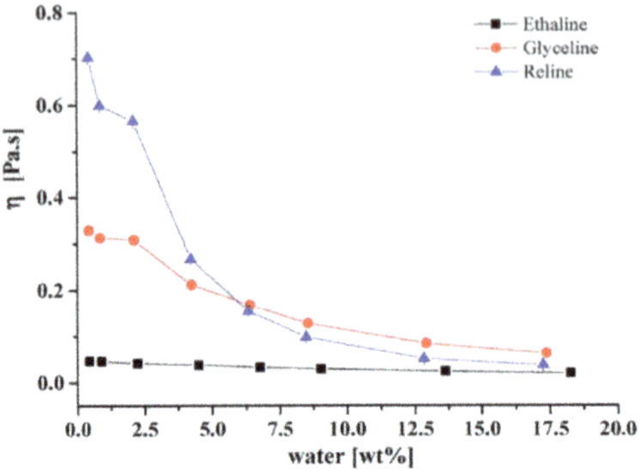

Fig. 2. Effect of Water Content on Viscosity for ChCl-based DESs at 20 °C [17]

It was observed that with the water content increasing, the viscosity decreased in all of the examined liquids. However, the decrease is not steady. The effect of the percentage of water on the viscosity of Reline was more than the other two liquids. Whereas, there is a small effect for Ethaline.

Abbott et al. showed the effect of ChCl concentration and temperature on the viscosity of the ChCl: glycerol mixture as shown in figure 3 [18]. The viscosity of pure glycerol (without ChCl) is around 1200 cP. It can be observed from the figure viscosity decreases by increasing temperature and ChCl concentration. A significant viscosity reduction of glycerol by the addition of ChCl was explained as due to the partial break of the strong hydrogen bond network of glycerol.

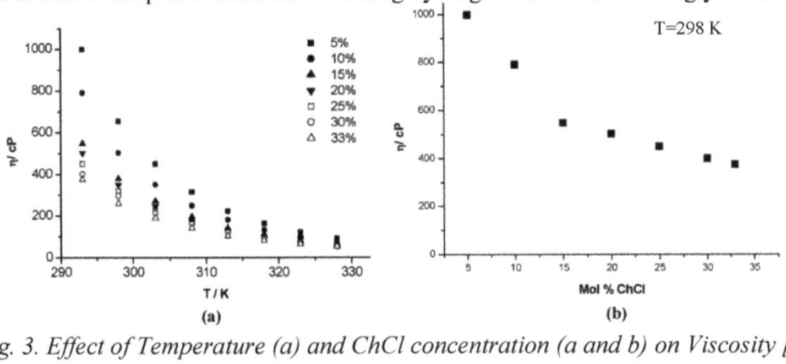

Fig. 3. Effect of Temperature (a) and ChCl concentration (a and b) on Viscosity [18]

Density.

Several studies were performed for the measurement of densities of DESs as a function of temperature [9, 12, 18]. Results showed that density decreases with increasing temperature which is the same behaviour as most of the liquid phase materials. The range of density of common DESs varies between 1.04-1.63 g/cm^3 at 25 °C and the salt/HBD molar ratio has a significant effect [4].

Conductivity.

Most of DESs have low ionic conductivities (lower than 2 mS cm^{-1} at 20 °C) due to their relatively high viscosity [4]. The viscosity of DES decreases with increasing temperature which increases ionic conductivity. Moreover, the ionic conductivity of DES increases with increasing water content due to the decrease in viscosity.

Applications of DESs for Efficient Separation Processes in the Oil and Gas Industry

Since the introduction of DES at beginning of the 21st century, their applications in areas are widening due to their advantages. The main advantages of DESs compared to ILs can be summarized as lower cost, non-toxicity, easy preparation, biodegradability and elimination of the purification step. Even though the main research area of DESs is chemistry [19], interest in other fields is also increasing including the oil and gas industry. The main applications of DESs are in the fields of extraction and separation processes, catalytic reactions and electroplating, material preparation, hydrometallurgy, biomass treatment and pharmaceutical applications [20, 21]. Important separation applications of DESs in oil and gas fields can be presented below:

Dearomatization

The separation of aromatic from aliphatic compounds is important to improve fuel properties and reduce environmental impacts. However, this is not an easy task due to very close boiling points and the formation of azeotropes. There are several processes such as liquid-liquid extraction (LLE), azeotropic distillation, extractive distillation and membrane processes. The most widely used method in industry is liquid-liquid extraction due to its low cost, simple operation and low energy consumption. In this method selection of extractants is vital. Organic solvents are the most commonly used chemicals but due to their volatility and toxicity, it is important to find novel solvents that will allow green and efficient separation [21].

ILs have been studied as an alternative to organic solvents by several researchers. Improved separation performance was achieved with most of the ILs compared to organic solvents. But due to the drawbacks of ILs as discussed before DESs started to be studied as an alternative to ILs for the removal of aromatics and to improve efficiency [22].

Kareem and his team investigated the separation of toluene and heptane with DES composed of tetra-butyl phosphonium-bromide as a salt and ethylene glycol or sulfolane as HBDs [23]. Tests were performed at different temperatures and different salt: HBD molar ratios. Results showed satisfactory performance of DESs for extracting aromatics and they exhibited higher selectivity at low concentrations of toluene compared to high ones.

Mulyono et al. performed the separation of different materials including ethylbenzene, toluene, benzene, and m-xylene (BTEX) aromatics from n-octane using a DES (tetrabutylammonium bromide + sulfolane) [24]. From the results benzene>toluene>m-xylene>ethylbenzene was the order of extraction performance on aromatics. Moreover, it was proven that the selective extraction of BTEX aromatics was possible from a mixture of aromatic and aliphatic compounds using DES.

Hou et al. investigated the separation of toluene from the toluene and n-hexane (or cyclohexane) mixtures using DES consisting of tetra-butyl phosphonium bromide (TBPB) and levulinic acid (LA) at different temperatures and molar ratios [25]. The optimal condition was achieved with a 6:1 molar ratio of levulinic acid: TBPB and a molar ratio of 6.4:1 of the DES: toluene at 20 °C. By distillation at 100 °C toluene in DES was completely recovered and DES was reused. It was concluded that the separation by DESs showed significance in terms of extraction rate and high selectivity.

Gouveia et al. evaluated the separation of toluene from n-heptane with a liquid-liquid extraction method using DESs [26]. DESs were based on various HBAs (cholinium chloride, tetrabutylammonium chloride, and benzyl-cholinium chloride) and levulinic acid as HBD with the 1:2 mole ratio of HBA: HBD. Improvement of the distribution coefficient led to the introduction of a more hydrophobic HBA and higher selectivity achieved by playing with the aromaticity of the DES.

Larriba et al. tested the extraction of aromatic hydrocarbons from reformer and pyrolysis gasoline using choline chloride-based DES [27]. In this study six choline chloride-based DESs (levulinic acid, malonic acid, ethylene glycol, phenylacetic acid, glycerol, and urea as HBDs) were tested. From the results, DES formed by choline chloride and levulinic acid (1:3) was selected as the most promising one.

Feng et al. studied the effective separation of aromatic hydrocarbons by pyridine-based DES [21]. The pyridine-based DESs were consisting of N-ethyl-pyridinium bromide and two HBDs (N-formyl morpholine and levulinic acid). By liquid-liquid extraction method, two ternary systems, toluene + n-heptane + DES and benzene + cyclohexane + DES were studied. N-formyl morpholine DES showed better performance in terms of separation compared to levulinic acid. The best separation for both DESs was at room temperature (20 °C) and a relatively low percentage of the aromatic concentration.

Desulfurization

Sulphur dioxide (SO_2) is a harmful gas to human health and the ecosystem. It is produced by the combustion of sulfur-containing fuels such as coal, oil or diesel. Exposure to SO_2 causes several health problems such as asthma, neurological disorders, wheezing and irritation of the skin are only some of them [28, 29]. Moreover, SO_2 damages the ecosystem as a result of acid rain. Due to these reasons, numerous regulations have been imposed to reduce SO_2 emissions. Maximum Sulphur content that is allowed in the fuel is one of these regulations. In recent years more strict policies are being taken [30].

Considering these circumstances, Sulphur removal from sour crudes is an important task to meet fuel standards. There are three Sulphur removal methods: Solvent extraction, catalytic hydro-desulphurization (HDS) and adsorption on molecular sieves. HDS has been the commonly used method to remove sulfur from crude oil but it is unable to remove polycyclic organic sulphides [30]. Due to these reasons and also due to the simplicity of the process, low energy consumption research on solvent extraction techniques is increasing. In this method, solvents are used to remove Sulphur compounds from the hydrocarbon stream based on the liquid-liquid extraction principle. Solvent selection is a key factor. Conventional solvents are alkanol amines. Past few years ILs have been widely researched. However, due to raised concerns about ILs such as toxicity and non-biodegradability of some ILs, in recent years DESs gained more attention as an alternative.

Li et al. reported the first paper on DESs as a solvent for desulfurization [31]. Desulfurization of fuels was carried out with ammonium-based DESs. Successful desulfurization of fuels was achieved from these DESs. Tetrabutylammonium chloride: polyethene glycol showed the optimum condition with an extraction efficiency of 82.83% for one cycle which is much higher than traditional solvents and ILs. Moreover amount of Sulphur in fuels decreased below 8.5 ppm.

Using $FeCl_3$-based DES, Gano et al. studied the desulfurization of simulated fuel and commercial diesel [32]. As a sulfur compound, dibenzothiophene and thiophene were present in the simulated fuel. Results showed 64% of extraction efficiency for dibenzothiophene and 44% for thiophene in a single-stage extraction. Due to the satisfactory results obtained for simulated fuel, the solvents were used for the desulfurization of real commercial diesel. Total sulfur removal from the diesel was 32%. Additionally, it was concluded also DESs could be regenerated and used repetitively without a significant decrease in their sulfur removal ability.

Makos and Boczkaj successfully applied DES (ChCl: phenol) for the desulfurization of model liquid fuel which contains thiophene, benzothiophene and dibenzothiophene [33]. The best conditions were the 1:4 mole ratio of ChCl: phenol, 2.5:1 volume ratio of DES: Fuel, 40 °C of temperature and 40 minutes of extraction time. In optimum conditions, the removal efficiency was 91.5% for thiophene, 95.4% for benzothiophene and 99.2% for dibenzothiophene in a one-step process. After three stages this value was 99.99%.

Purification of Biodiesel

Compared to fossil diesel, biodiesel is considered clean and renewable fuel which is less toxic and produces less amount of greenhouse gases [34]. After production biodiesel should be purified to pass standards. The glycerol is an undesirable by-product and must be removed before the biodiesel can be used as a fuel. It increases the viscosity of fuel and damages the injection system of diesel engines [15]. Glycerol is highly polar and unlikely to biodiesel. Liquid-liquid phase decantation is a common method for the separation of glycerol but a non-negligible amount remains and extra treatments are required [5].

Abbott et al. studied the removal of glycerol from biodiesel in two ways [35]. Initially, they added pure quaternary ammonium salt to glycerol-containing biodiesel to form a eutectic mixture with glycerol. This approach was not successful. It was stated this might be due to enthalpy formation. After that, they prepared DESs (ammonium salts: glycerol) and demonstrated the successful extraction of excess glycerol from biodiesel. For extraction of glycerol from the biodiesel the most effective molar ratio was 1:1 of glycerol: salt. Among the studied salts best results were achieved with choline chloride. The efficiency of glycerol removal was up to 99%. After studying the recovery of ammonium salts, it was observed that 25% of choline chloride could be recovered.

Shahbaz et al. presented the separation of glycerol from palm oil-based biodiesel using different combinations of DESs (ChCl: ethylene glycol and ChCl:2,2,2-trifluracetamide) [36]. All DESs showed successful results which removed all free glycerol with an efficiency of >99%. Moreover, Shahbaz et al. reported the removal of residual catalyst (KOH) from palm oil-based biodiesel using DESs.

Niawanti et al. performed a purification of biodiesel using choline chloride-based DESs. DES was a mixture of a 1:2 molar ratio of choline chloride: ethylene glycol. Best conditions were achieved when the molar ratio of crude biodiesel: DES was 1:4 with an efficiency of 96.60 %. The authors concluded that based on their study DES can be used for the purification of biodiesel from non-edible raw materials [37].

Salic et al. evaluated biodiesel purification in micro extractors using the water washing method and also using seven different DESs that are based on ChCl: glycerol and ChCl: ethylene glycol mixtures [38]. Three different sizes of micro extractors were used for both methods. Results from both processes compared to each other. DESs showed advantages over the water washing method such as higher efficiency, a lower amount of solvent usage, and less waste to be treated at the end of the process. It was possible to separate purified diesel and DES at the end of the micro extractor and recirculate DES. Figure 4 shows a graphical illustration of processes.

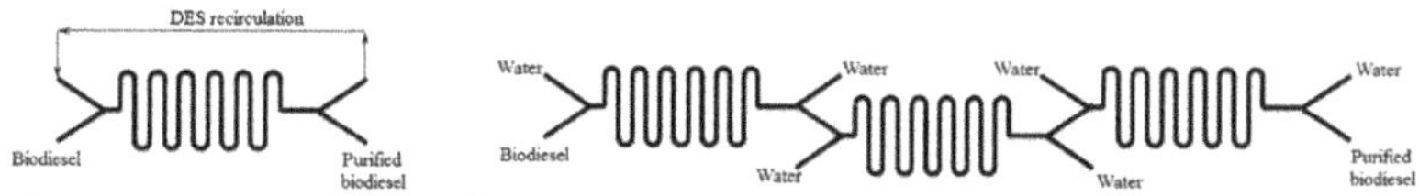

Figure 4. Glycerol Removal Using (a) DES and (b) Water Washing (Source: Salic et al., 2020)

CO_2 Capture

Global warming is one of the biggest issues in the world, especially in recent years. Great efforts are focused on reducing global warming. The emission of CO_2 gas is one of the major causes of global warming. Several technologies have been studied that are free of CO_2 emissions. But most of them are not yet at the stage of large-scale implementation. On the other hand, a large number of techniques and research are being conducted for CO_2 capture. Fossil fuels are the main cause of

CO_2 emissions which is the most important energy source. Hence, reducing CO_2 emissions from industry is essential [39, 40].

CO_2 capture using aqueous ethanolamine solutions such as monoethanolamine (MEA) and diethanolamine (DEA), has been introduced and studied broadly [41]. Even though studies showed efficient CO_2 capture, there are several drawbacks reported that need to be taken into account. The main drawbacks include corrosion problems due to the corrosive property of ethanolamine solutions, high cost due to high energy consumption, degradation, high solvent volatility and toxicity. Due to these drawbacks, ILs were proposed [42] but they have several limitations also as discussed in previous parts such as high cost and not being ecologically advantageous. DESs emerged recently as being free from most of these disadvantages [43].

As a first quantitative study, Li et al. presented DESs as a potential CO_2-capturing method [44]. Solubility of CO_2 in ChCl: urea DES was measured at different temperatures, pressures and molar ratios. Results showed solubility increases with increasing pressure and decreases with increasing temperature. Different molar ratios didn't have a significant impact on solubility.

Mirza et al. performed experiments and thermodynamic modelling on three different DESs to analyze the solubility of CO_2 at different temperatures and pressures [45]. Three different DESs were reline (ChCl:urea with 1:2 molar ratio), ethaline (ChCl:ethylene glycol with 1:2 molar ratio) and malinine (ChCl:malic acid:ethylene glycol with 1.3:1:2.2 molar ratio). The experimental study observed Henry's constants were between the ranges of 3.7-6.1 MPa. Thermodynamic modelling was used to correlate experimental data and results showed excellent agreement with a 1.6% deviation.

Ali et al. examined different types of DESs for the feasibility of the CO_2-capturing process [46]. The best performance was seen from methyl triphenylphosphonium bromide: monoethanolamine DES with a molar ratio of 1:6. Moreover, it was shown that the energy requirement with MEA-based DESs is lower than the energy requirement with the classical MEA process. However, a large solvent/feed ratio was required for all investigated DESs.

Zhang et al. also investigated CO_2 capture by DESs [47]. Results showed excellent CO_2 absorption capacity of DESs up to 1.00 mol CO_2/mol DES, and by changing the mole ratio of HBA and HBD, the absorption process could be adjusted. Moreover, high thermal stability and excellent recyclability were observed in the studied DESs.

Conclusion

In this paper, the applications of DESs in separation processes were reviewed. The DESs showed great potential in DES as a new-class green solvent for the oil and gas industry. The study proved the effectiveness of DES not only in improving oil recovery, but also in CO_2 capture, purification of biodiesel, desulfurization, and dearomatization. This study also introduces a way forward for current challenges with environmental sustainability. The new class solvents are not only green but also exhibit better performance than current similar solvents with added economical benefits.

Acknowledgement

The authors would like to thank ministry of higher education of Malaysia. This paper is under project FRGS/1/2019/TK02/HWUM/02/4.

References

[1] Andrew P. Abbott, Glen Capper, David L. Davies, Raymond K. Rasheed, and Vasuki Tambyrajah. "Novel solvent properties of choline chloride/urea mixtures." Chemical Communications, no. 1 (2003): 70-71. https://doi.org/10.1039/b210714g

[2] Zakaria Hamdi, Prashanth Nair Kumaran, and Oluwole Talabi. Investigation and Optimization of Deep Eutectic Solvent in Enhanced Oil Recovery Using Numerical Simulation.

in Abu Dhabi International Petroleum Exhibition & Conference. 2020. https://doi.org/10.2118/202633-MS

[3] Julio de Paula, Atkins'physical Chemistry. 2006: WH Freeman and Company New York.

[4] Tracy El Achkar, Sophie Fourmentin, and Hélène Greige-Gerges. "Deep eutectic solvents: An overview on their interactions with water and biochemical compounds." Journal of Molecular Liquids 288 (2019): 111028. https://doi.org/10.1016/j.molliq.2019.111028

[5] Qinghua Zhang, Karine De Oliveira Vigier, Sébastien Royer, and François Jérôme. "Deep eutectic solvents: syntheses, properties and applications." Chemical Society Reviews 41, no. 21 (2012): 7108-7146. https://doi.org/10.1039/c2cs35178a

[6] Daniel Carriazo, María Concepción Serrano, María Concepción Gutiérrez, María Luisa Ferrer, and Francisco del Monte. "Deep-eutectic solvents playing multiple roles in the synthesis of polymers and related materials." Chemical Society Reviews 41, no. 14 (2012): 4996-5014. https://doi.org/10.1039/c2cs15353j

[7] EFSA Panel on AdditivesProducts or Substances used in Animal Feed. "Scientific Opinion on safety and efficacy of choline chloride as a feed additive for all animal species." EFSA Journal 9, no. 9 (2011): 2353. https://doi.org/10.2903/j.efsa.2011.2353

[8] Ivo Mavrovic, A. Ray Shirley Jr., and G. R. "Buck" Coleman, Urea, in Kirk-Othmer Encyclopedia of Chemical Technology. p. 1-21.

[9] C. Florindo, F. S. Oliveira, L. P. N. Rebelo, Ana M. Fernandes, and I. M. Marrucho. "Insights into the Synthesis and Properties of Deep Eutectic Solvents Based on Cholinium Chloride and Carboxylic Acids." ACS Sustainable Chemistry & Engineering 2, no. 10 (2014): 2416-2425. https://doi.org/10.1021/sc500439w

[10] Seda Keskin, Defne Kayrak-Talay, Uğur Akman, and Öner Hortaçsu. "A review of ionic liquids towards supercritical fluid applications." The Journal of Supercritical Fluids 43, no. 1 (2007): 150-180. https://doi.org/10.1016/j.supflu.2007.05.013

[11] Hua ZhaoGary A. Baker. "Ionic liquids and deep eutectic solvents for biodiesel synthesis: a review." Journal of Chemical Technology & Biotechnology 88, no. 1 (2013): 3-12. https://doi.org/10.1002/jctb.3935

[12] M. Hayyan, M. A. Hashim, A. Hayyan, M. A. Al-Saadi, I. M. AlNashef, M. E. Mirghani, and O. K. Saheed. "Are deep eutectic solvents benign or toxic?" Chemosphere 90, no. 7 (2013): 2193-5. https://doi.org/10.1016/j.chemosphere.2012.11.004

[13] Katherine D. Weaver, Hye Jin Kim, Jiazeng Sun, Douglas R. MacFarlane, and Gloria D. Elliott. "Cyto-toxicity and biocompatibility of a family of choline phosphate ionic liquids designed for pharmaceutical applications." Green Chemistry 12, no. 3 (2010): 507-513. https://doi.org/10.1039/b918726j

[14] Burcu Gurkan, Henry Squire, and Emily Pentzer. "Metal-Free Deep Eutectic Solvents: Preparation, Physical Properties, and Significance." The Journal of Physical Chemistry Letters 10, no. 24 (2019): 7956-7964. https://doi.org/10.1021/acs.jpclett.9b01980

[15] Emma L. Smith, Andrew P. Abbott, and Karl S. Ryder. "Deep Eutectic Solvents (DESs) and Their Applications." Chemical Reviews 114, no. 21 (2014): 11060-11082. https://doi.org/10.1021/cr300162p

[16] Mukhtar A. Kareem, Farouq S. Mjalli, Mohd Ali Hashim, and Inas M. AlNashef. "Phosphonium-Based Ionic Liquids Analogues and Their Physical Properties." Journal of Chemical & Engineering Data 55, no. 11 (2010): 4632-4637. https://doi.org/10.1021/je100104v

[17] Cuiling Du, Binyuan Zhao, Xiao-Bo Chen, Nick Birbilis, and Haiyan Yang. "Effect of water presence on choline chloride-2urea ionic liquid and coating platings from the hydrated ionic liquid." Scientific Reports 6, no. 1 (2016): 29225. https://doi.org/10.1038/srep29225

[18] Andrew P. Abbott, Robert C. Harris, Karl S. Ryder, Carmine D'Agostino, Lynn F. Gladden, and Mick D. Mantle. "Glycerol eutectics as sustainable solvent systems." Green Chemistry 13, no. 1 (2011): 82-90. https://doi.org/10.1039/C0GC00395F

[19] Xiang Ge, Changdong Gu, Xiuli Wang, and Jiangping Tu. "Deep eutectic solvents (DESs)-derived advanced functional materials for energy and environmental applications: challenges, opportunities, and future vision." Journal of Materials Chemistry A 5, no. 18 (2017): 8209-8229. https://doi.org/10.1039/C7TA01659J

[20] A. N. El-hoshoudy, F. S. Soliman, E. M. Mansour, T. Zaki, and S. M. Desouky. "Experimental and theoretical investigation of quaternary ammonium-based deep eutectic solvent for secondary water flooding." Journal of Molecular Liquids 294 (2019): 111621. https://doi.org/10.1016/j.molliq.2019.111621

[21] Shanhua Feng, Jianjun Sun, Zhongqi Ren, Zhiyong Zhou, Fan Zhang, and Wei Liu. "Effective separation of aromatic hydrocarbons by pyridine-based deep eutectic solvents." The Canadian Journal of Chemical Engineering 97, no. 12 (2019): 3138-3147. https://doi.org/10.1002/cjce.23606

[22] A. B. Pereiro, J. M. M. Araújo, F. S. Oliveira, J. M. S. S. Esperança, J. N. Canongia Lopes, I. M. Marrucho, and L. P. N. Rebelo. "Solubility of inorganic salts in pure ionic liquids." The Journal of Chemical Thermodynamics 55 (2012): 29-36. https://doi.org/10.1016/j.jct.2012.06.007

[23] Mukhtar A. Kareem, Farouq S. Mjalli, Mohd Ali Hashim, Mohamed K. O. Hadj-Kali, Fatemeh Saadat Ghareh Bagh, and Inas M. Alnashef. "Phase equilibria of toluene/heptane with tetrabutylphosphonium bromide based deep eutectic solvents for the potential use in the separation of aromatics from naphtha." Fluid Phase Equilibria 333 (2012): 47-54. https://doi.org/10.1016/j.fluid.2012.07.020

[24] Sarwono Mulyono, Hanee F. Hizaddin, Inas M. Alnashef, Mohd A. Hashim, Anis H. Fakeeha, and Mohamed K. Hadj-Kali. "Separation of BTEX aromatics from n-octane using a (tetrabutylammonium bromide + sulfolane) deep eutectic solvent - experiments and COSMO-RS prediction." RSC Advances 4, no. 34 (2014): 17597-17606. https://doi.org/10.1039/c4ra01081g

[25] Yucui Hou, Zhongyi Li, Shuhang Ren, and Weize Wu. "Separation of toluene from toluene/alkane mixtures with phosphonium salt based deep eutectic solvents." Fuel Processing Technology 135 (2015): 99-104. https://doi.org/10.1016/j.fuproc.2014.11.001

[26] Andreia S. L. Gouveia, Filipe S. Oliveira, Kiki A. Kurnia, and Isabel M. Marrucho. "Deep Eutectic Solvents as Azeotrope Breakers: Liquid-Liquid Extraction and COSMO-RS Prediction." ACS Sustainable Chemistry & Engineering 4, no. 10 (2016): 5640-5650. https://doi.org/10.1021/acssuschemeng.6b01542

[27] Marcos Larriba, Miguel Ayuso, Pablo Navarro, Noemí Delgado-Mellado, Maria Gonzalez-Miquel, Julián García, and Francisco Rodríguez. "Choline Chloride-Based Deep Eutectic

Solvents in the Dearomatization of Gasolines." ACS Sustainable Chemistry & Engineering 6, no. 1 (2018): 1039-1047. https://doi.org/10.1021/acssuschemeng.7b03362

[28] Weiming Song, Jianan Zhou, Bao Wang, Shu Li, and Rijin Cheng. "Production of SO2 Gas: New and Efficient Utilization of Flue Gas Desulfurization Gypsum and Pyrite Resources." Industrial & Engineering Chemistry Research 58, no. 44 (2019): 20450-20460. https://doi.org/10.1021/acs.iecr.9b04403

[29] Jin Han, Kejian Li, Hanyun Cheng, and Liwu Zhang. "A Green Desulfurization Technique: Utilization of Flue Gas SO2 to Produce H2 via a Photoelectrochemical Process Based on Mo-Doped BiVO4." Frontiers in Chemistry 5 (2017): https://doi.org/10.3389/fchem.2017.00114

[30] Davannendran Chandran, Mohammad Khalid, Rashmi Walvekar, Nabisab Mujawar Mubarak, Swapnil Dharaskar, Wai Yin Wong, and Thummalapalli Chandra Sekhara Manikyam Gupta. "Deep eutectic solvents for extraction-desulphurization: A review." Journal of Molecular Liquids 275 (2019): 312-322. https://doi.org/10.1016/j.molliq.2018.11.051

[31] Changping Li, Dan Li, Shuangshuang Zou, Zhuo Li, Jingmei Yin, Ailing Wang, Yingna Cui, Zhilong Yao, and Qi Zhao. "Extraction desulfurization process of fuels with ammonium-based deep eutectic solvents." Green Chemistry 15, no. 10 (2013): 2793-2799. https://doi.org/10.1039/c3gc41067f

[32] Zaharaddeen S. Gano, Farouq S. Mjalli, Talal Al-Wahaibi, Yahya Al-Wahaibi, and Inas M. AlNashef. "Extractive desulfurization of liquid fuel with FeCl3-based deep eutectic solvents: Experimental design and optimization by central-composite design." Chemical Engineering and Processing: Process Intensification 93 (2015): 10-20. https://doi.org/10.1016/j.cep.2015.04.001

[33] Patrycja MakośGrzegorz Boczkaj. "Deep eutectic solvents based highly efficient extractive desulfurization of fuels - Eco-friendly approach." Journal of Molecular Liquids 296 (2019): 111916. https://doi.org/10.1016/j.molliq.2019.111916

[34] Samah E. E. Warrag, Cor J. Peters, and Maaike C. Kroon. "Deep eutectic solvents for highly efficient separations in oil and gas industries." Current Opinion in Green and Sustainable Chemistry 5 (2017): 55-60. https://doi.org/10.1016/j.cogsc.2017.03.013

[35] Andrew P. Abbott, Glen Capper, Katy J. McKenzie, and Karl S. Ryder. "Electrodeposition of zin*f* tin alloys from deep eutectic solvents based on choline chloride." Journal of Electroanalytical Chemistry 599 (2007): 288-294. https://doi.org/10.1016/j.jelechem.2006.04.024

[36] Kaveh Shahbaz, Farouq Mjalli, Mohd Hashim, and Inas Alnashef. "Using Deep Eutectic Solvents for the Removal of Glycerol from Palm Oil-Based Biodiesel." Journal of Applied Sciences 24 (2010): 3349-3354. https://doi.org/10.3923/jas.2010.3349.3354

[37] Helda Niawanti, Siti Zullaikah, and M. Rachimoellah. "Purification of biodiesel by choline chloride based deep eutectic solvent." AIP Conference Proceedings 1840, no. 1 (2017): 050006. https://doi.org/10.1063/1.4982280

[38] Anita Šalić, Ana Jurinjak Tušek, Martin Gojun, and Bruno Zelić. "Biodiesel purification in microextractors: Choline chloride based deep eutectic solvents vs water." Separation and Purification Technology 242 (2020): 116783. https://doi.org/10.1016/j.seppur.2020.116783

[39] Zakaria Hamdi, Raja Zuhaili Aimran Raja Zainal Raffik, Oluwole Talabi, Urval Patel, Eric Mackay, and Mahmood Bataee. CO2 Mineral Trapping in Basaltic Formation During CO2 Storage. in SPE Canadian Energy Technology Conference. 2022. https://doi.org/10.2118/208935-MS

[40] Z. Hamdi, M. Awang, M. Bataee, and M. Kamyab. Cold CO2 injection for balancing interfacial tension decrease and viscosity increase. in Society of Petroleum Engineers - SPE Asia Pacific Oil and Gas Conference and Exhibition 2016. 2016. https://doi.org/10.2118/182149-MS

[41] Mohammad R. M. Abu-Zahra, John P. M. Niederer, Paul H. M. Feron, and Geert F. Versteeg. "CO2 capture from power plants: Part II. A parametric study of the economical performance based on mono-ethanolamine." International Journal of Greenhouse Gas Control 1, no. 2 (2007): 135-142. https://doi.org/10.1016/S1750-5836(07)00032-1

[42] "Feasibility of using ionic liquids for carbon dioxide capture." International Journal of Environmental Technology and Management 4, no. 1-2 (2004): 105-115. https://doi.org/10.1504/IJETM.2004.004624

[43] Yizhak author aut http id loc gov vocabulary relators aut Marcus, Deep Eutectic Solvents. 1st ed. 2019. ed. 2019.

[44] Y.-J. Li, C.-S. Zhao, C.-R. Qu, L.-B. Duan, Q.-Z. Li, and C. Liang. "CO2 Capture Using CaO Modified with Ethanol/Water Solution during Cyclic Calcination/Carbonation." Chemical Engineering & Technology 31, no. 2 (2008): 237-244. https://doi.org/10.1002/ceat.200700371

[45] Nouman Rafique Mirza, Nathan J. Nicholas, Yue Wu, Kathryn A. Mumford, Sandra E. Kentish, and Geoffrey W. Stevens. "Experiments and Thermodynamic Modeling of the Solubility of Carbon Dioxide in Three Different Deep Eutectic Solvents (DESs)." Journal of Chemical & Engineering Data 60, no. 11 (2015): 3246-3252. https://doi.org/10.1021/acs.jced.5b00492

[46] Emad Ali, Mohamed K. Hadj-Kali, Sarwono Mulyono, and Inas Alnashef. "Analysis of operating conditions for CO2 capturing process using deep eutectic solvents." International Journal of Greenhouse Gas Control 47 (2016): 342-350. https://doi.org/10.1016/j.ijggc.2016.02.006

[47] Na Zhang, Zhaohe Huang, Haiming Zhang, Jingwen Ma, Bin Jiang, and Luhong Zhang. "Highly Efficient and Reversible CO2 Capture by Task-Specific Deep Eutectic Solvents." Industrial & Engineering Chemistry Research 58, no. 29 (2019): 13321-13329. https://doi.org/10.1021/acs.iecr.9b02041

Determination of green solvents' water activity by inference of freezing point depression

WENG CHOH, Cheong[1,a*], ZAKARIA Hamdi[1,b], ANTONIN Chapoy[2,c], BEHZAD Partoon[3,d]

[1] School of Energy, Geoscience, Infrastructure and Society, Heriot-Watt University, 62200 Putrajaya, Malaysia

[2] Institute of GeoEnergy Engineering, Heriot-Watt University, Edinburgh EH14 4AS, UK

[3] Biological and Chemical Engineering Department, Aarhus University, Hangøvej 2, 8000 Aarhus N, Denmark

[a]alex.cheong@hw.ac.uk, [b]z.hamdi@hw.ac.uk, [c]a.chapoy@hw.ac.uk, [d]behzad.partoon@bce.au.dk

Keywords: Green Solvents, Deep Eutectic Solvents, Water Activity, Freezing Point Depression

Abstract. Deep eutectic solvents (DESs) are novel green solvents with great potential in many industrial applications. Due to their high tunability, DESs can be synthesized and prepared into various concentrations by dilution with water. This work revolutionises the determination of selected DESs' water activity using inferencing their respective freezing point depression. Using the Hildebrand-Scott equation, the freezing point depression of the respective DESs in various concentrations is then correlated to the water activity. This inferential correlation provides an alternate but convenient way of determining a solution's water activity in absence of a water activity meter. The highest freezing point depression of almost -10 °C was observed on the choline chloride and ethylene glycol in 1:2 molar ratio DES at 50 wt% concentration, with corresponding water activity determined at 0.909.

Introduction

In the interest of sustainability, there has been growing interest around the world in the search for green solvents. An effective solvent not only acts as a key enabler in chemical processes, but the sustainability framework also prescribes the solvent to comply with major assessment criteria, ranging from conformance to green chemistry principles, the satisfaction of life cycle evaluations, as well as fulfilment of the "10R Framework" such as recycle, reuse, reduce etc associated with the modern circularity philosophy [1].

Deep eutectic solvents (DESs) of late have been classified as green alternatives to molecular solvents and ionic liquids (ILs) [2]. They are characterised as eutectic mixtures formed between quaternary ammonium, phosphonium or sulfonium salts and hydrogen bond donors (HBDs) with a freezing temperature lower than their respective individual components. Marcus [3] concluded that DESs have the advantage over other conventional industrial solvents in that they are relatively easy to be synthesised and highly tunable. Their synthesis ingredients are also easily available. Most importantly, the synthesis and production of these green solvents are highly cost-competitive, through relatively benign processes which support the sustainability framework compared to other common industrial organic solvents [4].

Deep Eutectic Solvents

A eutectic system is a mixture of chemical compounds or elements that demonstrates a single and homogenous chemical composition which freezes at a significantly lower temperature than its forming components [5]. The term DESs has been specified mainly to differentiate them from authentic ionic liquids (ILs) in light of different characteristics in some solvent properties, and the fact that some DES components are non-ionic. The great freezing point depression exhibited by the eutectic mixture does not occur in ILs but only in DESs as a unique form of change in the solvent's physical property. Abbot et al. noted that the freezing point depression of DESs can be as high as hundreds of degrees Celsius relative to its components [6].

DESs have a wide range of industrial applications with established performance. Fourmentin, Costa Gomes [7] presented several successful cases using DESs in innovative medical and pharmaceutical formulations in enhancing selective drugs' bioavailability. There are also growing interests in DESs for biomass treatments, metal processing, separation and extraction for pharmaceutical and medical applications. DESs have also been studied by Zhou et al. with prospective findings that when used as novel reaction mediums, they will help to improve enzymatic activity stabilization in the fields of bio-catalysis [8]. DESs have also been proven to offer cost-effective, reliable and efficient performance for CO_2 capture with excellent regeneration capacity [9]. Specific to the oil and gas industry, DESs have been identified as potential greener and more cost-effective alternatives compared to conventional surfactants in improving enhanced oil recovery (EOR) in the exploration and production of heavy oils [10, 11].

Selection and Synthesis of DESs and Derivatives

DESs are broadly classified into four categories depending on the nature and properties of the mixing components [12]. In this work, the focus is given on a few DES species from DES III, IV and the novel type DES V (also known as natural DES, or NADES) [13]. Table 1 below lists the classification of various conventional DES types, along with the general formula.

Table 1: Conventional DES classifications with general formula. [3]

Type	General Formula
I (quarternary salt + metal halide)	$Cat^{+}X^{-}zMCl_x$
II (quarternary salt + hydrated metal halide)	$Cat^{+}X^{-}zMCl_x \cdot yH2O$
III (quarternary salt + HBD)	$Cat^{+}X^{-}zRZ$
IV (metal halide + HBD)	$MCl_x{}^{+}RZ = MCl_{x-1}+\cdot RZ+MCl_{x+1}$

Fifteen DESs of various concentrations have been synthesized and are subject to freezing point depression measurement. The list of DESs prepared in this work is listed in Table 1 below, along with the respective Hydrogen Bond Acceptor (HBA) and Hydrogen Bond Donor(s) (HBDs) and their respective corresponding molar composition in each DES specimen.

Table 2: List of DESs in various concentrations synthesised for water activity determination.

DES ID	Description and Concentration	HBA: HBD molar composition
1	DES IIIa 2 wt%	choline chloride: ethylene glycol 1:2
2	DES IIIa 5 wt%	As above
3	DES IIIa 10 wt%	As above
4	DES IIIa 20 wt%	As above
5	DES IIIa 50 wt%	As above
6	DES IIIb 10 wt%	choline chloride: glycerol 1:2
7	DES IIIc 10 wt%	choline chloride: urea: glycerol 1:1:1
8	DES IIId 10 wt%	choline chloride: levulinic acid 1:2
9	DES IV 10 wt%	zinc chloride: ethylene glycol 1:4
10	DES V 10 wt% solubilised in Polysorbate 20	thymol: coumarin 2:1
11	DES V 10 wt% solubilised in Poly Suga Mulse	As above
12	DES Va 10 wt% solubilised in Polysorbate 20	thymol: coumarin: ethylene glycol 2:1:2
13	DES Va 10 wt% solubilised in Poly Suga Mulse	As above
14	DES Vb 10 wt% solubilised in Polysorbate 20	thymol: coumarin: glycerol 2:1:1
15	DES Vb 10 wt% solubilised in Poly Suga Mulse	As above

Water Activity

Water activity (a_w) of a solution is the ratio between the solution vapour pressure alone when subject to a completely undisturbed balance with the surrounding air media, and the vapour pressure of distilled water under an identical condition. A water activity of 0.90 means the vapour pressure is 90% of that of pure water. Water activity increases with higher temperatures. Mathematically, water activity (a_w) is defined as:

$$a_w = \frac{p}{p_0} \qquad \text{Eq. 1}$$

of which $a_w =$ water activity of solvent
$p =$ vapour pressure of the solvent
$p_0 =$ vapour pressure of pure water

Water activity is important in the evaluation of salting out effects of any aqueous two-phase system (ATPS). In addition, water activity also enables the determination of other critical thermodynamic properties of the solution, such as the solvent's vapour pressure and osmotic coefficients. [14]. The water activity of a solvent can be measured directly using an industrial-scale water activity meter. Alternately, water activity can be determined through experiments, and compared against the various Non-random Two Liquid (NTRL) models and derivatives available to examine the measured water activity accuracy [15]. The resultant water activity data can be utilised to assess the solute-solvent interactions and behaviours in an array of water activity and vapour pressure data.

This work prescribes an alternate empirical method using simple and easily available laboratory apparatus, to determine the DESs' water activity in absence of an industrial-scale water activity meter.

Materials Research Proceedings 29 (2023) 337-343 https://doi.org/10.21741/9781644902516-37

Experimental Procedures

The DESs freezing point depression is measured using a modified and simplified inflected temperature method adopted by Haghighi et al. [16]. This method requires a temperature bath for cooling control, thermometers to measure temperatures of the solution as well as the reference liquid, which in this case is ethylene glycol. For this work, the Pour and Cloud Point Apparatus (Model: Anton-Parr) is used to provide temperature bath control. The experimental setup is represented in Figure 1 below.

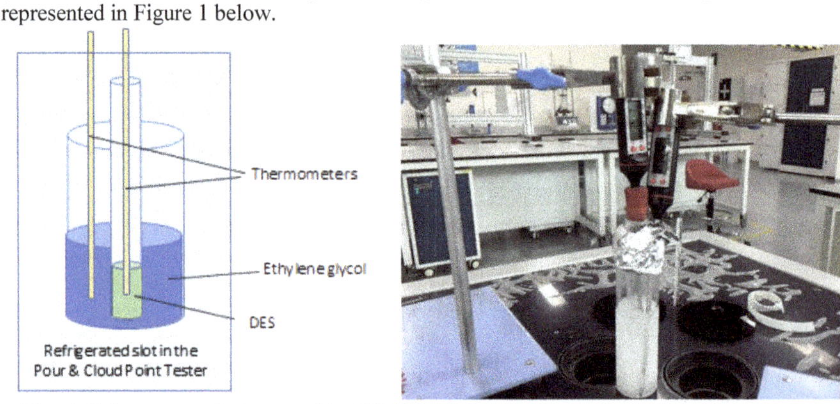

Fig. 1: Schematic and picture of the DES freezing point depression setup. The container was retracted from the refrigerated slot for a better illustration shown in the picture.

The Cloud and Pour Point Tester temperature control was set to 5°C. Upon temperature stabilisation, the Pour Point Tester container filled with 20-ml reference liquid i.e. ethylene glycol, and the test tube filled with 5-ml DES specimen were lowered down to the refrigerated slot. Both the reference liquid and DES specimen were installed with separate digital thermometers to record the respective temperature readings. The temperature control was set by reducing the refrigerated slot of the Cloud and Pour Tester gradually until such time that the temperature of the DES specimen increased upon reaching its freezing point. This phenomenon is indicated by a corresponding increase in the test tube content absolute temperature following the liberation of the exotherm from the DES specimen upon reaching its freezing point.

The experiment was then continued by adjusting the Cloud and Pour Point temperature control continuously with reducing temperature set point until such time that the absolute temperatures in both the reference liquid and DES specimen reached parity. When this was achieved, the Cloud and Pour Point Tester temperature control was reset to 5 °C and both thermometer readings were recorded every minute. A differential temperature plot between the reference liquid and DES specimen ($T_{DES} - T_{ref}$) was generated. The point of which an inflection on the differential temperature plot occurred corresponded with the inferred freezing point depression (ΔT_f) of the DES specimen.

Results and Discussions

Figure 2 below provided an example of an inflection temperature run on DES IIIa of 20 wt% concentration.

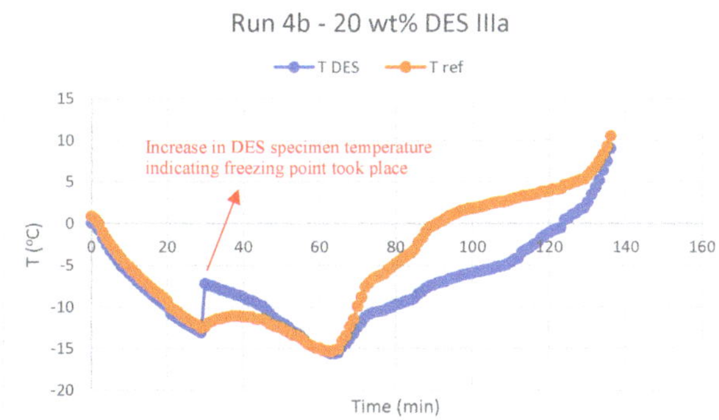

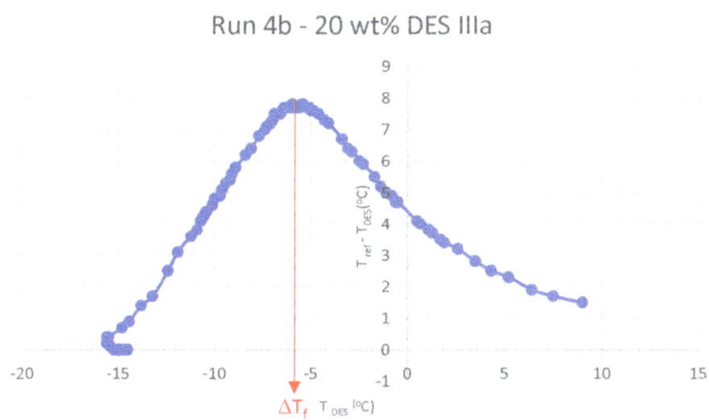

Fig. 2: (a) Absolute temperature plot, and (b) differential temperatures vs T_{DES} plot for DES IIIa 20wt%

From Figure 2(b) above, the inflection point occurred at $T_{DES} = -5.3°C$. This point signified the freezing point depression (ΔT_f) of the DES specimen tested in the experimental run. With this freezing point depression ascertained, the DES specimen freezing point (T_f) relative to 273.15K could then be calculated. Using Hildebrand & Scott's equation similar to previous studies, the respective DES specimen water activity could then be determined [17].

$$\ln a_w = \frac{-\Delta H_f}{R}\frac{(T_f - T)}{T_f T} + \frac{\Delta C_f}{R}\left[\frac{(T_f - T)}{T} - ln\left(\frac{T_f}{T}\right)\right] \qquad \text{Eq. 2}$$

of which T is the DES specimen freezing point in K, T_f signifies the water freezing point in K, ΔH is the latent heat of water which is 6,088 J/K/mol, ΔC_f is the change of specific heat of water equals 38.7 J/K/mol and R is the gas constant.

Table 3 below summarized the inference water activity of all DES specimens examined using the inflection temperature method.

From the runs generated, a higher concentration of the same DES yielded greater freezing point depression, which in return resulted in lower water activity. In terms of DESs III and IV at 10 wt% concentration, their empirical water activity appeared to be within the range between 0.97 and

0.98. For NADES DES V solubilized in either solubilizer of Polysorbate 20 and Poly Suga Mulse (PSM), the resultant freezing point depression of all these solutions yielded insignificant difference, with a corresponding water activity in the range of 0.95 to 0.96.

Table 3: Summary of freezing point depression and calculated water activity of the various DES specimens.

DES ID	Description and Concentration	Freezing Point Depression, ΔT_f (K)	Freezing Point, T_f (K)	Ln a_w	Water Activity, a_w
1	DES IIIa 2 wt%	-0.2	272.95	-0.0019	0.998
2	DES IIIa 5 wt%	-0.8	272.35	-0.0078	0.992
3	DES IIIa 10 wt%	-3.2	269.95	-0.0310	0.969
4	DES IIIa 20 wt%	-6.0	267.20	-0.0578	0.944
5	DES IIIa 50 wt%	-9.8	263.35	-0.0953	0.909
6	DES IIIb 10 wt%	-1.9	271.25	-0.0184	0.982
7	DES IIIc 10 wt%	-3.1	270.05	-0.0301	0.970
8	DES IIId 10 wt%	-2.2	270.95	-0.0213	0.979
9	DES IV 10 wt%	-3.4	269.75	-0.0330	0.968
10	DES V P20 10 wt%	-4.9	268.25	-0.0476	0.954
11	DES V PSM 10 wt%	-5.1	268.05	-0.0495	0.952
12	DES V(a) P20 10 wt%	-5.3	267.85	-0.0514	0.950
13	DES V(a) PSM 10 wt%	-4.5	268.65	-0.0437	0.957
14	DES V(b) P20 10 wt%	-4.6	268.55	-0.0446	0.956
15	DES V(b) PSM 10 wt%	-4.4	268.75	-0.0427	0.958

Conclusions

The water activity of the various DES specimens can be inferenced by determining its freezing point depression following the inflected temperature procedures. The experimental setup is relatively hassle-free using common equipment available in most research laboratories, and most importantly cost-effective relative to the investment in procuring a water activity meter. Further works may involve a comparison of the inference water activity values against measurements using industrial water activity meter, to determine the precision of such an inference method. The inference results can also be compared against modelling output, as a means for calibrating the model accuracy.

References

[1] Hessel, V., et al., Sustainability of Green Solvents - Review and Perspective. Green chemistry : an international journal and green chemistry resource : GC, 2021.

[2] Florindo, C., L.C. Branco, and I.M. Marrucho, Quest for Green-Solvent Design: From Hydrophilic to Hydrophobic (Deep) Eutectic Solvents. ChemSusChem, 2019. 12(8): p. 1549-1559. https://doi.org/10.1002/cssc.201900147

[3] Marcus, Y., Deep Eutectic Solvents by Yizhak Marcus. 1st ed. 2019. ed. 2019: Cham : Springer International Publishing : Imprint: Springer. https://doi.org/10.1007/978-3-030-00608-2

[4] Paiva, A., et al., Natural Deep Eutectic Solvents - Solvents for the 21st Century. ACS Sustainable Chemistry & Engineering, 2014. 2(5): p. 1063-1071. https://doi.org/10.1021/sc500096j

[5] Abbott, A.P., et al., Ionic Liquids Based upon Metal Halide/Substituted Quaternary Ammonium Salt Mixtures. Inorganic Chemistry, 2004. 43(11): p. 3447-3452. https://doi.org/10.1021/ic049931s

[6] Abbott, A.P., et al., Deep Eutectic Solvents Formed between Choline Chloride and Carboxylic Acids: Versatile Alternatives to Ionic Liquids. Journal of the American Chemical Society, 2004. 126(29): p. 9142-9147. https://doi.org/10.1021/ja048266j

[7] Fourmentin, S., M. Costa Gomes, and E. Lichtfouse, Deep eutectic solvents for medicine, gas solubilization and extraction of natural substances / Sophie Fourmentin, Margarida Costa Gomes, Eric Lichtfouse, editors. 1st ed. 2021. ed. 2021: Cham, Switzerland : Springer. https://doi.org/10.1007/978-3-030-53069-3

[8] Zhou, Y., et al., Natural deep eutectic solvents as green and biocompatible reaction medium for carbonic anhydrase catalysis. International journal of biological macromolecules, 2021. 190: p. 206-213. https://doi.org/10.1016/j.ijbiomac.2021.08.221

[9] Ahmad, N., et al., Understanding the CO2 capture performance by MDEA-based deep eutectics solvents with excellent cyclic capacity. Fuel (Guildford), 2021. 293: p. 120466. https://doi.org/10.1016/j.fuel.2021.120466

[10] Hamdi, Z.K., Prashanth Nair ; Talabi, Prashanth Nair Investigation and Optimization of Deep Eutectic Solvent in Enhanced Oil Recovery Using Numerical Simulation, in Abu Dhabi International Petroleum Exhibition & Conference. 2020, OnePetro. https://doi.org/10.2118/202633-MS

[11] Hadj-Kali, M.K., et al., Application of deep eutectic solvents and their individual constituents as surfactants for enhanced oil recovery. Colloids and surfaces. A, Physicochemical and engineering aspects, 2015. 487: p. 221-231. https://doi.org/10.1016/j.colsurfa.2015.10.005

[12] Zhang, Q., et al., Deep eutectic solvents: syntheses, properties and applications. Chemical Society Reviews, 2012. 41(21): p. 7108-7146. https://doi.org/10.1039/c2cs35178a

[13] van Osch, D.J.G.P., et al., A Search for Natural Hydrophobic Deep Eutectic Solvents Based on Natural Components. ACS Sustainable Chemistry & Engineering, 2019. 7(3): p. 2933-2942. https://doi.org/10.1021/acssuschemeng.8b03520

[14] Jimenez, Y.P., et al., Measurement and modelling of water activity, density and viscosity of the KClO4 + poly(ethylene glycol) + H2O system at various temperatures. Journal of molecular liquids, 2019. 286: p. 110922. https://doi.org/10.1016/j.molliq.2019.110922

[15] Majdan-Cegincara, R. and M.T. Zafarani-Moattar, Measurements and Correlation of Water Activity in Aqueous Solutions Containing Diphenhydramine-Hydrochloride Drug, (D+)-Galactose, (D−)-Fructose, (D+)-Lactose, and Sucrose at 298.15 K. Journal of chemical and engineering data, 2019. 64(12): p. 5671-5680. https://doi.org/10.1021/acs.jced.9b00694

[16] Haghighi, H., A. Chapoy, and B. Tohidi, Freezing Point Depression of Electrolyte Solutions: Experimental Measurements and Modeling Using the Cubic-Plus-Association Equation of State. Industrial & Engineering Chemistry Research, 2008. 47(11): p. 3983-3989. https://doi.org/10.1021/ie800017e

[17] Miyawaki, O., et al., Activity and Activity Coefficient of Water in Aqueous Solutions and Their Relationships with Solution Structure Parameters. Bioscience, Biotechnology, and Biochemistry, 1997. 61(3): p. 466-469. https://doi.org/10.1271/bbb.61.466

Fuel-synthesis wastewater treatment and purple non-sulfur bacteria biomass and pigments production: Effect of vitamin concentration

SULTAN Shaikh[1,a*], NAIM Rashid[1,b], GORDON McKay[1,c] and HAMISH R. Mackey[1,2,d*]

[1] Division of Sustainable Development, College of Science and Engineering, Hamad bin Khalifa University, Qatar Foundation, Doha, Qatar

[2] Department of Civil and Natural Resources Engineering, University of Canterbury, Private Bag 4800, Christchurch 8140, New Zealand

[a]sushaikh@hbku.edu.qa, [b]narashid@hbku.edu.qa, [c]gmckay@hbku.edu.qa, [d]hamish.mackey@canterbury.ac.nz

Keywords: Fuel Synthesis Wastewater, Purple Phototrophic Bacteria, Carotenoids, Bacteriochlorophylls, Vitamins, Micronutrients

Abstract. The effect of different concentrations of ATCC vitamin supplement on the production of purple non-sulfur bacteria biomass (suspended and biofilm), carotenoids (Crts) and bacteriochlorophylls (BChls) from fuel-synthesis wastewater (FSW) was investigated in this study. The results suggest that maximum COD removal (3465 ± 125 mg/L) from FSW was obtained at a vitamin media concentration of 10 mL/L. The best optimal condition for increased biomass production from suspended and biofilm culture is the addition of vitamin at concentrations of 20 mL/L and 10 mL/L, respectively. The optimum condition for maximum Crts and BChls production from suspended and biofilm culture was found as 5 mL/L and 0 mL/L, respectively. Hence, the biomass response to vitamins is complex and vitamin addition can be used to prioritize different end goals with purple-non sulfur bacteria treatment systems and biomass utilization.

Introduction

Photosynthetic bacteria (PSB) have been the focus of research since the 1960s for their potential use in treating different wastewaters including olive mill wastewater, poultry slaughterhouse wastewater, palm oil mill effluent wastewater, and dairy wastewater [1]. PSB have been found extremely effective at removing COD, phosphate, nitrate, and hydrogen sulfide from different wastewaters and have a high conversion ratio of organic carbon present in wastewater into valuable biomass [2]. PSB can attain high cellular concentrations of proteins, carotenoids (Crts), bacteriochlorophylls (BChls), biopolymers, antibacterial compounds, and pantothenic acid compared with many types of bacteria [3].

Among these products, Crts and BChls are of particular interest since they are useful pigments for numerous industries. They are applicable to food, drug, and cosmetic products. Crts, for instance, have been utilized as food coloring agents and cosmetic additives, while BChl is a potential chemical for photodynamic therapy [4–6]. These natural biochemicals are in growing demand in the modern industrial sector. To address this requirement, researchers are focusing on increasing the Crts and BChls content of PSB.

PSB are usually anoxygenic and have four different functional groups: purple non-sulfur bacteria (PNSB), purple sulfur bacteria, green non-sulfur bacteria, and green sulfur bacteria [8]. The most commonly used PSB are PNSB bacteria since they are heterotrophs allowing conversion of organic substrates in wastewater. The synthesis of pigments by bacterial culture in culture medium has already been reported [9]. However, the cost of culture medium contributes significantly to the cost of PNSB biomass and pigments production. One way to cut costs is to use non-hazardous wastewater instead of the culture medium as the substrate. Therefore, integrated application of PSB to treat relatively benign wastewaters with Crts and BChls production from the biomass is gaining attention [7]. However, few studies have examined the biological synthesis of PNSB cells and pigments from wastewater [2,7].

In this study fuel-synthesis wastewater (FSW) is selected as culture medium. FSW is the byproduct of the Fischer-Tropsch process where hydrogen and carbon monoxide are converted over a catalyst at high temperature and pressure to synthetic alkanes. It is a major industry in Qatar, the study location. FSW is transparent, having low levels of colloidal or settleable particles, and is deficient in nutrients and has a very high chemical oxygen demand (COD). Due to the synthetic conversion process and pure inputs, FSW contains very few chemicals apart from organics consisting of C, H, and O; mainly volatile fatty acids, alcohols and alkanes [10].

As PNSB also requires other nutrients like nitrogen, phosphate, trace elements, and vitamins for growth [11], these components were added to the FSW. Since no study has been conducted to optimize the vitamin composition for mixed culture PNSB biomass and pigments production, this study investigated the impact when the vitamin concentration was varied. The goal of this research is to treat FSW while producing maximum biomass and pigments. Key questions are a) What will be the effect of different vitamin concentrations on fuel synthesis wastewater treatment? b) What will be the effect of different vitamin concentrations on PNSB biofilm formation? c) What will be the effect of different vitamin concentrations on Crts and BChls production?

Materials and Methods

Microorganisms and Growth Media. This study used a mixed culture of bacteria dominant with PNSB. The substrate used in this experiment was fuel synthesis wastewater (FSW) with the addition of KH_2PO_4 (3.03 g/L), NH_4Cl (3.03 g/L), $NaHCO_3$ (4.29 g/L), ATCC trace minerals supplement (MD-TMS) (10 mL/L), and different concentrations of ATCC vitamin supplement (MD-VS) (0, 5,10, and 20 mL/L).

Experimental Setup and Culture Condition. In this study eight 250 mL DURAN bottles with a working volume of 235 mL were used as photobioreactors (PBRs) with four different ATCC vitamin (MD-VS) concentrations. To provide mixing and light all PBRs were placed in an incubating shaker (New Brunswick Innova 44, Canada) at 150 rpm, 35 °C temperature, and continuous light supply of 0.7 W/m^2. The initial pH of all conditions was 7.10 ± 0 with no control of pH during the study duration. A piece of green shade (25 cm^2) was suspended in each bottle as a supporting material for biofilm development. Green garden shade was used as it is porous, economical, thin, and readily available in the market and has previously been demonstrated in our group as an effective PNSB support stratum. The experiment was conducted for 30 days, and each condition was run in duplicate.

Analytical Methods. An UV-3600 plus spectrophotometer (Shimadzu, Japan) was used to assess the growth of PNSB in a suspended culture and at the end of the experiment volatile suspended solids (VSS) and total suspended solids (TSS) were determined using standard procedures [12]. PNSB biofilm biomass from the green shade at the end of experiment was released with distilled water and measured by TSS and VSS.

To extract the supernatant for COD, samples of effluent wastewater were centrifuged at 23,366 g for 10 minutes in a centrifuge (Sorvall LYNX 6000, Thermo Scientific, USA) and filtered using 0.45 μm polyethersulfone syringe filters (Nalgene). The COD was determined using Hach high range Test'N'tube kits and the USEPA Reactor Digestion Method 8000 [13].

Suspended and biofilm biomass Crts and BChls were extracted using acetone and acetone/methanol (7:2 v/v) solvent, respectively, and were quantified by measuring absorbance at 480 nm and 771 nm by UV-vis spectrophotometer as previously described [14].

Statistical Analysis. Given the uncertainties in equality of variance associated with small sample sizes, Welch's analysis of variance (ANOVA) was used to analyze statistical differences between samples. For similar reasons, the conservative Bonferroni test was used for post-hoc testing of significance. JASP software was used to conduct the analysis and α = 5% was used as the level of significance threshold.

Results and Discussion

Fuel Synthesis Wastewater Treatment. The COD removal was maximum (3465 ± 125 mg/L) at 10 mL/L vitamin concentration followed by 0, 20 and 5 mL/L. It shows that PNSB have the ability to remove enough COD (2355 ± 95 mL/L) even without the addition of vitamin (0 mL/L). A strange drop in COD removal was observed at 5 mg/L of vitamin solution, to less than half that of either the 0 or 10 mg/L conditions (Table 1). One possible explanation for the observed behavior is the selection pressure for different organisms at different vitamin concentrations and the resulting microbial community that is established.

Table 1. COD removal and biomass production at different vitamin concentrations. Alphabetical letters indicate significant differences from a Bonferroni post-hoc test conducted following a significant Welch's ANOVA test.

Vitamin concentration [mL/L]	COD Removal [mg/L]	Biomass [mg]	
		Suspended	Biofilm
0	2355 ± 95^{a}	200 ± 0.0^{a}	15 ± 5^{bcehi}
5	1065 ± 235^{bc}	75 ± 25^{ac}	12.5 ± 2.5^{bcehi}
10	3465 ± 125^{ade}	112.5 ± 62.5^{ace}	45 ± 20^{acehi}
20	2205 ± 235^{ace}	300 ± 25^{adfg}	10 ± 0.0^{bcehi}

PNSB Growth at Different Vitamin Concentration. At 20 mL/L vitamin concentration the biomass production from suspended culture was found to be higher (300 mg) than at any other concentration. However, when comparing the vitamin concentrations examined in biofilm cultures, the 10 mL/L concentration yields the highest biomass output of 45 mg. Except for the vitamin concentration of 0 mL/L in suspended culture, all other concentrations demonstrated an increase in biomass production with increasing vitamin concentration. However, in biofilm culture, biomass increases and decreases as the vitamin concentration increases (Table 1). The significant difference between all conditions is shown in table 1 which suggests that suspended biomass production from 20 mL/L vitamin concentration is different ($p<0.05$) from 5 mL/L and 10 mL/L but is similar to ($p>0.05$) 0 mL/L vitamin concentration. However, there is no significant difference between biomass of the biofilm cultures under the different vitamin concentrations.

Suspended culture biomass production can be confirmed from the absorbance value of the suspended culture. However, absorbance is not the accurate measurement for the biomass present in the suspended culture as PNSB cells get attached and detached from the green shade and can change the absorbance value. PNSB growth in the suspended culture was nearly the same (0.8 ± 0.0) across all vitamin concentrations on day 4. On day 8, PNSB growth was somewhat greater at a concentration of 0 and 20 mL/L compared to 5 and 10 mL/L of vitamin concentration. On day 12 and 30, the PNSB growth was found to be a maximum when the vitamin concentration was 20 mL/L, followed by 0 mL/L (Fig. 1a). Likewise, biofilm VSS can be confirmed from the visual observation of biofilm formation on green shade. Higher PNSB biofilm formation was observed at 10 mL/L vitamin concentration followed by 0, 5 and 20 mL/L vitamin concentration (Fig. 1b-d). Therefore, it is suggested that the best optimal condition for increased biomass production for suspended and biofilm growth is the addition of vitamin at concentrations of 20 mL/L and 10 mL/L, respectively.

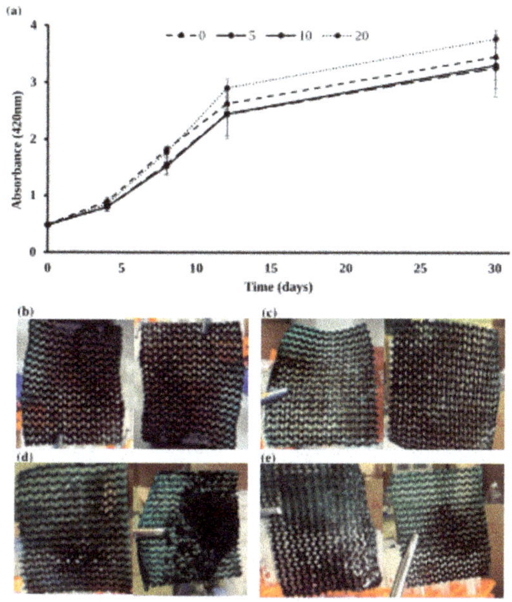

Fig. 1 (a) PNSB suspended growth at different vitamin concentrations. Biofilm formation on green shade at vitamin concentration of (b) 0 mL/L (c) 5 mL/L (d) 10 mL/L, and (e) 20 mL/L

Content of Carotenoids and Bacteriochlorophylls. The content of Crts and BChls in the suspended culture was found higher at a vitamin concentration of 5 mL/L and it was greater than in all other concentrations. However, from the biofilm culture, the maximum Crts content was observed from the condition where no vitamin was added. Except 0 mL/L vitamin concentration, the Crts concentration in the suspended culture decreased with an increase in vitamin concentration. Whereas, in the biofilm culture, the Crts concentration increases with an increase in vitamin concentration. However, BChls were always higher in suspended culture as compared to the biofilm culture. This was most noticeable with the 5 mg/L vitamin condition. Hence the optimum condition for maximum Crts and BChls production from suspended and biofilm culture was 5 mL/L and 0 mL/L. However, it should be noted that there was no significant difference between the biofilm and suspended culture values or between vitamin concentrations for both pigments (Table 2).

Table 2. PNSB Crts and BChls contents at different vitamin concentrations. Alphabetical letters indicate significant differences from a Bonferroni post-hoc test conducted following a Welch's ANOVA test.

Vitamin concentration [mL/L]	**Crts content [μg/g]**		**BChls content [μg/g]**	
	Suspended	**Biofilm**	**Suspended**	**Biofilm**
0	7.4 ± 1.4^a	15.4 ± 3.7^a	2.90 ± 0.2^a	2.60 ± 0.4^a
5	12.5 ± 6.9^a	5.1 ± 1.4^a	4.40 ± 2.3^a	1.24 ± 0.0^a
10	7.5 ± 5.5^a	6.8 ± 0.2^a	1.34 ± 0.5^a	1.05 ± 0.2^a
20	5.0 ± 0.9^a	10.5 ± 0.7^a	1.73 ± 0.3^a	1.68 ± 0.0^a

The Crts/BChls ratios are shown in Table 3. The Crts/BChls ratio of suspended culture for all conditions is in the range of 2.5 – 4.6. Biofilm cultures showed a slightly higher range of 4.1 to 6.5. However, there is no significant difference between all the vitamin concentration conditions and suspended and biofilm cultures. Wang et al. [2], while studying the influence of various concentrations of NaCl on the production of Crts and BChls by photosynthetic bacteria (PSB), discovered a ratio of 2.1 – 2.8. The ratio of PSB Crts/BChls was reported by Zhou et al. [15] to be between 1.15 and 1.42 when the authors treated wastewater using a variety of light sources. However, there is a lack of in-depth studies in the literature on PNSB Crts and BChls production from various substrates/wastewaters with different vitamin concentrations.

Table 3. Crts/BChls ratio of PNSB suspended and biofilm culture at different vitamin concentrations. Alphabetical letters indicate significant differences from a Bonferroni post-hoc test conducted following a Welch's ANOVA test.

Vitamin concentration [mL/L]	Crts/BChls	
	Suspended	**Biofilm**
0	2.5 ± 0.3^a	5.8 ± 0.4^a
5	2.8 ± 0.0^a	4.1 ± 1.2^a
10	4.7 ± 2.1^a	6.6 ± 0.8^a
20	2.9 ± 0.0^a	6.3 ± 0.7^a

Summary

It was demonstrated that culture media without addition of ATCC vitamin can be used to grow PNSB for COD removal as well as pigments production. The culture, without the addition of vitamin, removed around 55% of COD and produced Crts in the range of 7.4–15.4 μg/g and BChls in the range of 2.6–2.9 μg/g. Furthermore, it has also the potential to induce PNSB biofilm growth on green shade.

Acknowledgement

The authors would like to acknowledge the funding support by the Qatar National Research Foundation and Qatar Ministry of Environment and Climate Change for the financial support through the Food Security Program, project number MME01-0910-190029.

References

[1] H. Lu, G. Zhang, X. Dai, L. Schideman, Y. Zhang, B. Li, H. Wang, A novel wastewater treatment and biomass cultivation system combining photosynthetic bacteria and membrane bioreactor technology, Desalination. 322 (2013) 176–181. https://doi.org/10.1016/j.desal.2013.05.007

[2] H. Wang, A. Yang, G. Zhang, B. Ma, F. Meng, M. Peng, H. Wang, Enhancement of carotenoid and bacteriochlorophyll by high salinity stress in photosynthetic bacteria, Int. Biodeterior. Biodegrad. 121 (2017) 91–96. https://doi.org/10.1016/j.ibiod.2017.03.028

[3] H. Lu, G. Zhang, Z. Zheng, F. Meng, T. Du, S. He, Bio-conversion of photosynthetic bacteria from non-toxic wastewater to realize wastewater treatment and bioresource recovery: A review, Bioresour. Technol. 278 (2019) 383–399. https://doi.org/10.1016/j.biortech.2019.01.070

[4] C. Rudolf, H. Grammel, Fructose metabolism of the purple non-sulfur bacterium Rhodospirillum rubrum: Effect of carbon dioxide on growth, and production of

bacteriochlorophyll and organic acids, Enzyme Microb. Technol. 50 (2012) 238–246. https://doi.org/10.1016/j.enzmictec.2012.01.007

[5] Z. Aksu, A. Tuğba Eren, Carotenoids production by the yeast Rhodotorula mucilaginosa: Use of agricultural wastes as a carbon source, Process Biochem. 40 (2005) 2985–2991. https://doi.org/10.1016/j.procbio.2005.01.011

[6] C. Paliwal, I. Pancha, T. Ghosh, R. Maurya, K. Chokshi, S. V. Vamsi Bharadwaj, S. Ram, S. Mishra, Selective carotenoid accumulation by varying nutrient media and salinity in Synechocystis sp. CCNM 2501, Bioresour. Technol. 197 (2015) 363–368. https://doi.org/10.1016/j.biortech.2015.08.122

[7] Q. Zhou, P. Zhang, G. Zhang, Biomass and carotenoid production in photosynthetic bacteria wastewater treatment: Effects of light intensity, Bioresour. Technol. 171 (2014) 330–335. https://doi.org/10.1016/j.biortech.2014.08.088

[8] Q. Zhang, Z. Zhang, Biological Hydrogen Production From Renewable Resources by Photofermentation, 2018. https://doi.org/10.1016/bs.aibe.2018.03.001

[9] F.S. Kuo, Y.H. Chien, C.J. Chen, Effects of light sources on growth and carotenoid content of photosynthetic bacteria Rhodopseudomonas palustris, Bioresour. Technol. 113 (2012) 315–318. https://doi.org/10.1016/j.biortech.2012.01.087

[10] P.J. Boogaard, J.C. Carrillo, L.G. Roberts, G.F. Whale, Toxicological and ecotoxicological properties of gas-to-liquid (GTL) products. 1. Mammalian toxicology, Crit. Rev. Toxicol. 47 (2017) 121–144. https://doi.org/10.1080/10408444.2016.1214676

[11] S. Sali, H.R. Mackey, The application of purple non-sulfur bacteria for microbial mixed culture polyhydroxyalkanoates production, Rev. Environ. Sci. Biotechnol. 20 (2021) 959–983. https://doi.org/10.1007/s11157-021-09597-7

[12] APHA, Standard methods for the examination of water and wastewater, 2012. https://doi.org/10.5860/choice.49-6910

[13] HACH, Oxygen Demand, Chemical. USEPA reactor digestion method, HACH Method 8000. (2019) 1–6. https://www.hach.com/asset-get.download-en.jsa?id=7639983817

[14] S. Chumpol, D. Kantachote, T. Nitoda, H. Kanzaki, Administration of purple nonsulfur bacteria as single cell protein by mixing with shrimp feed to enhance growth, immune response and survival in white shrimp (Litopenaeus vannamei) cultivation, Aquaculture. 489 (2018) 85–95. https://doi.org/10.1016/j.aquaculture.2018.02.009

[15] Q. Zhou, P. Zhang, G. Zhang, Biomass and pigments production in photosynthetic bacteria wastewater treatment: Effects of light sources, Bioresour. Technol. 179 (2015) 505–509. https://doi.org/10.1016/j.biortech.2014.12.077

An integrated approach for treatment and biomass resource recovery from juice industry wastewater using purple non-sulfur bacteria

NAIM Rashid[1,a*], SULTAN Shaikh[1,b], HAMISH R Mackey[1,2,c] and GORDON McKay[1,d*]

[1] Division of Sustainable Development, College of Science and Engineering, Hamad bin Khalifa University, Qatar Foundation, Doha, Qatar

[2] Department of Civil and Natural Resources Engineering, University of Canterbury, Private Bag 4800, Christchurch 8140, New Zealand

[a]narashid@hbku.edu.qa, [b]sushaikh@hbku.edu.qa, [c]hamish.mackey@canterbury.ac.nz, [d]gmckay@hbku.edu.qa

Keywords: Purple Non-Sulfur Bacteria, Juice Wastewater, Biomass Production, Carbon Sources

Abstract. This study aims to treat juice industry wastewater and recover resources using purple non-sulfur bacteria (PNSB). Simulated juice wastewater was prepared and used as a feedstock for PNSB. Carbon sources, including sucrose, fructose, glucose, and tris-Na-citrate, were added to different bottles separately and inoculated with a mixed culture of PNSB. Tris-Na-citrate showed higher growth and biomass production (650±70 mg/L) than the other carbon sources. The PNSB growth was associated with the pH change; tris-Na-citrate showed different pH change behavior than the other carbon sources, i.e., it showed a lesser decrease in the medium pH than the others. The effect of substrate (using glucose) concentration on PNSB biomass production was determined in the following experiment. A substrate concentration of 400 mg/L (as COD) showed the highest growth, whereas 800 mg/L (as COD) returned the highest biomass production (776±1 mg/L). Total organic carbon removal (TOC) increased with an increase in substrate concentration. Inorganic carbon removal initially decreased but then increased over the growth period; however, it was also directly correlated with initial substrate concentration. Lastly, the effect of pH on TOC, IC, and COD removal was investigated. A pH of 6.0 or above (up to 8.0) was required to remove organic and inorganic carbon.

Introduction

The juice industry is widespread globally and generates a large volume of wastewater during packing, washing, and extraction. Juice industrial wastewater (JIW) is characterized by high COD, suspended solids, dissolved solids, and low pH [1]. Several treatment methods, including coagulation, flocculation, dissolved air flotation, and membrane filtration, are employed to treat JIW. Implementing these methods at an industrial scale is not currently cost-effective. In this perspective, biological treatments are desirable due to relatively low energy inputs. Recently, anaerobic treatment using PNSB has received considerable research attention [2-4]. They are considered favorable due to their unique metabolic features. They can utilize a variety of organic and inorganic carbon sources as a substrate, govern high substrate yield, and can switch their growth metabolism between photo-heterotrophy and chemo-heterotrophy, depending on environmental conditions. PNSB biomass can generate value-added bio-products such as single-cell protein, bioplastics, and photo-pigments. Hence, PNSB-based JIW treatment offers a dual benefit—wastewater treatment and biomass production [5, 6]. This study aims to utilization of JIW as a substrate for PNSB, their treatment performance, and biomass production. So far, no study has been reported that demonstrates the use of PNSB to treat JIW [7].

Materials Research Proceedings 29 (2023) 350-354 https://doi.org/10.21741/9781644902516-39

Methodology

Simulated juice wastewaters were prepared by dissolving glucose, fructose, sucrose, or tris-Na citrate individually to a theoretical chemical oxygen demand (COD) concentration of 1000 mg/L to create four different simulated juice wastewaters. Salts including potassium chloride (1.35 g/L), magnesium sulfate heptahydrate (9.93 g/L), sodium chloride (20 g/L), calcium chloride dihydrate (1.0 g/L), and sodium bio-carbonate (75 g/L) were added. Essential nutrients, trace metals, and vitamins were added per the ATCC-2 medium. The effect of glucose concentration as a sole substrate was also tested in a separate experiment. For this, glucose 0 to 1200 mg- as COD/L was added to the bottles. The remaining components of the simulated juice wastewater composition were the same. The solutions were transferred to 160 mL serum bottles and inoculated with a mixed microbial culture of PNSB. The initial pH of the solutions was noted but not adjusted during the treatment process. Using a crimper, the flasks were tightly sealed with a rubber stopper and an aluminum rim. The flasks were placed in an incubator shaker (Innova®44, Eppendorf), stirring continuously at 200 RPM with a temperature of 35 °C ± 2, under the illumination of 7.0 W/m^2 (Figure 1).

PNSB growth was monitored by measuring absorbance at 420 nm using a spectrophotometer (UV-3600 Plus, Shimadzu). Total suspended solids (TSS) were measured by the standard gravimetric method (APHA 2012), and COD was measured using HACH kits. Total organic carbon (TOC), total nitrogen, and inorganic carbon were measured using a TOC-L analyzer with a TNM-1 module (Shimadzu). Volatile fatty acids (VFAs) were measured by using ion chromatography (940 Professional IC Vario, Metrohm). The experimental treatments were analyzed by the student's t-test at a significance level of $p < 0.05$ using JASP (version 0.14.1) software to determine statistically significant differences between specific pairs.

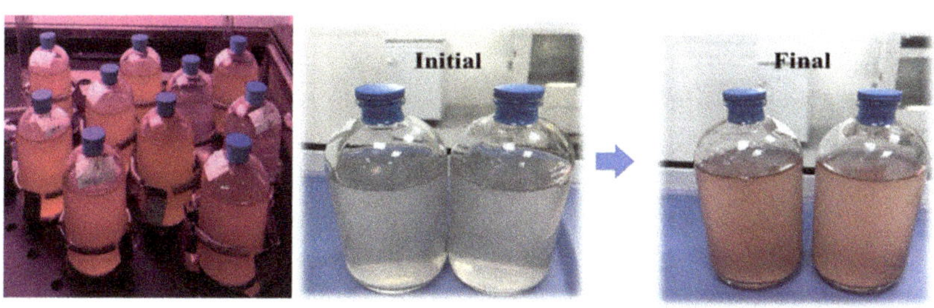

Fig. 1: An illustration of PNSB growth in a shaking incubator (left), and the initial and final growth culture of PNSB (right)

Results and Discussion

Effect of Different Carbon Sources. Juice wastewater is reported to contain one or multiple carbon sources, depending on the type of juice [8, 9]. In this experiment, PNSB growth was carried out using different carbon sources; each bottle contained only one kind of carbon source. A bottle without any carbon source was set as a control. The absorbance values indicate that the PNSB grew slowly for the first five days, and then a sharp increase in growth was observed. pH decreased significantly during the exponential growth period and then again increased. The same trend was found in all treatments except the control. Notably, PNSB fed with tris-Na-citrate showed the highest growth, and the pH drop in this treatment was relatively less than the other carbon sources. Tris-Na-citrate also returned the highest biomass production (total suspended solids, TSS). The biomass production in sucrose, fructose, glucose, tris-Na-citrate, and control were 575±35,

550±71, 600±5, 650±70, and 175±35 mg/L, respectively. Volatile fatty acids (VFAs) were also the highest (2096 ± 606 mg/L) while using tris-Na-citrate as a carbon source (Figure 2). High biomass productivity using Tris-Na-citrate can be linked to the pH changes. As Tris-Na-citrate showed less pH change than the other carbon sources, in those carbon sources, the pH was dropped to 6.0, which may not be favorable for PNSB.

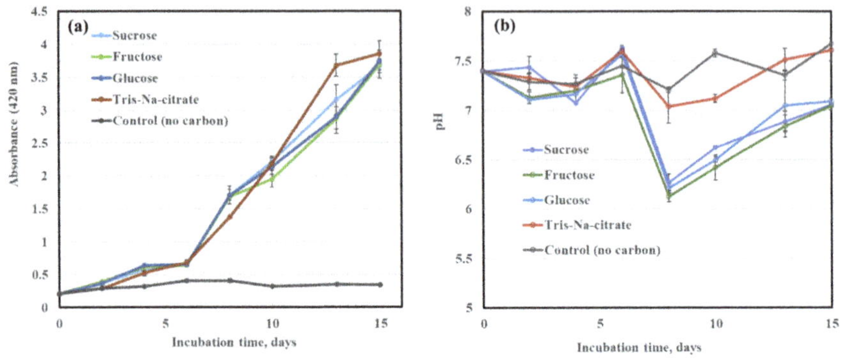

Fig. 2: (a) Effect of carbon different carbon sources on PNSB growth and (b) pH changes during the cultivation

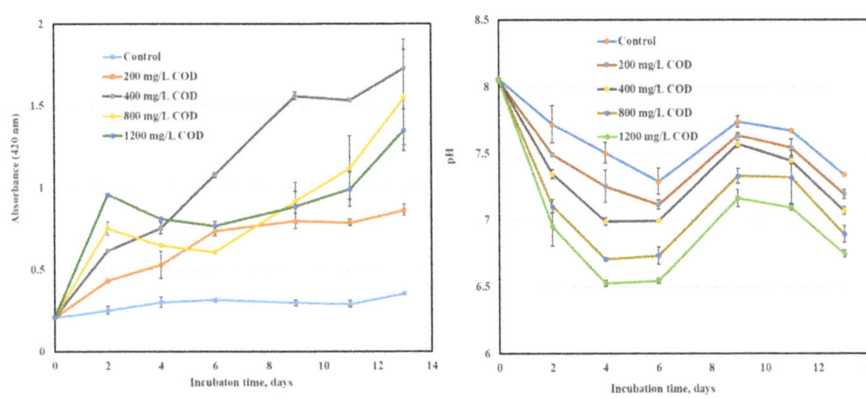

Fig. 3: (a) Effect of substrate (glucose) concentration, and (b) pH changes on PNSB cultivation

Effect of Substrate (Glucose) Concentration. The effect of substrate concentration on cell growth and medium pH was also investigated. The cells were grown under varying glucose concentrations (0-1200 mg/L as COD) at an initial pH of 8.0±0.1. Cell growth was pH-dependent, as the cells showed high growth when the pH was high, which decreased significantly as the pH dropped. However, the pH drop was larger at high substrate concentrations, linked with inorganic carbon production (Figure 3). A substrate concentration of 400 mg/L seems to show the highest growth performance; however, the absorbance values were not significantly ($p>0.05$) different than 800 mg/L and 1200 mg/L. The biomass production (TSS) was 555±21, 687±21, 729±4, 776±1 and 747±4 mg/L at 0, 200, 400, 800, 1200 mg/L substrate concentration, respectively. Based on TSS results, 800 mg/L can be considered the optimal substrate concentration. High glucose concentration showed an inhibitory effect, perhaps due to the reasons that it caused to drop in the pH, and the PNSB showed attenuated growth response.

Carbon Removal. Carbon removal is a crucial parameter in assessing the efficiency of a wastewater treatment system. TOC and IC removals were determined at different time points during the treatment process. The results indicate a continuous decrease in TOC over the treatment period. TOC removal corresponded to the initial substrate concentration. The highest TOC removal (233±46 mg/L) was found at a substrate concentration of 1200 mg/L, followed by 800, 400, 200 mg/L, and control, which were 208±39, 157±3, 103±10, and 26±11 mg/L, respectively (Figure 4). IC concentration gradually increased over the period of cultivation. This phenomenon was more pronounced at high initial substrate concentrations (800-1200 mg/L) than at the initial low substrate concentration (0-400 mg/L). IC removals were also high at higher substrate concentrations and vice versa.

COD, TOC, and IC removals were tested at different pH also in a separate experiment. The results indicate that the TOC removal at pH 6.0, 7.0, and 8.0 was almost the same, whereas, at pH 5.0, TOC removal was significantly lower than the others (data not shown). Similarly, COD removal at pH 5.0 was much lower than the other treatments. COD removal at pH 6.0, 7.0, and 8.0 was not significantly different. IC removal increased with an increase in the pH, the highest at pH 8.0 and the lowest at pH 5.0.

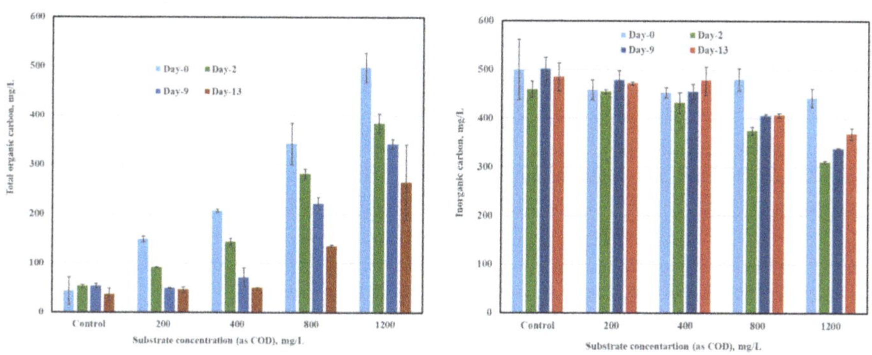

Fig. 4: (a) Total organic carbon, and (b) Inorganic carbon removal while using glucose as a substrate

Summary

The investigations of this study reveal that PNSB are suitable feedstock for the treatment of juice industry wastewater and produce a reasonable amount of biomass. The treatment performance depends on the pH, type of carbon sources, and substrate concentration. Further investigation should be carried out to characterize PNSB for producing value-added bio-products, including single-cell protein, photo-pigments, and lipids.

Acknowledgements

The authors would like to acknowledge the support of the Qatar National Research Fund (MME01-0910-190029) and Qatar Shell Research and Technology Centre for their financial and technical support, respectively.

References

[1] O.T. Can, COD removal from fruit-juice production wastewater by electrooxidation electrocoagulation and electro-Fenton processes, Desalination and Water Treatment 52 (2014) 65-73. https://doi.org/10.1080/19443994.2013.781545

[2] N. Rashid, U. Onwusogh, H.R. Mackey, Exploring the metabolic features of purple non-sulfur bacteria for waste carbon utilization and single-cell protein synthesis, Biomass Conversion and Biorefinery (2022). https://doi.org/10.1007/s13399-022-03273-8

[3] S.S. Tak, O. Shetye, O. Muley, et al. Emerging technologies for hydrogen production from wastewater, International Journal of Hydrogen Energy, (2022) https://doi.org/10.1016/j.ijhydene.2022.06.225

[4] B. Ramprakash, P. Lindblad, J.J. Eaton-Rye, et al., Current strategies and future perspectives in biological hydrogen production: A review. Renewable and Sustainable Energy Reviews, 168 (2022) 112773. https://doi.org/10.1016/j.rser.2022.112773

[5] R.J. TeKippe, Treatment and disposal of citrus fruit processing wastes, Journal (Water Pollution Control Federation), 44 (1972) 2001-2012.

[6] T. Hulsen, D.J. Batstone, J. Keller, Phototrophic bacteria for nutrient recovery from domestic wastewater, Water Res, 50 (2014) 18-26. https://doi.org/10.1016/j.watres.2013.10.051.

[7] G. Capson-Tojo, D.J. Batstone, M. Grassino, et al., Purple phototrophic bacteria for resource recovery: Challenges and opportunities, Biotechnology Advances, 43 (2020) 107567. https://doi.org/10.1016/j.biotechadv.2020.107567

[8] A. Tawfik, H. El-Kamah, Treatment of fruit-juice industry wastewater in a two-stage anaerobic hybrid (AH) reactor system followed by a sequencing batch reactor (SBR), Environmental Technology, 33 (2012) 429-436. https://doi.org/10.1080/09593330.2011.579178

[9] C. Amor, M. Lucas, A. Pirra, et al., Treatment of concentrated fruit juice wastewater by the combination of biological and chemical processes, Journal of Environmental Science and Health. Part A, Toxic/hazardous substances and Environmental Engineering, 47 (2012) 1809-1817. https://doi.org/10.1080/10934529.2012.689244

Physicochemical characterization of recovered palm oil from palm oil mill effluent by progressive freezing

AMRAN Nurul Aini[1,2,a] *, MOHAMED ANUAR Muhammad Athir[3,b] and RUSLAN Muhammad Syafiq[4,c]

[1]Chemical Engineering Department, Universiti Teknologi PETRONAS, 32610, Seri Iskandar, Perak, Malaysia

[2]HICoE-Centre for Biofuels and Biochemical Research (CBBR), Institute of Self-Sustainable Building (ISB), Universiti Teknologi PETRONAS, 32610, Seri Iskandar, Perak, Malaysia

[3]Renewable Energy & Green Technology, TNB Research Sdn. Bhd., 43000, Kajang, Selangor, Malaysia

[4]School of Chemical Engineering, College of Engineering, Universiti Teknologi MARA, Shah Alam 40450, Malaysia

[a]nurul.amran@utp.edu.my, [b]athir.anuar@tnb.com.my, [c]syafiqhazwan@uitm.edu.my

Keywords: Progressive Freezing, Recovered Palm Oil, Palm Oil Mill Effluent, Free Fatty Acid, DOBI

Abstract. Oil extraction rate (OER) is essential in measuring palm oil mills' performance and competence. One of the attempts to achieving high OER is exploring the possibility of recovering residual oil from palm oil mill effluent (POME). A recent technique called progressive freezing (PF) has been proposed to reclaim the residual oil from POME, forming a block of solid oil in the system. This article focuses on evaluating the quality of the recovered oil through its physicochemical characterization (free fatty acid (FFA), deterioration of bleachability (DOBI), moisture content and dirt analysis) before deciding on the subsequent usage or suitable treatment if needed. The results show that the residual oil contained 25 to 27% free fatty acid (FFA), 1.01 to 2.02% moisture content, 0.049 to 0.055 dirt and had deterioration of bleachability (DOBI) between 2.3 to 2.7. The recovered oil had been proposed to be suitable for recycle or utilization as a value-added product such as biodiesel production.

Introduction

Palm oil has progressively become vital in agriculture, accounting for 32% and 54% of global palm oil production for Malaysia and Indonesia [1], respectively. Due to the production of flexible, practical, and valuable palm oil products, this industry has become Malaysia's major income earner [2]. Palm oil is a resourceful vegetable oil used as a resource for food and non-food industries [3]. The worldwide demand for palm oil has progressively expanded since the price is lower than the other oils. This expansion of palm oil as a food ingredient has assisted in realizing sustainable food production and healthy food intake since it accelerated food guarantee, health, and profit for millions worldwide [4]. Technically, palm oil is extracted from the fruit's outer layer (mesocarp). For a ripe bunch, it is ordinarily orange-yellow. CPO is mainly utilized for food preparation purposes and is commonly used in South-East Asia (SEA), where restaurants use it due to the cheap price [5].

In 2010, worldwide palm oil utilization recorded around 71.1% of palm oil usage distribution came from food products such as margarine, salad, and cooking oil. The remaining palm oil is used for industry (24.2%) and energy usage (4.7%). Palm oil has not only been used to produce soap, cosmetic products and candles but also to produce electricity, heating and fuel. Additionally, a number of research papers uncovered that palm oil/kernel oil decides the advancement of biodiesel and oleochemical since they became significant feedstocks [6 – 8].

As palm oil production is projected to rise at the present pace, the increment of by-products is also expected. Indeed, it is proven by previous research that abundant biomass wastes have been generated by palm oil mills [9, 10] as shown in Table 1.

Table 1: Approximate by-products produced by palm oil mills (MPOB)

By-products	Million Tons	Moisture Content	Heat Value (kcal/kg)
EFB	9.6	65	3700
Shredded Fibre	5.8	42	4420
Palm Kernel Shell	3.7	7	3700
POME	21.0	95	-

Among the listed by-products above, the most plentiful is POME. POME is mainly produced through the separation of oil from the CPO solution. The main composition of POME is water (95-96%), followed by 4 - 5% total solid, 2 - 4 % suspended solid and 0.6 - 0.7 % oil [11]. The palm oil industry is concerned about the oil losses through POME since it contributes to reducing OER. Although the oil loss from POME is fewer than 1.0%, the annual oil loss was found to be significantly high and worth recovering.

In addition, POME can cause water pollution, leading to reduced dissolved oxygen in the water, which threatens marine life. Besides, the POME is acidic with an incredibly unpleasant odour. Thus, it is difficult for the palm oil industry to meet strict environmental regulations before releasing the POME to the water body. Accounting for approximately 452 palm oil mills in Malaysia, a substantial POME discharge could dangerously affect the environment [12]. Table 2 displays that POME holds almost 4000 to 6000 mg/L of oil and grease, which surpasses the permissible discharge limits which make direct disposal illegal. Moreover, POME's treatment's degradation processes could be restrained by the triacylglycerols and degradative presence inside POME [13]. Thus, removing and recovering residual oil is vital before specific treatment and disposal.

Table 2: Composition of POME

Parameter	POME (mg/L)	Range (mg/L)
Oil / grease	4000	4000-6000
Total Solid	40000	11500 – 79000
Suspended solid	18000	5000 – 54000
Total volatile solid	34000	9000 – 72000
Biochemical Oxygen Chemical Demand (BOD)	25000	10250 – 43750
Chemical Oxygen Demand (COD)	51000	15000 – 100000
Total nitrogen	750	180 – 1400

Recovering the residual oil from POME could suggest a great alternative in searching for appropriate POME treatment strategies. A reported study had proven that over 70% of COD was considerably reduced, and 78% of the oil was efficiently recovered from a POME originating from one mill in Thailand, which intensified the residual oil recovery rate [14]. Various separation methods can be applied to remove oil from palm oil mill effluent (POME). However, this report focuses on a recent innovative oil recovery process, progressive freezing (PF), which has never been investigated in Malaysia.

PF is a method used to concentrate a solution by solidifying one liquid element into a pure solid and sequentially separating the solid from the concentrated liquid. The most significant benefit of PF is that it can retain the thermally sensitive materials in the concentrate due to the involvement of low process temperatures and low energy conditions compared to evaporation [15]. The introduction of the PF method for residual oil recovery from POME may remove a high quantity of oil from the waste due to the separation driven by the difference in freezing or melting points of the oil and water. Therefore, this particular article aims to examine the physicochemical characteristics of residual oil (FFA content, DOBI, moisture content and dirt), which can define the final quality of the oil.

Materials and Methods

Materials. A Palm oil–water mixture model consisting of 6000 mg of CPO per litre of water was prepared as the primary raw material. The palm oil was first collected from the bulk storage tank (BST) of FELCRA Nasaruddin Belia Berhad, Bota Kanan, Perak. The sample was stored at a low temperature to avoid decomposition, oxidation, and free fatty acid content changes. Next, a 50% volume ratio (v/v) of ethylene glycol and water was mixed and utilized as the PF coolant. The coolant's composition was employed to facilitate the supercooling of the solution and to avoid the slushy form (viscous) of the ethylene glycol–water solution.

Experimental Procedure. The PF procedure was conducted through a lab-scale setup shown in Fig. 1. The refrigerated waterbath containing the coolant was operated to achieve the required cooling temperature between 4 to 12 ºC. As it reached the desired reading, the storage tank was filled with 2.5 L of the POME sample. The sample was then circulated inside the tubed crystallizer using a peristaltic pump for the specified circulation flowrate (2600 to 3400 mL/min) and freezing time (20 to 60 minutes). After the designated freezing time, the circulation was stopped, the remaining unfrozen liquid in the crystallizer was flushed out, and an adequate amount was collected for characterization purposes. Similarly, sample of the solidified oil was also collected and melted melted before being sent for FFA, DOBI, moisture content and dirt analysis.

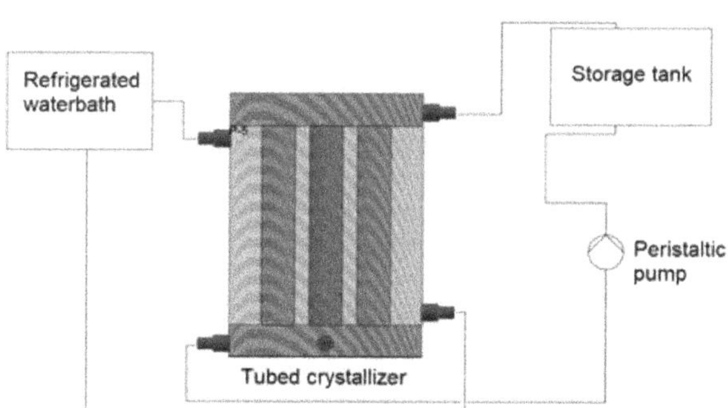

Fig. 1: Experimental setup for the PF process

The sodium hydroxide titration method was used to determine the FFA content, while DOBI was analyzed using a UV-VIS spectrophotometer. The absorbance of 446 nm signifies the carotene (pro-Vitamin A), while 269 nm reveals the secondary oxidation [16]. Table 3 shows the grades of crude palm oil based on the DOBI values. On the other hand, the moisture and dirt analysis were conducted through the standard heating and homogenization processes. Table 4 shows the quality guideline for CPO specifications released by palm oil refiners associations of Malaysia (PORAM).

Table 3: Crude palm oil grade at difference DOBI

DOBI	Grade
<1.68	Sludge palm oil or its equivalent
1.78-2.30	Poor
2.36-2.92	Fair
2.99-3.24	Good
>3.24	Excellent

Table 4: The quality limit for CPO specifications

Parameters	Quality Guideline
FFA	5.00% max
DOBI	2.3 min
Moisture Content	0.2% max
Dirt	0.02% max

Results and Discussion

Formation of Solid Palm Oil. After the PF process, the solid oil had been successfully formed on the tube crystallizer's wall, as shown in Fig.2. The amount of solid oil produced varied depending on the operating conditions.

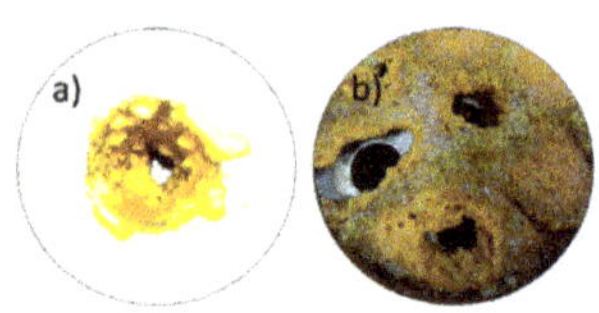

Fig. 2: Formation of solid oil on the crystallizer

Physicochemical Characteristics of Recovered Oil. Table 5 shows the physicochemical characteristics of the recovered oil by the PF process.

Table 5: Physicochemical sharacteristics of the recovered oil

Run	FFA (%)	DOBI	Moisture Content (%)	Dirt
1	27.24	2.5616	1.01	0.055
2	25.77	2.7059	1.74	0.062
3	26.51	2.3545	2.02	0.049
Previous study	27.67 ± 0.10%	1.94-2.43	2.20 ± 0.07%	-

Free Fatty Acid. As can be observed in Table 4.10, the minimum FFA value of 25.77 was obtained from the second run. The obtained values show that the quality of the residual oil was not meeting the specification limit of 5.00% maximum. A high value of FFA indicates a decline in oil quality owing to the hydrolysis of triglycerides in the residual oil [17]. FFA accumulation in the recovered oil can be due to the high moisture content or too long waiting time for the palm fruits to be processed [18]. Besides, nearly identical results had been obtained by previous researchers; for instance, a high amount of FFA was detected from the residual oil recovered by the soxhlet method (27.67 ± 0.10%) and novel polypropylene nanofiber (25%) [11,19]. Yet, this situation can be manipulated by processing the palm fruit or utilizing the recovered oil as value-added products. Residual oil with a high FFA content can produce biodiesel, which involves a simple production step. Additionally, the recovered oil can also be used to produce low-grade soap or fuel boilers.

DOBI. The extent of deterioration of palm oil can be determined by the DOBI analysis, which requires UV-VIS spectrophotometric analysis of absorbance ratio at A446 nm/A269 nm [20]. The absorbance of 446 nm corresponds to the carotene content and the absorbance of 269 nm corresponds to the secondary oxidation product. The minimum value for DOBI to indicate a reasonably good oil should be around 2.3 as it has been introduced in ISO, relevant to examine the fitness for refining a CPO. As can be seen in Table 5, all three runs fall within the fair grade range, and the lowest DOBI value was 2.3545. This value suggests an increase in secondary oxidation of the product content which could be caused by several considerations in the processing of oil palm fruit, such as the fruit ripeness, processing time and condition, supply process, storage, and contamination [20]. A quite similar result has been reported in the previous report at which the DOBI value of the residual oil obtained was yet in an satisfactory range of 1.94 to 2.43. Additionally, no substantial changes in the macronutrients and residual oil quality were reported, concluding that it can be utilized as a raw material for industrial purposes while enhancing the OER for the palm oil industry [21]. Nevertheless, the recovered oil for this study comprised a high FFA value; therefore, value-added products are preferable to recycling.

Moisture Content and Dirt. For CPO, the specification limits for moisture content and dirt are 0.2% and 0.02%, respectively. As shown in Table 5, the moisture and dirt content for all three runs exceeded these values; the lowest value obtained is 1.01%. The phenomenon might be due to the large quantities of non-toxic organic in the POME, supported by the obnoxious odour of the waste [11]. Ahmad et al. [22] had stated that just about 50% of water spent in the palm oil separation process was released as steam, while the remaining half resulted in POME. Consequently, the high water content in the residual oil will speed up the hydrolysis reaction, making the oil unsuitable for raw materials. Agreeing to earlier research, the water had been removed to reduce up to 2.20 ± 0.07% so that the utilization in the biodiesel production can be realized [19]. The removal was also crucial to avoid the reversible reaction of esterification; hence, the residual oil was pretreated by heat prior use [23].

On the other hand, dirt analysis test involves using hexane and a vacuum pump to wash the residual oil. Referring to Table 5, all three runs exceed the standard dirt limits for CPO production, where the nearest to the standard was 0.049. Nevertheless, the residual oil can be purified in the oil room to reduce the dirt percentage before further reuse or recycle [24]. So, it can be assumed that the residual oil that surpasses the specifications can be recycled or reused as value-added products, thus improving the oil extraction rate in the palm oil industry.

Conclusion

Through this study, PF process has successfully recovered the residual oil from POME by producing a single solid oil on the crystallizer's wall. The physicochemical representation had also been conducted to validate the quality of the recovered oil. The recovered residual oil shows free fatty acid (FFA) content from 25 to 27%, deterioration of bleachability (DOBI) from 2.3 to 2.7, moisture content from 1.01 to 2.02% and dirt content from 0.049 to 0.055. It can be concluded that the recovered oil can be either recycled or utilized as a value-added product such as biodiesel production.

Acknowledgement

The authors would like to acknowledge the funding supports from Ministry of Higher Education (MOHE) Malaysia through HICOE Grant (cost centre: 015MA0-052/015MA0-104/015MA0-136) to CBBR.

References

[1] H. Purnomo et al., Reconciling oil palm economic development and environmental conservation in Indonesia: A value chain dynamic approach, For. Policy Econ. 111 (2020). 102089. https://doi.org/10.1016/j.forpol.2020.102089

[2] I. Mukherjee, B. K. Sovacool, Palm oil-based biofuels and sustainability in southeast Asia: A review of Indonesia, Malaysia, and Thailand, Renew. Sust. Energ. Rev. 37 (2014) 1-12. https://doi.org/10.1016/j.rser.2014.05.001

[3] S. Kadandale, R. Marten, and R. Smith, The palm oil industry and non-communicable diseases, Bull. World Health Organ. 97 (2019) 118-128. https://doi.org/10.2471/BLT.18.220434

[4] W. Willett et al., Food in the Anthropocene: the EAT-Lancet Commission on healthy diets from sustainable food systems, The Lancet, 393 (2019) 447-492. https://doi.org/10.1016/S0140-6736(18)31788-4

[5] T. Silalertruksa et al., Environmental sustainability of oil palm cultivation in different regions of Thailand: Greenhouse gases and water use impact, J Clean. Prod. 167 (2017) 1009-1019. https://doi.org/10.1016/j.jclepro.2016.11.069

[6] P. Maneechakr and S. Karnjanakom, A combination of 2k factorial with Box-Behnken designs for FAME production via methanolysis of waste cooking palm oil over low-cost catalyst, J. Environ. Chem. Eng. 7 (2019) 103389. https://doi.org/10.1016/j.jece.2019.103389

[7] P. Alagi et al., Carbon Dioxide-Based Polyols as Sustainable Feedstock of Thermoplastic Polyurethane for Corrosion-Resistant Metal Coating, ACS Sustainable Chem. Eng. 5, (2017) 3871-3881. https://doi.org/10.1021/acssuschemeng.6b03046

[8] A. I. Matinja, N. A. Mohd Zain, M. S. Suhaimi, A. J. Alhassan, Optimization of biodiesel production from palm oil mill effluent using lipase immobilized in PVA-alginate-sulfate beads, Renew. Energy 135 (2019 1178-1185. https://doi.org/10.1016/j.renene.2018.12.079

[9] H.-M. Yoo, S.-W. Park, Y.-C. Seo, K.-H. Kim, Applicability assessment of empty fruit bunches from palm oil mills for use as bio-solid refuse fuels, J. Environ. Manage. 234 (2019) 1-7. https://doi.org/10.1016/j.jenvman.2018.11.035

[10] M. S. Umar, P. Jennings, T. Urmee, Sustainable electricity generation from oil palm biomass wastes in Malaysia: An industry survey, Energy 67 (2014) 496-505. https://doi.org/10.1016/j.energy.2014.01.067

[11] C. Hing Chung et al., Recovery of residual oil from palm oil mill effluent using polypropylene nanofiber: a field trial," MIJST 3 (2017) 276-294. https://doi.org/10.20319/mijst.2017.32.276294

[12] A. P. Ghulam Kadir, Oil palm economic performance in malaysia and R&D progress in 2019, JOPR, 2020. https://doi.org/10.21894/jopr.2020.0032

[13] A. Ahmad, Effect of Ozonation on Biodegradation and Methanogenesis of Palm Oil Mill Effluent Treatment for the Production of Biogas, Ozone: Sci. Eng. 41 (2019) 427-436. https://doi.org/10.1080/01919512.2019.1565987

[14] F. B. Ahmad, Z. Zhang, W. O. S. Doherty, and I. M. O'Hara, The prospect of microbial oil production and applications from oil palm biomass, Biochem. Eng. J. 143 (2019) 9-23. https://doi.org/10.1016/j.bej.2018.12.003

[15] M. J. Iskandar, A. Baharum, F. H. Anuar, and R. Othaman, Palm oil industry in South East Asia and the effluent treatment technology-A review, Environ. Technol. Innov. 9 (2018) 169-185. https://doi.org/10.1016/j.eti.2017.11.003

[16] M. Tapanwong, Effect of combination microwave and oven drying on the chemical properties of different ripeness crude palm oil, GEOMATE 18 (2020). https://doi.org/10.21660/2020.67.5567

[17] U. E. Shehu, T. Q. Chow, H. S. Hafid, M. N. Mokhtar, A. S. Baharuddin, N. M. Nawi, Kinetics of thermal hydrolysis of crude palm oil with mass and heat transfer in a closed system, Food Bioprod. Process. 118 (2019) 187-197. https://doi.org/10.1016/j.fbp.2019.09.009

[18] I. A. Kadir, L. S. Jurid, and S. I. Zubairi, Physico-chemical properties of recycled refined, bleached and deodorized (RBDPO) palm olein: which cycle should the olein considered spoilage, J. Teknol. 79 (2017). https://doi.org/10.11113/jt.v79.9829

[19] S. Suwanno et al., The production of biodiesel using residual oil from palm oil mill effluent and crude lipase from oil palm fruit as an alternative substrate and catalyst, Fuel. 195 (2017) 82-87. https://doi.org/10.1016/j.fuel.2017.01.049

[20] B. S. Baharin, R. A. Latip, Y. B. Che Man, R. A. Rahman, The effect of carotene extraction system on crude palm oil quality, carotene composition, and carotene stability during storage, J. Amer. Oil. Chem. Soc. 78 (2001) 851-855. https://doi.org/10.1007/s11746-001-0354-4

[21] N. S. H. Md Yunos et al., The Physicochemical Characteristics of Residual Oil and Fibers from Oil Palm Empty Fruit Bunches, BioResources. 10 (2014) 14-29. https://doi.org/10.15376/biores.10.1.14-29

[22] A. L. Ahmad, S. Sumathi, B. H. Hameed, Coagulation of residue oil and suspended solid in palm oil mill effluent by chitosan, alum and PAC, Chem. Eng. J. 118 (2006) 99-105. https://doi.org/10.1016/j.cej.2006.02.001

[23] S. R. Putri Primandari, Z. Yaakob, M. Mohammad, A. B. Mohamad, Characteristics of Residual Oil Extracted from Palm Oil Mill Effluent (POME), World Appl. Sci. J. 27 (2013) 1482-1484.

[24] C. Melches, H. Plate, J. Schürhoff, R. Buchfink, The Steps from Batchwise to Continuous Crystallization for a Fine Chemical: A Case Study, Crystals. 10 (2020) 542. https://doi.org/10.3390/cryst10060542

GHG accounting framework: Development of the GHG assessment indicator for the petroleum refinery sector

NABILA Farhana binti Jamaludin[1,2,a*], NUR 'Aifaa Syamimi Binti Shuhaimi[2,b], MUHAMMAD Fairis bin Hadipornama[3,c]

[1]Institute of Contaminant Management, Universiti Teknologi PETRONAS, 32610, Seri Iskandar, Perak, Malaysia

[2]Chemical Engineering Department, Universiti Teknologi PETRONAS, 32610, Seri Iskandar, Perak, Malaysia

[3] KINETICS Systems Malaysia Sdn. Bhd., 40300, Shah Alam, Selangor, Malaysia

[a]nabila.jamaludin@utp.edu.my, [b]nur_18001014@utp.edu.my, [c]Fairis@kinetics.net

Keywords: Carbon Accounting, GHG Accounting, Framework, Assessment, Petroleum Refinery

Abstract. The petroleum refineries sector is one of the highest-ranked sectors in terms of GHG emissions per facility. As a way forward in managing climate change, GHG assessment has become crucial in achieving net zero emissions. The objective of this study is to develop appropriate measurements for GHG assessment performance for the Malaysia petroleum refinery sector. The framework includes establishing indicators and parameters to quantify and assess carbon footprint and to obtain an approximate measure for carbon footprint accounting. The overall methodology for the GHG assessment framework is separated into two main parts with a total of eight steps for this study to be completed, however in this study the focus is only on the familiarization of plant, determination of parameters and indicators and greenhouse gases mapping which are part of the indicator development. With all information of the study being prepared, the next goal is to develop a comprehensive framework that can be used in the petroleum refinery process for carbon footprint accounting.

Introduction

Climate change is real, and Malaysia is not excluded from being impacted by this phenomenon. Rising temperatures, floods, droughts, and rising sea levels are among the examples of impact brought by climate change. Climate change occurs due to the emissions of greenhouse gases (GHG), for example carbon dioxide, methane, and nitrous oxide. As part of the effort to fight climate change, on July 30th 2021, the government of Malaysia has approved the unconditional Nationally Determined Contributions (NDC) target to reduce the intensity of greenhouse gas (GHG) emissions by 45% based on gross domestic product, also an initiative called the net zero carbon emission need to be achieved by 2030 [1] [2]. However, in order to meet the target, various initiatives, policies and instruments may need to be established. In light of this, GHG assessment and reporting from the industrial sector is expected to be one of the critical elements of the action plan. *Peter Drucker, the Management guru has famously said; "you can't manage what you can't measure"* [3]. Therefore, be it at the level of a country or an organisation, it is critical to determine the current status of GHG emissions and to establish a GHG baseline so that the relevant mitigation strategies and actions could be effectively designed to address the root causes of GHG emissions then to further with effective mitigation strategies. From the nature of the refining process, refinery process contributes to the increasing amount of CO_2 emissions. In 2020, CO_2 emissions in Malaysia was 262.2 million tonnes, and one of the top contributor of CO_2 emissions is the oil and gas industry [4]. With increasing demands and high market value of petroleum refinery products, the processes are further developed and improved to cater the demands and to obtain high profitability in business. Rapid development and growth have developed both positive and negative impacts on business, economy, social and the environment. Focusing on the environmental aspect, the development has caused negative impacts towards sustainability and emission of greenhouse gases (GHGs) which will lead to climate issues gradually. The emission of GHGs is one of the negative impacts due to the nature of the petroleum refinery process, it is important for companies in petroleum refinery sector to measure carbon level, or to conduct carbon footprint accounting. However, they were limited studies or framework focusing on the carbon accounting assessment of petroleum refineries in Malaysia [5]. One of the study calculated the carbon emissions of the products transported through pipeline [6]. While [7] has developed a well-to wheel method focusing on the bitumen processing area. The literature that is closely related to this study is summarized in Table 1. Therefore, this study is required to be conducted to establish a comprehensive framework on carbon accounting by identify relevant carbon footprint parameters and indicators to quantify the GHGs in the petroleum refinery process.

Table 1 Review carbon footprint research studies

Title and Author	Objectives	Application	Method	Research Gaps
Carbon footprint of oil products pipeline transportation [6]	To build a comprehensive Life Cycle Assessment (LCA) model of carbon footprint for oil products pipeline	Oil and Gas Industry: Oil products pipeline transportation	Life Cycle Assessment (LCA) model	This framework establish a Life Cycle Analysis (LCA) by studying the life cycle inventory of petroleum products pipeline system only not including the process of petroleum refining.
Statistically enhanced model of oil sands operations: Well-to-Wheel comparison of in situ oil sands pathways [7]	To investigation variability in well-to-wheel (WTW) GHG emissions	Oil and Gas Industry: Refinery with respect to bitumen	Well-to-Wheel method	This model have been developed for refinery process but only focusing on bitumen production
Assessing the carbon footprint of a university campus using a life cycle assessment approach [8]	To evaluate the carbon footprint of Clemson University's campus using a streamlined life cycle assessment approach	Public Buildings: University	Streamlined life cycle assessment approach	This model framework developed is suitable for public building such as universities only, however methodology of research can be used as reference.

Materials Research Proceedings 29 (2023) 362-369 https://doi.org/10.21741/9781644902516-41

Assessment of carbon footprint emissions and environmental concerns of solid waste treatment and disposal techniques; case study of Malaysia [9]	To alert Malaysian stakeholders on the uneven danger of carbon footprint emissions of waste technologies	Waste Management	2006 IPCC Methodology Assessed by 3 scenario: Scenario 1: Waste dumping in sanitary landfills equipped with gas recovery system Scenario 2: Anaerobic digestion of organics and recycling of waste Scenario 3: Waste Incineration	This framework developed is suitable for waste management, however methodology of research can be used as reference

Methodology

The overall framework of this GHG assessment is separated into a two-part process as shown in Fig. 1 with eight steps in total. Part 1 consist of three main steps of indicator development and Part 2 of the study framework consists of five main steps to quantify the carbon. The second step is to compile all related data obtained throughout the study, to quantify and assess the carbon footprint accounting with reference to IPCC guidelines and MYCarbon Malaysia. However, the methodology in this study only focuses on indicator development from steps 1 until 3 in Part 1. A structured methodology has been developed to select indicators as the applicability of the framework depends on the relevance and importance of these indicators. An overall framework of the development of petroleum refinery carbon footprint accounting is presented in Fig. 1.

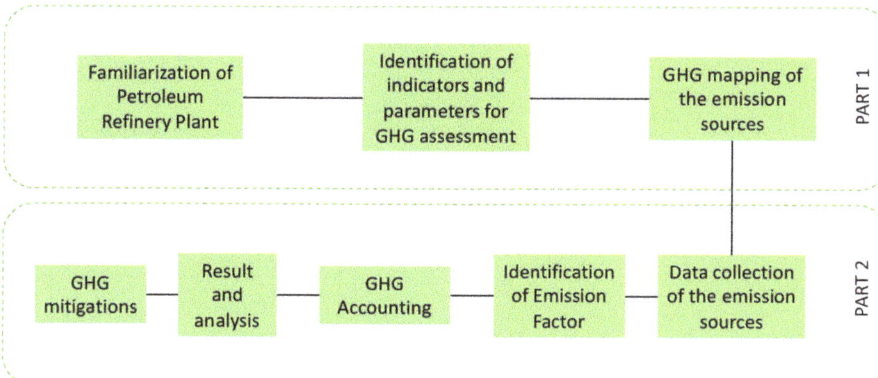

Figure 1 Development of Petroleum Refinery Carbon Footprint Accounting

Part 1: Development of Carbon GHG Footprint Accounting for Petroleum Refinery Process

The first part of developing petroleum refinery carbon footprint accounting first is to understand the overall process of the refinery, determine the parameters and indicators for carbon footprint accounting and map the emission sources to the respective parameters as shown in Fig. 2.

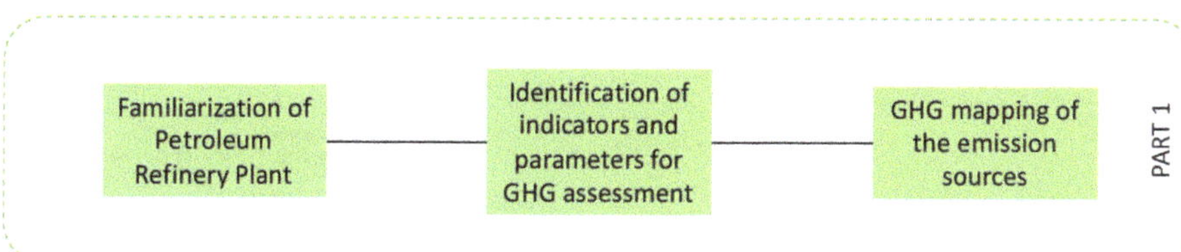

Figure 2 Part 1 of development of petroleum refinery carbon footprint accounting

Familiarization of Petroleum Refinery Plant

In the first step, a brief introduction of the petroleum refinery industry and the overall process of petroleum refining was explained. This step is crucial as it is important to understand the process involved, the main component of a petroleum refinery and the operational data that is required for the next step of the study. Moreover, this study is going to focus on the main process units in the petroleum refinery process. The main process units determined in the petroleum refinery process are the crude oil tankage, crude distillation unit, secondary process unit, product treatment unit and the final product tankage. The flowchart of the main process unit and overall process flow diagram of petroleum refinery processes can be referred to Fig. 3. and Fig. 4.In this study, the space boundary of the assessment is limited to the kerosene production plant as in Fig. 4.

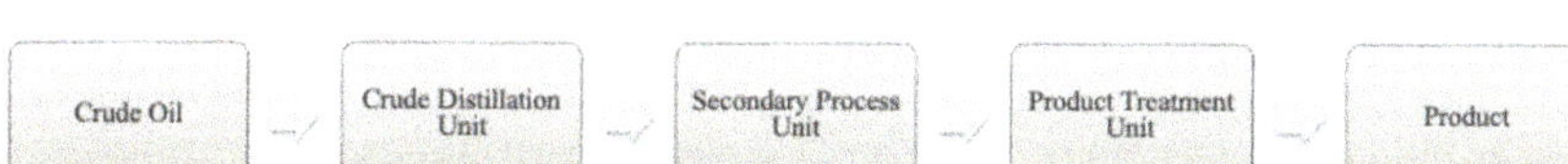

Figure 3 Flowchart of main process units in petroleum refinery process

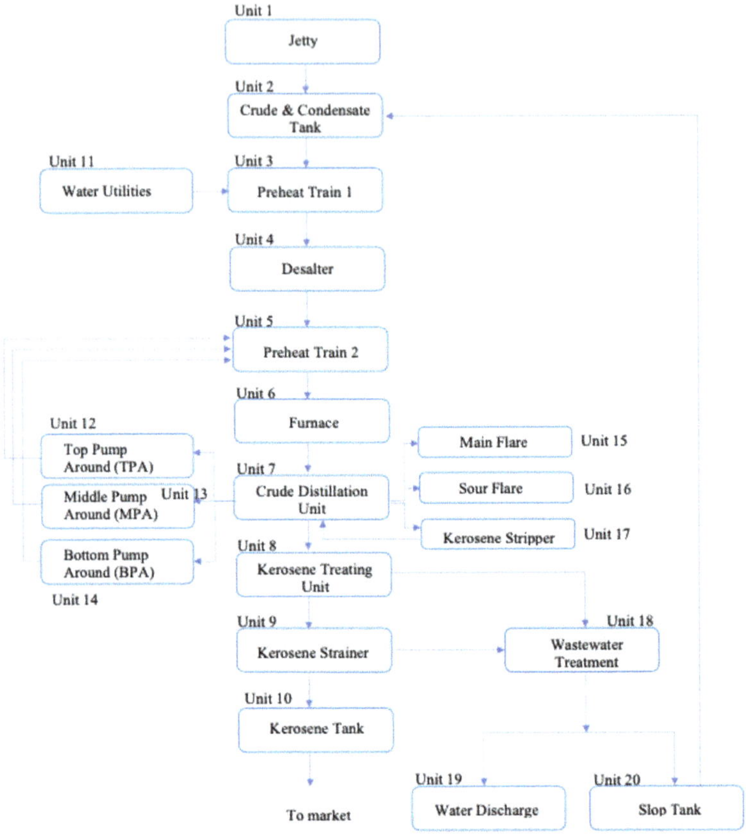

Figure 4 Overview of petroleum refinery processes

Selection of Indicators and Parameters of the GHGs Emission Source

The selection of indicators and parameters was based on the process flow diagram (PFD) of the industry illustrated in Figure 4 and is aligned with the IPCC guidelines. The final list of parameters and indicators is decided based on the input from the expert, subject to data availability. Five potential carbon footprint accounting parameters are identified which are water consumption for the process, water consumption, electric consumption, diesel consumption, wastewater and fugitive emission. The indicator of each parameter are shown in Table 2.

Mapping of Greenhouse Gases (GHGs)

Table 3 shows the mapping of greenhouse gases (GHGs) is conducted by determining which stream in the process unit emits GHGs and the streams with respect to the indicators and parameters. Then, the findings will be used in collecting the data for GHGs emissions.

Part 2: Carbon Footprint Accounting Analysis and Mitigations for Petroleum Refinery Process
The second part of the methodology, it consists of analyzing and suggesting proper mitigations to encounter the emissions produced by petroleum refinery processes as shown in Figure 5.

Collection of Monthly Consumption or Generation of Data for Each Mapped Stream
Based on the GHG mapping completed, the data of monthly consumption or data of generation for each mapped stream will be identified. Data collection will be done with reference to available petroleum refineries in Malaysia such as Petronas Penapisan Terengganu (PPT) Sdn. Bhd. And Malaysian Refining Company Sdn. Bhd. (MRCSB).

Accounting of Greenhouse Gases (GHGs)
Determination of emission factors from IPCC standard is the initial step to be conducted prior to any calculations. Based on the GHG parameter, the emission factor will be obtained from the IPCC standard. From the values obtained by referring to the IPCC standard, the accounting of greenhouse gas emissions can be completed. The calculations done will be tabulated in table form for easy reference and documentation.

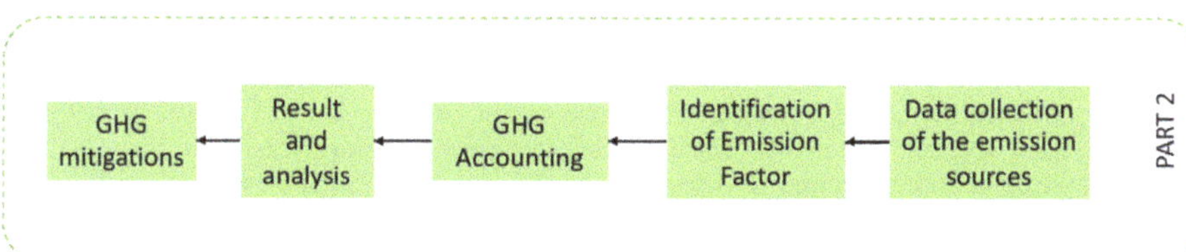

Figure 5 Part 2 of development of petroleum refinery carbon footprint accounting

Results and Discussion
The parameter has been established based on the IPCC guidelines and relevant literature on carbon assessment development. Based on the study, five parameters have been found in the petroleum refinery sector as shown in Table 2.

Table 2 List of parameters and indicators of the GHGs Emission Source

Parameter, p	Indicator, i
Water Consumption	Use of water
Electricity	Electricitiy consumption
Diesel Consumption	Diesel used for transport
Wastewater	Oily water sewer
Fugitive Emission	Fugitive emission

Based on the familiarization of the plant mapping, there are 19 units that contribute the GHG emissions. The name of the unit can be referred to the PFD in Fig. 4. Unit 1, inlet of raw material from the jetty has the most GHG parameters involved in the process, the parameters are wastewater, electric consumption and diesel consumption. While the electric consumption parameter involved with most of the units except units 11 and 19. From these results, it can be expected that the electric consumption will be the highest emitters compared to the other parameter. Unit 1 also expected the highest unit contribute to the total emissions. Based on this analysis, an effective mitigation study can be conducted to reduce emissions.

Table 3 Mapping of GHGs

Parameters	1	2	3	4	5	6	7	8	9	10	11	12	13	14	15	16	17	18	19	20
Water Consumption			/																	
Wastewater	/	/		/				/	/	/									/	
Electric Consumption	/	/	/	/	/	/	/	/	/	/		/	/	/	/	/	/	/		/
Diesel Consumption	/																			
Emission															/	/				

Conclusion

This study is aimed to establish indicators and parameters to quantify an approximate measure of carbon footprint accounting in the petroleum refinery process. There are five parameters have been identified associated with petroleum refineries, namely water consumption for the process, water consumption, electric consumption, diesel consumption, wastewater and emission. 19 unit operations have also been identified contributed to the carbon emission. This identification of the correct parameters and indicators is very important to ensure the measurement will reflect the actual emission level. With the aim objective of the study being stated above achieved, the way forward for this study is to do data collection related to the indicator, quantify the carbon footprint, thus analysed the result of the assessment and propose appropriate mitigation for the enhancement of the process.

Acknowledgment

The authors would like to offer their gratitude to Universiti Teknologi PETRONAS and Grant Research STIRF (015LA0-034) for providing a conducive work environment and state-of-the-art research facilities.

References

[1] UNFCCC, 'Malaysia nationally determined contribution update', 2021.

[2] Ministry of Energy and Natural Resources, 'Malaysia's Energy Transition Plan 2021-2040 featured at the Special Meeting Of Asean Ministers On Energy and The Minister Of Economy, Trade and Industry Of Japan', no. June, pp. 1–2, 2021.

[3] M. Mcmanus, 'Here's How The "Carbon Footprint" Should Really Be Used – It's Not For Shifting Blame : ScienceAlert', *Science Alert*, 2022. https://www.sciencealert.com/fossil-fuel-companies-have-co-opted-carbon-footprint-strategies-to-shift-the-blame-onto-us (accessed Sep. 27, 2022).

[4] Knoema, 'Malaysia CO2 emissions, 1970-2021 - knoema.com', 2020. https://knoema.com/atlas/Malaysia/CO2-emissions (accessed Sep. 27, 2022).

[5] H. L. C. Amanda, N. Z. Mahmood, and A. A. Abdul Raman, 'Carbon accounting initiatives: case study of a petroleum refinery in Malaysia to prepare for future carbon market', *Journal of Engineering Science and Technology*, vol. 5, no. 2, pp. 224–233, 2010.

[6] L. Huang, Q. Liao, J. Yan, Y. Liang, and H. Zhang, 'Carbon footprint of oil products pipeline transportation', *Science of The Total Environment*, vol. 783, 2021. https://doi.org/10.1016/j.scitotenv.2021.146906

[7] J. Guo, A. Orellana, S. Sleep, I. J. Laurenzi, H. L. MacLean, and J. A. Bergerson, 'Statistically enhanced model of oil sands operations: Well-to-wheel comparison of in situ oil sands pathways.', *Energy*, vol. 208, 2020. https://doi.org/10.1016/j.energy.2020.118250

[8] R. Clabeaux, M. Carbajales-Dale, D. Ladner, and T. Walker, 'Assessing the carbon footprint of a university campus using a life cycle assessment approach', *Journal of Cleaner Production*, vol. 273, 2020. https://doi.org/10.1016/j.jclepro.2020.122600

[9] A. Malakahmad, M. S. ' Abualqumboz, S. R. M. Kutty, and T. J. Abunama, 'Assessment of carbon footprint emissions and environmental concerns of solid waste treatment and disposal techniques; case study of Malaysia.', *Waste Management*, vol. 70, pp. 282–292, 2017. https://doi.org/10.1016/j.wasman.2017.08.044

Development of inherent safety assessment index for e-waste recycling process: flammability parameter

SYAZA IZYANNI Ahmad[1,2,a*], ENGKU SYIFA HAYANI Engku Saifuddin[1,b], MARDHATI Zainal Abidin[1,2,c], MUHAMMAD FIRDAUS Husin[3,d]

[1] Chemical Engineering Department, Universiti Teknologi PETRONAS, Perak, Malaysia

[2] Centre of Advanced Process Safety, Institute of Contaminant Management, Unversiti Teknologi PETRONAS, Perak, Malaysia

[3] School of Chemical Engineering, College of Engineering, Universiti Teknologi MARA, Cawangan Terengganu, Kampus Bukit Besi, Terengganu, Malaysia

[a]syaza.ahmad@utp.edu.my, [b] engku_21000527@utp.edu.my, [c]marhati.zainal@utp.edu.my, [d] firdaushusin@uitm.edu.my

Keywords: Flammability, Inherent Safety, Index Scoring, E-Waste Recycling Process, Logistic Function

Abstract. Nowadays, the usage of electronic devices such as computers and mobile phones are crucial in the daily life especially with the growing internet usage globally. The worldwide e-waste production is anticipated to be around 20 to 25 million tonnes per year and is expected to rise through the decade. Increasing amount of e-waste will become a major concern due to its harmful impact to the environment as well as human health. Investigations on the impact of e-waste recycling process in terms of flammability parameter is currently lacking. The objective of this work is to develop an inherent safety assessment index focusing on the flammability parameters assessment of e-waste recycling process. The inherent safety assessment index developed focused on the flammability value of every chemical used in e-waste recycling process particularly the hydrometallurgy and pyrometallurgy processes. Logistic function was used in developing the scores for flammability evaluation of e-waste processes. In this scoring index, higher flammability score indicates higher hazard. A simple case study was conducted to compare the flammability level of two e-waste recycling process, namely Process A and Process B. Process B has higher Total Flammability Score than Process A indicating it as more hazardous due to the existence of hydrochloric acid and cyanide in the process than Process A with only hydrochloric acid as its flammable chemical in terms of flammability level. This indicates that the inherent safety assessment index produced can be used to conduct preliminary evaluation on the flammability level of chemicals involved in an e-waste recycling process particularly the hydrometallurgy and pyrometallurgy processes. However, to achieve a more comprehensive inherent safety assessment, this index needs to be equipped with several others inherent safety assessment parameters for example explosiveness and toxicity.

Introduction

Nowadays, the usage of electronic devices such as computers and mobile phones are crucial in the daily life especially with the growing internet usage globally. These devices after some time will reach their end-of-life and degrade into a complicated waste matter which are referred to as e-waste [1]. Every year, an estimated 20 to 25 million tons of e-waste are generated globally [2], and the volume of e-waste will continue to increase in tandem with technological improvements [3,4]. Globally, 44.7 million metric tonnes of e-waste were produced in 2016 [5], an astounding rise from 20 million tonnes per year in 2006 [4]. The worldwide e-waste production is anticipated to be around 20 to 25 million tonnes per year and is expected to rise through the decade. Increasing amount of e-waste will become a major concern due to its harmful impact to the environment as well as human health [6] if it is not properly managed. Human health is impacted by e-waste through food chain in which the toxic substances end up entering the food chain and direct impact to workers who work in the e-waste pre-treatment plant. The impact of e-waste towards human health has been discussed by various works for example in the assessment of carcinogenic risk involving heavy metals in e-waste [7], detection of copper and lead in scalp hair samples [8] and detection of PCBs in human milk samples [9]. However, investigations on the impact of e-waste recycling process in terms of flammability parameter is currently lacking. The objective of this work is to develop an inherent safety assessment index focusing on the flammability parameters assessment of e-waste recycling process. The scopes involved in this work are;

1. The inherent safety assessment developed focus on the flammability value of every chemical used in hydrometallurgy and pyrometallurgy processes.
2. Logistic function is used in scores assignment for each flammability value.

There are many e-waste processing methods exist for example pyrometallurgy, hydrometallurgy, centrifugal separation and vacuum pyrolysis also surface passivation. However, this work only focuses on assessing inherent safety parameter of hydrometallurgy and pyrometallurgy processes. In hydrometallurgy process, e-waste collected will be dissolved in the appropriate solvent or liquid for the extraction of valuable metals through leaching in acidic or alkaline medium. Then further refining the target metal which is extracting it from the solutions using its chemical properties, either through currents and voltage such as electrolytic cells or simply through precipitation [10]. Pyrometallurgical processes have been used for the processing of metals from various waste materials over the last two decades. Smelting in furnaces, incineration, combustion, and pyrolysis are common e-waste disposal techniques. State-of-the-art smelters and refineries can extract precious metals efficiently and of isolating hazardous substances. These recycling facilities would close the loop for precious metals and reduce the effects of large volumes of e-waste on the environment. E-waste recycling is dominated by pyrometallurgical paths, while the steel industry embraces ferrous fractions for iron recovery and the secondary aluminium industry takes over aluminium fractions. Pyrometallurgical processes work with steps of release, separation/upgrading and purification that are basically like those of mechanical or hydrometallurgical routes. However, the release of precious metals is achieved not by leaching, crushing, or grinding, but by smelting in furnaces at high temperatures. E-waste/copper/lead scrap is fed into a furnace in these pyrometallurgical processes, whereby metals are collected in a molten bath and a slag phase is formed by oxides [11].

Development of Flammability Scoring Index for E-Waste Recycling Process

i. Brief Description on Flammability Parameter

Heikkila [12] defines flammability as how easily a material burns in air. In this work, flash point of a liquid is used to measure the flammability of the chemicals involved in the e-waste recycling process. The flash point of a liquid refers to the lowest temperature at which the liquid emits sufficient vapour to form an ignitable mixture with air [13]. Thus, chemicals with lower flash points present greater hazard risks compared to chemicals with higher flash points. Similar flammability value was used in the Numerical Descriptive Inherent Safety Technique for inherent safety assessment of flammability parameter in petrochemical industry [14], Inherent Benign-ness Index (IBI) [15], and Prototype Index for Inherent Safety (PIIS) [16].

ii. Brief Description of Logistic Function

The flammability scoring for the proposed inherent safety assessment index is developed through the application of logistic function. Equation 1 shows the general equation of logistic function

Materials Research Proceedings 29 (2023) 370-377 https://doi.org/10.21741/9781644902516-42

[17]. Equation 1 is supported by Equation 2 and 3. y variable in Equation 1 represents the flammability score for every flash point value, while x variable in Equation 1 indicates the flash point value to be evaluated. There are three main constants in Equation 1 namely C, B and A where C refers to the maximum score limit in which the y value will always be less than or equal to C. This feature of logistic function makes it suitable to be used in a scoring index. As an example, if the C value is set to 100, the maximum value for output y is 100. Equation 2 determines parameter B via the m value while Equation 3 determines parameter A via the k value.

$$y = \frac{C}{1 + Ae^{-Bx}} \tag{1}$$

$$B = \frac{4m}{C} \tag{2}$$

$$A = e^{Bk} \tag{3}$$

Research Methodology

a) Data Collection

Information on the types of chemicals used in hydrometallurgy and pyrometallurgy processes was conducted. Resources used from the available literatures such as journal papers and reports. Then, data on the flash point values for the chemicals identified were collected. Table 1 shows the chemicals identified in hydrometallurgy and pyrometallurgy processes along with their respected flash point.

Table 1 *Chemicals Involved in Hydrometallurgy and Pyrometallurgy Processes*

Chemicals	Flash Point (°C)
Carbon Monoxide	-191
Nickel	-25
Hexane	-22
Cyanide	-18
Acetone	-18
Thiourea	-9.3
Methanol	9
Lithium	18
Phosphorus	30
Acetic Acid	39
Hydrochloric Acid	48
Tri-N-Butyl Phosphate	145
Di-(2-ethylhexyl)phosphoric Acid	168.5
Ascorbic Acid	214.6
Raw and Modified Palm	250
Fructose	274.9
Sucrose	343.9
Zinc	680
Sodium Sulphide	950
Sodium Chloride	1413

Materials Research Proceedings 29 (2023) 370-377 https://doi.org/10.21741/9781644902516-42

b) Data Analysis

The data collected as shown in Table 1 was then divided into several range as shown in Table 2 and analyzed to identify the frequency and cumulative frequency of each range. Then, a cumulative curve of the flash point range versus cumulative frequency is plotted as shown in Figure 1. A linear trendline is plotted on the cumulative curve to identify its slope. According to Figure 1, the slope identified is 2.57. This value is used as the basis value in developing the flammability score using the logistic function in the next step.

Table 2 Frequency Analysis

Chemicals	**Flash Point (°C)**	**Flash Point Range (°C)**						
		< -50	**-49 to 0**	**1-50**	**51-100**	**101-150**	**151-200**	**>200**
Carbon Monoxide	-191	x						
Nickel	-25		x					
Hexane	-22		x					
Cyanide	-18		x					
Acetone	-18		x					
Thiourea	-9.3		x					
Methanol	9			x				
Lithium	18			x				
Phosphorus	30			x				
Acetic Acid	39			x				
Hydrochloric Acid	48			x				
Tri-N-Butyl Phosphate	145					x		
Di-(2-ethylhexyl) phosphoric Acid	168.5						x	
Ascorbic Acid	214.6							x
Raw and Modified Palm	250							x
Fructose	274.9							x
Sucrose	343.9							x
Zinc	680							x
Sodium Sulphide	950							x
Sodium Chloride	1413							x
Frequency		**1**	**5**	**5**	**0**	**1**	**1**	**7**
Cumulative Frequency		**2**	**7**	**12**	**12**	**13**	**14**	**21**

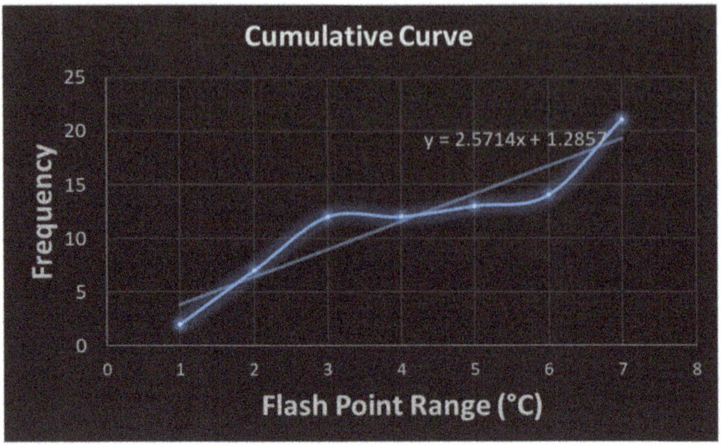

Figure 1 *Cumulative Curve*

c) Development of Flammability Scoring using Logistic Function

There are three values that need to be determined first before the logistic function as shown in Equation 1 can be used for score development. The first value is the C value which refers to the maximum score to be assigned in the logistic function. In this work, the C value is set as 100 indicating 100 as the highest score available for the flammability scoring index. The second value is the k-value which indicate the middle score of the scoring index. As 100 is the maximum score, 50 is taken as the k-value. The third value is the m value which can be used to determine coefficient B as shown in Equation 2. The slope value obtained from the cumulative curve in Figure 1 is used as the first m value to obtain coefficient B. After the B value was obtained, coefficient A value can be obtained using Equation 3 and lastly the final flammability score can be obtained using Equation 1. Equation 1 is used to produce a logistic curve, specific for flammability parameter of e-waste processing. The m value needs to be adjusted (either by reduction or addition, for this work the adjustment was done through reduction) so that a smooth logistic curve can be obtained as the final product.

Results and Discussion

Equation 4 shows the final logistic function developed while Figure 2 shows the final logistic curve developed for the flammability scoring of e-waste processing. In Equation 4, x_{FP} refers to the flash point of the chemical to be evaluated. If a user wanted to identify the flammability level of a chemical involved in e-waste processing, the user must insert the flash point value of the chemical into Equation 4 and can directly obtain its flammability score. This is similar to the approach taken by Ahmad et. al. [14] in the development of inherent safety assessment index for petrochemical processes. In addition, the user can also use the flammability score curve in Figure 2 to graphically obtain the score of the chemicals. Figure 2 can also be used in identifying the root-cause of flammability hazard in an e-waste recycling process as discussed by Ahmad et. al. [18]. In this scoring index, higher flammability score indicates higher hazard.

$$Flammability\ Score = 100 \times \frac{1}{1 + 2.7182e^{-0.02x_{FP}}} \quad (4)$$

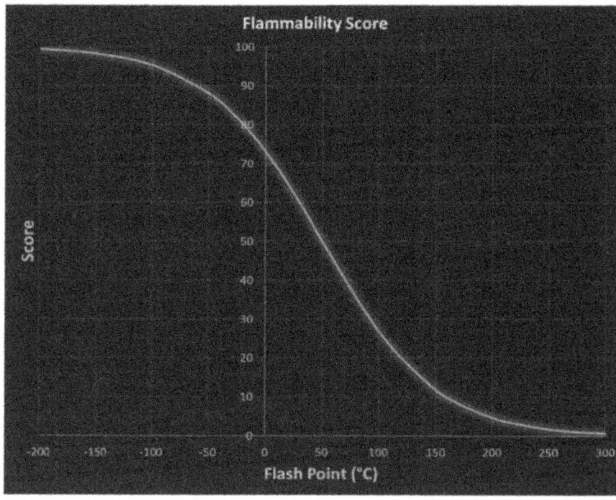

Figure 2 Logistic Curve for Flammability Scoring

If there are several chemicals involved in a process, the score for each chemical can be totalled up (Equation 5) to produce a score that can represented the whole process.

$$Total\ Flammability\ Score = \sum_i Flammability\ Score_i \tag{5}$$

Application of the Developed Index to a Case Study

A simple case study was conducted on two e-waste recycling process namely Process A and Process B to identify the flammability level of the chemicals involved in both processes. The purpose of the case study is to illustrate the usage of the scoring equation developed as shown in Equation 4. Process A involved four chemicals namely sulfuric acid, hydrogen peroxide, magnesium chloride, and hydrochloric acid while there are three chemicals involved in Process B which are nitric acid, hydrochloric acid, and cyanide. After identifying their flash point values, it was identified that sulfuric acid, hydrogen peroxide, and magnesium chloride in Process A while nitric acid in Process B are non-flammable which are assigned as having 0 flammability score. This is aligned with the score's assumption used in this scoring index in which higher flammability score indicates higher flammability hazard. The identified flash point values were inserted into Equation 4 to produce the flammability score for each chemical. Table 3 shows the chemicals involved in both processes as well as their flash point values and designated flammability scores.

Table 3 Chemicals involved in Process A and Process B

Process	Chemical Involved	Flash Point (°C)	Flammability Score
A	Sulfuric Acid	Non-flammable	0
	Hydrogen Peroxide	Non-flammable	0
	Magnesium Chloride	Non-flammable	0
	Hydrochloric Acid	48	51
B	Nitric Acid	Non-flammable	0
	Hydrochloric Acid	48	51
	Cyanide	-18	80

Individually, cyanide is deemed as the most hazardous in terms of flammability compared to the other chemicals used in Process A and Process B. Comparison can also be made between Process A and Process to identify which process possessed the most hazardous flammability level. This can be done by totalling the flammability scores using Equation 5. Table 4 shows the comparison between Process A and Process B in terms of their flammability level. According to Table 4, Process B has higher Total Flammability Score than Process A indicating it as more hazardous than Process A in terms of flammability level. Comparison of both processes indicates Process B as the most hazardous due to the existence of two flammable chemicals namely hydrochloric acid and cyanide.

Table 4 Comparison between Process A and Process B in terms of Flammability Level

Process	Chemical Involved	Flammability Score	Total Flammability Score
A	Sulfuric Acid	0	51
	Hydrogen Peroxide	0	
	Magnesium Chloride	0	
	Hydrochloric Acid	51	
B	Nitric Acid	0	131
	Hydrochloric Acid	51	
	Cyanide	80	

This inherent safety index developed specifically for flammability parameter needs to be equipped with several other inherent safety parameters, such as explosiveness, toxicity, and operating conditions, for a more comprehensive inherent safety evaluation. Furthermore, relationships between flash point and operating pressure of the process might also affect the inherent safety level of the process indicating its necessity for future investigations.

Conclusions

In conclusion, the inherent safety assessment index scoring equation produced can be used to conduct preliminary evaluation on the flammability level of chemicals involved in an e-waste recycling process particularly the hydrometallurgy and pyrometallurgy processes. Assessment on simple case study of two e-waste processing methods namely Process A and Process B were conducted. In process A, out of the four chemicals involved, hydrochloric acid is the most flammable represented by the highest flammability score of 51. Meanwhile, cyanide is identified as the most hazardous in Process B in terms of flammability parameter due to its highest score of 80. Comparison of both processes indicates Process B as the most hazardous due to the existence of two flammable chemicals namely hydrochloric acid and cyanide. However, to achieve a more comprehensive inherent safety assessment, this index needs to be equipped with several other inherent safety assessment parameters for example explosiveness and toxicity.

Acknowledgement

Special thanks to Universiti Teknologi PETRONAS (UTP) for the support received in completing this manuscript under STIRF research grant (cost center: 015LA0-022).

References

[1] P. Tanskanen, Management and recycling of electronic waste, Acta Materialia. 61 (2013) 1001-1011. https://doi.org/10.1016/j.actamat.2012.11.005

[2] M. Abul Hasan, T.A. Abbasi, R.P. Mahapatra, S. Ahmed, M.S. Abbasi, Ewaste management and hazardous effect on environment, In Proceedings of the 4th National Conference; INDIACom-2010 Computing for Nation Development. (2010)

[3] A. Ibitz, Environmental policy coordination in ASEAN: The case of waste from electrical and electronic equipment, Austrian Journal of South-East Asian Studies. 5(1) (2012) 30-51.

[4] B.H. Robinson, E-waste: An assessment of global production and environmental impacts, Science of the Total Environment. 408(2) (2009) 183-191. https://doi.org/10.1016/j.scitotenv.2009.09.044

[5] C.P. Balde, V. Forti, V. Gray, R. Kuehr, P. Stegmann, The Global E-Waste Monitor, United Nation University (UNU), International Telecommunication Union (ITU) & International Solid Waste Association (ISWA), Bonn/Geneva/Vienna. (2017)

[6] G. Song, J. Li, A Review on Human Health Consequences of Metals Exposure to E-Waste in China, Environ.Pollution. 196 (2015) 450-461. https://doi.org/10.1016/j.envpol.2014.11.004

[7] M. Xue, Y. Yang, J. Ruan, Z. Xu, Assessment of Noise and Heavy Metals (Cr, Cu, Cd, Pb) in the Ambience of the Production Line for Recycling Waste Printed Circuit Boards, Environ. Sci. Technol. 46 (2012) 494-499. https://doi.org/10.1021/es202513b

[8] T. Wang, J. Fu, Y. Wang, C. Liao, Y. Tao, G. Jiang, Use of Scalp Hair as Indicator of Human Exposure to Heavy Metals in an Electronic Waste Recycling Area, Environ. Pollution, 157 (2009) 2445-2451. https://doi.org/10.1016/j.envpol.2009.03.010

[9] G. H. Xing, J. K. Y. Chan, A. O. W. Leung, S. C. Wu, M. H. Wong, Environmental Impact and Human Exposure to PCBs in Guiyu, an Electronic Waste Recycling Site in China, Environ. International, 35 (2009) 76-82. https://doi.org/10.1016/j.envint.2008.07.025

[10] Chemistry: The Central Science. (Accessed: 14/05/2022) https://chem.libretexts.org/Courses/University_of_Missouri/MU%3A__1330H_(Keller)/23%3A_Metals_and_Metallurgy/23.3%3A_Hydrometallurgy 1212 (2020).

[11] M. Kaya, Recovery of Metals from Electronic Waste by Physical and Chemical Recycling Processes. World Academy of Science, Engineering and Technology. 10 (2) (2016) 232-243.

[12] A. M. Heikkila, Inherent Safety in Process Plant Design an Index-based Approach, PHd Thesis, Helsinki University of Technology (1999).

[13] D.A. Crowl, J.F. Louvar, Chemical Process Safety Fundamentals with Applications, 2nd Ed., Prentice Hall, New Jersey, 2002.

[14] S. I. Ahmad, H. Hashim, M. H. Hassim, Numerical Descriptive Inherent Safety Technique (NuDIST) for Inherent Safety Assessment in Petrochemical Industry, Process Saf. Environ. Protect. 92 (2014) 379-389. https://doi.org/10.1016/j.psep.2014.03.009

[15] R. Srinivasan, N.T. Nhan, A statistical approach for evaluating inherent benign-ness of chemical process routes in early design stages. Process Saf. Environ. Protect. 86 (2008) 163-174. https://doi.org/10.1016/j.psep.2007.10.011

[16] D.W. Edwards, D. Lawrence, Assessing the Inherent Safety of Chemical Process Routes: Is There a Relation Between Plant Costs and Inherent Safety? Trans IChemE, Vol. 71 (B) (1993) 252-258.

[17] R.J. Larsen, M.L. Marx, An Introduction to MathematicalStatistics and Its Applications, third ed. Prentice-Hall, NewJersey (2001).

[18] S. I. Ahmad, H. Hashim, M. H. Hassim, A graphical method for assessing inherent safety during research and development phase of process design, J. of Loss Prevent. In the Process Ind. 42 (2016) 59-69. https://doi.org/10.1016/j.jlp.2015.09.018

Assessment of enablers and disablers of environment-food-energy-water nexus in biomass value chain

SYAZMI ZUL Arif Hakimi Saadon[1,2,a], NORIDAH Osman[1,2,b *], SHARVEENA Shruthee[1,c], AHMAD SYAKIR Ismail[1,2d]

[1]Department of Chemical Engineering, Universiti Teknologi PETRONAS, Bandar Seri Iskandar, 32610 Perak, Malaysia

[2]HICOE – Center for Biofuel and Biochemical Research, Institute of Self-Sustainable Building, Department of Chemical Engineering, Universiti Teknologi PETRONAS, 32610, Seri Iskandar, Perak, Malaysia

[a]syazmi_19001700@utp.edu.my, [b]noridah.osman@utp.edu.my, [c]sharveena_17001674@utp.edu.my, [d]ahmad_18000240@utp.edu.my

Keywords: Biomass Value Chain, Environment, Food, Energy, Water

Abstract. The study aims to understand biomass value chains (BVC) in Malaysia, specifically those involving the use oil palm waste such as empty fruit bunch (EFB), palm oil mill effluent (POME) and oil palm trunks (OPT). This particular type of biomass source was selected as it is very abundant and is a major feedstock in existing biomass value chains in the country. The research uses survey methodology to obtain data related to BVCs by surveying relevant stakeholders. From the information obtained in Section 1 and Section 2 of the survey conducted, we were able to assess the enablers and disablers of environment-food-energy-water (EFEW) nexus in biomass value chain, which is the main research objective. Other than that, the raw data from survey is analyzed with SPSS statistical analysis methods to determine the relationship between technology, policies and role of government and private sector with the implementation of sustainable BVCs. Based on this, we found that the R value obtained was 0.759 with an R-square value of 0.575. From the Pearson correlations, we found that X3 (referring to role of private sector) showed the highest correlation and was the only statistically significant variable obtained based on the model. Finally, a comparative study on BVC's and fossil fuel value chain (FFVC) was also performed. From this comparison, we note that BVCs and FFVCs differ significantly in terms of carbon footprint, economic feasibility, technology, government and private role, policies, infrastructure, and volatility of resource.

Introduction

Biomass value chains (BVCs) refers to a network of infrastructure and technology designed to convert low value raw materials, such agricultural and forestry residue, to high value products [1]. BVCs typically undergo four main steps which are crop harvesting, biomass storage, logistical transportation, and biomass processing. Occasionally, pre-processing of biomass such as drying, and pelletizing may be necessary to increase the energy density of products. This is seen as being economical as it allows for easy storage and transportation.

The environment-food-energy-water (EFEW) nexus, on the other hand, is a systems-thinking approach that could holistically and critically address climate change [2]. The EFEW nexus provides us a useful framework to carefully assess the impacts of existing biomass value chains on these 4 interconnected and interdependent sectors (environment, food, energy, and water). Additionally, the EFEW nexus is considered to be able to anticipate unforeseen consequences towards our natural resources and seek common ground for various conflicting factors present in BVC's [3] . The enablers and disablers of EFEW refers to the mindsets, attitudes and structures that encourage or discourage implementation or adoption of sustainable practices, which in this case, within BVCs.

Therefore, this study aims to provide meaningful insight on the challenges of implementing sustainable BVCs within the context of EFEW by surveying local producers, processors, and distributors of biomass, as well as receive input from researchers and organizations. This will shed light onto the actual factors that enable and disable implementation of sustainable BVCs in Malaysia which can then be utilized to provide recommendations to sustainably optimize existing BVCs. In effect, biomass sources, specifically that from oil palm waste, will be able to better compete as a viable alternative source of energy and green products.

Materials and Methods

Methodology

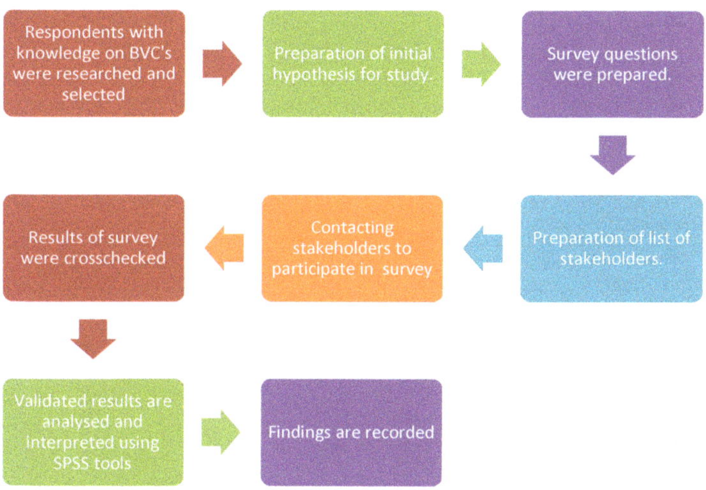

Selection of respondents

The respondents were selected by identifying groups or individuals that may have specific knowledge on biomass value chains. These include the stakeholders involved in actual biomass value chains in Malaysia. The stakeholders were identified as being the biomass producers, processing plants, researchers, organizations, and policy makers.

Table 1: Overview of relevant stakeholders in biomass sector

FUNCTION	NO. OF RESPONDENTS
FARMER	3
REPRESENTATIVE FARMERS' ORGANIZATION	1
CUSTOM HARVESTER	4
REPRESENTATIVE FROM INDUSTRY	3
POLICY MAKER	1
RESEARCHER	2
TOTAL	**14**

Survey and Interviews
A survey was prepared which consists of two main sections that included 50 rating type questions and essay type questions. The rating questions are divided into 7 individual parts. The data collected will then be analyzed through SPSS methods to derive the specific relationships and correlations between the variables and gauge if the initial hypothesis can be accepted or rejected. For interview section, a sample of the rating type or Likert scale questions is used in the survey.

Table 2: Breakdown in survey questionnaire

Section Type	Section Parts	Section Name	No. of questions
Ratings	Part 1	Biomass Industry in Malaysia	10
	Part 2	Biomass Value Chains in Malaysia	10
	Part 3	Challenges in Implementing Biomass Value Chains	10
	Part 4	Technology in Biomass Sector	5
	Part 5	Role of Government in	5
	Part 6	Biomass Industry	5
	Part 7	Role of Private Sector in Biomass Industry	5

Results and discussion
A total of 15 respondents from various segments of the biomass industry were recorded. The breakdown of respondents from the varying stakeholder segments in the biomass industry is as shown in Figure 1.

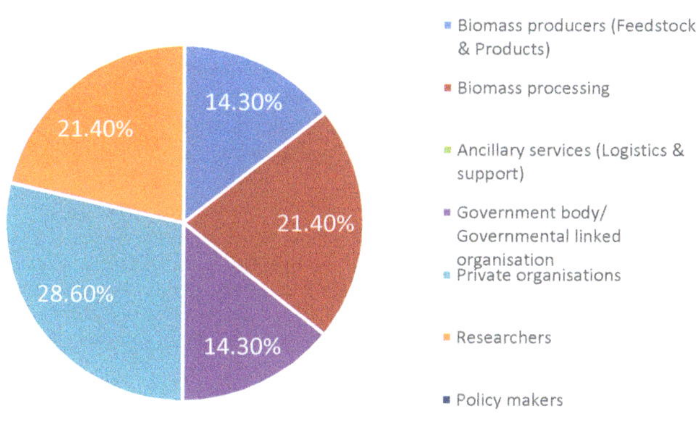

Figure 1: Stakeholder segments that participated in survey

Part 1: Biomass Industry in Malaysia

The percentage of responses in Part 1 for each question can be observed in Figure 2. The information collected from the data is grouped into two parts which are (i) Biomass sources and the extent of their use in Malaysia and (ii) Utilization of biomass as an energy source.

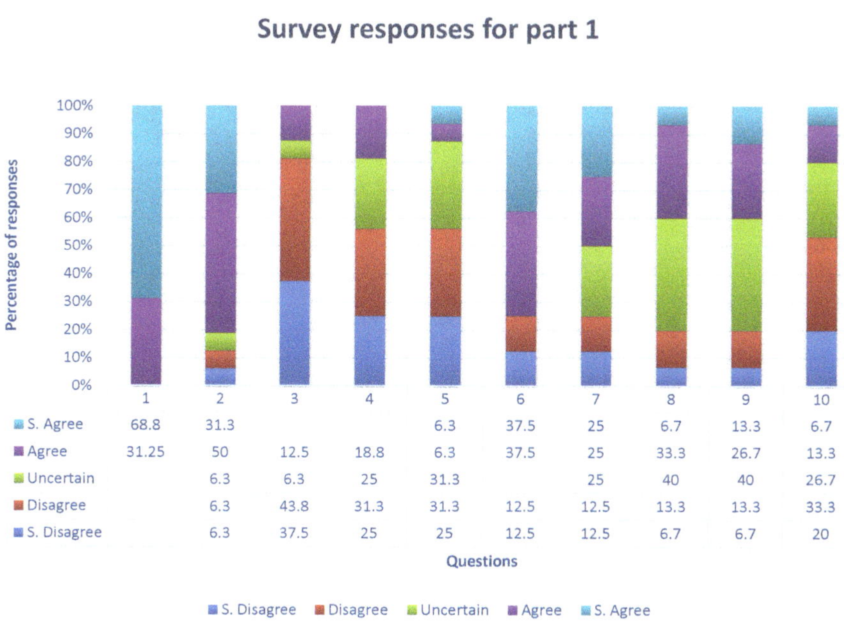

Figure 2: Responses for part 1 of survey

Materials Research Proceedings 29 (2023) 378-390 https://doi.org/10.21741/9781644902516-43

Biomass Sources and The Extent of Their Use in Malaysia

It is found that all respondents agrees that agricultural sector contribute most of biomass generated. In other words, municipal waste, urban garbage, and waste from animal breeding are less significantly utilized. Most of the respondents also agree that energy crops are not largely developed as a biomass source in Malaysia. As a whole, the respondents agree that the agricultural waste currently produced is not being effectively utilized as biomass.

Biomass As Energy Source

Most respondents agree on the fact that energy generation from biomass sources are not as widely developed as solar energy in Malaysia. However, energy from biogas and BioCNG plants are locally available through the national grid, although their capacity may not be very high. This can be corroborated from information obtained from the SEDA portal. Additionally, all experts agree that the biomass industry in Malaysia is focused on the production of solid biofuels and briquettes, which are mostly made for import and typically not consumed locally.

Part 2: Biomass Value Chains in Malaysia

The percentage of responses in Part 2 for each question can be observed in Figure 3.

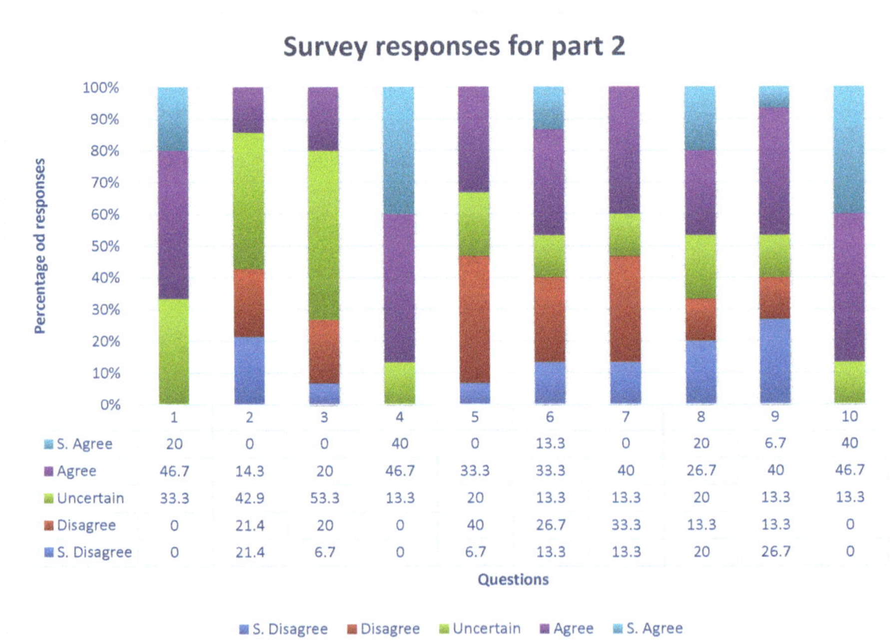

Figure 3: Responses for part 2 of survey

Opinion on Environmental Sustainability of Existing BVCs

There is some significant uncertainty among the respondents with regards to the environmental sustainability of BVCs currently implemented. This is gauged as a total of 42.9% and 53.3% of respondents were uncertain, when they were asked on whether or not existing BVCs in Malaysia pose concerns towards resource sustainability and negatively impacts the environment. This

Materials Research Proceedings 29 (2023) 378-390 https://doi.org/10.21741/9781644902516-43

indicates that the awareness on the adverse environmental effects of unsustainably implemented BVCs may not be very high among respondents.

Environmental Concerns Related to BVCs

A majority of respondents agree that BVCs are definitively more environmentally friendly compared to fossil fuel value chains. They also disagree that BVCs produce significant carbon footprint. This is consistent with various studies that have proven the impact of fossil fuels and their value chains on the environment.

Despite this, they do believe that there are some environmental concerns involved such as the water used for irrigation of crops and steam in processing plants may be used unsustainably. For the other environmental issues put forth, the results are less obvious. 46.7% agree on the issue of unsustainable energy use in biomass processing and water pollution from fertilizers and waste from biomass production. On the other hand, 46.7% disagree that competition for land between biomass cultivation and food crops is a major concern.

Part 3: Challenges in Implementing Biomass Value Chains in Malaysia

The percentage of responses in Part 3 for each question can be observed in Figure 4. The goal is to determine the difficulties and challenges in conducting and implementing BVC in Malaysia's biomass sector in an efficient and environmentally friendly way.

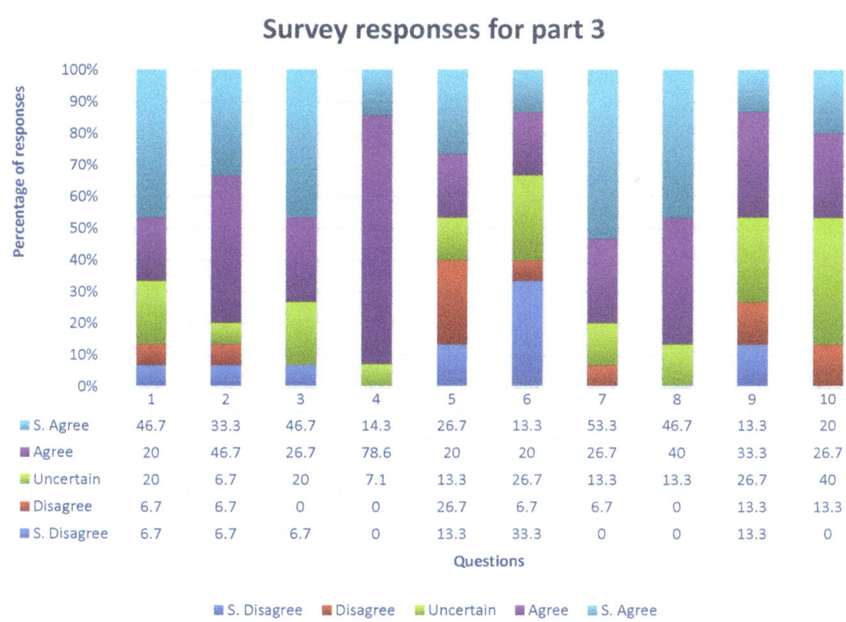

Figure 4: Responses for part 3 of survey

All the respondents agree that the implementing BVCs and being involved in the sector itself comes with several challenges. 66.7% believed that the uncertainties in biomass supply could be one of the main challenges. Additionally, 92.9% believe that involvement in the biomass sector may be hindered due to cost and economic reasons. Other than that, the lack of technology and insufficient policies are believed by most respondents to be a major challenge. This corresponds with our initial outcome hypothesis, whereby technology and policy are identified as being a factor that affects sustainable implementation of BVC.

The respondents mostly agreed that the nature of current pre-processing of biomass which utilizes large amount of heat power and the issue of scarcity and long distance between facilities is a major challenge that affects sustainability of BVC. However, results were more mixed when it came to other possible internal factors. 46.7% agree that harvesting/ collection activities require massive amounts of water, land, and energy, while 40% disagree. Additionally, 40% disagree that storage of biomass is not carbon free, while 33% agree. Besides that, 46.7% agree that the nature of biomass materials that are usually high volume and deteriorative further increases transportation needs that lead to environmental damage while 40% were uncertain.

Part 4: Technology in Malaysian Biomass Sector

The percentage of responses in Part 4 for each question can be observed in Figure 5.

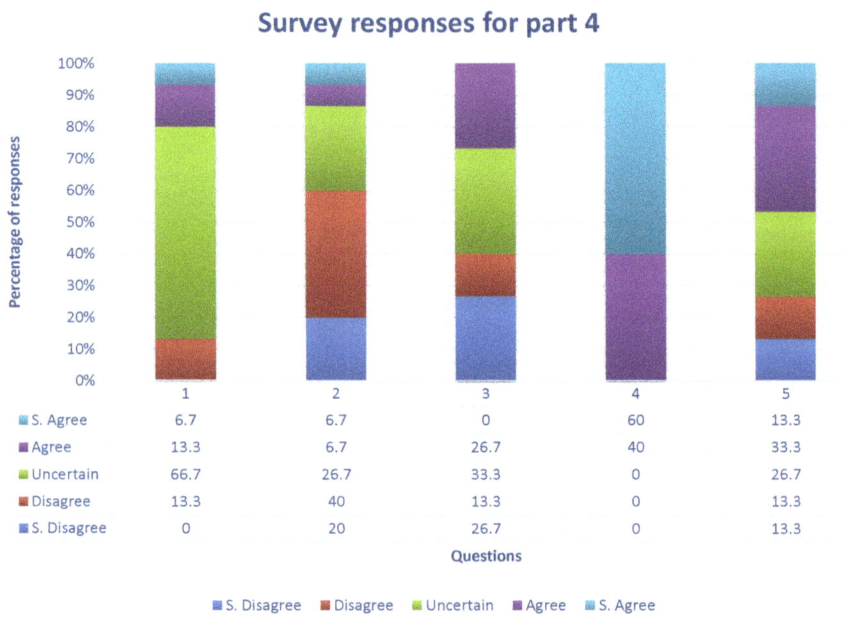

Figure 5: Responses for part 4 of survey

Based on the survey responses, it is evident that most of the respondents seem to be unaware of new technology currently being used or developed within the biomass industry. This is because 66.7% responded that they were uncertain on whether or not plenty of new technology was being utilized in BVCs in Malaysia. Despite this, they all agreed that there is no sufficient investment in developing technology in Malaysia and that government needs to play a greater role in creating incentives and putting forth initiatives to develop new technology for use in the biomass sector. They all also unanimously agree that the development of such technologies is essential in ensuring BVCs are implemented sustainably and in an environmentally friendly manner.

Interestingly, when asked if the technology used in Malaysia is comparable to technology used in other countries producing and processing similar biomass feedstock, 46.6% agreed. This indicates that the issue of lack of new technology being used and developed for more sustainable

BVCs is not just a problem locally, but also a major challenge in many other countries especially those within the region

Part 5: Role of Government in Malaysian Biomass Industry
The percentage of responses in Part 5 for each question can be observed in Figure 6.

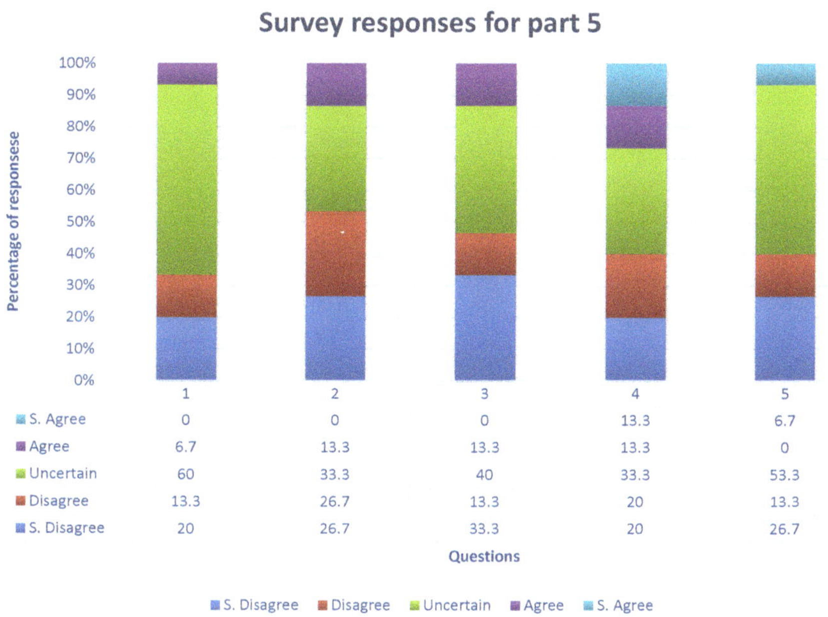

Figure 6: Responses for part 5 of survey

Government Involvement in Sustainable Optimization of BVCs
It was found that a majority of respondents were uncertain if current government efforts were sufficient or that more could be done on their part. They were also uncertain if the government had played a significant role in the growth of biomass sector within the past decade. However, most respondents agreed that at least when it comes to creating awareness of the environmental harms of unsustainable BVCs and investment in the sector, there is still some more room for improvement by the government. 53.4% also agreed that government ministries and agencies do not keep track of existing BVCs and monitor their impacts. This shows us that direct involvement by this entity within the biomass industry is still limited and should have better engagement and cooperation with industry stakeholders.

Part 6: Role of Private Sector in Biomass Industry
The percentage of responses in Part 6 for each question can be observed in Figure 7.

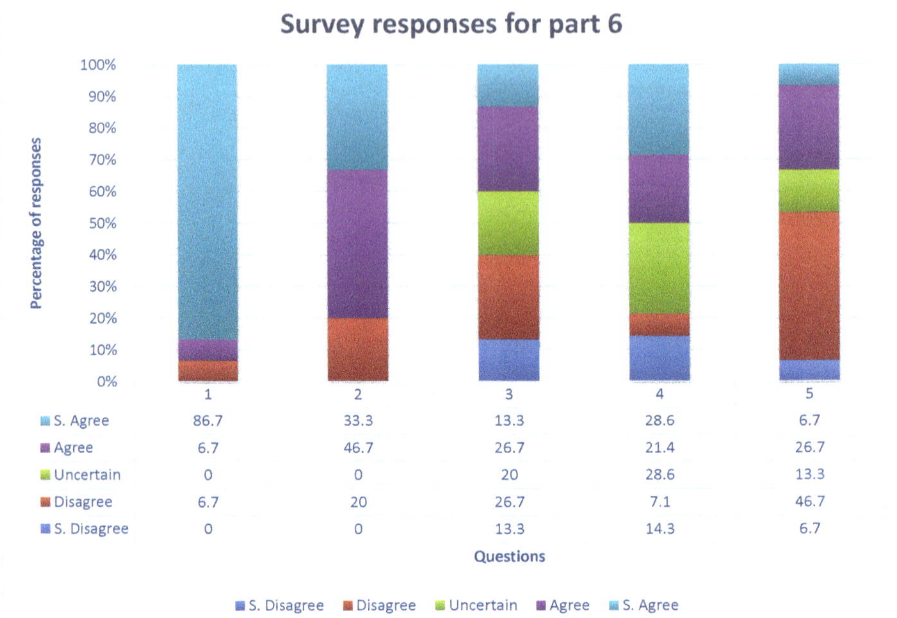

Figure 7: Responses for part 6 of survey

Private Sector Involvement in Sustainable Optimization of BVCs

It was found that the respondents all agree that private agencies and organizations play an important role in aiding the government to develop the biomass industry in Malaysia. 80% of respondents also agreed that currently the private sector provides many incentives and opportunities for the development of sustainable BVCs. Additionally, 53.4% of respondents agree that the private sector regularly invests in research and development for new technology that is able to reduce environmental impacts of existing BVCs. However, this may be limited to mostly the private organizations and not necessarily the privately operating biomass companies. This is determined as when asked if private companies were unconcerned about the environmental damage resulting from unsustainable BVCs, the results were split.

Other than that, 50% of respondents believe that biomass producers and processing plants are incentivized to implement more environmentally conscious practices in their BVCs. The results here prove to be inconclusive because another 50% disagreed or were uncertain on the matter. But if we take the results on a simple majority, we know that some incentives exist possibly from government efforts or initiatives. However, the efficacy of these incentives in creating more sustainable practices in BVCs is not known.

Part 7: Policies Related to Biomass Industry in Malaysia

The percentage of responses in Part 7 for each question can be observed in Figure 8.

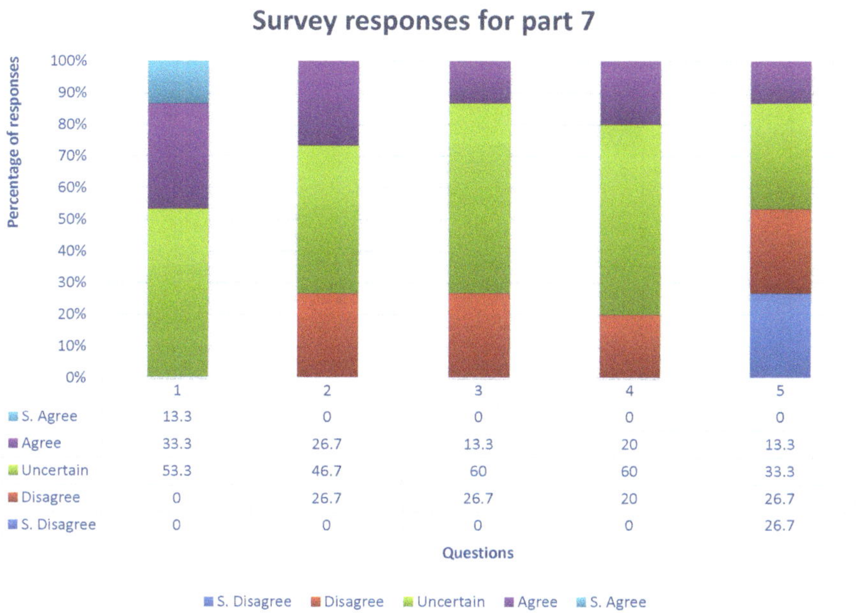

Figure 8: Responses for part 7 of survey

Policy Impact and Efficacy on BVCs

Based on the data collected, we can note that according to 53.4% of stakeholders, the existing environmental policies are not sufficient in ensuring sustainable BVC implementation. They believe that the policies require some revisions and updates to ensure they can effectively monitor and uphold environmental standards on existing BVCs being carried out.

Despite this, 53.3% of respondents were uncertain if policies enacted by the government have a significant enough effect in ensuring BVC related practices are in line with environmental regulations. 46.6% of respondents on the other hand, believed that policies were useful and necessary to deter unsustainable practices in the industry.

SPSS Results

Pearson correlation

The Pearson correlation matrix is shown in Table 3. Based on Table 3, we can observe that the role of private institutions (X3) shows the highest correlation at 0.410 followed by policies (X4) at 0.144, role of government (X2) at -0.005 and technology (X1) at -0.144. This tells us that based on the respondents surveyed, the sustainable implementation of biomass value chains in Malaysia can be improved most efficiently by encouraging a greater responsibility and having better policies on part of the private sector.

Table 3: Pearson correlation for identified dependent and independent variables

Variable	Y	X_1	X_2	X_3	X_4
Y	1	-0.144	-0.005	0.410	-0.008
X_1	-0.144	1	0.329	0.706	0.427
X_2	-0.005	0.329	1	0.315	0.411
X_3	0.410	0.706	0.315	1	0.223
X_4	-0.008	0.427	0.411	0.223	1

Model Summary

The Model Summary is shown in Table 4. From this, we note that the correlation coefficient, R=0.759. This value tells us about the correlation between the dependent variable and all the independent variables in the study. On the other hand, based on the R-square value, we know that 57.5% of the dependent variable can be explained by the variations in the independent variables.

$$y = \beta_0 + \beta_1 \cdot x_1 + \beta_2 \cdot x_2 + \beta_3 \cdot x_3 + \varepsilon \quad (1)$$

Table 4: Model summary

Model	R	R^2	Adjusted R^2	Std. Error of Estimate
1	0.759	-0.575	0.405	0.37219

The multifactorial regression model is shown in Table 5. the independent variable, X3 in the model is statistically significant (Sig. < 0.05). Additionally, the positive values on the beta coefficients confirms the convergent effect while the negative values indicate a divergent effect. As can be observed, the significance in values obtained do not pass the p-test as all except the X3 value showed significance greater than 0.05. This indicates that variables X1, X2 and X4 may not play a significant enough role in explaining the independent variable. Thus, the null hypothesis for these variables may have to be rejected except for that concerning variable X3.

Table 5: Multifactorial regression model

Model		Unstandardized Coefficient		Standardized Coefficients Beta	t	Significance
		B	Std. Error			
1	(constant)	2.392	0.806		2.968	0.014
	X_1	-0.823	0.274	-0.951	-3.007	0.130
	X_2	-0.710	0.144	-0.115	-0.493	0.633
	X_3	0.793	0.220	1.072	3.601	0.005
	X_4	0.233	0.275	0.207	0.849	0.416

It should also be noted that the sample size of this study is relatively small as only a total of 15 respondents had participated. Thus, this may have had an impact on the data analysis done and may explain the low significance and Pearson correlation values obtained.

We observed that variable, X3 which represents the role of private sector is the only statistically significant variable obtained through SPSS analysis. This tells us two things. It tells us that most of the respondents felt that the private sector plays a crucial role and are an important enabler for development of the local biomass sector in terms of environmental sustainability. It also tells us that most respondents believed that currently the private sector has been actively involved and is significantly contributing towards the implementation of BVCs in line with EFEW.

The current role of private sector can be seen through its many efforts in reducing carbon footprint within its BVCs and commitment towards sustainable practices. From a technical standpoint, private companies have started embracing process integration tools (PI) such as pinch analysis and mathematical modelling to reduce the carbon footprint produced during oil palm biomass processing. Other than that, implementations of good engineering practices and significant investment in new technology to address environmental concerns such as reduction of steam, electricity, and fuel consumption during processing. Notably, the private sector's active involvement in research and development of 'waste to wealth' conversion such as biogas capture, biochar and biofertilizers from biomass processing waste has also been an important enabler of EFEW nexus in BVCs.

When it comes to the other variables analyzed, the lack of significance obtained tells us that according to the respondents, there has not been sufficient contribution on the part of government, policies, and new technology to sustainably optimize existing BVCs. In other words, this means that the lack of active government involvement, effective policy and technology are disablers of EFEW nexus in BVCs. This tells us that improvements have to be made with regards to these factors in order to ensure local BVC implementation can reach optimum sustainability.

Conclusion

Biomass possesses great potential for development into a viable energy source environmentally friendly value-added product. However, existing biomass value chains often are too complex and inefficient in meeting performance requirements and are found to be detrimental to the environment. Thus, optimized biomass value chains that are in line with the energy-food-water-environment (EFEW) nexus is required to control the and minimize the impacts towards the environment. The EFEW nexus allows us to gauge this by providing meaningful insight on the interactions that BVCs have on our surrounding food, energy, and water resources. Understanding these interactions will allows the fulfilment of the set research objective, which is to identify and assess the enablers and disablers of EFEW in BVC. Through analysis of the survey questionnaire prepared we can come up with several findings that allowed us to understand the challenges and opportunities involved for implementation of EFEW nexus in BVC. Through analysis of results, we reflect insight into the perception of stakeholders on biomass value chains and its possible negative impacts on the environment. We were also able to gauge the challenges involved which allowed us to understand the enablers and disablers sustainable BVC implementation in line with EFEW nexus. From this, we find that most challenging part of BVC implementation according to the respondents is the issue of transportation and logistics, followed closely by cost, policy enforcement, technology, and inconsistency in supply. Analysis of SPSS survey results clearly visualize the relationship between the dependent variable, which is the implementation of sustainable BVC and the independent variables. From the Pearson correlations, we found that X3 (referring to role of private sector) showed the highest correlation and was the only statistically significant variable obtained based on the model. Thus, we are able to determine that role of private sector has significantly contributed to optimizing BVCs, making it an enabler, while the insufficiencies of government role, policies and technology have not significantly contributed towards this, making it a disabler.

Acknowledgements
The authors would like to acknowledge the funding support from the Ministry of Higher Education (MOHE) Malaysia through the Higher Education Center of Excellence (HICoE) grant to Center for Biofuel and Biochemical Research (cost centre: 015MA0-052) and supported by the Yayasan Universiti Teknologi PETRONAS grant (cost centre: 015LC0-205).

References

[1] J.F.D. Tapia, S. Samsatli, S.S. Doliente, E. Martinez-Hernandez, W.A.B.W.A.K. Ghani, K.L. Lim, H.Z.M. Shafri, N.S.N.B. Shaharum, Design of biomass value chains that are synergistic with the food–energy–water nexus: Strategies and opportunities, Food and Bioproducts Processing 116 (2019) 170-185. https://doi.org/10.1016/j.fbp.2019.05.006

[2] H. Hoff, Understanding the Nexus, The Water, Energy and Food Security Nexus, Stockholm Environment Institute, Stockholm, 2011.

[3] J. Liu, H. Mooney, V. Hull, S.J. Davis, J. Gaskell, T. Hertel, J. Lubchenco, K.C. Seto, P. Gleick, C. Kremen, S. Li, Systems integration for global sustainability, Science 347(6225) (2015) 1258832. https://doi.org/10.1126/science.1258832

A binder additional process in urea granule fertilizer by using adaptive fuzzy logic control

NORHIDAYAH Mohamad[1,a*], NOR AZLINA Ab Aziz[1,b] and NOR HIDAYATI Abdul Aziz[1,c]

[1]Faculty of Engineering & Technology, Multimedia University, Melaka, Malaysia

[a]norhidayah.mohamad@mmu.edu.my, [b]azlina.aziz@mmu.edu.my, [c]hidayati.aziz@mmu.edu.my

Keywords: Urea Fertilizer, Granulation, Ammonia, Fuzzy Logic, Fluidized Bed

Abstract. This paper presents the effect of binder feed rate addition towards ammonia gas released during urea granulation process. The binder feed rate as manipulated variable with few other constant parameters such as pressure and temperature of fluidized bed granulator. These parameters, binder flow rate and NH_3 emission were used to indicate the function ability of the designated fuzzy logic. The performance index of this study is then defined with percentage error from experimental value and actual value. An adaptive Fuzzy Logic Controller (FLC) is proposed to control the system conditions closed to the reference values. As binder flow rate increases the higher is the emission of NH3. The average of error percentage for whole project was 6.91%. The highest and lowest error in percentages are 81.5 and 0 respectively. The result shows that the proposed method can be efficiently implemented in the real-time determination and control of optimal conditions for granulation processes with efficient energy and to minimize the amount of ammonia gas (NH_3) release to the environment.

Introduction

Urea granule is used as fertilizers to supply additional nitrogen for paddy plant growth and sustaining the soil fertility [1]. The utilization of urea fertilizers on paddy fields has been well accepted for producing high rice grain yields and improved the nutrient availability. Urea fertilizer is also cost effective and having high Nitrogen content comparing to the other resources [2].

Granulation is a process for enlargement from small particles into larger diameter particles which called granules [3]. In fluidized bed granulation process, ammonia gas was distinguished. There are two contribution of ammonia gas emission which are binder solution added by spraying into powder particles and evaporation of the urea syntheses process [4]. Figure 1 shows the reaction process of fluid bed granulation where the ammonia gas released was detected during the mixing of urea powder with the urea binder solution. The ammonia gas was formed when the urea solution was heated by temperature under high pressure. The ammonia gas was released through the output ventilation from the fluid bed granulation's chamber.

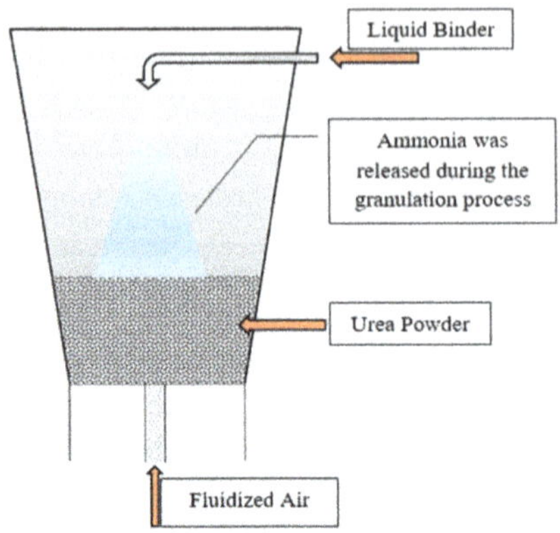

Fig. 1: Schematic of fluidized bed granulation process

In spite of this, during processing the urea fertilizer, a binder liquid solution was added to urea powders. The powder particles adhere to one another to form agglomerates, and the size of the particles is enlarged to form as a urea granule [5]. The binder fluid contains a carbon content that must be volatile due to unstable state of operating conditions. The involvements of heat, pressure and water existence was considered as the main factor of the gas released based on experimental observations.

A fuzzy logic control is constructed from a set of fuzzy If–Then rules using fuzzy logic principles and the learning algorithm adjusts parameters of the fuzzy logic control based on numerical information for better performance on tracking the desired outputs [6]. The objective of this paper is to apply adaptive fuzzy logic control in the fluidized bed granulation process and study the correlation between binder addition flow rate and ammonia gas emission

Ammonia (NH_3) emisssions

Urea melts due to process spraying and drying towards the completion of the granulation process. The air in the granulator contains urea dust and NH_3 [4]. The NH_3 emission due to evaporation section of the urea synthesis process which contribution from urea powder and the urea binder solution [5]

Ammonia (NH_3) reactions with urea

This research analyze of binder addition in fluidized bed granulation which focus on the emission of ammonia gas and condition that affect from it. Combination of ammonia (NH_3) and carbon dioxide (CO_2) can produce urea which also known as ammonia carbamide. Solution synthesis in which the reactions are as follows [6]:

$$2NH_2 + CO_2 \longrightarrow NH_2\,CO_2\,NH_4 \quad (1)$$

$$NH_2\,CO_2\,NH_4 \longrightarrow NH_2\,CO\,NH_4 + H_2O \quad (2)$$

The use of urea solution from solution synthesis process is produce solid urea by concentrated. Concentrate the urea "melt" by evaporation process of the solution [7]. According to F. T et. al [8], high temperature is to prevent the melt to solidify and the temperature able to breaks down to ammonia and biuret. Emission of ammonia from breakdown will transfer to the gas phase in the finishing step [9]. Besides that, another side reaction that occur is the hydrolysis reaction of urea [10]. Temperature is vital factor for the reaction which urea needs to be guard from expose [11]. The hydrolysis reaction is reversed reaction of the urea transformation [12]

Additionally, the fuzzy logic which include construction of the system that include variables, the set of rules require and also knowledge base use. An adaptive fuzzy logic was designed as a platform in Matlab-Simulink environment to control the granulation system conditions.

Methodology

A fuzzy logic has been designed to study relationship between one input and one output which are binder flow rate and NH_3 Emission. Mamdani inference system was chosen to develop this fuzzy logic controller. The characteristic of this fuzzy logic system was based on fuzzy inference system, a few fuzzy sets, input and output variables, type of membership functions, and definition of if-then rules which explained in the further part of the chapter. The total of 50 data set has been collected from the system which will be compare with actual data value. Each of data obtain will be compare with the actual data value for further analysis. Table 1 shows an experimental work conducted via Fuzzy Logic Toolbox in Mathworks Matlab software.

Table 1: Experimental Data for Fuzzy Values

No	Binder Flow rate (ml/min)	NH_3 Emission (ppm)	No	Binder Flow rate (ml/min)	NH_3 Emission (ppm)
1	0.2	0.0484	26	5.2	3.15
2	0.4	0.0968	27	5.4	3.29
3	0.6	0.145	28	5.6	3.44
4	0.8	0.194	29	5.8	3.53
5	1	0.242	30	6	3.68
6	1.2	0.29	31	6.2	3.82
7	1.4	0.581	32	6.4	3.92
8	1.6	0.726	33	6.6	4.07
9	1.8	0.871	34	6.8	3.92
10	2	1.02	35	7	4.16
11	2.2	1.21	36	7.2	4.21
12	2.4	1.5	37	7.4	4.31
13	2.6	1.65	38	7.6	4.4
14	2.8	1.79	39	7.8	4.45
15	3	1.94	40	8	4.5
16	3.2	2.03	41	8.2	4.55
17	3.4	2.13	42	8.4	4.6
18	3.6	2.18	43	8.6	4.65
19	3.8	2.27	44	8.8	4.65
20	4	2.37	45	9	4.69
21	4.2	2.42	46	9.2	4.74
22	4.4	2.52	47	9.4	4.79

23	4.6	2.61		48	9.6	4.79
24	4.8	2.71		49	9.8	4.79
25	5	2.52		50	10	4.79

An actual data collection of 50 sets contains of two different variables which are binder flow rate and NH_3 Emission. These data have been gathered from granulation process whereas fixed value of the parameters in the process are:

Inlet Temperature =65 °C
Spray Angle = 90 °C
Binder Viscosity = 0.236 μPa.s
Binder Atomizing Pressure = 0.5kg/ cm2
Binder Volume =250 ml
Nozzle Size = 5 mm

Around 50 sets of data were collected by using the designated fuzzy logic system in this project. The starting value of 0.2 for input variable and increase by 0.2 until reach the value of 10. These experimental values will be compared with actual value in the Table 4.3. The percentage error will be calculated based on an equation.

$$\%\ error = \frac{Experimental\ value - Actual\ value}{Actual\ value} \times 100\%$$

Result and Discussions

Based on the value obtained from designated fuzzy logic showed that as binder flow rate increase the higher NH_3 emission. This trend can be seen in Figure 2. Many factors that influenced the experimental value data for each set such as number of fuzzy sets, range of the fuzzy sets, specification of membership function, and defuzzification methods applied. A single input and a single output have been sets as parameters for designing fuzzy logic. The input and output variables are binder flow rate and NH_3 emission respectively. The measure of binder flow rate on a scale of 0.2 to 10 while NH_3 emission on a scale 0 to 4.84. The smallest of maximum was defuzzification method used in this fuzzy logic.

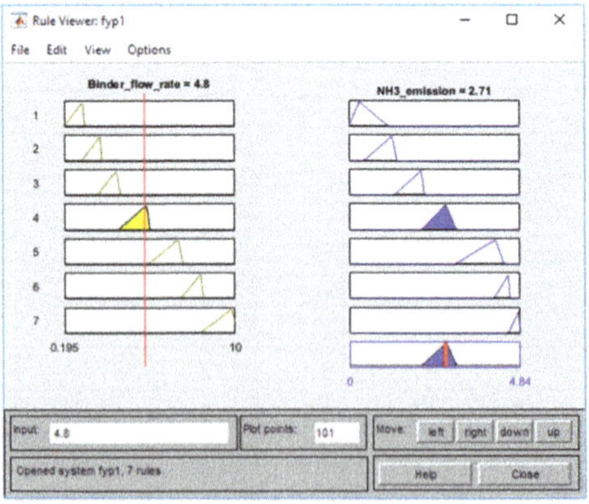

Fig. 2: Rule viewer of the designated fuzzy logic

The output for experimental value in this project was based on Figure 3 since the input value will correspond to their output. The Figure 3 shows that crisp value of both input and output variables in graphical presentation. It also can be used for finding the output crisp by choosing value for input variables in their respective axes.

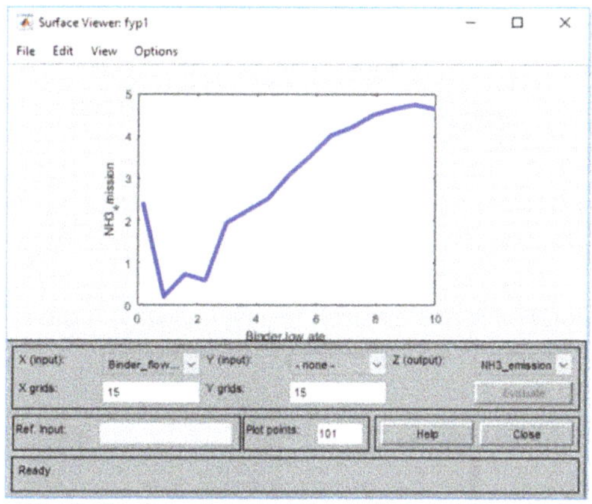

Fig. 3: Surface viewer of the designated fuzzy logic

According to the data collected result data in Figure 4, the NH_3 emission tends to increase with binder flow rate. The experimental data is closed to the actual data as discussed in this part. The influence of binder flow rate with the specific value of constant variables were used enhance the released of NH_3 gas during the granulation process. The impact of binder flow rate also improves completion time of granulation process and improved the size of granule [13].

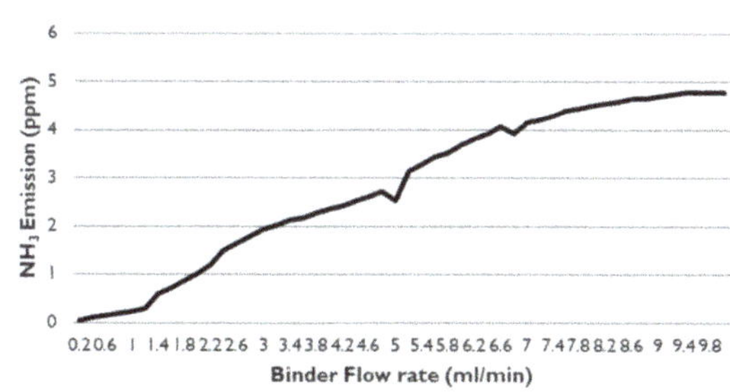

Fig. 4: Graph on NH_3 Emission vs Binder flow rate

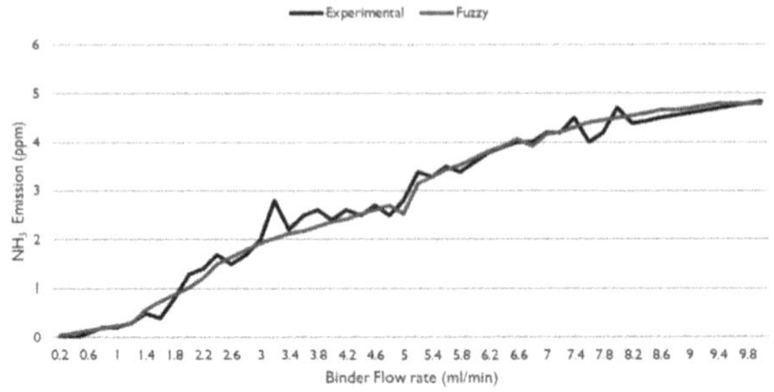

Fig. 5: Graph of comparison value of the NH_3 Emission

The trends of percentage errors are lessened along the input variables. Based on Figure 5 and Figure 6, around 6.91% of error in average in 50 sets of data.

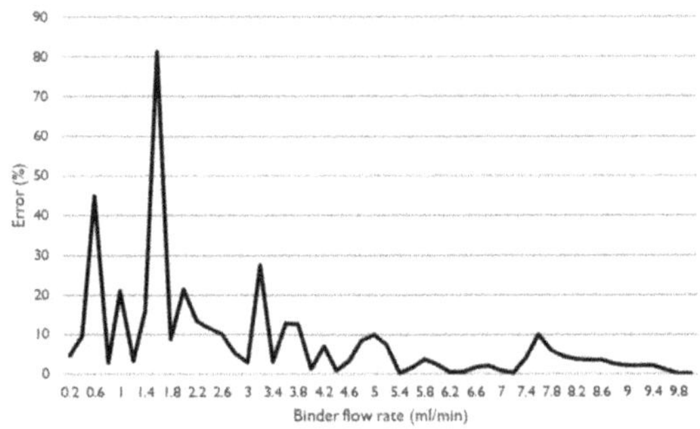

Fig. 6: Graph of percentage error of the NH_3 Emission

Based on data collection presented, the NH_3 emission tends to increase with binder flow rate. The experimental data is closed to the actual data as discussed in this part. The influence of binder flow rate with the specific value of constant variables were used enhance the released of NH_3 gas during the granulation process [14]. The impact of binder flow rate also improves completion time of granulation process and also improved granule size [15]. The defuzzification method which was small of maximum stand gives experimental value close to actual data. The selection of defuzzification method was a trial and error process to find better results [16].

Conclusion

An adaptive fuzzy logic controller has been successfully completed in this project. 50 sets of fuzzy values with different of binder flow rate which varies from 0.2 to 10 in ml/min. The results obtained by Mamdani inference system acceptable and close with experimental value thus the fuzzy designed applied was acceptable in this designing the project. As binder flow rate increases higher emission of NH_3. The average of percentage error for whole project was 6.91%. The highest and lowest error in percentage are 81.5 and 0 respectively

Extension of research with different properties and concentration of binder parameters are recommended. This project can be further studied by using other systems such as Takagi-Sugeno controller, Neural network or combination of fuzzy and neural which is Adaptive neuro fuzzy inference system (ANFIS) for better system comparison.

Acknowledgment
This research is funded by Multimedia University funding. The authors also would like to acknowledge the anonymous reviewers for their valuable comments and insights.

References

[1] Mohamad, N., and Shaaban, A., 2018. Modeling of Spray Angle and Nozzle Size to Gas Release in Processing Urea Fertilizers by Using Fluidized Bed Granulator. (July), pp. 2-3.

[2] Rahmanian, N., Naderi, S., Supuk, E., Abbas, R., and Hassanpour, A., 2015. Urea finishing process: Prilling versus granulation. Procedia Engineering, 102, pp. 174-181. https://doi.org/10.1016/j.proeng.2015.01.122

[3] D. M. Morkhade, " Comparative impact of different binder addition methods, binders and diluents on resulting granule and tablet attributes via high shear wet granulation," Powder technology, vol.320, pp. 114-124, 2017. https://doi.org/10.1016/j.powtec.2017.07.038

[4] C. Onwulata, Encapsulated and Powered Foods, CRC Press Taylor & Francis Group, 2005. https://doi.org/10.1201/9781420028300

[5] Patel, T. B., Patel, L. D., Patel, T. B., Makwana, S. H., and Patel, T. R., 2010. Influence Of Process Variables On Physicochemical Properties Of The Granulation Mechanism Of Diclofenac Sodium. 3(1), pp. 61-65.

[6] P. Roy, R. Khanna, D. Subbarao, "Granulation Time in Fluidized Bed Granulators," Powder Technology, vol.199, issue 1. pp. 95-99, 2010. https://doi.org/10.1016/j.powtec.2009.04.018

[7] W. L. Davies and W. T. Gloor, "Batch Production of Pharmaceutical Granulations in a Fluidized Bed I: Effects of process variables on physical properties of final granulation," Journal of Pharmaceutical Sciences, vol.60, issue. 12, pp. 1869-1874, 1971. https://doi.org/10.1002/jps.2600601224

[8] F. T. Van der Scheur, P. L. Goedendorp, A. J. Van der Goot and J. J. Olieman, "Fluidized Bed Agglomeration with Polyethylene Glycol Melt Binder: Effects of Bed temperature Droplet Size," in Proceedings of World Congress of Particle Technology, Kyto, 1998.

[9] P. G. Smith, Applications of Fluidization to Food Processing, John Wiley & Sons, 2008. https://doi.org/10.1002/9780470995426

[10] W. L. Davies and W. T. Gloor, "Batch Production of Pharmaceutical Granulations in a Fluidized Bed II: Effects of Various Binders and their Concentrations on Granulations and Compressed Tablets," Journal of Pharmaceutical Sciences, vol.61, issue. 4, pp. 618-622, 1972. https://doi.org/10.1002/jps.2600610428

[11] W. L. Davies and W. T. Gloor, "Batch Production of Pharmaceutical Granulations in a Fluidized Bed III: Binder dilution effects on granulation," Journal of Pharmaceutical Sciences, vol.62, pp. 170-171, 1973. https://doi.org/10.1002/jps.2600620144

[12] M. Gohel, R. Parikh, L. Baldaniya, B. S. Barot, H. Joshi, P. Parejiya, P. Mistry, A. Popat, T. Patel and R. Parmar, "Fluidized Bed Systems: A Review", 2007.

[13] M. Norhidayah, A. Shaaban, M. F. Dimin, M. Y. Norazlina, O. Rostam, "Optimization of Biodegradable Urea Production Process to Minimize Ammonia Release through Response Surface Method Experimental Design," Journal of Advanced research in Applied Sciences and Engineering Technology, vol. 2, no. 1. pp. 9-18, 2016.

[14] J. H. Meessen and H. Petersen, "Urea," Ullmann's Encyclopedia of Industrial Chmesitry, vol.5, pp. 333-365, 1996.

[15] J. Eliasson, Y. Eriksson, D. Holmström, F. Nilsson and F. Ojala, Reduction of Continuous Ammonia Emission in a Urea Plant Finishing Section, Lund University, 2009.

[16] R. Jang, C. Sun, E. Mizutani, Neuro-fuzzy and soft computation, Prentice-Hall,1997

The implementation of simple additive weighting (SAW) method in decision making process for paint & coating

MUHAMMAD DINI Zainuddin[1,a], MARDHATI Zainal Abidin[1,b*] and NOOR DIANA Abdul Majid[2,c]

[1]Chemical Engineering Department, Universiti Teknologi PETRONAS, 32610 Seri Iskandar, Perak, Malaysia

[2]Faculty of Industrial Sciences and Technology, Universiti Malaysia Pahang, 26300 Gambang, Pahang, Malaysia

[a]muhammad_17006248@utp.edu.my, [b]mardhati.zainal@utp.edu.my, [c]ndiana@ump.edu.my

Keywords: Multi-Criteria Decision Making, Simple Additive Weighting, Paint, Safety, Health

Abstract. New product development is a changing goal that becomes increasingly complex as a result of a variety of known and unknown circumstances. Different attributes need to be considered in selecting the best product formulation such as cost, safety, health, environment, performance and many others. A solid decision at the early stage of process design are all essential aspects in building a great product. Therefore, multi-criteria decision making is an important aspect during product development process. In this work, Simple Additive Weighting (SAW) has been used for the decision-making process to identify the best paint alternatives considering three attributes: safety, health and performance. The rank from the most favorable alternative to the less favorable alternative are 2K Formulation, Gloss 1K Formulation, Matte 1K Formulation with the score 0.73, 0.69 and 0.46 respectively. A comparison between the SAW results obtained in this work with simple index-based methodology shows that the results are in agreement with each other and both methodologies are suitable to be used at the early stage of process development as the methodologies are simple. Further investigation shows that the SAW result gives more flexibility in prioritizing the attributes compared to the simple index-based methodology based on predetermined weight value.

Introduction

The development and application of chemical products in a daily life are important for the technology revolution and economic progress around the world. However, chemical components in various designed items that have been launched to the market may expose consumers to a range of safety and health issues. Several cases have been recorded in the past for example lawsuits from consumers claimed the contamination of asbestos in baby powder causes on the development of ovarian cancer [1], the death from mesothelioma due the exposure asbestos-containing talc that was used in paint manufacturing [2] and melamine contamination in milk powder leading to fatalities [3]. As a result, it's vital to use a systematic procedure to develop the products that meet acceptable safety and health standards. Different criteria need to be considered in selecting the best product formulation such as cost, safety & health, and performance of the product. A solid product strategy that takes into account positioning possibilities, feasibility, and a flexible development methodology are all essential aspects in building a great product. To resolve the issue, multi-criteria decision making (MCDM) also referred to as multiple criteria decision analysis (MCDA) can be used. MCDM is a technique that involves the analysis of various available choices for the decision-making process. Among the common MCDM tools are Simple Additive Weighting (SAW), Weighted Product Method (WPM), Analytic Hierarchy Process (AHP), Analytic Network

Process (ANP), Technique for Order Preference by Similarity to Ideal Solutions (TOPSIS), and others. The comparisons between the methodologies are presented in Table 1 [4], [5] indicates the advantages and disadvantages for each methodologies. Various MCDM techniques have been applied for decision making process in chemical process industry. In the work by Sarder, T. & Khan, M.R. AHP and TOPSIS methodology has been implemented to identify the right chemical supplier considering six attributes: quality of chemicals, price, timely shipment, social responsibility, safety management facility and risk management [6]. Other applications of MCDM via Analytic Network Process (ANP) has been demonstrated by the work by Othman, M.R. to select sustainable biodiesel production considering economy, environment, societal and technical aspects [7]. In more recent work, AHP model that combines a deep residual network with multiway principal component analysis (MPIDRN-AHP) for Tennessee Eastman Process (TEP) operation evaluation [8].

In paint and coating industries, the selection of paint is determined by various factors such as desired properties, work requirements and limitations, safety and environmental restrictions, compatibilities, and costs [9]. The identification of the best paint alternatives will be challenging due to the conflicting attributes of the paint. A clear example demonstrated by the case of selection water-based and solvent-based paint. Water based paint offers a good option for health environment attributes as the paint have lower volatiles organic compound (VOCs) compared to the solvent-based paint. However, in terms of drying time, solvent based paint would be better as the paint is less susceptible to environmental conditions such as temperature and humidity during the curing process [10]. While there are a lot of discussions on the conflicting attributes of the paint, there is no standard framework for the decision-making process to select the best paint alternatives. Therefore, in this study, SAW will be used for the decision-making process to identify the best paint alternatives considering three attributes: safety, health and performance. A comparison between the SAW results with simple index-based methodology will be conducted in order to understand the advantages and disadvantages of SAW for decision making process.

Development of Simple Additive Weighting (SAW) Framework for Paint Selection

In this work, the multi-criteria decision-making process for chemical products using SAW will be presented. The process will be demonstrated using the alternatives of polyurethane coatings for wood flooring by [11]. The process steps are as follows:

Identification of Chemical Product Alternatives and the Attributes

The first step involving the identification of chemical product alternatives together with attributes of the alternatives. In this work, three alternatives of polyurethane coatings; 2K Formulation, Matte 1K Formulation and Gloss 1K Formulation are shown in Table 2. The attributes: safety, health and performance considered in this work together with the parameters are summarized in Table 3.

Table 1: Comparison of MCDM techniques [4], [5]

Technique	Type of normalization	Suitability	Inputs	Outputs	Preference function	Approach	Ranking scale	Best alternative
SAW	Normalized score	Choice problems, ranking problems	Option Weights and score	Complete ranking with scores	Addition of preference value for each alternative	Quantitative	Positive values	Max value
WPM	Normalized score	Choice problems, ranking problems	Option Weights and score	Complete ranking with scores	Multiplication of criterion ratio	Quantitative	Positive values	Max value
AHP	Vector normalisation	Choice problems,	Pairwise comparison	Complete ranking	Priorities of decision problem using	Qualitative	0 to 1	Max value

	(sum)	ranking problems, sorting problems (AHPsort)	on ratio scale (1–9)	with scores	hierarchy with a goal, decision criteria, and alternatives			
TOPSIS	Vector normalisation (square root of sum (L2 normalization	Choice problems, ranking problems	Ideal and anti-ideal option weights	Complete ranking with closeness score to ideal and distance to anti-ideal	Distance metric (Euclidean distance, Manhattan distance, Tchebycheff distance)	Qualitative and/or quantitative	0 to 1	Max value

Table 2: Alternatives of polyurethane coatings [11]

	2K Formulation	**Matte 1K Formulation**	**Gloss 1K Formulation**
Polyisocyanate	BAYHYDUR XP 2655	-	-
Dispersion	BAYHYDROL A 2651	BAYHYDROL UH 2874	BAYHYDROL UH 2874
Hardener solution	80% in PGDA	-	-
Matting agent	No	Yes	No
Mixing	Manual	Manual	Manual
VOC Content (g/L)	53	129	125
Bio-based content in paint (%)	45	51	46

Table 3: Attributes and parameters for MCDM

Safety parameters	Explosiveness (vol %), flammability (flash point), toxicity (TLV in ppm), chemical reactivity (NFPA rating)
Health parameters	Eye hazard, inhalation, ingestion, dermal
Performance parameters	Dry time, cure time, hardness, chemical resistance (water, ethanol, coffee, red wine, etc.), weathering artificial and natural tests of deterioration on external walls

Data Collection and Scoring Process

According to the parameters identified in Table 3, the data for each alternative will be collected. In this work, the safety and health data has been collected from the using Material Safety Data Sheet (MSDS) for BAYHYDUR XP 2655, BAYHYDROL A 2651 and BAYHYDROL UH 2874 while the data collected for performance originated from [11]. The scoring process for safety and health attributes was performed using index-based method, Product Safety and Health Index (PSHI) [10]. The scoring process for performance attribute was conducted using index-based method developed by M.Zainuddin [12]. In both indices, higher value indicates unfavorable situations compared to lower value. The sum of scores of each attribute will be used as the input for the next step.

Multicriteria Decision Making using SAW

The basic concept of the SAW method is to find the weighted sum of performance ratings on each alternative on all attributes. The first step in SAW method is normalizing the decision matrix (X) to a scale comparable to all existing alternative ratings.[13]. The formula for normalization for alternative i and attributes j given in Eq. 1 and Eq. 2:

$$R_{ij} = x_{ij} / \text{Max } x_{ij} \quad (1)$$

$$R_{ij} = \text{Min } x_{ij} / x_{ij} \quad (2)$$

With R_{ij} is the normalized performance rating of alternative i and W_j is the specified weight of attribute j, the preference value for each alternative (V_i) is given in Eq. 3. A larger value of V_i indicates that A_i's alternatives are preferred.

$$V_i = \sum W_j r_{ij} \quad (3)$$

Results & Discussion

The safety, health and performance assessment has been conducted using the methodology outlined in the previous section. The paint data for safety, health and performance together with the scores shown in Table 4 to Table 6. In general, higher score in the results indicated unfavorable situation i.e. poor performance, high potential of safety hazard and high potential of health hazard compared to the lower score. The comparison of the attributes for alternatives shown in Figure 1. Based on the results obtained, there are a conflict where if the safety and health aspects is the priorities of the decision, 2K Formulation is the best choice with the score 1 and 5 respectively. However, if the performance is the priorities of the decision, the Gloss 1K Formulation is the best choice while Gloss 1K Formulation is the worst choice.

Table 4: Safety Parameters Data and Scoring

Alternative	**2K Formulation**		**Matte 1K Formulation**		**Gloss 1K Formulation**	
Solvent Mixture	BAYHYDROL A 2651		BAYHYDROL UH 2874		BAYHYDROL UH 2874	
Properties and Score	Properties	Score	Properties	Score	Properties	Score
Explosiveness (vol%)	N/A	N/A	N/A	N/A	N/A	N/A
Flammability (Flash Point in C)	>96	1	>92	2	>92	2
Toxicity (TLV in ppm)	N/A	N/A	N/A	N/A	N/A	N/A
Chemical Reactivity (NFPA rating)	0	0	0	0	0	0
Total Score		**1**		**2**		**2**

Table 5: Health Parameters Data and Scoring

Alternative	**2K Formulation**		**Matte 1K Formulation**		**Gloss 1K Formulation**	
Solvent Mixture	BAYHYDROL A 2651		BAYHYDROL UH 2874		BAYHYDROL UH 2874	
Properties and Score	Properties	Score	Properties	Score	Properties	Score
Eye Hazard Statement	No Effect	0	H318 H319	2	H318 H319	2
Inhalation Volatility (Boiling Point in C)	96	1	92	1	92	1
Ingestion Hazard Statement	H302	1	H301	2	H301	2
Dermal Hazard Statement	H311	3	H310 H314	4	H310 H314	4
Total Score		**5**		**9**		**9**

Table 6: Performance Parameters Data and Scoring

Alternative	**2K Formulation**		**Matte 1K Formulation**		**Gloss 1K Formulation**	
Solvent Mixture	BAYHYDROL A 2651		BAYHYDROL UH 2874		BAYHYDROL UH 2874	
Dry Time Score	45	2	45	2	40	1
Cure Time Score	8 hrs, 12 min	3	2 hrs, 10 min	0	2hrs	0
Hardness Score	100	0	70	1	80	0
Chemical Resistance Score						
• Water	5	0	5	0	5	0
• Ethanol	5	0	4	1	5	0
• Coffee	5	0	5	0	5	0
• Red Wine	5	0	5	0	5	0
Total Score		**5**		**4**		**1**

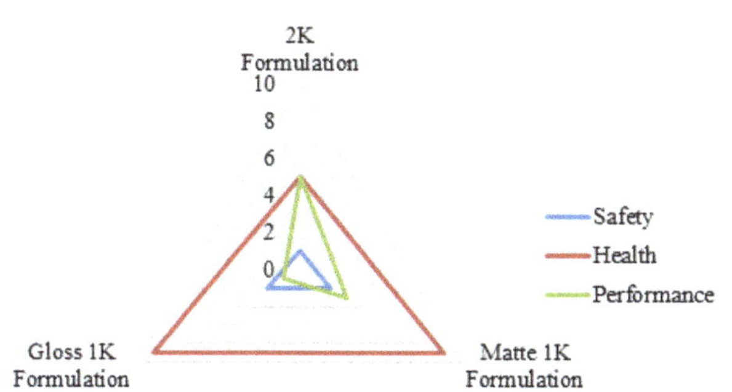

Figure 1: Paint Alternatives Evaluation Chart

To resolve the conflict, SAW was computed, and the results are presented in Table 7. A larger value of Vi indicates that the alternative is preferred. Based on SAW results, considering three attributes with equal weightage, the best choice is 2K Formulation followed by Gloss 1K Formulation and Matte 1K Formulation. The validation process has been done using a simple index-based method using sum of the score obtained in Table 4 to Table 6. Based on the score value, similar rank with SAW was obtained. However, in SAW methodology, more flexibility given to the assessor as the priorities can be assigned using pre-determine weight values. Higher weight value indicates more priority given to selected attribute. This flexibility has been demonstrated as the results presented in Table 8. The results show that, when more priorities given to performance attributes with weight value = 0.5, the best choice is Gloss 1K Formulation followed by 2K Formulation and Matte 1K Formulation.

Table 7 : Multicriteria Decision Making Results for Paint Alternatives

Alternative	Weight	2K Formulation	Matte 1K Formulation	Gloss 1K Formulation
Safety	0.33	1.00	0.50	0.50
Health	0.33	1.00	0.56	0.56
Performance	0.33	0.20	0.33	1.00
Final Value of Alternative (Vi)		**0.73**	**0.46**	**0.69**
Ranking		**1**	**3**	**2**
Simple Index Based Method		**11**	**14**	**12**
Ranking		**1**	**3**	**2**

Table 8 : Demonstration of Weight Impact in SAW for Paint Alternatives

Alternative	Weight	2K Formulation	Matte 1K Formulation	Gloss 1K Formulation
Safety	0.25	1.00	0.50	0.50
Health	0.25	1.00	0.56	0.56
Performance	0.50	0.20	0.33	1.00
Final Value of Alternative (Vi)		**0.60**	**0.43**	**0.76**
Ranking		**2**	**3**	**1**

Conclusion

In conclusion, in this work, the capability of SAW methodology to assist multicriteria decision making process for paint alternatives at the early stage of product development has been demonstrated. The rank obtained from the most favorable alternative to the less favorable alternative are 2K Formulation, Gloss 1K Formulation, Matte 1K Formulation with the score 0.73, 0.69 and 0.46 respectively. The validation process has been conducted and the results obtained using SAW is in agreement with the simple index-based methodology. Further investigation shows that the SAW result gives more flexibility compared to the simple index-based method in prioritizing the attributes based on predetermined weight value.

Acknowledgement

The authors would like to thank Universiti Teknologi PETRONAS for providing the facilities and supporting this project under the STIRF grant "Development of Predictive Model for Safe by Design (SbD) Assessment of Nanomaterials".

References

[1] O. Dyer, "Johnson and Johnson will discontinue talc based baby powder worldwide," *BMJ*, vol. 378, 2022, https://doi.org/10.1136/bmj.o2046

[2] T. Povtak, "Jury Awards $22M in Asbestos-Contaminated Paint Case," 2018. [Online]. Available: https://www.asbestos.com/news/2018/01/02/asbestos-contaminated-paint-verdict/

[3] "China's lethal milk scandal reverberates a decade later," *The Straits Times*, 2019.

[4] A. M. Ghaleb, H. Kaid, A. Alsamhan, S. H. Mian, and L. Hidri, "Assessment and Comparison of Various MCDM Approaches in the Selection of Manufacturing Process," *Adv. Mater. Sci. Eng.*, vol. 2020, p. 4039253, 2020, https://doi.org/10.1155/2020/4039253

[5] B. Zlaugotne, L. Zihare, L. Balode, A. Kalnbaļķīte, A. Khabdullin, and D. Blumberga, "Multi-Criteria Decision Analysis Methods Comparison," *Environ. Clim. Technol.*, vol. 24, pp. 454–471, 2020, https://doi.org/10.2478/rtuect-2020-0028

[6] T. Sarder and M. R. Khan, "Chemical Supplier Selection Using AHP and TOPSIS," *J. Bus. Manag.*, vol. 22, no. 11, pp. 50–54, 2020.

[7] M. R. Othman, "Sustainability assessment and decision making in chemical process design," 2011.

[8] Y. Wang and H. Li, "Complex Chemical Process Evaluation Methods Using a New Analytic Hierarchy Process Model Integrating Deep Residual Network with Multiway Principal Component Analysis," *Ind. Eng. Chem. Res.*, vol. 58, no. 31, pp. 13889–13899, Aug. 2019, https://doi.org/10.1021/acs.iecr.9b02857

[9] J. P. Guyer, "An Introduction to Coatings and Paints," 2010.

[10] R. Raslan, M. Hassim, N. Denny, N. Chemmangattuvalappil, and N. Norazahar, "Safety and health index development for formulated product design: Paint formulation," *E3S Web Conf.*, vol. 90, p. 3002, 2019, https://doi.org/10.1051/e3sconf/20199003002

[11] E. T. Rosales, B. V. Sanchez, M. Nakao, S. Reinstadtler, and A. Miller, "High-Performance Waterborne Wood Floor Coatings," *PCI Paint & Coating Industries*, 2020. https://www.pcimag.com/articles/106941-high-performance-waterborne-wood-floor-coatings.

[12] M. D. Zainuddin, "Development Of Multi-Criteria Decision-Making Index For Paint And Coating Industry," Universiti Teknologi PETRONAS, 2022.

[13] D. Putra and A. Punggara, "Comparison Analysis of Simple Additive Weighting (SAW) and Weigthed Product (WP) In Decision Support Systems," *MATEC Web Conf.*, vol. 215, p. 1003, 2018, https://doi.org/10.1051/matecconf/201821501003

Kinetic study of combined steam and CO_2 reforming of methane over Ni-Sr/MgO-ZrO_2 catalyst

AHMAD Salam Farooqi[1,2,a*], BAWADI Abdullah[1,2,b], FREDERIC Marias[3,d], AMANDA Yap Yi Tong[1,c] and UMAIR Ishtiaq[4,e]

[1]Department of Chemical Engineering, Universiti Teknologi PETRONAS, Seri Iskandar, 32610, Perak, Malaysia

[2]Centre of Contamination Control and Utilization (CenCoU), Universiti Teknologi PETRONAS, Seri Iskandar, 32610, Perak, Malaysia

[3]Universite de Pau et des Pays de l'Adour, E2S UPPA, LaTEP, Pau, France

[4]Department of Petroleum Engineering, Universiti Teknologi PETRONAS, Seri Iskandar, 32610, Perak, Malaysia

[a]ahmad.farooqi@utp.edu.my, [b]bawadi_abdullah@utp.edu.my, [c]amanda_18001492@utp.edu.my, [d]frederic.marias@univ-pau.fr, [e]umair.ishtiaq_g02875@utp.edu.my

Keywords: Reaction Kinetics, CSCRM, Bimetallic Catalyst, Activation Energy, Power Law

Abstract. Combined steam and CO_2 reforming of methane (CSCRM) has gained increasing attention as a potential solution to mitigate global warming and addressing the demand for alternative energy resources. In this study, the reaction kinetics of CSCRM over high-performance Ni-Sr/MgO-ZrO_2 bimetallic catalyst is investigated in a fixed bed reactor. The rate of reaction was analyzed at a temperatures ranging from 700-800°C and reactant partial pressures of CH_4, CO_2 and H_2O ranging from 5-50 kPa. The apparent activation energies for CH_4 and CO_2 consumption were found to be 20.94 kJ/mol and 27.53 kJ/mol respectively. The experimental results obtained were then fitted with a Power Law kinetic model and showed good agreement with R^2 values of 0.86-0.91.

Introduction

As economies thrive to grow with steady increase in global population, the demand for energy is conveniently anticipated to rise. According to recent findings, energy consumption at global scale is soaring at a rate of approximately 1-2% per year [1]. This ever increasing energy consumption globally is causing proliferation of greenhouse gas (GHSs) as these emissions lead to detrimental phenomenone of global warming. As a direct consequence, technological advacement in the H_2 production and synthesis gas (syngas) has gained momentum in recent years. Hydrogen is a fuel that is considered environmental friendly and suitable for use in fuel cells [2]. On the contrary, syngas is a combination of hydrogen and carbon monoxide that is used in a variety of petrochemical and metallurgical processes such as the production of methanol or oxo-chemicals. Syngas is used as a varied feedstock in the Fischer-Tropsch process, which is used to produce methanol and hydrocarbons [3]–[5].

Dry reforming of methane (DRM) is a method used for producing syngas that involves effective utilization of two greenhouse gases namely CH_4 and CO_2 [6], [7]. Steam methane reforming (SMR) technique, a two-step reforming method along with Autothermal reforming (ATR) technique are considered to be conventional reforming approaches widely utilized in industrial sector. Another different technique known as combined CO_2 and steam reforming (CSCRM) is a single-step reforming method. The enticing benefit of this method is that it reduces the amount of catalyst coking and modifies the ratio of hydrogen to carbon monoxide in syngas for methanol and hydrocarbons to be produced. Another potential approach for methane reforming to generate syngas is by combining hydrogen and carbon monoxide that is used as a source gas. This process is called Thermal Partial Oxidation (POX) [8]. However, this process has some disadvantages such as soot production and poor hydrogen yield [9], [10]. Therefore, the CSCRM method is considered to be energy efficient that utilizes waste flue gases, thereby rendering a viable approach for high quality syngas production [11], [12].

CSCRM reforming employs noble metals as catalayts that are widely recognized for their exceptional C-H bond cleavage, excellent catalytic characteristics and exhibit good resistance against coking. However, due to their exorbitant costs and rare availability, these precious metals are only rarely utilized in major commercial applications [13], [14]. These limitations prompted the non-nobel Nickel (Ni) metal which is affordable and easily accessible in vast quantities that made it the most promising candidate for the role of active metal replacement. Ni has been chosen above other transition metals for CSCRM reformation because of its catalytic performance, which is akin to that of noble metal. Even though Ni is frequently utilized in the reforming process, coke deposition and metal sintering are the two main downsides that might cause the catalyst to become inactive [15], [16]. In order to eviscerate catalyst deactivation, promoters (as pre-treatment statregy) mixed with oxides as support, improved catalyst synthesis technique and bimetallic systems are employed.

The kinetics and mechanistic of methane reforming over noble and non-noble catalysts had been extensively studied in the last decade. Oemar et al. [17] investigated the mechanism and kinetic modeling of La-doped Ni/SBA-15. According to the authors, the rate-determining step (RDS) of the DRM reaction in the 1% La promoted Ni/SBA-15 catalyst is the CH_4 decomposition. This step has low activation energy of 40.8 kJ mol^{-1} when compared to other supported Ni catalysts. During CH_4 decomposition, the widely dispersed Ni catalyst has elevated level of activity which results in high CH_4 conversion. According to Gallego et al. [18], the CH_4 adsorption and subsequent CH_4 breakdown were caused by the Ni active sites. In addition, Jun et al. [3] investigated the kinetics modelling of Ni-CeO_2/$MgAl_2O_4$ catalysts and reported that solely the H_2/CO ratio was impacted by CO_2 % under the Ni-based catalyst, despite the fact that both methane conversion and the H_2/CO ratio were considered to be affected by wall temperature.

In this work, the kinetics of the Ni-Sr/MgO-ZrO_2 bimetallic catalyst for CSCRM reactivity were investigated at temperatures ranging between 600 °C to 800 °C under typical atmospheric pressure. The effect of the reactant partial pressure was considered a vital aspect in comprehending CSCRM kinetics and reactor design. The obtained data was analyzed using the Power law kinetic model for reaction kinetics.

Methodology

The kinetic data of CSCRM reaction over the screened bimetallic catalyst was collected under differential conditions using a fixed bed reactor used for catalyst screening. 0.1g catalyst mass was taken. Before carrying out the kinetic experiments at each stage of the reaction temperature ranging from 700 to 800 °C, the screened bimetallic catalyst was decreased to include 50% hydrogen and nitrogen. The temperature of the reduction was brought up to 800 °C at a rate of 5 °C/min and then maintained at that level for two hours so that active metals could be generated in the catalyst. Then, a combined flow of CH_4, CO_2, H_2O, and N_2 with a total of 60 mL min^{-1} was introduced into the reactor. The output gas mixtures were sent via TCD online gas chromatography (model: (Agilent 7820) for examination. After reactant stabilization in a reactant stream for an hour, the first set of kinetic data was acquired. The change in the partial pressures of the reactants (CH_4, CO_2 and H_2O) alters the total pressure of the system. In order to maintain and keep the total pressure constant, nitrogen gas was introduced in the feed stream that acted as a balancing inert gas.

Based on the preliminary studies, the kinetic investigations were examined without transport limitations. The power law model (1) is fitted to the experimental kinetic data to determine the kinetic parameters. The effect of temperature on the kinetic constant is accounted by including the Arrhenius Equation (2) in the model.

$$r = k(P_{CH_4})^{\alpha}(P_{CO_2})^{\beta}(P_{H_2O})^{\gamma} \quad (1)$$

where r_i represents the reaction rates for the consumption of CH_4 and CO_2, as well as the formation rates of H_2 and CO. The partial pressure of CH_4, CO_2 and H_2O are denoted as PCH_4, PCO_2 and PH_2O respectively while α, β and γ are the corresponding reaction orders for CO_2, CH_4, and H_2O reactants respectively.

$$k = Ae^{\frac{-Ea}{RT}} \quad or \quad lnk = lnA - \frac{-Ea}{RT} \quad (2)$$

where A, R, T and Ea represent pre-exponential factor (frequency factor), universal gas constant, reaction temperature and activation energy respectively were used to explain the rate constant of the reaction.

In order to determine the kinetic parameters, including the order of reactions (*α, β* and *γ*) and the reaction rate constants (k_i), the rate of reactions for consumption as well as formation have been investigated and compared over a range of temperatures. The parameters were optimized using the Levenberg-Marquardt algorithm. For this purpose, a script was written in MATLAB (MathWorks, Inc.) using the FSOLVE function, where the objective function was the mean of absolute difference between the predicted and actual experimental data values (3). The summary of these kinetic parameters is shown in Table 1.

$$\sum_i^n \frac{|r_{pi} - r_{expi}|}{n} \quad (3)$$

Results and Discussion

This section presents the results and a discussion of the kinetic study of CSCRM reaction over the Ni-6Sr/MgO-ZrO_2 (6Sr) catalyst. The kinetic behaviors of the CSCRM over the optimum 6Sr catalyst were studied, where the effect of temperature (700-800 °C) and partial pressures of the reactants (CH_4, CO_2 and H_2O) were investigated. The experimental kinetic data of the CSCRM was fitted to Power-law kinetic model.

Effect of CH_4 Partial Pressure

The impact of CH_4 partial pressure at various temperatures was investigated by keeping the partial pressures of CO_2 and H_2O constant at 15 and 30 kPa respectively and altering the CH_4 partial pressure between 35 and 50 kPa as shown in Figure 1. At all temperatures, the consumption rates of CH_4 and CO_2 increased to the point at which the CH_4 partial pressure was 45 kPa. Beyond this point, the increase retarded down significantly at 50 kPa CH_4 partial pressure. Analyzing the effect of the rate of formation of H_2 and CO on CH_4 partial pressure, a similar behavior pattern was also observed (see Fig 1 c and d). The decrease in catalytic efficiency in an environment with a high concentration of CH_4 may be attributed to an increase in the rate at which CH_4 is fissured into carbonaceous species, that encapsulates the active Ni metallic sites. [19]. Therefore, the amount of extra methane in the feed did not have an impact on the reactant conversion and yield product.

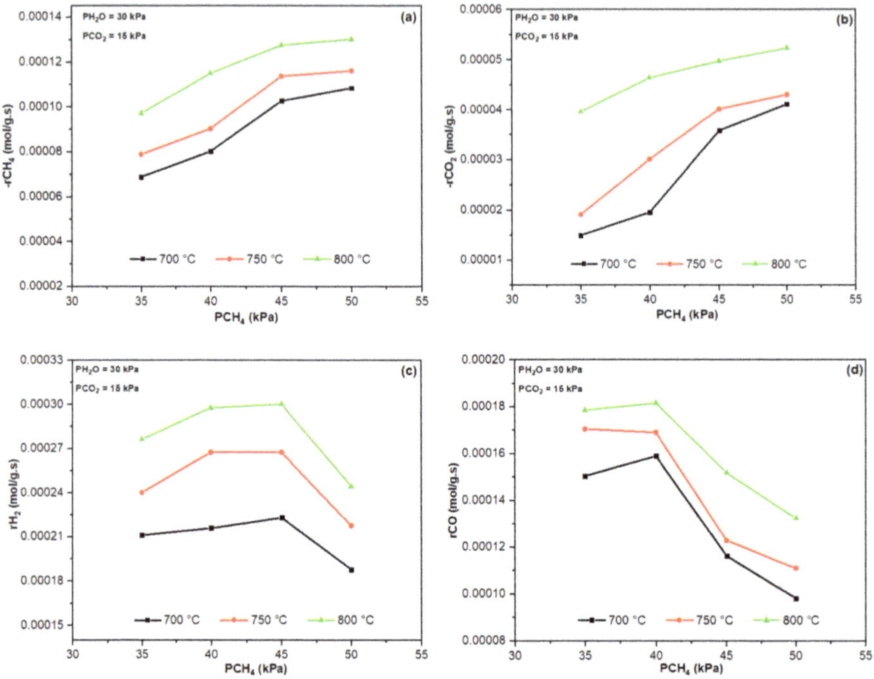

Figure 1: Effects of CH_4 partial pressure on the consumption rates of (a) CH_4 (b) CO_2 and formation rates of (c) H_2 (d) CO

Effect of CO_2 Partial Pressure

To investigate the effect of partial pressure of CO_2 on the rate of consumption and generation of products, the partial pressures of PCO_2 was varied from 5 to 20 kPa while the partial pressures of PCH_4 and PH_2O were held constant at 45 and 30 kPa respectively. Figure 2 (a-b) illustrates that there was a considerable rise in the consumption rates of both CH_4 and CO_2 as the pressure of PCO_2 raised from 5 to 20 kPa. It is assumed that the increase in consumption rates might be due to the amplification of the simultaneous DRM and WGS response over SRM. This finding was in line with studies that were done in the past [20], [21]. In addition, the effect of the partial pressure of CO_2 on the formation of H_2 and CO was also investigated. In the range of 5 to 20 kPa, a rise in the partial pressure of CO_2 caused a reduction in the rate at which H_2 was produced, while there was an increase in the rate for CO formation. The results are shown in Figure 2 (c-d) for temperatures of 700, 750, and 800 °C. The results shown here depict that hydrogen gas was used up at the RWGS reaction, whereas all the DRM, SRM, and RWGS processes result in the formation of extra carbon monoxide.

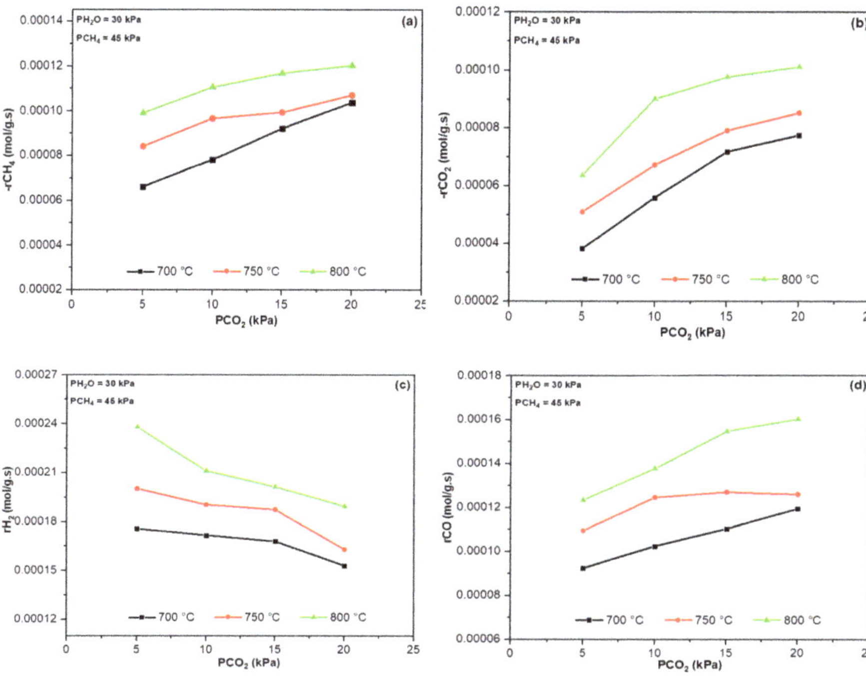

Figure2: Effects of CO_2 partial pressure on the consumption rates of (a) CH_4 (b) CO_2 and formation rates of (c) H_2, (d) CO

Effect of H_2O Partial Pressure

Apart from CO_2 reactant, H_2O is another oxidizing agent playing an important role in CSCRM reaction. The role of H_2O partial pressure in CSCRM reaction was investigated at fixed PCH_4 = 45 kPa and PCO_2 = 15 kPa while PH_2O was varied from 15 to 30 kPa as shown in Figure 3. The CH_4 consumption rate increased substantially with rising PH_2O from 10 to 30 kPa whilst a decrease in CO_2 consumption rate was observed at 700, 750 and 800 °C respectively as shown in Figure 3 (a-b). CH_4 comsumption was directly propotional to PH_2O formation presumely due to the improvement of steam gasification of deposited carbon produced from CH_4 decomposition. In the thermodynamic study of CSCRM using Gibbs free energy minimization approach, Ozkara-Aydnolu [22] found that regardless of reaction temperature, the conversion of CH_4 increased noticeably with growth in H_2O:CH_4 ratio from 0:1 to 3:1. The decline in CO_2 consumption rate and increase in CH_4 consumption rate could suggest that SRM was more preferable than DRM reaction. Although both CO_2 and H_2O possess oxidizing properties, CH_4 would likely react with H_2O instead of CO_2 owing to the greater stable nature of CO_2 molecules as to H_2O species. The rate of the generation of H_2 increased as the partial pressure of H_2O in the atmosphere rose (see Fig. 3 c). In addition, the introduction of steam led to an increase in the rate of H_2 production through SRM and WGS reactions, which in turn caused the rate of CO formation to first accelerate before eventually slowing down (see Fig. 3 d).

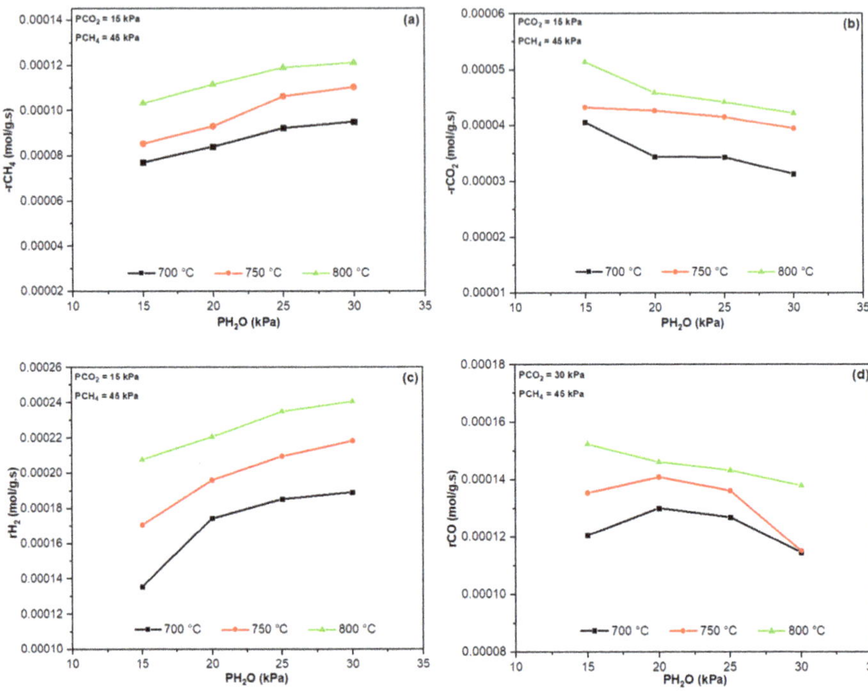

Figure 3: Effects of H_2O partial pressure on the consumption rates of (a) CH_4 (b) CO_2 and formation rates of (c) H_2, (d) CO

Power Law Kinetic Modelling

During the kinetic evaluation of a Ni/CeO_2-Al_2O_3 catalyst at a reaction temperature between 700 and 800 °C, the values of apparent activation energy (E_a) for the rate of consumption of CH_4 and CO_2 were estimated to be 20.94 kJ/mol and 27.53 kJ/mol respectively. These values were in agreement with those reported by Karemore et al. [6]. The authors indicated that this tendency may be ascribed to the greater sensitivity of CO_2 to temperature fluctuations than CH_4. As CH_4 also interacted with steam during the CSCRM reaction, its consumption rate was found to be greater than that of CO_2. E_a values for the production of H_2 and CO were 19.70 kJ/mol and 23.05 kJ/mol respectively. This explains the reason for higher H_2 production than the yield of CO. This tendency can also be explained based on the impact of the water gas shift reaction ($CO + H_2O \rightarrow CO_2 + H_2$) on the rate of CO formation [23]. According to the findings of Kathiraser et al. [24], the activation energy required for the methane reforming process over various Ni-based catalysts can range from 26 kJ/mol to 243 kJ/mol.

***Table 1**: Estimated kinetic parameters for the power-law model*

Reaction species	α	β	γ	A (mol/g.s.kPa $^{(\alpha+\beta+\gamma)}$)	Ea (kJ/mol)	R^2
CH_4	1.247	0.234	0.286	0.000021	20.94	0.910
CO_2	1.815	0.111	0.508	0.000017	27.53	0.864
H_2	-0.638	0.016	0.262	0.014	19.70	0.905
CO	1.137	0.167	-0.079	0.078	23.05	0.893

Figure 4 illustrates the adequacy of the power law model by showing the parity plots of the observed rates and the projected rates of the kinetic model. In each instance, the rates that were determined via experimentation had an R^2 value that is more than 0.85. The experimentally obtained rates were close to the rates that were anticipated using the Power Law kinetic model. It was observed that the predicted rates data fit the Power Law model and were close to the experimental rates, indicating that the kinetic data obtained from the power-law model was adequate to represent the kinetic performance of the Ni-Sr/MgO-ZrO_2 catalyst in CSCRM reforming, although this model could not explain the reaction mechanism.

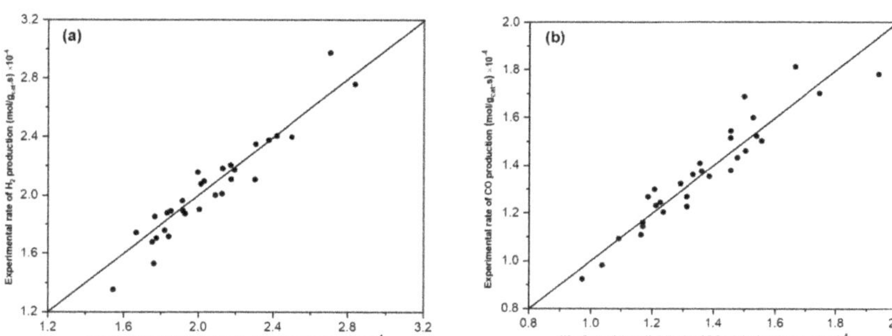

Figure 4: Comparison of the experimental rate with kinetic model predicted rate based on power law model (a) H_2 production (b) CO production

Conclusion

This work investigated the kinetics of the bimetallic catalyst Ni-Sr/MgO-ZrO_2 during the CSCRM reforming under reactant partial pressures ranging from 10 to 50 kPa. The study focused on the reactant impact of the partial pressures (CH_4, CO_2, and H_2O) on the CSCRM process at atmospheric pressure in the temperature range of 700-800°C. The Power Law model was utilized to predict the kinetic modelling of the CSCRM reaction over the ideal Ni-Sr bimetallic catalyst. The apparent activation energy for CH_4 consumption was determined to be 20.94 kJ/mol, while the activation energy for CO_2 consumption was found to be 27.53 kJ/mol. The results showed good agreement between the experimental rates and predicted rates with an R^2 value above 0.86.

Acknowledgment

The authors express their gratitude to Yayasan Universiti Teknologi PETRONAS (YUTP) for funding the research (Grant no. 015LC0-343). Additionally, the authors acknowledge Universiti Teknologi PETRONAS Malaysia and the Erasmus+ Program for providing the opportunity to study abroad as exchange student at Université de Pau et des Pays de l'Adour in France.

References

[1] K. Jabbour, "Tuning combined steam and dry reforming of methane for 'metgas' production: A thermodynamic approach and state-of-the-art catalysts," *J. Energy Chem.*, vol. 48, pp. 54–91, 2020, https://doi.org/10.1016/j.jechem.2019.12.017

[2] A. Midilli and I. Dincer, "Hydrogen as a renewable and sustainable solution in reducing global fossil fuel consumption," *Int. J. Hydrogen Energy*, vol. 33, no. 16, pp. 4209–4222, 2008, https://doi.org/10.1016/j.ijhydene.2008.05.024

[3] H. J. Jun, M.-J. Park, S.-C. Baek, J. W. Bae, K.-S. Ha, and K.-W. Jun, "Kinetics modeling for the mixed reforming of methane over Ni-CeO2/MgAl2O4 catalyst," *J. Nat. Gas Chem.*, vol. 20, no. 1, pp. 9–17, 2011, https://doi.org/10.1016/S1003-9953(10)60148-X

[4] A. S. Farooqi *et al.*, "Syngas production via dry reforming of methane over Nibased catalysts," *IOP Conf. Ser. Mater. Sci. Eng.*, vol. 736, no. 4, p. 042007, 2020, https://doi.org/10.1088/1757-899X/736/4/042007

[5] S. Aslam, A. S. Farooqi, M. Y. A. Rahman, and S. A. M. Samsuri, "Titanium-Based Vacancy-Ordered Double Halide Family in Perovskite Solar Cells," *Phys. status solidi*, vol. 219, no. 8, p. 2100671, Apr. 2022, https://doi.org/10.1002/pssa.202100671

[6] A. L. Karemore, R. Sinha, P. Chugh, and P. D. Vaidya, "Mixed reforming of methane over Ni–K/CeO2–Al2O3: Study of catalyst performance and reaction kinetics," *Int. J. Hydrogen Energy*, vol. 46, no. 7, pp. 5223–5233, 2021, https://doi.org/10.1016/j.ijhydene.2020.11.221

[7] A. S. Farooqi *et al.*, "Syngas Production via Bi-Reforming of Methane Over Fibrous KCC-1 Stabilized Ni Catalyst," *Top. Catal.*, 2022, https://doi.org/10.1007/s11244-022-01713-3

[8] M. B. Bahari, H. D. Setiabudi, T. D. Nguyen, A. A. Jalil, N. Ainirazali, and D.-V. N. Vo, "Hydrogen production via CO2CH4 reforming over cobalt-supported mesoporous alumina: A kinetic evaluation," *Int. J. Hydrogen Energy*, vol. 46, no. 48, pp. 24742–24753, 2021, https://doi.org/10.1016/j.ijhydene.2020.04.130

[9] P. Kulchakovsky, Syngas Production By Thermal Partial Oxidation of Methane With Technical Oxygen. 2016.

[10] A. S. Farooqi *et al.*, "Combined H2O and CO2 Reforming of CH4 Over Ca Promoted Ni/Al2O3 Catalyst: Enhancement of Ni-CaO Interactions," in *Lecture Notes in Mechanical Engineering*, 2021, pp. 220–229, https://doi.org/10.1007/978-981-16-3641-7_26

[11] A. S. Farooqi *et al.*, "A comprehensive review on improving the production of rich-hydrogen via combined steam and CO2 reforming of methane over Ni-based catalysts," *Int. J. Hydrogen Energy*, vol. 46, no. 60, pp. 31024–31040, 2021, https://doi.org/10.1016/j.ijhydene.2021.01.049

[12] A. S. Farooqi, M. Yusuf, N. A. M. Zabidi, K. Sanaullah, and B. Abdullah, "CO2 conversion technologies for clean fuels production," in *Carbon Dioxide Capture and Conversion*, S. Nanda, D.-V. N. Vo, and V.-H. B. T.-C. D. C. and C. Nguyen, Eds. Elsevier, 2022, pp. 37–63.

[13] H. S. Whang *et al.*, "Enhanced activity and durability of Ru catalyst dispersed on zirconia for dry reforming of methane," *Catal. Today*, vol. 293–294, pp. 122–128, 2017, https://doi.org/10.1016/j.cattod.2016.12.034

[14] A. S. Farooqi *et al.*, "Hydrogen-rich syngas production from bi-reforming of greenhouse gases over zirconia modified Ni/MgO catalyst," *Int. J. Energy Res.*, vol. 46, no. 3, pp. 2529–2545, Mar. 2022, https://doi.org/10.1002/er.7325

[15] J. Estephane *et al.*, "CO2 reforming of methane over Ni–Co/ZSM5 catalysts. Aging and carbon deposition study," *Int. J. Hydrogen Energy*, vol. 40, no. 30, pp. 9201–9208, 2015, https://doi.org/10.1016/j.ijhydene.2015.05.147

[16] A. Awan *et al.*, "Green synthesis of molybdenum-based nanoparticles and their applications in energy conversion and storage: A review," *Int. J. Hydrogen Energy*, 2021, https://doi.org/10.1016/j.ijhydene.2021.10.076

[17] U. Oemar, Y. Kathiraser, L. Mo, X. K. Ho, and S. Kawi, "CO2 reforming of methane over highly active La-promoted Ni supported on SBA-15 catalysts: mechanism and kinetic modelling," *Catal. Sci. Technol.*, vol. 6, no. 4, pp. 1173–1186, 2016, https://doi.org/10.1039/C5CY00906E

[18] G. Sierra Gallego, C. Batiot-Dupeyrat, J. Barrault, and F. Mondragón, "Dual Active-Site Mechanism for Dry Methane Reforming over Ni/La2O3 Produced from LaNiO3 Perovskite," *Ind. Eng. Chem. Res.*, vol. 47, no. 23, pp. 9272–9278, Dec. 2008, https://doi.org/10.1021/ie800281t

[19] N. Kumar, M. Shojaee, and J. J. Spivey, "Catalytic bi-reforming of methane: From greenhouse gases to syngas," *Curr. Opin. Chem. Eng.*, vol. 9, pp. 8–15, 2015, https://doi.org/10.1016/j.coche.2015.07.003

[20] W. J. Jang *et al.*, "Combined steam and carbon dioxide reforming of methane and side reactions: Thermodynamic equilibrium analysis and experimental application," *Appl. Energy*, vol. 173, pp. 80–91, 2016, https://doi.org/10.1016/j.apenergy.2016.04.006

[21] O. Omoregbe *et al.*, "Syngas production from methane dry reforming over Ni/SBA-15 catalyst: Effect of operating parameters," *Int. J. Hydrogen Energy*, vol. 42, no. 16, pp. 11283–11294, 2017, https://doi.org/10.1016/j.ijhydene.2017.03.146

[22] Ş. Özkara-Aydnolu, "Thermodynamic equilibrium analysis of combined carbon dioxide reforming with steam reforming of methane to synthesis gas," *Int. J. Hydrogen Energy*, vol. 35, no. 23, pp. 12821–12828, 2010, https://doi.org/10.1016/j.ijhydene.2010.08.134

[23] N. M. Schweitzer, J. A. Schaidle, O. K. Ezekoye, X. Pan, S. Linic, and L. T. Thompson, "High Activity Carbide Supported Catalysts for Water Gas Shift," *J. Am. Chem. Soc.*, vol. 133, no. 8, pp. 2378–2381, Mar. 2011, https://doi.org/10.1021/ja110705a

[24] Y. Kathiraser, U. Oemar, E. T. Saw, Z. Li, and S. Kawi, "Kinetic and mechanistic aspects for CO2 reforming of methane over Ni based catalysts," *Chem. Eng. J.*, vol. 278, pp. 62–78, 2015, https://doi.org/10.1016/j.cej.2014.11.143

Dry reforming of methane over Ni/KCC-1 catalyst for syngas production

SITI NUR SABIHA Suhaimi[1,a], SYED MUHAMMAD WAJAHAT Ul Hasnain [1,b], AHMAD Salam Farooqi[1,2,c], HERMA Dina Setiabudi[3,d], SYED Anuar Faua'ad Syed Muhammad[4,e], SHAHRUL Ismail[5,f], BAWADI Abdullah[1,2,g*]

[1]Department of Chemical Engineering, Universiti Teknologi PETRONAS, Seri Iskandar, 32610, Perak, Malaysia

[2]Centre of Contamination Control and Utilization (CenCoU), Universiti Teknologi PETRONAS, Seri Iskandar, 32610, Perak, Malaysia

[3]Faculty of Chemical and Process Engineering Technology, College of Engineering Technology, Universiti Malaysia Pahang, 26300 Gambang, Kuantan, Pahang, Malaysia

[4]School of Chemical and Energy Engineering, Faculty of Engineering, Universiti Teknologi Malaysia, UTM Skudai, 81310 Skudai, Johor, Malaysia

[5]School of Ocean Engineering, Universiti Malaysia Terengganu, 21030 Kuala Nerus, Terengganu, Malaysia

[a]siti.nur_24455@utp.edu.my, [b]syed_22000103@utp.edu.my, [c]ahmad.farooqi@utp.edu.my, [d]herma@ump.edu.my, [e]syed@utm.my, [f]shahrul.ismail@umt.edu.my, [g]bawadi_abdullah@utp.edu.my

Keywords: Catalyst Development, Ni-Based Catalysts, Syngas Production, KCC-1

Abstract. Dry reforming of methane (DRM) is a trendy topic of investigation as a means of reducing global warming. However, the adoption of DRM for a commercial purpose is still a question due to the deactivation and sintering of catalysts. The performance of the 5Ni/KCC-1 catalyst by using the ultrasonic-assisted impregnation method was examined in this study. The micro-emulsion method and ultrasonic-assisted impregnation method were used to prepare KCC-1 support and 5Ni/KCC-1 catalyst respectively. The catalyst was characterized by N_2 adsorption-desorption and field emission scanning electron microscopy (FESEM) techniques. FESEM morphology shows that KCC-1 support experienced a well-defined fibrous morphology in a uniform microsphere which can promote high catalytic activity. The results show that the catalyst has optimum performance with higher reactant conversions and H_2/CO ratio when operated at 850°C in a tubular furnace reactor as compared to 750°C.

Introduction

The recent global temperature rise is closely related to increased anthropogenic greenhouse gas emissions over the past century. The most common greenhouse gases are methane (CH_4) and carbon dioxide (CO_2) and they are the major developer of the recent problems of climate change [1]. This has resulted in greenhouse gas (GHG) accumulation such as CH_4 and CO_2 that induces a global climate change that triggers extreme natural phenomena effects such as heavy floods in some areas and severe droughts in other parts of the world that have intensified environmental pressure, thereby impacting biodiversity [2], [3].

Emissions of methane from natural causes such as termites, grasslands, wildfires, wetlands and lakes while human activities such as the refining of oil and gas from coal mines and agricultural activities [4]. This has resulted in greenhouse gas (GHG) accumulation such as CH_4 and CO_2 that induces a global climate change that triggers extreme natural phenomena effects such as heavy floods in some areas and severe droughts in other parts of the world that have intensified environmental pressure, thereby impacting biodiversity [5], [6].

Fossil fuels are the most important source of CO_2 emissions, with the most significant contribution to this energy paradigm [7]. In addition, human activities in industries such as textiles, aluminum, pulp and paper, refineries, cement, iron and steel, and landfills also result in excess atmospheric CO_2 pollution due to the conversion of raw materials. According to the United States Environmental Protection Agency, there is inadequate natural removal of CO_2 from forests and oceans to remove excess CO_2 from the atmosphere [8]. Renewable energy may serve as alternatives to fossil fuels that alleviate this challenge, such as hydropower, wind and solar power, but these sources depend on geographical, nocturnal and seasonal variations. In green energy science, reducing greenhouse gas emissions has become one of the most important challenges in minimizing temperature increases and coping with global warming [9], [10]. The conversion of two greenhouse gases, methane and carbon dioxide, into a value-added product called synthetic gas, can be accomplished by methane steam reforming (SRM), methane partial oxidation (POM) and methane dry reforming (DRM). Synthetic gas is a combination of two gases containing hydrogen (H_2) and carbon monoxide (CO), known as syngas [11], [12].

DRM is used due to its, lower cost, environmentally friendly, and effective energy utilization. In the DRM process, the catalyst is the main existence as it is responsible for changing the reaction rate, and improving syngas production without being consumed in it [13]. However, coke deposition and catalyst sintering at high temperatures are the main challenges in this reaction [14]. These two problems can cause the catalyst to be deactivated. Therefore the discovery of the extremely active and stable catalyst used in DRM is of great importance [15]. The addition of mesoporous silica particle (KCC-1) into Ni-based catalyst is being studied in this project. The high accessibility to the active sites of the bulky mass reactant is one of the advantages of this fibrous morphology, thus increasing the reaction rate and the formation of the material [16]. Oxygen vacancy has recently been regarded as one of the reactive sites in the methanation process and the rate of action is significantly improved by adsorbing and activating the carbon-oxygen bond [17].

Thus, the primary aim of this research is to study the influence of KCC-1 support over Ni catalyst on the catalytic performance during the DRM reaction. The freshly prepared catalyst was studied by BET technique. The DRM reaction was conducted at 700 °C and 800 °C under atmospheric pressure by feeding CH_4, CO_2 at a stoichiometric ratio of 1:1 respectively.

Materials and methods

Catalyst Synthesis. The catalyst support KCC-1 was synthesized by a micro-emulsion method. For this, tetraethyl orthosilicate, butanol, and toluene are stirred in a beaker for 30 minutes at a temperature 30 °C while urea, cetyl tri ammonium bromide (CTAB, Aldrich), and water were stirred in another beaker for 30 minutes at temperature 30 °C. Both solutions were then mixed, and stirred for 45 minutes at 30 °C. After that, the solution was heated in a Teflon-sealed hydrothermal reactor for 5 hours at 120 °C. The obtained solution is centrifuged, rinsed, and oven-dried for 12 h at a temperature 100 °C and calcined for 5 hours at 550 °C.

In the next step, the Ni was impregnated onto the support KCC-1 by the ultrasonic-assisted impregnation method. The necessary amount of $Ni(NO_3)_2 \cdot 6H_2O$ was dissolved in diluted water to make the aqueous solution and then mixed with the calculated amount of the support. The resulting slurry was then immersed in the ultrasonic cleaner bath and heated slowly at 80°C until nearly all the water had evaporated. The solid residue was dried overnight at 110°C followed by calcination at 550°C for 3 h.

Catalyst Characterization. At 77 K, a gas adsorption analyzer (Tristar 3020, Micromeritics) was used to measure the textural properties of catalysts using an N_2 physisorption isotherm. The BET method by relative pressure range (p/po) of 0.05 – 0.20 was used to calculate particular surface areas. While the amount of N2 absorbed in a relative pressure range (p/po) was used to calculate pore volumes. The morphologies of the catalyst were observed by using the FESEM technique. XRD was used to determine the composition and structure of the catalyst. Powder X-ray diffraction patterns with a Bruker D8B Advance X-ray diffractometer are reported on the calcined catalysts in which Cu-Kα at a wavelength of 1.54Å was used. With an exposure time of 3 seconds and a step size of 1°, the XRD had a scanning range of 10-90°.

Catalyst Evaluation. The catalyst was tested in the tubular reactor to perform a DRM reaction. The catalyst was placed in the catalyst tube, which was made from quartz with 10 mm internal diameter. To hold the catalyst in a catalyst tube, 100 mg of catalyst was packed between two glass wool. the catalyst was purged with nitrogen at 30 mL min^{-1} at a temperature 750 °C. After 1 hour of purging, hydrogen gas was added to the catalyst bed with the same flow rate 30 mL min^{-1} at the same temperature of 750°C. This means a flow of H_2 and N_2 into the catalyst bed is 50% each to reduce the catalyst for catalyst activation. A reduction process was carried out for 3 h. After that, the catalyst bed was purged back with 20 mL min^{-1} to remove all hydrogen gas presence during reduction. Reactants were fed into the catalyst bed with an equimolar rate at 60 mL min^{-1} in a ratio of 1:1:1 for CH_4:CO_2:N_2 at a temperature of 750°C and atmospheric pressure. The catalyst activity was observed for 6 hours on stream. Online gas chromatography was used to continuously monitor and analyzed the excess reactant and product gases of hydrogen and carbon monoxide. The same procedure was repeated for the temperature at 850°C. The conversions of the gases were calculated by using the following equation:

$$CH_4\ (\%) = \left(\frac{F_{CH4,in} - F_{CH4,out}}{F_{CH4,in}}\right) \times 100 \quad (1)$$

$$CO_2\ (\%) = \left(\frac{F_{CO2,in} - F_{CO2,out}}{F_{CO2,in}}\right) \times 100 \quad (2)$$

$$\frac{H_2}{CO} = \frac{F_{H2,out}}{F_{CO,out}} \quad (3)$$

Where F_{in} and F_{out} are the molar flowrates (mol.s^{-1}) of the gases going in and out of the reactor.

Results and Discussion

N_2 Physisorption Analysis. The table below shows the N_2 adsorption-desorption isotherms and pore size distributions of 5Ni/KCC-1 catalysts. The 5Ni/KCC-1 catalyst performed type IV isotherm due to the average pore diameter is 8.68 nm, which falls under the range of mesoporous material (2-50 nm). The type IV isotherm occurs at pressures below the gas saturation pressure. It shows the formation of a monolayer followed by the formation of multilayers at lower pressures. The 5Ni/KCC-1 catalyst experienced H3 hysteresis loops as it can be thought of as a middle ground between adsorption and desorption limit forms. The relative pressure (p/po) of 5Ni/KCC-1 is also high which is from 0.8 to 1.0 causing the aggregates of plate-like particles to produce slit-shaped pores resulting the type H3 hysteresis loop. The average pore diameter of the 5Ni/KCC-1 catalyst is 8.68 showing that this catalyst a has high value of pore diameter. The surface area and pore volume of this catalyst are 456.6031 m_2/g and 0.991964 cm^3/g respectively. This can cause an increasing activity of KCC-1 due to its fibrous morphology, big pore size and large surface area.

Materials Research Proceedings 29 (2023) 416-424 https://doi.org/10.21741/9781644902516-47

Table 1 Textural properties of 5Ni/KCC-1 catalyst

Sample	BET Surface Area (m^2/g)	Pore Volume (cm^3/g)	Average Pore Diameter (nm)	Isotherm Type
5Ni/KCC-1	456.6031	0.991964	8.68614	IV

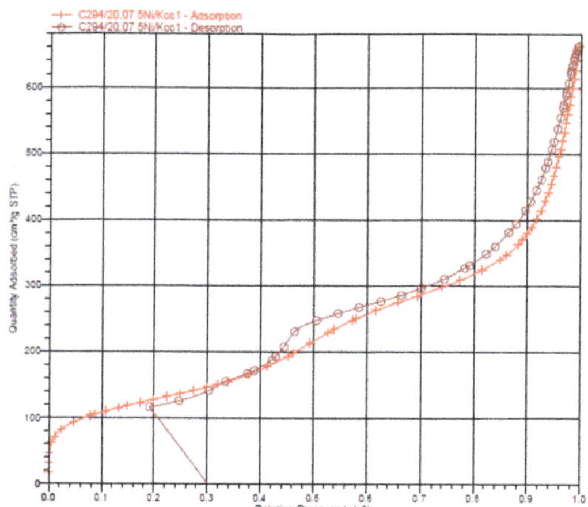

Fig. 1 N_2 adsorption-desorption isotherm curves

FESEM analysis. The morphologies of the Ni/ KCC-1 sample was studied using a Field Emission Scanning Electron Microscope (FESEM). The magnification of sample testing are 10000, 20000, 30000 and 100000. The distribution of Ni metals on the fibrous KCC-1 support can be seen via FESEM micrographs as shown in Figure 2. The dispersion of Ni grains is represented by the white spots and it sticks on the wall of the KCC-1 without uniform shape. The FESEM imaging also shows that Ni metal is well dispersed on the KCC-1 support without the agglomeration. While the formation of the fibrous KCC-1 structures is expected to be a well-defined fibrous morphology in a uniform microsphere in the range of size 202.7-790.1 nm. As shown in Figure 2, mesoporous silica KCC-1 also has a structure of big pores and dendrimers that are radially symmetric molecules with a well-defined, homogenous and monodisperse structure that has a typically symmetric core. This is due to the distance between each specific dendrimer, the reactant was able to diffuse easily, allowing for high catalytic performance [18]. Additionally, the ultrasonic-assisted impregnation technique showed that the fibrous KCC-1 structure was not demolished when loaded with Ni metal.

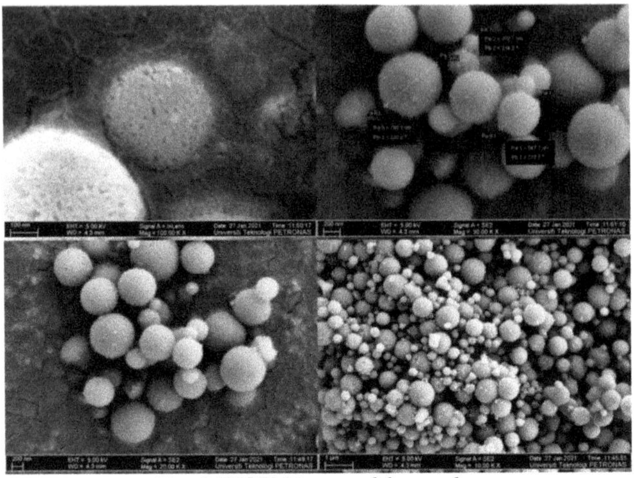

Fig. 2 FESEM image of the catalyst

H_2-TPR Analysis. The metal-support interaction was investigated using the H_2-TPR of the 5Ni/KCC-1. The reduction peak temperatures and the catalysts' reducibility are shown in the Figure 3. The 5Ni/KCC-1 H_2-TPR profile shows a broad reduction peak between temperature 100 – 200 °C (region IV) in a time frame of 74 – 108 minutes. The lower peak is attributed to the reduction of Ni^{2+} at the catalyst's surface sites or to the reduction of Ni metal that has not been affected by the KCC-1 support while the higher peak refers to a decrease in complex Ni metal species that interact strongly with KCC-1 support. This can be related to the calcination temperature. Three reduction peaks were observed in the TPR profile of the 5Ni/KCC-1 catalyst at 658 °C, 426 °C and 406 °C respectively to the first, second and third peak. The first peak (higher peak) should be linked to the possibility that this is due to the reduction of NiO to Ni, which was affected by the KCC-1 support and the surface lattice of an oxide removal in NiO crystalline. The second peak (region II) is due to the reductions of Ni species with weak 5Ni/KCC-1 interactions as it is unaffected by the KCC-1 support. While the third peak (region III) could be the reductions of Ni species with strong 5Ni/KCC-1 interactions. This indicated that Ni species in low Ni loading catalysts could interact strongly with KCC-1, which was highly stable during the reduction process. Furthermore, the hydrogen consumption of 5Ni/KCC-1 shows multiple reading which are 316.90, 360.54 and 119.39 µmol/g respectively on the first, second and third peak. This means 5Ni/KCC-1 catalyst could be attributed to a reduction in surface NiO with Ni species that have weak and strong interactions with the KCC-1 support. The reason of the high peak, which is strong interaction, is due to high surface area of the catalyst and can have high catalytic performance [19].

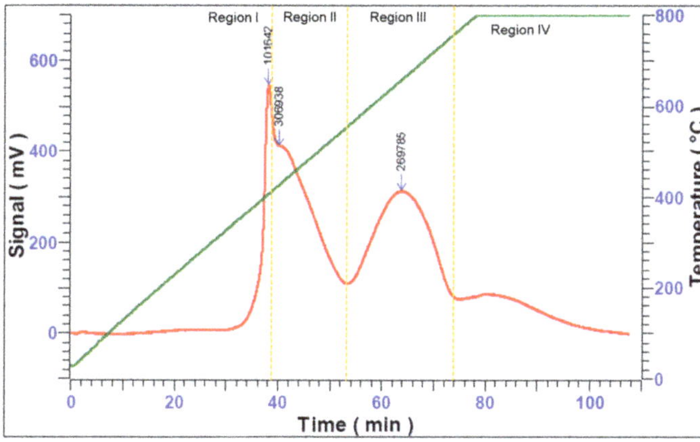

Fig. 3 TPR Profile of 5Ni/KCC-1 Catalyst

Catalyst Performance on DRM. The catalyst was tested for DRM on tubular reactor at different reaction temperatures of 750°C and 850°C to evaluate the effect of support morphology and preparation technique on catalyst efficiency in terms of operation and stability. The DRM performance of the 5Ni/KCC-1 catalyst was investigated for 6 h time on stream at 750°C and 850°C. Catalysts tested at 850°C had higher CH_4 and CO_2 conversions than those tested at 750°C. The 5Ni/KCC-1 catalyst achieved 85.51% and 81.52% of CH_4 and CO_2 conversions within 6 h at 850°C. A reaction temperature of 750°C caused CH_4 and CO_2 conversions to fall and decrease to 49.78% and 44.26 % respectively. It shows that carbon formation is more intense at lower temperatures, and at the higher reaction temperature, there is a strong interaction between Ni and KCC-1 that is responsible to prevent sintering and coke deposition [20].

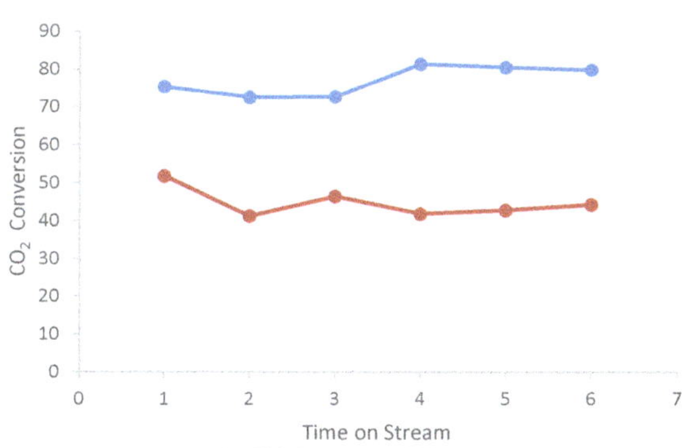

Fig. 4 Conversion of CO_2 at 750°C and 850°C

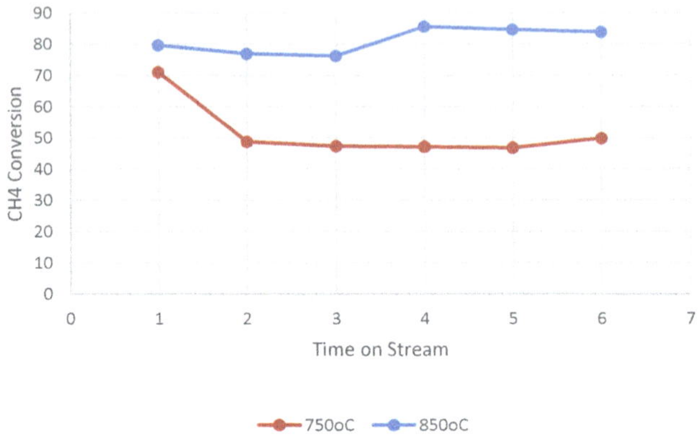

Fig. 5 Conversion of CO_2 at 750°C and 850°C

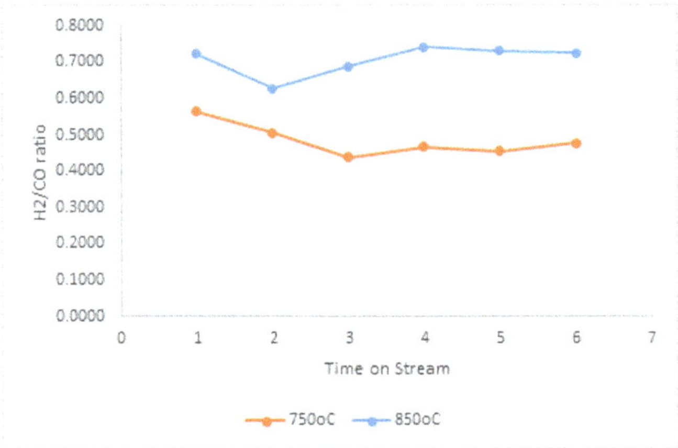

Fig. 6 H_2/CO ratio at different reaction temperature

Conclusions

In the current study, DRM was carried out over 5Ni/KCC-1 catalyst. The catalyst was prepared by ultrasonic assisted impregnation method and was tested for DRM at different temperatures. The catalyst characterization was performed to study the surface area, morphologies, behavior, and active sites of the catalyst. The characterization shows that 5Ni/KCC-1 has a BET surface area and large pore volume which are 456.6031 m^2/g and 0.991964 cm^3/g respectively. FESEM morphology shows that KCC-1 support experienced a well-defined fibrous morphology in a uniform microsphere which can promote high catalytic activity. The catalyst shows higher conversion when operated at 850°C which is 85.51% and 81.52% of CH_4 and CO_2 as compared to the lower reaction temperature. It is proved that conversion increases by increasing the reaction temperature due to less coke formation and high catalyst stability.

Acknowlegement

The authors would like to thank the Collaboration Research Fund (CRF) grant no: 015MD0-010 for providing the financial support and thanks to Universiti Teknologi PETRONAS for providing the research facilities.

References

[1] B. Abdullah, N. A. Abd Ghani, and D. V. N. Vo, "Recent advances in dry reforming of methane over Ni-based catalysts," *J. Clean. Prod.*, vol. 162, pp. 170–185, 2017, https://doi.org/10.1016/j.jclepro.2017.05.176

[2] A. S. Farooqi *et al.*, "Combined H_2O and CO_2 Reforming of CH_4 Over Ca Promoted Ni/Al_2O_3 Catalyst: Enhancement of Ni-CaO Interactions," in *Lecture Notes in Mechanical Engineering*, 2021, pp. 220–229, https://doi.org/10.1007/978-981-16-3641-7_26

[3] B. M. Al–Swai *et al.*, "Low-temperature catalytic conversion of greenhouse gases (CO_2 and CH_4) to syngas over ceria-magnesia mixed oxide supported nickel catalysts," *Int. J. Hydrogen Energy*, vol. 46, no. 48, pp. 24768–24780, 2021, https://doi.org/10.1016/j.ijhydene.2020.04.233

[4] X. Li, "Diversification and localization of energy systems for sustainable development and energy security," *Energy Policy*, vol. 33, no. 17, pp. 2237–2243, 2005, https://doi.org/10.1016/j.enpol.2004.05.002

[5] A. S. Farooqi *et al.*, "Hydrogen-rich syngas production from bi-reforming of greenhouse gases over zirconia modified Ni/MgO catalyst," *Int. J. Energy Res.*, vol. 46, no. 3, pp. 2529–2545, Mar. 2022, https://doi.org/10.1002/er.7325

[6] S. A. Farooqi, A. S. Farooqi, S. Sajjad, C. Yan, and A. B. Victor, "Electrochemical reduction of carbon dioxide into valuable chemicals: a review," *Environ. Chem. Lett.*, 2023, https://doi.org/10.1007/s10311-023-01565-7

[7] J. S. Yu, J. M. Park, J. H. Kwon, and K. S. Park, "Roles of Al_2O_3 coating layer on an ordered mesoporous Ni/m-Al_2O_3 for combined steam and CO_2 reforming with CH_4," *Fuel*, vol. 3, no. July 2022, pp. 37–38, 2023, https://doi.org/10.1016/j.fuel.2022.125702

[8] "Inventory of U.S. Greenhouse Gas Emissions and Sinks | US EPA." .

[9] T. Saelee *et al.*, "Experimental and computational investigation on underlying factors promoting high coke resistance in NiCo bimetallic catalysts during dry reforming of methane," *Sci. Rep.*, vol. 11, no. 1, pp. 1–17, 2021, https://doi.org/10.1038/s41598-020-80287-0

[10] S. Ullah *et al.*, "Sulfur enriched cobalt-based layered double hydroxides for oxygen evolution reactions," *Int. J. Hydrogen Energy*, 2021, https://doi.org/https://doi.org/10.1016/j.ijhydene.2021.09.201

[11] A. Abdulrasheed, A. A. Jalil, Y. Gambo, M. Ibrahim, H. U. Hambali, and M. Y. Shahul Hamid, "A review on catalyst development for dry reforming of methane to syngas: Recent advances," *Renewable and Sustainable Energy Reviews*, vol. 108. pp. 175–193, 2019, https://doi.org/10.1016/j.rser.2019.03.054

[12] A. S. Farooqi *et al.*, "Syngas Production via Bi-Reforming of Methane Over Fibrous KCC-1 Stabilized Ni Catalyst," *Top. Catal.*, vol. 66, no. 1, pp. 235–246, 2023, https://doi.org/10.1007/s11244-022-01713-3

[13] A. S. Farooqi *et al.*, "A comprehensive review on improving the production of rich-hydrogen via combined steam and CO_2 reforming of methane over Ni-based catalysts," *Int. J.*

Hydrogen Energy, vol. 46, no. 60, pp. 31024–31040, 2021, https://doi.org/10.1016/j.ijhydene.2021.01.049

[14] J. Gao, Z. Hou, J. Guo, Y. Zhu, and X. Zheng, "Catalytic conversion of methane and CO_2 to synthesis gas over a La_2O_3-modified SiO_2 supported Ni catalyst in fluidized-bed reactor," *Catal. Today*, vol. 131, no. 1–4, pp. 278–284, 2008, https://doi.org/10.1016/j.cattod.2007.10.019

[15] M. A. Salaev, L. F. Liotta, and O. V. Vodyankina, "Lanthanoid-containing Ni-based catalysts for dry reforming of methane: A review," *Int. J. Hydrogen Energy*, vol. 47, no. 7, pp. 4489–4535, 2022, https://doi.org/10.1016/j.ijhydene.2021.11.086

[16] G. Torres-Sempere *et al.*, "Recent advances on gas-phase CO_2 conversion: Catalysis design and chemical processes to close the carbon cycle," *Curr. Opin. Green Sustain. Chem.*, vol. 36, 2022, https://doi.org/10.1016/j.cogsc.2022.100647

[17] Y. Tian, X. Ma, X. Chen, and C. Zhang, "Effect of Ni-Co bimetallic core-shell catalyst for coke resistance in CO_2 reforming of biomass Tar," *J. Anal. Appl. Pyrolysis*, vol. 164, p. 105539, Jun. 2022, https://doi.org/10.1016/j.jaap.2022.105539

[18] S. N. Bukhari, C. C. Chong, H. D. Setiabudi, Y. W. Cheng, L. P. Teh, and A. A. Jalil, "Ni/Fibrous type SBA-15: Highly active and coke resistant catalyst for CO_2 methanation," *Chem. Eng. Sci.*, vol. 229, p. 116141, 2021, https://doi.org/10.1016/j.ces.2020.116141

[19] A. S. Al-Fatesh *et al.*, "Greenhouse gases utilization via catalytic reforming with Sc promoted Ni/SBA-15," *Fuel*, vol. 330, no. August, p. 125523, 2022, https://doi.org/10.1016/j.fuel.2022.125523

[20] Y. Zhang *et al.*, "Combined steam and CO_2 reforming of methane over Co–Ce/AC-N catalyst: Effect of preparation methods on catalyst activity and stability," *Int. J. Hydrogen Energy*, vol. 47, no. 5, pp. 2914–2925, 2022, https://doi.org/10.1016/j.ijhydene.2021.10.202

Materials Research Proceedings 29 (2023) 425-429 https://doi.org/10.21741/9781644902516-48

Drying kinetics and statistical modeling of thin-layer drying of purple non-sulfur bacterial biomass

ASIFA Anwar[1,a], GORDON McKay[1,b] and HAMISH Mackey[1,2,c,*]

[1] Division of Sustainable Development, College of Science and Engineering, Hamad bin Khalifa University, Qatar Foundation, Doha, Qatar

[2] Department of Civil and Natural Resources Engineering, University of Canterbury, Private Bag 4800, Christchurch 8140, New Zealand

[a]asan39545@hbku.edu.qa, [b]gmckay@hbku.edu.qa, [c]hamish.mackey@canterbury.ac.nz

Keywords: Biomass, Drying Kinetics, Modeling, Phototrophic Purple Bacteria, Thin-Layer Drying

Abstract. Thin-layer drying studies have been done for various materials, including vegetables, fruit, grains, wood, plants, biomass, and others. The present study is concerned with the investigation of the drying kinetics of harvested photosynthetic non-sulfur bacteria (PNSB) biomass. Four different drying models, i.e., Logarithmic, Newton, Page, and Henderson and Pabis were fitted to the experimental data obtained from drying a thin-layer of biomass in a laboratory oven at 60, 80, 100, and 120 °C. The results were assessed on statistical parameters such as the Akaike information criterion, and residual standard error. The results showed that Logarithmic model fitted the overall best for all the states of the thin layer at 60 °C and 80 °C, while Page model gave the best fit for 100 °C and 120 °C. The current research is aimed to study the drying behavior of the biomass and to aid in optimizing the drying procedure of the biomass.

Introduction

Aquaculture is a growing industry that requires a constant supply of protein for the cultivation of aquatic organisms. Recently single-cell protein (SCP) has emerged as an attractive option due to its high rate of productivity and lack of competition with arable land. Purple non-sulfur bacteria are one group of organisms that shows strong potential and can be integrated with photo-anaerobic wastewater treatment. The recovery of PNSB from such a system for SCP involves many steps, including preparation of media, fermentation, harvesting, and in the end, drying of the biomass [1]. After biomass harvesting, the dry solid content of the achieved product is still relatively low. Consequently, drying is carried out to remove/lower the moisture content. Generally, drying is defined as preserving a product by reducing the product's water content to limit microbiological or chemical deterioration [2]. A few of the most preferred biomass drying methods are convective, vacuum, solar, spray and freeze drying. Freeze-drying is recently a very popular method based on sublimation, preserving the product's actual characteristics and nutritional value. However, it is an expensive and time extensive method [3]. Thermal drying methods such as conventional oven drying, and solar drying are comparatively cheaper. Therefore, in the present study, the drying of PNSB biomass was done using a laboratory oven. The study focused on the impacts of temperature on drying and on understanding the multiple stages the biomass undergoes before it becomes dehydrated through fitting various mathematical drying models. It examines moisture movements occurring in the internal and external surfaces of the biomass based on a set of equations [4].

Materials and Methods

PNSB Cultivation. In this study, wastewater from a gas-to-liquids reforming process was used as a medium for the growth of photosynthetic non-sulfur bacteria. A 4 L anaerobic lab-scale bioreactor was utilized for cultivating the PNSB under a white LED light source.

PNSB Harvesting. PNSB biomass was harvested when the organic content of the wastewater was almost exhausted. The biomass from the wastewater was harvested by centrifuging at 8000 rpm for 10 minutes. For the thin layer drying experiment, approximately 13.7 to 13.8 grams of the harvested PNSB biomass was evenly spread over a circular aluminum petri dish.

PNSB Drying. The drying of harvested biomass was performed in a laboratory oven. The centrifuged biomass was placed in petri dishes was dried at 60, 80, 100, and 120 °C, with each condition duplicated. The mass of PNSB paste in each petri dish was measured before drying using a digital weighing balance. Later, the instantaneous weight of the biomass was measured every 15 minutes until a constant weight was achieved. The mass balance-time data was converted to moisture ratio-time data for its use in equations of the drying models.

Modeling of Drying Curves

Four moisture ratio models have been employed in the study, enlisted in Table 1. These models are selected based on their successful application in previous literature.

Table 1: Drying models, their expressions, and successful application on various materials

Drying Model	Analytical Expression	Parameters
Henderson and Pabis	MR = a exp (-kt)	a k, t
Logarithmic	MR = a exp (-kt) + c	a, k, c, t
Newton	MR = exp (-kt)	k, t
Page	MR = exp ($-kt^y$)	k, y, t

In Table 1, a, c, k, and y are model constants where t represents the time in minutes. The governing factor that forms the model equations is the dimensionless moisture ratio (MR), which is calculated as:

$$MR = \frac{Mt-Me}{Mi-Me} \quad (1)$$

Where Mt is the moisture content at time t, Me is the equilibrium moisture content, and Mi is the initial moisture content. Me is relatively smaller than Mt and Mi. Therefore, Eq. 1 can be simplified as follows,

$$\mathrm{MR} = \frac{\mathrm{Mt}}{\mathrm{Mi}} \quad (2)$$

Statistical Analysis of Modeling Results. The experimental drying data were fitted to the four selected models using R software (R version 4.2.1). Statistical parameters such as residual standard error (RSE) and Akaike information criterion (AIC) were used for the goodness of fit of the chosen models to the experimental data. The lower the AIC and RSE values, the better the model will fit [5].

Results and Discussion

Drying Kinetics. In the study, the initial moisture content of the selected samples of harvested PNSB biomass varied between 70% and 80%. There are three generic phases in the drying of biomass. The first is the warm-up phase, in which the temperature of the biomass increases until the bulk temperature reaches the oven environment's temperature. The next phase is the phase of maximum drying rate. This phase is isenthalpic. Water evaporation occurs on the biomass's surface, covered by a layer of free non-bonded water molecules. In the last reduced drying phase, water molecules entrapped within the biomass, i.e., with intercellular connections, tend to evaporate until the mass of the biomass becomes constant [6].

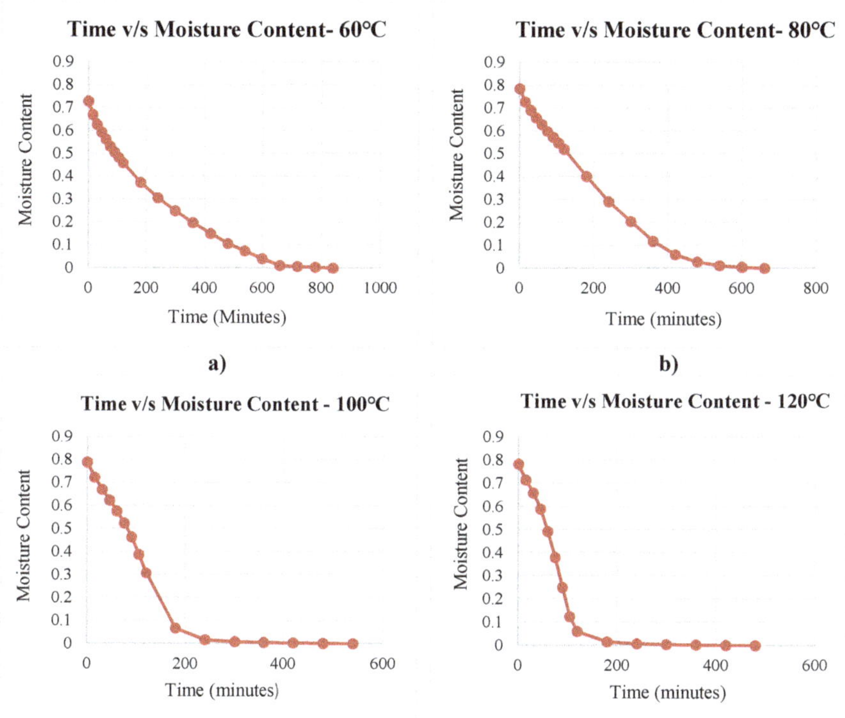

c) d)

Figure 1: Graphs illustrating the variation of mass of the samples with time for different temperatures (a represents for 60°C, b for 80 °C, c for 100 °C, and d for 120 °C)

The drying kinetics show that drying of thin-film of the PNSB biomass is the slowest for the lowest temperature (60 °C). The drying process took the longest time (approx. 840 minutes) for the moisture content to be removed completely from the samples at 60 °C. In comparison, drying of thin film took the least time at 120 °C, i.e., 480 minutes. At 100 °C, the drying rate was only slightly slower, in comparison to 80 °C, which took twice as long to dry completely. The analysis indicates that the drying time decrease as the temperature increase, which portrays an inverse relationship between the two parameters. At higher temperatures, the heat and mass transfer rate increases, resulting in quicker moisture evaporation from inside the film to the outside.

In Figure 1, the drying curve is initially more upwards concave at 60 °C, with a steeper, more linear initial response as the temperature increases. There was a slower decrease in drying rate at lower temperatures (60 °C and 80 °C). However, at the higher temperatures of 100 °C and 120 °C, the drying rate was much quicker with a short or unnoticeable warming up period, resulting in only maximum and reduced drying phases being observed.

Statistical Analysis of Drying Models. Modeling was done to analyze the model that best predicts the experimental data. The two best-fitted models of the selected ones (in ascending order) are given in Table 2:

Table 2: Statistical model fitting to the acquired drying data

Temperature [°C]	Model	Checks	
		AIC	RSE
60	Logarithmic	-208.5684	0.01906
	Newton	-170.0247	0.03085
80	Logarithmic	-127.7947	0.03833
	Page	-125.2065	0.04025
100	Page	-110.5012	0.04048
	Logarithmic	-73.57193	0.07106
120	Page	-110.7946	0.03576
	Logarithmic	-55.67538	0.08826

The lower the values of the statistical parameters, the better the model fit is. It can be concluded from Table 2 that the Logarithmic model performed the best at lower temperatures such as 60 °C and 80 °C. After which, the Page model displayed the best fit at 100 °C and 120 °C.

As seen in Figure 1, the drying curves at 60 °C and 80 °C display a more logarithmic fit than at 100 °C and 120 °C, which justifies the best fit of the Logarithmic model. The slope of the curve is more time-dependent at lower temperatures. On the contrary, heat plays a major role at higher temperatures along with time, showing a much quicker decrease in the mass. Page model has its dependency on other external factors besides time. Therefore, this explains the best fitting of the Page model at 100 °C and 120 °C. The results are found to be concurrent with previous studies on microwave drying of microalgae (*Chlorella vulgaris*) [7], air drying of some food materials such as tomatoes [8] and banana slices [9], and thyme leaves [10].

Summary

This study has investigated the drying kinetics of PNSB biomass cultivated in gas-to-liquid process wastewater. The three phases in a drying process are warming, maximum drying, and reduced drying. All the phases were perceivable at 60 °C and 80 °C, while only the latter two mentioned phases were apparent at 100 °C and 120 °C. Four widely studied models were applied to the experimental data, and statistical analysis was done. Based on the acquired results for the examined models, the Logarithmic model was the best to define the drying behavior of the considered biomass film at 60 °C and 80 °C. Likewise, Page model predicted the best results for 100 °C and 120 °C.

Acknowledgements

The authors would like to thank the support of the Qatar National Research Fund and Ministry of Environment and Climate Change through the Food Security Call, grant number MME01-0910-190029.

References

[1] M. Adedayo, E. Ajiboye, J. Akintunde, and A. Odaibo, "Single cell proteins: as nutritional enhancer," *Adv Appl Sci Res,* vol. 2, no. 5, pp. 396-409, 2011.

[2] I. Doymaz and M. Pala, "The thin-layer drying characteristics of corn," *Journal of food engineering,* vol. 60, no. 2, pp. 125-130, 2003. https://doi.org/10.1016/S0260-8774(03)00025-6

[3] S. Huang *et al.*, "Spray drying of probiotics and other food-grade bacteria: A review," *Trends in food science & technology,* vol. 63, pp. 1-17, 2017. https://doi.org/10.1016/j.tifs.2017.02.007

[4] A. R. Celma, F. Cuadros, and F. López-Rodríguez, "Convective drying characteristics of sludge from treatment plants in tomato processing industries," *Food and Bioproducts Processing,* vol. 90, no. 2, pp. 224-234, 2012. https://doi.org/10.1016/j.fbp.2011.04.003

[5] D. A. Kenny, "Measuring model fit," ed, 2015.

[6] D. Jung, S. Couturier, J. Pannejon, J. Jossent, and D. Chenu, "A Solar Sludge Drying Model: Study of Drying Phenomena—Validation of an Industrial Solar Dryer," *Optimization of a Rapid DNA Extraction and Purification Protocol for Wastewater Biosolids.* p. 17, 2013.

[7] A. R. C. Villagracia *et al.*, "Microwave drying characteristics of microalgae (Chlorella vulgaris) for biofuel production," *Clean technologies and environmental policy,* vol. 18, no. 8, pp. 2441-2451, 2016. https://doi.org/10.1007/s10098-016-1169-0

[8] I. Doymaz, "Air-drying characteristics of tomatoes," *Journal of Food engineering,* vol. 78, no. 4, pp. 1291-1297, 2007. https://doi.org/10.1016/j.jfoodeng.2005.12.047

[9] İ. Doymaz, "Evaluation of mathematical models for prediction of thin-layer drying of banana slices," *International Journal of Food Properties,* vol. 13, no. 3, pp. 486-497, 2010. https://doi.org/10.1080/10942910802650424

[10] O. Y. Turan and E. Fıratlıgil, "Modelling and characteristics of thin layer convective air-drying of thyme (Thymus vulgaris) leaves," *Czech Journal of Food Sciences,* vol. 37, no. 2, pp. 128-134, 2019. https://doi.org/10.17221/243/2017-CJFS

Optimizing graphene nanoplatelets (GNPs) dispersion for improved scale inhibitor squeeze treatment

UMAIR Ishtiaq[1,a], ALI SAMER Muhsan[2,b*], SYED MOHAMMAD Mahmood[1,c], ALI AREF Ali Alzanam[1,d], AHMAD SALAM Farooqi[3,e]

[1]Department of Petroleum Engineering, Universiti Teknologi PETRONAS, Seri Iskandar 32610, Perak, Malaysia

[2]Department of Mechanical Engineering, Universiti Teknologi PETRONAS, Seri Iskandar 32610, Perak, Malaysia

[3]Department of Chemical Engineering, Universiti Teknologi PETRONAS, Seri Iskandar 32610, Perak, Malaysia

[a]umair.ishtiaq_g02875@utp.edu.my, [b]ali.samer@utp.edu.my, [c]mohammad.mahmood@utp.edu.my, [d]ali_19000986@utp.edu.my [e]ahmad.farooqi@utp.edu.my

Keywords: Graphene Nanoplatelets, Nanomaterial dispersion, Squeeze treatment, UV-vis spectrometry, Scale Inhibitor

Abstract. The formation of mineral scale has been a major nuisance in oil field operations as it leads to multifarious flow assurance issues. Conventional squeeze treatment commonly employed to inhibit detrimental scale formation lacks in various aspects to effectively mitigate it thereby making this treatment less effective and at times unviable as production operations are moving into deeper and tighter formations. Employing nanotechnology to improve the efficiency of squeeze treatment is presented in this paper by the formulation of high and stably dispersed of Graphene Nanoplatelets (GNPs) nanofluid that can render high active surface area for PPCA scale inhibitor chemical to effectively non-covalently adsorb to the highly dispersed GNPs platform. GNPs dispersion is characterized and measured using UV-Vis spectrophotometry analysis by the parameter of percentage extractability. The concentration of 40mg/L of GNPs corresponds to the highest value of 40% extractability that depicts optimum concentration of GNPs for its highest dispersion. The presence of benzene ring in Triton X-100 surfactant molecular scheme acting as a dispersing agent has proved to be the dominant factor that promotes strong bonding between the surfactants hydrophobic tail with GNPs surface creating sufficient steric hinderance thereby giving high and stable GNPs dispersion. Furthermore, sonication process is essential for effective GNPs dispersion process.

Introduction

Scale formation and deposition in the reservoir rock matrix is a common flow assurance predicament that leads to multiple problems during hydrocarbon production. Scale mineral is formed by precipitation due to chemical reactions as ions that are dissolved in produced water react with each other to form an insoluble solid [1]. The driving forces that are responsible for scale formation are changes in temperature, pressure, alkalinity and incompatible mixing of formation and sea water as oil and gas fields use sea water or brine injection for pressure maintenance and for primary/secondary oil recovery [1-2]. Brine is the primary cause of calcite and sulfate scales. The most common types of scales witnessed are sulfates and carbonates scales of calcium, barium and strontium [1-3]

Uninhibited scale formation and its subsequent deposition in the reservoir formation is problematic for oil and gas operations that adversely affecting its economy. Solid scale deposition in the producing reservoir matrix can lead to irrevocable permanent formation damage by building-

Materials Research Proceedings 29 (2023) 430-438 https://doi.org/10.21741/9781644902516-49

up in the near well-bore region plugging reservoir pore throat and perforations thereby eventually cutting-off production. If not remediated, continuous scale deposition can lead to permanent formation damage as shown in Figure 1.0. Scale build-up can narrow down or completely choke production tubing and flow lines. Pipeline plugging due to scale deposition can result in drastic hype in operational pressure drop thereby reducing well productivity which is economically unwarranted and pose acute safety hazard in oil field operations [4]. It may also lead to perilous scenarios by interfering with running and operation of downhole equipment which can cause equipment damage leading to operational failure [4-5]. Scale formation also gives impetus to corrosion phenomenon.

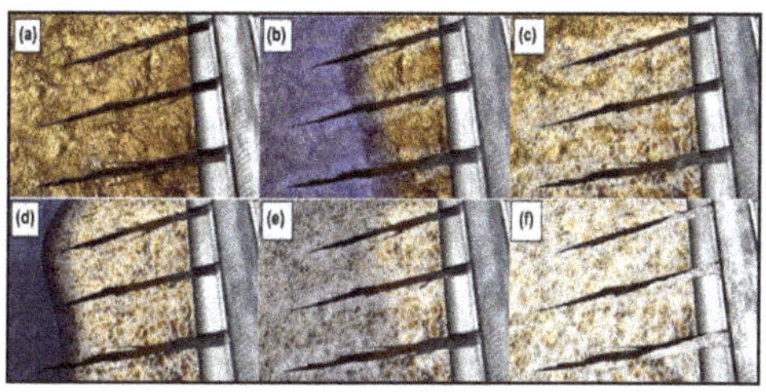

Fig 1 Scale deposition process (a-f) in the near wellbore region depicting formation damage [5]

Scale removal is a common well-intervention operation with a wide range of mechanical, chemical and scale inhibitor treatment options. There are various scale inhibitor delivery methods such as continuous injection of scale inhibitor chemical into the well bore via macroni string, injection into the gas lift system [6] and slow dissolution of the inhibitor chemical placed in rat hole [7-8], but the most commonly used and effective method for scale control is scale inhibitor squeeze treatment [9].

Downhole conventional squeeze treatment employed in many reservoirs use acid phosphonate inhibitors and various types of polymers [10]. In this technique, commercial scale inhibitor chemical solution such as Phosphino polycarboxylic acid (PPCA) [11] is pumped downhole (squeezed) at high injection pressures in order to partially adsorb by temperature activated precipitation onto the reservoir rock followed by 12 to 24 hours well shut-off. The adsorbed and retained chemical on the reservoir rock acts as a scale inhibitor chemical storage as shown in Figure 2.0.

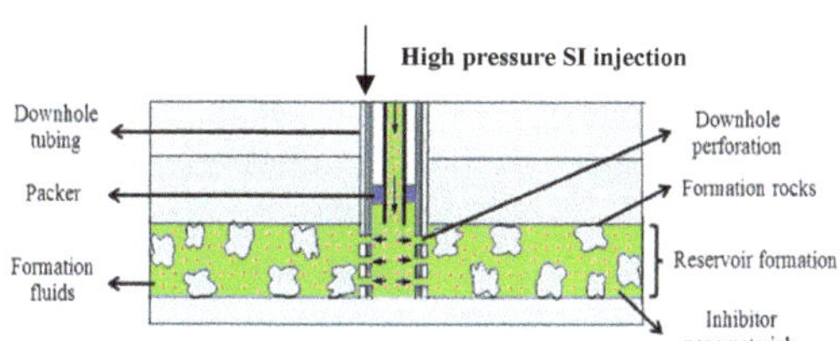

Fig 2 Scale inhibitor (SI) injection in downhole region [12]

During production, flow of reservoir formation water desorbs and dissolves this adsorbed inhibitor chemical which flows into the production and distribution system to inhibit scale formation process. Scale inhibitor chemical adsorption and retention onto the producing formation rock is of paramount importance which determines the effectiveness (success) of the employed squeeze treatment. Squeeze treatment 'lifetime' can be defined as the volume of water produced by a well before the adsorbed scale inhibitor concentration dwindles below the critical threshold concentration value referred as minimum inhibitor concentration (MIC) or minimum effective dose (MED) [13-14]. Scale inhibitor concentration below MIC/MED makes it incapable to prevent or mitigate scale formation phenomenon.

Squeeze treatment is an exorbitant process that requires large volumes of chemical with powerful pumping requirements for its injection into the reservoir. Well must be shut-off for about 12-24 hours which causes production loss and consequently translates into lesser revenue generation. In addition, due to high injection pressures, a significant quantity of scale inhibitor chemical is lost due to vaporization into the permeable reservoir formation matrix. The efficiency of squeeze treatment needs to be improved by enhancing the adsorption and retention of the chemical onto the producing reservoir rock [4-15].

In light of recent interest in nanotechnology application in oil and gas industry, research efforts are being oriented towards exploring the role of nanomaterials to improve and enhance conventional squeeze treatment process widely applied for scale control [16]. By analysing the adsorption potential of conventional squeeze treatment, it can be deduced that lack of availability of active surface for scale inhibitor chemical adsorption is one of the primary reasons for lower efficiency with short squeeze lifetimes. Furthermore, weak interactive bonding between the scale inhibitor chemical molecules and reservoir rock surface is also mainly responsible for premature inhibitor flush-out when water starts to leach out of the reservoir matrix at certain flow rates. In order to effectively inject nanomaterial into the reservoir to mitigate the negative outcomes and improve conventional squeeze treatment, formulation of highly stabilized GNPs nanofluid is proposed. GNPs nanofluid should be resistant to agglomeration and flocculation which can be achieved by adopting an effective dispersion strategy by optimally dispersing GNPs into the solvent solution.

Graphene Nanoplatelets (GNPs)

GNPs are nanoparticles with an average thickness of 3-10 nm that are comprised of short stacks of platelet-shaped graphene sheets in a planar form shown in Figure 3.0. They are good conductors of heat and electricity. Hydrogen or covalent bonding capability can be added through functionalization at sites on the edges of the platelets. Due to their unique nanoscale size, shape and composition material, they can be used to improve properties of a wide range of polymeric materials and composites. They are emerging as a multi-dynamic coating material rendering high surface area for various applications [17].

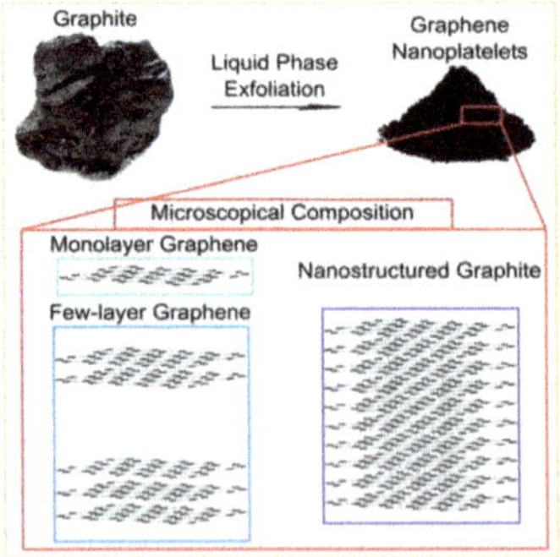

Fig 3 Schematic of the manufacture of GNPs starting from natural graphite [17]

Materials and Methods

Materials

GNPs were procured in powder form from Sigma Aldrich. Triton X-100 and PPCA were supplied by Merck. Probe Sonicator (model: Q-Sonica 500 with 12mm probe diameter) was used to mechanically disperse the GNPs nanofluid solution. Ultra-visible (UV-vis) spectrophotometer (model: Shimadzu UV-1900) was used to determine concentrations of the prepared solutions.

Preparation of dispersed GNPs nanofluid

Highly dispersed GNPs enable to formulate stabilized nanofluid that offer high active surface area to chemically bind scale inhibitor chemical for enhanced squeeze treatment along with superior rheological characteristics for its injection into the reservoir. GNPs are not readily soluble and flocculate in PPCA scale inhibitor solution as shown in Figure 4.0

Fig 4 GNPs flocculation in PPCA solution

In order to determine optimum GNPs concentration for maximum dispersion, three concentrations of at 30, 40 and 50 mg/L GNPs labelled as sample A, B and C respectively were prepared in 18 MΩ cm deionized water while the concentration of Triton X-100 was kept constant at 5ml/L. These prepared solutions were sonicated for 10 mins for homogenous mixture of the components and was left static under the influence of gravity for 24 hours for any unfunctionalized and undispersed GNPs to precipitate and settle at the bottom of the beaker. Table 1 shows concentrations of A, B and C prepared samples.

Table 1 Concentration of GNPs prepared solutions

Sample	GNPs	Triton X-100
A	30mg/L	5ml/L
B	40mg/L	5ml/L
C	50mg/L	5ml/L

Characterization of GNPs dispersion

UV-Vis Spectroscopy

The dispersions of GNPs by Triton X-100 surfactant were characterized using UV-Vis spectrophotometry operating between the range of 200 to 800nm. To characterize the dispersion and measure the concentration of GNPs in a dispersed format, UV-Vis spectroscopy absorbance values were recorded at 650nm wavelength at which Beer-Lambert law is well obeyed as reported in previous studies [18-19].

Probe Sonication

Mechanical technique of sonication is applied to the prepared solutions for 10 mins to homogenously dispersed GNPs into the solvent solution. High shear force created by ultrasonic cavitation breaks-up molecular agglomeration by overcoming Van der Waals forces of attraction.

Results and Discussion

Percentage (%) Extractability

Bundled GNPs are not active in the UV-Vis region and only individual GNPs absorb in this region. The degree of GNPs dispersion in a solution can be evaluated by the parameter of percentage extractability [20] which can be calculated as:

$$\% \text{ extractability} = c_1/c \times 100 \quad (1)$$

Where c_1 = concentration of GNPs recovered in solution (after dispersion) and c = original concentration of GNPs initially added in the solution to be dispersed.

Nanomaterial dispersion depends on the dispersing power of the dispersing agent such as surfactant, solution, sonication process and concentration of nanomaterial. Characterization of GNPs in Triton X-100 surfactant solution using UV-Vis spectroscopy is performed by recording the absorbance value for each dispersed GNPs solution at wavelength of 650nm as shown in Figure 5.0. Absorbance in the wavelength range till 380nm is deliberately ignored due to the presence of turbulence in the dispersed GNPs solution [21-22]. Therefore, it is cogent to take the absorbance at the wavelength of 650nm as the solution is at relative stability and the Beer-Lambart law is well obeyed by GNPs as absorbance is minimally affected by the ambient conditions of GNPs. The absorbance value for each GNPs is shown in Table 2.0.

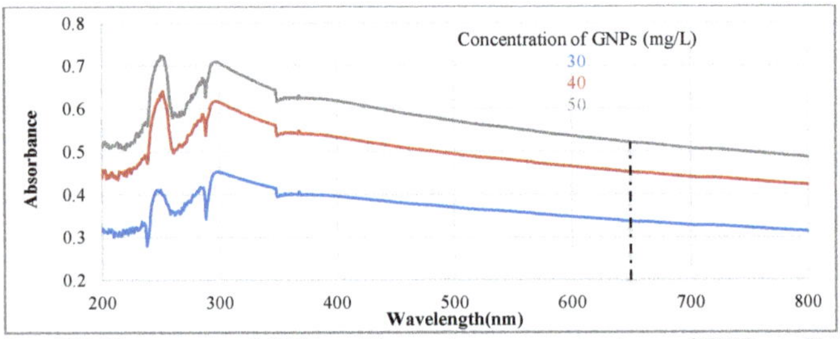

Fig 5 depicts UV-Vis spectra of GNPs with varying concentrations of GNPs in Triton X-100 surfactant solution

Table 2 Absorbance value for each graphene nanoplatelets concentration

Concentration (mg/L)	Absorbance
30	0.33
40	0.45
50	0.52

The correlation between absorbance and GNPs solution can be developed by plotting straight line graph validating assumption of Beer-Lambart law as shown in Figure 6.0.

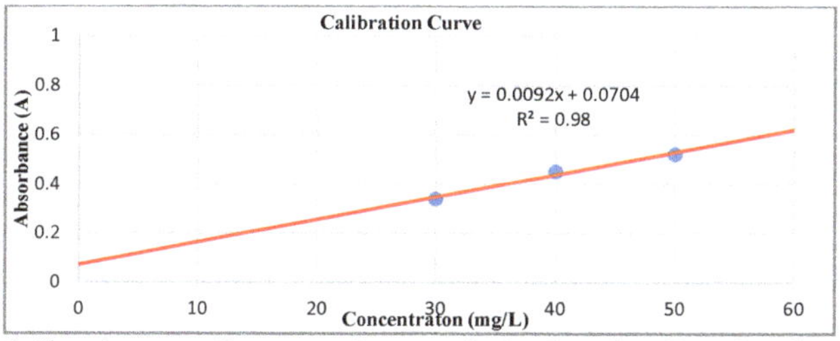

Fig 6 Absorbance vs Concentration correlation based on Beer-Lambert law

From this above graph, the concentration of a sample can be determined from its absorbance value by simple interpolation method. This is done by drawing a horizontal line from the y-axis (sample's absorbance) until it intersects the best fit line. Then a vertical line is drawn from that point to the x-axis. The x-axis value is the concentration of the sample. Absorbance using Beer-Lambert law can be defined by equation (2)

$$A = \varepsilon \, l \, c_1 \quad (2)$$

where **A** = absorbance value of GNPs, **ε** = specific extinction coefficient of GNPs at 650nm, **l** = cuvette path length (constant value of 1.0) and **c_1** = concentration of GNPs after dispersion.

From equation, concentration of GNPs after dispersion (c_1) is calculated which is used to evaluate % extractability values for each GNPs nanofluid solution as shown in table 3.0. The calculated % extractability concentration vs concentration follows a Gaussian trend as shown in Figure 7.0 that depicts a linear increase with increase in GNPs concentration until a maximum extractability limit is achieved. This limit is presumably the optimum amount of Triton X-100 surfactant to sufficiently coat the GNPs surface in a homogenous manner thereby rendering its highest dispersion. Decrease in % extractability at higher GNPs concentration indicates presence of insufficient Triton X-100 surfactant to effectively disperse GNPs. Also, this decrease indicates agglomeration of GNPs as higher GNPs concentration giving rise to the presence intrinsic Van Der Waals forces of attraction. The highest dispersion of GNPs solution was shown to be at 40mg/L followed by a sharp decline at 50 mg/L as shown in Figure 7.0. Figure 8.0 shows homogenously dispersed stable GNPs solutions left under the influence of gravity for 48 hours.

Table 3 %Extractability and Absorbance values of GNPs concentrations

Concentration (mg/L)	Absorbance	%Extractability
30	0.33	39
40	0.45	40
50	0.52	36

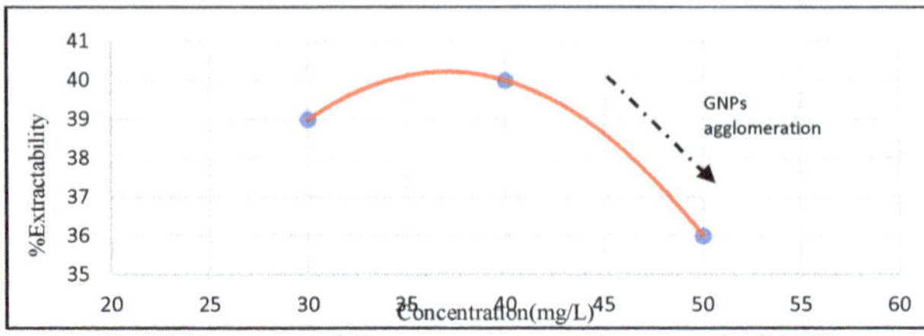

Fig 7 % extractability vs concentration trend of GNPs

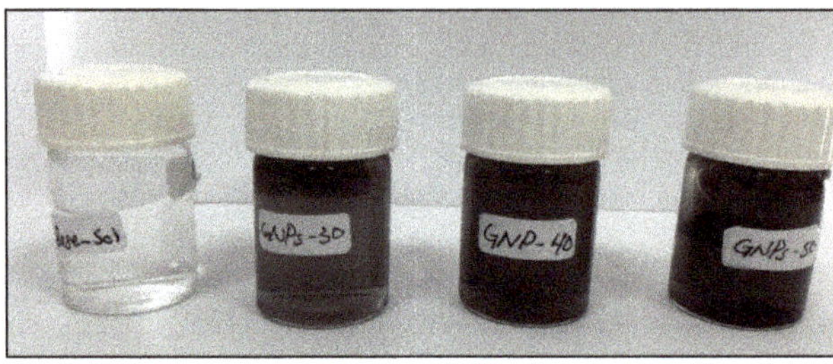

Fig 8 Different concentrations of GNPs dispersed stable solutions

Conclusion

These preliminary results demonstrate that surfactant molecules generally orient themselves in such a manner that their hydrophobic tail attaches with the GNPs surface while their hydrophilic tail bonds with the aqueous phase thereby reducing GNPs/water interfacial tension. The parameter of % extractability is a reliable measure that depicts the degree of dispersion of a solution. The concentration of 40mg/L GNPs gives the highest % extractability value of 40 depicting the best and stable dispersion of GNPs solution. Further increase in GNPs concentration leads to its agglomeration. The dispersing power of the surfactant depends on its bonding strength with the GNPs surface and thus creating sufficient steric hinderance [20] that can overcome the agglomeration potential of GNPs due to the presence of Van Dar Waals forces of attraction therefore rendering high degree of stable nanomaterial dispersion. The presence of benzene ring structure in Triton X-100 surfactant is a dominant factor that adsorb more firmly to the graphitic surface of GNPs due to $\pi - \pi$ stacking type interaction. In conclusion, the dispersing power of surfactant is inherent that mainly depends on its chemical structure organization.

Acknowledgement

The authors profoundly acknowledge Universiti Teknologi Petronas (UTP), Malaysia for the required research support. This research work was supported under YUTP-FRG 015LC0-233 Research grant.

References

[1] Crabtree, M., Eslinger, D., Fletcher, P., Miller, M., Johnson, A., King, G. (1999). Fighting Scale-Removal and Prevention, Oilfield Rev. 11. 30-45

[2] Amiri, M., Moghadasi, J., & Pordel, M. (2012). The effect of temperature, pressure, and mixing ratio of the injection water with formation water on strontium sulfate scale formation in the Siri oilfield. Petroleum Science and Technology, 30(7), 635-645 https://doi.org/10.1080/10916466.2010.489088

[3] Pons-Jiménez, M., Hernández-Altamirano, R., Cisneros-Dévora, R., Buenrostro-González, E., Oviedo-Roa, R., Martínez-Magadán, J. M., & Zamudio-Rivera, L. S. (2015). Theoretical and experimental insights into the control of calcium sulfate scales by using random copolymers based on itaconic acid. Fuel, 149, 66-77. https://doi.org/10.1016/j.fuel.2014.09.064

[4] Al Salami, A.R., Monem, A.A., 2010.Downholeand top side scale challenge Removal, prevention and inhibition strategy, [SPE 137906], in the Abu Dhabi International Petroleum Exhibition & Conference. Society of Petroleum Engineers, Abu Dhabi, p.7.

[5] Muhsan, A., Ishtiaq, U., Rozali, A., Mohamed, N., & Albarody, T. (2018, December). Nanocarbon-based enhanced squeeze treatment for improved scale management. In IOP Conference Series: Materials Science and Engineering (Vol. 458, No. 1, p. 012044). IOP Publishing. https://doi.org/10.1088/1757-899X/458/1/012044

[6] Poggesi, G., Hurtevent, C., & Brazy, J. L. (2001, January). Scale Inhibitor Injection Via the Gas Lift System in High Temperature Block 3 Fields in Angola. In International Symposium on Oilfield Scale. Society of Petroleum Engineers. https://doi.org/10.2118/68301-MS

[7] Hsu, J. F., Al-Zain, A. K., Raju, K. U., & Henderson, A. P. (2000, January). Encapsulated scale inhibitor treatments experience in the Ghawar Field, Saudi Arabia. In International Symposium on Oilfield Scale. Society of Petroleum Engineers. https://doi.org/10.2118/60209-MS

[8] Bourne, H., Heath, S. M., McKay, S., Fraser, J., Stott, L., & Müller, S. (2000, January). Effective treatment of subsea wells with a solid scale inhibitor system. In International Symposium on Oilfield Scale. Society of Petroleum Engineers. https://doi.org/10.2118/60207-MS

[9] Ghorbani, N., Wilson, M. C. T., Kapur, N., Fleming, N., & Neville, A. (2014, May). Carbon nanotubes: A new methodology for enhanced squeeze lifetime CNTs. In SPE International Oilfield Scale Conference and Exhibition. Society of Petroleum Engineers. https://doi.org/10.2118/169763-MS

[10] Hasson, D., Shemer, H., & Sher, A. (2011). State of the art of friendly "green" scale control inhibitors: a review article. Industrial & Engineering Chemistry Research, 50(12), 7601-7607. https://doi.org/10.1021/ie200370v

[11] Farooqui, N. M., & Sorbie, K. S. (2016). The use of PPCA in scale-inhibitor precipitation squeezes: solubility, inhibition efficiency, and molecular-weight effects. SPE Production & Operations, 31(03), 258-269. https://doi.org/10.2118/169792-PA

[12] Shen, D., Zhang, P., Kan, A.T., Fu, G., Alsaiari, H.A. and Tomson, M.B. 2008. Control Placement of Scale Inhibitors in the Formation with Stable Ca-DTPMP Nanoparticle Suspension and its Transport Porous Media, in SPE International Oilfield Scale Conference. Society of Petroleum Engineers: Aberdeen, UK. https://doi.org/10.2118/114063-MS

[13] Ghorbani, N., Wilson, M. C. T., Kapur, N., Fleming, N., Tjomsland, T., & Neville, A. (2017). Adsorption of polyphosphinocarboxylic acid (PPCA) scale inhibitor on carbon nanotubes (CNTs): A prospective method for enhanced oilfield scale prevention. Journal of Petroleum Science and Engineering, 150, 305-311. https://doi.org/10.1016/j.petrol.2016.12.016

[14] Ghorbani, N., Wilson, M., Kapur, N., Fleming, N., & Neville, A. (2012, January). Using Nanoscale Dispersed Particles to Assist in the Retention of Polyphosphinocarboxylic Acid (PPCA) Scale Inhibitor on Rock. In SPE International Oilfield Nanotechnology Conference and Exhibition. Society of Petroleum Engineers. https://doi.org/10.2118/156200-MS

[15] Ishtiaq, U., Muhsan, A. S., Rozali, A. S., Mohamed, N. M., & Badri Albarody, T. M. (2020). Graphene oxide/carbon nanotubes nanocoating for improved scale inhibitor adsorption ability onto rock formation. Journal of Petroleum Exploration and Production Technology, 10(1), 149-157. https://doi.org/10.1007/s13202-019-0689-7

[16] Mady, M. F., & Kelland, M. A. (2020). Review of nanotechnology impacts on oilfield scale management. ACS Applied Nano Materials, 3(8), 7343-7364. https://doi.org/10.1021/acsanm.0c01391

[17] Cataldi, P., Athanassiou, A., & Bayer, I. S. (2018). Graphene nanoplatelets-based advanced materials and recent progress in sustainable applications. Applied Sciences, 8(v9), 1438. https://doi.org/10.3390/app8091438

[18] Abedi, M., Fangueiro, R., & Correia, A. G. (2020). An effective method for hybrid CNT/GNP dispersion and its effects on the mechanical, microstructural, thermal, and electrical properties of multifunctional cementitious composites. Journal of Nanomaterials, 2020. https://doi.org/10.1155/2020/6749150

[19] Mehrali, M., Sadeghinezhad, E., Latibari, S. T., Kazi, S. N., Mehrali, M., Zubir, M. N. B. M., & Metselaar, H. S. C. (2014). Investigation of thermal conductivity and rheological properties of nanofluids containing graphene nanoplatelets. Nanoscale research letters, 9(1), 1-12. https://doi.org/10.1186/1556-276X-9-15

[20] Rastogi, R., Kaushal, R., Tripathi, S. K., Sharma, A. L., Kaur, I., & Bharadwaj, L. M. (2008). Comparative study of carbon nanotube dispersion using surfactants. Journal of colloid and interface science, 328(2), 421-428. https://doi.org/10.1016/j.jcis.2008.09.015

[21] Wang, B., Jiang, R., Song, W., & Liu, H. (2017). Controlling dispersion of graphene nanoplatelets in aqueous solution by ultrasonic technique. Russian Journal of Physical Chemistry A, 91(8), 1517-1526. https://doi.org/10.1134/S0036024417080040

[22] Nazari, B., Ranjbar, Z., Moghaddam, A. R., Momen, G., & Ranjbar, B. (2019). Dispersing graphene in aqueous media: Investigating the effect of different surfactants. Colloids and Surfaces A: Physicochemical and Engineering Aspects, 582, 123870. https://doi.org/10.1016/j.colsurfa.2019.123870

Materials Research Proceedings 29 (2023) 439-445 https://doi.org/10.21741/9781644902516-50

Preliminary study for catalytic gasification of water hyacinth for syngas production

APRIL LING KWANG Chee[1,a], BRIDGID LAI FUI Chin[1,b*], SHAHARIN ANWAR Sulaiman[2,c], YEE HO Chai[3,4,d], AGUS Saptoro[1,e], HADIZA AMINU Umar[2,5,f], MAN KEE Lam[3,4,g], CHUNG LOONG Yiin[6,7,h]

[1]Department of Chemical and Energy Engineering, Faculty of Engineering and Science, Curtin University Malaysia, CDT 250, 98009 Miri, Sarawak, Malaysia

[2]Department of Mechanical Engineering, Universiti Teknologi PETRONAS, 32610, Seri Iskandar, Perak Darul Ridzuan, Malaysia

[3]Chemical Engineering Department, Universiti Teknologi PETRONAS, 32610, Seri Iskandar, Perak Darul Ridzuan, Malaysia

[4]HICoE-Centre for Biofuel and Biochemical Research, Institute of Self-Sustainable, Building, Universiti Teknologi PETRONAS (UTP), 32610 Seri Iskandar, Perak Darul Ridzuan, Malaysia

[5]Mechanical Engineering Department, Bayero University Kano, PMB 3011 Kano, Kano State Nigeria

[6]Department of Chemical Engineering and Energy Sustainability, Faculty of Engineering, Universiti Malaysia Sarawak (UNIMAS), Kota Samarahan, 94300, Malaysia

[7]Institute of Sustainable and Renewable Energy (ISuRE), Universiti Malaysia Sarawak (UNIMAS), 94300, Kota Samarahan, Sarawak, Malaysia

[a]aprilchee@postgrad.curtin.edu.my, [b]bridgidchin@curtin.edu.my, [c]shaharin@utp.edu.my, [d]yeeho.chai@utp.edu.my, [e]agus.saptoro@curtin.edu.my, [f]hadiza_16000717@utp.edu.my, [g]lam.mankee@utp.edu.my, [h]clyiin@unimas.my

Keywords: Gasification, Biomass, Water Hyacinth, Syngas, Downdraft Gasifier

Abstract. Water hyacinth being one of the top invasive aquatic plants has brought upon various challenges towards the humanity and the environment. The magnitude of the menace of uncontrollable growth and spread of water hyacinth has sparked the interest of researchers in identifying its potential as a biomass feedstock for biofuel production. Biomass gasification is deemed as a promising green technology which is capable of converting biomass into value-added commodity. Conversion of such large quantity of biomass into biofuel via gasification does not only help to promote sustainable resource utilization but also facilitates the reduction of global carbon impacts and engender socioeconomic development. The addition of catalysts to the gasification process could enhance the formation of gaseous products where the gas composition may be altered. This study aims to present the preliminary study on the gasification performance of water hyacinth biomass in a lab scale fixed-bed downdraft gasifier (67 mm diameter and 750 mm height), with the use of air as the gasifying agent in a batch feeding of 50 grams for each run. The results showed that temperature has a substantial effect on the gasification of water hyacinth whereby hydrogen produced was raised from 2.92 vol.% to 11.19 vol.%. Further gasification tests are expected for the optimization of the main process parameters such as biomass particle size and catalyst loading.

Introduction

Water hyacinth, with its scientific name known as *Eichhornia crassipes*, is regarded as a potential and economically viable feedstock for biofuel production via thermochemical conversion. It was reported that water hyacinth comprises of higher cellulose and hemicellulose contents in comparison to algae [1]. Water hyacinth has been perceived to be one of the worst aquatic weeds due to the tremendous challenges brought upon the urban management as a result of its abundance and rapid reproduction rate. In fact, water hyacinth has been a major encumbrance to fishing and hydroelectric schemes while also causing disruption to navigation and irrigation systems [2]. Biomass gasification would be an appropriate alternative solution for managing such large amount of aquatic plant biomass and help to eradicate the excessive amount of water hyacinth present in the ecosystem. Meanwhile, such initiative could also facilitate the paradigm shift to alternative renewable energy sources and possibly reducing the heavy reliance on the dwindling supply fossil fuels for most industrial activities [3,4,5,6,7,8,9,10,11]. Utilization of conventional fossil fuel resources often times contribute a huge share to the net greenhouse gas emission. Hence, exploration of potential alternative renewable and sustainable energy resources is much needed not only for reducing the global carbon footprint but also be able to meet the ever-increasing energy demand by the rapid population growth.

Gasification process involves the conversion of organic or carbonaceous materials into gaseous products such as carbon monoxide, hydrogen, methane and carbon dioxide via high temperatures for wide application such as power generation, liquid fuels, production of hydrogen, ethanol, and methanol [12]. Partial oxidation reaction is involved where the amount of oxygen/air, steam or combination of these two oxidizing agents is controlled and introduced into the system. Gasification outweighs other thermochemical conversions in terms of syngas production as it is capable of transforming biomass into syngas with low net CO_2 emissions. There are some studies on the gasification of water hyacinth in the literature. [13] studied on the supercritical water gasification of water hyacinth and examined its potential against biomethanation. Results showed that gasification offers better conversion and energy efficiency of 64.8%. However, it was lack of in-depth study on the thermal degradation behaviour and kinetic parameters related to the gasification of water hyacinth.

Most researchers investigating on conversion of water hyacinth have mostly employed the pyrolysis process. Study conducted by [13] involved catalytic pyrolysis of water hyacinth using KCl, MgO and CaO catalysts whereby the optimum conditions for attaining maximum syngas were studied. [14] reported on the maximum bio-oil yield of 44.9 wt% achieved at 350°C with heating rate of 30°C/min and particle size of less than 0.5 mm via pyrolysis of water hyacinth. [15] found that the optimum concentration of ferric chloride solution for maximum syngas production from pyrolysis of water hyacinth at 540°C was 2M with syngas composition of 42% H_2 and 22% CO. Notably, these studies have converted water hyacinth into valuable commodity via pyrolysis but yet to investigate on the kinetic parameters of the thermal degradation of water hyacinth. However, [16] had evaluated on the bioenergy potential and kinetic parameters of the pyrolysis of water hyacinth. Nonetheless, there was no studies reported on the catalytic gasification of water hyacinth biomass for syngas enriched hydrogen production and hence this preliminary study intends to investigate on the effect of temperature on the gasification of water hyacinth.

This research involves gasification of water hyacinth biomass for producing syngas enriched hydrogen. It could be a solution to the major issue of the abundance of water hyacinth especially in Malaysia and thereby possibly alleviating the negative impacts of water hyacinth brought to both human and local ecosystems. For instance, the removal of water hyacinth which occurred to be nutrient extraction tool in Lake Victoria was found to have reduced algal blooms as nutrients level in its surrounding was decreased. Furthermore, the rate of disease spreading could be reduced as schistosomiasis and malaria were found to be associated to the spread of water hyacinth [17,18].

Materials Research Proceedings 29 (2023) 439-445 https://doi.org/10.21741/9781644902516-50

Moreover, production of syngas by gasification of water hyacinth could be used for multiple applications such as electricity generation as it can be directly used in engine, steam or gas turbines as well as fuel cells. Syngas could also be employed to produce chemicals for commercialization such as ammonia, methanol and Fischer Tropsch diesel [19].

This study may offer a solution by utilizing the wet biomass found in ASEAN countries and thereby encourage the development of renewable and sustainable energy while also resolving the negative ecological impact brought by the aquatic weeds such as water hyacinth. Furthermore, this could be part of the initiatives for attaining a circular bioeconomy approach in phytoremediation through gasification or pyrolysis of biomass. For instance, production of biochar may be applied for multiple applications such as conductor or electrode in fuel battery, adsorbents to remove the aquatic pollutants and catalysts in green industry. Furthermore, gasification technology is deemed to possess better carbon sequestration potential (146–264 kg C/ton) when compared to combustion (~87 kg C/ton) and anaerobic digestion (146 kg C/ton) [20]. This would aid in achieving low carbon footprint as well as reduction in greenhouse gas emissions.

Methodology and Methods

Water hyacinth (WH) was extracted from the lake of Curtin University, Malaysia. WH biomass was dried under sunny days for 36-48 hours and further dried in oven at 105°C for 4 hours. The dried biomass was then grounded and sieved to particle size of 4 mm for preliminary gasification experiment.

Gasification Experimental Setup

The gasification system consists of a gasifier, a temperature controller, an air compressor and a cleaning system. The gasification experiment was conducted in a lab-scale gasifier which operates in allothermal mode. The gasifier is composed of a vertical cylindrical casing of 750 mm height and 67 mm internal diameter, which is enclosed within a ceramic electric heater of 1.25 kW. Prior to the 50 g batch feeding of each experimental run, the gasifier has to be cleaned. The temperature microcontroller which employs the Type-K thermocouple is responsible for the temperature adjustment. The gasification temperature used for this study is range from 600 to 800°C. An air compressor is connected to the gasifier for air supply which is regulated by a rotameter. A constant air flow rate of 2 L/min was supplied to the gasifier. A 4.0 μm ceramic filter is affixed for filtering possible particulate matter in the gas sample prior to entering the gas analyzer. The gaseous products exit from the bottom of the gasifier and flows through a tube connected to the cooling and filtering unit and enters the gas analyzer (X-STREAM X2GP model). The gas analyzer provides information on the gas constituents in terms of volume percentage.

Results and Discussion

The operating temperature employed to gasify water hyacinth biomass for investigation purposes range from 600 to 800°C. As illustrated in Table 1, the gaseous products generally increase alongside with the increment of temperature except for carbon dioxide (CO_2) content.

Table 1: Gas composition from gasification of water hyacinth at different temperatures

Temperature (°C)	CO [vol.%]	CO_2 [vol.%]	CH_4 [vol.%]	H_2 [vol.%]	H_2+CO [vol.%]	H_2:CO
600	4.59	14.6	4.56	2.92	7.51	0.389
700	9.10	22.48	5.07	5.74	14.84	0.631
800	15.06	20.48	8.63	11.19	26.25	0.426

A huge leap in hydrogen (H_2) and carbon monoxide (CO) content, from 2.92 to 11.19 vol.% and 4.59 to 15.06 vol.% respectively, is observed when temperature is raised from 600 to 800°C. Study conducted by [21] have shown similar trend in the hydrogen content which was increased from 7.28 to 14.59 vol.% when temperature was increased from 700 to 900°C. It was reported that the gasification of cellulose with temperature increasing beyond 930°C was found to cause the decline in hydrogen (H_2) production [22]. However, the desirable gasification temperature range for maximizing hydrogen production is also subjected to the gasifier, gasification agent and biomass feedstock used. [22] reported that a reduction in H_2 composition was observed from the gasification of rice husk and rice straw while corn cob exhibited greater amount of H_2 production, when pure steam was employed as the gasifying agent.

Due to the increasing trend of carbon monoxide (CO) and hydrogen (H_2), the syngas produced, which is mainly the sum of CO and H_2, from the gasification process shows significant increase from 7.51 to 26.25 vol.%. The proportional relationship between the H_2+CO composition and temperature could be attributed to the endothermic reactions (primary and secondary water gas reactions, Eq. 1-2, and Boudourd reaction, Eq. 3) which may have prevailed over the methanisation reaction (Eq. 4) [23-25]. This is because high temperatures favours the right side of the equilibrium reactions for endothermic reactions and hence promotes the formation of H_2 and CO.

$$C + H_2O \leftrightarrow CO + H_2 \qquad \Delta H_r = 131.3\ kJ/mol \qquad (1)$$

$$C + 2H_2O \leftrightarrow CO_2 + 2H_2 \qquad \Delta H_r = 90.2\ kJ/mol \qquad (2)$$

$$C + CO_2 \leftrightarrow 2CO \qquad \Delta H_r = 172.5\ kJ/mol \qquad (3)$$

$$C + 2H_2 \leftrightarrow CH_4 \qquad \Delta H_r = -74.9\ kJ/mol \qquad (4)$$

Based on the tabulated results in Table 1, it can be deduced that the temperature of 700°C could be the ideal gasification temperature for high quality syngas due to the high H_2:CO ratio of 0.631 which is a significant increment of 62.2% compared to the H_2:CO ratio at 600°C.

Pertaining to the composition of carbon dioxide (CO_2), it shows an increase from 14.6 to 22.48 vol.% when temperature was raised from 600 to 700°C which later decreases to 20.48 vol.% when the gasification temperature was further increased to 800°C. It was presumed that the decline in CO_2 content would be due to the prevalence of Boudourd reaction (Eq. 3) at higher temperature which consumes the CO_2 produced from partial oxidation (Eq. 5), secondary water gas (Eq. 2) and water gas shift (Eq. 6) reactions [24,26,27].

$$C + O_2 \leftrightarrow CO_2 \qquad \Delta H_r = -394.1\ kJ/mol \qquad (5)$$

$$CO + H_2O \leftrightarrow CO_2 + H_2 \qquad \Delta H_r = -41.2\ kJ/mol \qquad (6)$$

With regards to methane (CH_4) content, Fig. 1. shows a rather relatively minor increment in the CH_4 composition as the temperature increases when compared to the other gases. Regardless, the increasing trend is also observed in the study conducted by [23] whereby grapevine pruning waste was gasified using air as the gasifying agent which is also employed in the present study. Despite so, it can be construed that CH_4 production is not greatly affected by the gasification temperature but rather due to the pyrolysis process as minor changes in CH_4 composition was observed in several studies [24,28,29] . The slight change is attributable to the formation and displacement of CH_4 involving exothermic reactions (methanation, Eq. 7 [30], methane steam reforming reaction, Eq. 8 [24], and combustion reaction, Eq. 9 [30]) which are favoured at low and high temperatures respectively [23,24,31].

$$C + 2H_2 \leftrightarrow CH_4 \quad \Delta H_r = -74.9 \text{ kJ/mol} \tag{7}$$

$$CH_4 + H_2O \leftrightarrow CO_2 + 3H_2 \quad \Delta H_r = 206.2 \text{ kJ/mol} \tag{8}$$

$$CH_4 + 2O_2 \leftrightarrow CO_2 + H_2O \quad \Delta H_r = -802.3 \text{ kJ/mol} \tag{9}$$

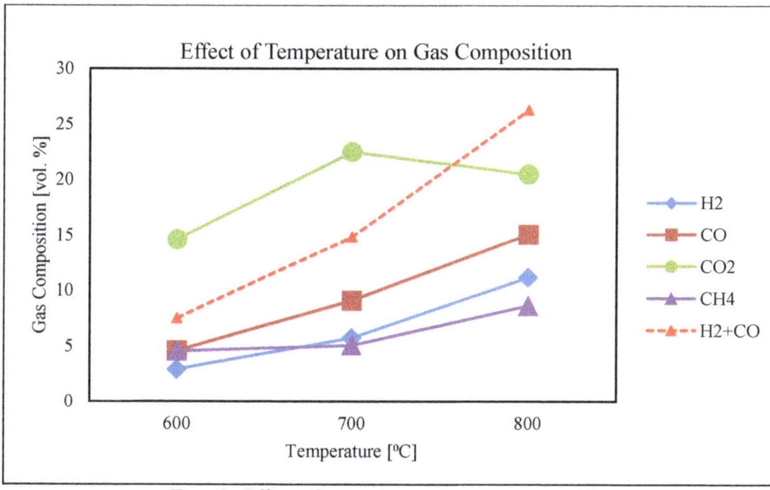

Fig. 1: Effect of temperature on gas composition

Acknowledgements

The authors would like to thank Curtin University Malaysia and Biomass Processing Laboratory, Centre of Biofuel and Biochemical (CBBR) of Universiti Teknologi PETRONAS (UTP) for the technical support. In addition, we would like to thank Curtin University Malaysia for providing the financial support on the Curtin Malaysia Research Grant (Sustainability) grant entitled, 'Conversion of water hyacinth biomass to syngas enriched hydrogen production via catalytic gasification process'.

References

[1] Q. Chen, X. Lu, J. Liu, J. Yu, J. Gu, P. Ning, R. Miao, Q. Guan, Recycling water hyacinth by gasification in supercritical water, Fresenius Environ. Bull. 26 (2017) 2163–2170.

[2] D. Güereña, H. Neufeldt, J. Berazneva, S. Duby, Water hyacinth control in Lake Victoria: Transforming an ecological catastrophe into economic, social, and environmental benefits, Sustain. Prod. Consum. 3 (2015) 59–69. https://doi.org/https://doi.org/10.1016/j.spc.2015.06.003

[3] B.L.F. Chin, A. Gorin, H.B. Chua, F. Twaiq, Experimental invesigtation on tar produced from palm shells derived syngas using zeolite HZSM-5 catalyst. J Energy Inst. 89(4) (2016) 713-724. https://doi.org/10.1016/j.joei.2015.04.005

[3] A.L.K. Chee, B.L.F. Chin, S.M.X. Goh, Y.H. Chai, A.C.M. Loy, K.W. Cheah, C.L. Yiin, S.S.M. Lock, Thermo-catalytic co-pyrolysis of palm kernel shell and plastic waste mixtures using bifunctional HZSM-5/limestone catalyst: kinetic and thermodynamic insights, J. Energy Inst. 107 (2023) 101194. https://doi.org/10.1016/j.joei.2023.101194

[4] Y.H. Chai, S. Yusup, W.N.A. Kadir, C.Y. Wong, S.S. Rosli, M.S.H. Ruslan, B.L.F. Chin, C.L. Yiin, Valorization of tropical biomass waste by supercritical fluid extraction. Sustainability 13 (2021) 233, https://doi.org/10.3390/su13010233

[5] H.Y. Lim, S. Yusup, A.C.M. Loy, S. Samsuri, S.S.K. Ho, A.S.A. Manaf, S.S. Lam, B.L.F. Chin, M.N. Acda, P. Unrean, E. Rianawati. Review on conversion of lignin waste into value-added resources in tropical countries. Waste and biomass valorization 12 (2021) 5285-5302, https://doi.org/10.1007/s12649-020-01307-8

[6] D.K.W. Gan, B.L.F. Chin, A.C.M. Loy, S. Yusup, M.N. Acda, P. Unrean, E. Rianawati, Z.A. Jawad, R.J. Lee. An in-situ thermogravimetric study of pyrolysis of rice hull with alkali catalyst of CaCO3. IOP Conf. Ser.: Mater. Sci Eng. 458 (2018) 012085. https://doi.org/10.1088/1757-899X/458/1/012085

[7] V.S.Y. Soon, B.L.F. Chin, A.C.R. Lim. Kinetic study on pyrolysis of oil palm frond. IOP Conf. Ser.: Mater. Sci. Eng. 121 (2016) 012004. https://doi.org/10.1088/1757-899X/121/1/012004

[8] Y.H. Chai, M. Mohamed, Y.W. Cheng, B.L.F. Chin, C.L. Yiin, S. Yusup, M.K. Lam, A review on potential of biohydrogen generation through waste decomposition technologies. Biomass Conv. Bioref. (2021) https://doi.org/10.1007/s13399-021-01333-z

[9] Y.H. Chan, A.C.M. Loy, K.W. Cheah, S.Y.W. Chai, L.H. Ngu, B.S. How, C. Li, S.S.M. Lock, M.K. Wong, C.L. Yiin, B.L.F. Chin, Z.P. Chan, S.S. Lam, Hydrogen sulfide (H2S) conversion to hydrogen (H2) and value-added chemicals: progress, challenges and outlook. Chem. Eng. J. 458 (2023) 141398. https://doi.org/10.1016/j.cej.2023.141398

[10] Y.H. Chan, S.S.M. Lock, M.K. Wong, C.L. Yiin, A.C.M. Loy, K.W. Cheah, S.Y.W. Chai, C. Li, B.S. How, B.L.F. Chin, Z.P. Chan, S.S. Lam, A state-of-the-art review on capture and separation of hazardous hydrogen sulfide (H2S): recent advances, challenges and outlook. Environ. Pollution 314 (2022). 120219. https://doi.org/10.1016/j.envpol.2022.120219

[11] S.W. Hii, B.L.F. Chin, F.R.S.A. Majing, H.Y. Lim, A.C.M. Loy, C.L. Yiin, S. Yusup, A.T. Quitain, M.N. Acda, P. Unrean, E. Rianawati, Chapter 12 – Isoconversional kinetic and thermodynamic analysis of catalytic pyrolysis for palm oil wastes. Value-Chain of Biofuels (2022), 277-300. https://doi.org/10.1016/B978-0-12-824388-6.00025-7

[12] Y. Matsumura, Evaluation of supercritical water gasification and biomethanation for wet biomass utilization in Japan, Energy Convers. Manag. 43 (2002) 1301–1310. https://doi.org/10.1016/S0196-8904(02)00016-X

[13] Z. Hu, X. Ma, L. Li, Optimal conditions for the catalytic and non-catalytic pyrolysis of water hyacinth, Energy Convers. Manag. 94 (2015) 337–344. https://doi.org/https://doi.org/10.1016/j.enconman.2015.01.087

[14] M.A. Rahman, Pyrolysis of water hyacinth in a fixed bed reactor: Parametric effects on product distribution, characterization and syngas evolutionary behavior, Waste Manag. 80 (2018) 310–318. https://doi.org/10.1016/j.wasman.2018.09.028

[15] T.K. Tran, N. Kim, H.J. Leu, M.P. Pham, N.A. Luong, H.K. Vo, The production of hydrogen gas from modified water hyacinth (Eichhornia Crassipes) biomass through pyrolysis process, Int. J. Hydrogen Energy. 46 (2021) 13976–13984. https://doi.org/10.1016/j.ijhydene.2020.08.225

[16] H. Huang, J. Liu, H. Liu, F. Evrendilek, M. Buyukada, Pyrolysis of water hyacinth biomass parts: Bioenergy, gas emissions, and by-products using TG-FTIR and Py-GC/MS analyses, Energy Convers. Manag. 207 (2020) 112552. https://doi.org/10.1016/j.enconman.2020.112552

[17] C.C. Gunnarsson, C.M. Petersen, Water hyacinths as a resource in agriculture and energy production: A literature review, Waste Manag. 27 (2007) 117–129. https://doi.org/https://doi.org/10.1016/j.wasman.2005.12.011

[18] N. Minakawa, G.O. Dida, G.O. Sonye, K. Futami, S.M. Njenga, Malaria vectors in Lake Victoria and adjacent habitats in Western Kenya, PLoS One. 7 (2012). https://doi.org/10.1371/journal.pone.0032725

[19] Ö. Tezer, N. Karabağ, A. Öngen, C.Ö. Çolpan, A. Ayol, Biomass gasification for sustainable energy production: A review, Int. J. Hydrogen Energy. 47 (2022) 15419–15433. https://doi.org/https://doi.org/10.1016/j.ijhydene.2022.02.158

[20] F. Li, X. He, A. Srishti, S. Song, H.T.W. Tan, D.J. Sweeney, S. Ghosh, C.H. Wang, Water hyacinth for energy and environmental applications: A review, Bioresour. Technol. 327 (2021) 124809. https://doi.org/10.1016/j.biortech.2021.124809

[21] A.M. Ali, M. Inayat, A.A. Zahrani, K. Shahzad, M. Shahbaz, S.A. Sulaiman, H. Sadig, Process optimization and economic evaluation of air gasification of Saudi Arabian date palm fronds for H2-rich syngas using response surface methodology, Fuel. 316 (2022) 123359. https://doi.org/10.1016/j.fuel.2022.123359

[22] W. Jangsawang, A. Klimanek, A.K. Gupta, Enhanced yield of hydrogen from wastes using high temperature steam gasification, J. Energy Resour. Technol. Trans. ASME. 128 (2006) 179–185. https://doi.org/10.1115/1.2134733

[23] M. Lapuerta, J.J. Hernández, A. Pazo, J. López, Gasification and co-gasification of biomass wastes: Effect of the biomass origin and the gasifier operating conditions, Fuel Process. Technol. 89 (2008) 828–837. https://doi.org/10.1016/j.fuproc.2008.02.001

[24] L. Emami Taba, M.F. Irfan, W.A.M. Wan Daud, M.H. Chakrabarti, The effect of temperature on various parameters in coal, biomass and CO-gasification: A review, Renew. Sustain. Energy Rev. 16 (2012) 5584–5596. https://doi.org/10.1016/j.rser.2012.06.015

[25] X. Zhang, Y. Han, Y. Li, Y. Sun, Effect of Heating Rate on Pyrolysis Behavior and Kinetic Characteristics of Siderite, Minerals. 7 (2017). https://doi.org/10.3390/min7110211.

[26] J. Gil, P. Aznar, M.A. Caballero, E. France, with Steam - Oxygen Mixtures . Product Distribution for, Energy & Fuels Fuels. 11 (1997) 1109–1118.

[27] H. Zhang, X. Guo, Z. Zhu, Effect of temperature on gasification performance and sodium transformation of Zhundong coal, Fuel. 189 (2017) 301–311. https://doi.org/10.1016/j.fuel.2016.10.097

[28] W.J. Lee, S.D. Kim, B.H. Song, Steam Gasification of Coal with Salt Mixture of Potassium and Nickel in a Fluidized Bed Reactor, Korean J. Chem. Eng. 18 (2001) 640–645. https://doi.org/10.1007/BF02706380

[29] W.J. Lee, S.D. Kim, B.H. Song, Steam gasification of an australian bituminous coal in a fluidized bed, Korean J. Chem. Eng. 19 (2002) 1091–1096. https://doi.org/10.1007/BF02707238

[30] S. Pang, Fuel flexible gas production: Biomass, coal and bio-solid wastes, in: J. Oakey (Ed.), Fuel Flex. Energy Gener., Woodhead Publishing, Boston, 2016: pp. 241–269. https://doi.org/10.1016/B978-1-78242-378-2.00009-2

[31] M. Pohořelý, M. Vosecký, P. Hejdová, M. Punčochář, S. Skoblja, M. Staf, J. Vošta, B. Koutský, K. Svoboda, Gasification of coal and PET in fluidized bed reactor, Fuel. 85 (2006) 2458–2468. https://doi.org/10.1016/j.fuel.2006.04.022

Prediction of the conductivity and compatibility of the selected ionic liquids (ILs) with Nafion™ using COSMO-RS

RUWAIDA ASYIKIN Abu Talip[1,a], WAN ZAIREEN NISA Yahya[1,2,b*]

[1]Chemical Engineering Department, Universiti Teknologi PETRONAS, 32610 Seri Iskandar, Perak, Malaysia

[2]Centre of Research in Ionic Liquids, Universiti Teknologi PETRONAS, 32610 Seri Iskandar, Perak, Malaysia

[a]ruwaida.abutalip@utp.edu.my, [b]zaireen.yahya@utp.edu.my

Keywords: Ionic Liquids, Nafion™, Ionic Conductivity, Compatibility, COSMO-RS

Abstract. Proton exchange membrane (PEM) electrolysis is one of the waters splitting techniques available for producing green hydrogen. As such, improvement of the membrane ion conductivity will result in improvement of hydrogen production. Ionic liquids have recently been reported to enhance ionic conductivity of PEM. Herein, a screening method to select suitable ionic liquids for the development of efficient proton exchange membrane. COnductor-like Screening MOdel for Realistic Solvents (COSMO-RS) was used to predict the ionic conductivity as well as the compatibility of the ions with the Nafion™ through the interpretation of σ-profile as well as interaction energy of the selected cations and anions. It was found that the anions namely of trifluoromethanesulfonate and nitrate with the cation of ammonium and imidazolium may be the best candidate for the ILs to be incorporated to Nafion™ for polymer electrolyte membrane (PEM) as the combination gives high ionic conductivity with considerable high interaction towards Nafion™. It is to be highlighted that the ionic liquids mainly interact with Nafion™ through the anion as implied by the high interaction energy of the anion towards Nafion™ compared to the cation.

Introduction

It is recorded in the Global Energy Statistical Yearbook 2022 that the domestic consumption of electricity in the Asia region for 2021 is 12,165 terawatt-hours (TWh)[1] pinpointing the demand for electricity that keeps on rising over the years. At present, coal remains the main source to generate electricity globally. Not only it is a non-renewable source, the usage of coal to generate electricity has contributed to greenhouse gas production. Due to this, it is of paramount importance to make use of a better option which is a renewable source. Hydrogen source for power generation particularly green hydrogen is an attractive choice of source as no greenhouse gas (GHG) is being released and with no carbon footprint as the only byproduct that is released through the process is water vapor [2].

Proton exchange membrane (PEM) electrolysis is one of the waters splitting techniques available for producing green hydrogen. The advantage of the techniques lies in the ease of fabrication where the PEM electrolyzer can be built according to the desired system. The most commonly used membrane for hydrogen production is perfluorosulfonic acid (PFSA) polymer membrane particularly Nafion™ due to its good ionic conductivity and excellent physicochemical properties [2, 3]. The former properties are vital as a high ionic conductivity is required to have efficient hydrogen production. The inclusion of ionic liquids (ILs) into the membrane is also another way to achieve this. The presence of ILs is reported to improve the proton conductivity of the membrane. The ionic liquids composed of cations such as ammonium, imidazolium, and triazolium with a variety of anions such as tetrafluoroborate, chloride, trifluoromethanesulfonate, formate and nitrate have been reported to improve the conductivity [4-14].

As there are up to at least a million possible of ionic liquids due to various combination of cations and anions available, a prior prediction must be done to look for suitable ILs to be included in the membrane. COSMO-RS is a versatile tool for predicting the thermophysical properties of pure or mixed fluids. COSMO-RS which is based on quantum chemical calculations using the molecular structure and confirmation information can be used to gain a better understanding of the behavior of a molecule. In this study, COSMO-RS was used to predict the ionic conductivity as well as compatibility of ILs with the Nafion™ through the interpretation of σ-profile as well as interaction energy prediction of the selected cations and anions.

Methods

Structure optimization generation of the σ-profile, of the cations, anions, and Nafion™ structure

The method to optimize the structure in this study is adapted from [15-22]. COSMO-RS was used to determine the σ-profile of a molecule by performing quantum chemical calculations using molecular structure and confirmation information. The sigma (σ) profile can be defined as a probability of surface charge distribution on a molecular surface. To generate the σ-profile and σ-surface, T-Molex version 21.0.0 was used to optimize the structure of cations and anions at the DFT level. Firstly, the 3D molecular structure of the target molecule was built. The step was then followed by the geometry optimization through DFT calculation using parameterization of B3LYP with triple –ξ valence polarized basis set (TZVP). The optimized structures were then exported to COSMOthermX version 19.0.4 where the σ-profiles and σ-surface of the structures were generated.

COSMO-RS ionic conductivity prediction

The resulting optimized structures that were in form of COSMO file was used as the input in COSMO-RS to predict the properties of ionic conductivity of different combination of anions and cations at room temperature. The ionic conductivity (x=κ) is determined by equation (1):

$$\ln (x(T)x_0) = d^* + e^*(\ln(r^*_m)) + f^* (\ln/\sigma)) + E_{diel} / E_{diel0} + g^* (E/(RT)) + i^* (E_{diel}(T_0))/(R^*T^2) + d \quad (1)$$

where the field without the dielectric, $E_{diel\,0}$ = 1 kJ/mol, initial temperature, T_0 = 1 K. The term r^*_m is molecular radius, σ is symmetry number and R is gas constant. For ionic conductivity the regression parameter values are d= 8.784179, e= -15.668627, f= -1.141751, g= 0.025274, h= -0.054461 and i= 19.107936. The parameter values were obtained through the correlation of the experimental data to develop the model.

Estimation of the interaction energy of ions pairs of ionic liquids towards Nafion™

COSMO-RS was further used to estimate the ion interaction energy towards Nafion™ in which the Nafion™ and the ionic liquid molecule were treated as an equimolar mixture. To simplify quantum chemical calculation of the Nafion™ polymer, a dimer structure of two repeating units is used to represent the Nafion™ in this study as shown in Figure 1. The interaction energy namely misfit (E_{MF}), hydrogen-bond (E_{HB}), and van der Waals (E_{vdW}) were quantified using Equations (2), (3), and (4) respectively. Equation (5) on the other hand is the total interaction energy (E_{INT}).

$$E_{HB} = \alpha_{eff} c_{HB}\, min(0; \sigma_{donor} + \sigma_{HB}) \times max\,(0; \sigma_{acceptor} + \sigma_{HB}) \quad (2)$$

$$E_{MF} = \alpha_{eff} \frac{\alpha}{2} (\sigma + \sigma')^2 \quad (3)$$

$$E_{vdW} = \alpha_{eff} (\tau_{vdW} + \tau_{vdW}') \quad (4)$$

$$E_{INT} = E_{MF} + E_{HB} + E_{vdW} \quad (5)$$

Where α_{eff}, c_{HB} and σ_{HB} denote the effective contact area, the degree of hydrogen bond, and the lower limit of hydrogen bond respectively; σ_{donor}, and $\sigma_{acceptor}$ are the screening charge densities of hydrogen bond donor and acceptor, respectively; σ and σ' represent the screening charge densities of two different parts; τ_{vdW} and τ_{vdW}' are the specific van der Waals interaction parameter for each element.

Results and Discussions

COSMO-RS is a model to screen ionic liquids with specific properties. It can predict physicochemical properties such as viscosity and ionic conductivity henceforth helping the selection process of potential ionic liquids to be done effectively. This allows to leverage in terms of cost as it will avoid unnecessary synthesis reactions taking part as the properties of the ionic liquids can be predicted beforehand. Nevertheless, it should be bear in mind that, COSMO-RS merely serves as a tool to give a prediction of the physicochemical properties of an ionic liquid. The data predicted served as a guideline to narrow down the overwhelming number of simple ionic liquids. Hence, it is necessary for the prediction to be followed by experimental work [15-19].

The key properties for an ideal IL to be incorporated in the membrane for PEM is its ionic conductivity [4, 8]. The selection of the cation and anion was based on the previously incorporated ILs into the membrane of the Nafion™. In this study, the combination of the cations of imidazolium, boronium, ammonium, and phosphonium (Figure 2) with the anion of nitrate, tetrafluoroborate, formate, and trifluoromethylsulfonate (Figure 3) were predicted. The conductivities of the different combinations of anions and cations are summarized in Figure 4. Apart from that, the length of the alkyl chain of the imidazolium was also varied.

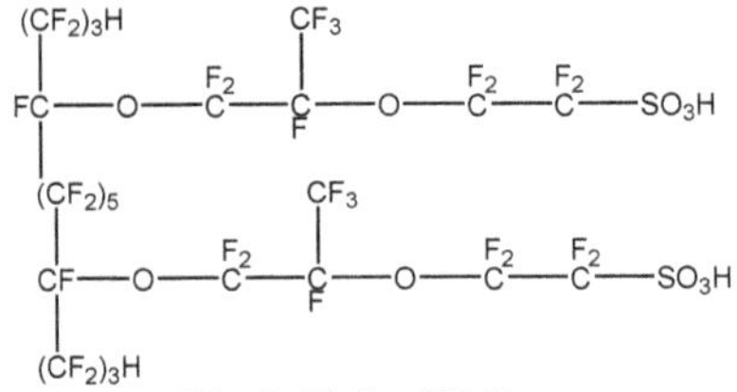

Fig. 1. Nafion™ dimer

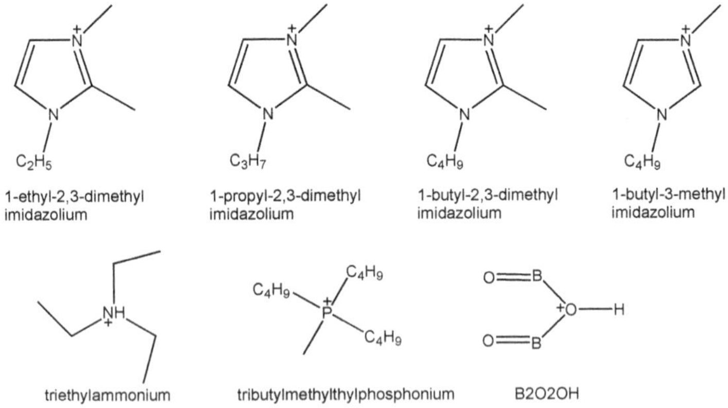

Fig. 2. Different cations structures of the ILs used in this study

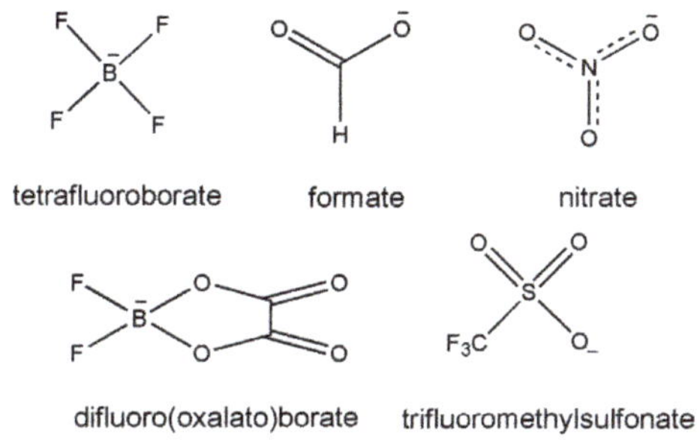

Fig. 3. Different anions structures of the ILs used in this study.

Based on the predicted conductivities, the variation of the anion has a greater effect on the ionic conductivity of the resulting ILs [18, 23]. The combination of nitrate anion followed by trifluoromethanesulfonate with all cations has resulted in the highest conductivity compared to other anions. This is in accord with the study reported in [12] where the incorporation of nitrate-based ILs into the Nafion™ membrane has resulted in a greater conductivity of the membrane compared to the usage of formate-based ILs. The variation in length of the alkyl chain attached to the imidazolium-based showed that the longer alkyl chain caused the ionic conductivity to be lower as it induced greater intermolecular attractions and impedes ion mobility as reported in [18, 23-26]. The phosphonium and boron-based ILs showed significantly lower conductivities compared to the imidazolium and ammonium-based ILs. This explains the prominent usage of imidazolium and ammonium cation-based ILs to improve the conductivity of the proton exchange membrane such as Nafion™ as reported in quite a significant number of studies [4-9, 11-13, 27-30].

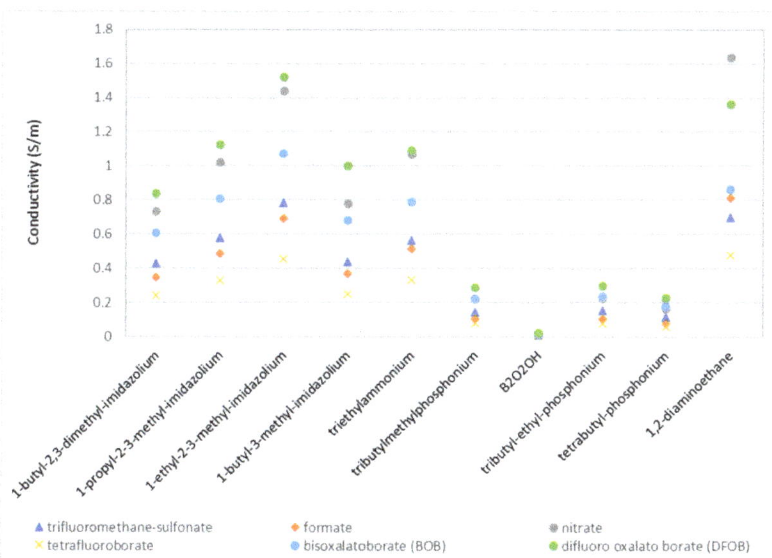

Fig. 4. Predicted conductivities of the cations and anions included in the study

Materials Research Proceedings 29 (2023) 446-455 https://doi.org/10.21741/9781644902516-51

The incorporation of σ-profile probability distribution in this study aids in explaining the compatibility of the cations and anions towards Nafion™. There are three different regions in a graph of σ-profile distribution that represents the nature of a molecule namely hydrogen bond donor (HBD), hydrogen bond acceptor (HBA), and nonpolar (NP) region [15-18]. The σ-profile probability distribution of Nafion™ is mostly nonpolar and with a very slight hydrogen bond acceptor in nature. This is attributable to the repeating fluorinated backbone unit (Figure 5). The σ-profile of the anion showed the compatibility of the anion namely nitrate, formate and triflate, with Nafion™. All three have σ-profile probability distribution in the non-polar region. The non-polar molecule has great affinity towards the non-polar molecule [17, 18] suggesting the compatibility of the anions with the polymer. Moreover, the slight distribution of σ-profile of Nafion™ in the hydrogen bond donor region may add up to the affinity of the three anions namely nitrate, formate and triflate, towards Nafion™ that are mostly of HBA in nature.

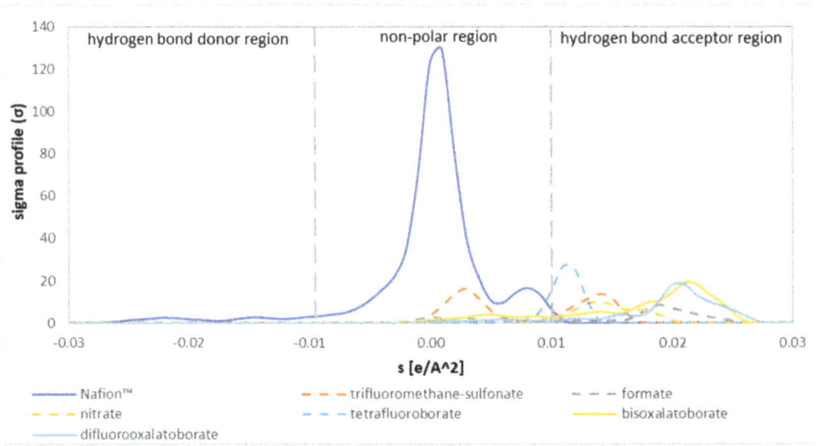

Fig. 5. Sigma profile distributions of Nafion™ and anions included in the study

As for the cation, the cation boronium and phosphonium may have a great affinity towards the Nafion™ structure as their σ-profile distributions are mostly in HBA in nature for the former and partially HBA and non-polar in nature for the latter (Figure 6). On the other hand, the cation, imidazolium, and ammonium may interact to the Nafion™ molecule through their nonpolar area of the cations as their σ-profile distributions mainly are in the nonpolar region.

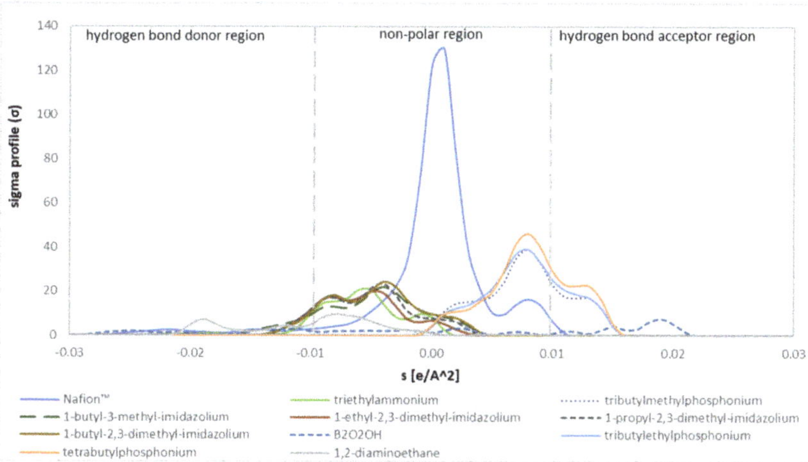

Fig. 6. Sigma profile distributions of Nafion™ and cations included in the study

The COSMO-RS was further used to predict the interaction energy of the ionic liquids towards Nafion™. The effect of the anion and cation was individually assessed. In the decreasing order, the interaction energy of the anions (kcal/mol) is formate > nitrate > trifluoromethane-sulfonate > tetrafluoroborate (Figure 7). If we correlate the data with the σ-profile, the nature of the formate anion that has the greatest HBA character among the anions included in this present study has caused the interaction energy of formate towards Nafion™ to be the highest. This also suggested the mode of interaction of the ILs towards the Nafion™ dimer is through the anion. Although the distribution of the σ-profile of the Nafion™ molecule is highly concentrated at the non-polar region attributable to the fluorinated backbone but the main mode of the interaction between the anion and the molecule is through the acidic area of the molecule that is the sulphonic acid group.

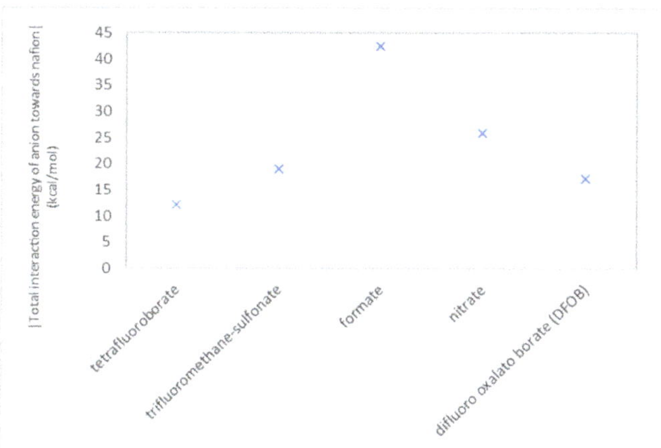

Fig. 7. Interaction energies prediction of anions included in the study towards Nafion™

This is further supported by the interaction energy of the cations. The boron-based cation that has the highest affinity towards the Nafion™ has the most prominent distribution in the HBA region pinpointed the main interaction of the cation towards the Nafion™ molecule is through the HBD area (Figure 8). As such, the interaction of phosphonium towards the Nafion™ is the second highest after boronium as the distribution of phosphonium cation lies partially in HBA and non-polar. The cation of imidazolium and ammonium showed slightly lower interaction energy compared to the phosphonium cation in which the mode of interaction of the cations is through the alkyl chain of the cations that contributed to the non-polar nature.

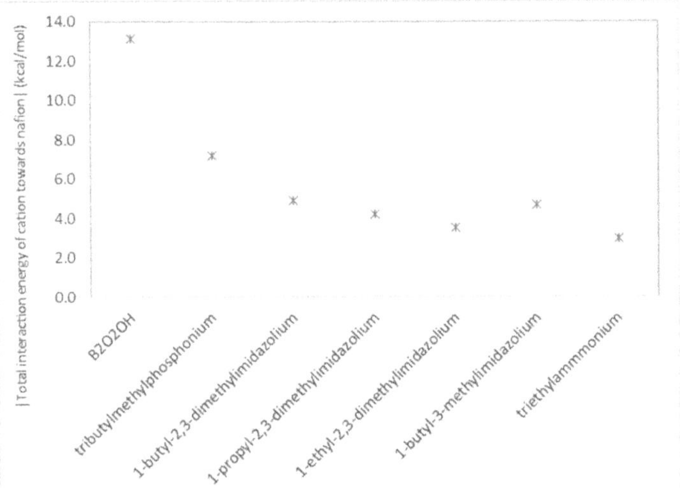

Fig. 8. Interaction energies prediction of cations included in the study towards Nafion™

Conclusion

The combination of imidazolium and nitrate anion gives out the best conductivity followed by the combination of the same cation with trifluoromethanesulfonate. The σ-profile showed the non-polar nature of the Nafion™ dimer. The imidazolium and ammonium cation lies mostly in the non-polar region whilst boronium and phosphonium have some distributions in the hydrogen bond acceptor region. The interaction energy suggested that the main interaction of the ionic liquids is through the anion as the anion has significantly high interaction energy towards the Nafion™ compared to the cation. Furthermore, as the anions are hydrogen bond acceptor in nature and has high affinity towards hydrogen bond donor, this revealed that the main interaction mode of the Nafion™ is through the sulfonic group that contributed to the HBD nature of the polymer. Taking into consideration the prediction through COSMO-RS, the anion of trifluoromethane sulfonate, nitrate with the cation of ammonium and imidazolium may be the best candidate for the ILs to be incorporated into Nafion™ for polymer exchange membrane (PEM) as the combination gives considerably high ionic conductivity with considerable high interaction towards Nafion™. This result may act as the basis for the selection of suitable anions and cations to be incorporated into Nafion™ experimentally.

Acknowledgement

This research was funded by Petroleum Research Fund (PRF) (015MD0-132). The authors also acknowledge technical support and facilities from the Chemical Engineering Department and Centre of Research in Ionic Liquids of Universiti Teknologi PETRONAS.

References

[1] Enerdata. "Electricity domestic consumption." https://yearbook.enerdata.net/electricity/electricity-domestic-consumption-data.html (accessed 10 February, 2020).

[2] M. F. Ahmad Kamaroddin et al., "Membrane-based electrolysis for hydrogen production: A review," Membranes, vol. 11, no. 11, p. 810, 2021. https://doi.org/10.3390/membranes11110810

[3] P. Millet, "Hydrogen production by polymer electrolyte membrane water electrolysis," in Compendium of hydrogen energy: Elsevier, 2015, pp. 255-286. https://doi.org/10.1016/B978-1-78242-361-4.00009-1

[4] L. Zanchet et al., "3-Triethylammonium propane sulfonate ionic liquids for Nafion-based composite membranes for PEM fuel cells," Journal of Materials Science, vol. 55, no. 16, pp. 6928-6941, 2020. https://doi.org/10.1007/s10853-020-04454-4

[5] F. Lu et al., "Preparation and characterization of nonaqueous proton-conducting membranes with protic ionic liquids," ACS Applied Materials & Interfaces, vol. 5, no. 15, pp. 7626-7632, 2013. https://doi.org/10.1021/am401940y

[6] M. Doyle, S. K. Choi, and G. Proulx, "High-temperature proton conducting membranes based on perfluorinated ionomer membrane-ionic liquid composites," Journal of the Electrochemical Society, vol. 147, no. 1, p. 34, 2000. https://doi.org/10.1149/1.1393153

[7] V. Di Noto, E. Negro, J.-Y. Sanchez, and C. Iojoiu, "Structure-relaxation interplay of a new nanostructured membrane based on tetraethylammonium trifluoromethanesulfonate ionic liquid and neutralized nafion 117 for high-temperature fuel cells," Journal of the American Chemical Society, vol. 132, no. 7, pp. 2183-2195, 2010. https://doi.org/10.1021/ja906975z

[8] R. Sood, C. Iojoiu, E. Espuche, F. Gouanvé, H. Mendil-Jakani, and S. Lyonnard, "Influence of different perfluorinated anion based Ionic liquids on the intrinsic properties of Nafion®," Journal of membrane science, vol. 495, pp. 445-456, 2015. https://doi.org/10.1016/j.memsci.2015.07.006

[9] Y. Li et al., "More sustainable electricity generation in hot and dry fuel cells with a novel hybrid membrane of Nafion/nano-silica/hydroxyl ionic liquid," Applied Energy, vol. 175, pp. 451-458, 2016. https://doi.org/10.1016/j.apenergy.2016.03.075

[10] S. Sekhon, J.-S. Park, E. Cho, Y.-G. Yoon, C.-S. Kim, and W.-Y. Lee, "Morphology studies of high temperature proton conducting membranes containing hydrophilic/hydrophobic ionic liquids," Macromolecules, vol. 42, no. 6, pp. 2054-2062, 2009. https://doi.org/10.1021/ma8027112

[11] C. Schmidt, T. Glück, and G. Schmidt-Naake, "Modification of Nafion membranes by impregnation with ionic liquids," Chemical Engineering & Technology: Industrial Chemistry-Plant Equipment-Process Engineering-Biotechnology, vol. 31, no. 1, pp. 13-22, 2008. https://doi.org/10.1002/ceat.200700054

[12] J. T. E. Goh, A. R. Abdul Rahim, M. S. Masdar, and L. K. Shyuan, "Enhanced performance of polymer electrolyte membranes via modification with ionic liquids for fuel cell applications," Membranes, vol. 11, no. 6, p. 395, 2021. https://doi.org/10.3390/membranes11060395

[13] Y. Zhang, J. Li, L. Ma, W. Cai, and H. Cheng, "Recent developments on alternative proton exchange membranes: strategies for systematic performance improvement," Energy Technology, vol. 3, no. 7, pp. 675-691, 2015. https://doi.org/10.1002/ente.201500028

[14] M. Martinez et al., "Proton-conducting ionic liquid-based Proton Exchange Membrane Fuel Cell membranes: The key role of ionomer-ionic liquid interaction," Journal of Power Sources, vol. 195, no. 18, pp. 5829-5839, 2010. https://doi.org/10.1016/j.jpowsour.2010.01.036

[15] A. Klamt, "Conductor-like screening model for real solvents: a new approach to the quantitative calculation of solvation phenomena," The Journal of Physical Chemistry, vol. 99, no. 7, pp. 2224-2235, 1995. https://doi.org/10.1021/j100007a062

[16] A. Klamt and F. Eckert, "COSMO-RS: a novel and efficient method for the a priori prediction of thermophysical data of liquids," Fluid Phase Equilibria, vol. 172, no. 1, pp. 43-72, 2000. https://doi.org/10.1016/S0378-3812(00)00357-5

[17] N. I. M. F. Hilmy, W. Z. N. Yahya, and K. A. Kurnia, "Eutectic ionic liquids as potential electrolytes in dye-sensitized solar cells: Physicochemical and conductivity studies," Journal of Molecular Liquids, vol. 320, p. 114381, 2020. https://doi.org/10.1016/j.molliq.2020.114381

[18] R. A. A. Talip, W. Z. N. Yahya, and M. A. Bustam, "Understanding the physicochemical and transport properties of pyrazolium based ionic liquids bearing iodide and triiodide anions," Journal of Molecular Liquids, vol. 346, p. 118270, 2022. https://doi.org/10.1016/j.molliq.2021.118270

[19] Z. K. Koi, W. Z. N. Yahya, R. A. A. Talip, and K. A. Kurnia, "Prediction of the viscosity of imidazolium-based ionic liquids at different temperatures using the quantitative structure property relationship approach," New Journal of Chemistry, vol. 43, no. 41, pp. 16207-16217, 2019. https://doi.org/10.1039/C9NJ03436F

[20] M. R. Islam and C.-C. Chen, "COSMO-SAC sigma profile generation with conceptual segment concept," Industrial & Engineering Chemistry Research, vol. 54, no. 16, pp. 4441-4454, 2015. https://doi.org/10.1021/ie503829b

[21] M. S. Khan, C. S. Liew, K. A. Kurnia, B. Cornelius, and B. Lal, "Application of COSMO-RS in investigating ionic liquid as thermodynamic hydrate inhibitor for methane hydrate," Procedia engineering, vol. 148, pp. 862-869, 2016. https://doi.org/10.1016/j.proeng.2016.06.452

[22] T. Aissaoui, Y. Benguerba, and I. M. AlNashef, "Theoretical investigation on the microstructure of triethylene glycol based deep eutectic solvents: COSMO-RS and TURBOMOLE prediction," Journal of Molecular Structure, vol. 1141, pp. 451-456, 2017. https://doi.org/10.1016/j.molstruc.2017.04.009

[23] R. A. A. Talip, W. Z. N. Yahya, and M. A. Bustam, "Viscosity and Ionic Conductivity of Imidazolium based Ionic Liquids bearing Triiodide Anion," in E3S Web of Conferences, 2021, vol. 287: EDP Sciences, p. 02015. https://doi.org/10.1051/e3sconf/202128702015

[24] J.-f. Liu, G.-b. Jiang, and J. Å. Jönsson, "Application of ionic liquids in analytical chemistry," TrAC Trends in Analytical Chemistry, vol. 24, no. 1, pp. 20-27, 2005. https://doi.org/10.1016/j.trac.2004.09.005

[25] W.-L. Yuan, X. Yang, L. He, Y. Xue, S. Qin, and G.-H. Tao, "Viscosity, conductivity, and electrochemical property of dicyanamide ionic liquids," Frontiers in chemistry, vol. 6, p. 59, 2018. https://doi.org/10.3389/fchem.2018.00059

[26] R. Khalil, N. Chaabene, M. Azar, I. B. Malham, and M. Turmine, "Effect of the chain lengthening on transport properties of imidazolium-based ionic liquids," Fluid Phase Equilibria, vol. 503, p. 112316, 2020. https://doi.org/10.1016/j.fluid.2019.112316

[27] J. Hao et al., "Development of proton-conducting membrane based on incorporating a proton conductor 1, 2, 4-triazolium methanesulfonate into the Nafion membrane," Journal of Energy Chemistry, vol. 24, no. 2, pp. 199-206, 2015. https://doi.org/10.1016/S2095-4956(15)60301-1

[28] A. K. Mishra, T. Kuila, D.-Y. Kim, N. H. Kim, and J. H. Lee, "Protic ionic liquid-functionalized mesoporous silica-based hybrid membranes for proton exchange membrane fuel cells," Journal of Materials Chemistry, vol. 22, no. 46, pp. 24366-24372, 2012. https://doi.org/10.1039/c2jm33288d

[29] M. D. Bennett, Electromechanical transduction in ionic liquid-swollen Nafion™ membranes. Virginia Polytechnic Institute and State University, 2005.

[30] A. P. Sunda, "Ammonium-based protic ionic liquid doped Nafion membranes as anhydrous fuel cell electrolytes," Journal of Materials Chemistry A, vol. 3, no. 24, pp. 12905-12912, 2015. https://doi.org/10.1039/C5TA02315G

Detection and characterization of chloride induced stress corrosion cracking on SS304 under perlite thermal insulation

SURESH Nuthalapati[1, a] K.E. Kee[1,b*], SRINIVASARAO Pedapati[1,c]

Department of Mechanical Engineering, Universiti Teknologi PETRONAS, Malaysia

suresh_20001751@utp.edu.my[a], keekokeng@utp.edu.my[b], srinivasa.pedapati@utp.edu.my[c]

Keywords: SS304, Sensitization, Grain Size, Perlite Thermal Insulation, Pitting Corrosion, Crack Evaluation

Abstract. Austenitic stainless steel (ASS) type SS304 has been extensively employed in various sectors because of its good structural strength and corrosion resistance. These steels are susceptible to chloride-induced stress corrosion cracking (CISCC), particularly by the effect of chloride corrosive environment under thermal insulation. Corrosive environment formed by external sources or condensation due temperature differences under thermal insulation as a result of inadequate maintenance or unfavourable climatic conditions. It causes localised corrosion and leads to catastrophic failure under the action of tensile stress. The objective of the research was to detect the chloride-induced stress corrosion cracking on SS304 under thermal insulation. This study simulated a real corrosive environment using water containing chloride ions on U-bend samples to estimate the susceptibility and assess CISCC under the drip test method. SS304 as-received (AR) and sensitized (SEN) over-stressed U-bend samples were tested as per ASTM C692 standard. Samples were exposed to 0.1, 1.0, and 3.5 wt. % of NaCl concentrations under perlite thermal insulation and tested at 90°C. Under high chloride concentrations, SEN samples were susceptible more and showed little evidence of crack initiation. The rest of the concentrations showed no evidence of crack, but they showed tiny localized corrosion near the dripping zone. The characteristics of the material structure and the corrosion mechanism were described in a pictorial view.

Introduction

Austenitic stainless steels (ASSs) are widely utilized in various industries such as oil, gas, petrochemical, food, nuclear etc. because of their remarkable corrosion resistance and favourable mechanical properties [1], [2]. However, ASSs are more susceptible to CISCC in a chloride-containing environment. This environment can damage the passive film, resulting in pitting and leads to cracking under mechanical loading [3]. Generally, the stress corrosion cracking (SCC) failures induced by presenting chlorides from external sources such as rain, coastal fog, leaks etc., and thermal insulation are likely to develop. These are the major sources to form chlorides on the metal surface to initiate cracks [4]. Sodium chloride (NaCl) is the most frequent offender due to its high solubility and extensive distribution. This is most common but not much aggressive. Under thermal insulation, the concentration of chloride deposits on the metal surface is still controversial. But it is mostly affected by the temperature transformations on the metal surface. Although it was difficult to estimate the necessary chloride concentration for SCC, but it was observed remarkably low levels of chloride (10ppm) in the solution [5].

Thermal insulation is essential for economical and safety concerns in various industries. The continuous use of insulated components like pipes or pressure vessels requires an assessment to define their damages. Predominantly those industries located close to unfavourable environments, ASSs might be more susceptible to corrosion such as pitting, stress corrosion cracking, crevice corrosion and intergranular attack [6]. Furthermore, CISCC occurs due to the generation of moisture under thermal insulation which means the interface between insulation and the surface of pipe or equipment. There is a temperature differential that leads to condensation when the material surface temperature below the atmospheric dew point. The condensed chloride ions with other dissolved particles were unfree in the annular space. Then, over time the metal surface will corrode, particularly for surfaces without coating [7].

Under the insulation, CISCC is possible in the presence of tensile stress when the chlorides or other contaminants are present in the range of 50°C-150°C temperature [8]. No cracking will take place if the tensile stress is removed or significantly decreased. The majority of CISCC cracks are quite small and exhibit considerable crack branching in a transgranular path that is mostly uninterrupted by grain boundaries. The possible cracks propagate along the grain boundary indicating that the steel is in a sensitised state when loses its integrity [9]. The ASSs becomes sensitive as a result of sensitization or heat-affected area from the perspective of the welding or heat treatment applications. When these applications allow holding or decreasing the temperature range between 900°C and 480°C, leads to chromium depletion at the grain boundary due to the formation of chromium carbides [10], [11]. It could be affected to increase the microhardness slightly as well grain growth [12].

Perlite thermal insulation has been used in various insulation applications because of its unique properties, and it is non-flexible granular type material [13]. In a recent experimental work under perlite thermal insulation, SS304L did not show any evidence of CISCC [14]. Despite numerous studies done on the field of SCC under thermal insulation, there are still many issues and difficulties. SCC under thermal insulation remained one of the biggest challenges in various industries [15]. Many researchers focused on SCC detection and its mechanism without thermal insulation, only few worked on insulated components. As result, there is no clear understanding of the CISCC vulnerability, mechanism and characteristics. SCC mechanism has been assessed in different testing methods with various corrosive solutions [16]. The studies reported that static tensile stress combined with active corrosion causes fractures, which eventually lead to catastrophic damage to the sensitive component [17], [18]. These unexpected failures cause dangerous explosions leads to economic loss for the oil, gas and other chemical industries. According to a 2016 NACE international study, the global cost of corrosion was expected to be 2.5 trillion dollars, or 3.4 percent of the world's GDP in 2013 [19]. Consequently, periodical inspection is required to detect the damages or fractures under the insulation so as can forecast the potential hazards.

In the present study, the effect of the combined action of chloride concentration and tensile stress on the SCC susceptibility has been studied systematically for SS304 AR and SEN U-bend samples using drip test apparatus simulating the corrosive environment. Microstructures of different test samples were also studied to reveal its damage mechanism.

Material and Methods

Sample Preparation:

AR and SEN (SS304) samples were used for drip test with a hot fluid circulating pipe-spool. Based on the pipe-spool external diameter, samples were designed and cut as dimensions of 3^{t}x 25^{w}x 220^{l} in millimeters from the long bar. AR flat samples were heat treated by soaking in muffle furnace for 3 hours at constant temperature of 649°C and cooled in furnace itself as per ASTM C 692 standard [20]. Both samples were prepared for their microstructure followed by grinding-polishing and electrolytic etching according to ASTM E3 [21] and ASTM A262 [22] respectively. The composition elements of SS304 AR and SEN samples were presented in Table 1 as identified by Hitachi table top SEM (TM3000) with energy dispersive X-ray (EDX) spectroscopy (Bruker Quantax 70) system according to ASTM E572 [23]. For the tensile test used a sub-sized flat specimen with 3mm thickness as per ASTM E08M [24]. Also, Vickers microhardness mechanical test was performed on 3 mm thickness of 30x30 mm square plate using Leco microhardness tester (LM 247AT) according to the ASTM E384 standard [25]. The hardness tester was equipped with diamond pyramidal indenter with 1Kgf load at maximum. The mechanical properties of AR and SEN samples are presented in Table 2 in accordance with ASTM A240 [26].

Table 1: Chemical compositions of SS304 As received and sensitized

Elements	As received	ASTM A240	Sensitized	ASTM A240
C	0.091 ± 0.002	>0.08	0.12 ± 0.003	>0.08
Mn	1.83 ± 0.1	<2.00	1.94 ± 0.1	<2.00
P	0.058 ± 0.03	>0.045	0.037 ± 0.05	<0.045
S	0.036 ± 0.02	>0.030	0.042 ± 0.02	>0.030
Si	0.48 ± 0.05	<0.75	1.09 ± 0.04	>0.75
Cr	18.23 ± 0.2	18.00-20.00	18.92 ± 0.1	18.00-20.00
Ni	8.45 ± 0.04	8.00-10.50	8.90 ± 0.06	8.00-10.50
N	0.16 ± 0.02	>0.10	0.08 ± 0.01	<0.10

Table 2: Mechanical properties of SS304 AR and SEN samples

Mechanical Properties	As-received	ASTM A240	Sensitized	ASTM A240
Yield strength (MPa)	363.38	>205	346.26	>205
Tensile strength (MPa)	687.51	>515	681.05	>515
Micro-Hardness	206.3HB (217.2HV)	>201HB (HV≈1.05HB)	215.1HB (226.5HV)	>201HB (HV≈1.05HB)

The test samples were prepared according to the ASTM G30 [27]. Geometrical dimensions were depicted in a graphical representation of mechanical loading assembly using two-stage method to make U-bend as shown in Figure 1 (a). A stressed U-bend accomplished by allowing the elastic strain to relax at the first stage and followed by the second stage through tightening nut and bolt until legs held in parallel with each other. And a decent approximation of the strain (ε) was derived on an investigated area of U-bend using $\varepsilon = t/2R$, where, t is thickness and R is radius of the sample. Then, an appropriate value of the maximum stress was determined from the stress-strain curve of the test material. Furthermore, calculated the leg deflection as shown in Figure 1 (b) to produce the desired elastic stress using equation 1 [28].

$$\text{Deflection } (\delta) = \left[\frac{12S(2R+t)}{(L+R)\,(8R+t)(t)(E)}\right]\cdot\left[\frac{L^3}{3}+R\left(\frac{\pi L^2}{2}+\frac{\pi R^2}{4}+2LR\right)\right] \ldots\ldots\ldots (1)$$

Where δ is the leg deflection (in), S is the applied stress (psi), E is the modulus of elasticity (psi), R is the radius of bend (in), t is the thickness of sample (in) and L is the length of the straight section (in).

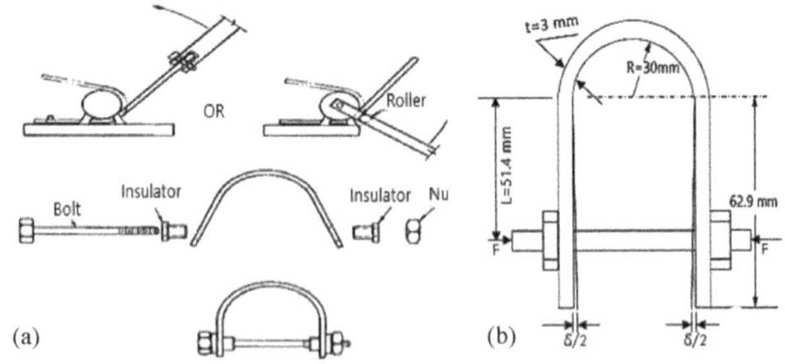

Figure 1: (a) Schematic view of a two-stage method for U-bend making [27], (b) Schematic view of leg deflection

Equipment setup:

The drip test apparatus consists of pipe-spool, oil bath circulating tank with temperature control setup, solution tank, peristaltic pump, drippers and temperature sensors for temperature readings. AR and SEN samples were placed over the two inch diameter pipe-spool as shown in Figure 2 according to the ASTM C692 [20]. The samples with pipe-spool heated by circulating silicone oil which was heated at 90°C in an oil bath circulating tank. The pipe-spool inlet and outlet temperatures were observed through the given indicators. The perlite thermal insulation blocks as per ASTM C610 [29], selected and prepared according to the ASTM C585 standards [30]. It has a good physical properties such as density (50-150 Kg/m^3), thermal conductivity (0.057W/mK at 80 Kg/m^3), service temperature (-250-1000°C) and thickness (25-300 mm) at 20°C [31]. The prepared blocks were enclosed over the samples along the pipe-spool.

Sodium chloride (NaCl) solution was prepared in weight percentage as 0.1, 1.0, 3.5 with deionized water according to ASTM D1193 standard [32]. The prepared solution dripped over the insulated material by peristaltic pump which was calibrated for 21 ml per two hours. Three AR and three SEN insulated samples were tested in one attempt under three different NaCl concentrations. The test matrix as shown in Table 3, was prepared in accordance with the aforesaid test requirements in order to execute the test.

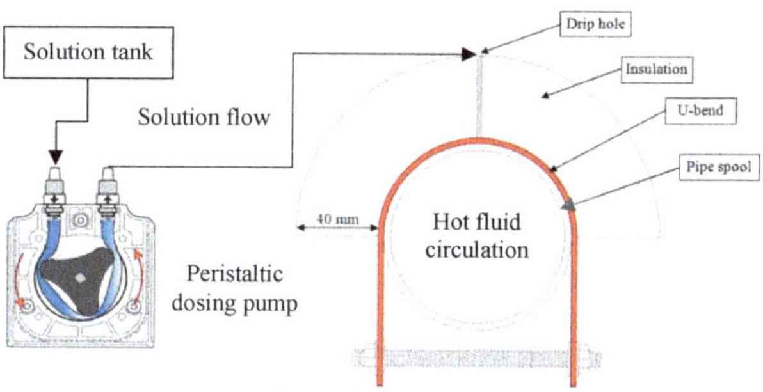

Figure 2: Line diagram of auto drip control system set up for SCC test

Materials Research Proceedings 29 (2023) 456-471 https://doi.org/10.21741/9781644902516-52

Table 3: Test process and its parameters

Test Duration	68 days					
Temperature	90 °C					
Insulation Type	Perlite					
Sample type	SS304 As-received			SS304 Sensitised		
NaCl concentration (wt.%)	3.5	1.0	0.1	3.5	1.0	0.1
Test No	AR1	AR2	AR3	SEN1	SEN2	SEN3
Evaluation	Visual, DPT, Stereo microscope, Optical microscope					

Finally, Aluminium sheet was used to cladding along the insulated pipe-spool and tied with binding wires evenly then run the test at 90°C. During the test maintained neutral pH by adding hydrochloric acid (HCl) reagent according to the ASTM G44 [33]. Each sample has been observed periodically by using die penetrant test (DPT) and stereo microscope. Following the observation, each sample was stressed by leg deflection and then sustained the test. After completion of the test, results were analysed by using an optical microscope with suitable magnifications followed by grinding-polishing and etching. Also, used TESCAN CLARA's FESEM (Field Emission Scanning Electron Microscopy) to characterize the material structure. In addition, Vickers hardness was used to analyse the micro hardness at the centre of an investigated area by applying 1kgf load to the given indenter. The results were compared to those from the preceding tests.

Result and Discussion

Formation of Localised Corrosion:

Table 4 summarizes the periodical inspection results for as-received (AR) and sensitized (SEN) insulated SS304 samples as per test conditions. Upon removal of perlite thermal insulation block for inspection, the sample surfaces were slightly moist around the dripping zone with stains and specks of salt deposits that built-up with time as shown in Figure 3 (a). All the U-bend samples tested at 90°C under perlite thermal insulation revealed brown rusting stains around the dripping zones except AR3 sample. For U-bend samples exposed to higher salinity i.e. 3.5 wt.% NaCl, colonies of pits were found distributed in the vicinity around the dripping zones as shown in Figure 3 (b). At the inspection time after 972 hours, only stains but no pits were observed for samples exposed to 0.1 and 1 wt% NaCl. Pitting was only found on samples under 3.5 wt% NaCl. It is well established that the presence of chloride can rupture the passive oxide film and lead to initiation of new pits [34]. At the end of 1632 hours, all samples showed varying levels of localized corrosion without cracks, except for SEN1 sample under 3.5 wt% NaCl. The severity of localized corrosion can be related to the concentration levels of NaCl and the type of samples, whereby higher NaCl concentration resulted in more severe localized corrosion. Also, the heat-sensitized samples showed more susceptibility than as-received sample to localized corrosion and stress corrosion cracking [28], [35].

Table 4: Summarized results under the test conditions

Temperature (°C)	90					
Insulation type	Perlite					
Sample Type	SS304 As-received (AR)			SS304 Sensitized (SEN)		
NaCl Concentration (%)	3.5	1.0	0.1	3.5	1.0	0.1
Test No	AR1	AR2	AR3	SEN1	SEN2	SEN3
After 72 h.	X	X	X	X	X	X
After 272±5 h.	Stains	X	X	Stains	Stains	X
After 472±5 h.	pits	Stains	X	pits	Stains with brown rust	Stains
After 672±5 h.	pits	Stains	X	pits	Stains with brown rust	Stains
After 972±5 h.	pits	Stains	X	pits	Stains with brown rust	Stains
After 1632± 5h.	pits	Stains with brown rust	Stains	SCC	Stains with brown rust	Stains with brown rust

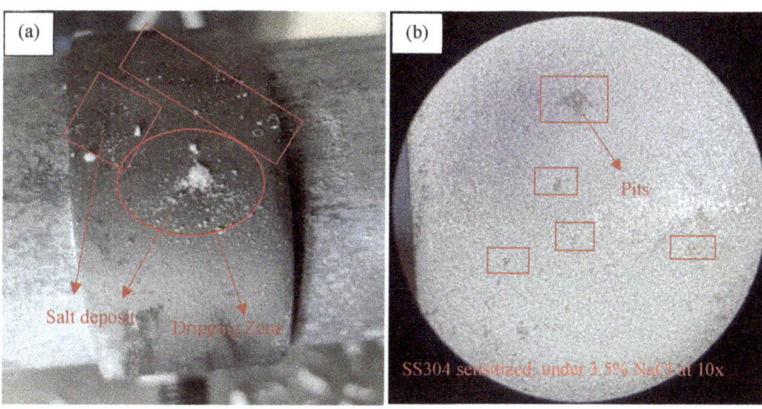

Figure 3: (a) Salt deposit observed onSEN1 sample, (b) Stereo microscope image of corroded sample

Figure 4 illustrates the progressive evolution, state of pitting for AR1 and SEN1 samples. These visuals focused on a specific pit from both samples and observed periodically. The pit area was observed to grow based on the direction of passive film dissolution around the long-standing pit. It is believed that the pit growth will eventually lead to crack formation as time progresses, as evidenced by SEN1 sample after 1632 hours as shown in Figure 4 (d). However, the AR1 sample did not produce any crack under the same test condition. This indicates the heat-sensitized SS304 is more susceptible to cracking under high chloride concentration.

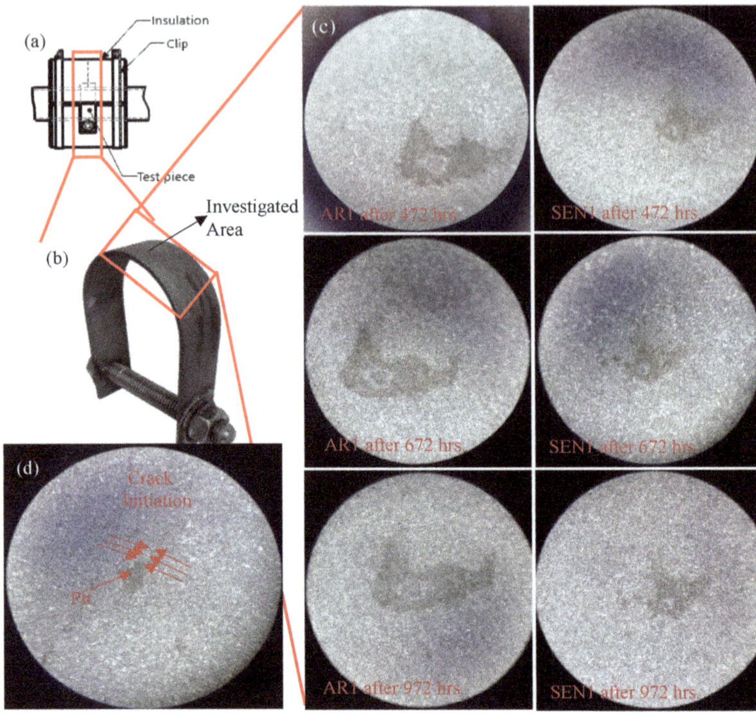

Figure 4: (a) Schematic view of U-bend covered by thermal insulation, (b) SS304 testing U-bend sample and (c) stereo microscopic images of SS304 as received (AR1) and sensitized (SEN1) samples pit variation under 3.5 wt% NaCl solution, (d) Crack initiation.

Vickers hardness observation:

Figure 5 compares the Vickers micro-hardness (HV) test results for both AR and SEN taken before test and after test. The measurement points were taken at center of U-bend apex area, and the HV reading were taken at 1mm intervals by applying 1 kg force in 10 seconds dwell period.

Before the test, the SEN sample had slightly higher average HV than AR, most likely due to carbide formation. These may impede dislocation movement after heat sensitization treatment at 649°C. Generally, the hardness depends on the types of heat treatment process and crystal lattice dislocation at grain boundaries in relation to flow stress in the Hall-Petch relation [36]–[39]. The average HV measured after test were found to be higher than before test by 21% and 30% for AR1 and SEN1, respectively. The rise of hardness may be attributed to the 'semi-quenching' state of the samples. In the present work, the samples were heated at constant temperature (90°C) and then quenched periodically by aqueous solution dripping to create periodic dry/wet cycle in the insulated sample. This condition would favours the formation of hard carbide compounds on the surface, thereby contributing to micro-hardness enhancement.

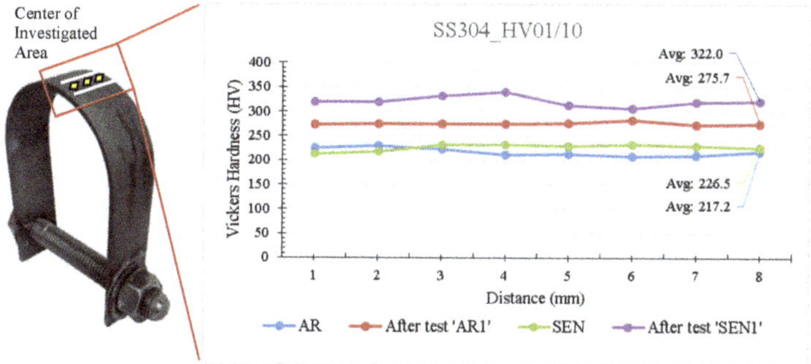

Figure 5: Micro-hardness (HV) comparison of before and after the test

Metallographic observation:

Metallography is a study of materials microstructure and can influence its physical properties, and employed in failure analysis to investigate the mechanism of corrosion and oxidation [40].

Optical microscopic observation:

Figure 6 (a), (b) represents the austenite phase with a homogeneous matrix of both samples before the test. Expected that the SEN sample shown chromium carbides (Cr_{23} C_6) in the grain boundaries, but does not completely encircle the grains. After the test, there is no much difference in structure of AR1 as shown in Figure 6 (c) but seems to be a grain size has been increased slightly. Figure 6 (d) shown, the SEN1 sample's grain boundary thickness has been increased, also observed small particles on grains supposed to be σ (Fe-Cr) particles. As of the literature, the σ-phase transition happens from the δ-ferrite, because it is a ferritic stabilizer which accumulates the chromium became rich Cr region. When there is no δ -ferrite present in the stainless steels, the σ -phase can potentially precipitate from γ –austenite [41].

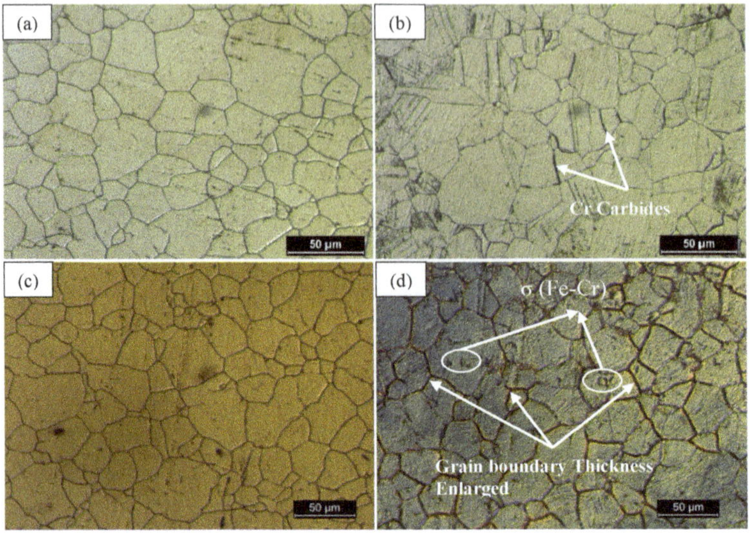

Figure 6: Photomicrographs of (a) AR, (b) SEN before the test, (c) AR1 and (d) SEN1 after the test (500x magnification)

Grain size is an important material characteristic, and it is denoted by ASTM 'G' number. For grain size, both samples can be determined by applying line intercept technique according to ASTM E112 [42]. Note that a large ASTM 'G' number indicates a smaller grain size and vice versa. Figure 7 summarizes the calculated ASTM 'G' number and average grain diameter (in µm) for four different sample conditions. Results shows that average grain diameter of SEN sample has slightly larger than AR sample, after the test also it was increased gradually. It can be assumed that the grain boundary migration happened if there is no delta-ferrite precipitation along the grain boundary. Under this circumstances, the previous grain boundary migrated quickly and formed new boundary as like double Y-shape, produced large grains [43]. Due to larger the grain size, the distance dislocation is longer that can take long time to move compared to shorter distance in case of small grains then the material loss its ductility. This can be assumed that, the austenite grain size is dependent on the volumetric density of nuclei induced by deformation [44]. Furthermore, while increasing the area of grain boundary it would be distribute more carbides along it and increases the sensitivity causes inter granular cracking corrosion when applied tensile stress [45], [46]. Due to overage of experimental work, occurred crack on SEN1 sample as shown in Figure 8 and it was a greater depth and intergranular. Finally, during the test the factors such as strain (0.05), temperature (90°C) in long range and periodically used chloride (3.5wt.%) drips (rapid quenching) attributed to enhance the sensitization gradually leads to metal deformation occurred.

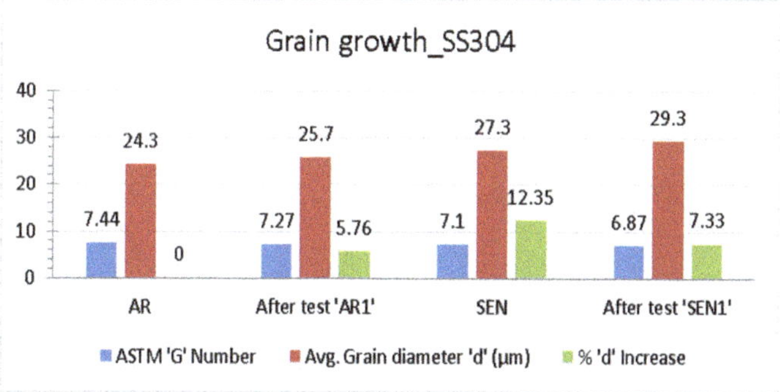

Figure 7: ASTM G number and average grain growth for as received and sensitized samples

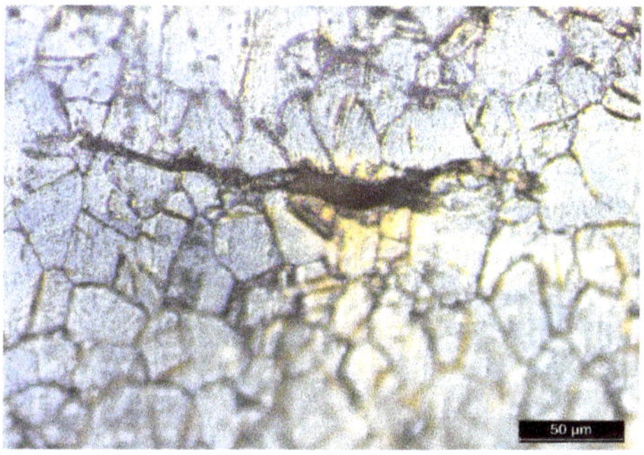

Figure 8: IGSCC on SS304 SEN1 sample under higher chloride concentration

FESEM observation:

For detailed investigation, the visuals taken by using FESEM for both SEN1 and AR1 shown in Figure 9. As the visuals characterized by seems to be an equiaxed grains with less carbide particles shown in Figure 9 (a), (d). Figure 9 (b), (e) representing brittle phase, there is almost dispersed the carbide particles along the grain boundary and it shown as a grooved grain structure due to the long range of experiment. Also observed the micro-fissures in the grain boundary as it might be developed by tensile stress and hence occurs the grain incoherence and it would supposed to intergranular cracking corrosion. In other side, Figure 9 (c) represents the flake carbide compounds along the grain boundary and the grain interior micro voids. Moreover, in case of tensile stress and temperature increased the flake carbides would disperse in the grain boundary. Moreover, as compared to SEN1 sample AR 1 structure has less plane defects (might be voids or stacking fault) with smooth grain surface.

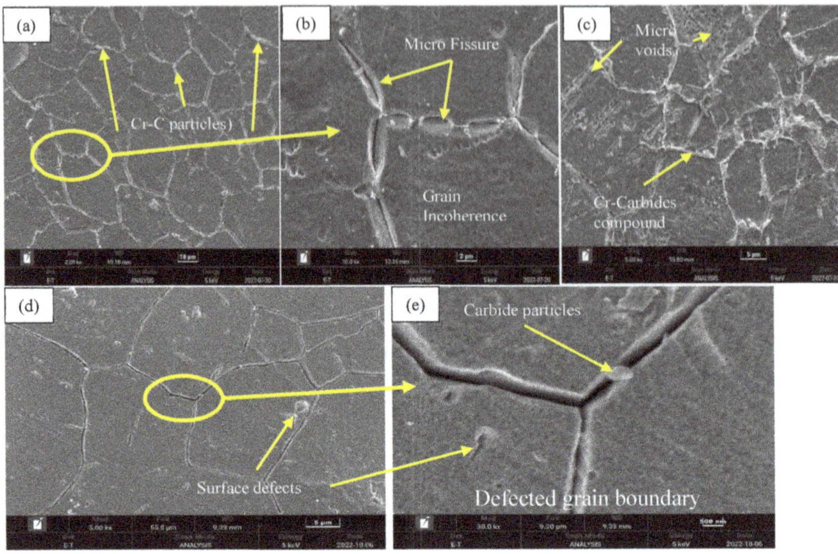

Figure 9: FESEM visuals and its characteristics of SEN1 (a), (b), (c) and AR1 (d), (e) tested samples

Finally observed the intergranular stress corrosion cracking (IGSCC), the crack seems to be more depth. It was distressed by the overage of experimental work and formed a lot of residual debris in the crack as shown in Figure 10. The metal dissolution in the crack interfaces causes by the applied parameters and those responsible for the development of metalloid debris along the crack. The initiation of crack takes place perpendicular to the applied tensile stress and the direction of dissemination slightly inclined. Here, the significant mechanism is the breakdown of interatomic bonding or decohesion along the grain boundary influenced by the mechanical load/tensile stress. At the crack tip, the strained interatomic bonds weaken by the absorbed chloride ions, hence the crack propagated and it depends on the loading condition, chloride concentration, exposure duration and temperature [47].

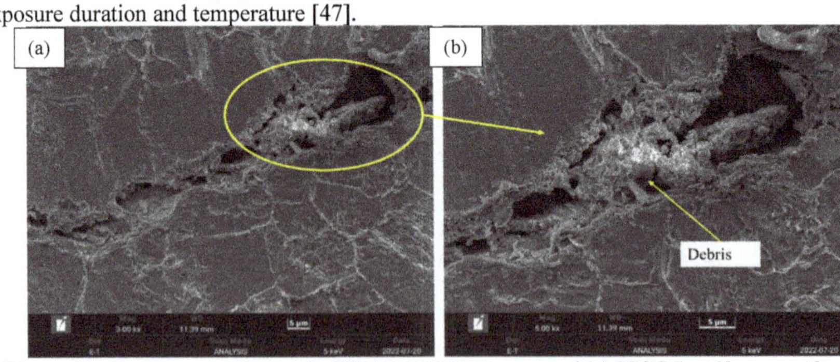

Figure 10: Overage tested sample (SEN1) FESEM images of (a) IGSCC at 3kx magnification and (b) IGSCC with debris at 5kx magnification

Materials Research Proceedings 29 (2023) 456-471 https://doi.org/10.21741/9781644902516-52

The energy dispersion spectrum (EDS) analysis was performed on the matrix surface (spectrum 22) and at grain boundary (spectrum 23) of SEN1 and AR1 as shown in Figure 11, contains the major elements of Fe, Cr, Ni and O. The phase of grain boundary in the SEN1 sample seems to be quite fragile. This phase exhibit frequently in austenitic stainless steels when used at temperatures more than 600 °C and owing to the corrosion products produced by stainless steels when Fe, Cr, and O are present [48]. From the spectrum 22, at the vicinity of grain boundary as rich in Cr and Fe than spectrum 23 and vice versa in AR1 sample. Assumed that where the chromium oxide is formed it might be initiate the pitting corrosion and leads to SCC when exposed to the residual or applied stress. Spectrum 23 at the grain boundary of SEN1, appears to be more sensitised and heavily oxidised than AR1. Sensitisation causes as result of chromium carbide precipitates at grain boundary. Chromium carbide precipitation creates chromium depleted zone in the vicinity of grain boundary. The precipitate happens continuously, the grain boundary becomes brittle and incoherence as leads to intergranular corrosion. Also, the cracks by residual stress propagates easily along the boundary [49]. In this case, observed sensitised samples are more possible to root the intergranular crack in the practice of drip test under the test parameters. The sensitised metal finally exposed as a result of crack propagation that are created at the oxide/metal interface.

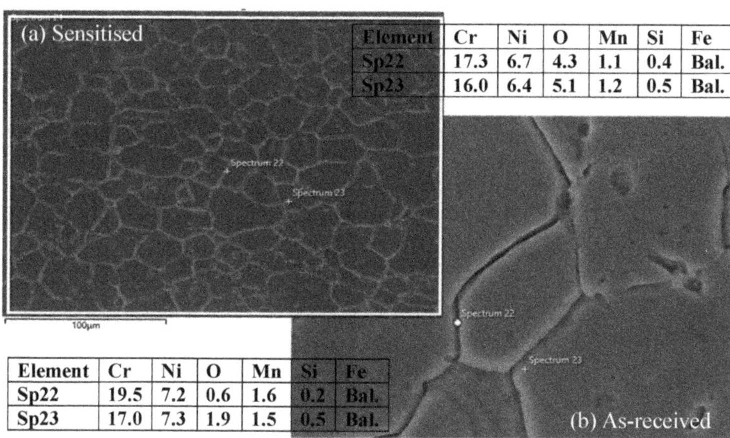

Element	Cr	Ni	O	Mn	Si	Fe
Sp22	17.3	6.7	4.3	1.1	0.4	Bal.
Sp23	16.0	6.4	5.1	1.2	0.5	Bal.

Element	Cr	Ni	O	Mn	Si	Fe
Sp22	19.5	7.2	0.6	1.6	0.2	Bal.
Sp23	17.0	7.3	1.9	1.5	0.5	Bal.

Figure 11: EDS spectrum analysis of element compositions of (a) SEN1 (b) AR1 tested samples

Conclusions

This work investigated the characteristics of CISCC on SS304 AR and SEN U-bend samples subjected to various chloride concentrations under perlite thermal insulation at 90°C. The key findings can be summed up as follows:

- All the samples observed periodically and firstly detected the pitting corrosion on AR1 and SEN1 after 472 hours. SCC was detected on heat-sensitised sample under 3.5 wt% of NaCl solution after 1632 hours, while other sample were crack free.
- The sensitised sample had more CISCC susceptibility than as-received sample at higher chloride concentration. Grain size and micro-hardness has been increased in SEN1 than AR1 sample.

- Micro-fissures were detected in sensitised specimen's grain boundary by the dissolution of carbides in the action of tensile stress at 90°C under higher NaCl concentration.
- In the extended test range, the sensitised sample experienced an intergranular crack with significant quantities of debris under 3.5 wt% of chloride concentration.

According to the practice of this work, physical inspection is a risk-based process and would be a difficult assignment. As result there will be a financial and production losses. For further enhancement, advise improving the design parameters and approach to non-destructive techniques (NDT) such as radiography, ultrasonic, eddy current testing, and acoustic emission technique.

Ethics Statement

The authors conducted and published this research in accordance with generally accepted standards of ethical behaviour.

Acknowledgement

The author would like to thank Ministry of Higher Education Malaysia (MOHE) for providing financial assistance under Fundamental Research Grant Scheme (FRGS/1/2/2020/TK0/UTP/03/2) and Universiti Teknologi PETRONAS for providing required facilities to conduct this research work.

References

[1] T. Michler, "Austenitic Stainless Steels," *Ref. Modul. Mater. Sci. Mater. Eng.*, no. June 2015, pp. 1–6, 2016, https://doi.org/10.1016/b978-0-12-803581-8.02509-1.

[2] S. M. Elsariti and Haftirman, "Behaviour of stress corrosion cracking of austenitic stainless steels in sodium chloride solutions," in *Procedia Engineering*, 2013, vol. 53, pp. 650–654, https://doi.org/10.1016/j.proeng.2013.02.084

[3] J. E. TRUMAN, "The influence of chloride content, pH and temperature of the test solution on the occurrence of stress corrosion cracking with austenitic stainless steel," *Corros. Sci. Pergamon Press. Print. Gt. Britain*, vol. 17, no. August 1976, pp. 737–746, 1977.

[4] NACE SP0198-2016, Control of Corrosion Under Thermal Insulation and Fireproofing Materials, vol. 2nd Ed., no. 21084. 2016.

[5] A. Bahadori, Thermal Insulation Handbook for the Oil, Gas, and Petrochemical Industries. 2014.

[6] F. De Vogelaere, "Corrosion Under Insulation," *Process Saf. Progress, Wiley InterScience-AIChE*, vol. 28, no. 1, pp. 30–35, 2009, https://doi.org/10.1002/prs.10276

[7] E. O. Eltai, F. Musharavati, and E. S. Mahdi, "Severity of corrosion under insulation (CUI) to structures and strategies to detect it," *Corros. Rev.*, no. 1988, pp. 1–12, 2019, https://doi.org/10.1515/corrrev-2018-0102

[8] B. R. Sanders, M. Production, J. R. Davis, and M. Park, *ASM Handbook- Volume 13C Corrosion : Environments and Industries*, vol. 13C. 2006.

[9] A. Almubarak, W. Abuhaimed, and A. Almazrouee, "Corrosion Behavior of the Stressed Sensitized Austenitic Stainless Steels of High Nitrogen Content in Seawater," *Int. J. Electrochem.*, vol. 2013, pp. 1–7, 2013, https://doi.org/10.1155/2013/970835

[10] M. Dahmen, K. D. Rajendran, and S. Lindner, "Sensitization of Laser-beam Welded Martensitic Stainless Steels," *Phys. Procedia*, vol. 78, no. August, pp. 240–246, 2015, https://doi.org/10.1016/j.phpro.2015.11.034.

[11] R. Parrott and H. Pitts, Chloride stress corrosion cracking in austenitic stainless steel. 2011.

[12] M. A. and E.-N. T. J. Bernard-Maxmillan Sim , Sai-Hong Tang, "Analyzing the Effects of Heat Treatment on SMAW Duplex Stainless Steel Weld Overlays," *Materials-MDPI*, pp. 1–12, 2022, https://doi.org/https://doi.org/10.3390/ma15051833

[13] İ. Demir, S. Başpınar, and E. Kahraman, "Production of insulations and construction materials from expanded perlite," *Springer Int. Publ.*, vol. 6, no. November, pp. 24–32, 2018, https://doi.org/10.1007/978-3-319-63709-9_3.

[14] J. W. Lee and K. E. Kee, "Experimental Study of Chloride-Induced Stress Corrosion Cracking (Ciscc) for Austenitic Sus 304L Under Thermal Insulation," *Platf. - A J. Eng.*, vol. 4, no. 2, pp. 31–43, 2020.

[15] Q. Cao *et al.*, "A Review of Corrosion under Insulation: A Critical Issue in the Oil and Gas Industry," *Metals (Basel).*, vol. 12, no. 4, pp. 1–24, 2022, https://doi.org/10.3390/met12040561

[16] U. K. Chatterjee, "Stress corrosion cracking and component failure: Causes and prevention," *Sadhana*, vol. 20, no. 1, pp. 165–184, 1995, https://doi.org/10.1007/BF02747288

[17] A. Bahadori, "Fundamentals of Corrosion in the Oil , Gas , and Chemical Industries," in *Corrosion and Materials Selection: A Guide for the Chemical and Petroleum Industries*, First Edit., John Wiley & Sons, Ltd, 2014, pp. 1–16.

[18] S. H. Khodamorad, N. Alinezhad, D. Haghshenas Fatmehsari, and K. Ghahtan, "Stress corrosion cracking in Type.316 plates of a heat exchanger," *Case Stud. Eng. Fail. Anal.*, vol. 5–6, pp. 59–66, 2016, https://doi.org/10.1016/j.csefa.2016.03.001

[19] U. NACE International, Houston, Texas, "International Measures of Prevention , Application , and Economics of Corrosion Technologies Study," 2016.

[20] A. C692-08, "Standard Test Method for Evaluating the Influence of Thermal Insulations on External Stress Corrosion Cracking Tendency of Austenitic Stainless," 1987.

[21] ASTM and E03-21, "Standard Guide for Preparation of Metallographic Specimens 1," 2001.

[22] A. International and A262-02a, "Standard Practices for Detecting Susceptibility to Intergranular Attack in Austenitic," 2002.

[23] ASTM and E572, "Standard Test Method for Analysis of Stainless and Alloy Steels by X-ray," *ASTM Int.*, vol. 03, no. June, pp. 1–10, 2003.

[24] ASTM Standard E8/E8M-13a, "'Standard Test Methods for Tension Testing of Metallic Materials,'" *ASTM Int.*, vol. i, pp. 1–27, 2013, [Online]. Available: http://www.astm.org/Standards/E8.htm.

[25] E384 and ASTM, "Microindentation Hardness of Materials 1," in *ASTM Standards*, vol. 14, 2002, pp. 1–24.

[26] ASTM A240, "ASTM A 240/A 240M – 04. Standard Specification for Chromium and Chromium-Nickel Stainless Steel Plate, Sheet, and Strip for Pressure Vessels and for General Applications," *ASTM Int.*, vol. i, p. 12, 2004, [Online]. Available: http://www.ussa.su/gosts2/ASTM_A240.PDF

[27] A. International and G30-97, "Standard Practice for Making and Using U-Bend Stress-Corrosion Test," in *ASTM Standards*, vol. i, no. Reapproved, 2009, pp. 1–7.

[28] A. . Dana, "Stress- Corrosion Cracking of Insulated Austenitic Stainless Steel," *ASTM Bull.*, pp. 46–52, 1957.

[29] C610 and ASTM, "Standard Specification for Molded Expanded Perlite Block and Pipe Thermal," in *ASTM Standards*, vol. 04, 2000, pp. 1–4.

[30] C585 and ASTM, "Standard Practice for Inner and Outer Diameters of Rigid Thermal Insulation for Nominal Sizes of Pipe and Tubing (NPS System) 1," in *ASTM Standards*, vol. 90, no. Reapproved, 2004, pp. 1–9.

[31] AAAMSA, THERMAL INSULATION Incorporating the Architectural Glass Industry Incorporating the Architectural Glass Industry, no. April. 2001.

[32] D1193 and ASTM, "Reagent Water 1," *ASTM Int.*, vol. 51, no. 7916, pp. 1–3, 2001.

[33] G44 and ASTM, "Standard Practice for Exposure of Metals and Alloys by Alternate Immersion in," *ASTM Int.*, vol. 11, no. Jan, pp. 1–4, 2000.

[34] P. Street, "Metastable pitting corrosion of stainless steel and the transition to stability," *Philos. Trans. R. Soc. London. Ser. A Phys. Eng. Sci.*, vol. 341, no. 1662, pp. 531–559, 1992, https://doi.org/10.1098/rsta.1992.0114.

[35] S. Caines, F. Khan, J. Shirokoff, and W. Qiu, "Journal of Loss Prevention in the Process Industries Experimental design to study corrosion under insulation in harsh marine environments," *J. Loss Prev. Process Ind.*, vol. 33, pp. 39–51, 2015, https://doi.org/10.1016/j.jlp.2014.10.014

[36] P. Wang, Y. Zhang, and D. Yu, "Microstructure and mechanical properties of pressure-quenched SS304 stainless steel," *Materials (Basel).*, vol. 12, no. 2, pp. 1–9, 2019, https://doi.org/10.3390/ma12020290

[37] P. R. Rios and A. F. Padilha, "Precipitation From Austenite," *Ref. Modul. Mater. Sci. Mater. Eng.*, no. February 2015, pp. 1–8, 2019, https://doi.org/10.1016/b978-0-12-803581-8.02522-4.

[38] G. Yin, Y. Li, J. Sun, and J. Chen, "Effect of Heat Treatment Temperature on Mechanical Properties of the AISI 304 Stainless Steel," *Int. J. Innov. Res. Sci. Eng. Technol.*, vol. 3, no. 2, pp. 9516–9520, 2014, https://doi.org/10.7521/j.issn.0454-5648.2014.03.17

[39] R. W. Armstrong, "60 years of hall-petch: Past to present nano-scale connections," *Mater. Trans.*, vol. 55, no. 1, pp. 2–12, 2014, https://doi.org/10.2320/matertrans.MA201302.

[40] M. M. Louwerse and A. C. Graesser, "Macrostructure," in *Metallography: Principles and Practice (#06785G)*, 2006, pp. 426–429.

[41] C.-C. Hsieh and W. Wu, "Overview of Intermetallic Sigma (σ) Phase Precipitation in Stainless Steels," *ISRN Metall.*, vol. 2012, no. 4, pp. 1–16, 2012, https://doi.org/10.5402/2012/732471

[42] ASTM E112, "Standard Test Methods for Determining Average Grain Size," 2010.

[43] Y. Komizo, "Correlation of delta-ferrite precipitation with austenite grain growth during annealing of steels," *Philos. Mag. Lett.*, vol. 91, no. July, pp. 491–497, 2011, https://doi.org/10.1080/09500839.2011.587464

[44] P. Kral *et al.*, "Creep Resistance of S304H Austenitic Steel Processed by High-Pressure Sliding," *Materials (Basel).*, vol. 15, no. 1, 2022, https://doi.org/10.3390/ma15010331.

[45] M. A. Mohtadi-Bonab, "Effects of different parameters on initiation and propagation of stress corrosion cracks in pipeline steels: A review," *Metals (Basel).*, vol. 9, no. 5, pp. 1–18, 2019, https://doi.org/10.3390/met9050590

[46] S. K. Pradhan, T. S. Prithiv, and S. Mandal, "Through-thickness microstructural evolution during grain boundary engineering type thermomechanical processing and its implication on sensitization behavior in austenitic stainless steel," *Mater. Charact.*, vol. 134, no. July, pp. 134–142, 2017, https://doi.org/10.1016/j.matchar.2017.10.014

[47] Woodhead Publishing Limited, *Stress corrosion cracking: Theory and practice*. Woodhead Publishing Limited, 2011.

[48] P. P. Psyllaki, G. Pantazopoulos, and A. Pistoli, "Degradation of stainless steel grids in chemically aggressive environment," *Eng. Fail. Anal.*, vol. 35, pp. 418–426, 2013, https://doi.org/10.1016/j.engfailanal.2013.04.016

[49] M. W. A. Rashid, M. Gakim, Z. M. Rosli, and M. A. Azam, "Formation of Cr23C6 during the sensitization of AISI 304 stainless steel and its effect to pitting corrosion," *Int. J. Electrochem. Sci.*, vol. 7, no. 10, pp. 9465–9477, 2012.

Current resource assessment in Brunei's coastal waters

AHMAD FARID Dato Hj Ali[1,a*] and ROSLYNNA Rosli[1,b]

[1]Universiti Teknologi Brunei, Tungku Highway, Gadong BE1410, Brunei Darussalam

[a]ahmadfaridali955@gmail.com, [b]roslynna.rosli@utb.edu.bn

Keywords: Tidal Current, Tidal Resource, Resource Assessment, Tidal Energy, Brunei

Abstract. Current Resource Assessment plays a major part in Tidal Energy Resource Assessment. The paper looks at analysing the total current data collected from BSP at Fairley Oilfield from 1st January 2019 up to 31 December 2019. The total current data and its direction was measured using ADCP, and is placed at an area with a depth of about 60m. The data is categorized into 3 depths; current speed and direction for overall depth, current speed and direction for top half of the sea, current speed and direction for bottom half of the sea. According to the current speed for the overall depth, the current speed was recorded to have an average speed of 0.27m/s and a maximum speed of 1.4m/s. For the current speed and direction at the top half of the sea, the current speed was recorded to have an average speed of 0.33m/s and a maximum speed of 2.24m/s. Finally, for current speeds at the bottom part of the sea, the current speed has an average of 0.22m/s and a maximum speed of 1.66m/s. The direction for all three categorizes were consistent in which the flow is recorded to be coming in from the East-North East to East, majority of the time.

Introduction

Coastal water management places an important part in the economic development of a country. Unfortunately, due to climate change, damages and breakdown of a country's coastline has been witnessed around the world [1]. Not only that, the affects towards marine life could also be seen, with marine animals losing their sense of direction and the extinction and reduction of marine life in general due to climate change [2]. Tidal currents are generated due to the rise and fall of the sea tide [3]. According to Newton's Theory, tides can be defined by its harmonic periods. However, even though tides can be predictable, due to the presences of bathymetry, coastlines, island and other forces that will affect the movement and flow of total currents, it is very difficult to exactly workout the behaviour of the total currents [4].

The total current is referred to the flow of water in rivers, sea or ocean, that is caused by factors including, Tidal Current, Ocean Current, Wind, Salinity, Temperature and etc. For areas with low Tidal Range, such as Brunei Darussalam, the Tidal Currents are not dominants and does not play a major role in Total Currents. However, in some areas, with high tidal range, the tidal current is dominant and follows a clear pattern with the tides. This would mean that the Tidal Currents are predictable and reliable. Even though some researcher would refer to the flow of the water as Tidal currents, it is impossible to collect data for tidal currents only, as other factors such as wind, temperature and etc. will affect the tidal current. Therefore, total current is more suitable to be mentioned.

There are many potentials in harnessing energy from tidal currents. 70% of the world is covered in water and it is estimated that, the potential energy generated by the movement of the water globally can go up to 3000GW [5]. With that being said, tidal energy is still considered a new form of renewable energy. Even though there has been implementation of tidal energy in the past. Not many studies have been done on tidal energy resource and further development of the tidal energy converters still needs to be made [6].

To work out the feasibility of tidal energy in a specific location, an in-depth resource assessment needs to be done. This resources assessment includes, the tide and tidal current resource assessment, bathymetry assessment and etc. [7,8]. The feasibility of tidal energy will determine includes the practicality and economic impact in harnessing energy through tides.

As mentioned earlier, Tidal turbines are still considered in the testing stages and much more development is required. With the existing tidal extraction devices available, the minimum cut in speed for the tidal extraction device is reported to be 0.5m/s [9]. Therefore, it is advisable that the average speed of the currents within the area to be at least 0.5m/s. However, there are a lot of concept designs that have been simulated which suggest that the cut in speeds to be lower to support low velocity currents. There are also many types of tidal turbine deployment suggested, such as Gravity Structure, Piled Structure or Floating Structure [10]. These deployment methods would be dependent on the site specification available at the area.

Most studies that have been done on tidal energy resource assessment have been seen at locations with high tidal currents with high tidal range at the oceans or seas, whereas less energetic currents have also been studied, however in shallower areas such as rivers for easy excess and installation. However not many studies on tidal energy have been done on many parts of the world yet [11]. Tidal current energy is proportional to the Fluid Density, Area and the cube of its velocity. The velocity plays a major role in the power output and therefore, a slight change in velocity will see a significant impact on the power output. The power output would also be determined by the size of the turbine, as channels and river that would usually have higher current flow may be limited to the size of the turbine generator compared to the sea or lake that will allow a bigger size turbine generator [12].

Brunei Darussalam is located in South East Asia, in Borneo Island. The country, faces the South China Sea and has a coastline of 161km, which means the country would have an opportunity to extract energy through the currents. Not only that, about 85% of the residence, resides in the coastal area in which majority of activities such as social, cultural and economic activities are concentrated. However, as it stands, there have been no proper study on the currents in Brunei Darussalam to suggest the possibility of Tidal Energy in Brunei Darussalam [13].

Methodology

The total currents at the coastal waters of Brunei Darussalam were measured using an ADCP, which was set up by Brunei's Shell Petroleum (BSP). The ADCP was placed near one of BSP's oilfield, Fairley Oilfield. Fairley Oilfield, is located west of Brunei Darussalam, and is located 45km away from the shoreline. The ADCP was placed on the seabed which has a depth of about 63m. The data collected was from 1st January 2019 up to 31 December 2019. The data was measured with 10mins interval at depths 3.5m, 7.5m, 11.5m, 15.5m, 19.5m, 23.5m, 27.5m, 31.5m, 35.5m and 39.5m. On top of that the direction of the currents were also recorded by the ADCP for each depth. The direction of the current, which was measured by the ADCP, was measured using bearing system.

The investigation is done by looking at the effects of the depth towards the velocity of the currents and its directions in the Coastal Waters of Brunei Darussalam. This study will present how the current behaves in the Coastal waters of Brunei Darussalam, which includes looking at the maximum and average velocities of the current, the average direction of the currents and how it all behaves as the depth changes. Consistency in the water velocity and direction will also be touched upon to see the stability of the water at different depths and time. These assessments play a major role in assessing the current resource for energy extraction and device determination.

The data are categorized into three depths; the data for overall depth of the sea, the data at the top half of the sea & the data at the bottom half of the sea. The water velocities are compared for the three categorize depths to see how different depths affect the water velocity. The direction in which the water flow is also investigated to see whether the incoming flow changes direction or remains constant.

Results & Discussion

Fig. 1 shows the direction of the current compared to the 3 categorized depths of the sea. As can be seen in Fig. 1, for all the categorize, the highest percentage of the direction in which the current flows are flowing towards the North East to East direction, with about 18% of the flow flowing towards the East and 14% flowing towards the East-North East for the total average current for the entire depth of the sea. For the top half of the sea, about 14% of the flow is recorded to be flowing towards the East-North East and about 13% of the flow is recorded to be flowing towards the South East. For the Bottom half of the sea, 23% of the flow is recorded to be flowing towards the East-North East Direction, whereas the other incoming flow is recorded to be heading towards the North East and East direction with a total of about 22% combined.

For all categories, it can be seen that majority of the water flow would be flowing towards the East-North East to East direction, with only a small percentage of the water flow changing its direction. However, comparing to all three categorizes the currents flowing at the Bottom half of the sea, shows the most consistent in which the current flows towards the East-North East majority of the time. This would make it favorable for energy extraction using currents, as the device does not need to consider the changing direction of currents.

Fig. 2 shows the average current speed for all depths at the coastal waters of Brunei Darussalam for the whole of 2019. As can be seen in Fig. 3, there are times in which the device stopped recording, due to device malfunctioning or maintenance where being done on it. From the data collected, the average velocity is recorded at 0.27m/s only, with the highest recorded value at 1.43m/s. However, it can be seen that there are fluctuations in the water velocity which suggest that the water velocity is not stable and also, in general, the middle of the year provides a higher water velocity.

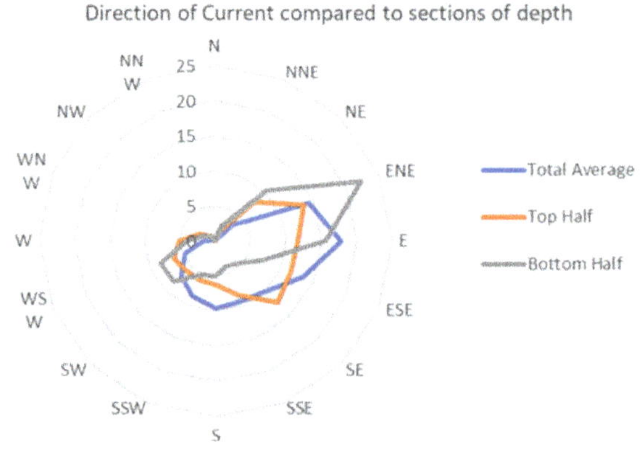

Fig. 1. Direction of the currents at different depths

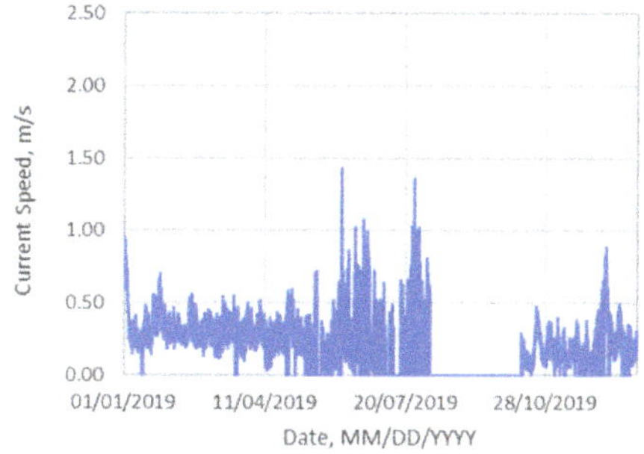

Fig. 2. Total Average Current at the Coastal Waters of Brunei Darussalam

According to Fig. 2, there are no evidence in which the total currents follow a pattern, which means that predicting the water velocity in Brunei would be very difficult. The current speed also rarely exceeds, 1m/s for the whole year of 2019, with sudden spikes appearing that do exceed 1.0m/s. These sudden spikes are not consistent and may be due to a lot of factures, such as bad weather in the South China Sea. Fig. 3 shows the average current speed for 2019, for the top half of the sea at the coastal waters of Brunei Darussalam. The average current flow was found to be 0.3m/s for the whole year of 2019 with the highest recorded value of 2.24m/s.

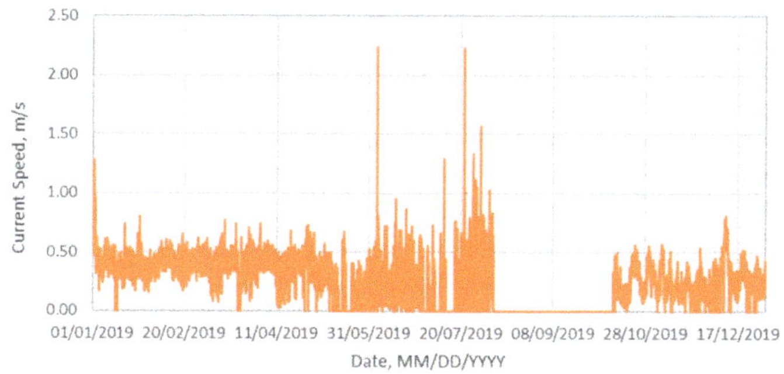

Fig. 3. Average Current at the top half at the Coastal Waters of Brunei Darussalam

From the data analysis from Fig. 3, it is clear that the average speed is higher compared to the average current speed for the whole depth of the sea. This could be due to less friction from the sea bed which would affect the water flow, and also, the bathymetry of the sea would have less impact on the current speed. One of the main reasons why, the currents at the top half of the sea are travelling faster is also due to the influence on the wind at the South China Sea, which would accelerate the water flow. Similar to the data for the overall depth, it can be seen that the data is not stable and that fluctuations form. In fact, it can be seen that there are more fluctuations at the top half of the sea.

Fig. 4 shows the average current at the bottom half of the sea at the coastal waters of Brunei Darussalam. The average current speed for the bottom half of the sea is recorded at 0.22m/s, whereas the highest recorded value is at 1.7m/s. Again, similar to Fig. 2 & 3, spikes are form at the same time as when the other recordings recorded the spikes. The average current speed is lower than the average current speed at the top half of the sea and overall depth of the sea. However, if we compare the highest recorded value for the current at the Bottom half of the sea, and the overall depth of the sea, the bottom half of the sea has a higher value with 1.66m/s, whilst the overall depth of the sea has a value of 1.4m/s. Fig. 5 shows the average current speed throughout the year of 2019, against the depth of the sea. As can be seen in the figure, as the water gets deeper, the velocity of the water reduces drastically. The figure shows that at the surface of the water, the water is flowing more than 0.6m/s. However, after 10m depth into the sea, the water reduces drastically to about 0.23m/s and fluctuates slightly up to about 40m in depth. However, pass 40m deep, the velocity decreases even more as by 60m, which is few meters from the seabed, the velocity falls to zero.

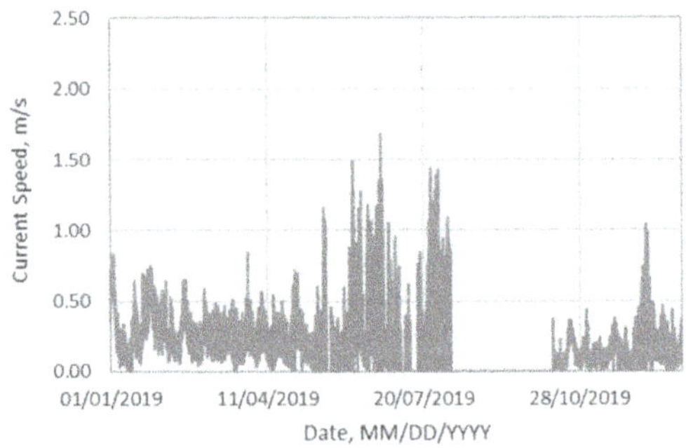

Fig. 4. Average Current at the bottom half at the Coastal Waters of Brunei Darussalam

According to the research, the water flows much faster at the top of the surface as, wind accelerated the flow of the water. However, as it goes deeper, bathymetry and friction affect the flow causing the flow to slow down. Fig. 6 represents that satellite projection of the water flow at the South China Sea for 15th June 2019 at 08:00 Local time. Brunei Darussalam is located in the Borneo Island and to the north of Brunei is the South China Sea. It shows that there are circulation form in the South China Sea, which would result in the overall velocities to be slow. On top of that, there are other countries surrounding Brunei Darussalam, which would result in the water flow to reduce its energy, hence why the water flow in Brunei Darussalam is considerably low.

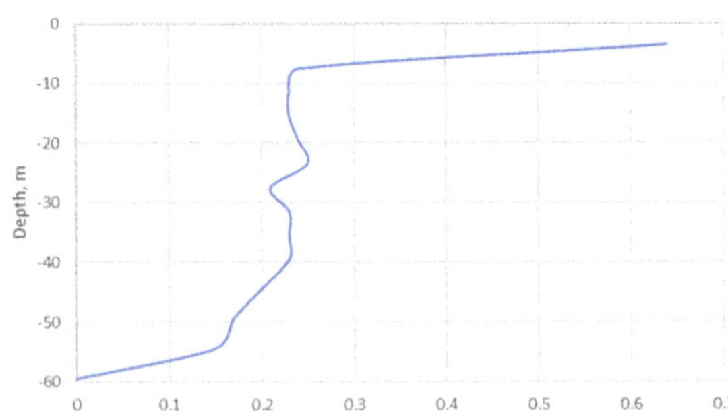

Fig. 5. The Average Current Speed vs the Depth of the Sea

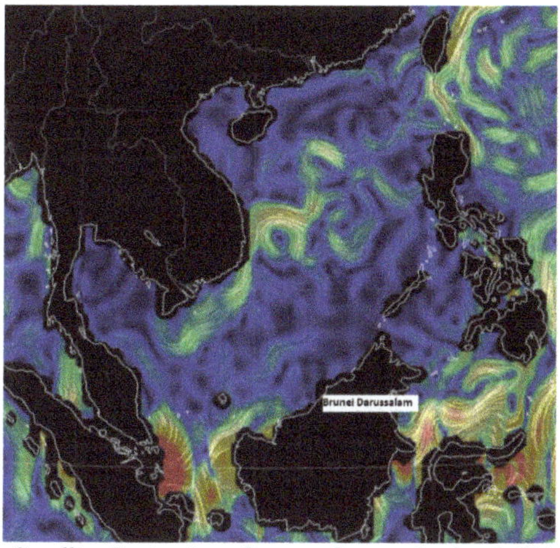

Fig. 6. Satellite Projection of water flow at the South China Sea

Conclusion

According to existing Tidal turbines, current tidal turbine designs would have a minimum cut in speed of about 0.5m/s [14]. However, in order to produce energy economically, the average velocity should be around 1.5-3.5m/s [15]. With that being said, even though Brunei has a good location and that the currents in Brunei Darussalam shows promising results. It is not sufficient to generate electricity through tidal currents in Brunei Darussalam economically. Furthermore, more research needs to be done at different parts of the Coastal waters of Brunei Darussalam, to fully determine the feasibility of Tidal Energy and examine the Total Currents Resource in Brunei Darussalam.

From the data collected, the overall depth, the current speed was recorded to have an average speed of 0.27m/s and a maximum speed of 1.4m/s. For the current speed and direction at the top half of the sea, the current speed was recorded to have an average speed of 0.33m/s and a maximum speed of 2.24m/s. Finally, for current speeds at the bottom part of the sea, the current speed has an average of 0.22m/s and a maximum speed of 1.66m/s. The direction for all three categorizes were consistent in which the flow is recorded to be flowing towards the East-North East to East, majority of the time. Therefore, when comparing which area provides the highest average and maximum velocity, the top half of the sea would provide the highest velocity, however with the least consistent direction. With that being said, even though the bottom half of the sea provides the lowest average velocity, it is also the most consistent in the direction of the current flow.

Acknowledgment

The research contributed for this paper has been conducted with the help of my main supervisor, who is also the second author of this paper, Dr Roslynna Rosli. The tide data that has been used in this paper as well has been obtained with the permission of Brunei Shell Petroleum, BSP, and Ministry of Development of Brunei Darussalam's Survey Department and without their permission, the research and study would not be able to be conducted. The Univesiti Teknologi of Brunei, UTB, should also be mentioned in this acknowledgment as they provide the facilities in order to do this research.

References

[1] Horstman E. M., Wijnberg K. M., Smale A. J., & Hulscher S. J. M. H., 2009. Long-term Coastal Management Strategies: Useful or Useless? Journal of Coastal Research, 233-237.

[2] Richardson A. J., Brown C. J., Brander, K. Bruno, J. F., Buckley L., Burrows M. T., … Poloczanska, E. S., 2012. Climate change and marine life. Biology Letters, 8(6), 907-909. https://doi.org/10.1098/rsbl.2012.0530

[3] David A. J., Adam D., Deborah I., Eric W. P., 2021. Tides and Coastal Geomorphology: The Role of Non-Stationary Processes. Treatise on Geomorphology, 161-198. https://doi.org/10.1016/B978-0-12-818234-5.00166-8

[4] Bryden I. G., Couch S. J., Owen A., and Melville G., 2007. Tidal current resource assessment. Proceedings of the Institution of Mechanical Engineers, Part A: Journal of Power and Energy, 221(2), 125-135. https://doi.org/10.1243/09576509JPE238

[5] Lemonis G., 2004. Wave and Tidal Energy Conversion. Encyclopedia of Energy, 385-396. https://doi.org/10.1016/B0-12-176480-X/00344-2

[6] Boretti A., 2020. Trends in tidal power development. E3S Web of Conferences, 173, 01003. https://doi.org/10.1051/e3sconf/202017301003

[7] Liu X., Chen Z., Si Y., Qian P., Wu H., Cui L., & Zhang D., 2021. A review of tidal current energy resource assessment in China. Renewable and Sustainable Energy Reviews, 145, 111012. https://doi.org/10.1016/j.rser.2021.111012

[8] Uihlein A., and Magagna D. 2016. Wave and tidal current energy - A review of the current state of research beyond technology. Renewable and Sustainable Energy Reviews, 58, 1070-1081. https://doi.org/10.1016/j.rser.2015.12.284

[9] Lewis M., O'Hara Murray R., Fredriksson S., Maskell J., de Fockert A., Neill S. P., and Robins P. E., 2021. A standardised tidal-stream power curve, optimised for the global resource. Renewable Energy, 170, 1308-1323. https://doi.org/10.1016/j.renene.2021.02.032

[10] Purvi C., Pathik P., Saurin M. S., 2015. Tidal Stream Turbine-Introduction, current and future Tidal power stations, National Conference on Innovative and Emerging Technologies.

[11] Burić M., Grgurić S., Mikulčić H., and Wang, X. 2021. A numerical investigation of tidal current energy resource potential in a sea strait. Energy, 234, 121241. https://doi.org/10.1016/j.energy.2021.121241

[12] Li D., Wang S., & Yuan P., 2010. An overview of development of tidal current in China: Energy resource, conversion technology and opportunities. Renewable and Sustainable Energy Reviews, 14(9), 2896-2905. https://doi.org/10.1016/j.rser.2010.06.001

[13] Geronimo S. and Chua T. E., 1992. The Coastal Resources of Brunei Darussalam, Department of Fishery, Ministry of Industry and Primary Resources. ISBN 971-8709-1 1-8

[14] Orhan K., and Mayerle R. 2017. Assessment of the tidal stream power potential and impacts of tidal current turbines in the Strait of Larantuka, Indonesia. Energy Procedia, 125, 230-239. https://doi.org/10.1016/j.egypro.2017.08.199

[15] Khare V., Khare C., Nema S., and Baredar P. 2019. Prefeasibility Assessment of a Tidal Energy System. Tidal Energy Systems, 115-188. https://doi.org/10.1016/B978-0-12-814881-5.00003-X

Design selection of a tidal turbine for low velocity application

NURJAH BAZILAH Haji Jemluddin[1,a] and ROSLYNNA Rosli[1,b*]

[1]Universiti Teknologi Brunei, Tungku Highway, Gadong BE1410, Brunei Darussalam

[a]bazilahjemluddinii@gmail.com, [b]roslynna.rosli@utb.edu.bn

Keywords: Tidal Turbine, Turbine Design, Turbine Blade, Design Matrix, Low Velocity

Abstract. Tidal energy is the sole renewable energy associated with tidal movements due to the gravitational and centrifugal forces between the Earth, the Moon, and the Sun. Although tidal energy is not yet widely used, it has the potential to generate energy for the future. It is a clean, renewable energy source and more predictable than other renewable sources of energy. Various existing tidal turbines have been developed in Europe, where tidal velocity is much higher than the tidal velocity in Brunei Darussalam, which is around 0.5 m/s. Therefore, this study investigates the design of a tidal turbine that can be effectively implemented in Brunei Darussalam. There are five conceptual designs, and each of them was evaluated with the evaluation matrix with a three-bladed commercial Horizontal Axis Tidal Turbine as the datum, obtaining the final design, which is Design 1, has the highest scoring point.

Introduction

Although tidal energy is not yet widely used, it has the potential to generate energy for the future [1]. It is a renewable energy source that produces energy from the ebb and flow of tides. Tides are caused by the combined action of gravitational forces exerted by the Moon and the Sun on Earth. As the Earth rotates, the position of a given area relative to the moon changes, creating the tides. Thus, a periodic succession of high and low tides. Two high tides and two low tides occur during the day. The high and low tide follows a sinusoidal curve. The difference between ebb tide and low tide is called the tidal range. When near a full moon or new moon, the range of tide is relatively high, known as spring tides, whereas nearly half-moon, the range of tide is relatively small, known as neap tides.

Several technologies can generate tidal energy, including tidal barrages, tidal lagoons, tidal fences, and tidal turbines. Both tidal barrages and tidal lagoons utilise the potential energy generated by the change in tidal height between high and low tides. It generates electricity via turning turbines similar to a hydropower. Tidal fences are turbines that operate like turnstiles, while tidal turbines are driven similarly to wind turbines but underwater. Both tidal fences and tidal turbines generate electricity by extracting the kinetic energy from the tidal currents. Ocean currents are 832 times denser than air and exert more force on the turbine, so ocean currents are able to generate more energy than winds at smaller size [2].

Khan et al [3] categorised the tidal technologies system into two classes: turbine and non-turbine. The turbine systems consist of axial (horizontal), vertical, crossflow, venturi, and gravitational vortex, whereas the non-turbine consists of flutter vane, piezoelectric, vortex-induced vibration, oscillating hydrofoil, and sail [3]. Recent technologies of tidal energy converters also include Archimedes Screw, Tidal Kite and Hydro-Spinna turbine. When the tide flows across the turbine, the turbine extracts the kinetic energy from the tides, reducing the kinetic energy of the tides and their velocity. Factors that affect tidal turbines are the orientation of the turbine, the number and blade profile used in the turbine, the twist and the tapper of the blade, and the rotor solidity of the turbine. According to Betz's limit, the theoretical maximum efficiency of a tidal or wind turbine is 59.3%. However, in practice, this limit cannot be reached, and the common efficiency is 35 to 45%.

Several methods have been developed to predict the performance of the turbine blade, one of them the Blade Element Momentum Theory (BEMT). Computational Fluid Dynamics (CFD) is another way to study different airfoils [4,5,6,7]. It can optimise the blades and perform turbine analysis [4,8]. Another method using a combination of Reynolds Average Navier-Stokes (RANS) equations and Shear-Stress Transportation (SST) type k – omega (ω) turbulence can perform with higher accuracy in investigating the performance of the turbine. Experimentations of the turbine blade in testing facilities to compare the CFD and the experimental data to validate the CFD model.

Designs of Turbine

Five conceptual designs were generated, based on existing tidal turbines that were known for low velocity applications. The first design is based on a Horizontal Axis Tidal Turbine with three-blades where the leading and trailing edge has a geometrical shape of one-third of cardioid. The rotation of the turbine is anticlockwise as the leading edge takes the lead. The shape of the hub is a semi-circle, and the blades are connected to a cylindrical shaft. This design is inspired by the Hydro-Spinna having the cardioid blades [1]. The illustration of the first design is in Fig. 1.

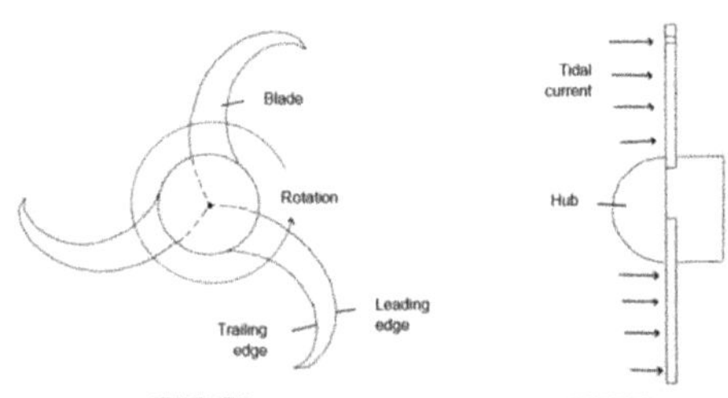

Fig. 1 First design.

The second design is also a Horizontal Axis Tidal Turbine design illustrated in Fig 2. This design is similar to the first design with a modification of extended blades in the axial direction. The turbine rotates in an anticlockwise direction.

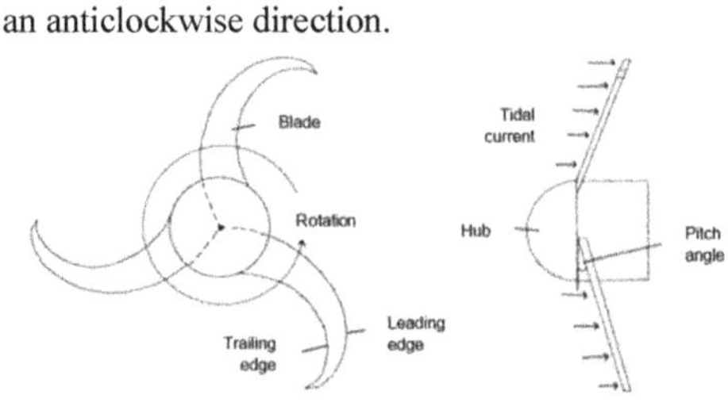

Fig. 2 Second design.

The third design is based on Vertical Axis Tidal Turbine with three blades shown in Fig. 3. Each blade's cross-sectional shape is one-third of cardioid, and the blades are mounted with a cylindrical shaft. The rotation of the turbine is anticlockwise. The Savonius Vertical Axis Tidal Turbine inspires this design, it provides uniform torque distribution along the whole rotation of the turbine.

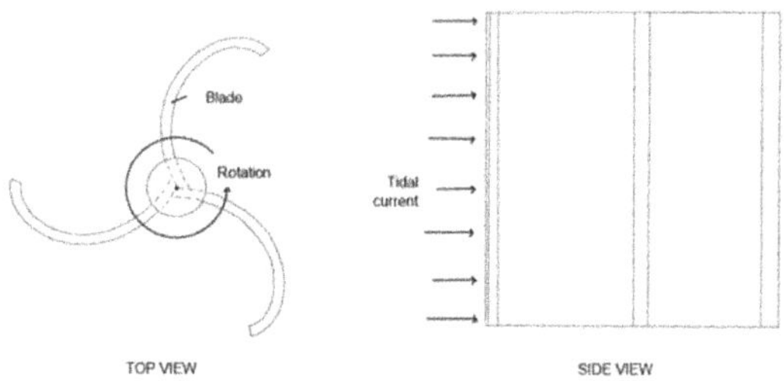

Fig. 3 Third design.

The fourth design is based on the Vertical Axis Tidal Turbine, and it is similar to the third design with twisted physical features presented in Fig. 4. This feature of the helical design had been proven to give better performance than the traditional Savonius Vertical Axis Tidal Turbine.

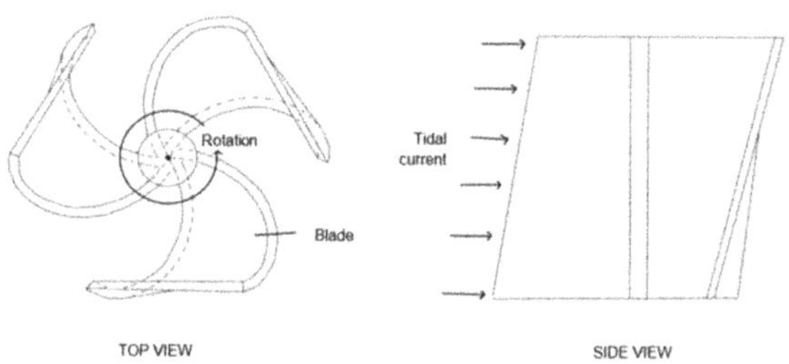

Fig. 4 Fourth design.

The fifth design is based on a Horizontal Axis Tidal Turbine with three blades which the leading and trailing edge are one-third of cardioid and circle respectively shown in Fig.5. The rotation of the turbine is anticlockwise. The hub is a semi-circle shaped and the blades are attached to a cylindrical shaft. This design was also motivated by the Hydro-Spinna turbine.

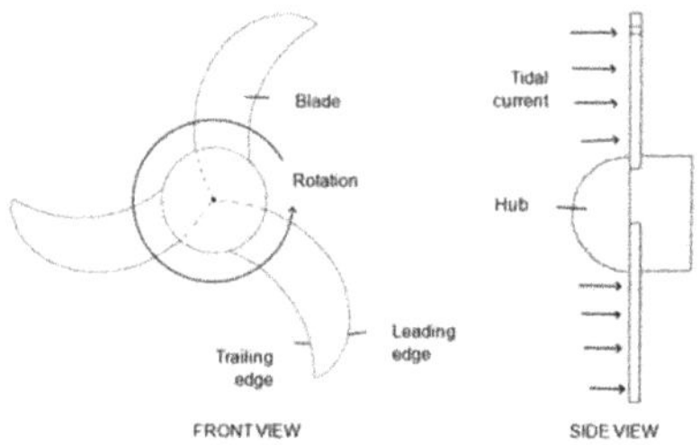

Fig. 5 Fifth design.

Design Selection of Turbine

The method of evaluation matrix is used to determine the final design by comparing all the preliminary designs with the datum. The criteria are the detailed requirements needed to be fulfilled, creating the successful product tabulated in Table 1 with the description. In finalising the design, each preliminary design was compared with the datum, a typical three-bladed commercial Horizontal Axis Tidal Axis, as illustrated in Fig. 6 by having the highest points due to its high efficiency and design popularity in real life deployment. The scoring system is based on the effectiveness of design according to properties. The design will gain a point if it is more effective than the datum and vice versa. The designs can also obtain no point with similar properties as the datum. This point system of the preliminary designs is tabulated in Table 2.

Fig. 6 Typical Horizontal Axis Tidal Turbine [9]

Table 1. Criteria of the design selection matrix

No.	Criteria	Description
1	Robust	• The turbine is able to sustain the tidal force. • The turbine is able to withstand other marine conditions. • The turbine is able to maintain its structural integrity.
2	Efficiency	• The power coefficient of the turbine is maximised.
3	Shelf life	• The turbine is able to last for more than 20 years with proper scheduled maintenance every six month.
4	Maintenance	• The blades are easy to maintain and repair.
5	Safety	• The blades should have no or less sharp edges.
6	Environment	• The turbine has little impact on the marine environment.
7	Fabrication	• The turbine should be easy to fabricate.
8	Transportation	• The turbine is easy to transport from one place to another.
9	Installation	• The turbine is easy to install. • The turbine is easy to assemble and disassemble.
10	Performance	• The turbine is able to be performed in 0.5 m/s of water velocity.
11	Storage	• The turbine is easy to store.

Materials Research Proceedings 29 (2023) 479-486 https://doi.org/10.21741/9781644902516-54

Based on the calculated point from Table 2, the design having the highest points in the evaluation matrix is design 1. This design is more robust, safer, and easier to fabricate. In Fig. 7 illustrated the final design of the turbine in 3D model and its blade profile. The cross-sectional profile of the blade is that of the NACA 0018. The final design was verified using a previously validated numerical analysis in ANSYS Fluent. Fig. 8 shows the numerical domain that was used to investigate the power and thrust coefficient of the turbine using RANS and SST k-ω turbulence model. The solver used is Pressure-Based with absolute velocity formation in steady time. In calculating the C_P and C_T from the data obtained from the simulation uses the following mathematical expressions in Equations 1, and 2. The power and thrust coefficient performance will be plotted against the tip speed ratio (TSR) of the turbine. C_P is the ratio of the power generated by the turbine to the power available in the flow while C_T is the thrust ratio on the turbine to the axial load available in the flow.

$$C_P = \frac{Q\Omega}{0.5\,\rho A_s U^3} \tag{1}$$

$$C_T = \frac{T}{0.5\,\rho A_s U^2} \tag{2}$$

$$TSR = \frac{\Omega \times \mathrm{r}}{U} \tag{3}$$

where Q is the torque generated, Ω is the angular speed, ρ is the density, r is the radius of the turbine, A_S is the area of the turbine and U is the velocity of the water.

All simulations are performed at a tidal velocity of 3 m/s, resulting with a Reynolds Number (Re) of about $8.39\text{x}10^5$ based on the diameter of the turbine. The calculated Re is over 2000, indicating that the flow is turbulent.

Table 2. Evaluation Matrix

Design	**1**	**2**	**3**	**4**	**5**	**6**
Robust	+	+	+	+	+	
Efficiency	+	=	-	-	+	D
Shelf life	=	=	=	=	=	
Maintenance	=	=	-	-	=	
Safety	+	+	-	-	+	A
Environment	=	=	=	=	=	
Fabrication	+	-	-	-	+	
Transportation	=	=	-	-	=	T
Installation	=	-	-	-	=	
Performance	=	=	=	=	=	
Storage	=	=	-	-	=	U

Total (+)	3	2	1	1	4	
Total (-)	0	2	7	7	0	
Total (=)	7	7	3	3	8	M
Total	+3	0	-6	-6	+4	

Note: + (+1 point): Property is better than datum. - (-1 point): Property is worse than datum.
= (0 point): Property is comparable to datum.

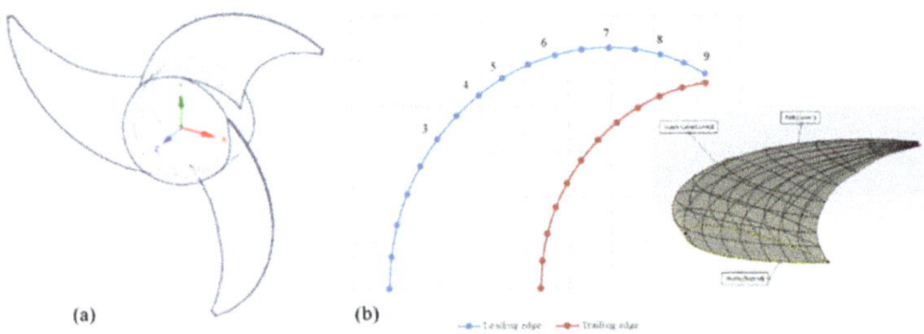

Fig. 7 (a) The 3D model of the finalised design turbine and (b) its blade profile with the NACA0018 cross sectional profile

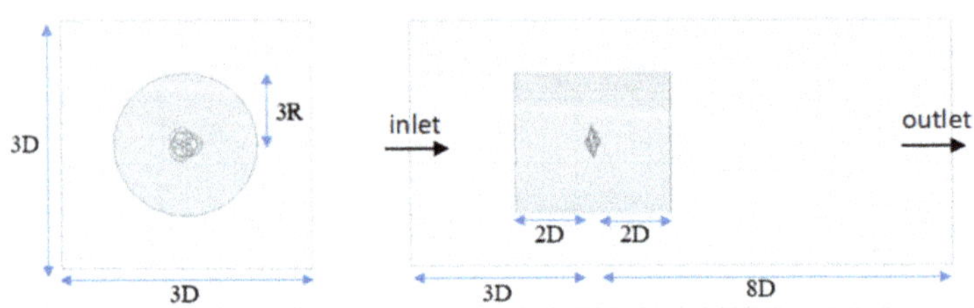

Fig. 8 The numerical domain model for the performance analysis (a) front view (b) side view.

Results and Discussion

The initial performance of the turbine was investigated at a pitch angle of 0°. The numerical analysis produced showed that the turbine performed optimally at a $C_P = 0.24$ at a TSR = 4 whereas the thrust coefficient of the turbine reached a maximum value of 0.29.

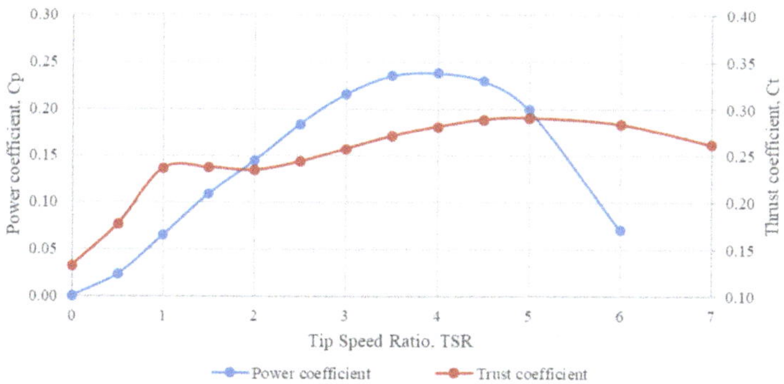

Fig.9 Power and thrust coefficient performance of the turbine against TSR.

The full range of the operational TSR is between 0 < TSR < 6.5. In comparison, a generic horizontal axis tidal turbine has the optimal C_P of more than 0.4 with the optimal TSR at around 6 and a full range of 0 < TSR < 12 [10]. It supports that the turbine operates optimally at a lower TSR albeit at a lower C_P.

Summary

This study shows the procedure for designing a tidal turbine for low velocity application by generating five conceptual designs based on the literature reviews. Appropriate criteria for the matrix were carefully selected to represent the production, installation, and operational conditions of the turbine. The finalised design i.e. Design 1 was further analysed using numerical model was conducted to investigate the performance of the turbine. It was found that the turbine operates optimally at a C_P of 0.4 at TSR = 4.

References

[1] R. Rosli, R. Norman and M. Atlar, Experimental investigations of the Hydro-Spinna turbine performance. Renewable Energy. 99 (2016) 1227-1234. https://doi.org/10.1016/j.renene.2016.08.034

[2] F. O. Rourke, F. Boyle and A. Reynolds, Tidal energy update 2009. Applied Energy. 87 (2010) 398-409. https://doi.org/10.1016/j.apenergy.2009.08.014

[3] M. J. Khan, G. Bhuyan, M. T. Iqbal and J. E. Quaicoe, Hydrokinetic energy conversion systems and assessment of horizontal and vertical turbines for river and tidal application: A technology status review. Applied Energy. 86 (2009) 1823-1835. https://doi.org/10.1016/j.apenergy.2009.02.017

[4] C. Bai, F. Hsiao, M. Li, G. Huang and Y. Chen, Design of 10kW Horizontal Axis Wind Turbine (HAWT) Blade and Aerodynamic Investigation Using Numerical Simulation. Procedia Engineering 67 (2013) 279-287. https://doi.org/10.1016/j.proeng.2013.12.027

[5] F. B. Hsiao, C. J. Bai and W. T. Chong, The Performance Test of Three Different Horizontal Axis Wind Turbine (HAWT) Blade Shapes Using Experimental and Numerical Methods. Energies. 6 (2013) 2784-2803. https://doi.org/10.3390/en6062784

[6] C. J. Bai and F. B. Hsiao, Code Development for Predicting the Aerodynamic Performance of a HAWT Blade with Variable-Speed Operation and Verification by Numerical Simulation, in 17th National Computational Fluid Dynamics (CFD) Conference, Taoyuan, Taiwan, 2010.

[7] C. J. Bai and F. B. Hsiao, Using CFD Computation for Aerodynamic Performance Design and Analysis of Horizontal Axis Wind, in 15th National Computational Fluid Dynamics (CFD) Conference, Kaohsiung, Taiwan, 2010.

[8] N. S. Tachos, A. E. Filios, D. P. Margaris and J. K. Kaldellis, Computational Aerodynamics Simulation of the NREL Phase II Rotor. The Open Mechanical Engineering Journal. 3 (2009) 9-16. https://doi.org/10.2174/1874155X00903010009

[9] Information on https://www.andritz.com/products-en/hydro/products/tidal-current-turbines

[10] A.S Bahaj, A.F. Molland, J.R. Chaplin and W.M.J. Batten, Power and thrust measurement of marine current turbines under various hydrodynamic flow conditions in a cavitation tunnel and a towing tank. Renewable Energy. 32 (2007) 407-426. https://doi.org/10.1016/j.renene.2006.01.012

Influence of ultra-fine friction grinding on microstructured Sago starch particles and suspension films

AFIQAH Nabihah Ahmad[1,a], NAMASIVAYAM Navaranjan[1,b], HIROSHI Uyama[2,c], SYAZANA Abdullah Lim[1,d*]

[1] Food Science and Technology Programme, School of Applied Sciences and Mathematics, Universiti Teknologi Brunei, Tungku Highway, Gadong, BE1410, Brunei Darussalam

[2] Department of Applied Chemistry, Graduate School of Engineering, Osaka University, 2-1 Yamadaoka, Suita, Osaka 565-0871, Japan

[a]afiqahahm95@gmail.com, [b]nava.navaranjan@utb.edu.bn, [c]uyama@chem.eng.osaka-u.ac.jp, [d]syazana.lim@utb.edu.bn

Keywords: Green Fabrication, Sago, Starch, Micro-Particles, Mechanical Grinder

Abstract. Microstructured sago (*Metroxylon sagu*) possesses an excellent potential to be used for numerous applications. In this paper, sago starch microparticles (SMPs) were fabricated using an ultra-fine grinder without any chemical treatments or purification steps thus making it a "green" and an eco-friendly procedure. The effects on increasing the numbers of passages through the grinder on the resulting SMPs and as suspension films were investigated. We observed that after 7 cycles the average diameter was reduced to a microscale region of 0.811 ± 0.754 μm (811 nm). The highest tensile strength was shown to be the highest after 3 cycles of 2.308 ± 0.210 MPa and repeated cycles beyond 7-cycled SMPs process had resulted in damaged microparticles. Furthermore, water vapor permeability was observed in SMPs with higher number of grinding cycles, with 7-cycled SMPs being 15.522 ± 0.184 g mm/m^2 h Pa and 10-cycled SMPs, 19.763± 0.233 g mm/m^2 h Pa. Our results demonstrated that the final particle size, water vapor permeability and tensile strength of the SMPs suspensions films were affected by the number of passages through the mechanical grinder.

Introduction

After cellulose, starch is the second most abundant carbohydrate in nature [1]. Due to their abundance, cost-saving, biocompatibility and renewability, starch and its derivatives have emerged as important biomaterials that can be exploited for various applications, most notably in biomedical and industrial sectors. Some of the promising applications are as drug carriers [2], biodegradable packaging materials [3] and fat replacers in food materials [4]. Microparticles have also wide applications in novel food products development since structural changes to micro-size levels can affect physicochemical and sensory properties qualities of food thus improving the quality of processed foods.

Several methods to fabricate starch microparticles have been explained in the literature by adipic acid-modified dry preparation technique, aqueous-alcoholic treatment, molecular rearrangement of short-chain glucans, rapid ultrasound, electrostatic spray and supercritical CO_2, premix membrane emulsification, reactive extrusion, inverse crosslinking-emulsion method and alcohol–alkaline treatment. However, these methods involve chemicals treatments or addition of chemical reagents and purification steps to obtain a final product, which could be undesirable for certain applications. In this study, a "green" process was employed to obtain microparticles of sago starch. The process involved with an ultra-fine friction grinder or known as "super mass colloider (SMC)" without using any chemical treatments or reagents. The ultra-fine particles obtained *via* this method permits the use of concentrated suspensions without prior filtration [5].

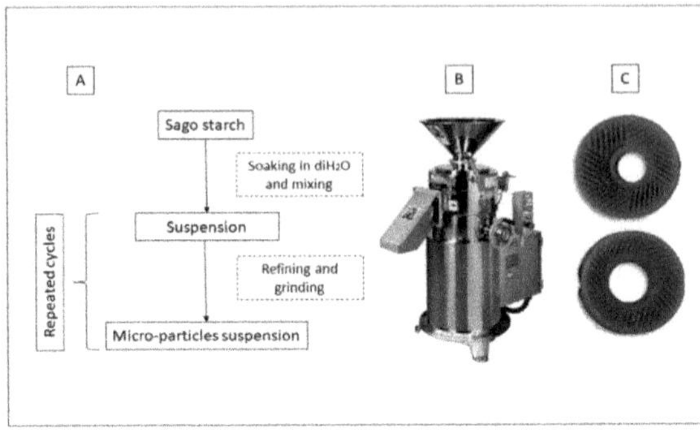

Figure 1. (a) Schematic flow of sago starch microparticles fabrication; (b) Super Masscolloider (Masuko Sangyo, Saitama, Japan); (c) grinding stones.

As seen in Figure 1(c), SMC comprises of two grinding discs or rotary grinder with a grove design to ensure an accurate flow of the feed toward the grinding area. The gap distance amongst the two discs can be modified vertically by displacing the rotary disc in 1/100 mm increment. Furthermore, higher compression, shearing, and rolling friction forces can be achieved by employing a smaller gap resulting in closer contact of rotary grinders. The ceramic and non-porous grinding stones avoid bacterial growth and cracks linked with thermal stress that can be normally found in conventional grinding. The energy consumption of this technique is also comparable with homogenization and micro-fluidization [7]. Conversely, the particle size acquired after ultra-fine grinding is larger than that obtained after other methods such as high-pressure homogenization. Various sources have been described and reported in literature using this technique to obtain micro- and nano materials. Such sources include cellulose, jute fibers, date palm, carrotsbamboo fibres, turmeric and sugar palm.

Sago (*Metroxylon sagu*) palm is a type of starch native to South East Asia [8] with its stem possesses the highest starch content as compares with maize, rice and cassava [9]. Sago palm is known to be exceptionally resistant to unfavourable weather environments portraying its important part in food security issues. The chemical and physical characteristics of sago starch as a biomaterial, however, was less known to researchers in the past. Consequently, there have been a recent rise in scientific and commercial interest in the investigation and development of sago starch particularly modifications in chemical and physical structure, molecular structure, physicochemical and functional properties [9]. At present, research in sago starch potential is still in its infancy compared to its starch counterparts.

Experimental

Materials

Only sago starch powder - acquired from NZA Enterprise, a local sago production company in Brunei Darussalam and deionized water were used to produce sago micro-particles (SMPs) in this work.

Fabrication of micro-particles

Figure 1 shows the production steps of SMPs. Sago starch (2 wt. %) powder were mixed with deionized water and stirred for 2 days in room temperature. The sago homogenates were processed using SMC (Masuko Sangyo, Saitama, Japan) that consists of two grinding stones that rotates at a high speed. Procedures were repeated 10 times to obtain SMPs in smaller sizes and samples produced were denoted based on their cycles as x-SMPs, whereby, each batch of grinding was counted as one pass.

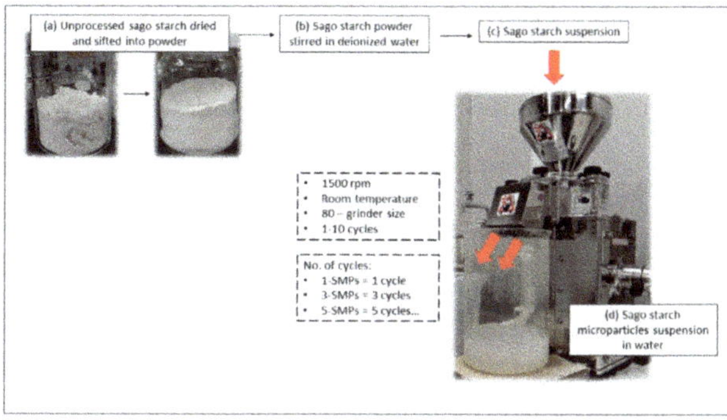

Figure 2. Stages of sago-microparticles production using ultra-fine grinder machine.

Preparation of sago micro-particles suspension films
Films of the sago suspensions after every cycle were prepared using solvent casting method in which 40 mL aliquots of the suspension were poured onto teflon plates and placed in an oven (Eyela Vacuum Oven, VDS-201SD) at 30 ± 1 °C for 15 hours. Prior to any testing, the film specimens were preconditioned in a climate chamber at 25 °C and 50% RH for at least 48 hours.
Micro-particles analysis

Scanning electron microscopy
The particles microstructure and fracture surfaces were observed a scanning electron microscope (SEM) instrument (SU3500, Hitachi, Tokyo, Japan). A drop of undiluted microparticles suspension was deposited on a glass plate attached on metal stubs, formerly covered with double sided adhesive. All specimens were sputter-coated with gold using Eiko Sputter Coater under vacuum condition.

Particle size determination
Dynamic light scattering method was performed using particle size analyzer device (ELSZ-2000 Series, Otsuka Electronics, Japan) to determine the distribution of particles. The undiluted microparticles dispersions were sonicated prior to the measurement and immediately measured to prevent sample precipitation.

Characterization of SMPs suspension films
Thickness, water solubility and swelling kinetics
The thickness of suspension films was carried out using a digital micrometer (Mitutoyo, Japan) with an accuracy of ±0.001 mm. The reported results were the mean value of ten measurements for each sample expressed in millimeters (mm).

Total water solubility of SMPs suspension films were defined as the weight of dissolved film after being immersed in distilled water. Film specimens (20 mm by 20 mm) were prepared dried at 100 ± 2°C for 24 hours in a laboratory oven and weighed to determine the initial dry mass. Film specimens were then immersed in performed in 50 mL of distilled water for five hours with constant agitation. The resultant pieces of films were taken out, filtered and dried at 100 ± 2°C for 24 hours.

Prior to determining the swelling kinetics of SMPs suspension films, the specimens were prepared by cutting into 10 mm by 10 mm squares and dried in an oven at 60°C for 24 hours. The dried film specimens were immersed in beakers containing distilled water and weighed for every 2 hours. The test was carried out in triplicates and the swelling ratio (SR) was calculated by the following equation:

$$SR\ (\%) = \frac{W_s - W_d}{W_d} \times 100 \tag{1}$$

where W_s and W_d are the swollen and dry weight of films at time t, respectively.

Water vapor permeability

Prior to water vapor permeability (WVP) analysis, the films were first conditioned in a desiccator at 25 °C for 48 hours. SMPs suspension films were cut in circular shape about 30 mm in diameter and positioned over the test cup, which was prefilled with anhydrous calcium chloride, leaving 5 mm to the top. The suspension films were then fastened with melted paraffin. The assembly was placed in a chamber conditioned at 25 °C and 100% RH and weight increments of the cup were measured and plotted at intervals. The WVP was calculated as follows:

$$WVP = \frac{m \times d}{(A \times t \times P)} \tag{2}$$

where d is the thickness of films (m), m is the weight increment of the cup (g), A is the area exposed (m^2), t is the permeation time (h), and P is the difference in partial pressure of water vapor across the film (Pa). All specimens were tested in triplicates.

Tensile test

Film specimens were cut into dumbbell shape strips and dried in oven at 50 °C prior to the investigation to eliminate any leftover moisture. The samples were stretched at 0.500 mm/s with 100 N load. Tensile strength (MPa), Young's Modulus (MPa) and elongation at break (%) were assessed using Universal Testing Machine (EZ Graph, Shimadzu, Kyoto, Japan).

Statistical analysis

All data in this study were expressed as means ± standard deviations of triplicates. Statistical significance of differences between mean values was assigned by one-way ANOVA with Tukey's test at a confidence level of $P < 0.05$ using OriginPro software (OriginLab Corporation, USA).

Results and discussion

Morphological and Mechanical studies of SMPs

The surface morphology of microparticles in Figure 3 shows that increased number of cycles produced particles of finer sizes. Clear differences can be seen in Fig. 3(a), whereby 0-SMPs exhibited an average diameter of 25.140 ± 1.853 μm. Minimal polydispersed microparticles were also obtained by ultra-fine grinder after 7 cycles where average diameter was reduced further to a microscale region of 0.811 ± 0.754 μm (811 nm) in Fig. 3(b).

Materials Research Proceedings 29 (2023) 487-492 https://doi.org/10.21741/9781644902516-55

An increment in grinding cycles of SMPs had resulted in lower tensile strength, Young's modulus as well as elongation at break ($P < 0.05$). The highest tensile strength was observed in 3-SMPs of 2.308 ± 0.210 MPa and it gradually decreases with increasing grinding cycles.

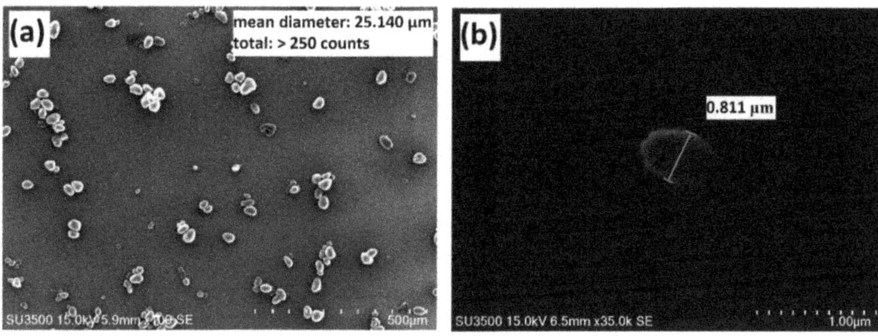

Figure 3. SEM images of (a) sago starch at 0 cycle at x100 and (b) single sago micro-particle after 7 cycles at x35.0k.

Consequently, repeated cycles beyond 7-SMPs process had resulted in damaged microparticles as shown in Fig 4.

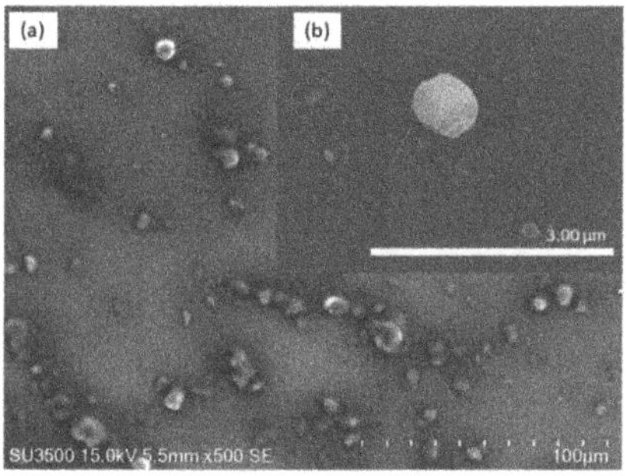

Figure 4. (a) Destroyed SMPs after 10 complete cycles. (b) Breakdown of smaller particles from a single microparticles.

Particle size distribution

The result by dynamic light scattering exhibited a slightly broad particle size distribution with the highest intensity of micro-particles recorded at 11.338% for 811.561 µm.

Physical properties of SMPs suspension films

Thickness of suspension films ranged between 0.087 mm to 0.093 mm and there were no significant differences detected in these film suspensions. In terms of total solubility in water after 24 hours immersion, it can be deduced that water solubility was significantly higher after 10 passages. Increasing the grinding cycles had caused the suspension films to become more susceptible to water uptake and retention. The increase in water vapor permeability was observed in SMPs with higher number of grinding cycles, with 7-SMPs being 15.522 ± 0.184 g mm/m^2 h Pa and 10-SMPs, 19.763± 0.233 g mm/m^2 h Pa.

Conclusion

This work reports on green fabrication of sago starch microparticles using a rapid, simple, and eco-friendly ultra-fine grinding technique. Repeated cycles had proven to have major drawbacks where the fabricated microparticles were damaged resulting in cracks and fissures and decreased SMPs suspension films total water solubility, water vapor solubility and mechanical properties. Nonetheless, damaged granules still exhibited great potentials to be employed in several applications such as drug carriers in drug delivery systems and food processing especially considering their renewable source, nontoxicity and cost-effectiveness.

References

[1] A. Blennow, T.H. Nielsen, L. Baunsgaard, R. Mikkelsen and S.B Engelsen, Starch phosphorylation: A new front line in starch research, Trends Plant Sci 7(2002) 445-450 https://doi.org/10.1016/S1360-1385(02)02332-4

[2] J. Yang, F. Li, M. Li, S. Zhang, J. Liu, C. Liang, Q. Sun and L. Xiong, Fabrication and characterization of hollow starch nanoparticles by gelation process for drug delivery application, Carbohydr Polym 173 (2017) 223-232. https://doi.org/10.1016/j.carbpol.2017.06.006

[3] A.Mittal, S.Garg and S. Bajpai, Thermal decomposition kinetics and properties of grafted barley husk reinforced PVA/starch composite films for packaging applications. Carbohydr Polym (2020) https://doi.org/10.1016/j.carbpol.2020.116225 https://doi.org/10.1016/j.carbpol.2020.116225

[4] S. Punia, A.K. Siroha, K.Sandhu and M. Kaur, Rheological and pasting behavior of OSA modified mungbean starches and its utilization in cake formulation as fat replacer, Int J Biol Macromol, 128 (2019) 230-236. https://doi.org/10.1016/j.ijbiomac.2019.01.107

[5] J. Zhu, Y. Sun, W. Sun, Z. Meng, Q. Shi, X. Zhu, H. Gan, R. Gu, Z. Wu, and G. Dou, Calcium ion-exchange cross-linked porous starch microparticles with improved hemostatic properties. Int J Biol Macromol 134 (2019) 435-444. https://doi.org/10.1016/j.ijbiomac.2019.05.086

[6] J. Velásquez-Cock, P. Gañán, P. Posada, C. Castro, A. Serpa, C.H. Gomez, J.L. Putaux, and R. Zuluaga, Influence of combined mechanical treatments on the morphology and structure of cellulose nanofibrils: Thermal and mechanical properties of the resulting films. Ind Crops Prod. 85 (2016) 1-10. https://doi.org/10.1016/j.indcrop.2016.02.036

[7] R. Shamsudin, C.S. Ling, N.M. Adzahan and W.R. Daud, Rheological properties of ultraviolet-irradiated and thermally pasteurized Yankee pineapple juice, J. Food Eng. Journal of Food Engineering, 116 (2013) 548-553. https://doi.org/10.1016/j.jfoodeng.2012.12.031

[8] S.M. Sapuan, J. Sahari, M.R. Ishak and M.L. Sanyan In: Sugar palm biofibers, biopolymers, and biocomposites, CRC Press/Taylor & Francis Group.Boca Raton, FL, Florida, 2018 https://doi.org/10.1201/9780429443923

[9] F. Zhu, Recent advances in modifications and applications of sago starch, Food Hydrocoll, 96 (2019) 412-423. https://doi.org/10.1016/j.foodhyd.2019.05.035

Towards green extraction: A study on solvent selection and methods for antioxidant activities of *Diplazium esculentum* Retz. and *Stenochlaena palustris*

HAZIMAH Sharifulazar[1,3,a], AIDA Maryam Basri[1,3,b*], MASAYOSHI Arai[2,3,c]

[1] Food Science and Technology, School of Applied Sciences and Mathematics, Univerrsiti Teknologi Brunei, Tungku Link Road, Gadong, BE1410, Brunei Darussalam

[2] Laboratory of Natural Products for Drug Discovery, Graduate School of Pharmaceutical Sciences, Osaka University, 1-6 Yamadaoka, Suita, Osaka, 565-0871, Japan

[3] Osaka University ASEAN Campus Brunei Darussalam, Jalan Kianggeh, BSB111, Brunei Darussalam

[a]hazimah.sharifulazar@gmail.com, [b]aida.basri@utb.edu.bn, [c] araim@phs.osaka-u.ac.jp

Keywords: Aqueous, Green Extraction, Antioxidant, *Diplazium Esculentum* Retz., *Stenochlaena Palustris*

Abstract. Green extraction method by combination of stirring method with no presence of heat and the use of aqueous as solvent were highlighted in this study. Various solvents (aqueous, chloroform, ethyl acetate, hexane and methanol) and two extraction methods (stirring and Soxhlet) were used to study their effect on the yield, qualitative phytochemical content, and antioxidant properties of *Diplazium esculentum* Retz. and *Stenochlaena palustris*. Stirring extraction method in aqueous has shown to provide highest yield in both plant species with *D. esculentum* at 8.88% and *S. palustris* at 9.40%. Saponin was also seen present in both aqueous extracts qualitatively. In the case of FRAP (Ferric Reducing Antioxidant Power) assay, aqueous extract of *D. esculentum* (DEA) had the highest value with 687.57 ± 0.01 µg Fe(II)E/ml, while stirring method in various solvents had shown to produce high antioxidant activities compared to Soxhlet method. This study revealed that aqueous extracts by stirring method is a promising method for extraction of plant materials and at the same time leading towards a green environment.

Introduction.

Plants are known to contain bioactive components which are very beneficial for health. These bioactive components include phenolic, flavonoid, alkaloid and families of terpenoid that contributes to antioxidant properties through various mechanisms [1]. *Diplazium esculentum* Retz. and *Stenochlaena palustris* are two medicinal ferns shown to have antioxidant activities when the extraction was done *via* Soxhlet extraction method in methanol solvent [2,3]. Traditional extraction methods such as Soxhlet extraction, maceration and decoction tend to be time-consuming and requires a large volume of solvent such as methanol and hexane which are harmful and toxic. The methods also involve presence of heat that can cause degradation of bioactive compounds. Hence, the need to reduce the use of hazardous substances, and at the same time reduces the consumption of energy for extraction of plant materials is very crucial.

Green extraction process of plant materials has been a promising approach that aims to save the environment, energy and promote sustainability. This includes the use of green and sustainable solvents that is non-toxic to health, has low environmental impact and provides low cost and energy. Water is potentially considered as a green solvent as it is recognised as safe and environmental-friendly. The use of water as an extraction solvent combined with a suitable extraction method such as microwave assisted extraction and ultrasound-assisted extraction methods have been shown to enhance extraction processes and improve efficiency [4]. However, there have been very limited studies on the combination of water as extraction solvent and the use of stirring method to extract plant materials. Extraction method by stirring has low cost, uses less energy and no heat is required, thus contributing towards green environment.

This study used various solvents (water/aqueous, chloroform, ethyl acetate, hexane and methanol) and two extraction methods (stirring and Soxhlet) to study their effect on the yield, qualitative phytochemical content, and antioxidant properties of *D. esculentum* Retz. and *S. palustris*. This paper will focus on green extraction process, hence the discussion will highlight on the effect of extraction method by stirring with no presence of heat and the use of aqueous as the extraction solvent.

Materials and Methods.

Reagents and chemicals. All chemicals used were of analytical grade and without further purification. Ascorbic acid, gallic acid, quercetin, sodium acetate buffer, 2,4,6-tris(2-pyridyl)-s-triazine (TPTZ), Folin–Ciocalteu's phenol reagent, single reagent 2,20-azino-bis(3-ethylbenziazoline-6-sulfonic acid) (ABTS), potassium peroxodisulfate, chloroform, ethyl acetate, hexane and methanol were purchased from Merck, Sigma-Aldrich (Darmstadt, Germany). Free radical 2,2-diphenyl-1-picrylhydrazyl (DPPH) powder was from Alfa Aesar, Thermo Fisher Scientific (Heysham, UK).

Plant materials. *Diplazium esculentum* Retz. (Voucher specimen AMB.DE.1) and *Stenochlaena palustris* (Voucher specimen AMB.SP.1) were identified and authenticated by Brunei National Herbarium (BRUN), under Forestry Department, Sg. Liang, Brunei.

Preparation of dried plant powder [5]. Plant powders were prepared following previous method with modifications. Fresh leaves samples of *D. esculentum* Retz. and *S. palustris* were rinsed, cleaned, and dried in an oven at 45°C for 3–4 days. The dried samples were then ground to powder using an electrical blender (Philips HR 2056/90TT, Netherlands) until fine (approximately 1000 μm).

Preparation of plant extracts by stirring method. In a 1:10 ratio of powdered sample and respective solvents (aqueous, chloroform, ethyl acetate, hexane, methanol), the sample was left to stir for 24 hours, and filtered. It was then concentrated using rotary evaporator and stored at 4 ± 1°C until further use.

Preparation of plant extracts by Soxhlet method [5]. 80g of dry powdered sample in 250ml of respective solvents (chloroform, ethyl acetate, hexane and methanol) for approximately 6 hours or until colourless. The extracts were then concentrated using rotary evaporator and stored at 4 ± 1°C until further use.

Preliminary qualitative phytochemical screening of plant extracts [6].

Phytochemical	Method
Saponins	Extracts were diluted with 20ml of distilled water and shaken for 15 minutes. Presence of saponins can be confirmed by formation of foam on top of the solution.
Steroids	To 5ml of extract in a test tube, 2ml of chloroform and 2ml of concentrated H_2SO_4 are added by the walls of the test tube. Presence of steroids confirmed by red colour in lower layer with yellow colour and green fluorescent in the sulfuric acid layer.
Flavonoids	10mg of extract were mixed with a few drops of diluted NaOH. Presence of flavonoids confirmed when a yellow colour disappears or turns colourless with the addition of a few drops of diluted H_2SO_4.
Alkaloids	10mg of extract was dissolved in 2ml of Wagner's solution. Presence of alkaloid is confirmed when appearance of reddish-brown coloured precipitate is observed.
Tannins	10 mg of extract was dissolved in 45% ethanol, and then boiled for 5 minutes before adding 1ml of 15% ferric chloride solution. Presence of tannins confirmed when there is appearance of dark blue or greenish black colour.

Materials Research Proceedings 29 (2023) 493-499 https://doi.org/10.21741/9781644902516-56

DPPH free radical scavenging ability antioxidant assay [7]. 0.1mM solution of DPPH was prepared by dissolving 0.394g of DPPH in 10ml of methanol. 10mg of plant extract/ascorbic acid was diluted in 10ml of solvent as stock solution and as positive standard, respectively. Serial dilutions were prepared in 10, 20, 30, 40, 50, 100, 250, 500 and 750 ug/ml. To 1ml of sample, added was 1ml of DPPH solution. The mixture was left to incubate in the dark for 30 minutes. A colour change from violet of the stable DPPH compound to yellow of the reduced DPPH compound can be seen. After incubation, the absorbance was read at 517.0nm using a UV-Vis spectrophotometer (JENWAY) 6850UV/Vis. Spectrophotometer). Blank sample was prepared by replacing the sample extract with the mother solvent added with DPPH solution. The readings were done in triplicates. DPPH % scavenging ability was calculated using the formula: % inhibition = [A0 – A1] / A0] x 100, where A0 is the absorbance of control and A1 is the absorbance of the extract or standard. IC_{50} value is calculated from the formula obtained in the % inhibition against concentration graph.

Ferric reducing antioxidant power assay [8]. FRAP reagent was prepared by adding 10mM of 2,4,6-Tris(2-pyridyl)-1,3,5-triazine (TPTZ) dissolved in 40mM of hydrochloric acid, 20mM of ferrous chloride ($FeCl_3$) and 300mM of sodium acetate buffer (pH 3.6) in ratio of 1:1:10. Under dark condition, 100ul of sample, 900ul of distilled water and 2ml of FRAP reagent were mixed and left to incubate for 3minutes at room temperature. A colour change into intense blue suggests a reduction of the ferric tripyridyltriazine complex. Using a spectrophotometer, absorbance was read at 593nm. The ferric reducing antioxidant ability was measure using the formula: FRAP value of sample = Absorbance (sample) x FRAP value of standard / absorbance (standard).

ABTS antioxidant assay [9]. ABTS working solution was prepared by 1:1 ABTS to 2.46 mM potassium persulfate solution. The mixture was left in the dark for 16-18hr in room temperature. The mixture was then diluted with ethanol to reach absorbance of 0.700-0.900nm before use. In a test tube, 20ul of sample with 80ul of ethanol was combined and added 2ml of ABTS working solution. A colour change from dark blue to colourless suggest a reaction. The absorbance was read at 734nm. The percentage of inhibition was measured and compared to the standard Gallic acid using the formula: ABTS·+ scavenging effect (%) = ((AB–AA)/ AB) ×100, where AB is absorbance of ABTS radical + ethanol and AA is absorbance of ABTS radical + sample/standard The reading was taken in triplicates.

Determination of total phenolic content (TPC) [10]. 10mg of extract diluted in 10ml of extract solvent; methanol and chloroform, to make concentration of 1mg/ml stock solution. Stock solution was then diluted to prepare concentrations of 500, 250, 100, 50, 10 ppm. A mixture of 0.5ml of each concentration, 2.5ml of 10% Folin-Ciocalteu's reagent in water and 2.5% ml 7% $NaHCO_3$ were prepared. The samples were left to incubate at 45°C for 45 minutes. After incubation, absorbance was read at 765nm. When phenolic compounds are introduced, the Folin-Ciocalteu's reagent produces a blue colour. The samples were prepared in triplicates. Positive control of the assay was gallic acid. The total phenolic content was calculated using the formula: A = cV / m, where A is the total phenolic content, c is the concentration of gallic acid obtained from the curve (mg/ml), V is the volume of extract in ml and m is the weight of plant extract (g).

Determination of total flavonoid content (TFdC) [10]. 0.5ml of extract or standard solution in a test tube and adding 0.1ml of 10% aluminium chloride followed by 0.1ml of 1M potassium acetate. In the same test tube, 1.5ml of 80% methanol was added followed by 2.8ml of distilled water. They were then incubated for 30 minutes in the dark. Absorbance was read at 415nm. Standard solution was made from Quercetin. The flavonoid content was measured using the formula: A = (c x v) /m, where A is total flavonoid content in mg/g Quercetin equivalent, c is concentration of quercetin in mg/ml obtained from the standard graph equation, v is the volume of extract and m is the mass of extract. Blank was calculated as mean absorbance at 0.0959.

Determination of total flavonol content (TFC) [11]. 1ml of extract or standard solution in a test tube. In the same test tube, 1ml of 2% aluminium chloride was added to each test tube, followed by 3ml of sodium acetate. The mixture was incubated for 2hrs after shaking gently, in room temperature. Absorbance was read at 440nm using spectrophotometer. Blank was calculated as mean absorbance at 0.0032.

Data analysis. The experiments were performed in triplicate. All analyses were performed with Analysis ToolPak extension in Excel 2016 and results presented as mean ± standard deviation. A regression correlation using Pearson's used to observe any correlation between assays. ANOVA test was used to see any significant difference between the type of solvents used for the crude extracts. Statistical analyses were considered significant when $p < 0.05$.

Results and Discussion.

Extraction yields can be used to indicate the effects of extraction in respective solvents. Table 1 shows the percentage yield for all extracts of both *D. esculentum* and *S. palustris* which was conducted using two different methods: 24-hour stirring method (s) with no presence of heat and Soxhlet method. DEA: *D. esculentum* aqueous extract; DEC: *D. esculentum* chloroform extract; DEE: *D. esculentum* ethyl acetate extract; DEH: *D. esculentum* hexane extract; DEM: *D. esculentum* methanol extract; SPA: *S. palustris* aqueous extract; SPC: *S. palustris* chloroform extract; SPE: *S. palustris* ethyl acetate extract; SPH: *S. palustris* hexane extract; SPM: *S. palustris* methanol extract. Interestingly, aqueous extracts by stirring method of both plant species have shown to produce the highest yield compared to other solvents, with 8.88% and 9.40% for *D. esculentum* and *S. palustris* aqueous extracts, respectively. The higher polarity of water and the kinetics of stirring method may have caused higher and faster diffusion or partition rate of the solute to move from the solid matrix into solvent [12]. This reveals that the use of water as a sole extractant with no presence of heat is possible to give a higher yield of aqueous extracts.

Table 1. The percentage yield of D. esculentum (DE) and S. palustris (SP) extracts by stirring and Soxhlet method in different solvents of aqueous (A), chloroform (C), ethyl acetate (E), hexane (H) and methanol (M).

Extract	Yield (%)	Extract	Yield (%)
		Stirring method	
DEA	8.88	SPA	9.40
DEC	0.77	SPC	1.17
DEE	1.60	SPE	3.90
DEH	4.70	SPH	1.84
DEM	5.95	SPM	0.78
		Soxhlet method	
DEC	3.34	SPC	0.69
DEE	1.33	SPE	0.56
DEH	2.00	SPH	0.54
DEM	0.70	SPM	0.99

Preliminary phytochemical screening on plant extracts were observed qualitatively to indicate the presence of known antioxidants which are saponins, steroids, flavonoids, alkaloids and tannins. There is no aqueous extract for both plant species using Soxhlet method as it is not efficient due to high temperature of 100°C is required for evaporation. Nevertheless, extraction by Soxhlet method has shown the presence of more phytochemicals qualitatively when compared to extraction by stirring method, as seen in Table 2. However, this does not confirm the complete absence of the phytochemical in that extract and their antioxidant potential, hence, we proceed to the analysis of antioxidant activities and specific phytochemical content for further evidence.

Materials Research Proceedings 29 (2023) 493-499 https://doi.org/10.21741/9781644902516-56

Table 2. Phytochemical screening of D. esculentum and S. palustris extracts, (+) = present; (-) absent.

Property	Saponins		Steroids		Flavonoids		Alkaloids		Tannins	
Solvents	*D. esculentum*	*S. palustris*	*D. esculentum*	*S. palustris*	*D. esculentum*	*S. palustris*	*D. esculentum*	*S. palustris*	*D. esculentum*	*S. palustris*
Stirring method										
Aqueous	+	+	-	-	-	-	-	-	-	-
Chloroform	+	+	-	-	-	-	-	-	-	-
Ethyl acetate	-	-	-	-	-	-	-	+	-	-
Hexane	-	-	+	+	-	-	-	-	-	-
Methanol	+	+	-	-	-	-	-	+	-	-
Soxhlet method										
Chloroform	+	+	-	+	+	+	+	+	-	-
Ethyl acetate	+	+	+	+	-	+	+	+	-	-
Hexane	-	-	+	-	-	-	-	-	-	-
Methanol	-	+	-	-	+	+	+	+	-	-

In vitro antioxidant activities of plant extracts. There are various colorimetric methods that can be used to analyse *in vitro* antioxidant activities. These methods undergo different reaction mechanisms, hence the need to analyse antioxidant activities of samples using more than one colorimetric method to obtain promising results. There are six different antioxidant assays used in this study, which are 2, 2–diphenyl–1–picrylhydrazyl (DPPH), 2,2'-azino-bis 3-ethylbenzothiazoline-6-sulphonicacid (ABTS), ferric reducing-antioxidant power (FRAP), total phenolic content (TPC), total flavonoids content (TFdC) and total flavonol content (TFC).

Based on the results in Table 3, it is interesting to observe that aqueous extract and stirring method revealed high antioxidant activities and phytochemical content in some assays conducted. This is particularly in FRAP assay whereby aqueous extract of *D. esculentum* (DEA) had the highest FRAP value compared to the other solvents with 687.57 $\pm$ 0.01 μg Fe(II)E/ml, which also correlates with its high TPC value of 29.06 ± 4.21 mg GAE/g showing that phenolic content has relation towards ferric reducing potential. Water is also known to be a good solvent for phenolic acids which could be the reason for the high TPC value in DEA [13].

When comparing between the two extraction methods, stirring method has shown to produce high antioxidant activity and phytochemical content. This can be seen in DPPH assay where DEC (stirring) extract gave the highest IC_{50} value of 42.19 ± 0.407 μg/ml, the highest value in ABTS assay is seen by SPE (stirring) extract of 84.72 ± 0.001 μg GAE/ml, the highest value in FRAP assay is DEA (stirring) extract with 687.57 $\pm$ 0.01 μg Fe(II)E/ml, DEM (stirring) extract gave the highest TPC value of 31.87 ± 5.32 mg GAE/g and DEA at 29.06 ± 4.21 mg GAE/g, and lastly, highest TFC value is by SPC (stirring) extract with 63.77 ± 0.00 mg QE/g. This shows that stirring method without the presence of heat allows better permeability for compounds in plant materials to be extracted into the solvent matrix as compared to Soxhlet method [12].

Conclusion

Extraction method by combination of stirring method with no presence of heat and the use of aqueous as solvent is a promising method towards green environment. This method is non-toxic, has low environmental impact, provides low cost and energy and can prevent the degradation of bioactive compounds as there is no heat involved. Aqueous extracts of *D. esculentum* and *S. palustris* produced the highest yield compared to other solvents, when using the stirring method instead of Soxhlet method. In addition, *D. esculentum* and *S. palustris* have shown to have good antioxidant activities that can be applied into many areas such as functional food and pharmaceutical sectors.

Acknowledgements
We would like to thank Miss Hazirah Abd Hamid for the preparation of plant voucher specimens at Brunei National Herbarium (BRUN), and this research was funded by UTB/GSR/2/2018(1).

Table 3 In vitro antioxidant activities and total phenolic (TPC), flavonoid (TFdC) and flavonol (TFC) contents of D. esculentum and S. palustris extracts

Extract	DPPH IC_{50} (µg AAE/ml)	ABTS µg GAE/ml	FRAP µg Fe(II)E/ml	TPC mg GAE/g	TFdC mg QE/g	TFC mg QE/g
			Stirring method			
DEA	953.56 ± 0.112 [14]	42.49 ± 0.001	687.57 ± 0.01	29.06 ± 4.21 [14]	2.74 ± 0.01	51.54 ± 0.22
DEC	42.19 ± 0.407	35.67 ± 0.000	47.46 ± 0.01	13.97 ± 12.89	44.79 ± 0.10	3.67 ± 0.04
DEE	1022.85 ± 0.128	41.72 ± 0.001	31.33 ± 0.00	6.39 ± 7.97	57.10 ± 0.17	5.52 ± 0.04
DEH	187.55 ± 0.296	36.39 ± 0.000	48.55 ± 4.7E-05	0.35 ± 15.40	33.71 ± 0.12	16.47 ± 0.04
DEM	65.98 ± 0.918	48.66 ± 0.001	118.25 ± 0.00	31.87 ± 5.32	40.26 ± 0.10	6.05 ± 0.00
SPA	838.26 ± 0.199 [14]	65.93 ± 0.000	92.70 ± 0.00	27.18 ± 0.46 [14]	12.92 ± 0.19	-
SPC	2387.34 ± 0.276	50.72 ± 0.000	40.16 ± 4.7E-05	6.70 ± 17.06	53.94 ± 0.18	63.77 ± 0.00
SPE	372.41 ± 0.223	84.72 ± 0.001	63.51 ± 0.00	5.98 ± 5.01	46.86 ± 0.24	1.80 ± 0.01
SPH	1238.28 ± 0.235	48.72 ± 0.001	48.52 ± 0.00	24.35 ± 21.21	53.47 ± 0.22	48.56 ± 0.04
SPM	127.16 ± 0.390	47.58 ± 0.000	56.33 ± 0.00	4.81 ± 12.71	64.95 ± 0.18	5.38 ± 0.03
			Soxhlet method			
DEC	2372.43 ± 0.066^{*b}	43.63 ± 0.001^{*b}	39.04 ± 0.00^{*b}	18.77 ± 10.69	28.77 ± 0.12^{*b}	2.84 ± 0.01^{b}
DEE	61.72 ± 0.264^{*b}	46.37 ± 0.001^{*b}	142.36 ± 0.00^{*b}	14.66 ± 3.77*	60.23 ± 0.21^{b}	59.75 ± 0.01^{*b}
DEH	476.18 ± 0.413^{*b}	48.41 ± 0.001^{*b}	37.39 ± 0.00^{*b}	20.21 ± 5.77*	24.66 ± 0.32^{b}	27.12 ± 0.11^{b}
DEM	1890.28 ± 0.081^{*b}	51.98 ± 0.001^{*b}	80.51 ± 0.00^{*b}	5.41 ± 5.48	34.26 ± 0.10^{b}	47.98 ± 0.00^{*b}
SPC	415.86 ± 0.115^{*d}	51.77 ± 0.001^{*d}	42.38 ± 0.00^{*d}	19.87 ± 3.09*	15.92 ± 0.10^{*d}	13.86 ± 0.01^{*d}
SPE	433.00 ± 0.245^{*d}	46.60 ± 0.001^{*d}	104.96 ± 0.00^{*d}	26.21 ± 6.47*	60.05 ± 0.13^{d}	4.06 ± 0.03^{d}
SPH	275.34 ± 0.086^{d}	46.91 ± 0.000^{*d}	34.97 ± 0.00^{*d}	30.64 ± 5.88	48.35 ± 0.08^{d}	18.36 ± 0.00^{*d}
SPM	777.10 ± 0.144^{d}	47.28 ± 0.001^{d}	110.55 ± 0.00^{*d}	13.85 ± 13.16	90.35 ± 0.17^{*d}	38.31 ± 0.02^{d}
Standard	Ascorbic acid 14.04 ± 0.141	Gallic acid 100ug/ml 173.90 ± 0.023	Quercetin (100ug/ml) 100.15 ± 0.00	-	-	-

Conflict of interest.
There is no conflict of interests.

References

[1] J. Oh, H. Jo, A. R. Cho, S.-J.Kim & J. Han, Antioxidant and antimicrobial activities of various leafy herbal teas. Food Control. 31(2) (2013) 403-409. https://doi.org/10.1016/j.foodcont.2012.10.021

[2] A. Semwal, M.S. Farswan, K. Upreti, K. Bhatt, & K. Upadhyaya, Evaluation of antioxidant activity of some pteridophytes. Int. J. Herb. Med. 1(1) (2013).

[3] A.H. Ndanusa, D. Cicuzza, & M.M. Siddique, Analysis of the phytochemical contents and anti-oxidative properties of Stenochlaena palustris. Int. Food Res. J. 27(5) (2020).

[4] S. Carpentieri, F. Soltanipour, G. Ferrari, G. Pataro, F. Donsì, Emerging Green Techniques for the Extraction of Antioxidants from Agri-Food By-Products as Promising Ingredients for the Food Industry. Antioxidants. 10, (2021) 1417. https://doi.org/10.3390/antiox10091417

[5] N. N. Azwanida, A review on the extraction methods use in medicinal plants, principle, strength and limitation. Med. Aromat. Plants, 4:3 (2015).

[6] G. E. Trease and W. C. Evans, Pharmacology. 11th ed. London: Bailliere Tindall Ltd (1989)

[7] S. Akter, M.M Hossain, I. Ara & P. Akhtar, Investigation of in vitro antioxidant, antimicrobial and cytotoxic activity of Diplazium esculentum (Retz). Sw. Int. J. Adv. Pharm. Biol. Chem. 3(3) (2014).

[8] N. S. Rajurkar & S. M. Hande, Estimation of phytochemical content and antioxidant activity of some selected traditional Indian medicinal plants. Indian J. Pharm. Sci. 73(2) (2011). https://doi.org/10.4103/0250-474X.91574

[9] P. Tipduangta, J. Julsrigival, K. Chaithatwatthana, N. Pongterdsak, P. Tipduangta & S. Chansakaow, Antioxidant properties of Thai traditional herbal teas. Beverages, 5(3), (2019) 44. https://doi.org/10.3390/beverages5030044

[10] M.S. Stankovic, Total phenolic content, flavonoid concentration and antioxidant activity of Marrubium peregrinum L. extracts. Kragujevac J. Sci, 33, (2011) 63-72.

[11] N. Wuttisin, T. Nararatwanchai & A. Sarikaputi, Total phenolic, flavonoid, flavonol contents and antioxidant activity of Inca peanut (Plukenetia volubilis L.) leaves extracts. Food Research, 5(1) (2021). https://doi.org/10.26656/fr.2017.5(1).346

[12] QW. Zhang, LG. Lin & WC. Ye, Techniques for extraction and isolation of natural products: a comprehensive review. Chin. Med. 13, 20 (2018). https://doi.org/10.1186/s13020-018-0177-x

[13] M. Flórez, P. Cazón, M. Vázquez, Antioxidant Extracts of Nettle (Urtica dioica) Leaves: Evaluation of Extraction Techniques and Solvents. Molecules 2022, 27, 6015. https://doi.org/10.3390/molecules27186015

[14] A. Samad, S. Makhbar, H. Sharifulazar, A. M. Basri & S. A. Lim, The effects of Diplazium esculentum Retz. and Stenochlaena palustris incorporated with sodium alginate as edible coating on packaged figs (Ficus carica L.): A preliminary study. J. Food Process Preserv. 46(7), (2022). https://doi.org/10.1111/jfpp.16611

Modelling Malaysia's energy transition pathways: The role of renewable energy policy measures

KAZEEM ALASINRIN Babatunde[1,a*], MOAMIN A Mahmoud[2,b*], NAZRITA Ibrahim[3,c] and FATHIN FAIZAH Said[4,d]

[1]Institute of Informatics and Computing in Energy, Universiti Tenaga Nasional, Kajang 43000, Malaysia

[2]Institute of Informatics and Computing in Energy, Department of Computing, College of Computing and Informatics, Universiti Tenaga Nasional, Kajang 43000, Malaysia

[3]Institute of Informatics and Computing in Energy, Department of Informatics, College of Computing and Informatics, Universiti Tenaga Nasional, Kajang 43000, Malaysia

[4]Center for Sustainable and Inclusive Development Studies (SID), Faculty of Economics and Management, Universiti Kebangsaan Malaysia (UKM), Bangi 43600, Selangor, Malaysia

[a]alasinrin@uniten.edu.my, [b]moamin@uniten.edu.my, [c]nazrita@uniten.edu.my, [d]fathin@ukm.edu.my

Keywords: Agent-Based Model, Electricity Sector, Renewable Energy Policies, Energy Decarbonization, Simulation

Abstract. Coal's rising prominence in the power industry has raised concerns about future CO2 emissions and energy reliability. As of 2017, it is estimated that Malaysia's existing natural gas production can only be maintained for another 40 years. Consequently, the carbon intensity of electricity production has increased due to the increasing share of coal-fired plants and electricity infrastructure inefficiencies. To sum it up, energy industries were the highest emitters of CO2 emissions, with a 54 percent share. In response to these challenges, the government implemented series of Renewable Energy (RE) policy measures. Whether these policies are sufficient in driving Malaysian energy decarbonization is yet to be answered. In the study, we simulate different scenarios from 2015 to 2050 with an agent-based model to explore the roles of renewable energy policies toward emission reduction in the energy sector. The simulation results reveal that when all renewables initiatives were implemented, the share of RE increased to 16 percent, and emissions intensity fell by 26 percent relative to its level in 2005, albeit with increasing absolute carbon emissions. This milestone is still far below the government's 45 percent reduction target. The simulation results demonstrated that renewable energy policies alone are less effective in driving Malaysian electricity toward desired low-carbon pathways. Furthermore, it is evidenced that no single policy can achieve the emission reduction target. Therefore, a combination of energy efficiency and renewable energy policy measures is unavoidable to decarbonize the electricity sector in Malaysia.

Introduction

Energy markets are developing across the globe, from vertically integrated monopolies to liberalized market structures that encourage high competitiveness among electricity producers and allow customers freedom of services and choice [1]. As part of the continuous transformation, the liberalization of the distribution, transmission, and most critically producing sectors creates potential for new companies in the industry. Economic theory teaches that if market mechanisms function efficiently, this will provide greater productivity in terms of higher quality services and goods at lower costs. Malaysia's energy industry has been partly liberalized and labeled a quasi-competitive one [2]. However, due to the conclusions of several scientific studies indicating traditional fossil fuel power systems are not ecologically and economically sustainable, sustainable electricity generation and consumption are the subject of policy debates among environmental economists and policymakers [3–5]. The fear of depletion of natural resources such as coal, oil, and natural gas and global warming has made the green economy an urgent issue.

Renewable energies are the backbone of any energy shift toward net zero. As countries across the globe progressively turn away from carbon-emitting fossil fuels, knowing the present role renewable energy plays in the energy transition is crucial to maintaining a smooth pathway to net zero [6,7]. Renewable sources, notably solar and wind power, play a significant role in energy decarbonization. According to a recent international Energy Agency report, renewable energy capacity expansion driven by solar energy reached another annual record of about 290,000 megawatts (MW) in 2021. This exceptional capacity expansion is 3 percent greater than in 2020. More than half of all renewable energy expansion in 2021 came from solar energy alone, followed by hydropower and wind [8]. In the next two decades, renewable energy is projected to be the fastest-growing low-carbon energy source worldwide, accounting for almost 66 percent of global investment in electricity generation until 2040 [9].

Significant transformations will be needed in all economic sectors, including resource extraction, manufacturing, transportation, and hospitality, to ensure a timely transition to a low-carbon economy [10,11]. Renewable energy is essential for low-carbon transitions, yet despite increases in renewable energy use, the pace of decarbonization is still low. Energy transition initiatives have been affected by policy misalignments and lack of leadership [12,13]. Malaysia's economy, from the industrial, commercial, transportation to agricultural sectors, is largely supported by the country's electricity sector. A significant level of economic damage can result from interruptions in the energy supply, as was seen during the 1992 widespread blackout in Peninsular Malaysia. As a result, no sector is more important to the economy than power. However, given the current population and economic growth rate, the rapidly rising demand for power [14] and the growing proportion of coal-fueled power in the energy mix raises grave concern [15].

Consequently, the carbon intensity of electricity production has increased due to the increasing share of coal-fired plants and electricity infrastructure inefficiencies [16–18]. To sum it up, energy industries were the highest emitters of CO2 emissions, with a 54 percent share. Meanwhile, emissions of CO2 by the energy industry were mainly due to the fuel used by the power sector [19]. Thus, a significant reduction in electricity-related CO2 emissions must be the core of the climate change policy thrust of the government. Reducing carbon emissions and the national transformation agenda (TN2050) pose unique and tremendous problems for Malaysia's energy transition. No wonder the decoupling of carbon emissions from economic growth has been challenging, with the government prioritizing energy security, albeit low-carbon sources. Opening two locks with one key, the electricity sector plays a significant role in addressing economic and environmental issues. Therefore, implementing the Renewable Energy Policy for the smooth transformation of the power sector is inevitable. However, it is less clear what their potential impacts will be on Malaysia's power sector. Thus, exploring their effects on the electricity generation mix, electricity price, and carbon emissions within an established framework could contribute to formulating policies that could address greenhouse gas emissions while sustaining long-term energy security in Malaysia. The agent-based simulation model provides such a framework [20].

Agent-based Modelling

A model often used to study energy decarbonization is to simulate dynamically different agents' (e.g., electricity producers and consumers, government, electricity market, fuel market) behavior by ABM. The literature on energy and technology has been on a steady rise in using agent-based models to model complex emergent phenomena like energy transition. To answer why an ABM is a valuable tool for modeling the low-carbon pathways, we must first consider why this is the case. Individual behavior at the micro level is captured by an ABM, which then predicts the emerging behavior (like a transition) on a macro scale [21, 22]. However, linear models are inadequate for capturing the complex dynamics of decarbonization. Transition behavior is both an economically rational process and a highly complex system with different heterogeneous actors, preferences, social network effects, time, and learning processes from experiences. An ABM provides a simple platform for simulating and, therefore, analyzing such complicated decision systems, taking into account autonomous actors, a changing set of parameters, the influence of social interactions, and the interdependence of individual agents. As a result, there are numerous possibilities for technological development throughout the time that analytical tools and cross-sectional investigations cannot capture. The unpredictable and non-linear character of technological change can be accommodated by a social simulation (i.e., ABM) [23-27]. Agent-based simulation models (ABM) are employed to analyze energy transition policies' implications on the electricity sector in Malaysia for the next four decades, as recommended by Babatunde et al. (2017) [28] and Wooldridge et al. (1995) [29].

Agent description and interaction

Actors, markets, and environments are the three autonomous and active agents that comprise the model structure. Producers of electricity, the government, and customers are the key actors. The model's key components are the electric power market and the fuel market, where commerce is enabled, and players engage with each other by acquiring fuels and trading electric power, as shown in Fig. 1. Agent actions are updated yearly based on the design. Therefore, the simulation time step or unit is a year.

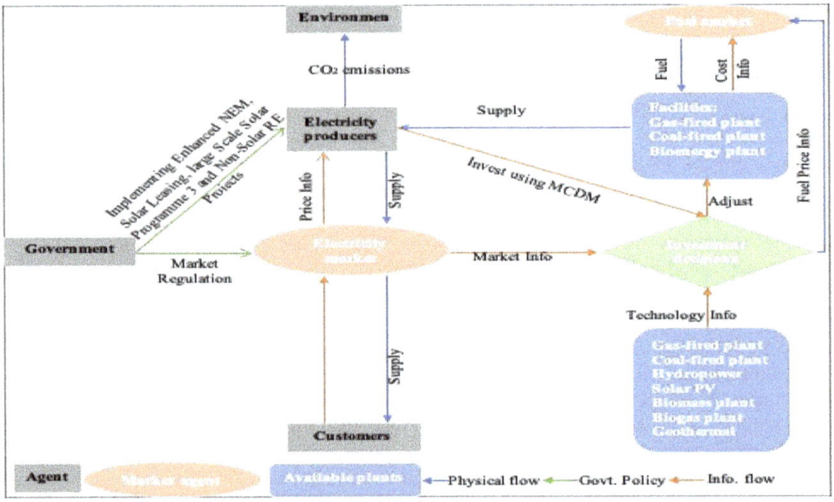

Fig. 1. Model structure

There are six different kinds of agents in our concept. Rectangles represent agents, and ovals show two markets. The physical movement of goods (fuel inputs and energy), as well as CO2 emissions, are shown by the blue arrows. The flow of cost and price information, governmental actions, and investment decisions by electricity producers are represented by peach arrows. Government agents oversee the electricity market and create rules. Based on legally binding contracts from the electricity market, energy producers buy fuel from the global market and ensure a steady electricity supply to the market. After the generation phase, emissions are discharged into the atmosphere, and producers determine whether it is necessary to shut down power plants or build new ones. If it is decided to build a new one, the producer uses the Multi-Criteria Decision Making (MCDM) technique to choose the preferred power plant. Finally, the customer receives electricity from the electrical market. Last but not least, the first rounded blue rectangular box displays different types of electricity generation plants that are now accessible in Malaysia. In contrast, the last blue rounded box lists the carbon-emitting generating technologies.

Policy Scenarios

Based on the existing energy market structure, we run a series of economically and politically significant scenarios to examine the implications of renewable energy policy and its influences on the electrical sector's transition to low-carbon pathways. Four scenarios are defined to examine the effects of energy policies on Malaysia's electricity sector in various economic situations until 2050. First, the scenarios vary in accordance with the state of the economy ('Initial-value' and 'Economic-trend'). The agents' virtual environment is designed using scenarios (Table 1). Consequently, a scenario is characterized by well-defined limiting conditions on government regulations and the input data of an energy system.

Table 1 Number of Scenarios Considered

Scenarios	Government policy	Economic	Conditions
	Renewable Energy	Initial-Value (INVA)	Economic-Trend (ECOT)
Business-As-Usual (BaU)	No	INVA-1	ECOT-2
Policy implementation	Yes	INVA-3	ECOT-4

Results and Discussion

The decarbonization of power will necessitate a substantial shift in the existing policy framework. It begins with the alignment of energy policy with emission reduction goals, followed by creating a favorable and supportive investment environment to boost private investment in technologies that may shift power generation toward renewable energies. Without articulating renewable energy policies, cost-effective decarbonization of the power sector would be difficult, if not impossible. Finally, long-term market mechanisms should be considered when providing a level playing field for sustainable energy technologies.

Business-As-Usual Scenario

This section focuses on baseline scenarios since they provide the basis for evaluating the impacts of policy scenarios. These scenarios aim to elucidate what carbon emissions and intensity, electricity price, and portfolio will likely be in 2050 without government interventions. We developed projections for electricity transition indicators under two baseline scenarios. The first baseline, an Initial-Value (INVA) scenario, assumes that the initial values of parameters remain unchanged throughout the simulation period. While the second baseline, Economic Trend (ECOT) scenario, considers the trend of economic parameters such as rising prices of fuels and falling costs of renewable energy through the simulation period.

Renewable Energy Policy

This is a case when the government implements renewable policies and programs as it is currently through a feed-in tariff, net energy metering, large-scale solar, self-consumption initiative, and other RE incentives such as green income tax exemption and green investment tax allowance.

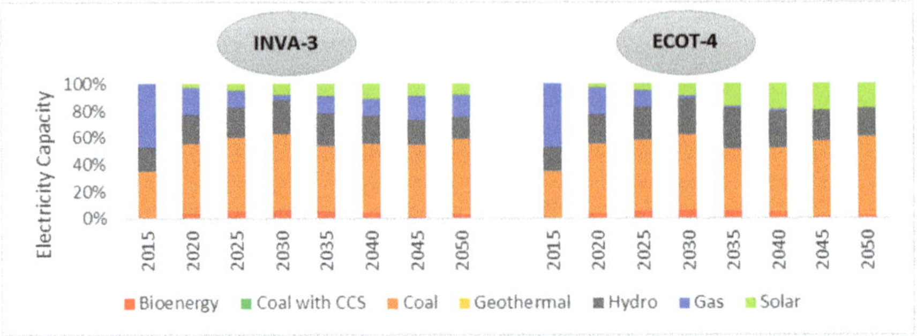

Fig. 2. Electricity Capacity by Technology under Renewable Energy Policy Consideration

Installed Capacity Levels by Technology

As presented in Fig. 2. the generation capacity for INVA-3 and ECOT-4 scenarios is expected to increase from 25064 MW in 2015 to a value between 72425 MW and 73071 MW by 2050 for INVA and ECOT scenarios. In the INVA scenario, the share of gas-fired plants falls from 47% in 2015 to about 16% by 2050, just as coal-fired counterparts increase from 35% in 2015 to 56% by 2050. Generators fueled by renewable technology from almost 0% to 3% and 9% for bioenergy and solar PV between 2015 and 2050. However, the generation fueled by natural gas will be retired from the electricity system by 2044, while coal-fired power plants will increase from 86,96.9 MW (35%) in 2015 to 42,400 MW (59%) by 2050. An imminent increase in capacity is observed in renewable deployment from 0% in 2015 to 2% and 18% for bioenergy and solar PV by 2050 (see Fig. 2).

Electricity Carbon Intensity

More than 45% of CO2 emissions from energy combustion are from electricity generation. The reason for this is the sector's over-reliance on non-renewable energy sources such as coal and gas for its electricity production. In the INVA scenario, about 72% of the electricity is generated through fossil fuels compared with 59% in the ECOT scenario by 2050 against 82% in the reference year. Intensity generally increases as more fossil fuels are used to generate electric power. In the INVA RE policy scenario, carbon intensity improves from 0.549 in 2015 to 0.447 (18.6% decrease) by 2040 before deteriorating again to 0.507 (13.4%) by 2050, making a small amount of 7.7% between 2015 to 2050 (see Table 2). At the same time, carbon intensity in the ECOT scenario experiences noticeable improvement throughout the simulation period. Although the RE policy could not fully drive the sector towards achieving a 45% emissions intensity target of 0.291, it reduces the intensity by 31% and 26% by 2035 and 2050, respectively (Table 2). When this result is viewed from the government's ambitious unconditional emission intensity target of 45 percent against Gross Domestic Product (GDP) by 2030 based 2005 level, the renewable energy policies only achieved a 26 percent reduction. This indicates that renewable energy policies and programs remain integral to the policy consideration for electricity decarbonization.

Table 2 Electricity Carbon Intensity Under Renewable Energy Policy Scenario

	BaU Scenarios		Renewable Energy Policy Scenarios	
	Initial Value	Economic Trend	Initial Value	Economic Trend
Operating Year	INVA-1	ECOT-2	INVA-3	ECOT-4
2015	0.549	0.549	0.549	0.549
2020	0.681	0.693	0.472	0.479
2025	0.750	0.766	0.452	0.440

2030	0.770	0.783	0.441	0.409
2035	0.787	0.787	0.415	0.364
2040	0.810	0.804	0.447	0.385
2045	0.800	0.732	0.507	0.375
2050	0.824	0.758	0.545	0.390

Electricity Carbon Emissions

Fig. 3 displays emissions from the power sector under renewable energy scenarios. Emissions under the baseline (INVA-1) scenario increase from 93 Mt in 2015 to 423 Mt in 2050, while the increase would be capped at 360 Mt for the ECOT-2 scenario with an average of 199 Mt, 268 Mt, and 317 Mt for the period between 2021 to 2030, 2031 to 2040 and 2041 2050 respectively. But with renewable energy policies in place, emissions increase slower than the baseline scenario. Under INVA-3, emissions only increased from 93 Mt in 2015 to 243 Mt in 2050 with an average of 178 Mt compared to 246 Mt in the BaU scenario, while ECOT-4 projected the best in terms of emissions trajectory at 173 Mt by 2050.

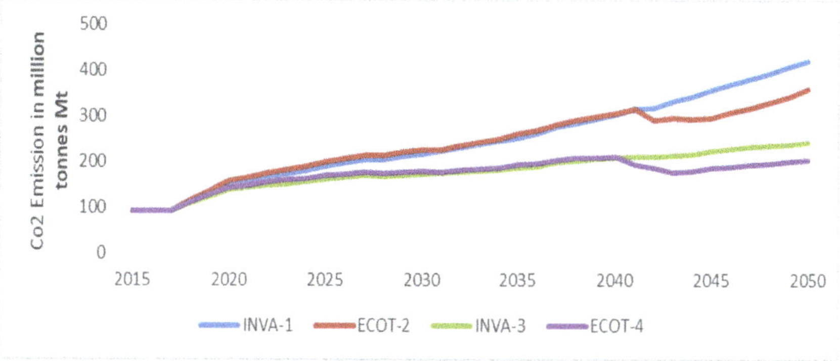

Fig. 3. CO2 Emissions under Renewable Energy Policy

Conclusion and Policy Implications

The reason for considering renewable energy scenarios is to see how renewable target achievement will impact the decarbonization process. One of the government's renewable energy policy initiatives is to increase the share of power generation from renewable sources. Scenarios INVA-3 and ECOT-4 capture these effects. When RE policies are implemented, more electricity is generated from RE sources such as hydro, solar PV, and bioenergy. Specifically, in scenario 3, the electricity system witnesses the injection of solar PV complemented by hydroelectric power with a combined share of 1/5 of electricity generation. The RE situation becomes better in scenario 4 with 2 percent hydro, 12 percent solar PV, and 2 percent bioenergy share of power generation by 2050. The deployment of RE sources has come at the expense of fossil fuel power generation. The RE influence, on one hand, and the increasing cost of natural gas electricity generation, on the other hand, are responsible for the early elimination of gas-fired plants from the generation mix as of 2044 with a reduced role for coal-powered plants. The improvement in the share of RE electricity generation is assumed to result from production subsidy, which reduces the investment cost of all RE sources except hydro, which is considered a matured technology.

The simulation results reveal that when all RE policy initiatives are well-articulated, and renewable energy share of electricity generation is at least as presented above, emissions intensity will fall by 26 percent relative to its level in 2005, albeit with increasing absolute carbon emissions.

However, this milestone is still far below the government's 45 percent reduction target. As a result, RE deployment effects on electricity prices fluctuate around baseline values with insignificant effects in both scenarios.

Successful energy policies and an ambitious emissions reduction target can't get off the ground without well-articulated renewable energy policy measures. However, renewable energy policies alone have proven less effective in driving Malaysian electricity toward desired low-carbon pathways. The simulation results demonstrated that renewable energy policies alone are less effective in driving Malaysian electricity toward desired low-carbon pathways. Furthermore, it is evidenced that no single policy can achieve the emission reduction target. Therefore, a combination of energy efficiency and renewable energy policy measures is unavoidable to decarbonize the electricity sector in Malaysia. The energy transition will be achieved if energy policymakers and related government agencies with a climate change plan can review the existing energy policy instruments to achieve emissions reduction targets.

References

[1] V. Ahlqvist, P. Holmberg, T. Tangerås, A survey comparing centralized and decentralized electricity markets, Energy Strateg. Rev. 40 (2022). https://doi.org/10.1016/j.esr.2022.100812

[2] S. Ahmad, R.M. Tahar, Selection of renewable energy sources for sustainable development of electricity generation system using analytic hierarchy process: A case of Malaysia, Renew. Energy. 63 (2014) 458–466. https://doi.org/10.1016/j.renene.2013.10.001

[3] M. Kalkuhl, O. Edenhofer, K. Lessmann, Renewable energy subsidies: Second-best policy or fatal aberration for mitigation?, Resour. Energy Econ. 35 (2013) 217–234. https://doi.org/10.1016/j.reseneeco.2013.01.002

[4] C. Chong, W. Ni, L. Ma, P. Liu, Z. Li, The use of energy in Malaysia: Tracing energy flows from primary source to end use, Energies. 8 (2015) 2828–2866. https://doi.org/10.3390/en8042828

[5] S. S.A., N.A. Omar, M.S. Bin Ahmad, H.R. Siddiquei, S. Mohd. Nor, Renewable Energy in Malaysia: Strategies and Development, Environ. Manag. Sustain. Dev. 2 (2013) p51-66. https://doi.org/10.5296/emsd.v2i1.3197

[6] M.J. Pickl, The renewable energy strategies of oil majors – From oil to energy?, Energy Strateg. Rev. 26 (2019) 100370. https://doi.org/10.1016/j.esr.2019.100370

[7] G. Falchetta, A. Adeleke, M. Awais, E. Byers, P. Copinschi, S. Duby, A. Hughes, G. Ireland, K. Riahi, S. Rukera-Tabaro, F. Semeria, D. Shendrikova, N. Stevanato, A.́e Troost, M. Tuninetti, A. Vinca, A. Zulu, M. Hafner, A Research Agenda for Planning and Financing Nexus Development Objectives in Rural Sub-Saharan Africa, Renew. Stategy Rev. 43 (2022). https://doi.org/10.2139/ssrn.4024065

[8] IEA, Renewables 2021: Analysis and forecast to 2026, Int. Energy Agency Publ. Int. (2021) 167. www.iea.org/t&c/%0Ahttps://webstore.iea.org/download/direct/4329

[9] EIA, Annual Energy Outlook 2018 with projections to 2050, Washington, D.C., 2018. https://doi.org/DOE/EIA-0383(2012) U.S

[10] U. Bhattarai, T. Maraseni, A. Apan, Assay of renewable energy transition: A systematic literature review, Sci. Total Environ. 833 (2022) 155159. https://doi.org/10.1016/J.SCITOTENV.2022.155159

[11] K.A. Babatunde, F.S. Fathin, M.N. Nor Ghani, A.B. Rawshan, Reducing Carbon Dioxide Emissions from Malaysian Power Sector: Current Issues and Future Directions, J. Kejuruter. 1 (2018) 59–69.

[12] S. Zaman, Z. Wang, S.F. Rasool, Q. uz Zaman, H. Raza, Impact of critical success factors and supportive leadership on sustainable success of renewable energy projects: Empirical evidence from Pakistan, Energy Policy. 162 (2022) 112793. https://doi.org/10.1016/j.enpol.2022.112793

[13] K.A. Babatunde, F.F. Said, N.G. Md Nor, R.A. Begum, M.A. Mahmoud, Coherent or conflicting? Assessing natural gas subsidy and energy efficiency policy interactions amid CO2 emissions reduction in Malaysia electricity sector, J. Clean. Prod. 279 (2021) 123374. https://doi.org/10.1016/j.jclepro.2020.123374

[14] F.F. Said, K.A. Babatunde, N.G.M. Nor, M.A. Mahmoud, R.A. Begum, Decarbonizing the Global Electricity Sector through Demand-Side Management: A Systematic Critical Review of Policy Responses, J. Ekon. Malaysia. 56 (2022) 71–91. https://doi.org/10.17576/JEM-2022-5601-06

[15] Ministry of Natural Resources and Environment, Malaysia Biennial Update Report to the United Nations Framework Convention on Climate Change (UNFCCC), Putrajaya, 2015. https://doi.org/10.1017/CBO9781107415324.004

[16] B.W.W. Ang, B. Su, Carbon emission intensity in electricity production: A global analysis, Energy Policy. 94 (2016) 56–63. https://doi.org/10.1016/j.enpol.2016.03.038

[17] S. Hinchliffe, R. V. Diemen, C. Heuberger, N. Macdowell, A Study on Transitions in Electricity Systems Towards 2030 Examining: Australia, China, India, Malaysia, Singapore, South Africa and the United Kingdom, 2015.

[18] B. Tranberg, O. Corradi, B. Lajoie, T. Gibon, I. Staffell, G.B. Andresen, Real-time carbon accounting method for the European electricity markets, Energy Strateg. Rev. 26 (2019). https://doi.org/10.1016/j.esr.2019.100367

[19] MESTECC, Third National Communication (TNC) and Biennial Update Report to the UNFCCC, Putrajaya, 2018. https://www.mestecc.gov.my/web/documents/jqhRxaP3ff20SNzCaBfiyqiOdsLhuBa9Jyz5D0C_gQ0

[20] É.J.L. Chappin, I. Nikolic, N. Yorke-Smith, Agent-based modelling of the social dynamics of energy end use, Elsevier Inc., 2020. https://doi.org/10.1016/B978-0-12-818567-4.00029-6

[21] E.J.L. Chappin, L.J. De Vries, J.C. Richstein, P. Bhagwat, K. Iychettira, S. Khan, Simulating climate and energy policy with agent-based modelling : The Energy Modelling Laboratory (EMLab), Environ. Model. Softw. 96 (2017) 421–431. https://doi.org/10.1016/j.envsoft.2017.07.009

[22] Mahmoud, M. A., Md Nasir, N. R., Gurunathan, M., Raj, P., & Mostafa, S. A. (2021). The current state of the art in research on predictive maintenance in smart grid distribution network: Fault's types, causes, and prediction methods—a systematic review. Energies, 14(16), 5078.

[23] P. Hansen, X. Liu, G.M. Morrison, Agent-based modelling and socio-technical energy transitions: A systematic literature review, Energy Res. Soc. Sci. 49 (2019) 41–52. https://doi.org/10.1016/j.erss.2018.10.021

[24] X. Wang, K. Yang, Economic load dispatch of renewable energy-based power systems with high penetration of large-scale hydropower station based on multi-agent glowworm swarm optimization, Energy Strateg. Rev. 26 (2019) 100425. https://doi.org/10.1016/j.esr.2019.100425

[25] Mahmoud, M. A., Ahmad, M. S., Ahmad, A., Yusoff, M. Z. M., Mustapha, A., & Hamid, N. H. A. (2013, May). Obligation and Prohibition Norms Mining Algorithm for Normative Multi-agent Systems. In KES-AMSTA (pp. 115-124).

[26] , M. A., Ahmad, M. S., Ahmad, A., Mustapha, A., Yusoff, M. Z. M., & Hamid, N. H. A. (2013). Building norms-adaptable agents from potential norms detection technique (PNDT). International Journal of Intelligent Information Technologies (IJIIT), 9(3), 38-60.

[27] Mahmoud, M. A., Ahmad, M. S., Ahmad, A., Mohd Yusoff, M. Z., & Mustapha, A. (2011, July). Norms detection and assimilation in multi-agent systems: a conceptual approach. In Knowledge Technology Week (pp. 226-233). Springer, Berlin, Heidelberg.

[28] K.A. Babatunde, R.A. Begum, F.F. Said, Application of computable general equilibrium (CGE) to climate change mitigation policy: A systematic review, Renew. Sustain. Energy Rev. 78 (2017) 61–71. https://doi.org/10.1016/j.rser.2017.04.064

[29] M. Wooldridge, C. Street, M. Manchester, N.R. Jennings, Q. Mary, W. College, Intelligent Agents : Theory and Practice, Knowl. Eng. Revie. 10 (1995) 1–62.

Keyword Index

About the Editors

Associate Professor Ir Dr Yeong Yin Fong is currently serving as Associate Professor at the Department of Chemical Engineering, Universiti Teknologi PETRONAS, Malaysia. For the past 12 years, AP Ir Dr Yeong Yin Fong is actively working on the synthesis and characterization of polymeric, inorganic, hybrid and mixed matrix membranes (flat sheet, tubular and hollow fiber) for CO2 separation from small molecule gases. She is expert in advanced synthesis of high performance polymeric and hybrid organic-inorganic membranes, and familiar with the specific needs, relevant facilities and basic procedures to physically and chemically characterize the membrane materials. Besides, she is also expert in design, fabrication and operation of membrane texting facilities. To date, she has authored/co-authored more than 70 research articles in these fields.

Ts. Dr. Lam Man Kee is a senior lecturer in the Chemical Engineering Department at Universiti Teknologi PETRONAS (UTP) and a key core member of the HICoE-Centre for Biofuel & Biochemical Research at UTP. He received his BEng (Hons) (2008), MSc (2010), and PhD (2014) in Chemical Engineering from Universiti Sains Malaysia (USM). His research interests include microalgae cultivation, biofuel production, life cycle energy and carbon assessment, and the social-economic impact of biofuels. He has published over 130 papers in ISI-indexed journals with an H-index of 48 and cumulative citations of 7500 (SCOPUS database). Currently, he serves as the Deputy Editor-in-Chief for the International Journal of Biomass & Renewables (UTP Press) and is an Editorial Board Member for the Journal of Advanced Chemical Engineering (Omics International). In 2021 and 2022, Dr. Lam was recognized as one of the global Top 2% Scientists in a study conducted by Stanford University. He was awarded the Leaders in Innovation Fellowships Programme by the Royal Academy of Engineering (UK) under the Newton-Ungku Omar Fund in 2020 and was named a Top Peer Reviewer by Publon (Web of Science Group) in 2019. As of today, he has obtained 13 research grants as Principal Investigator, totaling RM 1.38 million. He has also delivered several short courses to academia and industry, such as "Biomass to Fuels & Chemicals," "Cultivation of Microalgae and Larvae for Biofuel Production," and "Writing and Publishing in High Impact Journals: Tricks & Tips."

Ts Dr Norwahyu Jusoh joined Universiti Teknologi PETRONAS in June 2018 as a lecturer. She is a researcher with more than 8 years' experience in the field of CO2 capture technologies. She demonstrates great consistency in publishing her research in more than 50 leading scientific journals, filed 2 patents and won more than 20 awards national and internationally. She is one of the key developers for the high pressure system for acid gas and heavy hydrocarbon vapor separation as well as the inventor for the development of novel hybrid composite membrane for CO2 separation. Her active involvement in multiple research projects and great networking and linkages with national and international institutions have contributed prominent impact in the area of contaminant management.

Ts Dr Chong Yang Chuah received his B. Eng. in Chemical and Biomolecular Engineering from Nanyang Technological University (NTU) Singapore in 2015. He received his Ph.D. degree from NTU in 2019 after obtaining the prestigious Nanyang President's Graduate Scholarship (NPGS) in 2015. He worked as a research fellow at Singapore Membrane Technology Centre (SMTC) from 2019 to 2021 before working at Universiti Teknologi PETRONAS (UTP) Malaysia as a lecturer in Chemical Engineering Department. His research area mainly focuses on the development of porous materials and membranes in gas separation. At the current stage, he has published more than 40 peer-reviewed journal papers with > 1,800 citations.

www.ingramcontent.com/pod-product-compliance
Ingram Content Group UK Ltd.
Pitfield, Milton Keynes, MK11 3LW, UK
UKHW020659041026
15590UKWH00001B/368

9 781644 902509